Lecture Notes in Electrical Engineering

Volume 539

Lecture Notes in Electrical Engineering (LNEE) is a book series which reports the latest research and developments in Electrical Engineering, namely:

- Communication, Networks, and Information Theory
- Computer Engineering
- Signal, Image, Speech and Information Processing
- Circuits and Systems
- Bioengineering
- Engineering

The audience for the books in LNEE consists of advanced level students, researchers, and industry professionals working at the forefront of their fields. Much like Springer's other Lecture Notes series, LNEE will be distributed through Springer's print and electronic publishing channels.

More information about this series at http://www.springer.com/series/7818

Bruno Andò · Francesco Baldini
Corrado Di Natale · Vittorio Ferrari
Vincenzo Marletta · Giovanna Marrazza
Valeria Militello · Giorgia Miolo
Marco Rossi · Lorenzo Scalise
Pietro Siciliano
Editors

Sensors

Proceedings of the Fourth National
Conference on Sensors,
February 21–23, 2018, Catania, Italy

 Springer

Editors
Bruno Andò
University of Catania
Catania, Italy

Francesco Baldini
IFAC-CNR
Sesto Fiorentino, Florence, Italy

Corrado Di Natale
University of Rome Tor Vergata
Rome, Italy

Vittorio Ferrari
Department of Information
Engineering (DII)
University of Brescia
Brescia, Italy

Vincenzo Marletta
University of Catania
Catania, Italy

Giovanna Marrazza
Department of Chemistry
University of Florence
Sesto Fiorentino, Florence, Italy

Valeria Militello
University of Palermo
Palermo, Italy

Giorgia Miolo
University of Padova
Padua, Italy

Marco Rossi
Sapienza University of Rome
Rome, Italy

Lorenzo Scalise
Università Politecnica delle Marche
(UNIVPM)
Ancona, Italy

Pietro Siciliano
IMM-CNR
Lecce, Italy

ISSN 1876-1100   ISSN 1876-1119   (electronic)
Lecture Notes in Electrical Engineering
ISBN 978-3-030-04323-0   ISBN 978-3-030-04324-7   (eBook)
https://doi.org/10.1007/978-3-030-04324-7

Library of Congress Control Number: 2018961731

This Springer imprint is published by the registered company Springer Nature Switzerland AG
The registered company address is: Gewerbestrasse 11, 6330 Cham, Switzerland

Preface

This book gathers scientific contributions presented at the 4th National Conference on Sensors held in Catania, Italy from 21 to 23 February 2018. The conference has been organized by a partnership of the major scientific societies and associations involved in the research area of sensors, the Italian Society of Chemistry (SCI), the Italian Association of Electric and Electronic Measures (GMEE), the Italian Association of Ambient Assisted Living (AITAAL), the Italian Society of Optics and Photonics (SIOF), the Italian Association of Sensors and Microsystems (AISEM), the Italian Society of Pure and Applied Biophysics (SIBPA), the Italian Association of Photobiology (SIFB), the Association Italian Group of Electronics (GE), and the Association NanoItaly.

The fourth edition of the conference has confirmed a large participation with approximately 70 oral presentations, 80 poster presentations, and over 150 delegates. The driving idea of the first conference, to gather scientists having different expertise and with different cultural backgrounds, dealing with all the different aspects of sensors, has proved to be indeed successful again.

In this perspective, the book represents an invaluable and up-to-the-minute tool, providing an essential overview of recent findings, strategies, and new directions in the area of sensor research. Further, it addresses various aspects based on the development of new chemical, physical, or biological sensors, assembling and characterization, signal treatment, and data handling. Lastly, the book applies electrochemical, optical, and other detection strategies to the relevant issues in the food and clinical environmental areas, as well as industry-oriented applications.

Catania, Italy	Bruno Andò
Florence, Italy	Francesco Baldini
Rome, Italy	Corrado Di Natale
Brescia, Italy	Vittorio Ferrari
Catania, Italy	Vincenzo Marletta
Florence, Italy	Giovanna Marrazza
Palermo, Italy	Valeria Militello
Padua, Italy	Giorgia Miolo
Rome, Italy	Marco Rossi
Ancona, Italy	Lorenzo Scalise
Lecce, Italy	Pietro Siciliano

Contents

Part I
Chemical Sensors

Low Temperature NO$_2$ Sensor Based on YCoO$_3$ and TiO$_2$ Nanoparticle Composites

Tommaso Addabbo⬤, Ada Fort⬤, Marco Mugnaini⬤
and Valerio Vignoli⬤

Abstract Chemical sensors based on metal oxides have been widely explored and used in the literature and have found different application fields as a function of their operating characteristics like selectivity, sensitivity, stability over time etc. Recently, some papers started to diffuse the idea that innovative chemical sensors could be obtained using two different metal oxides combined together providing enhanced sensing capabilities. In this paper the authors propose a new sensor based on perovskite support modified by a TiO$_2$ based compound in order to test enhanced sensing performance. Moreover, the present work aims to show that nanocomposites obtained introducing in a matrix of a given metal oxide a second nano-structured metal oxide, which can act either as a catalyst or as a structure modifier, can provide improved sensitivity, selectivity and stability.

Keywords Chemical sensors · Metal oxide sensors · Hetero-junctions

1 Introduction

Recently many research works have shown that in general all types of nano-composites are very promising for the development of resistive gas sensors [1]. It was established that the surface-related properties important for gas sensor applications

T. Addabbo · A. Fort (✉) · M. Mugnaini · V. Vignoli
Dipartimento di Ingegneria dell'informazione e Scienze Matematiche, Università di Siena,
Via Roma 56, 53100 Siena, Italy
e-mail: ada@dii.unisi.it

T. Addabbo
e-mail: addabbo@dii.unisi.it

M. Mugnaini
e-mail: mugnaini@dii.unisi.it

V. Vignoli
e-mail: vignoli@dii.unisi.it

© Springer Nature Switzerland AG 2019
B. Andò et al. (eds.), *Sensors*, Lecture Notes in Electrical Engineering 539,
https://doi.org/10.1007/978-3-030-04324-7_1

such as electronic, catalytic, mechanical, and chemical ones can be highly modified and tuned to the specific applications by combining different nano-structured materials [1–3]. The most traditional composites used in gas sensors are those obtained by the addition of noble metals to metal oxide matrices, but recently also many other composites were proposed and tested, such as films consisting of polymers mixed with metals or metal oxides, or carbon nanotubes mixed with polymers, or again composites based on fullerenes and graphene, etc.

In this context, composites based on the combination of two different nano-structured metal oxides, $Me_I O$ and $Me_{II} O$, have proven to be extremely interesting. Nanocomposites obtained introducing in a matrix of $Me_I O$ a nano-structured $Me_{II} O$, which can act either as a catalyst or as a structure modifier, can provide improved sensitivity, selectivity and stability. Metal Oxide-Metal Oxide structures can be implemented in several ways. They can be created during the process of either synthesis or deposition of initial material, or can be formed by various layer by layer techniques, or alternatively they can be generated by mixing already synthetized materials in certain proportions.

Anyhow it must be underlined that both in thin and thick film deposition, controlling the chemical compositions, surface morphology microstructure and phase state is still a challenge since, in general, the sensing characteristic are controlled independently by three factors (which are Receptor function, Transducer function and Utility factor [1]) which describe and participate in the generation of the sensor signature. The Receptor function is strictly linked to the material redox properties, to the stochiometry, and to the adsorption/desorption parameters. The Transducer function mainly depends on the carrier mobility, concentration and on the grain size, whereas the Utility factor is linked to the film thickness and geometrical factors [1–3]. Therefore any sensor has its peculiar characteristic depending on the combination of these factors which are really difficult to be controlled during the fabrication process.

2 Sensors and Characterization

2.1 Sensor Preparation

In this paper we propose a NO_2 sensor based on a composite obtained exploiting an already-prepared metal oxide matrix of $YCoO_3$ pervoskite. Perovskite powders were prepared by means of a sol-gel method, described in detail in [4], both in stoichiometric form and in Pd doped and defective versions. The powders were mixed to organic vehicles to prepare a paste which was screen printed on Alumina substrates across two electrodes. A transparent n-TiO_2 coating produced by Italvernici

Fig. 1 Chemical sensor layout. On side (**a**) the sensing film and the temperature sensors are designed and implemented through screen printing technique. On the (**b**) side the heater is realized by means of the same technique

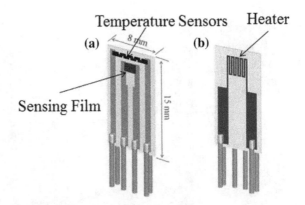

(ITALVERNICI-FELCE150) [5] has been used for wet impregnation of the printed layer by drop casting (2 μL doses) using a micropipette; the as-obtained film was heated at 320 °C for 24 h. The n-TiO$_2$ coating used in this study is based on crystalline anatase (nanoparticle dimensions in the range 25–55 nm) diluted in water with a concentration of 32×10^{-4} mol/L to get a transparent paint with 2.0 cPs viscosity. The obtained nano-composite consists of highly dispersed TiO$_2$ in the frame-work of the perovskite matrix. The sensing support is a screen-printed thick-film alumina circuit hosting the sensing layer and a temperature sensor on one side and a heater on the other side (Fig. 1).

2.2 Sensor Characterization

The sensors were characterized by means of a system which allows for accurately controlling the temperature of each sensor, as well as the gas mixture flow and composition, also in terms of humidity, as shown in Fig. 2 and described in [4–8].

In detail, with reference to Fig. 2, the system exploits gas reservoirs which feed, by means of a flowmeter bench, a measurement chamber equipped with 8 chemical sensors. The chamber is maintained at a constant temperature in an incubator. The front end electronic boards (one for each sensor, housed inside the measurement chamber) are connected to a NI PXI rack where a host processor board and ADC and DAC boards are used to control the measurement process and to acquire the measurement data. A personal computer is used to set the measurement parameters and to process and display the measurement data.

The sensors were tested under a constant flow (200 mL/min) of mixtures of NO$_2$ with a carrier gas that was N$_2$ or air, humid and dry.

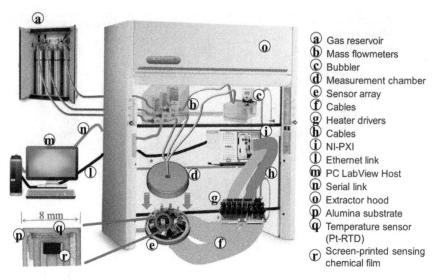

Ⓐ Gas reservoir
Ⓑ Mass flowmeters
Ⓒ Bubbler
Ⓓ Measurement chamber
Ⓔ Sensor array
Ⓕ Cables
Ⓖ Heater drivers
Ⓗ Cables
Ⓘ NI-PXI
Ⓛ Ethernet link
Ⓜ PC LabView Host
Ⓝ Serial link
Ⓞ Extractor hood
Ⓟ Alumina substrate
Ⓠ Temperature sensor (Pt-RTD)
Ⓡ Screen-printed sensing chemical film

Fig. 2 Chemical sensing system used for characterization

Figures 3 and 4 show the responses to NO_2 of sensors obtained with different $YCoO_3$ based materials, before and after the impregnation with n-TiO_2, as a function of temperature and as a function of time, respectively.

The response is defined as $(R - R_0)/R_0$, where R_0 is the baseline resistance of each individual sensors measured at the same temperature in the carrier gas (nitrogen or air), whereas R is the value of the sensor resistance after 4 min of exposition to the test mixture.

The results presented in this paper show that the introduction of TiO_2 (n-type semiconductor) in the matrix of $YCoO_3$ (p-type) highly improves the sensitivity toward NO_2 at low temperature (in fact the sensors could be used also at room temperature, as shown in Fig. 5).

The insertion of TiO_2 nano-particles on the surface of the $YCoO_3$ larger grains produced hetero-junctions (see Fig. 6), which due to microstructure of the layer contribute only marginally to the electronic conduction in the sensing film, which is instead mainly determined by the behavior of the homo-junctions at the $YCoO_3$ grains boundaries. Nevertheless, the large effect observed could be explained both by the enhancement of the depleted region at the surface of the $YCoO_3$ grains, which can both modify the height of the Schottky barriers at the homo-junctions and favor the adsorption of oxidizing gases, and by the large reactivity of TiO_2 toward NO_2 also at low temperature.

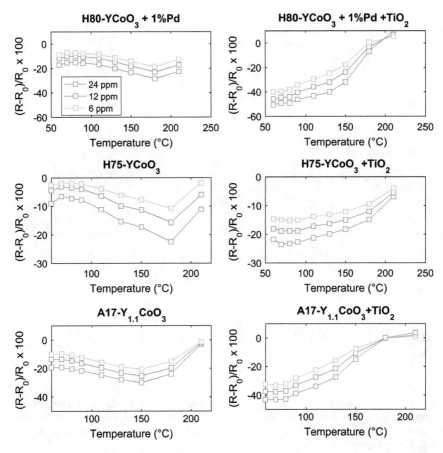

Fig. 3 Responses of 3 different materials based on YCoO₃ to NO₂ as a function of working temperature for different NO₂ concentrations (as per legend). Leftmost plots: without TiO₂. Rightmost plots: with TiO₂ nanoparticles. The carrier gas is nitrogen, the flow is 200 mL/min. The values of R are obtained after 4 min of exposition to the target mixtures

3 Conclusions

In this paper the authors presented a new sensor obtained by means of an heterojunction based on both a perovskite material and TiO₂. The introduction of TiO₂ (n-type semiconductor) in the matrix of YCoO₃ (p-type) highly improves the sensitivity toward NO₂ at low temperatures. The insertion of TiO₂ nano-particles on the surface of the YCoO₃ larger grains produced hetero-junctions, which slightly modify the material conductive properties. Nevertheless, the electronic conduction in the sensing film is still mainly determined by the behavior of the homo-junctions at the YCoO₃ grains boundaries. As a matter of fact, it seems that the depletion

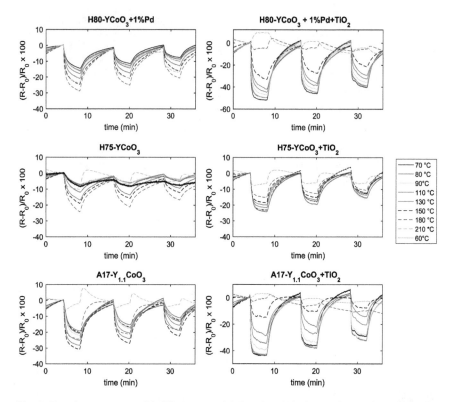

Fig. 4 Transient responses of 3 different materials based on $YCoO_3$ to NO_2 as a function of time at different temperature (as per legend). The carrier gas is nitrogen, the flow is mL/min with the following protocol: 4 min N_2, 4 min N_2 + 24 ppm NO_2, 8 min N_2, 4 min N_2 + 12 ppm NO_2, 8 min N_2, 4 min N_2 + 6 ppm NO_2, 4 min N_2

region of newly induced junctions is affected by the TiO_2 nano-particles favoring the adsorption of oxidizing gases.

The proposed sensors seem interesting in all the possible applications where temperature may play an important role in terms of power consumption requirement due to the fact that exploitable sensor responses can be obtained even starting from ambient temperature.

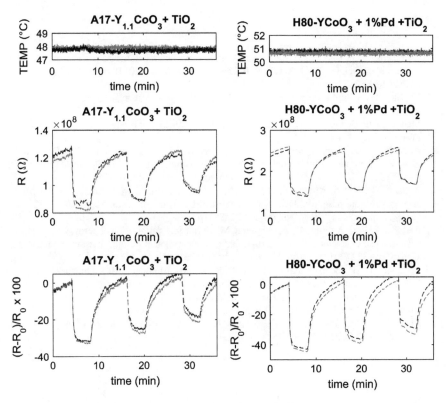

Fig. 5 Transient responses of 3 different materials based on YCoO₃ to NO₂ as a function of time. The heater wasn't driven

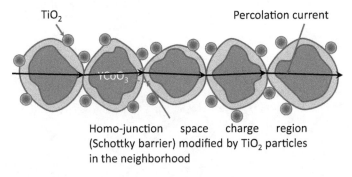

Fig. 6 Schematic of the structure and of the conduction mechanism of the proposed composite

References

1. Korotcenkov, G., Cho, B.K.: Metal oxide composites in conductometric gas sensors: achievements and challenges. Sens. Actuators B Chem. **244**, 182–210 (2017). https://doi.org/10.1016/j.snb.2016.12.117
2. Moseley, P.T.: Solid state gas sensors. In: Tofield, B.C. (ed.) Adam Hilger Series on Sensors, Bristol and Philadelphia (1987)
3. Madou, M.J., Morrison, S.R.: Chemical Sensing with Solid State Devices. Academic Press, San Diego (1989)
4. Addabbo, T., Bertocci, F., Fort, A., Gregorkiewitz, M., Mugnaini, M., Spinicci, R., Vignoli, V.: Gas sensing properties and modeling of YCoO3 based perovskite materials. Sens. Actuators B Chem. **221**, 1137–1155 (2015). https://doi.org/10.1016/j.snb.2015.07.079
5. Addabbo, T., Fort, A., Mugnaini, M., Vignoli, V., Baldi, A., Bruzzi, M.: Quartz-crystal microbalance gas sensors based on TiO2 nanoparticles. IEEE Trans. Instrum. Meas. **67**(3), 722–730 (2018). https://doi.org/10.1109/TIM.2017.2785118
6. Bertocci, F., Fort, A., Vignoli, V., Mugnaini, M., Berni, R.: Optimization of perovskite gas sensor performance: characterization, measurement and experimental design. Sensors (Switzerland), **17**(6), No. 1352 (2017). https://doi.org/10.3390/s17061352
7. Addabbo, T., Bertocci, F., Fort, A., Gregorkiewitz, M., Mugnaini, M., Spinicci, R., Vignoli, V.: Gas sensing properties of YMnO3 based materials for the detection of NOx and CO. Sens. Actuators B Chem. **244**, 1054–1070 (2017). https://doi.org/10.1016/j.snb.2017.01.054
8. Addabbo, T., Bertocci, F., Fort, A., Mugnaini, M., Vignoli, V.: WO3 nanograined chemosensor: a model of the sensing behavior. IEEE Trans. Nanotechnol. **15**(6), 1–11 (2016). https://doi.org/10.1109/TNANO.2016.2558099

Effect of Humidity on the Hydrogen Sensing in Graphene Based Devices

Brigida Alfano, Ettore Massera, Tiziana Polichetti, Maria Lucia Miglietta and Girolamo Di Francia

Abstract In this work, we investigate the effect of humidity variations on the sensing performance of Pd-graphene (GR) based devices. Palladium nanoparticles are directly synthetized onto GR sheets by microwave irradiation; the optimal palladium coverage results into a sensitive and fast hydrogen device. The dynamic conductance changes exposed to different hydrogen concentrations from 2.5 to 0.2% are displayed at room temperature, using humidified air as carrier gas at different Relative Humidity (RH) levels. The results show how the sensing curves in low humidity conditions have higher sensitivity with respect to humid environment. On the other hand, dry conditions negatively affect the sensing layer stability over time while humid conditions preserve the material.

Keywords Graphene · Metal decoration · Hydrogen detection

1 Introduction

Graphene is a two-dimensional material made of carbon atoms, so far called "wonder material" for its exceptional physical characteristics. Since its discovery in 2004, researchers are enthusiastically studying graphene to exploit its outstanding properties in numerous applications [1].

Graphene could represent a powerful sensing material to design smaller and lighter sensor respect to conventional ones. Moreover, employing its characteristic electronic structure, the device based on graphene could be able to detect rapidly a single molecule also at room temperature [2].

Pristine graphene is particularly sensitive towards nitrogen dioxide [3, 4].

In this regard, the functionalization with metal and metal oxide nanoparticles proves to be an effective way to extend the range of analytes to which the material is sensitive [5–9].

B. Alfano (✉) · E. Massera · T. Polichetti · M. L. Miglietta · G. Di Francia
ENEA, P.le E. Fermi 1, 80055 Portici (Naples), Italy
e-mail: brigida.alfano@enea.it

© Springer Nature Switzerland AG 2019
B. Andò et al. (eds.), *Sensors*, Lecture Notes in Electrical Engineering 539,
https://doi.org/10.1007/978-3-030-04324-7_2

The possibility of functionalization further improves the prospects of graphene-based electronics for a real industrial application. In this scenario, the effect of metal nanoparticles decoration onto graphene surface has been addressed in our latest works [10]. We have fabricated a chemiresistor based on graphene, functionalized with palladium nanoparticles, highly sensitive towards hydrogen gas [11].

Since the functionalization has a direct effect on sensing behavior, it is crucial to determine the optimum coverage surface to obtain the most performing device [12].

The fabricated device follows linearly the variations in hydrogen concentration, showing repeatable responses when exposed to cyclic tests.

Herein, a further investigation about the dynamic behavior of sensing device is displayed, with the aim to approach the real world operating conditions. First of all, the effect of humidity was considered and then the history of device.

1.1 Material Preparation

As reported in our previous work [13], pristine graphene was synthesized by Liquid Phase Exfoliation (LPE) method: natural graphite flakes, dispersed in a hydro-alcoholic solution, were exfoliated by means of ultrasound treatment. The surface of pristine graphene was decorated according to the process described in our previous works [11, 12].

1.2 Material Characterization

A drop of the freshly-sonicated dispersions of graphene decorated with Pd nanoparticles (PdNPs) was casted on n-doped Silicon substrate and dried on hot plate for Scanning Electron Microscopy characterization.

The graphene flakes, with an average lateral size of 300 nm, are randomly overlapped (Fig. 1). These are constituted by few layers, as evidenced in Raman spectroscopy data reported in our previous works [11, 12]. In the SEM image (Fig. 1), PdNPs appear as bright spots all over the graphene surface with average size of 30–40 nm.

1.3 Sensing Characterization

For the electrical and sensing characterization, few microliters of suspension were drop casted onto rough alumina transducers with five pairs of gold interdigitated electrodes (fingers 350 μm wide, 4650 μm long and 350 μm spaced).

For each device, the achievement of the ohmic contact between GR-PdNPs films and gold electrodes was proven by the linear response of I–V measurements (Fig. 2c).

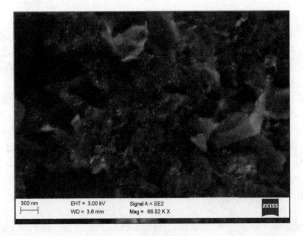

Fig. 1 SEM image Pd nanoparticles onto graphene sheets

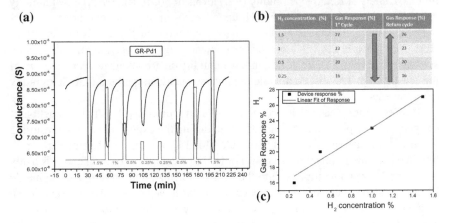

Fig. 2 a The dynamic conductance changes of GR-Pd device exposed to different hydrogen concentrations from 1.5 to 0.25% and the return cycle; **b** gas responses according to H_2 concentration for the first cycle and for return cycle; **c** IV characteristic

The as-fabricated devices were characterized by recording their conductance change when exposed to hydrogen gas. In particular, we have investigated the performances of the devices in terms of the percent conductance change consequent to a gas exposure, in absolute value, which we regard as gas response (Eq. 1).

$$\text{Gas Response} (\%) = \frac{|G_{max} - G_0|}{G_0} * 100 \tag{1}$$

where G_0 and G_{max} are the conductances of the devices before and after the exposure to a gas, respectively.

The device is able to follow the changes of hydrogen gas concentration from 0.25 to 1.5% (the specific concentrations are labeled under each peak in Fig. 2a) in humid atmosphere (RH 50%). By considering the responses recorded during the measurement cycle, it can be observed as the response values at the same hydrogen concentration are identical within the error of measurement (±0.2%).

This chemiresistor presents interesting sensing properties: in controlled environment the device follows linearly the hydrogen concentration and the sensing responses are repeatable. It is worth noting that these performances have been realized working at room temperature. However, for the application in a real environment, it is crucial to know how the device behaves when external conditions change. In actual environment, the most important interferent for any sensing device, even for high-temperature operating devices, is water. As so, in the first step, the relative humidity (RH) was changed in the test chamber in the range 0–50% RH.

Figure 3 shows the dynamic conductance changes of the device exposed to different hydrogen concentrations at room temperature under different humidity conditions, namely at Relative Humidity (RH) levels of 50, 10 and 0%. In low humidity conditions, from 0% RH to 10% RH, upon hydrogen exposure and after the sensing phase, the conductance signal shows a continuous drift towards higher values (Fig. 3b, c), while at 50% RH, the conductance signal drops slightly after every sensing phase (Fig. 3a). Basically, in low humidity environment an *over-recovery* of the sensing signal is observed while the reverse occurs in medium to high humidity environments.

The desorption phase of hydrogen from Pd structures involves the interaction of oxygen with the palladium hydride (formed during the exposure) with the restoring of Pd and formation of a water molecule [14, 15]. As can be seen in Fig. 3a, in humid environment this process does not lead to a complete reaction and it is likely that some H_2 remains trapped into the Pd structures so determining its volumetric expansion also concurring to the lower conductance values. On the contrary, it can be hypothesized that in low humidity conditions, where the reaction between oxygen and adsorbed hydrogen is favored, even the hydrogen previously trapped reacts with oxygen. This mechanism could be at the basis of the *over-recovery* phenomenon observed in Fig. 3b, c).

These results suggest that environmental conditions can affect sensing performances, and those performances may degrade more or less quickly based on the "history" of the device. This is clear if we refer to Fig. 3d, where the sensitivity curves recorded at different humidity levels and the subsequent measurement cycles in dry conditions are summarized. As can be seen, the sensitivity in dry environment appears higher, but the curves also lose the linearity as the overall exposure of device to hydrogen increases. This can be ascribed to the aforementioned volumetric expansion of the Pd nanoparticles, which is at the basis of the well-known issue of the metallic embrittlement by hydrogen [16]. Of course, this modification of the material causes a parallel loss of sensing properties.

In other words, the operation of the device in real environmental conditions, characterized by the constant presence of humidity, preserves this kind of sensitive film from a fast deterioration even though this leads to a decrease in sensitivity.

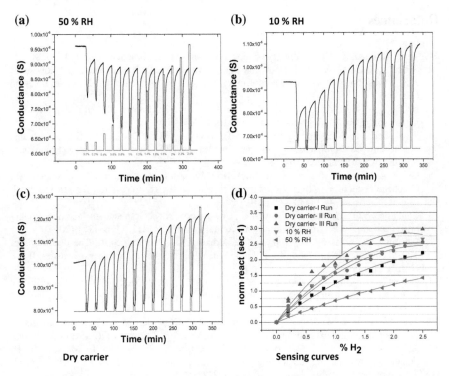

Fig. 3 The dynamic conductance change of device exposed to different hydrogen concentrations from 2.5 to 0.2% changing RH levels **a** 50% RH, **b** 10% RH, **c** Dry carrier, at RT; **d** Humidity effect on sensing curves. To overcome the uncompleted recovery and the changing baseline, we have used the correlation between the maximum rate of the relative response and hydrogen concentration (reactivity), method discussed in our previous work [4]

2 Conclusions

In this work, we have investigated the effect of humidity variations on the sensing performance of Pd-graphene (GR) based devices.

The experiments have shown that exists a loss of response linearity at high concentrations of hydrogen as a consequence of cyclic H_2 exposures, which cause morphological changes. Tests in environmental conditions (50% RH) have shown that despite a smaller sensitivity with respect to dry conditions, the presence of humidity preserves the sensing film.

References

1. Yang, T., Zhao, X., He, Y., Zhu, H.: Graphene-based sensors. Graphene 157–174 (2018)
2. Schedin, F., Geim, A.K., Morozov, S.V., Hil, E.W., Blake, P., Katsnelson, M.I., Novoselov, K.S.: Detection of individual gas molecules adsorbed on graphene. Nat. Mater. **6**, 652–655 (2007)
3. Ko, G., Kim, H.Y., Ahn, J., Park, Y.M., Lee, K.Y., Kim, J.: Graphene-based nitrogen dioxide gas sensors. Curr. Appl. Phys. **10**(4), 1002–1004 (2010)
4. Ricciardella, F., Massera, E., Polichetti, T., Miglietta, M.L., Di Francia, G.: A calibrated graphene-based chemi-sensor for sub parts-per-million NO_2 detection operating at room temperature. Appl. Phys. Lett. 104(18) (2014)
5. Gutés, A., Hsia, B., Sussman, A., Mickelson, W., Zettl, A., Carraro, C., Maboudian, R.: Graphene decoration with metal nanoparticles: towards easy integration for sensing applications. Nanoscale **4**(2), 438–440 (2012)
6. Chatterjee, S.G., Chatterjee, S., Ray, A.K., Chakraborty, A.K.: Graphene–metal oxide nanohybrids for toxic gas sensor: a review. Sens. Actuators B Chem. **221**, 1170–1181 (2015)
7. Esfandiar, A., Ghasemi, S., Iraji zad, A., Akhavan, O., Gholami, M.R.: The decoration of TiO_2/reduced graphene oxide by Pd and Pt nanoparticles for hydrogen gas sensing. Int. J. Hydrog Energy (37) (2012)
8. Chung, M.G., Kim, D.H., Seo, D.K., Kim, T., Im, H.U., Lee, H.M., Kim, Y.H., et al.: Flexible hydrogen sensors using graphene with palladium nanoparticle decoration. Sens. Actuators B Chem. (169) 387–392 (2012)
9. Zhao, Z., Knight, M., Kumar, S., Eisenbraun, E.T., Carpenter, M.A.: Humidity effects on Pd/Au-based all-optical hydrogen sensors. Sens. Actuators B Chem. **129**(2), 726–733 (2008)
10. Alfano, B., Polichetti, T., Mauriello, M., Miglietta, M.L., Ricciardella, F., Massera, E., Di Francia, G.: Modulating the sensing properties of graphene through an eco-friendly metal-decoration process. Sens. Actuators B Chem. **222**, 1032–1042 (2016)
11. Alfano, B., Polichetti, T., Miglietta, M.L., Massera, E., Schiattarella, C., Ricciardella, F., Di Francia, G.: Fully eco-friendly H_2 sensing device based on Pd-decorated graphene. Sens. Actuators B Chem. **239**, 1144–1152 (2017)
12. Alfano, B., Massera, E., Polichetti, T., Miglietta, M.L., Di Francia, G.: Effect of palladium nanoparticle functionalization on the hydrogen gas sensing of graphene based chemi-resistive devices. Sens. Actuators B **253**, 1163–1169 (2017)
13. Fedi, F., Miglietta, M.L., Polichetti, T., Ricciardella, F., Massera, E., Ninno, D., Di Francia, G.: A study on the physicochemical properties of hydroalcoholic solutions to improve the direct exfoliation of natural graphite down to few-layers graphene. Mater. Res. Express **2**(3), 035601 (2015)
14. Sun, Y., Wang, H.H., Xia, M.: Single-walled carbon nanotubes modified with Pd nanoparticles: unique building blocks for high-performance, flexible hydrogen sensors. J. Phys. Chem. C **112**(4), 1250–1259 (2008)
15. Chung, M.G., Kim, D.H., Seo, D.K., Kim, T., Im, H.U., Lee, H.M., Kim, Y.H., et al.: Flexible hydrogen sensors using graphene with palladium nanoparticle decoration. Sens. Actuators B Chem. (169) 387–392 (2012)
16. Zhao, Z., Knight, M., Kumar, S., Eisenbraun, E.T., Carpenter, M.A.: Humidity effects on Pd/Au-based all-optical hydrogen sensors. Sens. Actuators B Chem. **129**(2), 726–733 (2008)

A Networked Wearable Device for Chemical Multisensing

Tiziana Polichetti, Maria Lucia Miglietta, Brigida Alfano, Ettore Massera, S. De Vito, Girolamo Di Francia, A. Faucon, E. Saoutieff, S. Boisseau, N. Marchand, T. Walewyns and L. A. Francis

Abstract The present contribution illustrates the early stage activities of the project CONVERGENCE FLAG-ERA H2020. The project is aimed at improving the quality of healthcare during active life by preventing the development of diseases through earlier diagnosis of cardiovascular and/or neurodegenerative diseases, and meets the growing desire of consumers for a deeper awareness of their conditions; indeed, the extensive availability of smartphones and tablets and the technology therein incorporated enable the monitoring and transmission of vital parameters from the body of a patient to medical professionals. CONVERGENCE extends this concept, aiming to create a wireless and multifunctional wearable system, able to monitor, in addition to key parameters related to the individual physical condition (activity, core body temperature, electrolytes and biomarkers), even the chemical composition of the ambient air (NO_x, CO_x, particles). Herein is summarized the project activity, which involves ENEA group together with CEA (Commissariat à l'Energie Atomique, France) and UCL (Université catholique de Louvain, Belgium).

Keywords Wearable · Environmental monitoring · Sensor · Internet of things

T. Polichetti (✉) · M. L. Miglietta · B. Alfano · E. Massera · S. De Vito · G. Di Francia
Department of Energy Technologies, ENEA, Italian National Agency for New Technologies, Energy and Sustainable Economic Development, Department of Energy Technologies (DTE) P.le E. Fermi 1, 80055 Portici (Naples), Italy
e-mail: tiziana.polichetti@enea.it

A. Faucon · E. Saoutieff · S. Boisseau
Université Grenoble Alpes, CEA, LETI, MINATEC Campus, 38054 Grenoble, France

N. Marchand · T. Walewyns · L. A. Francis
Electrical Engineering Department (ELEN), ICTEAM Institute, Université catholique de Louvain (UCL), Place du Levant 3, 1348 Louvain-la-Neuve, Belgium

© Springer Nature Switzerland AG 2019
B. Andò et al. (eds.), *Sensors*, Lecture Notes in Electrical Engineering 539,
https://doi.org/10.1007/978-3-030-04324-7_3

1   Introduction

Medical technologies include a variety of devices, such as pacemakers, defibrilla-
tors, deep-brain stimulators, insulin pumps, which aim at improving the quality of
healthcare through non-invasive treatment of pathological conditions and/or earlier
diagnosis. The extensive availability of smartphones and tablets is directing consumers towards wearable technologies, capable of monitoring and transmitting vital
parameters from the body of a patient to medical professionals, thus preventing the
development of diseases [1–4]. CONVERGENCE, a FLAG-ERA H2020 project,
broadens this concept by realizing a wireless and multifunctional wearable system,
able to monitor, in addition to key parameters related to the individual physical condition (activity, core body temperature, electrolytes and biomarkers), even the chemical
composition of the ambient air (NO_x, CO_x, particles).

In this contribution we focused on the early stage of the project activities, which
involves our group together with CEA (Commissariat à l'Energie Atomique, France)
and UCL (Université catholique de Louvain, Belgium) and regards the implementation of a demonstrator composed by gas sensors integrated into a test platform
realized by CEA. In the project framework, one of the tasks is to develop sensors for
the detection of NO_2 based on graphene, a material widely known in the scientific
literature for the specificity towards this pollutant [5–7]. The provided sensors, based
on bare and ZnO nanoparticles decorated-graphene, are chemiresistive devices able
to detect nitrogen dioxide in the range 100–1000 ppb. They were connected with an
electronic board, developed for sensor data acquisition in real time; data were directly
sent by Bluetooth on smartphones. Finally, a microarray platform made of micro-
interdigitated electrodes, developed by UCL, will allow the miniaturization of the
demonstrator in a post-CMOS process [8]. Detection limit below 100 ppb (≥ 10 ppb),
selectivity, ultra-low-power consumption ($<20\,\mu W$ continuously), and low-cost fabrication (<1 €/die), are the key challenges towards Internet-scale chemical sensing
applications in environmental monitoring.

2   The Environmental Sensors

In line with the project requirements, ENEA labs worked on the synthesis of the
functionalized material and its optimization, realizing chemiresistive devices based
on bare and ZnO nanoparticles decorated-graphene, able to detect nitrogen dioxide at
room temperature in the range 100–1000 ppb; pristine graphene (GR) was prepared
by sonication assisted graphite exfoliation. Graphite flakes were dispersed into a mixture of ultrapure water and i-propanol (at 1 mg/ml) and sonicated in an ultrasonic bath
for 48 h. Afterwards, graphite crystallites were removed by centrifuging for 45 min
at 500 rpm. The concentration of the graphene suspension was 0.1 ± 0.1 mg/ml. The
preparation of ZnO decorated graphene (GZnO) was performed by freeze drying of
graphene suspension. 2.5 mg of graphene powder, mixed with 4 mg of ZnO (Ø 14 nm)

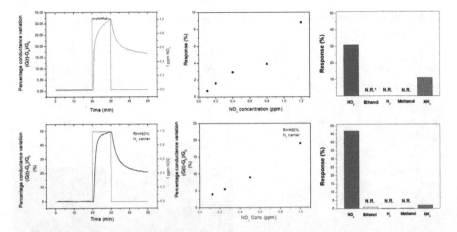

Fig. 1 Sensing responses to a single pulse of 1000 ppb NO_2 and relative sensitive curves of chemiresistors based on graphene (panel a) and ZnO decorated graphene (panel c). The selectivity towards 1 ppm NO_2, 50 ppm ethanol, 1% hydrogen (H_2), 50 ppm methanol and 250 ppm ammonia (NH_3) of chemiresistors based on graphene (b) and ZnO decorated graphene (d). *N.R. = No response

and microwave irradiated for 5 min at 1000 W. The resulting powder was dissolved in ethanol and the sensitive material was dispensed on interdigitated commercial macroscopic electrodes realized on alumina substrate for a preliminary test towards NO_2, whose results are illustrated in the Fig. 1.

As can be inferred from the graphs of Fig. 1, both preparations exhibit a high degree of specificity towards nitrogen dioxide, besides being able to detect this analyte in the range 100–1000 ppb in agreement with the requirements of the project.

3 Test Platform with NO_2 Gas Sensors

Four devices, prepared starting from pristine and functionalized graphene, were sent to the CEA for the connection with an electronic board, specifically designed for sensor data acquisition in real time. The platform is compatible with different kinds of sensors, i.e. able to monitor both vital parameter (activity, core body temperature, electrolytes and biomarkers) and the chemical composition of the ambient air (NO_x, CO_x, particles); data were directly sent by Bluetooth on smartphones. Table 1 summarizes the main features of the prepared devices.

The sensors were plugged onto the sensor platform that converts through an ADC Voltage Divider Bridge the analog electric signal into a measurable signal, as shown in Fig. 2.

Table 2 displays the main electrical parameters measured after the connection to the platform. The resistance values, both measured by a multimeter and by the plat-

Table 1 Summary of the basic resistance values of the sensors produced by ENEA and the corresponding sensitivity towards NO$_2$

	Sample name	R (kΩ)	Sensitivity to NO$_2$
Pristine graphene	ENEA 1	0.46	37% @ 300 ppb
Pristine graphene	ENEA 2	0.4	31% @ 1 ppm
Pristine graphene	ENEA 3	1.9	23% @ 1 ppm
ZnO NP decorated graphene	ENEA 4	88	50% @ 1 ppm

Fig. 2 NO$_2$ sensors (ENEA) tests

Table 2 Summary of the main electrical parameter checked after the connection to the Leti-CEA platform

	Multimeter measures (Ω)	Platform measures (Ω)	Error (%)	Value converted by the ADC (V)
ENEA 2	593	598	0.8	145
ENEA 3	1984	2009	1.26	477

form, are in agreement with an error of about 1%, thus indicating that the connection to the platform did not introduce any contact resistance.

After these characterizations, devices ENEA 3 and ENEA 4, connected to the CEA platform were tested into ENEA sensor test chamber with a measurement protocol that includes 20 min in an inert environment, 10 min of exposure to the analyte and the restoration phase in an inert environment for 20 min. As can be seen in Fig. 3,

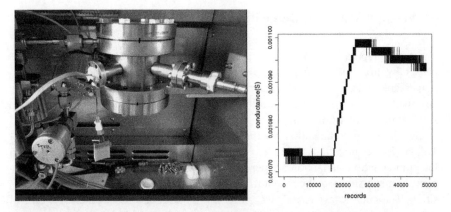

Fig. 3 Pristine graphene-based sensor, named ENEA3, installed on LETI board, exposed to 300 ppb of NO$_2$ for 10 min

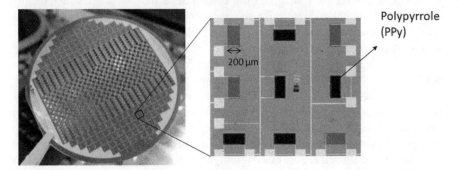

Fig. 4 Example of a multi-pixel platform; each transducer consuming less than 20 μW in continuous operation

ENEA3 device installed on LETI board and exposed to 300 ppb NO$_2$, exhibits a variation of 3%, so demonstrating that the platform is able to follow and visualize in real time on a Smartphone the signal variation consequent to the exposure to such analyte.

These preliminary tests allowed identifying some parameters on which to optimize the sensor devices. In particular, for an optimal Analog to Digital Conversion, the coupling of the sensor device resistance with that of the Voltage Divider Bridge (67 kΩ) should be realized so as those values result as close as possible. According to this, the further development of sensing devices will be carried out in such a way as to obtain a basic resistance contained in the identified range, in order to work with the maximum ADC input dynamic voltage range, that will allow a limit of NO$_2$ detection as low as 50 ppb.

4 Sensing Material Deposition on UCL Microarrays

The same preparations based on bare and ZnO-functionalized graphene were dispensed by drop casting on a multi-pixel resistive CMOS-compatible platform functionalizable with gas sensing materials operating at room temperature, specially designed by UCL looking at ultra-low-power low-cost environmental monitoring. An example of platform is depicted in Fig. 4.

Besides, a communication platform has also been implemented for Internet of Things (IoT) applications through LoRaWAN protocol on dedicated networks. Data integration and visualization are performed in partnership with Opinum S.A. company. Figure 5 shows the dies sent to the ENEA lab for the sensitive materials deposition. The microarrays are of two types: 2×2 pixel^2 and 3×3 pixel^2, the size of sensitive surfaces goes down to 300×200 μm^2. This should highlight miniaturization capabilities towards a versatile low-cost multi-gas sensing microsystem.

In this first phase we tested the drop casting deposition on 2×2 structures, by dispensing the materials onto microstructures by using a microsyringe. The 3×3 arrays were instead put aside and destined to inkjet deposition to be carried out in a second phase. Figure 6 displays a photo of the 2×2 microarrays on which the sensing materials were dispensed. The resistance values of the devices resulted to fall into the $k\Omega$ range, therefore suitable for the subsequent bonding and encapsulation.

The main critical issues emerged during these tests indicate the need to further refined the deposition technique, in order to avoid both the overlapping of material between two adjacent devices and also the contact between the microsyringe tip and the substrate, which can cause scratches on the interdigitated structure. We are confident that all the abovementioned issues can be avoid on 3×3 array kept for inkjet deposition; indeed, such approach allows to deposit the sensing material based-ink in controlled way also avoiding contact between the dispensing medium and the sample surface, thus resulting totally safe.

5 Conclusions

In this document we presented the first steps of the CONVERGENCE project, which involves ENEA group together with CEA (Commissariat à l'Energie Atomique, France) and UCL (Université catholique de Louvain, Belgium).

The interfacing of graphene-based gas sensors with LETI board and the subsequent exposure to 300 ppb NO_2, demonstrated the ability of the platform to follow and visualize in real time the sensing signal; an adequate sizing of the coupling of the sensor device resistance with that of the Voltage Divider Bridge will allow a limit of NO_2 detection as low as 50 ppb.

Deposition of bare and ZnO-functionalized graphene on a multi-pixel resistive CMOS-compatible platform has produced devices with electrical resistance values suitable for the subsequent bonding and encapsulation; some weaknesses emerged

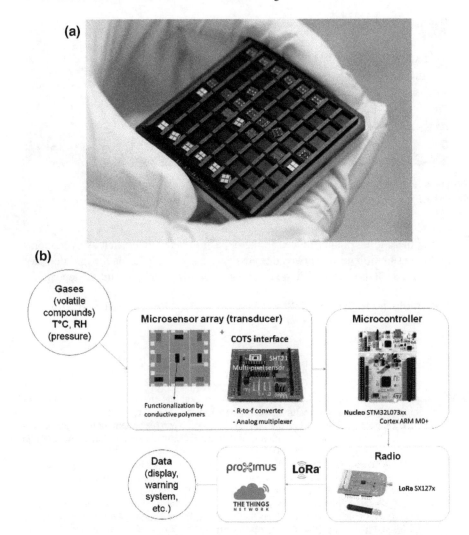

Fig. 5 **a** Multi-pixel array platforms based on interdigitated microelectrodes sent to ENEA for pristine graphene and GZnO deposition and further characterizations after encapsulation and bonding. **b** Full IoT system implementation

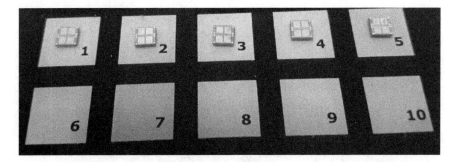

Fig. 6 Photograph of the bare and functionalized graphene-based devices realized by drop casting on UCL microarrays

during these tests, mainly related to the deposition technique, such as overlapping of the sensing film between two adjacent devices and scratches on the interdigitated structure. Such drawbacks will be overcome by utilizing inkjet printing deposition.

Acknowledgements CONVERGENCE is funded by the Flag ERA (ERANET—JTC2016).

References

1. Chen, M., Ma, Y., Song, J., Lai, C.F., Hu, B.: Smart clothing: connecting human with clouds and big data for sustainable health monitoring. Mob. Netw. Appl. **21**(5), 825–845 (2016)
2. Surrel, G., Rincón, F., Murali, S., Atienza, D.: Low-power wearable system for real-time screening of obstructive sleep apnea. In: Conference: 2016 IEEE Computer Society Annual Symposium on VLSI (ISVLSI), pp. 230–235 (2016)
3. Sopic, D., Aminifar, A., Aminifar, A., Atienza, D.: Real-time classification technique for early detection and prevention of myocardial infarction on wearable devices. In: 13th IEEE Biomedical Circuits and Systems Conference. No. EPFL-CONF-230328 (2017)
4. Magno, M., Pritz, M., Mayer, P., Benini, L.: DeepEmote: towards multi-layer neural networks in a low power wearable multi-sensors bracelet. In: 7th IEEE International Workshop on Advances in Sensors and Interfaces (IWASI), pp. 32–37 (2017)
5. Schedin, F., Geim, A.K., Morozov, S.V., Hill, E.W., Blake, P., Katsnelson, M.I., Novoselov, K.S.: Detection of individual gas molecules adsorbed on graphene. Nat. Mater. **6**(9), 652–655 (2007)
6. Ricciardella, F., Massera, E., Polichetti, T., Miglietta, M.L., Di Francia, G.: A calibrated graphene-based chemi-sensor for sub parts-per-million NO_2 detection operating at room temperature. Appl. Phys. Lett. **104**(18), 183502 (2014)
7. Singh, E., Meyyappan, M., Nalwa, H.S.: Flexible graphene-based wearable gas and chemical sensors. ACS Appl. Mater. Interfaces **9**(40), 34544–34586 (2017)
8. Marchand, N., Walewyns, T., Lahem, D., Debliquy, M., Francis, L.A.: Ultra-low-power chemiresistive microsensor array in a back-end CMOS process towards selective volatile compounds detection and IoT applications. In: Proceedings of International Symposium on Olfaction and Electronic Nose (2017)

High Performance VOC$_s$ Sensor Based on ɣ-Fe$_2$O$_3$/Al-ZnO Nanocomposites

N. Zahmouli, S. G. Leonardi, A. Bonavita, M. Hjiri, L. El Mir, Nicola Donato and G. Neri

Abstract In this study, ternary ɣ-Fe$_2$O$_3$/Al-ZnO nanocomposites (NC) ware prepared using the solvothermal sol-gel process and a successive supercritical drying in ethanol. SEM analysis of the ternary NC samples showed clearly that they are formed by very small nanoparticles in the nanometer range. XRD highlighted the presence of the characteristic diffraction peaks of ɣ-Fe$_2$O$_3$ and ZnO phases in all samples. Conductometric sensors were fabricated and tested for the monitoring of acetone in air. Results obtained have demonstrated that the ternary composite-based sensors display higher response to acetone and ethanol compared to that obtained with Al-ZnO and ɣ-Fe$_2$O$_3$ ones.

Keywords ɣ-Fe$_2$O$_3$ · Al-ZnO · Composite nanoparticles · Sol-gel · Acetone Selectivity

1 Introduction

Metal oxide-based chemical sensors have been widely used for the detection of toxic pollutant, combustible gases and volatile organic compounds (VOCs), due to their high sensitivity, small size, low power consumption and easy fabrication. To minimize the damage caused by atmospheric pollution, controlling and alarm

N. Zahmouli (✉) · M. Hjiri · L. El Mir
Laboratory of Physics of Materials and Nanomaterials Applied at Environment,
Faculty of Sciences of Gabes, University of Gabès, 6072 Gabès, Tunisia
e-mail: nassim.zahmouli1989@gmail.com

N. Zahmouli · S. G. Leonardi · A. Bonavita · N. Donato · G. Neri
Department of Engineering, University of Messina, Messina 98166, Italy

L. El Mir
Department of Physics, College of Sciences, Al Imam Mohammad Ibn Saud Islamic University (IMSIU), Riyadh 11623, Saudi Arabia

M. Hjiri
Faculty of Sciences, Physics Department, King Abdulaziz University, Jeddah, Saudi Arabia

© Springer Nature Switzerland AG 2019
B. Andò et al. (eds.), *Sensors*, Lecture Notes in Electrical Engineering 539,
https://doi.org/10.1007/978-3-030-04324-7_4

systems are needed. Previously, we reported the high performances of Al(III)-doped ZnO nanomaterials for CO sensing [1]. Here, we extended the study to a ternary γ-Fe_2O_3/Al-ZnO composite, using Fe(III) as a dopant for monitoring low concentration of ethanol and acetone in air.

ZnO is one of more investigated metal oxide for VOCs. Al-doping introduces defects into the ZnO structure and moreover enhance the electrical conductivity of the resulting Al-ZnO binary nanocomposite. γ-Fe_2O_3 is a ferromagnetic material and is widely used as a magnetic recording medium, which has been also investigated, pure or doped, as gas-sensitive material [2, 3]. In this study, we are aimed to increase the sensing properties of Al-ZnO by doping it further with iron. γ-Fe_2O_3/Al-ZnO with different Fe loading were therefore synthesized by the same sol gel route used to synthetize the binary Al-ZnO nanocomposite [4]. Preliminary sensing tests obtained in the monitoring of acetone and highlighting the performances of the ternary composites are reported here for the first time.

2 Experimental

2.1 Samples Preparation

Pure Al-ZnO, γ-Fe_2O_3 and the ternary γ-Fe_2O_3/Al-ZnO nanocomposite (γ-Fe_2O_3/Al-ZnO NCs) were prepared using the solvothermal sol-gel process and supercritical drying in ethanol as follows.

(i) *Al-ZnO*. Zinc oxide doped with 3% of aluminum was prepared using a sol–gel route with 20 g of zinc acetate dihydrate [$Zn(CH_3COO)_2 \cdot 2H_2O$; 99%] as a precursor in 140 ml of methanol. After 10 min of magnetic stirring at room temperature, an adequate quantity of aluminum nitrate-9-hydrate corresponding to [Al/Zn] ratio of 0.03 is added. After 15 min of magnetic stirring, the solution was placed in an auto-clave and dried in supercritical ethyl alcohol (Tc = 243∘C; Pc = 63.6 bar [1]).

(ii) *γ-Fe_2O_3*. For the preparation of pure γ-Fe_2O_3, 16 g of iron III precursor (acetylacetonate) [$C_{15}H_{21}FeO_6$] are poured into 32 ml of ethanol. After stirring for 15 min, 220 ml of ethanol are added for drying, then the solution is poured into an autoclave and heated under supercritical conditions (Tc = 243∘C, Pc = 63.6 bar).

(iii) *Al-ZnO/γ-Fe_2O_3*. For obtaining Al-ZnO/γ-Fe_2O_3 different quantities of 3 atomic% Al/ZnO nanoparticles previously prepared were added to the poured solution and stirred for 15 min. The ratio between Fe and ZnO varied in the range 0–100%. The as-obtained hybrid nanocomposites were annealed at 400 °C for 2 h in air in order to stabilize their microstructures.

2.2 Morphological, Microstructural and Sensing Properties

Samples morphology was observed through scanning electron microscope (SEM) operating at 30 kV while the microstructure was analyzed by the X-ray diffraction (XRD) technique. XRD measurements were performed using the CuK-α radiation ($\lambda = 1.5406$ Å) of a Bruker-AXS D5005 diffractometer.

Conductometric devices for sensing tests were fabricated by printing films (about 20 μm thick) of the synthesized samples dispersed in water on alumina substrates (6×3 mm^2) with Pt interdigitated electrodes and a Pt heater located on the backside. The fabricated sensors based were used for monitoring low concentration of acetone in air at different temperatures.

3 Results and Discussion

3.1 Metal Oxides Characterization

The morphological properties of the samples have been investigated by SEM. The typical fine grained morphology of some of the synthesized powder samples is shown in Fig. 1. γ-Fe₂O₃/Al-ZnO composites with 33% of γ-Fe₂O₃, display smaller grain size and reveal different morphology.

XRD analysis has been performed in order to acquire information about the phase present in these samples. XRD patters are shown in Fig. 2, highlighting the presence of peaks indexed according to hexagonal wurtzite structure for Al-ZnO and to cubic structure for γ-Fe₂O₃, respectively. The average crystallite size (d), as calculated by the full width at half maxima (FWHM) of the most intense diffraction peaks of Al-ZnO, indicate that in the ternary samples Al-ZnO grains reduce its dimension from 26 nm to 16 nm. The particle size decrease with the introduction of γ-Fe₂O₃ is in according with the results acquired by SEM analysis.

Fig. 1 From left to right: SEM images of Al-ZnO, Al-ZnO/γ-Fe₂O₃ (33%), γ-Fe₂O₃. The samples were annealed at 400 °C for 2 h in air

Fig. 2 X-ray diffraction spectra of Al-ZnO, Al-ZnO/ɣ-Fe₂O₃ (33%) and ɣ-Fe₂O₃ samples after annealing at 400 °C for 2 h

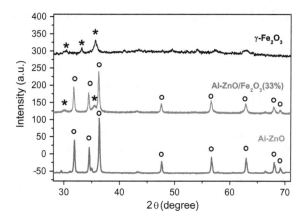

Fig. 3 3D Response to 10 ppm of acetone at different temperatures of Al-ZnO, 3%Al-ZnO/ɣ-Fe₂O₃ (33%), 3%Al-ZnO/ɣ-Fe₂O₃ (56%), 3%Al-ZnO/ɣ-Fe₂O₃ (83%) and ɣ-Fe₂O₃ sensors

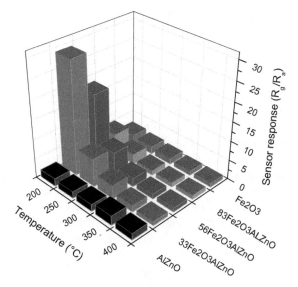

3.2 Acetone Sensing Tests

In order to find the optimal working temperature, we carried out preliminary sensing tests with all sensors fabricated in an interval temperature between 200 and 400 °C. Figure 3 reports the sensor response to acetone as target gas with these sensors. Al-ZnO/Fe₂O₃ (33%) sensor resulted the most responsive device at the optimal temperature of 200 °C.

Figure 4 shows the dynamic response to different concentrations of acetone for Al.ZnO/ɣFe₂O₃ (33%) and the related calibration curve. According to behavior shown, the ternary nanocomposite sensor shows very high response to acetone being very sensitive in the ppm range.

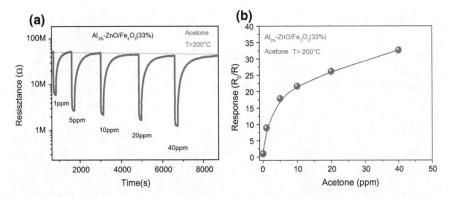

Fig. 4 **a** Dynamic response to different concentrations of acetone for the ternary composite sensor. **b** Calibration curve

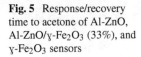

Fig. 5 Response/recovery time to acetone of Al-ZnO, Al-ZnO/γ-Fe₂O₃ (33%), and γ-Fe₂O₃ sensors

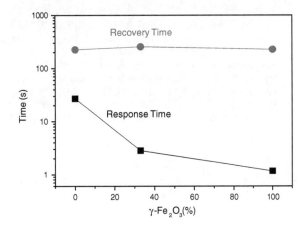

On the basis of the characterization results above reported, the good improvement of the sensitivity to acetone can be attributed to the increase in the surface area and the electron modification due to the introduction of iron [3].

The response/recovery time for these sensors is shown in Fig. 5. Very interestingly, the response time is strongly reduced by increasing the γ-Fe₂O₃ loading in the composite, whereas the recovery time remain approximatively the same.

Therefore, the introduction of γ-Fe₂O₃ into Al-ZnO based sensor increases not only the sensitivity but also the speed of the sensor response, helping in the developing sensors with better dynamic properties.

4 Conclusions

In this study, we proposed a novel ternary Al-ZnO/ɣ-Fe$_2$O$_3$ nanocomposite for the fabrication of high performance acetone sensors. Fe-doping induces some changes in the morphological and microstructural properties of base Al-ZnO binary composite. These modifications have been correlated with the sensing performances obtained in the monitoring of acetone with the ternary nanocomposite-based sensors. These sensors showed better sensing characteristics such as fast response times and sensitivity. In summary, results demonstrated that Al-ZnO/ɣ-Fe$_2$O$_3$ nanocomposites synthesized by the sol-gel route are highly promising sensing materials for acetone detection.

References

1. Hjiri, M., Zahmouli, N., Dhahri, R., Leonardi, S.G., El Mir, L., Neri, G.: Doped-ZnO nanoparticles for selective gas sensors. J. Mater. Sci. Mater. Electron. **28**, 9667–9674 (2017)
2. Jing, Z.: Ethanol sensing properties of ɣ-Fe$_2$O$_3$ nanopowder doped with Mg by nonaqueous medium method. Mater. Sci. Eng. B **133**, 213–217 (2006)
3. Jing, Z.: Synthesis, characterization and gas sensing properties of undoped and Zn-doped-ɣ-Fe$_2$O$_3$-based gas sensors. Mater. Sci. Eng. A **441**, 176–180 (2006)
4. Hjiri, M., El Mir, L., Leonardi, S.G., Pistone, A., Mavilia, L., Neri, G.: Al-doped ZnO for highly sensitive CO gas sensors. Sens. Actuators B **196**, 413–420 (2014)

Electrochemical Sensor Based on Molybdenum Oxide Nanoparticles for Detection of Dopamine

S. Spadaro, Enza Fazio, Martina Bonsignore, N. Lavanya, C. Sekar, S. G. Leonardi, F. Neri and G. Neri

Abstract Water nanocolloids of molybdenum oxide were synthesized by using a laser writing of a solid molybdenum target by a focused picosecond pulsed laser beam. The molybdenum oxide nanoparticles are then used to fabricate modified screen-printed carbon paste electrode. Morphology and compositional-structural properties of the samples were investigated by Scanning Transmission Electron Microscopy and X-ray diffraction spectroscopy. The sensors tested show enhanced electro-catalytic behavior for dopamine detection (also in presence of KCl, NaCl, glucose, uric acid, ascorbic acid and folic acid), in phosphate buffered saline (pH = 7). Under the optimal conditions, the peak current of dopamine increases linearly with the concentration in the 10–500 μM range, with the lowest detection limit of 43 nM. All these data indicate an excellent selectivity of this type of sensor towards main interferents, made it as a potential candidate for the detection of dopamine in pharmaceutical and clinical preparations.

Keywords Molybdenum oxide nanoparticles · Pulsed laser ablation
Electrochemical sensors · Dopamine

1 Introduction

Nanotechnology involves the creation and manipulation of materials at the nanoscale to obtain innovative products able to exhibit unique properties. In particular, new type of analytical tools for biotechnology and medical field involving the use of metal

S. Spadaro · E. Fazio (✉) · M. Bonsignore · F. Neri
Department of Mathematical and Computational Sciences, Physical Science
and Earth Science, Messina University, 98166 Messina, Italy
e-mail: enfazio@unime.it

N. Lavanya · C. Sekar
Department of Bioelectronics and Biosensors, Alagappa University, Karaikudi 630003, India

S. G. Leonardi · G. Neri
Department of Engineering, Messina University, 98166 Messina, Italy

© Springer Nature Switzerland AG 2019
B. Andò et al. (eds.), *Sensors*, Lecture Notes in Electrical Engineering 539,
https://doi.org/10.1007/978-3-030-04324-7_5

oxide nanomaterials, have received huge attention in recent times [1]. For these applications, the choice of green techniques is much appreciated to grant the production of contaminant free nanoparticles. Over the last decade, pulsed laser ablation in liquid (PLAL) is gradually becoming an irreplaceable technique to synthesize metal oxide nanostructures which represent a new type of analytical tools for biotechnology and life science [2]. This green technique ensures the control of physical-chemical properties of the synthesized nanostructures by changing ablation parameters and the absence of by-products, very useful issues in biological applications.

Dopamine (DA) is a neurotransmitter located in the ventral tegmental area of the midbrain, the substantia nigra pars compacta, and the arcuate nucleus of the hypothalamus of the human brain. Its detection is important in understanding neural behavior and in developing therapeutic intervention technologies for neurological disorders. Monitoring of extracellular DA concentration can serve as a clinically relevant biomarker for specific diseases states as well as a gateway to monitor treatment efficacy [3]. Among other analytical techniques used in biomedical diagnostics, electrochemical sensors show some advantages such as low cost, ease of operation and fast response with high accuracy and sensitivity. In Parkinson's disease for example, electrochemical sensors can be integrated with therapeutic interventions such as deep brain stimulation systems, to enhance the ability to continuously monitor DA and other neurotransmitters that are prone to fluctuations. Despite the advantages, there are some drawbacks limiting the widespread use of DA electrochemical sensors, such as the overlap of voltammetric responses due to several interfering substances present in biological systems and the formation of a passivating polymeric film on the electrode surface [4]. Recently, modified electrodes with assembled monolayers, polymer or metal oxides/graphene composites were developed to overcome the above limitations [5]. In this contest, the nanostructured molybdenum oxide particles could act as efficient electron redox mediators. Being characterized by three oxidation states and a high chemical stability, they can readily participate in redox reactions, contributing to the electrochemical sensing of some biological compounds such as DA. Their application in bio-sensing and specifically for DA sensing is still limited.

In this work, molybdenum oxide nanocolloids different in shape and size distribution as well as in surface chemical bonding coordinations were analyzed. Their electrochemical response was tested. Then, a systematic study was made on the sample with the best performance, ultimately proposing an electrochemical sensor based on a MoO_2 nanocomposite synthesized at room temperature (RT) in water using a picosecond pulsed laser source. The electrochemical sensor proposed enables the determination of DA level without any "engineered" electron transfer mediator, being the molybdenum oxide nanoparticles directly deposited on commercial carbon electrodes. The potentiality lies in the intrinsic properties of the molybdenum oxide nanostructures such as the high surface to volume ratio and the participation to surface oxidation processes.

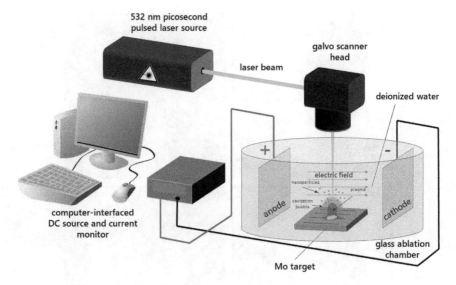

Fig. 1 Scheme of external field-assisted laser ablation setup

2 Materials and Methods

High purity (99.9%) Mo solid target in a 20 mL of deionized water were ablated using the 532 nm line of a laser source, operating at 100 kHz repetition rate with a pulse width of 6–8 ps. The target was irradiated with a typical laser power of 2.5 W and an irradiation time of 30 min. The laser beam was focused to a spot of about 75 μm in diameter on the surface of the target with a galvanometric scanner having a telecentric objective with a focal length of 163 mm. During the ablation process two Pt electrodes were immersed in the liquid in such a way to face their basal planes (see Fig. 1). The two electrodes were polarized by applying a DC potential (20 V/cm) and, between them, the plume was left to develop during the ablation. The experiments were performed at different water temperature (RT and 80 °C) and its values were monitored both before and after the ablation. During the overall process, only a moderate increase was found (less than 10 °C in 30 min), compatible with the thermal energy release due to the ablation phenomena.

Sample morphology was observed through a scanning transmission electron microscope (STEM) operating at 30 kV while chemical environment of the atomic species was analyzed by the X-ray diffraction (XRD) techniques. XRD measurements were performed using the CuK-α radiation ($\lambda = 1.5406$ Å) of a Bruker-AXS D5005 diffractometer. The surface Mo chemical bonding fractions was measured by means of X-ray photoelectron spectroscopy (XPS), using a Thermo Scientific spectrometer equipped with a conventional Al-Kα X-ray source (1486.6 eV) and a concentric hemispherical analyzer. The electrochemical sensors were fabricated depositing some drops (10 μL) of the nanocolloids onto commercial screen-printed

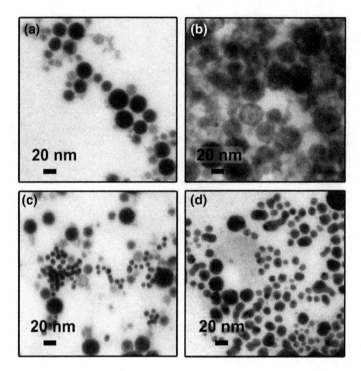

Fig. 2 STEM images of the samples prepared at RT (**a, b**) and 80 °C (**c, d**) without and applying the electric voltage, respectively

electrodes consisting of carbon working, carbon counter and silver pseudo-reference electrodes. The so prepared samples were dried at RT to obtain the MoO/SPCE (screen printed carbon paste electrodes). Cyclic voltammetry (CV), square wave voltammetry (SWV) and amperometric measurements were carried out in pH 7 Phosphate Buffered Solution (PBS) in presence and in absence of DA.

3   Results and Discussion

Figure 2 shows electron microscopy images of the colloids prepared in water at RT and T = 80 °C, without and by applying an external DC electric voltage.

We observe that some molybdenum oxide nanoparticles have a mean size of 20 nm when the ablation process was carried out at RT (Fig. 2a, b). However, overlapped spherical nanoparticles are obtained by applying the electric field (Fig. 2b). Nevertheless, smaller nanoparticles (size less than 10 nm) characterize colloids obtained at 80 °C with respect to the ones obtained at RT (Fig. 2c) while nearly oblong nanos-

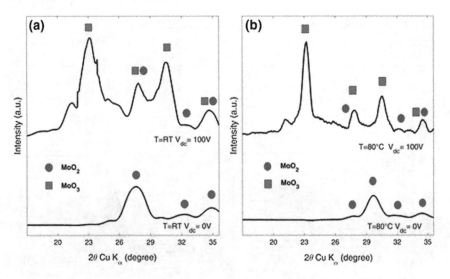

Fig. 3 XRD spectra of the samples prepared at RT (**a**) and 80 °C (**b**) without or applying the electric voltage. All spectra are shifted to highlight spectral differences

tructures mixed with spherical shaped ones are obtained at 80 °C and applying the DC electric field (Fig. 2d).

Figure 3 shows XRD spectra with the proper reflections assignments of the samples prepared at RT and 80 °C, by applying, or not, the electric voltage during the ablation process. XRD patterns show mixed phases which can be attributed to MoO_2 and MoO_3 coordinations. Moreover, the reflections (110), (040), (021), (111), (002) and (112) observed at about 23°, 28° and 30° match with the orthorhombic structure of MoO_3 (International Centre for Diffraction Data, JCPDS file no. 05-0508). In absence of the electric voltage, the samples show only the MoO_2 coordination. On the other hand, in presence of the electric field, the more crystalline MoO_3 phase dominates as indicated by the nearly lack of the reflections due to MoO_2 phase and the narrowing of the main characteristic contributions of MoO_3.

The surface chemical bonding nature of the samples was investigated by studying the modifications, induced by the synthesis parameters, on the Mo 3d X-ray photoemission peak lineshape. It is quite well established that the hot water and the electric voltage induce an increase of the MoO_3 bonding structure while, in RT & 0 V conditions, the sub-stoichiometric coordinations, mainly the MoO_2 configuration, were found (see Table 1).

Figure 4a shows CV responses of 50 μM DA in pH 7 PBS at the bare SPCE and molybdenum oxide modified SPCE. We outline that the modified electrode did not show any response in the absence of DA (not shown). In the CVs, both cathodic and anodic peaks of the modified electrode shifted negatively, relative to those of the screen-printed carbon electrode. The oxidation and reduction currents of DA on all the investigated electrodes were comparable. However, using the molybdenum

Table 1 Mo bonding fractions percentage estimated from XPS fitting procedure

Synthesis parameters	Mo (%)	MoO₂ (%)	MoO₃₋ₓ (0 < x < 1) (%)	MoO₃ (%)
RT & 0 V	0.4	68.7	22.7	8.2
80 °C & 0 V	0.3	60.2	25.5	14.0
RT & 100 V	0.0	58.4	26.2	15.4
80 °C & 100 V	0.0	57.9	28.9	13.2

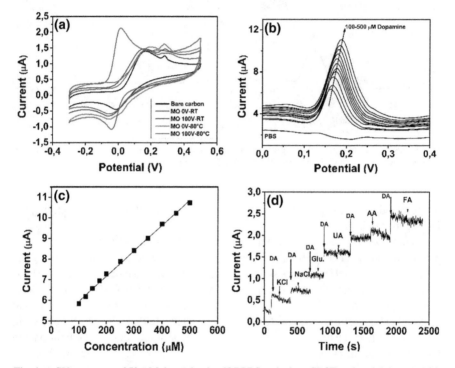

Fig. 4 **a** CV responses of 50 μM dopamine in pH 7 PBS at the bare SPCE and molybdenum oxide modified screen printed carbon paste electrodes; **b** SWVs for different concentration of dopamine; **c** Calibration curve in the range of concentration 100–500 μM of dopamine; **d** Selectivity test

oxide nanocolloids prepared at RT in absence of the electric field, the redox reaction shifted at more negative potentials and a well-defined redox peak corresponding to DA with improved peak current (2.2 μA) is achieved.

This sample shows the biggest nanoparticles mean size (see Fig. 2), i.e. a low surface to volume ratio, that cannot explain the observed CV behaviour. It is more likely that the electrochemical response and selectivity toward DA is affected by the different Mo-O bonding configurations (MoO₂ or MoO₃ phases) as observed by XRD and XPS analyses. It is well known [6] that the oxygen deficient, i.e. sub-stoichiometric MoO₃₋ₓ films, contain excess metal atoms which act as doping centers.

Generally, stoichiometric MoO_3 is an insulator with a band gap of about 3.1 eV. When the oxide is reduced, by the introduction of impurity donor atoms or by oxygen deficiency, electronic donor levels are created near the bottom of the conduction band and, therefore, the reduced oxides behave as semiconductors. These "doping centers" control the optical and the electrical field response of the nanomaterials by affecting the electronic mobility. As reported in a previous paper [7], our samples prepared at 80 °C have an optical bandgap of approximately 3.0 eV, while the RT prepared colloids exhibit smaller values due to the presence of the occupied gap states within their forbidden gap. Hence, from all the collected data, it seems that the nanoparticles made of MoO_2 configuration might favour the electrons transfer kinetic, then the electrocatalytic reaction of the DA on the electrode [8].

Owing to the best performance obtained for the sample prepared at RT in absence of electrical field, its electrochemical behavior toward detection of DA has been further investigated. Figure 4b shows the SWV recorded for the fabricated sensor in absence and in presence of different concentrations of dopamine ranging from 100 μM to 500 μM in PBS. When DA was present in solution an intense anodic peak was observed which shifted with increasing of concentration. In addition, in the whole investigated range the anodic peak current (i_a) increased linearly with the concentration of DA (c) according to the equation i_a (μA) $= 0\ 0.012\ c$ (μM) $+ 4.7$, $R^2 = 0.996$ (Fig. 4c). The selectivity of the sensor against common interfering species was also investigated by amperometric experiments. Figure 4d shows the chronoamperometric curve recorded at an applied potential of 0.1 V during step increases of 50 μM of dopamine and in the presence of 10 fold excess of K^+, Na^+, glucose, uric acid (UA), ascorbic acid (AA) and folic acid (FA), demonstrating that the molybdenum oxide modified-electrode is also highly selective towards the determination of dopamine in the presence of potential interferents.

4 Summary

Molybdenum oxide NPs with tunable surface physical-chemical properties were synthesized by a picosecond pulsed laser source and then used to fabricate modified screen-printed carbon paste electrodes. A combination of XPS, XRD and STEM analyses has shown that the synthesis parameters influence the nanoparticles size distribution and the Mo-O bonding configurations which, in turn, affects the dopamine electrochemical response of the colloids. The best performance, also in terms of selectivity, was exhibited by the sample prepared at RT in absence of electric voltage. This sample shows nanoparticles with the biggest mean size, i.e. a low surface to volume ratio, made of MoO_2 configuration. On the overall, from the collected data, it emerges that nanoparticles with the MoO_2 surface bonding configuration favour the electrons transfer kinetic and, ultimately, the electrocatalytic reaction of the dopamine on the electrode.

References

1. Tuli, H.S., Kashyap, D., Bedi, S.K., Kumar, P., Kumar, G., Sandhu, S.S.: Molecular aspects of metal oxide nanoparticles (MO-Nps) mediated pharmacological effects. Life Sci. **143**, 71–73 (2015)
2. Braydich-Stolle, L., Hussain, S., Schlager, J.J., Hofmann, M.-C.: In vitro cytotoxicity of nanoparticles in mammalian germline stem cells, toxicological. Sciences **88**(2), 412 (2005)
3. Xu, H., Zuo, P., Wang, S., Zhou, L., Sun, X., Hu, M., Liu, B., Wu, Q., Dou, H., Liu, B., Zhu, F., Teng, S., Zhang, X., Wang, L., Li, Q., Jin, M., Kang, X., Xiong, W., Wang, C., Zhou, Z.: Striatal dopamine release in a schizophrenia mouse model measured by electrochemical amperometry in vivo. Analyst **140**(11), 3840–3845 (2015)
4. Jackowska, K., Krysinski, P.: New trends in the electrochemical sensing of dopamine. Anal. Bioanal. Chem. **405**, 3753–3771 (2013)
5. Li, Y., Liu, J., Liu, M., Yu, F., Zhang, L., Tang, H., Ye, B.C., Lai, L.: Fabrication of ultra-sensitive and selective dopamine electrochemical sensor based on molecularly imprinted polymer modified graphene@carbon nanotube foam. Electrochem. Commun. **64**, 42–45 (2016)
6. Inzani, K., Nematollahi, M., Vullum-Bruer, F., Grande, T., Reenaas, T.W., Selbach, S.M.: Electronic properties of reduced molybdenum oxides. Phys. Chem. Chem. Phys. **19**, 9232–9245 (2017)
7. Spadaro, S., Bonsignore, M., Fazio, E., Cimino, F., Speciale, A., Trombetta, D., Barreca, F., Saija, A., Neri, F.: Molybdenum oxide nanocolloids prepared by an external field-assisted laser ablation in water. In: EPJ Web of Conferences, vol. 167, p. 04009 (2018)
8. Gao, F., Cai, X., Wang, X., Gao, C., Liu, S., Gao, F., Wang, Q.: Highly sensitive and selective detection of dopamine in the presence of ascorbic acid at graphene oxide modified electrode. Sens. Actuators B Chem. **186**, 380–387 (2013)

Sensing Properties of Indium, Tin and Zinc Oxides for Hexanal Detection

A. Malara⊙, L. Bonaccorsi ⊙, A. Donato⊙, P. Frontera⊙, A. Piscopo⊙, M. Poiana⊙, S. G. Leonardi⊙ and G. Neri⊙

Abstract The properties of ZnO, In_2O_3, and SnO_2 have been investigated as possible sensing layer in resistive sensors for monitoring hexanal in food applications. Sensors performances were tested at different temperatures (100–350 °C) and analyte concentrations (50–100 ppm). Results showed the ability of the analyzed metal oxides sensors, each one characterized by its own features and operating conditions, to detect hexanal. Moreover, the different oxides response has been also related to their Gibbs free energy of formation. According to preliminary results both indium and zinc oxides show promising sensing characteristics compared to tin oxide.

Keywords Hexanal · Metal oxides · Resistive sensors

1 Introduction

Hexanal is one of the major volatile compound released during food storage due to lipids oxidation and is considered an important indicator of food quality in packaging [1–3]. In addition, hexanal has been proposed as an odor reference standard for sensory analysis of drinking water [4]. Despite its importance as a quality marker in the food industry, hexanal has not been extensively studied yet. Tin oxide-based resistive sensors have been previously proposed for hexanal detection [5]. According to the best authors' knowledge, indium and zinc oxides have instead not been tested so far. Therefore, we started a study with the aim to investigate the sensing properties of In_2O_3 and ZnO as possible sensing layer in resistive sensors for monitoring hexanal in food applications and compare their performances with the most investigated SnO_2-based sensor.

A. Malara (✉) · L. Bonaccorsi · A. Donato · P. Frontera
Dipartimento DICEAM, Università Mediterranea, Loc. Feo di Vito, 89060 Reggio Calabria, Italy
e-mail: angela.malara@unirc.it

A. Piscopo · M. Poiana
Dipartimento di Agraria, Università Mediterranea, Loc. Feo di Vito, 89060 Reggio Calabria, Italy

S. G. Leonardi · G. Neri
Dipartimento di Ingegneria, Università di Messina, C.da Di Dio, 98166 Messina, Italy

© Springer Nature Switzerland AG 2019
B. Andò et al. (eds.), *Sensors*, Lecture Notes in Electrical Engineering 539,
https://doi.org/10.1007/978-3-030-04324-7_6

39

2 Experimental

2.1 Samples Preparation and Characterization

Metal oxide powders have been prepared by precipitation from aqueous solution of nitrate precursors (0.68 M) hydrolyzed with an aqueous potassium carbonate solution (1 M). The precipitates were then filtered, washed with deionized water, dried at 110 °C for 12 h and then calcinated at 500 °C for 2 h in air.

Powder samples were characterized by XRD analysis (Bruker, D2 Phaser) in the 2θ range $10°$–$80°$ (Cu $K_{\alpha1} = 1.54056$ Å) and their morphology studied by Scanning Electron Microscopy SEM (Phenom ProX). The Brunauer–Emmett–Teller (B.E.T.) surface areas of the prepared powders were determined from nitrogen adsorption–desorption isotherms at 77 K (ChemiSorb 2750 Micromeritics).

2.2 Sensor Preparation and Testing

The procedure to prepare a metal oxide sensor was as follows: a paste was obtained by mixing the oxide powder with a proper quantity of ethanol and deposited on an alumina planar substrate (3 mm × 6 mm) supplied with interdigitated Pt electrodes and a heating element on the back side. Before sensing tests, the sensor was conditioned in air for 2 h at 400 °C in order to stabilize the deposited film.

Measurements were performed positioning the sensor in a stainless steel testing cell and flowing a mixture of dry air and hexanal vapor at different concentrations for a total gas stream of 100 sccm. The hexanal vapor was obtained by bubbling dry air in liquid hexanal maintained at a controlled temperature by a refrigerated circulating bath (temperature range $-5/-15 \pm 0.01$ °C). All air fluxes were measured by Brooks mass flow controller systems. The sensors resistance data were collected in the four-point mode by an Agilent 34970A multimeter while a dual-channel power supplier instrument (Agilent E3632A) allowed to control the sensor temperature.

Sensor response S to hexanal was defined by the sensor resistance ratio:

$$S = \frac{R_{(air)}}{R_{(air+hexanal)}}$$

Measurements were performed varying sensors temperature in the range of 100–350 °C and for three hexanal concentrations: 50, 150 and 300 ppm.

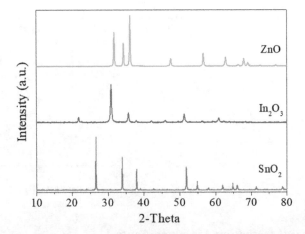

Fig. 1 XRD diffractograms of the synthesized ZnO, In$_2$O$_3$ and SnO$_2$ powders after calcination

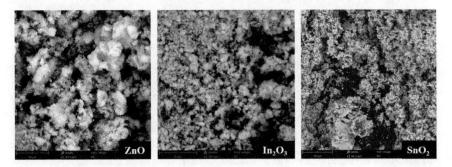

Fig. 2 SEM images of the synthesized powders after calcination

3   Results and Discussion

3.1   Metal Oxides Characterization

The metal oxides crystallinity was measured by XRD after calcination (T $=500$ °C). All synthesized powders resulted with no impurities or amorphous phases, as shown in diffractograms of Fig. 1.

The morphological analysis carried out by scanning electron microscopy demonstrated the formation of very small crystalline particles in all cases, as consequence of the synthesis method used (Fig. 2). The B.E.T. surface area, however, showed some differences among samples. Indium oxide, indeed, was the powder with the highest specific surface area equal to 20.3 m^2/g, followed by zinc oxide with a specific surface area of 9.6 m^2/g and tin oxide with the lowest value of 8.4 m^2/g.

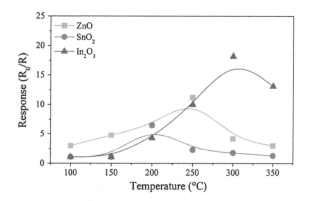

Fig. 3 Sensor responses versus temperatures at 50 ppm hexanal

3.2 Hexanal Detection Measurements

In this work, a preliminary study on the sensing properties of three metal oxides largely used in resistive sensors, ZnO, In_2O_3 and SnO_2, have been carried out to evaluate the best response to the hexanal molecule under controlled conditions. Zinc oxide, in particular, has not been tested for hexanal detection so far while SnO_2 is the most studied [5].

In Fig. 3 the responses of the tested sensors are compared at different temperatures for a hexanal concentration of 50 ppm in air.

The results demonstrate that both In_2O_3 and ZnO can detect hexanal in air with higher response than tin oxide. Some differences, however, exist indeed indium oxide has given the highest response at 300 °C, while ZnO can detect hexanal also at lower temperature, as low as 100 °C, with appreciable response.

The hexanal sensing mechanism is based on the aldehyde oxidation on the metal oxides surface due to the adsorbed oxygen and the consequent electrons release:

$$C_6H_{12}O + O_{(ads)}^- \rightarrow C_5H_{11}COOH + e^-$$

The different oxides response has been related to their Gibbs free energy of formation, ΔG, calculated at the different sensing temperatures according to the equation [6]:

$$\Delta G(T_f) = \frac{T_f}{T_0} G^0(T_0) - T_f(T_f - T_0) \frac{\Delta H^0}{T_f^2}$$

being G^0 and ΔH^0 (enthalpy of formation) at the standard temperature $T_0 = 298.15$ K. In Fig. 4 the Gibbs energy of the three metal oxides is plotted against temperature. For all metal oxides, the Gibbs energy shows the same trend, being negative from room temperature up to $T \approx 350$ °C, when the metal oxides become less stable.

Fig. 4 Metal oxides Gibbs free energy of formation versus temperature

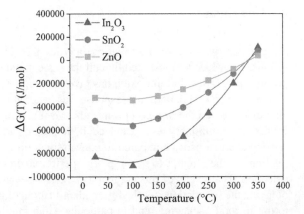

Fig. 5 Sensor responses versus hexanal concentration at the temperature of maximum sensitivity

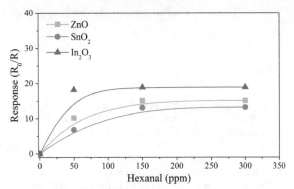

Indium oxide shows the highest stability while ZnO the lowest (Fig. 4). The difference in among Gibbs energy of the three metal oxides is more evident at low temperatures and decreases approaching 350 °C. Zinc oxide, indeed, can be considered "less stable" compared to In_2O_3 and SnO_2 and was the sensor able to detect hexanal in air at 100 °C (see Fig. 3). On the opposite, indium oxide, which is the more stable oxide, showed the highest sensitivity but at 300 °C.

In Fig. 5, the sensor response versus concentration is shown for all sensors investigated and operating at the temperature of maximum sensitivity. It can be observed that all the analyzed samples exhibit a similar trend in the range of concentration investigated. In particular, sensors are highly sensitive at low hexanal concentration but saturate at higher concentration. Sensing results are also in agreement with specific surface area values. Indeed, the more sensitive sensor, In_2O_3, is characterized by a high specific surface area, whereas SnO_2 and ZnO sensing behavior reflects a low specific surface area value.

4 Conclusions

The comparative study on the sensing properties of In_2O_3, ZnO and SnO_2 in detecting hexanal in air has demonstrated that all the three metal oxides gave a response with peculiar properties. Indium oxide has given the highest response but at the highest temperature (300 °C), tin oxide response was better at low temperature (200 °C) and zinc oxide sowed the best compromise between sensing temperature and response. From these preliminary results, indeed, both indium and zinc oxides show promising sensing characteristics compared to conventional tin oxide and merit further investigations in the attempt to develop high performance resistive sensors for hexanal detection. The results of this comparative study are particularly important in terms of laying the foundations for the future identification of the better metal oxide sensor to use in food packaging, and in particular as an intelligent packaging, that could monitor the quality variation of different food products during their shelf life.

References

1. Spadafora, N.D., Amaro, A.L., Pereira, M.J., Müller, M.J., Pintado, M., Rogers, H.J.: Multi-trait analysis of post-harvest storage in rocket salad (Diplotaxis tenuifolia) links sensorial, volatile and nutritional data. Food Chem. **211**, 114–123 (2016)
2. Jayasena, D.D., Ahn, D.U., Nam, K.C., Jo, C.: Flavour chemistry of chicken meat: a review. Asian-Australas. J. Anim. Sci. **26**, 732–742 (2013)
3. Schindler, S., Krings, U., Berger, R.G., Orlien, V.: Aroma development in high pressure treated beef and chicken meat compared to raw and heat treated. Meat Sci. **86**, 317–323 (2010)
4. Abd Wahab, N.Z., Nainggolan, I., Nasution, T.I., Derman, M.N., Shantini, D.: Highly response and sensitivity chitosan-polyvinyl alcohol based hexanal sensors. MATEC Web Conf. 1072 (2016)
5. Huang, K., Zhu, C., Yuan, F., Xie, C., J.: Nanoscale SnO_2 flat-type coplanar hexanal gas sensor arrays at ppb level. Nanosci. Nanotechnol. **13**, 4370–4374 (2013)
6. Korotcenkov, G.: Metal oxides for solid-state gas sensors: what determines our choice? Mater. Sci. Eng. B **139**, 1–23 (2007)

On-Glass Integration of Thin Film Devices for Monitoring of Cell Bioluminescence

D. Caputo, N. Lovecchio, M. Nardecchia, L. Cevenini, E. Michelini, M. Mirasoli, A. Roda, A. Buzzin, F. Costantini, A. Nascetti and G. de Cesare

Abstract This paper reports the development of a miniaturized lab-on-glass, suitable for the on-chip detection of living cell bioluminescence and their on-chip thermal treatments. The glass substrate hosts, on one side, hydrogenated amorphous silicon diodes, working as both temperature sensors and photosensors, and, on the other side, transparent thin films acting as heating sources. The main challenge of the work is the determination of the correct fabrication recipes in order to satisfy the compatibility of different microelectronic steps. The measured uniformity of temperature distribution, sensitivity of the temperature sensors, reverse dark current and spectral response of the photosensors demonstrate the successful technological integration and the suitability of the developed lab-on-glass to control the cell temperature and detect the BL emission with high sensitivity.

Keywords Lab-on-chip · Amorphous silicon · Bioluminescence · Cells
Indium tin oxide · Photosensor · Thin film heater

1 Introduction

The development of cost-effective, on-site analytical methods is an urgent need for diagnostics, monitoring food safety and detecting environmental pollution. Many laboratory procedures can be simplified through lab-on-chip (LoC) systems

D. Caputo (✉) · N. Lovecchio · M. Nardecchia · A. Buzzin · F. Costantini · G. de Cesare
Department of Information Engineering, Electronics and Telecommunications,
Sapienza University of Rome, Rome, Italy
e-mail: domenico.caputo@uniroma1.it

M. Nardecchia · F. Costantini · A. Nascetti
School of Aerospace Engineering, Sapienza University of Rome, Rome, Italy

L. Cevenini · E. Michelini · M. Mirasoli · A. Roda
Department of Chemistry G. Ciamician, Alma Mater Studiorum - University of Bologna,
Bologna, Italy

© Springer Nature Switzerland AG 2019
B. Andò et al. (Eds.) *Sensors*, Lecture Notes in Electrical Engineering 539,
https://doi.org/10.1007/978-3-030-04324-7_7

45

exploiting microfabricated sensors and heaters with microfluidics [1]. Moreover, to obtain highly valuable information, especially related to bioactivity and toxicology, living cells can be used as sensing elements and integrated in LoC. Indeed whole-cell biosensors have shown the potential to complement both laboratory-based and on-field analytical methods for the detection of general stress conditions, cyto- and genotoxic compounds, organic xenobiotics and metals [2–4].

Additionally, bioluminescence (BL), thanks to its peculiar features can be easily implemented in LoC as detection technique [5, 6]. BL reactions are characterized by high quantum yield emission and low background which results in high detectability and sensitivity of BL-based assays. In addition, BL does not require an external excitation light source and therefore it is an ideal detection principle for miniaturized and low weight systems.

The recent availability of new BL probes that emit at different wavelengths, such as new reporter genes or luciferase mutants, could also enable multiplexing detection relying on the spatial and spectral resolution of BL signals [7].

In this framework, we integrate on a single glass substrate different thin film technologies [8] in order to develop a lab-on-chip system suitable for on-chip thermal treatments of living cells and simultaneous on-chip detection [9] of their bioluminescence emission [10]. The main challenge of the work is the determination of the correct fabrication recipes of several electronic devices requiring different technological specifications.

2 System Structure and Operation

The developed system integrates on a single glass substrate different thin film technologies in order to obtain a multifunctional platform suitable for on-chip thermal treatments and on-chip monitoring of cell bioluminescence.

In particular, a 5×5 cm^2 glass substrate includes:

- transparent thin film resistors acting as heating sources, in order to provide the required temperature to the cells;
- hydrogenated amorphous silicon (a-Si:H) diodes, acting as temperature sensors, to monitor the temperature distribution on the active area of the heater;
- a-Si:H diodes, acting as photosensors, to detect the bioluminescence provided by the analytes during the biomolecular recognition.

Figure 1 depicts the integrated thin film devices and their combination with the biological sample, which is contained in a Polydimethylsiloxane (PDMS) microfluidic network. The presented structure reduces the distance between the detector and the bioluminescence source, minimizing the optical losses and system dimensions.

In order to perform the thermal treatment of the cells, an Indium Tin Oxide (ITO) layer is driven by a custom-made electronics, applying a voltage that increases the sample temperature by conduction heat transfer. The temperature is stabilized at the set-point (37 °C) using a software Proportional-Integral-Derivative algorithm, which

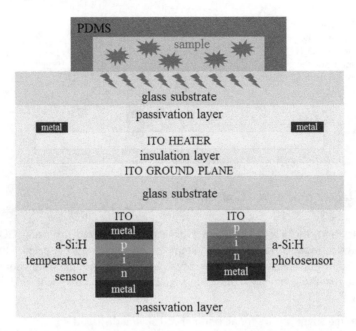

Fig. 1 Cross section of the LoC: the bottom glass substrate hosts the transparent heater on the top side and the a-Si:H sensors on the bottom side, while the top substrate hosts the PDMS microfluidic network

takes, as input, the temperature inferred by a-Si:H diode [11]. This is possible since the voltage across the diode varies linearly with the temperature when it is biased with a constant current. In order to correctly infer the temperature, a metal layer covering the ITO contact of the temperature sensors is used to shield the light radiations (see Fig. 1).

During the thermal control, the detection of the BL signal emitted by the cells is performed by the amorphous silicon photosensors, thanks to the transparency of the thin film heater that is crucial for the correct transmission of the emitted light.

The preliminary characterization of a first prototype highlighted a crosstalk between the heater and the a-Si:H sensors. In particular, it has been found that the biasing of the heater affects the current of the thin film sensors, because the leakage current through the glass resulting from the voltage applied to the heaters interferes with the current flowing through the sensors. A ITO transparent ground plane has then been inserted between the thin film heaters and the glass to ensure an electrical path of this current to ground. Furthermore, an insulation layer has been deposited on the ground plane before the heater fabrication.

Fig. 2 Top view of the
heater geometry. The yellow
region represents the ITO
active layer, while the grey
regions are the metal
electrical contacts

2.1 Heater and a-Si:H Sensor Design

Geometry and thickness of the heater have been optimized, by using COMSOL
Multiphysics, to satisfy requirements of temperature uniformity of treated active area,
cell spatial displacement and transparency to light. Basing on the simulation results,
we found that the geometry reported in Fig. 2 satisfies the biological specifications
because:

1. the round shape allows for an optimal cell spatial distribution;
2. the temperature distribution has a standard deviation equal to ± 0.3 °C.

The material of the active area (yellow region in Fig. 2) is a 250 nm-thick ITO,
which has a transmittance around 90% at wavelengths above 400 nm. Each heater
has a circular shape with an active area of $0.18 \, \text{cm}^2$. The electrical pads (gray regions
in Fig. 2) are a 30/150/30 nm-thick Cr/Al/Cr stacked structure.

The a-Si:H diodes are p-i-n stacked layers, whose thickness and energy gap have
been designed [12] to match the photosensor responsivity with the BL emission
spectrum, which is about 460 nm, and minimizing the reverse dark current.

3 Fabrication and Characterization

Heaters and thin film sensors have been fabricated using standard thin film microelectronic technologies, and a careful tuning of sequence and parameters of the
technological steps has been carried out to achieve the compatibility of the different
processes. In particular, the a-Si:H diodes have been deposited by Plasma Enhanced
Chemical Vapor Deposition in a three-chamber high vacuum system (GSI, Denver,
CO, USA). The ITO film has been instead deposited by magnetron sputtering (MRC,
Orangebourg, New York, NY, USA).

The fabrication steps of the whole structure are the following:

- Heater layer:

1. deposition by magnetron sputtering of a 20 nm-thick ITO layer, which acts as
   transparent ground plane for the heaters;

2. deposition by spin coating of a 5 μm-thick SU-8 3005 (from MicroChem, MA, USA) insulation layer;
3. deposition by magnetron sputtering of a 250 nm-thick ITO layer and its patterning by photolithography and dry etching processes for the fabrication of the heater central part;
4. vacuum evaporation of 30/150/30 nm-thick Cr/Al/Cr stacked layer acting as electrical pads of the heaters and its patterning by photolithography and wet etching processes;
5. deposition by spin coating of a 5 μm-thick SU-8 3005 passivation layer;

- Sensors layer:

1. deposition by magnetron sputtering of a 100 nm-thick ITO layer, which acts as transparent bottom contact of the diodes;
2. patterning of the ITO layer by photolithography and wet etching processes;
3. deposition by Plasma Enhanced Chemical Vapor Deposition (PECVD) of the a-Si:H 10/150/30 nm-thick p-type/intrinsic/n-type stacked structure;
4. deposition by vacuum evaporation of a 50 nm-thick chromium layer, which acts as top contact of the sensors;
5. wet etching of the chromium and dry etching of the a-Si:H layers for the mesa patterning of the diodes;
6. deposition by spin coating of a 5 μm-thick SU-8 3005 (from MicroChem, MA, USA) passivation layer and its pattering for opening via holes over the diodes;
7. deposition by magnetron sputtering of a 150 nm-thick titanium/tungsten alloy layer and its patterning for the definition of the top contacts and of the connection to the pad contacts;
8. deposition by spin coating of a 5 μm-thick SU-8 3005 passivation layer.

Figure 3 reports a picture of the fabricated chip.

The heaters, characterized using a thermo-camera FLIR A325 and a custom made electronic board, show a very good temperature uniformity according to the numerical analysis (Fig. 4). The a-Si:H photosensors show dark current below 10^{-10} A/cm^2 (which leads to a dark current noise of few fA) and responsivity around to 200 mA/W in the BL emission spectrum range. The temperature sensors present a sensitivity of 3.2 mV/°C when biased at a forward constant current of 8 μA/cm^2. The performances of the a-Si:H sensors are comparable with those achievable with the state-of-the-art crystalline silicon sensors.

From these results, we infer that the integrated devices fulfill the requirements for a wide range of cell-based BL assays.

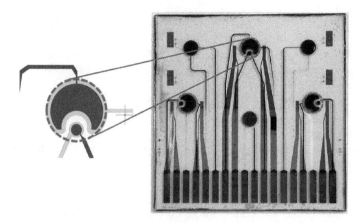

Fig. 3 Picture of the fabricated LoC, hosting heaters on the top glass side and sensors on the bottom side. The expanded area highlights the bottom glass moon-like shaped photosensors and circle-shaped temperature sensors

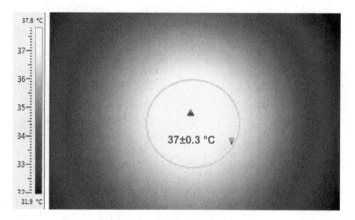

Fig. 4 Temperature distribution measured using a FLIR A325 thermocamera. The thermal energy is provided by the thin film heater driven by a custom electronic board. The green circle specifies the area of the photosensor aligned with the heater

4 Conclusions

In this work, an integrated lab-on-chip system designed to monitor and control the activity of living cells has been presented. The developed system-on-glass is based on thin film microelectronic technologies and integrates, on the same glass substrate, thin film heaters and amorphous silicon sensors in order to achieve a compact system to control the cell temperature and detect the BL emission with high sensitivity.

References

1. Mark, D., Haeberle, S., Roth, G., Von Stetten, F., Zengerle, R.: Microfluidic lab-on-a-chip platforms: requirements, characteristics and applications. In: Microfluidics Based Microsystems, pp. 305–376. Springer, Dordrecht (2010). https://doi.org/10.1007/978-90-481-9029-4_17
2. Waggoner, P.S., Craighead, H.G.: Micro-and nanomechanical sensors for environmental, chemical, and biological detection. Lab on a Chip 7(10), 1238–1255 (2007). https://doi.org/10.1039/B707401H
3. Wongkaew, N., He, P., Kurth, V., Surareungchai, W., Baeumner, A.J.: Multi-channel PMMA microfluidic biosensor with integrated IDUAs for electrochemical detection. Anal. Bioanal. Chem. 405(18), 5965–5974 (2013). https://doi.org/10.1007/s00216-013-7020-0
4. Mirasoli, M., Nascetti, A., Caputo, D., Zangheri, M., Scipinotti, R., Cevenini, L., de Cesare, G., Roda, A.: Multiwell cartridge with integrated array of amorphous silicon photosensors for chemiluminescence detection: development, characterization and comparison with cooled-CCD luminograph. Anal. Bioanal. Chem. 406(23), 5645–5656 (2014). https://doi.org/10.1007/s00216-014-7971-9
5. Pires, N.M.M., Dong, T., Hanke, U., Hoivik, N.: Recent developments in optical detection technologies in lab-on-a-chip devices for biosensing applications. Sensors 14(8), 15458–15479 (2014). https://doi.org/10.3390/s140815458
6. Pires, N., Dong, T., Hanke, U., Hoivik, N.: Integrated optical microfluidic biosensor using a polycarbazole photodetector for point-of-care detection of hormonal compounds. J. Biomed. Optics 18(9), 097001 (2013). https://doi.org/10.1117/1.JBO.18.9.097001
7. Branchini, B.R., Southworth, T.L., Fontaine, D.M., Kohrt, D., Welcome, F.S., Florentine, C.M., Henricks, E.R., DeBartolo, D.B., Michelini, E., Cevenini, L., Roda, A., Grossel, M.J.: Red-emitting chimeric firefly luciferase for in vivo imaging in low ATP cellular environments. Anal. Biochem. 534, 36–39 (2017). https://doi.org/10.1016/j.ab.2017.07.001
8. Petrucci, G., Caputo, D., Lovecchio, N., Costantini, F., Legnini, I., Bozzoni, I., Nascetti, A., de Cesare, G.: Multifunctional system-on-glass for Lab-on-chip applications. Biosens. Bioelectron. 93, 315–321 (2017). https://doi.org/10.1016/j.bios.2016.08.060
9. Costantini, F., Sberna, C., Petrucci, G., Reverberi, M., Domenici, F., Fanelli, C., Manetti, C., de Cesare, A., Nascetti, A., DeRosa, M., Caputo, D.: Aptamer-based sandwich assay for on chip detection of Ochratoxin A by an array of amorphous silicon photosensors. Sens. Actuators B Chem. 230, 31–39 (2016). https://doi.org/10.1016/j.snb.2016.02.036
10. Mirasoli, M., Guardigli, M., Michelini, E., Roda, A.: Recent advancements in chemical luminescence-based lab-on-chip and microfluidic platforms for bioanalysis. J. Pharm. Biomed. Anal. 87, 36–52 (2017). https://doi.org/10.1016/j.jpba.2013.07.008
11. Lovecchio, N., Petrucci, G., Caputo, D., Alameddine, S., Carpentiero, M., Martini, L., Parisi, E., De Cesare, G., Nascetti, A.: Thermal control system based on thin film heaters and amorphous silicon diodes. In: 6th IEEE International Workshop on Advances in Sensors and Interfaces (IWASI) 2015, pp. 277–282. Springer (2015). https://doi.org/10.1109/IWASI.2015.7184977
12. Caputo, D., Forghieri, U., Palma, F.: Low-temperature admittance measurement in thin film amorphous silicon structures. J. Appl. Phys. 82(2), 733–741 (1997). https://doi.org/10.1063/1.365607

Yeast-DMFC Device Using Glucose as Fuel: Analytical and Energetic Applications. Preliminary Results

Mauro Tomassetti, Emanuele Dell'Aglio, Riccardo Angeloni, Mauro Castrucci, Maria Pia Sammartino and Luigi Campanella

Abstract We carried out a preliminary study, using a Direct Catalytic Methanol (or Ethanol) Fuel Cell (DMFC) in association with yeast cells (Saccharomices Cerevisiae), both to evaluate the possibility of analytically determining glucose solutions of unknown concentration, and to determine the power (in μW) obtainable from a solution with a fixed glucose concentration. In this first research, after having experimentally verified the actual functioning of the system, we optimized the operating conditions, working at room temperature, such as the current intensity as a function of the yeast concentration in solution and the contact time between cells and glucose, carrying out the measurement in potentiostatic mode. It was therefore possible both to identify a short linearity range of the method for analytical applications and to evaluate the performance of this system from an energy perspective, using the experimental power curve. We lastly have also tried to obtain a short calibration curve, to be used for analytical purposes. Finally we also measured the current intensity obtainable using a weighed amount of glucose as fuel and estimated the power achieved from the system.

Keywords Direct catalytic fuel cell · Coupled yeast cells · Glucose as fuel
Analytical and energetic purposes

1 Introduction

As part of the studies, based fuel cell, devoted to evaluating energetic possibilities, recently conducted by some authors of our department, which recently performed several researches by different kind of fuel cell producing power, for energetic purposes [1–7], our research group carried out studies, in order to both develop new possible analytical methods and convert chemical energy into electrical energy, using

M. Tomassetti (✉) · E. Dell'Aglio · R. Angeloni · M. Castrucci · M. P. Sammartino
L. Campanella
Department of Chemistry, University of Rome "La Sapienza", P.le A. Moro 5, 00185 Rome, Italy
e-mail: mauro.tomassetti@uniroma1.it

© Springer Nature Switzerland AG 2019
B. Andò et al. (eds.), *Sensors*, Lecture Notes in Electrical Engineering 539,
https://doi.org/10.1007/978-3-030-04324-7_8

53

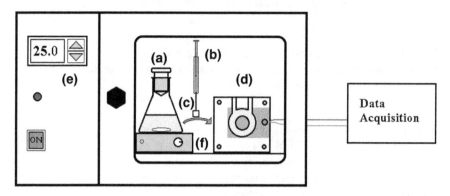

$$Glucose \xrightarrow{Yeast} 2\,Pyruvate \rightarrow 2\,Acetylaldehyde \rightarrow 2\,Ethanol$$

Fig. 1 Scheme of reactions chain producing ethanol from glucose

Fig. 2 Experimental apparatus: **a** flask containing glucose solution and yeast; **b** graduated syringe; **c** small filter; **d** catalytic fuel cell; **e** thermostatic apparatus; **f** magnetic stirrer

a Direct catalytic Methanol (or ethanol) Fuel Cell (DMFC). First of all applications to real food-grade (alcoholic drinks) [8, 9], or pharmaceutical samples (tinctures and two particular antibiotics) [10] have been investigated. We are currently studying the possibility of both analytically determining glucose and using it for the production of electrical energy. In practice, from glucose, ethanol can be obtained by means of yeast cell (Saccaromices cerevisiae) [11] using the following reactions (Fig. 1).

The produced ethanol supplies the DMFC device, giving electrical energy (Fig. 2).

2 Materials and Methods

The solutions used for the experiments were made by dissolving in 30 mL of distilled water, 0.34 g Glycine (Fluka, assay > 99%), 0.6 g of commercial yeast, (available from Conad shop [Rome, Italy]) and a weighed amount of glucose (D(+)-Glucose Monohydrate, Fluka, assay ≥ 99%). Glycine was used to make an isotonic solution. A Mettler PM460 balance (Columbus, Ohio, USA) was used for weighing all solid products.

A 50 mL glass flasks, suitably closed with a glass stopper and containing 30 mL of yeast-glucose-glycine solution, was kept in an incubator (Fig. 2) at fixed temperature of 25 ± 1 °C, for 24 h; using a magnetic stirrer, set at 100 rpm, to keep the yeasts in suspension. At the end of this time, 2 mL of solution was taken from the flask with a graduate syringe, equipped with small filter, and placed in a DMFC, H-TEC Model F111, fuel cell (50 × 50 × 40 mm and weighing 100 g), obtained from Fuel Cell Store (College Station, TX, USA). The Cell was made of Plexiglas©, while the electrode end plate was made of Pt-Ru black catalyst, assembled with a Nafion™ membrane. For potentiostatic format measurement, a Palmsens mod. EmStat potentiostat (Houten, The Netherlands) was used, connected to the fuel cell, supplied with PSTrace Software ver. 4.6 data interface on a Compaq Presario PC. For each measurement, the current was recorded until the steady state, at which the supplied current value was read.

The fuel cell was connected to Emstat, using anode as working electrode, while the cathode acts as reference and counter electrode.

Before the current measurement, the Emstat automatically measured the Open Circuit Voltage (OCV) value, for a time of about 200 s, and then the anode potential was set to a value of the Optimized Applied Potential (OAP = OCV − 100 mV) [8, 10].

In all cases, the fuel cell, before the measurement, was carefully washed with 0.5% water-ethanol solution and then several times with distilled water.

3   Results and Discussion

First of all we have experimentally verified that a glucose solution, inserted in the DMFC cell, does not give any significant signal; but, if we put the glucose solution in contact with yeast cells (Saccharomyces cerevisiae), we observed the source of an e.m.f. between the DMFC cathode and anode, from which electrical current can be generated. In the first results obtained with the present research, we have optimised, by operating in isotonic glycine solution and at room temperature, the quantity of yeast (Fig. 3) and the time necessary (Fig. 4) for the best production of ethanol and to maximize e.m.f. under the chosen operating conditions. Finally we recorded the cell response to the increasing glucose concentration in solution, using the optimized conditions as resulted by observing the curves reported in Figs. 2 and 3, (i.e. 0.6 g of yeast and after 24 h). The so obtained calibration curve is shown in Fig. 5.

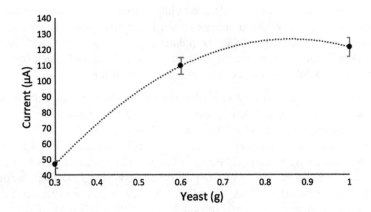

Fig. 3 Recorded current intensity increasing yeast mass (g) in solution

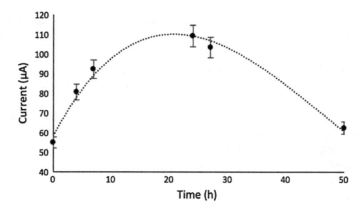

Fig. 4 Recorded current intensity as a function of contact time between yeast cells and glucose

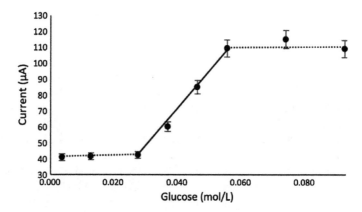

Fig. 5 Behaviour of current as a function of glucose concentration

In addition, we have carried out a preliminary estimation of the power (as watts) obtainable by the fuel cell, under the established operating conditions, extrapolating it from the power curve (see both the polarization curve in Fig. 6 and the power curve in Fig. 7). At the present state of the research and in the adopted experimental conditions, the results obtained can be summarized as follows:

(a) From an analytical point of view, the width of the linearity range of the method is slightly less than 1/2 decade (between about 28 and 56 mmol/L of glucose). The minimum detection limit is of about 22 mmol/L of glucose.

(b) From the energetic point of view, comparing μA obtained (checking a glucose concentration approximately equal to the one which corresponds to the upper end of the calibration curve) with those achieved by a suitable standard of EtOH concentration and taking into account the curve of power obtained, it can be concluded that, from a glucose solution of 0.056 mol/L and, after 1 h, about

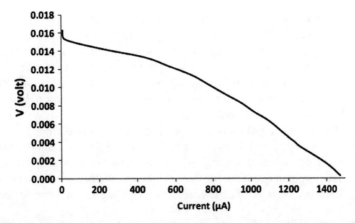

Fig. 6 Polarization curve for a glucose concentration of 56 mmol/L

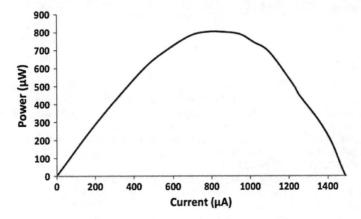

Fig. 7 Power curve for a glucose concentration of 56 mmol/L

0.004 mol/L of EtOH have been obtained. This glucose concentration generates, about 110 μA, i.e. it supplies a power of about 200 μW.

4 Conclusions

First results, using the DMFC-yeast system, are reported in the present article. Preliminary research encouraged us to move forward in this research by applying the method to real samples containing glucose, after confirming the optimized conditions. It is also clear that the research will be extended to the use of other types of yeasts and substrates (i.e. other carbohydrates), as well as to the extent of the chemical metabolism of the utilized yeasts. In addition further research will be carried out by varying the operating conditions with especially regard to the operating temperature. It is indeed foreseeable that, by carefully increasing the thermostating temperature of the measurement, it will probably be possible to shorten the measurement times. On the other hand, the present research was on purpose carried out at 25 °C, that is at room temperature (even if in thermostat), precisely in order to check whether the system can operate even at room temperature, without thermostation, however in a closed environment, i.e. in a room where the temperature variations was in all cases approximately ≤ 1 °C. In fact, as already observed in previous research [8, 10], when the used fuel was only ethanol, also in the present research it was found that the measurement is still possible by operating, at room temperature, without thermostatation; in fact, in these operating conditions the reproducibility does not worsen more than 1.5 times, compared to when the measurement is conducted in a thermostated mode.

Acknowledgements This work was funded by the University of Rome "La Sapienza", Center "Protezione dell'Ambiente e dei Beni Culturali (CIABC)" and "Istituto per lo Studio dei Materiali Nanostrutturati (ISMN)" of CNR.

References

1. Zeppilli, M., Lai, A., Villano, M., Majone, M.: Anion vs cation exchange membrane strongly affect mechanisms and yield of CO_2 fixation in a microbial electrolysis cell. Chem. Eng. J. **304**, 10–19 (2016). https://doi.org/10.1016/j.cej.2016.06.020
2. Zeppilli, M., Ceccarelli, I., Villano, M., Majone, M.: Reduction of carbon dioxide into acetate in a fully biological microbial electrolysis cell. Chem. Eng. Trans. **49**, 445–450 (2016). https://doi.org/10.3303/CET1649075
3. Villano, M., Ralo, C., Zeppilli, M., Aulenta, F., Majone, M.: Influence of the set anode potential on the performance and internal energy losses of a methane-producing microbial electrolysis cell. Bioelectrochemistry **107**, 1–8 (2016). https://doi.org/10.1016/j.bioelechem.2015.07.008
4. Brutti, S., Scipioni, R., Navarra, M.A., Panero, S., Allodi, V., Giarola, M., Mariotto, G.: SnO2-Nafion® nanocomposite polymer electrolytes for fuel cell applications. Int. J. Nanotechnol. **11**, 882–896 (2014). https://doi.org/10.13140/2.1.1079.3601
5. Navarra, M.A., Abbati, C., Croce, F., Scrosati, B.: Temperature-dependent performances of a fuel cell using a superacid zirconiadoped nafion polymer electrolyte. Fuel Cells **3**, 222–225 (2009). https://doi.org/10.1002/fuce.200800066
6. Bollella, P., Fusco, G., Stevar, D., Gorton, L., Ludwig, R., Ma, S., Boer, H., Koivula, A., Tortolini, C., Favero, G., Antiochia, R., Mazzei, F.: A glucose/oxygen enzymatic fuel cell based on gold nanoparticles modified graphene screen-printed electrode. Proof-of-concept in human saliva. Sens. Actuators B Chem. **256**, 921–930 (2018). https://doi.org/10.1016/j.snb.2017.10.025
7. Carbone, M., Gorton, L., Antiochia, R.: An overview of the latest graphene-based sensors for glucose detection: the effects of graphene defects. Electroanalysis **27**, 16–31 (2015). https://doi.org/10.1002/elan.201400409
8. Tomassetti, M., Angeloni, R., Merola, G., Castrucci, M., Campanella, L.: Catalytic fuel cell used as an analytical tool for methanol and ethanol determination. Application to ethanol determination in alcoholic beverages. Electrochim. Acta **191**, 1001–1009 (2016). https://doi.org/10.1016/j.electacta.2015.12.171

9. Tomassetti, M., Angeloni, R., Castrucci, M., Martini, E., Campanella, L.: Ethanol content determination in hard liquor drinks, beers, and wines, using a catalytic fuel cell. Comparison with other two conventional enzymatic biosensors: correlation and statisitcal data. Acta Imeko. **7**(2), 91–95 (2018). ISSN: 2221-870X (in press)
10. Tomassetti, M., Merola, G., Angeloni, R., Marchiandi, S., Campanella, L.: Further development on DMFC device used for analytical purpose: real applications in the pharmaceutical field and possible in biological fluids. Anal. Bioanal. Chem. **408**, 7311–7319 (2016). https://doi.org/10.1007/s00216-016-9795-2
11. https://en.wikipedia.org/wiki/Ethanol_fermentation

YCoO₃ Resistive Gas Sensors for the Detection of NO₂ in 'Resistance Controlled Mode'

Tommaso Addabbo, **Ada Fort**, **Marco Mugnaini** and **Valerio Vignoli**

Abstract In this paper the unconventional measurement technique, 'resistance controlled mode', which consists in driving the sensor heater in order to maintain constant the sensing-film resistance, and reading as a sensor output the temperature of the film, is applied to the detection of NO_2 gas with $YCoO_3$ based resistive gas sensors. The technique is discussed and its results are compared to those obtained with conventional measurements of resistance at constant temperature, showing that it has similar performance in terms of resolution and better in terms of speed.

Keywords Gas sensors · Metal oxides · Gas measurements

1 Introduction

Gas sensors find application in many different fields, including environmental monitoring and industrial process control, food quality assessment and biomedicine. In this framework, resistive gas sensors based on metal oxides are a very popular technology, mainly due to their very large sensitivity and low-cost. Nevertheless, there are still some major issues concerning their performance, including the lack of

T. Addabbo · A. Fort (✉) · M. Mugnaini · V. Vignoli
Dipartimento di Ingegneria dell'Informazione e Scienze Matematiche, Università di Siena,
Via Roma 56, 53100 Siena, Italy
e-mail: ada@dii.unisi.it

T. Addabbo
e-mail: addabbo@dii.unisi.it

M. Mugnaini
e-mail: mugnaini@dii.unisi.it

V. Vignoli
e-mail: vignoli@dii.unisi.it

© Springer Nature Switzerland AG 2019
B. Andò et al. (eds.), *Sensors*, Lecture Notes in Electrical Engineering 539,
https://doi.org/10.1007/978-3-030-04324-7_9

selectivity and hence a large cross-sensitivity to influence quantities, that can result in unexpected drifts. For these reasons there is still a lot of research activity concerning these devices, including both the development of new sensing materials and the improvement of the measurement techniques and measurement systems. In this context the idea of using a non-conventional measurement method based on a feedback loop, which allows for operating at a constant resistance value by conveniently changing the sensor temperature has already been proposed [1, 2]. In this context the temperature becomes the sensor output. The performance of the technique are not fully explored and understood. The idea at the basis is to keep the surface of the sensor in a fixed chemical state, with a fixed amount of chemisorbed species, by changing the temperature. This could avoid saturation of the surface, extend the measurement range and grant a more linear response. Moreover, in conventional measurements (fixed temperature variable resistance), the sensor resistance spans over a large range of values in different working conditions. This large variation, that can be of several orders of magnitude and can be therefore challenging from the front end circuit design point of view, corresponds to a much lower sensing film temperature variation, due to the exponential relationship between resistance and temperature, with clear advantages in terms of measurement management.

Actually the dependency of the sensor resistance on temperature is related both to an increase of carriers energy and to the variation of the reaction rates and equilibrium constants, so it is not easy to model the sensor behavior. The aim of this paper is to explore the applicability of the unconventional measurement technique named hereafter 'resistance controlled mode'. In particular the behavior of $YCoO_3$ sensors for NO_2, which have been studied by the authors in the conventional working mode [3, 4], is analyzed.

2   Measurement Technique and Measurement System

The technique consists in driving the heater, on the basis of the sensing film temperature, in order to maintain constant the sensing film resistance also while varying the test gas mixture composition. The measurement system and the sensor structure are shown in Fig. 1. The sensor is obtained depositing a film based on $YCoO_3$ powders on an alumina substrate. The substrate is equipped with a couple of interdigitated electrodes for the sensing film, a temperature sensor (Pt RTD) and finally a heater on the backside, all realized by screen-printing.

To perform measurements in 'resistance controlled mode' a PI digital control system was used, which exploits the acquisition of both the chemical sensing film resistance (R_{chem}) and the chemical sensing film temperature (T_s), and which sets the

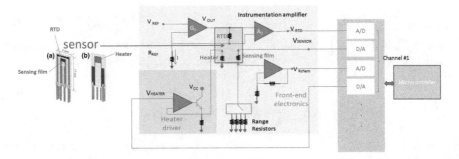

Fig. 1 Measurement system

voltage across the heater resistance, thus setting the desired power. To this purpose the hardware shown in Fig. 1 was developed [1]. The control implements the following strategy:

$$e_k = R_{chem_k} - R_{chem_{des}} \tag{1}$$

$$P_{k+1} = P_k + t_c \alpha e_k + \beta(e_k - e_{k-1}) \tag{2}$$

$$V_{heater_{k+1}} = \sqrt{P_{k+1} R_{heater}} \tag{3}$$

where e_k is the error at the k-th sampling time (difference between the measured chemical film resistance value and the desired value), P_k is the power delivered to the sensor at the k-th time step, by generating the voltage $V_{heaterk}$, R_{heater} is the heater resistance, t_c is the sampling time, whereas α and β are the integral and proportional gains, respectively.

Note that relationship between the measured resistance and the temperature, under some assumptions acceptable for the sensors used in this paper (p-type sensing material, depleted surface), is of this kind:

$$R_{chem} = R_0 \, e^{\frac{q^2 N_s^2 (T_s.[Gas_1].[Gas]_2...)}{2kT_s \epsilon p}} \tag{4}$$

where k is the Boltzmann constant, p is the density of carriers in the bulk (holes in this case), ε is the dielectric constant, q is the charge of the electron, $[Gas]_i$ represents the i-th gas concentration, whereas N_s represents the density of the unoccupied acceptor surface states. In other words qN_s is the density of the positive surface charge (for a p-type material with a depleted surface), which includes the contribution of chemisorbed gas molecules. For instance considering a sensor exposed to a mixture of air and NO$_2$ we can assume:

$$N_s\left(T, \left[O_{2gas}\right], \left[NO_{2gas}\right]\right) = N_i - \left[O_{ads}^-\right] - \left[NO_{2ads}^-\right] \tag{5}$$

where N_i is the positively ionized intrinsic surface defects density, whereas $[X_{ads}]$ indicated the surface density of the adsorbed X species. Moreover we can write:

$$\left[O_{ads}^-\right] = K_{ox}(T)\left[O_{2gas}\right][S_o] \tag{6}$$

$$\left[NO_{2ads}^-\right] = K_{NO_2}(T)\left[NO_{2gas}\right]\left[S_{NO_2}\right] \tag{7}$$

where K_{ox} and K_{NO_2} represent the equilibrium constants for the two chemisorption reactions; for both an Arrhenius form is expected, so that:

$$K_{ox}(T) = K_{0ox}e^{-\frac{E_{ox}}{kT}}, \quad K_{NO_2}(T) = K_{0NO_2}e^{-\frac{E_{NO_2}}{kT}} \tag{8}$$

So N_s depends obviously on the gas concentrations but also, heavily, on the temperature T_s because this latter influences the chemical reaction equilibrium and rate constants. So when operating at constant R_{chem} and changing the gas concentration, the temperature T_s will vary so as to maintain the exponential factor constant, causing also a different N_s equilibrium point. Note that in Eqs. (6)–(8) provide a coarse approximation of the chemical behavior and in general the exact relationship between N_s and the gas concentration is not known; moreover N_s is not a monotonic function of T_s given a certain gas concentration. For these reasons the response of the sensor in this mode of operation is very difficult to be predicted.

3 Experimental Results

Some results obtained using defective and doped YCoO$_3$ materials [3] with the proposed system are reported hereafter. The data are obtained exploiting a gas delivery system able to maintain constant the gas flow in the test chamber during the measurement (in this paper the flow is 200 mL/min) while varying the gas mixture composition. Usually the test mixtures are composed by a carrier gas that can be air of nitrogen, and a target gas (NO$_2$).

In Fig. 2 the transient responses of 3 different materials based on YCoO$_3$ to a pulse of 6 ppm of NO$_2$ are shown as a function of time; the carrier gas is air. The figures show the efficiency of the 'resistance control'.

Figure 3 shows the different dynamics found in similar conditions for the two working modes. It can be seen how the 'chemical resistance control' mode provides faster responses with respect to the traditional temperature controlled mode.

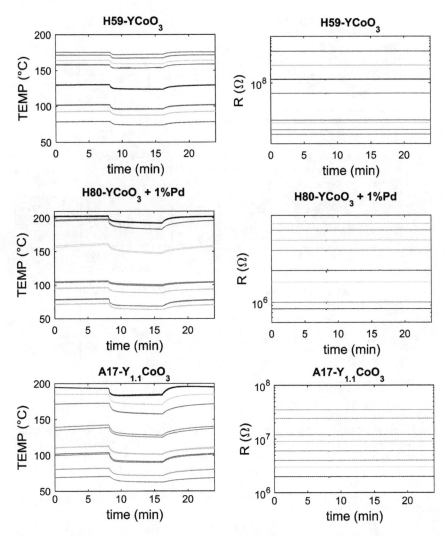

Fig. 2 Transient responses of 3 different materials based on YCoO$_3$ to a pulse of 6 ppm of NO$_2$ as a function of time. Left, measured T_s. Right, measured (and controlled) R_{chem}. The figure shows the efficiency of the resistance control

Finally Fig. 4 shows the response of the three examined materials to different NO$_2$ concentrations: it can be seen that the 'controlled chemical resistance mode' requires a sensitive temperature measurement system, because the temperature variations are of some degrees, but in any case the detection limit (about 1 ppm of NO$_2$) results similar to the one obtained in the traditional 'temperature controlled' mode.

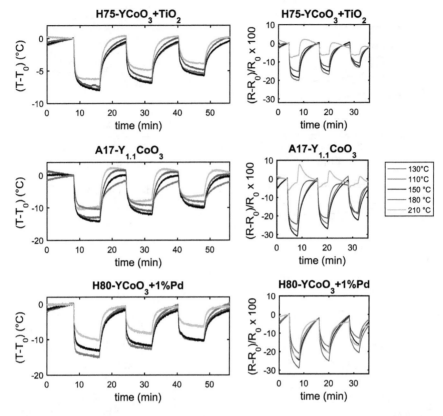

Fig. 3 Transient responses of 3 different materials based on $YCoO_3$ to NO_2 as a function of time. The carrier gas is nitrogen, the flow is 200 mL/min with the following protocol: t1 N_2, t2 N_2 + 24 ppm NO_2, t3 N_2, t4 N_2 + 12 ppm NO_2, t5 N_2, t6 N_2 + 6 ppm NO_2, t7 N_2. Left: measurements performed in 'chemical resistance control mode', at different resistance values (corresponding to the baseline temperature value in the legend). Right: measurements at different (constant) temperatures (as per legend). The figure shows the different dynamics found in similar condition for the two working modes

4   Conclusions

In this paper the unconventional measurement technique 'resistance controlled mode', which consists in driving the sensor heater in order to maintain constant the sensing-film resistance, and reading as a sensor output the temperature of the film, is applied to the detection of NO_2 gas with $YCoO_3$ based resistive gas sensors. The technique proved to provide results similar to those obtained with conventional measurements of resistance at constant temperature, in fact it was shown that it has

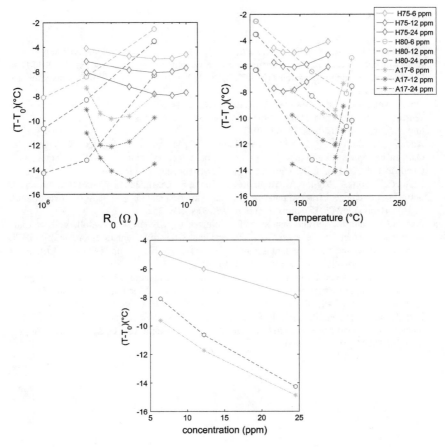

Fig. 4 Responses of 3 different materials based on YCoO3 to NO2, in 'controlled chemical resistance mode' and for different NO2 concentrations (as per legend). Left: as a function of controlled resistance (R_{chem}). Right: as a function of the corresponding baseline temperature. Bottom: temperature responses as a function of gas concentration in the optimum working conditions. The carrier gas is nitrogen, the flow is 200 mL/min

similar performance in terms of resolution and better in terms of speed. The transient responses induced by gas composition sudden variations obtained by the two techniques have different shapes, as expected. The study of the transient responses [5–7] could be exploited to further the knowledge about the chemical reactions contributing to the senor response.

References

1. Addabbo, T., Fort, A., Mugniani, M., Parri, L., Vignoli, V.: An unconventional type of measurement with chemoresistive gas sensors exploiting a versatile measurement system. In: 2017 New Generation of Circuit and Systems (NGCAS), pp. 113–116. IEEE Press (2017). https://doi.org/10.1109/ngcas.2017.34
2. Dominguez-Pumar, M., Kowalski, L., Calavia, R., Llobet, E.: Smart control of chemical gas sensors for the reduction of their time response. Sens. Actuators B Chem. **229**, 1–6 (2016). https://doi.org/10.1016/j.snb.2016.01.081
3. Addabbo, T., Bertocci, F., Fort, A., Gregorkiewitz, M., Mugnaini, M., Spinicci, R., Vignoli, V.: Gas sensing properties and modeling of $YCoO_3$ based perovskite materials. Sens. Actuators B Chem. **221**, 1137–1155 (2015). https://doi.org/10.1016/j.snb.2015.07.079
4. Bertocci, F., Fort, A., Vignoli, V., Mugnaini, M., Berni, R.: Optimization of perovskite gas sensor performance: characterization, measurement and experimental design. Sensors (Switzerland), 17 (6), art. no. 1352 (2017). https://doi.org/10.3390/s17061352
5. Fort, A., Rocchi, S., Serrano-Santos, M.B., Spinicci, R., Ulivieri, N., Vignoli, V.: Electronic noses based on metal oxide gas sensors: The problem of selectivity enhancement. In: 21st IEEE Instrumentation and Measurement Technology Conference, vol. 1, pp. 599–604. IEEE Press (2004). https://doi.org/10.1109/imtc.2004.1351121
6. Addabbo, T., Bertocci, F., Fort, A., Gregorkiewitz, M., Mugnaini, M., Spinicci, R., Vignoli, V.: Gas sensing properties of $YMnO_3$ based materials for the detection of NOx and CO. Sens. Actuators B Chem. **244**, 1054–1070 (2017). https://doi.org/10.1016/j.snb.2017.01.054
7. Fort, A., Mugnaini, M., Pasquini, I., Rocchi, S., Vignoli, V.: Modeling of the influence of H_2O on metal oxide sensor responses to CO. Sens. Actuators B Chem. **159**(1), 82–91 (2011). https://doi.org/10.1016/j.snb.2011.06.052

Monitoring Shelf Life of Carrots with a Peptides Based Electronic Nose

Sara Gaggiotti, Flavio Della Pelle, Vania Masciulli, Corrado Di Natale and Dario Compagnone

Abstract Monitoring and control of vegetable ripeness is a necessary and challenging issue in the food industry; in fact, the state of ripeness during harvest, storage, and market distribution defines the quality of the final product which is approved by customer preferences. Conventional methods used to determine the shelf life of vegetable are based on chemical, microbiological, physical and sensory indices. The majority of the classical methods are time-consuming and require skilled personnel. The aim of this work was to demonstrate that a methodology based on ZnO-peptide based QCMs array of gas sensors are useful to predict the shelf life of carrots. Samples of blanched carrots were stored at different temperatures (4, 25 °C and −18 °C) and analyzed after one month in gas-chromatography and with the sensor array. The results, analysed using principal component analysis (PCA) indicated that the sensors are able to clearly discriminate the different temperatures of storage.

Keywords Gas sensors · Shelf-Life · Carrots · Peptide · ZnO

1 Introduction

Aroma is one of the most important sensory properties in food, in particular, the aroma changes due to the treatment of samples, temperature change and storage time.

S. Gaggiotti (✉) · F. Della Pelle · V. Masciulli · C. Di Natale · D. Compagnone (✉)
Faculty of Bioscience and Technology for Food, Agriculture, and Environment,
University of Teramo, TE 64100 Teramo, Italy
e-mail: sgaggiotti@unite.it

D. Compagnone
e-mail: dcompagnone@unite.it

C. Di Natale
Department of Electronic Engineering, University of Roma Tor Vergata, RM 00133 Rome, Italy

© Springer Nature Switzerland AG 2019
B. Andò et al. (eds.), *Sensors*, Lecture Notes in Electrical Engineering 539,
https://doi.org/10.1007/978-3-030-04324-7_10

Carrot consumption, both fresh and processed, has increased over the past years due not only to the nutritional and health benefits this vegetable provides, but also to the introduction of new carrot-derived products. The ready-to-eat fresh-cut products are one of the major growing segments in food markets. The rapid growth is due to the new lifestyles and to the health-consciousness of the consumers [1]. The characteristic aroma and flavor of carrots are mainly due to volatile constituents. In particular, terpene secondary metabolites, which are synthesized during carrot root development, have a direct effect on quality and stability of the product. Terpenes constitute, anyway, the largest class of plant secondary metabolites, represent the major components of floral scents and essential oils of herbs and are important in determining the quality and nutraceutical properties of horticultural food products [2].

The VOCs composing the aroma of the vegetables are produced by metabolic activities during the stages of maturation, harvesting, post-harvest, and storage.

The most used technique for the determination and monitoring of VOCs is gas chromatography frequently coupled with mass spectrometry detection. However, cost of the instrumentation, sample pretreatment and the necessity to use skilled personnel does not allow the frequent use of such approach in food companies. Gas sensors arrays (electronic noses), in this respect can represent a feasible alternative. The simplicity of the approach together with the use of basic multivariate statistic can provide useful information for the rapid monitoring of the volatile components of vegetables [3, 4], and in general, of food [5].

In this work we report how a peptides based sensor array can detect changes during storage of food samples. Peptides have been recently used by our group as molecular binding elements for volatiles. They proved to be useful for a nice discrimination of VOCs originated from food when assembled either onto Au [6–8] or ZnO nanoparticles and deposited onto quartz crystal microbalances.

As a case study, a thermally sterilized carrots sample were selected. Sterilized samples were stored at three different temperatures: 25, 4 and $-18\ °C$. An array composed of six gas sensors modified with zinc oxide nanoparticles functionalized with different peptides (WHVSC, IHRIC, TGKFC, KSDSC, LAWHC and LGFDC) was used. The analysis of the headspace of the carrot samples for the identification of volatile compounds was carried out using GC-MS to obtain the conventional characterization of the aromatic pattern and to compare the data.

2 Materials and Methods

ZnONPs-Peptides gas sensor array. All reagents used were purchased from Sigma-Aldrich (Italy). The six peptides (IHRIC, LAWHC, TGKFC, KSDSC, LGFDC, and WHVSC) used for ZnONPs functionalization were purchased from Espikem (Italy, purity > 85%). ZnONPs were synthesized following the procedure used in another work [9]. The functionalization of ZnONP with six peptides was obtained by adding 100 μL of 10^{-3} M aqueous solution of the peptide to 900 μL of H_2O/Ethanol 9: 1 v/v containing 1 mg of ZnONPs. The suspensions of ZnONP-Peptides were incubated over-night at 4 °C.

The QCM sensors modification was achieved by drop casting five μL of the ZnONPs-peptide suspension on each side of the crystal and drying at room temperature. Cysteine was used as a spacer for the binding to ZnO nanoparticles. Analyses with the gas sensors were carried out with a "Ten 2009" Electronic nose (Tor Vergata Sensors Group, Rome, Italy), equipped with six Quartz Crystal Microbalances (QCMs). Carrier gas was N_2 used at a flow rate of 2 L/h.

Gas chromatography-mass spectrometry (GC-MS). A Clarus SQ8S GC-MS (Perkin Elmer) was used to analyse the headspace of all samples. Sampling of the volatile compounds was performed by solid phase micro-extraction (SPME). The sample was kept for 20 min at 40 °C and then exposed to the fiber (SPME Fiber Assembly 50/30um DVB/CAR/PDMS, Stableflex 24 Ga, Manual Holder, 3pk Grey) for 20 min at fixed temperature. The fiber was inserted in the desorption chamber where GC analysis was carried out with the following temperature gradient: the column was kept 6 min at 40 °C then the temperature was raised up to 250 °C for 50 min. Chromatograms obtained were analyzed in the mass spectrometer library.

Samples. All samples of carrots used in this study were from Abruzzo (Italy). Samples were prepared according to the procedure indicated in the literature [10]. The carrots were carefully washed, peeled and cut into standardized cylindrical pieces about one cm thick. To avoid enzymatic reactions during processing, storage and thawing, the packaged carrots were blanched at 95 °C for 8 min in a water bath. 3 g of carrots were then placed in 20 ml gas-tight vials and stored at three different temperatures: 25, 4 and −18 °C.

The analysis of the headspace was carried out after 1 day and after one month of storage with GC-MS and the gas sensor array. The data obtained were analyzed by the unsupervised multivariate technique principal component analysis (PCA) using Excel (XLSTAT).

3   Results and Discussion

SPME-GC-MS was used to detect the main components in the aroma profile of the different carrots samples.

Eight volatile compounds were detected and identified, the amount of these compounds for typical samples, expressed as relative abundance is reported in Table 1. All molecules found were already reported in the literature [10]. Moreover a decrease of the relative amount of terpinolene was observed confirming that this molecule can be a shelf-life marker [10].

The sensors outputs of the array for all the samples are reported in Table 2. The peptides TGKFC, KSDSC, LAWHC, have shown the higher response (Hz). The signals were very stable and reproducible (inter-sample RSD ≤ 15) demonstrating the ability of the ZnO-peptide gas sensors to work with samples containing high

Table 1 Relative abundance of GC-MS analysis of carrots headspace at different temperatures

Sample	α-Pinene (%)	β-Myrcene (%)	Limonene (%)	γ-Terpinene (%)	Terpinolene (%)	Caryophyllene (%)	Linalol (%)	Tetraydrofuran (%)
25day1	11	5	3	7	23	16	n.d	n.d
25day30	44	27	8	12	19	0	n.d	n.d
4day1	22	7	4	9	31	14	n.d	n.d
4day30	44	n.d	5	4	7	n.d	n.d	n.d
−18day1	22	14	7	7	54	7	n.d	n.d
−18day30	n.d	18	9	n.d	2	n.d	14	6

Table 2 ZnONPs-peptides QCMs ΔF response on carrots samples

Sample	ZnO_WHVSC	ZnO_IHRIC	ZnO_TGKFC	ZnO_KSDSC	ZnO_LAWHC	ZnO_LGFDC
25day1	92	212	480	370	324	110
25day1	93	212	486	370	331	114
25day1	91	201	475	390	324	110
4day1	89	191	444	454	313	107
4day1	89	203	444	465	315	105
4day1	89	193	448	465	318	106
−18day1	93	200	475	441	327	110
−18day1	89	204	464	450	321	108
−18day1	89	210	461	453	320	108
25day30	73	193	254	425	313	83
25day30	73	196	290	410	315	82
25day30	72	188	314	410	318	80
4day30	70	181	293	301	363	83
4day30	70	180	295	302	358	88
4day30	72	183	290	300	358	83
−18day30	69	170	202	205	315	75
−18day30	71	182	201	204	315	75
−18day30	73	183	200	205	320	78

amount of water [11]. Normalized ΔF signals were the analyzed with a multivariate statistical method (PCA).

The PCA reported in Fig. 1 clearly shows that the sensor array was able to separate the samples at different temperatures and times of storage.

The biplot of the first two principal components of the PCA model is reported that corresponds to 93.75% of the total variance of the data. In the biplot it is possible to observe the blue circle that includes all samples analyzed in day 1 at all temperature (25, 4, −18 °C). The green circles, corresponds to day 30. All the different temperatures are clearly discriminated. Moreover the loadings evidence that all the sensors are contributing to the separation of the classes with partial overlap of the contribution of IHRIC and LAWHC peptides.

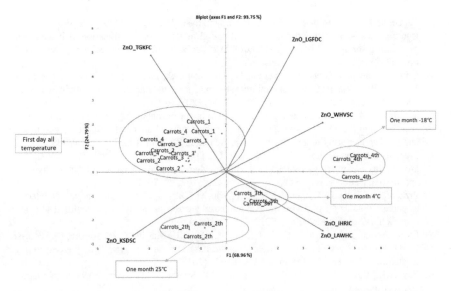

Fig. 1 Loading and score plots of ZnONPs-peptides assay on carrots. Blue circle = day 1 all temperatures; green circles = day 30 at different temperatures (25, 4, −18 °C) after one month of storage

4 Conclusions

In this study, we used ZnO nanoparticles functionalized with peptides for the realization of an array of QCMs gas sensors to follow the evolution of headspace of carrots stored at different temperatures.

The results obtained show that the sensor array was able to discriminate the different aroma patterns generated in the samples. Evaluation of the headspace with GC-MS confirmed that the amount of the shelf-life marker terpinolene decreased in the headspace. The reported approach appears useful for the evaluation of the shelf life of food. Current studies at shorter times on incubation are currently in progress to accurately assess the shelf-life of the product.

References

1. Gamboa-Santos, J., Soria, A.C., Pérez-Mateos, M., Carrasco, J.A., Montilla, A., Villamiel, M.: Vitamin C content and sensorial properties of dehydrated carrots blanched conventionally or by ultrasound. Food Chem. **136**(2), 782–788 (2013)
2. Yahyaa, M., Tholl, D., Cormier, G., Jensen, R., Simon, P.W., Ibdah, M.: Identification and characterization of terpene synthases potentially involved in the formation of volatile terpenes in carrot (Daucus carota L.) roots. J. Agric. Food Chem. **63**(19), 4870–4878 (2015)

3. Sanaeifar, A., Mohtasebi, S.S., Ghasemi-Varnamkhasti, M., Siadat, M.: Application of an electronic nose system coupled with artificial neural network for classification of banana samples during shelf-life process. In: Control, Decision and Information Technologies (CoDIT), pp. 753–757 (2014)
4. Compagnone, D., Faieta, M., Pizzoni, D., Di Natale, C., Paolesse, R., Van Caelenberg, T., Beheydt, B., Pittia, P.: Quartz crystal microbalance gas sensor arrays for the quality control of chocolate. Sens. Actuators B Chem. **207**, 1114–1120 (2015)
5. Santonico, M., Pittia, P., Pennazza, G., Martinelli, E., Bernabei, M., Paolesse, R., D'Amico, A., Compagnone, D., Di Natale, C.: Study of the aroma of artificially flavoured custards by chemical sensor array fingerprinting. Sens. Actuators B Chem. **133**, 345–351 (2008)
6. Pizzoni, D., Compagnone, D., Di Natale, C., D'Alessandro, N., Pittia, P.: Evaluation of aroma release of gummy candies added with strawberry flavours by gas-chromatography/mass-spectrometry and gas sensors arrays. J. Food Eng. **167**, 77–86 (2015)
7. Pizzoni, D., Mascini, M., Lanzone, V., Del Carlo, M., Di Natale, C., Compagnone, D.: Selection of peptide ligands for piezoelectric peptide based gas sensors arrays using a virtual screening approach. Biosens. Bioelectron. **52**, 247–254 (2014)
8. Compagnone, D., Fusella, G.C., Del Carlo, M., Pittia, P., Martinelli, E., Tortora, L., Paolesse, R., Di Natale, C.: Gold nanoparticles-peptide based gas sensor arrays for the detection of foodaromas. Biosens. Bioelectron. **42**, 618–625 (2013)
9. Zak, A.K., Razali, R., Majid, W.A., Darroudi, M.: Synthesis and characterization of a narrow size distribution of zinc oxide nanoparticles. Int. J. Nanomed. **6**, 1399 (2011)
10. Kebede, B.T., Grauwet, T., Magpusao, J., Palmers, S., Michiels, C., Hendrickx, M., Van Loey, A.: An integrated fingerprinting and kinetic approach to accelerated shelf-life testing of chemical changes in thermally treated carrot puree. Food Chem. **179**, 94–102 (2015)
11. Mascini, M., Gaggiotti, S., Della Pelle, F., Di Natale, C., Qakala, S., Iwuhoa, E., Compagnone, D., et al.: Peptide modified ZnO nanoparticles as gas sensors array for volatile organic compounds (VOCs). Front. Chem. **6**, 105 (2018)

An Innovative Optical Chem-Sensor Based on a Silicon Photomultipliers for the Sulfide Monitoring

Salvatore Petralia🄳, Emanuele Luigi Sciuto, Maria Anna Messina🄳,
M. Francesca Santangelo, Sebania Libertino and Sabrina Conoci🄳

Abstract The monitoring of pollutants such as sulfide anion species S^{2-} and HS^- is receiving a growing interest since they can cause acute and chronic toxicity including neurological effects and at high concentrations, even death. This study describes a new approach for optical detection of sulfide species in water samples. The method uses a silicon microchip with reagent-on-board and an integrated silicon photomultiplier (SiPM) device. The sulfide species are detected by the fluorescence signal emitted upon the reaction with N,N-dimethyl-phenylenediamine sulfate in the presence of iron(3+), leading to the formation of the fluorescent methylene blue (MB) species. A comparison with conventional fluorimetric detection method has been also carried out. Data show a very good linear correlation, proving the effectiveness of the method.

Keywords Chemical sensor · Fluorimetry · Silicon photomultiplier

S. Petralia (✉) · S. Conoci
STMicroelectronics, Stradale Primosole 50, 95121 Catania, Italy
e-mail: salvatore.petralia@st.com

E. L. Sciuto
Department of Physics and Astronomy, University of Catania, 95100 Catania, Italy
e-mail: emanueleluigi.sciuto@imm.cnr.it

M. A. Messina
Azienda Ospedaliero Universitaria Policlinico Vittorio Emanuele, Via S. Sofia 78, Catania, Italy

M. A. Messina
Centro Speleologico Etneo, Via Valdisavoia 5, Catania, Italy

M. Francesca Santangelo · S. Libertino
CNR-IMM Sede, Strada VIII Z.I. 5, 95121 Catania, Italy

© Springer Nature Switzerland AG 2019
B. Andò et al. (eds.), *Sensors*, Lecture Notes in Electrical Engineering 539,
https://doi.org/10.1007/978-3-030-04324-7_11

1 Introduction

The water is a primary good for the entire world's population, therefore the monitoring of the water quality is a crucial task for all states. In this context, the European Union has funded various research projects mainly focused on the development of miniaturized chemical sensor for water monitoring and on the water treatment technology (H2020-Water).

The development of nanotechnologies has permitted a fast growing of chemical sensors and their integration processes on miniaturized analyzer system [1, 2]. In example smart sensors integrated on multiparametric probes have been developed for the water pollution monitoring such as inorganic charged species (nitrate, ammonium, sulfide, nitrite etc.), heavy metals (Pb, Cd, Cr, Hg etc.), organic pollution (dioxin, polycyclic aromatic hydrocarbon) and biological pollution (fecal coliform, fecal streptococci, etc.) [3]. Sulfide species (H_2S, S^{2-} and HS^-) are contaminants typically produced by human and animal wastes or by bacterial species. They can cause acute toxicity at high concentrations, including death and neurological disorders [4, 5]. High concentrations of sulfide species are also find on groundwater especially in the presence of sulfidic-hot-spring typically located inside the caves [6].

Conventional methods for sulfide species sensing, normally, imply very labor-consuming sample preparation and tricky detection equipment [7].

Here we report the use of a miniaturized Silicon Photomultipliers as optical sensors for the colorimetric quantification of sulfide species on water samples. Due to the use of micro-sized components, that sensor provided a new low-cost, portable and easy-to-use method for in situ measurements [8, 9]. The analytical approach employed to monitor sulfide species (H_2S, HS^- and S^{2-}) consists on reaction with N,N-dimethyl-phenylenediamine and $FeCl_3$. The fluorescence intensity of methylene blue (MB) produced by this reaction is proportional to sulfide amount in water samples. The fluorescence signal was carried out with an excitation light of 632.8 nm measuring the emission light at 690 ± 10 nm. The groundwater samples collected in a sulfidic-spring located in Monte Conca Sinkhole were analyzed by our Silicon photomultipliers and for comparison by conventional optical absorption method. The experimental result reports for the proposed optical sensor a linear range for sulfide concentration from 0 to 10 mg L^{-1} with a detection limit value of about 0.5 mg L^{-1}. A very good correlation (R^2 0.99331) between the two methods was found. In conclusion thanks to the low reagent demand (a volume of about 20 μL), a short response time and excellent selectivity, the proposed optical sensor is environmentally friendly and it is a potential candidate for practical applications [10].

2 Materials and Methods

2.1 Chemicals

All chemicals employed in the experiments were purchased from Signal Aldrich (Merck KGaA, Darmstadt, Germany) and used as received. The N,N-dimethyl-phenylenediamine sulfate solution was prepared dissolving 0.2 g of solid in 100 mL of deionized water containing 10 mL of sulfuric acid. The $FeCl_3$ reagent was prepared dissolving 0.8 g of solid in 50 mL of deionized water containing 2 mL of sulfuric acid.

2.2 Silicon Photomultipliers

The system employed is composed of a SiPM detector, a laser excitation light and (c) and a miniaturized silicon-plastic device with the reagent-on-board (see Fig. 1).

The SiPM system is formed of 25 pixels, each one electrically and optically insulated by optical trenches fabricated all around each pixel and filled with silicon oxide. For the measurements, it has been mounted in a 32-pin open package (Fig. 1c), through which it is possible to bias it and collect the output signal.

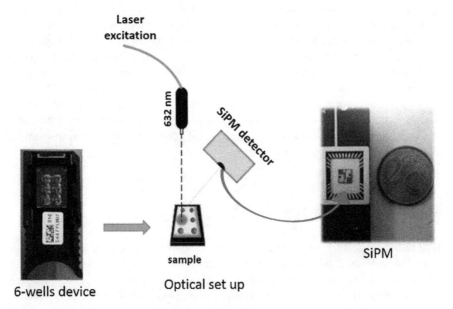

Fig. 1 Details for the device with reagent-on-board, the optical set up and SiPM

A Keithley 236 source meter was used to bias the SiPM and collect the signal. The signal was recorder by a PC and home-made software. The off-line analysis was performed using a home-made MATLAB™ (MathWorks, Natick, MA, USA) routine.

(b) The excitation light at 632.8 nm is provided by a He–Ne fiber-coupled laser. A bandpass filter, centered at 690 nm, is interposed between the sample and the SiPM sensor in order to reduce the optical noise due to the laser radiation reflected from the surface. The measurement system is automated by a customized software.

The plastic device was made of silicon and polycarbonate material. It contains six round-shaped micro-chambers (3 mm in diameter) of 20 μL each. For the measurements, the SiPM detector is placed at 45° with respect to both the laser and the plastic chip (see Fig. 1).

2.3 Sulfide Water Samples Collection

The sulfidic-water-samples were collected from a sulfidic spring located in the Monte Conca cave (Caltanissetta, Italy). The sulfidic spring is located in the final gallery of the cave at about −150 mt from the enter level. To reach the sulfidic spring properly speleological techniques were used. Figure 2a reports a speleologist during the sulfidic-water collection.

In details, eight samples from different cave sites featured by different sulfide concentration were collected. Each sample, was divided in two aliquots. Aliquots were stabilized with 1.5 mL of zinc acetate and stored at room temperature for the sulfide amount measurement by conventional fluorimeters method. The second aliquot was dispensed on miniaturized chips containing all the reagent on board for the assay (N,N-dimethyl-pphenylenediamine sulfate and $FeCl_3$) to form the methylene blue.

Quantitative studies were performed using a calibration curve obtained by standard sulfide solutions at concentration from 0 to 10 mg L^{-1}. Each standard solution reacts as described into scheme 1 to form methylene blue and the fluorescence emission was measured by SiPM and by conventional fluorimeter (excitation wavelength 633 nm, wavelength emission 690 nm).

3 Results and Discussion

The analytical performances of the system were evaluated measuring analytical water samples containing different sulfide concentrations from 0 to 10 mg L^{-1}. Figure 3 shows the calibration curve. The finding data showed a linear trend in the range between 0 and 10 mg L^{-1} (Fig. 3), as expected samples exhibiting sulfide concentration greater than 10 mg L^{-1}, a quenching of the fluorescence signal occur (see inset Fig. 3).

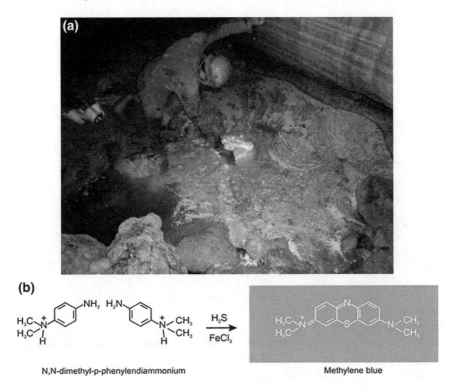

Fig. 2 **a** Sample collection into the cave, **b** Cline's procedure for sulfide detection

Data were interpolated by a linear fit giving the equation $Y = 3.8 \times 10^{-6} + 7.6 \times 10^{-6} X$ (R^2 0.98838). This leads to a sensitivity value of about 7.6 μA mg^{-1} L. The limit of detection (LoD) has been estimated considering the current value corresponding to 3-times the current measured without sulfide analyte. A LoD value of about 0.5 mg L^{-1} was obtained.

The working linear range for the proposed detection system is better than those obtained by other techniques reported in literature and based on the same colorimetric reaction such as spectrophotometry (LoD: 0.05–2 mg L^{-1} [11]), fluorimetry (LoD: 0.75–1.5 mg L^{-1} [12]), amperometry (LoD: 0.1–4.8 mg L^{-1} [13], electrochemistry (LoD: 0.9 mg L^{-1} [14]) and liquid chromatography (LoD: from 6.7×10^{-5} to 1.5×10^{-4} mg L^{-1} [15]).

To evaluate the performances of our system sulfidic water samples collected by a sulfidic spring were employed. At this scope three samples were collected and analyzed. Aliquots with a volume of 20 μL after collection were loaded on the silicon chip pre-loaded with NPS and FeCl$_3$ dried reagents. The SiPM current responses were measured and recorded, and the sulfide estimated by the linear regression equation (Fig. 3). For comparison the same aliquots sample stabilized after collection as reported in Material and method section, were analyzed by conventional fluorimeter,

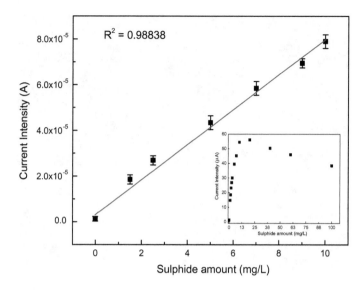

Fig. 3 Calibration curve of SiPM current versus sulfide amount in solution. In the inset, the calibration curve extended curve to the range 0–100 mg L^{-1}

Table 1 Sulfide concentration in sulfidic water samples measured using the approach proposed in this paper (SiPM) and the standard fluorimetry method

Sample	SiPM current intensity (θ A)	Sulfide mg L^{-1} by SiPM	Sulfide mg L^{-1} by fluorimetry
1	45.0±2.0	5.5±0.5	5.9±0.5
2	68.4±3.0	8.5±0.7	8.9±0.4
3	72.6±2.0	9.0±0.7	9.4±0.5

the sulfide amount calculated by an internal calibration curve obtained by standard solution of methylene blue (data not showed). All data are shown in Table 1 (Columns 1–3) and range between 5.5 and 9.0 mg L^{-1}.

By comparing the results reported in Columns 3 and 4 of Table 1, it is possible to note that the data are quite comparable.

4 Conclusions

The monitoring of Sulfide species is relevant for water quality, since these species can cause acute and chronic toxicity including neurological effects and, at high concentrations, even death. Existing procedures for these measurements are normally connected with very labor-consuming preparation of test samples and complicated measuring equipment that it is difficult to provide practically outside the laboratory.

In this study, we proposed a new strategy for sensitive optical detection of sulfide species using an integrated silicon photomultiplier coupled with a properly strategy applied in a silicon microchip with reagents on board. The sulfide species are detected by the fluorescence signal from MB species formed upon the Kline reaction method. The results herein presented proved that the system is able to measure the sulfide concentration in a linear range from 0 to 10 mg L^{-1} with a sensitivity value of about 7.6 μA mg^{-1} L and a detection limit of 0.5 mg L^{-1}. These indicate that the proposed method is reliable and thanks to its miniaturized components such as the SiPM detector and the silicon microchip, it can be certainly considered a very promising for the development of a portable easy-to-use system for fast and effective monitoring of sulfide species.

References

1. Petralia, S., Castagna, M.E., Cappello, E., Puntoriero, F., Trovato, E., Gagliano, A., Conoci, S.: A miniaturized silicon based device for nucleic acids electrochemical detection. Sens. Biosens. Res. **6**, 90–94 (2015)
2. McDonagh, C., Burke, C.S., MacCraith, B.D.: Optical chemical sensors. Chem. Rev. **108**, 400–422 (2008)
3. Libertino, S., Conoci, S., Scandurra, A., Spinella, C.: Sens. Actuators B Chem. (179), 240–251 (2013)
4. Banna, M.H., Imran, S., Francisque, A., Najjaran, H., Sadiq, R., Rodriguez, M., Hoorfar, M.: Crit. Rev. Env. Sci. Technol. **44**, 1370–1421 (2014)
5. Pandey, S.K., Kim, K., Tang, K.: A review of sensor-based methods for monitoring hydrogen sulfide. Trends Anal. Chem. 32 (2012)
6. Mukhopadhyay, S.C., Mason, A. (eds.): Smart Sensors for Real-Time Water Quality Monitoring. Springer, Berlin, Germany (2013). ISBN 978-3-642-37006-9
7. Nikolaev, I.N., Litvinov, A.V.: Procedure for measuring low concentrations of H_2 and H_2S above a water surface. Meas. Techn. 47(5) (2004)
8. Doujaiji, B., Al-Tawfiq, J.A.: Ann. Saudi. Med. (30), 76–80 (2010)
9. Santangelo, M.F., Sciuto, E.L., Lombardo, S., Busacca, A.C., Petralia, S., Conoci, S., Libertino, S.: Siphotomultipliers for bio-sensing applications. J. Sel. Top. Quantum Electron. **22**, 335–341 (2016)
10. Petralia, S., Sciuto, E.L., Santangelo, M.F., Libertino, S., Messina, M.A., Conoci, S.: Sulfide species optical monitoring by a miniaturized silicon photomultiplier. Sensors (18), 727 (2018)
11. Kuban, V., Dasgupta, P.K., Marx, J.N.: Anal. Chem. **64**, 36–43 (1992)
12. Spanziani, M.A., Davis, J.L., Tinani, M., Carroll, M.K.: Analyst (122), 1555–1557 (1997)
13. Lawrence, N.S., Davis, J., Jiang, L., Jones, T.G.J., Davies, S.N., Compton, R.G.: Electroanalysis (12), 1453–1460 (2000)
14. Lawrence, N.S., Davis, J., Marken, F., Jiang, L., Jones, T.G.J., Davies, S.N., Compton, R.G.: Sens. Actuators B Chem. (69), 189–192 (2000)
15. Tang, D., Santschi, P.H.J.: Chromatogr. A. **883**, 305–309 (2000)

Samarium Oxide as a Novel Sensing Material for Acetone and Ethanol

S. Rasouli Jamnani, H. Milani Moghaddam, S. G. Leonardi, Nicola Donato and G. Neri

Abstract Self-assembly structured Sm_2O_3 nanomaterials were prepared by a simple and cost effective hydrothermal method and subsequent annealing at 800 °C in air. The size, shape and phase composition of the structures synthesized by changing the hydrothermal processing time were characterized by SEM-EDX, XRD, micro-Raman and photoluminescence (PL) analysis. Differences in particle aggregation and coalescence were reported upon varying the timing of hydrothermal process (from 24 to 36 h). Sensors performances of the synthesized Sm_2O_3 nanostructures have been investigated for the monitoring of two important volatile organic compounds (VOCs) such as ethanol and acetone.

Keywords Sm_2O_3 · Gas sensor · Ethanol · Acetone

1 Introduction

Nowadays, rare-earth oxide materials have gained widespread attentions in many technological fields, due to their unique electrons configuration [1]. Among them, samarium oxide (Sm_2O_3), shows distinct optical properties, special magnetic and high permittivity, finding various applications in semiconductor glass, solar cells and nano-magnets [1, 2]. In gas sensing field, samarium is largely used as a dopant for improving the performance of metal oxide semiconductor-based conductometric gas sensors [3], but the applications as pure Sm_2O_3 are still very rare [4]. Therefore, we decided to investigate the sensing properties of samarium oxide towards the monitoring of some organic vapor species.

In the present work, Sm_2O_3 nanoparticles have been synthesized by a simple hydrothermal-calcination method. Two samples, A and B, were synthesized by

S. Rasouli Jamnani (✉) · H. Milani Moghaddam
Department of Physics, University of Mazandaran, Babolsar, Iran
e-mail: khorshid_adham@yahoo.com

S. G. Leonardi · N. Donato · G. Neri
Department of Engineering, Messina University, 98166 Messina, Italy

© Springer Nature Switzerland AG 2019
B. Andò et al. (eds.), *Sensors*, Lecture Notes in Electrical Engineering 539,
https://doi.org/10.1007/978-3-030-04324-7_12

changing the reaction time (24 and 36 h, respectively), maintaining all other synthesis conditions the same. The as-prepared samples were annealed at high temperature (800 °C) in air, resulting in the final self-assembly structured Sm_2O_3 nanomaterials. These Sm_2O_3-based nanomaterials have been used for the first time for developing high performance conductometric sensors for detecting ethanol and acetone.

2 Experimental

Hierarchical self-assembly structured Sm_2O_3 nanomaterials were prepared by an hydrothermal method as follows. Equimolecular quantity of samarium nitrate hexahydrate and citric acid monohydrate were dissolved in 30 ml distilled water under magnetic stirring. The mixture was transferred into a 100 ml Teflon-lined stainless autoclave, sealed and maintained at 180 °C for 24 h (Sample A) and 36 h (Sample B). The autoclave was then cooled down to room temperature. The obtained light yellow coloured precipitate was separated by centrifugation, washed with distilled water and ethanol several times, dried in air at 100 °C and calcinated at 800 °C for 2 h.

Sample morphology was observed by a Field Effect Scanning Electron Microscopy (FESEM) operating at 15 kV while the microstructure was analyzed by X-ray diffraction (XRD) technique. XRD measurements were performed using a Bruker-AXS D5005 diffractometer equipped with CuK α radiation ($\lambda = 1.5406$ Å).

Conductometric devices for electrical and sensing tests were then fabricated by printing films (about 20 μm thick) of the Sm_2O_3 samples dispersed in water on alumina substrates (6×3 mm) with Pt interdigitated electrodes and a Pt heater located on the backside. The fabricated sensors based on Sm_2O_3 samples were used for monitoring low concentration of ethanol and acetone in air.

3 Results and Discussion

Annealed Sm_2O_3 samples were characterized by FESEM and XRD to investigate their morphology and microstructure, respectively. From SEM analysis, both samples result composed by self-assembled Sm_2O_3 grains to form micrometer hierarchical spheres (Fig. 1). The size of the microspheres is in the range 1–10 μm. Further, many of these spheres are linked one each other. As a difference between these two samples, it can be noted that in sample A the contact area between the microsphere is limited compared to sample B. Thus, in the latter sample, large ensembles of microspheres with extended grain boundaries are formed.

Results from XRD showed that primary grains are single phase Sm_2O_3 with cubic crystal structure (Fig. 2). Characterization results also highlighted that increasing the hydrothermal reaction time leads to an increase of average grains size (from 22 to 28 nm) and an increased grains coalescence degree.

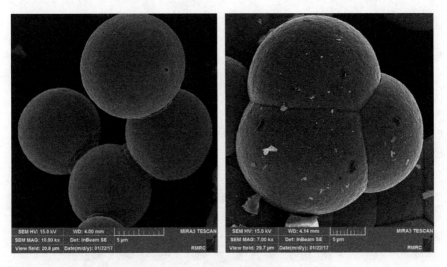

Fig. 1 FESEM images of sample A (left) and sample B (right)

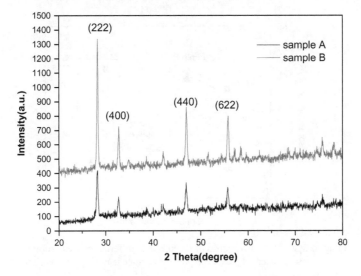

Fig. 2 XRD pattern of sample A and B

Results reported in Fig. 3 indicate that Sm_2O_3-sensors are able to detect few ppm of ethanol and acetone in a well reliable way. Furthermore, data obtained also reveal clearly that sample B display higher sensitivity toward both ethanol and acetone compared to sample A. To the best of our knowledge, this is the first report of Sm_2O_3-based sensors for monitoring these volatile organic compounds. Tests are in progress with Sm_2O_3 having different morphology with aim to elucidate the role of morphological characteristics on the sensing properties.

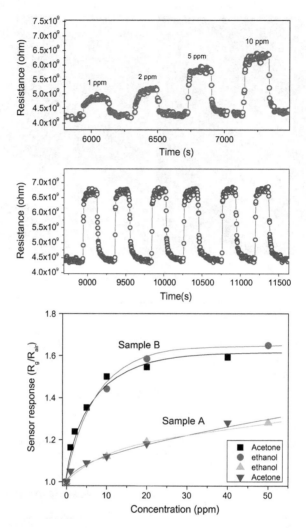

Fig. 3 From up to down: sensor response of sample B to different concentration of acetone (up graph); sensor response to repeated pulses (20 ppm) of acetone in air; calibration curves of samples A and B to acetone and ethanol

4   Conclusion

Self-assembly hierarchical Sm_2O_3 nanomaterials were prepared by a simple and effective method. It was found that Sm_2O_3 particle morphology is substantially affected by hydrothermal reaction time. Therefore, two different Sm_2O_3 conductometric sensors were fabricated and tested, for the first time, for detecting acetone and ethanol. Sensor B, based on large ensembles of microspheres with extended grain boundaries synthesized by adopting longer hydrothermal reaction time, showed the highest response. Further, this sensor exhibits high selectivity and response repeatability and stability. The good sensing performances suggested that hierarchical structured Sm_2O_3 nanomaterials may be promising sensing material for high performance VOCs sensors.

References

1. Atwood, D.A.: The Rare Earth Elements: Fundamentals and Applications. Wiley (2013)
2. Guria, A., Dey, K., Sarkar, S., Patra, B.K., Giri, S., Pradhan, N.: Tuning the growth pattern in 2D confinement regime of Sm_2O_3 and the emerging room temperature unusual superparamagnetism. Sci. Rep. **4**, 6514 (2014)
3. Michel, C.R., Martínez-Preciado, A.H., Parra, R., Aldao, C.M., Ponce, M.A.: Novel CO_2 and CO gas sensor based on nanostructured Sm_2O_3 hollow microspheres. Sens. Actuators B **202**, 1220–1228 (2014)
4. Hastir, A., Kohli, N., Singh, R.C.: Comparative study on gas sensing properties of rare earth (Tb, Dy and Er) doped ZnO sensor. J. Phys. Chem. Solids **105**, 23–34 (2017)

Crowdfunding for Increased Awareness Crowd-Sensing: A Technical Account

S. De Vito, Girolamo Di Francia, E. Esposito, G. Fattoruso, S. Fiore,
F. Formisano, Ettore Massera, M. Salvato and A. Buonanno

Abstract This work presents the results of the crowdfunding campaign devised for MONICA, an air quality monitoring portable device. The initiative is strongly focused on the increased awareness and involvement of citizens in the air pollution issue solution. Specifically, MONICA is an architecture composed by a portable device based on an array of commercial electrochemical sensors calibrated in lab, an Android App for smartphone, a web portal (MENA) and a NOSQL backend. This infrastructure is able to manage data communication/storage and map visualization of personal exposure to air pollutants. Two associated calibration procedures are depicted, one based on in-lab recordings while the second, based on the emerging on field calibration paradigm, will refine the performance of the node. The successful, both in financial and participatory terms, campaign has reached a crowdfunded contribution of 8000 € (145% of expected 6000 €) by 102 supporters. Among them, 44 users, have opted to become part of a small fleet of human-sensors able to produce air quality data during their daily mobility routine.

Keywords Air quality · Microsensors device · Crowdfunding

1 Introduction

Air pollution has both acute and chronic effects on human health, affecting different systems and organs [1]. Particulate air pollution, mostly, endangers people illness ranging from cardiovascular and respiratory problems to cancer [2, 3]. One topic of concern among population is due to the stream of scientific data about air pollution at regional scale and the subsequent impacts on public health and economy

S. De Vito (✉) · G. Di Francia · E. Esposito · G. Fattoruso · F. Formisano · E. Massera · M. Salvato
A. Buonanno
DTE-FSN-DIN, ENEA, C.R. Portici, P.le E. Fermi 1, 80055 Portici (NA), Italy
e-mail: saverio.devito@enea.it

S. Fiore
FSN-TECFIS, ENEA, C.R. Portici, P.le E. Fermi 1, 80055 Portici (NA), Italy

© Springer Nature Switzerland AG 2019
B. Andò et al. (eds.), *Sensors*, Lecture Notes in Electrical Engineering 539,
https://doi.org/10.1007/978-3-030-04324-7_13

[4]. OECD estimates that pollutants emission could cause 6 to 9 million premature deaths a year by 2060 and cost 1% of global GDP—around USD 2.6 trillion annually (USD 300 per person)—as a result of sick days, medical bills and reduced agricultural output [5]. Knowing pollutant concentration distribution may help population, including those who suffer from specific pathologies (autoimmune, COPD, asthma) and elderlies, to reduce their exposure while keeping an adequate active life [6]. In response, EU has issued an extensive legislation that sets health-based objectives and assessment standards for several air pollutants (see [7, 8]). However, given the sparseness of the resulting regulatory monitoring networks, targeted AQ information conflicts with current EU, not reflecting the spatial-temporal variability of AQ at road scale. The lack of local AQ information limits citizens 'awareness, reducing the adoption of AQ friendly lifestyle and exacerbating the distance between citizens and urban authority. Furthermore, air quality data were often communicated with an over-simplistic approach, determining a lack of understanding and ultimately low levels of engagement in citizens. For these reasons, in the last decade the interest in low cost mobile smart sensors devices, based on solid state chemical sensors, has grown. First of all, these portable solutions allow to draw up an estimate of personal exposure to air pollutants, secondly they help to provide, along with pervasive fixed nodes and certified analyzers, the data stream needed for obtaining high resolution air quality maps. Currently, several research groups are active in the development of these technologies (see [9]) as well as private companies that are beginning to ship low cost air quality monitoring nodes for indoor or outdoor applications. However, their accuracy must be seriously screened and taken into account (see [10]) because of most of COTS products available nowadays are sold without no information about it and consequently without any warranty. In this scenario, users perception can be misled causing false alarms, conflicts with concerning authorities, false expectations, etc.

Bridging the gap among citizens, urban and air quality authority, several participatory gas sensing has been funded in the framework of citizen science. Some initiatives focus on adopting intelligent transport systems and/or equipping them with AQ analyzers for sharing gathered data (Zurich Open Sense). Instead, the concept of citizens' observatory for urban AQ has been technically explored in several US and European RIA initiatives. The main focus has been on the development of novel fixed and mobile components for environmental monitoring to be also used by citizens, as well as innovative DSSs for policy makers helping them to identify measures for AQ improvement and to quantitatively assess their impact on citizens health. Relevant examples are EU projects as FP7-ENV-CITI-SENSE (https://co.citi-sense.eu/default.aspx) or COST-EUNetAir (http://www.eunetair.it/) and, respectively, LIFE15 ENV/PT/674 Index-Air (http://www.lifeindexair.net/). Starting from this, ENEA has designed a crowdfunding and a related crowd sensing campaign for allowing pervasive and cooperative air quality monitoring network, validated by analyzers operated by citizens. Main objectives are increasing awareness about possibilities and limitations of these new technologies in citizens and institutional stakeholders along with testing functionality and acceptability of the MONICA^{TM} device. We feel that along with accuracy and functional testing, the interaction with citizens have been of great help for a deeper understanding our technical role as researchers in this rapidly changing scenario. Aim of this contribution is hence to provide a keynote account of the lesson learned during this two steps campaign.

2   Crowdfunding Campaign

Monica 2.0 research project arises from the need to improve citizen's awareness of the air quality issue. In order to speed up the development of the project and to boost citizen's involvement in the initiative, a crowdfunding campaign has been launched on the Eppela web site [11] and linked on ENEA website [12]. This innovative self-financing strategy seemed useful for two aims. First of all, it would offer the opportunity to bridge the gap between researchers, citizens and regulatory monitoring authority (ARPAC). On the other hand, it appeared a good financial instrument for a non-funded project. The campaign lasted three months, including preparation and set up, ending on 17 December 2016. Our goal was fixed at achieving of a funding equal to 6000 €. The raised funds have been used for the development of a first fleet of 10 multisensor units and for their calibration. The campaign, holding the main principles of citizen science, was carried out arousing people curiosity such as to galvanize their active participation in the project. So, funders have been solicited by using a media coverage campaign showing the functionality of bike mounted MONICA devices and related smartphone applications. According to own interest in the project, citizens could finance it with 7 differently weighted stakes, receiving a corresponding reward in return (see Table 1). Anyway, all funders was awarded having the access to the anonymized data recorded by premium crowd-funders and participate in a newsletter campaign being informed of the development status and introduced to the limits of certified analyzers and smart sensing devices. Instead, "Smog Hunters" have had the opportunity to participate in the functional test campaign receiving a MONICA device and the related smartphone app for a one month duration at their premises. Acting as a fleet of mobile sensors, they are building their own city AQ pollution exposure map, focused on their daily routes. These maps will foster the possibility to make informed choices about their mobility routed choosing paths with reduced pollutant exposure, depending on the time of the day thy will be following them.

The campaign has been advertised on national press, national television and radio programs, and social networks. In particular, scientific television and radio programs have shown their interest towards the project, helping the campaign to reach the expected success. Moreover, the citizen's involvement was kept high in the campaign by means of feedback and suggestion questionnaires. Personal acknowledgments to each funder were published on the Eppela web page.

Table 1 Available stakes and related premiums during crowdfunding campaign

Stakes (€)	Qualification/Reward
5	Smog enemy: updated periodically by newsletter on project developments
10	Smog mapper: informed about developed tests
20	Smog tracer: accessibility to air quality maps produced by Smog Hunters
45	Smog hunter: received Monica 2.0 at home for 1 month
100	Smog researcher: visitor of ENEA lab for 1 day
200	Smog patron: supporter of the project
300	Smog master: ambassador of the project

3 Monica 2.0 Multi-sensor Device

MONICA2.0 is a portable pervasive air quality multi sensing architecture based on a mobile microsensor node with on board computational intelligence.

3.1 Sensor Node

The sensor array is composed by three electrochemical sensors from Alphasense, Ltd. [13]. They are targeted to CO (CO-A4), NO2 (NO2-A43F) and O3 (OX-A431) gaseous pollutants. Electrochemical sensors generally show enhanced stability and specificity with respect to MOX sensors although when operating in the required low ppb range they have experimentally shown cross interference from temperature and specific non target pollutants (with specific reference to NO2-O3 cross interference). Together they represent significant threats to human health with particular regard to long-term exposure (CO, NO2) and short term exposure (O3). Out of the box, the CO sensor (CO-A4) is accredited of a noise level of 20 ppb equivalent and a sensitivity of 220–375 nA/ppm at 2 ppm. Conversely, the NO2 sensor should show a noise level of 15 ppb and a sensitivity of -175 to -450 nA/ppb at 2 ppm. Finally, the O3 sensor datasheet information shows a noise level of 15 ppb and a sensitivity of -200 to -650 nA/ppm at 1 ppb. Additionally, a temperature and humidity sensor (sensirion SHT75) has been included in the sensor array to provide the needed information to correct chemical sensor responses for the environmental influences on their responses. Sensors have been mounted on a $10 \times 10 \times 15$ cm (500 g) box and the air flow toward the sensitive end of their case can be controlled by two small 3×3 cm fans (see Fig. 1).

Fig. 1 Different views of Monica 2.0 prototype, including a bike mounting device

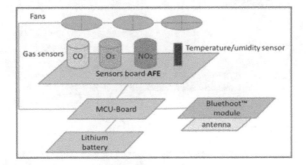

Fig. 2 Mounting scheme of sensor array

3.2 Data Acquisition System

The sensing array has been connected to a microcontroller ADC unit (16 bit) via the Alphasense analog front end (AFE 810-0020-00) providing sensor polarization and current to voltage conversion (see Fig. 2).

The current microcontroller platform is a STM32 Nucleo board from ST microelectronics and in particular its L-432KC version. The firmware controls data acquisition and conversion providing a duty cycle with a configurable sampling rate period of 1–75 s for the entire sensor array. Furthermore, oversampling techniques have been implemented to perform noise reduction. Data transmission facility has been provided by 2.0 EDR Bluetooth™ shield. System is powered by lithium battery which guarantees more than thirty hours of autonomy.

3.3 Back-End

Monica backend allows for data storage and visualization. Once data recorded by Android app, they are formatted in a timestamped JSON document and are sent as soon as IP connectivity is available to a single server NOSQL repository. Specifically, data storage relies on MONGODB2.4 server from which session data can be retrieved

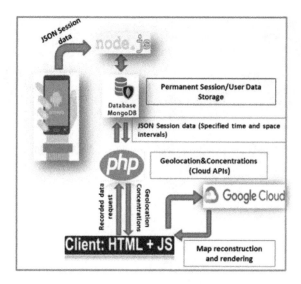

Fig. 3 Monica 2.0 Backend schema

Fig. 4 Start page of Android™ app (on the left) and Mena web portal

whenever needed by the user that provided it and visualized again (see Fig. 3). Further, a JavaScript/Php engine, including Google API Maps, allows to visualize, on phone and web app interface, session user data in a graph and on 2D route map (see Fig. 4).

3.4 Front-End

Monica 2.0 device is able to communicates data to users in two different ways, by means of an Android™ app for smartphone and/or by MENA (Monica Environmental and Network Analysis) web app. (see Fig. 4).

Both allow to visualize all users mobility sessions in different formats: raw data, graphical trend of pollutant concentration or 2D-route map (see Fig. 5). Specifically, the Android app allows for user geo-referencing using features provided by the operating systems and ultimately by the smartphone sensor. The position information is then stored in an array together with the pollutant concentrations and environmental data. Collectively, these information can be used to reconstruct the followed colored-encoded route and the related user exposure to pollutant. The resulting color code ranges from green (clean air) to black (heavily polluted air), as in the legend Fig. 6. The color code reflects the value of a cumulative air pollution index (AQI) based on the norm of the pollutant concentrations vector. The green color indicates a good air quality according to synthetic index value. Specifically, concentration of target gases are collectively below 20% of the 8-h concentration threshold set by EU in the 2008 Air quality directive. The map is actually generated averaging the relevant location based recordings found.

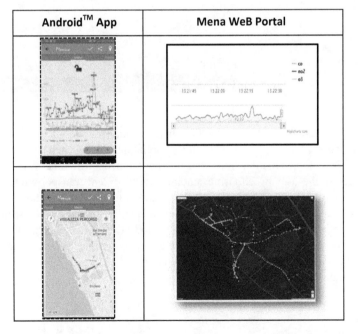

Fig. 5 On the top, trend of pollutants during an user session on app (left) and web interface (right). On the bottom, colored 2D route map on smartphone app (left) and web interface (right)

Map Legend   ×

European Air Quality Index

Gas	20%	40%	60%	80%	100%	>100%
CO (ppm)	(0 - 1)	(1 - 2)	(2 - 4)	(4 - 6)	(6 - 8)	
NO2 (ppb)	(0 - 21)	(21 - 53)	(53 - 106)	(106 - 212)	(212 - 319)	
O3 (ppb)	(0 - 40)	(40 - 60)	(60 - 90)	(90 - 120)	(120 - 300)	

Fig. 6 Gas concentration color code as adopted in the AQ map presentation by the proposed system

Fig. 7 Instantaneous and averaged pollutant concentration recorded during a mobility user session in Portici (NA)

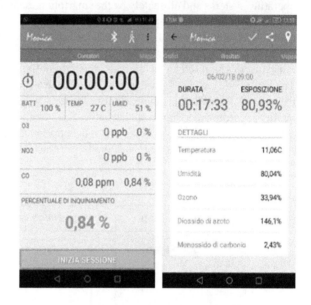

During operational phase, the Android app provides for showing to the users the instantaneous and averaged concentration of pollutants during each mobility session (see Fig. 7). The acquisition time, that reflects the sensor module sampling time, is defined by the user by choosing among three different transportation methods (on foot, by bike, by car). Selecting the car symbol, the position acquisition time is set each 10 s; using feet option will set the acquisition period at 30 s while using the bike option will cause the acquisition period at 20 s. If needed, the user can select a specific acquisition period by using an ad hoc graphical component. Vector averaged components contains single pollutant concentration normalized to the 8 h concentration limit in EU.

These information can be shared among all the user and are used to build a 2D map-route of air pollution in order to provide a detailed view to the benefit of all the contributing users that were not able to obtain a MONICA device via web interface. Otherwise, the DB storage allows to the users to consult data recorded and color-coded route on a map even if mobility sessions have it already available on the user's phone.

3.5 Calibration Procedures

During operational phase, Monica app provides for a refined on board calibration function execution as well as backend connectivity via Wi-Fi or 4G networks. The on board calibration will be based on a machine learning component operating in prediction mode. Two calibration methodologies have been devised for the sensing system. The first is based on lab recorded responses dataset and aims to develop a linear calibration algorithm for single gases with temperature correction. The results of the calibration also permit to validate in-factory sensor calibration parameters extracting temperature/humidity dependencies of sensibility and zero gas output. The calibration system is based on STM32 MCU Nucleo board and performs signal acquisition at 30 s sampling period, 5 points moving average noise filtering and raw data storage. Raw data can be then downloaded via usb connection. Eventually, least square sensor parameters extraction can be performed in an off-line fashion. During response recording time, the sensor board is located in a climatic chamber in which the gas is carried by the carrier synthetic air. The set point is varied according to a specified recipe spanning the predefined range of relevant concentrations. Humidity and temperature may be also changed. If needed, quaternary mixtures can be generated mixing different analytes for cross interference assessments. Figure 8 shows the response of the CO sensor to 5 different concentrations (from 0 to 15 ppm) of the target analyte at three different temperatures (20, 23, 34 °C), being T considered as a primary interferent for EC sensors. Results of the univariate (CO_WE) linear calibration procedure confirm a significant influence of the temperature on sensors sensitivity. Preliminary results for CO sensor confirm a significant influence of the temperature on sensor responses (see Fig. 9).

In order to deal with cross-sensitivity and instability sensors issues, a second nonlinear multivariate calibration has been performed by means of a feed-forward neural network cross-validated with leave-one-out strategy (Fig. 10).

A comparison among the performance indicator related to two calibration methods shows a mean absolute error (MAE) in CO estimate equal to 0.2 ppm for the multivariate approach against to a MAE of 0.8 ppm, estimating the same analyte with the linear univariate approach. A further planned procedure will refine these calibrations with on field recorded data. It will be based on dynamic non-linear multivariate calibration concept using a data driven computational intelligence approach [14]. Our previous findings have shown that the use of these specific electrochemical

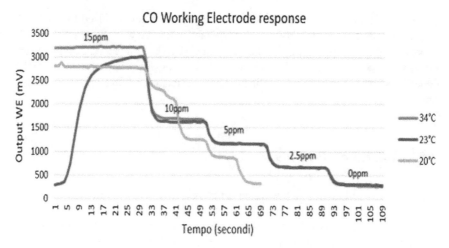

Fig. 8 CO sensor response in a climatic chamber at 3 different temperatures

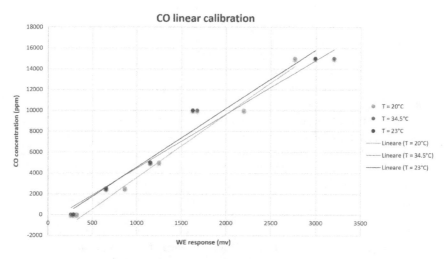

Fig. 9 Three different linear univariate calibration curves for CO computed from the in lab recorded data

sensors together with on field approach lead to an overall performance may come close or even meet the data quality objectives set by European Union.

4 Crowd Sensing Campaign

At the end of the campaign, Monica project has been funded for 8730 €, reaching 145% of expected 6000 €. With tenth of thousands of contacts, the campaign fully accomplished and outperformed the set funding goals with 102 contributors. Among

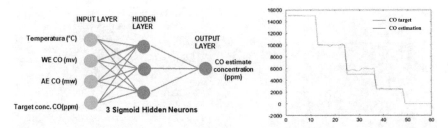

Fig. 10 On the left, FFNN architecture. On the right, CO estimates performed by a FFNN

Table 2 Features of funders mobility sessions

MONICA	Time period	Location	Sampling time	No. sessions	No. samples
1	16/1–22/2/2018	Rome	30	31	2656
2	29/1–26/2/2018	Bologna	20	19	1129
3	19/2–16/3/2018	Padova	20	29	1246
4	21/2–16/3/2018	Segrate (MI)	30	20	774
5	19/2–15/3/2018	Novate Milaese (MI)	30	27	3836
6	14/3–27/3/2018	Sanremo (IM)	10	18	6150

them, Smog Hunters are 44, they would constitute a small fleet of sensors having entitled to be owner of MONICA device for 30 days and relate Android app. After a calibration refinement conducted with on field data, the Monica's shipping is begun on January 2018. Nowadays, the 10 devices have been shipped to Italian crowd funders, returning to ENEA lab at the end of the 30 day period to be shipped again to remaining users (see Table 2). A detailed instruction manual is delivered to them together with MONICA system. This operative campaign is allowing to test, on field, the calibrated platform as well as back-end. Furthermore, their data are contributing to fill up a significant air pollution database on which building up maps, at high time-space resolution, including their routine routes in their cities (see Fig. 11). At the end of the trial period, a feedback questionnaire has been distributed to the participants. This is allowing to us to verify the acceptability and the ease of use of the device by non-technician personnel as well as the impact of the acquired air quality information on their daily routines. This active interaction with citizens have been of great help for a deeper understanding MONICA weakness in the operative urban environment such as to ameliorate the overall system at the user side yet.

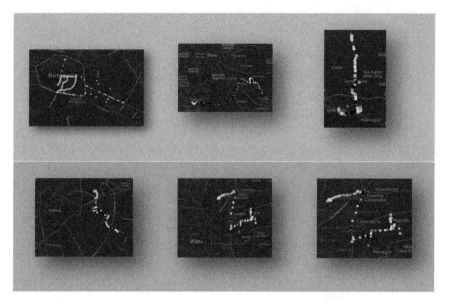

Fig. 11 Examples of pollutant concentration (color encoded) route maps of Smog Hunters

5   Conclusions

In this work, we presented the MONICA crowdfunding campaign aimed to perform a direct involvement of the citizens in air pollution measurement by a personal air quality monitoring device. The multisensor device, along with the supporting infrastructure functionalities and characteristics, has been described. The in-lab calibration tests, along with preliminary on field calibration results, have shown the capability to support qualitative measurements of NO2, CO and O3. The crowd sensing data gathered during the first shipping are summarized, allowing to build a consistent dataset. The next step will include a further on field calibration refinement and validation procedure co-locating the prototypes with an external ARPAC certified reference analyzer [15] at the aim to generate of map at high spatial-temporal resolution yet.

References

1. Kampa, M., Castanas, E.: Human health effects of air pollution. Environ. Pollut. **151**, 362–367 (2008). https://doi.org/10.1016/j.envpol.2007.06.012
2. Seaton, A., Godden, D., MacNee, W., Donaldson, K.: Particulate air pollution and acute health effects. The Lancet **345**(8943), 176–178 (1995)
3. Pope, C.A., Dockery, D.W.: Health effects of fine particulate air pollution: lines that connect. Air Waste Manag. Assoc. **56**, 709–742 (2006)
4. Baklanov, A., et al.: Atmos. Environ. **126**, 235–249 (2016). https://doi.org/10.1016/j.atmosenv.2015.11.059

5. http://www.oecd.org
6. Flag ERA Convergence Project. https://www.flagera.eu/wp-content/uploads/2016/02/FLAG-ERA_JTC2016_Project_flyer_Convergence_v0.3.pdf
7. Directive 2008/50/EC of the European Parliament and of the Council of 21 May 2008 on ambient air quality and cleaner air for Europe. https://eur-lex.europa.eu/legal-content/EN/TXT/?uri=CELEX:32008L0050
8. Air Quality Guidelines Global Update (2005). www.euro.who.int/__data/assets/pdf_file/0005/78638/E90038.pdf
9. Mead, I., et al.: Atmos. Environ. **70**, 186–203 (2013)
10. Lewis, A., Edwards, P.: Nature **535**, 29–31. https://doi.org/10.1038/535029a
11. https://www.eppela.com/it/projects/9652-monica-il-tuo-navigatore-personale-antismog
12. http://www.citizenscience.enea.it/regala-la-ricerca/
13. Alphasense company website. www.alphasense.com. Accessed Apr 2018
14. Esposito, E., et al.: Dynamic neural network architectures for on field stochastic calibration of indicative low cost air quality sensing systems. Sens. Act. B Chem. **231**, 701–713 (2016). ISSN 0925-4005
15. Marco, S.: The need for external validation in machine olfaction: emphasis on health-related applications. Anal. Bioanal. Chem. **406**, 3941 (2014)

Part II
Biosensors

Nickel Based Biosensor for Biomolecules Recognition

Salvatore Petralia⑩, Emanuele Luigi Sciuto⑩, Salvo Mirabella,
Francesco Priolo, Francesco Rundo and Sabrina Conoci⑩

Abstract A novel electrochemical device based on Nickel oxide sensing species is described. The miniaturized device contains three integrated metal microelectrodes with the working active electrode made of Ni zero-valence. It has been proved to be sensitive and versatile in the detection of glucose on saliva. The findings here reported pay the way to future development of versatile portable sensors addressing easy-to-use and low-cost system.

Keywords Glucose sensing · Electrochemical measurement · Silicon device

1 Introduction

The integration of new sensing materials with miniaturized is a fascinating fields of material science. It stimulates scientist for a wide range of areas such as molecular diagnostic, food industry, pollution environment, pharmaceutics [1–5].

In this framework, the development of new systems for the non-invasive biomolecules monitoring, such as glucose, aminoacids, protein etc. is receiving great attention due to its high impact to various diseases including diabetes and metabolic disorder (phenylketonuria) etc. [6–8]. In this context the glucose sensors devices are diffuse in different fields including food control, healthcare and industry. The traditional approaches to measure biomolecules levels are mainly based on enzymatic methods. In this scenario the non-enzymatic sensing based on the catalytic oxidation of biomolecules by nanomaterials represents the newest generation of glucose sensor technology [9–11]. Nickel is one of the most studied material since it allows the

S. Petralia (✉) · F. Rundo · S. Conoci
STMicroelectronics, Stradale Primosole 50, 95121 Catania, Italy
e-mail: salvatore.petralia@st.com

E. L. Sciuto · S. Mirabella · F. Priolo
Department of Physics and Astronomy, CNR-IMM-MATIS, University of Catania,
95123 Catania, Italy
e-mail: emanueleluigi.sciuto@imm.cnr.it

© Springer Nature Switzerland AG 2019
B. Andò et al. (eds.), *Sensors*, Lecture Notes in Electrical Engineering 539,
https://doi.org/10.1007/978-3-030-04324-7_14

direct electro-oxidation of molecules by means of the active species NiOOH/Ni(OH)$_2$ under alkaline medium. More in details, the sensing mechanism is based on the oxidation of analytes by the active NiOOH species, the produced Ni(OH)$_2$ species are reconverted to NiOOH by applying a specific potential (~0.7 V). A nanosized layer of sensing species (NiOOH) on top of Ni$^{(0)}$ working electrode is formed by Cyclic Voltammetry experiments in NaOH 0.1 M [12–14]. In this contribution we present a silicon miniaturized three planar electrodes device integrating Ni zero-valence layer (thickness 10 nm) on working electrode. The device exhibited good response towards glucose detection on human blood and saliva samples. The sensing experiment indicates a strong dependence of the sensitivity from the pH values, in particular the sensitivity increases with the increasing of pH value. This biosensor pave the way to future development of versatile, easy-to-use, low-cost and portatile multiparametric chemical sensors.

2 Materials and Methods

2.1 Chemicals

Sodium phosphate (PBS powder, pH 7.4), NaOH pellets (purity 97%), glucose, Potassium Chloride, PBS tablet were purchased by Sigma-Aldrich and used as received.

2.2 Electrochemical Measurements

Cyclic voltammetry (CV) and chronoamperometry measurements were performed by a Verstat 4 (Princenton Applied Research), The CV experiments were executed with a scan rate 10 mV/s and a voltage range −0.1 V/+1.0 V. The nickel layer was activated by about 60 sweeps of cyclic voltammetry in NaOH 0.1 M. The chrono amperometry were performed at a fixed voltage of 0.48 V.

2.3 Saliva Sample and Pre-treatment

The saliva sample after collection was treated with a precipitation buffer (COPAN) to remove the protein interferences. In detail a volume of 200 μL of fresh saliva was mixed with same volume of precipitation-buffer, after 5 min a volume of 20 μL of the supernatant was analyzed with EC-device. For comparison chrono-amperometry measurements were performed for the precipitation-buffer.

3   Results and Discussion

The electrochemical device (EC-device) was manufactured using the VLSI technology on a 6″ silicon wafer substrate. It is composed by 4 electrochemical cells, each one containing three planar microelectrodes, a working electrode (WE) in nickel, a counter (CE) and a reference (RE) electrodes made in gold. The EC-device contains four reaction chambers each with a volume of 22 μl in volume (Fig. 1).

The analytical performances for the EC-devices were evaluated measuring various amount of glucose on treated saliva sample. At this scope 10 μL of treated saliva was mixed with 10 μL of NaOH 0.1 M and chrono-amperometry measurement were recorded. After signal stabilization various glucose amount were added (from 20 to 250 μM) and the signal registered.

The Fig. 2 shows the experimental results of current intensity versus glucose amount, inset an example of chronoamperometric signal.

The data showed a limit of detection (estimated considering the current value corresponding to 3-times the current measured without glucose) of about 10 μM. These results encourage the development of miniaturized sensor integrating the Ni-device to be used in PoC format by not specialized end-users.

The future effort will be focused to improve the selectivity in complex biological matrices, the strategy will based on two approaches: implementation of algorithms in the analysis of sensing responses and integration of nanostructured Nickel species.

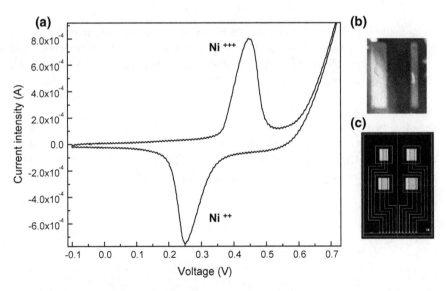

Fig. 1 Details of measurement and device: **a** CV curves at NaOH 0.1 M, **b** three planar electrodes (from left to right: RE in gold, WE in Ni and CE in gold) and **c** EC-device layout

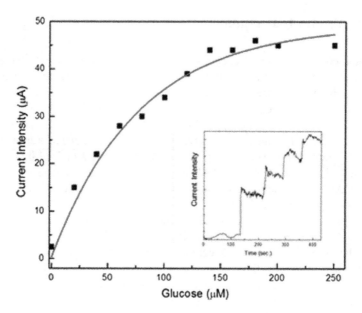

Fig. 2 Current intensity versus glucose amount in solution. Inset chronoamperometric signal

4 Conclusions

A miniaturized electrochemical device based on Nickel sensing material is here presented. The device is composed by three integrated metal microelectrodes with an active working electrode made of a Nickel. The sensing aptitude is based on the reversible conversion of Ni^{3+} (NiOOH) to Ni^{2+} ($Ni(OH)_2$) interacting to redox active glucose molecules. The results herein presented proved that the system is able to measure the glucose amount on saliva after an easy precipitation of protein, with a detection limit of $10\ \mu M$. These data indicate that the proposed method is reliable and thanks to its miniaturized components such as the silicon microchip, it can be certainly considered a very promising for the development of a portable easy-to-use system for fast and effective monitoring of glucose on human specimen.

References

1. Petralia, S., Sciuto, E.L., Santangelo, M.F., Libertino, S., Messina, M.A., Conoci, S.: Sulfide species optical monitoring by a miniaturized silicon photomultiplier. Sensors (18), 727 (2018)
2. Petralia, S., Sciuto, E.L., Di Pietro, M.L., Zimbone, M., Grimaldi, M.G., Conoci, S.: Innovative chemical strategy for PCR-free genetic detection of pathogens by an integrated electrochemical biosensor. Analyst **42**, 2090–2093 (2017)
3. Petralia, S., Conoci, S.: PCR technologies for point of care testing: progress and perspectives. ACS Sens. **2**, 876–891 (2017)

4. Petralia, S., Castagna, M.E., Cappello, E., Puntoriero, F., Trovato, E., Gagliano, A., Conoci, S.: A miniaturized silicon based device for nucleic acids electrochemical detection, 2015. Sens. Bio.-Sens. Res. **6**, 90–94 (2015)
5. Zeng, Y., Zhu, Z., Du, D., Lin, Y., J.: Electroanal. Chem. (781) 14–154 (2016)
6. Petralia, S., Sciuto, E.L., Messina, M.A., Scandurra, A., Mirabella, S., Priolo, F., Conoci, S.: Miniaturized and multi-purpose electrochemical sensing device based on thin Ni oxides. Sens. Actuators B Chem. **263**, 10–19 (2018)
7. Messina, M.A., Melim, C., Conoci, S., Petralia, S.: A facile method for urinary phenylalanine measurement on paper-based lab-on-chip for PKU therapy monitoring. Analyst (142) 2090 (2017)
8. Donlon, J., Sarkissian, C., Levy, H.L., Scriver, C.R.: Hyperphenylalaninemia: Phenylalanine Hydroxylase Deficiency. Scriver's Online Metabolic and Molecular Bases of Inherited Disease (2015). https://doi.org/10.1036/ommbid.97
9. Kun, T., Prestgard, M.: A review of recent advances in nonenzymatic glucosesensors. Mater. Sci. Eng. C (41) 100–118 (2014)
10. Iwu, K.O., Lombardo, A., Sanz, R., Scirè, S., Mirabella, S.: Facile synthesis of Ninanofoam for flexible and lowcost nonenzymatic glucose sensing. Sens. Actuators B **224**, 764–771 (2016)
11. Zhang, W., Du, Y., Wang, M.L.: Non Invasive Glucose Monitoring using Saliva nano-biosensor. Sens. Biosens. Res. **4**, 23–29 (2015)
12. Luo, L., Li, F., Zhu, L., Ding, Y., Zhang, Z., Deng, D., Lu, B.: Nonenzymatic glucose sensor based on nickel(II)oxide/ordered mesoporous carbon modified glassy carbon electrode. Colloids Surf. B **102**, 307–311 (2013)
13. Fleischmann, M., Korinek, K., Pletcher, D.: The oxidation of organic compounds at a nickel anode in alkaline solution. J. Electroanal. Chem., 31–39 (1971)
14. Petralia, S., Mirabella, S., Strano, V., Conoci, S.: A miniaturized electrochemicalsystem based on nickel oxide species for glucose sensing applications. Bionanoscience **7**, 58–63 (2017)

Electrochemical DNA-Based Sensor for Organophosphorus Pesticides Detection

Giulia Selvolini⬡, Ioana Băjan, Oana Hosu⬡, Cecilia Cristea⬡, Robert Săndulescu and Giovanna Marrazza⬡

Abstract In this work, we propose an electrochemical DNA-based sensor for sensitive detection of organophosphorus pesticides. To improve the sensitivity of the DNA-based sensor, polyaniline film and gold nanoparticles were progressively electrodeposited on the graphite screen-printed electrode surface by cyclic voltammetry. Gold nanoparticles were then employed as platform for the immobilization of thiol-tethered DNA oligonucleotide sequence complementary to the selected biotinylated DNA aptamer for profenofos detection. Streptavidin-alkaline phosphatase enzyme conjugate was then added to trace the affinity reaction through the hydrolysis of 1-naphthyl phosphate to 1-naphthol, which was then detected by differential pulse voltammetry. A decrease of the signal was obtained when the pesticide concentration was increased, making the sensor work as signal off sensor.

Keywords Pesticide · Aptamer · DNA · Competitive assay · Biosensor

1 Introduction

Organophosphorus pesticides (e.g. phorate, profenofos, isocarbophos) are highly toxic substances that nowadays are still used outside EU in some harvesting protocols. This class of compounds, like some nerve agents, acts on the enzyme acetylcholinesterase as neuromuscular inhibitors, affecting normal functions in insects, but also in humans and many other animals [1].

Conventional analytical methods for organophosphorus pesticides are based on chromatographic techniques, which provide sensitive and selective detection. Despite these advantages, chromatographic techniques require highly skilled technicians for

G. Selvolini · O. Hosu · G. Marrazza (✉)
University of Florence, Via della Lastruccia 3-13, 50019 Sesto Fiorentino, Italy
e-mail: giovanna.marrazza@unifi.it

I. Băjan · O. Hosu · C. Cristea · R. Săndulescu
University of Medicine and Pharmacy "Iuliu Hatieganu", 4 Louis Pasteur Street, 400439 Cluj Napoca, Romania

© Springer Nature Switzerland AG 2019
B. Andò et al. (eds.), *Sensors*, Lecture Notes in Electrical Engineering 539,
https://doi.org/10.1007/978-3-030-04324-7_15

operation and they are not suitable for screening analysis. Therefore, there are continuing developments in rapid and cost-effective devices for environmental monitoring including in situ analysis [2, 3]. In this perspective, biosensor development for pesticide analysis becomes urgent and proposes itself as an easily alternative to conventional techniques for screening analysis [4]. The main difficulties about the use of biosensors for pesticides detection concern the possibility to obtain antibodies for targets with high toxicity. In this way, the use of synthetic receptors such as DNA aptamer has recently become an interesting alternative to the antibodies in affinity biosensors technology [5, 6].

In this work, we report preliminary experiments using an electrochemical DNA aptasensor for profenofos organophosphorus pesticide detection based on a competitive assay format. The method combines the portability of graphite screen-printed electrochemical cells and of a computer-controlled instrument. Herein, the proposed aptasensor exploits the use of oligonucleotides coupled with a built-in electrochemical platform consisting of a metallic nanoparticles/polymeric film nanocomposite for disposable and cost-effective screening in situ analysis.

The DNA aptamer used in this work was selected from a library of aptamers, designed by SELEX, that proved itself to show the highest ability to recognize profenofos [7].

2 Materials and Methods

Perchloric acid (HClO$_4$), sulfuric acid (H$_2$SO$_4$), aniline, streptavidin-alkaline phosphatase enzyme, di-sodium hydrogen phosphate (Na$_2$HPO$_4$), sodium di-hydrogen phosphate di-hydrate (NaH$_2$PO$_4$·2H$_2$O), tetrachloroauric acid (HAuCl$_4$), 6-mercapto-1-hexanol (MCH), KCl, MgCl$_2$, BSA, di-ethanolamine (DEA) and profenofos have been purchased from Sigma-Aldrich (Italy). The DNA sequences (oligo-SH: 5'-(SH)-(CH$_2$)$_6$-CCGATCAAGAATCGCTGCAG-3'; apt-BIO: 5'-(biotin)-TEG(trietylene glycol)-AAGCTTGCTTTATAGCCTGCAGCGATTCTTGATCGGAAAAGGCTG AGAGCTACGC-3') were purchased from Eurofins Genomics (Germany). Immobilization buffer: 0.5 M phosphate buffer, pH 7. Detection buffer: 0.1 M DEA buffer, 0.1 M KCl, 1 mM MgCl$_2$, pH 9.6. Milli-Q water was used for all preparations. Electrochemical measurements were carried out with a portable potentiostat/galvanostat PalmSens electrochemical analyser (PalmSens, the Netherlands), and the results analysed by PSTrace 2.3 software.

The graphite screen-printed electrodes (GSPEs) were modified with polyaniline (PANI) [8] and gold nanoparticles (AuNPs). Sensors were then modified by self-assembly of a mixed monolayer of thiolated DNA capture probe (oligo-SH) and MCH [9]. Then, a solution containing a proper concentration of biotinylated DNA aptamer (apt-BIO) and the target pesticide was dropped on the sensor surface and the competitive reaction was allowed to proceed. The biotinylated hybrids formed onto the developed aptasensor surface were coupled with a streptavidin-alkaline phosphatase conjugate and the enzymatic product thus formed (1-naphthol) was detected by differential pulse voltammetry (DPV).

3 Results

The primary surface modification of the graphite screen-printed working electrodes was obtained by electrodepositing via CV a polyaniline (PANI) layer, followed by gold nanoparticles electrodeposition (AuNPs) carried out via CV too, in accordance with previously reported studies. Cyclic voltammetry provides detailed information on electrode surface changes. Thus, it was used to characterize the layer-by-layer formation of PANI during the electropolymerization process (Fig. 1a, b). Cathodic and anodic peak current height was recorded at different number of cyclic scans. After

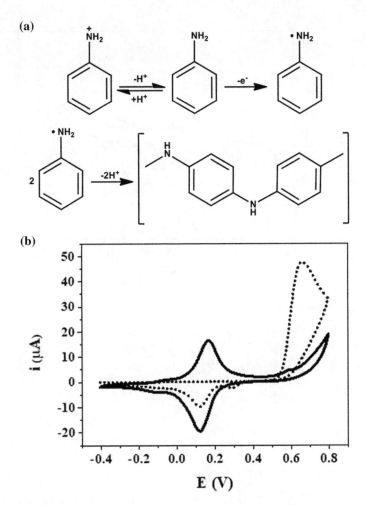

Fig. 1 **a** Aniline polymerization mechanism. **b** Aniline electropolymerization (first and last scans shown)

the 10th cycle, no further increase of peak current height was obtained; consequently, ten cyclic scans were then selected for further experiments.

Then, gold nanoparticles were electrodeposited on polyaniline modified graphite screen-printed electrodes (PANI/GSPEs) using CV in accordance with the optimized procedure previously reported (Fig. 2).

After the electrodeposition of PANI and AuNPs, the electrode conductivity increases of around 40% in comparison to bare GSPE surface.

Preliminary experiments were performed for profenofos pesticide (Fig. 3) detection by competitive assay. By analyzing a 5 μM profenofos solution, a decrease of 50% in the current peak height was obtained with respect to the blank value.

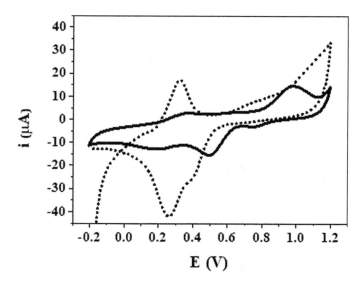

Fig. 2 Gold nanoparticles electrodeposition (first and last scans shown)

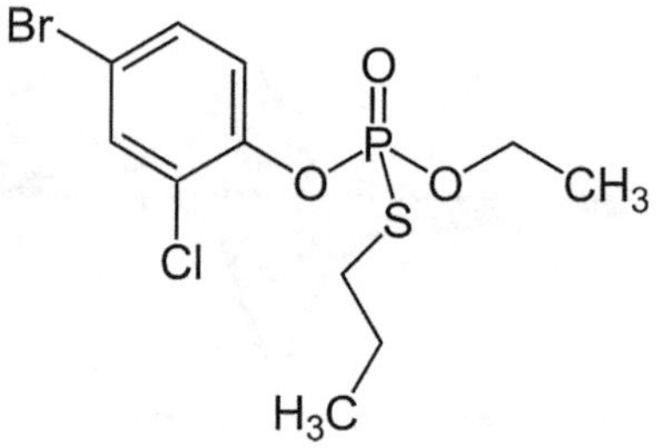

Fig. 3 Profenofos pesticide structure

4 Conclusions

The realized platform, based on graphite screen-printed electrodes (GSPEs) modified with polyaniline (PANI), gold nanoparticles (AuNPs) and an aptamer can contribute to profenofos detection as a valid and innovative analytical approach. The developed aptasensor allows the direct determination of profenofos, thanks to an easy to achieve, rapid and quite low-cost method.

These preliminary results encourage the application in the immediate future of this platform for pesticide detection, testing the developed assay for multiscreening analysis.

References

1. Pesticide Exposure in Womb Linked to Lower IQ. https://www.webmd.com/baby/news/20110421/pesticide-exposure-in-womb-linked-to-lower-iq#1. Last accessed 13 April 2018
2. Aceña, J., Stampachiacchiere, S., Pérez, S., Barceló, D.: Advances in liquid chromatography–high-resolution mass spectrometry for quantitative and qualitative environmental analysis. Anal. Bioanal. Chem. 407, 6289–6299 (2015)
3. Liang, H.C., Bilon, M., Hay, M.T.: Analytical methods for pesticide residues in the water environment. Water Env. Res. 87(10), 1923–1937 (2015)
4. Verma, N., Bhardwaj, A.: Biosensor technology for pesticides-a review. Appl. Biochem. Biotechnol. 175, 3093–3119 (2015)
5. Rapini, R., Marrazza, G.: Electrochemical aptasensors for contaminants detection in food and environment: recent advances. Bioelectrochemistry 118, 47–61 (2017)
6. Rapini, R., Cincinelli, A., Marrazza, G.: Acetamiprid multidetection by disposable electrochemical DNA aptasensor. Talanta 161, 15–21 (2016)
7. Wang, L., Liu, X., Qiang, Z., Zhang, C., Liu, Y., Tu, K., Tu, J.: Selection of DNA aptamers that bind to four organophosphorus pesticides. Biotechnol. Lett. 34, 869–874 (2012)
8. Saberi, R.-S., Shahrokhian, S., Marrazza, G.: Amplified electrochemical DNA sensor based on polyaniline film and gold nanoparticles. Electroanalysis 25(6), 1373–1380 (2013)
9. Ravalli, A., Rossi, C., Marrazza, G.: Bio-inspired fish robot based on chemical sensors. Sens. Actuators B Chem. 239, 325–329 (2017)

A Novel Lab-on-Disk System for Pathogen Nucleic Acids Analysis in Infectious Diseases

Emanuele Luigi Sciuto⊙, Salvatore Petralia⊙ and Sabrina Conoci⊙

Abstract The miniaturization of Real Time PCR (qPCR) systems is a crucial point towards the development of "genetic point-of-care" (PoC) that are able to offer sample-in-answer-out diagnostic analysis. Centralized laboratories and specialized staffs are needed for conventional DNA analysis. To solve this issue, we propose an innovative easy-to-use PoC technology based on a Lab-on-Disk miniaturized system, integrating nucleic acids extraction process based on Mags-Beads technology and detection based on qPCR. Lab-on-Disk system is composed by a polycarbonate disk with reagent-on-board for DNA extraction and a qPCR silicon-chip. A customized reader integrating electronic and optical modules was developed for driving the polycarbonate disk. Here we present results in the detection of Hepatitis B Virus (HBV) genome.

Keywords Pathogen DNA extraction/detection · Microfluidics · Magnetic beads QPCR

1 Introduction

The development of diagnostic technologies able to perform complete DNA analysis in a miniaturized-integrated-automated way is one of the fascinating fields in biomedical research. The scientific innovation is moving towards the decentralization of molecular diagnostics from the hospital's core laboratory. The aim is to allow massive diagnostic screening and better facing the threat of genetic and infectious diseases. Its clinical utility could be much more relevant, for example, in developing

E. L. Sciuto (✉)
Dipartimento di Fisica ed Astronomia, Università degli Studi di Catania, Catania, Italy
e-mail: emanueleluigi.sciuto@imm.cnr.it

S. Petralia (✉) · S. Conoci
STMicroelectronics, Stradale Primosole 50, Catania, Italy
e-mail: salvatore.petralia@st.com

© Springer Nature Switzerland AG 2019
B. Andò et al. (eds.), *Sensors*, Lecture Notes in Electrical Engineering 539,
https://doi.org/10.1007/978-3-030-04324-7_16

countries where infectious disease diagnosis is still a challenge due to poor clinical laboratory infrastructures and cost constraints [1].

Based on the above considerations, multidisciplinary research teams have spent significant efforts to develop innovative technologies and new chemical strategies for DNA analysis, leading to the introduction of the "genetic Point-of-Care" (PoC) [2, 3]. These systems must integrate and automate all steps necessary for the molecular diagnostics, including the DNA sample preparation (extraction and purification) and detection (identification and quantification).

The DNA preparation from biological samples represent one of the major issues in genetic PoC approach, since several starting materials can be employed (blood, urine, saliva etc.) and, consequently, complex architectures and protocols are needed.

Strategies for DNA extraction evolved from liquid–liquid purification method, such as isolation by precipitation with phenol-chloroform [4, 5], to most advanced liquid–solid purification systems, mostly, based on silica in the form of micro-filter mounted in a plastic column or layers covering magnetic beads [6, 7].

However, despite the improvements in terms of yield quality and protocol safety, DNA purification methods still have many widespread limitations.

One of those is the volume and the amount of biological sample required to trigger all procedure. From liquid-liquid to solid-phase extraction, in fact, hundreds of microliters to milliliters volumes are used; this implies invasive biological samplings from patients for diagnosis.

Another common drawback is the miniaturization and integration of the purification technology in a single portable device. A lot of stuff is required to perform the whole experiment of genetic material isolation, frequently with a complex architecture to manage the fluidic steps, which implies an increase of the design complexity, high costs and a laboratory with dedicated and specialized staff.

A solution to all the drawbacks, above reported, is brought by the microfluidic technology [8]. Microfluidic platforms are extremely attractive thanks to the number of advantages, they present, compared to macroscopic equivalents. The small volumes, required for the experimental setup, reduce sample and reagent consumption. Moreover, reduced size of device implies the possibility of miniaturization and integration, so the possibility of working outside the laboratory.

The literature reports few examples of miniaturized sample preparation devices; many of these are developed using plastic materials and include a complex microfluidic network that manages fluid movement, mixing, splitting etc. [9–17]. Plastic materials for integrated microfluidics have the advantage of low cost but the miniaturization, integration and automation required by some of the extraction steps (i.e. lysis) remain the main limitations towards the development of a point-of-care device.

Another issue in PoC is the detection of genetic material.

The DNA analysis is, commonly, based on the Real Time PCR (qPCR) method. In literature, various PCR based PoC have been reported for DNA detection of different pathogen microorganisms. Fernandez-Carballo et al. in 2016 presented a portable and low-cost point-of-care (PoC) system based on continuous flow PCR for quantitative detection of Chlamydia trachomatis and Escherichia coli [18]. Hsien et al. described a sequence-specific electrochemical DNA technology (E-DNA) able

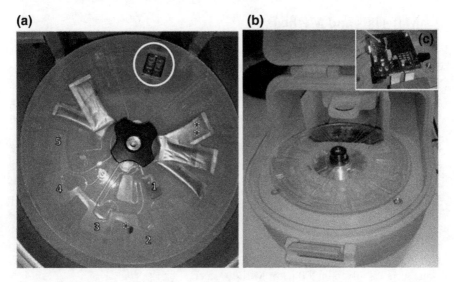

Fig. 1 **a** Lab-on-Disk: sample (1), lysis (2), binding (3), washing (4) and elution (5) chamber; magnetic beads (*); buffer stickpacks ( ); qPCR chip (yellow ring). **b** Disk-reader. **c** Detail of electronic board for qPCR detection

to detect up to 100 copies per ml of S. typhimurium by loop-mediated isothermal amplification, 10 TCID50 for H1N1 influenza virus, and 300 copies (in 50 μL) of Salmonella enterica [19]. Other research teams reported miniaturized devices based on PCR using innovative transduction methods for the detection of pathogen species such as: Escherichia coli (by using piezoelectric-excited cantilever sensors) [20]; Neisseria meningitidis and HBV (using electrochemical transduction) [21–23]; Staphylococcus aureus (by means of graphene oxide based fluorescent probes) [24]; Mycobacterium tuberculosis (using surface plasmon resonance) [25]. Example of genetic SNP (Single Nucleotide Polymorphisms) application are, also, reported [26].

However, among all systems, reported, the miniaturization of the qPCR amplification system, together with the possibility of a sample-in answer-out diagnostic analysis, enabling shorter analysis times, reducing reagent consumption, minimizing risk of sample contamination and enhancing the assay performance (such as sensitivity, specificity and limit of detection), are common drawbacks still to be solved [27].

In this sense, we introduced an innovative Lab-on-Disk microfluidic technology, shown in Fig. 1a, for the complete analysis of pathogen DNA in infectious diseases.

A polycarbonate module, containing the DNA extraction reagent-on-board for the Magnetic-Beads based purification, and a silicon chip, for the qPCR detection of extracted DNA, compose the disk. This is introduced, then, into a specific disk-reader (Fig. 1b) to perform the rotations and amplification protocols.

The disk is able to perform sensitive and high throughput analysis of Hepatitis B Virus genome from a biological sample, overpassing all limitations towards the genetic PoC approach.

2 Materials and Methods

2.1 Chemicals and Reagents

Stock solutions of Hepatitis B virus (HBV) clone complete genome (ref product 05960116), consisting of the HBV genome 3.2 kbps and a plasmid PBR322 vector 3.8 kbps in TE (Tris 10 mM, EDTA 1 mM, pH = 8), and the HBV real time PCR kit (ref product FO2 HBV MMIX KIT 48) were purchased from CLONIT and used according to the Instructions for Use.

2.2 Extraction Experiments

To test the extraction efficiency of polycarbonate module, 10^6 copies/μl of HBV clone were dissolved in 200 μl of Milli-Q water.

Once loaded the sample, a series of rotations allowed the beads to move through all reaction chambers (Fig. 1a), so that the HBV DNA can be adsorbed and purified. To quantify the eluted DNA, a qPCR amplification and cycles threshold (Ct) analysis was performed using the Q3 platform, developed by STMicroelectronics [27, 28]. For the qPCR experiment, 5 μL of both starting and eluted DNA were pipetted into the qPCR chip (described in Sect. 2.3), and preloaded with 10 μL of the HBV kit master mix. The thermal protocol used for qPCR reaction is reported in Table 1. The same experimental conditions have been applied for Magazorb and Qiagen extraction yield analysis.

2.3 Real Time Amplification on the Chip

A miniaturized silicon chip, integrating temperature sensors and heaters for the qPCR amplification and detection of purified DNA (developed by STMicroelectronics [27, 28]), has been assembled to the disk platform.

The chip was thermally and optically driven by an electronic miniaturized board, inside the disk-reader (Fig. 1c). The fluorescence signals were collected by the CCD detector inside the board and analyzed by a smart detection software [27].

	Step	Temperature (°C)	Time (s)
Table 1 Real time PCR thermal protocol	Initial denaturation	99	600
	Denaturation	99	15
	Annealing	62	60
	Extension	50	1

For the qPCR detection test, a negative sample (1 μL of water + 14 μL of HBV master mix) and three positive controls (1 μL of 1×10^6, 1×10^5 and 1×10^3 cps μL^{-1} + 14 μL of HBV master mix) of HBV clone genome were loaded on chip. Thermal protocol was the same reported in Table 1.

The HBV samples were amplified, in parallel, by the commercial Applied Biosystems 7500 as comparison.

3 Results and Discussions

3.1 Module for DNA Extraction

Extraction data from HBV purification on Lab-on-Disk, compared to those from gold standard Magazorb and Qiagen kit, are reported in Table 2.

As shown, the disk platform is able to perform a complete purification process of HBV clone DNA, with an eluate of about 10^3 cps/μl (29.92 Ct).

The protocol for Lab-on-Disk extraction is extremely simplified, since no more than 1 step is required for the experiment (as described in table), with an elution yield comparable to the Magazorb and Qiagen one (the 2 Ct gap is due to a partial DNA retention on beads after the elution in Lab-on-Disk).

3.2 Module for DNA Detection

Results of DNA amplification and fluorescence imaging from qPCR on integrated chip in Lab-on-Disk are shown in Fig. 2a, b, respectively.

In order to estimate the detection efficiency, the qPCR data of 10^6-10^5-10^3 cps/μl HBV DNA amplification on chip have been compared to the gold standard Applied Biosystems 7500 and the Ct values, from quantification, are reported in Table 3.

These data show a detection efficiency improvement (about 1 Ct for all HBV samples) in the case of integrated silicon chip, respect to the commercial instrument. These enhancements in sensitivity and time, since less PCR cycles are required for

Table 2 Extraction data comparison

KIT	Protocol step	Sample volume (μl)	Elution buffer volume (μl)	Eluate volume (μl)	qPCR Ct
Prep disk	1	200	200	200 ± 5	29.92
Magazorb	15	200	200	200 ± 5	27.98
Qiagen	11	200	200	200 ± 5	27.91

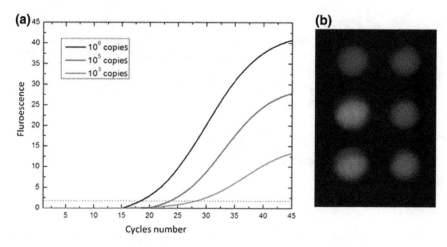

Fig. 2 **a** Lab-on-Disk amplification plot. **b** Fluorescent imaging of qPCR chip

Table 3 Detection data comparison

DNA (copies/μl)	Lab-on-disk (Ct)	Applied biosystems (Ct)
10^6	19.4 ± 0.2	19.9 ± 0.2
10^5	23.9 ± 0.3	23.9 ± 0.2
10^3	28.5 ± 0.5	29.9 ± 0.4

the DNA quantification, result from a perfect synergy between the specific chip architecture and the smart detection software used for the analysis.

4 Conclusions

In this work, we introduce a Lab-on-Disk system for the extraction and detection of Hepatitis B Virus genome. The system is composed by a reader and a polycarbonate disk integrating all reagents for Mag-Beads purification and a miniaturized silicon chip for qPCR detection.

DNA extraction test revealed that the system is able to perform a complete purification of HBV genome in one-step, since no actions of operator are required. In parallel, detection test on integrated qPCR chip in Lab-on-disk, showed that the system is also able to quantify HBV DNA with an improvement of sensitivity of about 1 Ct, respect to the commercial qPCR platforms, which means an optimization in time and material consuming.

These results, together with the miniaturization and integration of all stuff inside the portable disk-reader, are important features towards the identification of Lab-on-Disk as new Point-of-Care technology suitable for infectious disease diagnosis.

Acknowledgements The authors acknowledge HSG-IMIT for the disk design development and CLONIT for providing HBV sample for the extraction and detection tests.

References

1. Yager, P., Domingo, G.J., Gerdes, J.: Point-of-care diagnostics for global health. Annu. Rev. Biomed. Eng. **10**, 107–144 (2008)
2. Mabey, D., Peeling, R.W., Ustianowski, A., Perkins, M.D.: Diagnostics for the developing world. Nat. Rev. Microbiol. **2**, 231–240 (2004)
3. Petralia, S., Conoci, S.: PCR Technologies for point of care testing: progress and perspectives. ACS Sens. **2**, 876–891 (2017)
4. Sambrook, J., Russel, D.: Molecular Cloning: A Laboratory Manual, vol. 3, 3rd edn. Cold Spring Harbor Laboratory Press, New York, NY, USA (2001)
5. Chomczynski, P., Sacchi, N.: The single-step method of RNA isolation by acid guanidinium thiocyanate-phenolchloroform extraction: twenty-something years on. Nat. Protoc. **1**(2), 581–585 (2006)
6. Padhye, V.V., York, C., Burkiewiez, A.: Nucleic acid purification on silica gel and glass mixture. United States patent US 5658548, Promega Corporation (1997)
7. Berensmeier, S.: Magnetic particles for the separation and purification of nucleic acids. Appl. Microbiol. Biotechnol. **73**(3), 495–504 (2006)
8. Chen, L., Manz, A., Day, P.J.R.: Total nucleic acid analysis integrated on microfluidic devices. Lab Chip **7**, 1413–1423 (2007)
9. Petralia, S., Sciuto, E., Conoci, S.: A novel miniaturized biofilter based on silicon micropillars for nucleic acid extraction. Analyst **142**, 140–146 (2017)
10. Benitez, J.J., et al.: Microfluidic extraction, stretching and analysis of human chromosomal DNA from single cells. Lab Chip **12**, 4848–4854 (2012)
11. Breadmore, M.C., et al.: Microchip-based purification of DNA from biological samples. Anal. Chem. **75**, 1880–1886 (2003)
12. Chen, X., Cui, D., Liu, C., Li, H., Chen, J.: Continuous flow microfluidic device for cell separation, cell lysis and DNA purification. Anal. Chim. Acta **584**, 237–243 (2007)
13. Cady, N.C., Stelick, S., Batt, C.A.: Nucleic acid purification using microfabricated silicon structures. Biosens. Bioelectron. **19**, 59–66 (2003)
14. Hwang, K.-Y., Lim, H.-K., Jung, S.-Y., Namkoong, K., Kim, J.-H., Huh, N., Ko, C., Park, J.-C.: Bacterial DNA sample preparation from whole blood using surface-modified Si pillar arrays. Anal. Chem. **80**, 7786–7791 (2008)
15. Hegab, H.M., Soliman, M., Ebrahim, S., Op de Beeck, M.: In-flow DNA extraction using on-chip microfluidic amino-coated silicon micropillar array filter. J. Bionsens. Bioelectron. 4, 1–6 (2013)
16. Petralia, S., Ventimiglia, G.: A facile and fast chemical process to manufacture epoxy-silane coating on plastic substrate for biomolecules sensing applications. BioNanoScience **4**, 226–231 (2014)
17. Petralia, S., Castagna, M.E., Motta, D., Conoci, S.: Miniaturized electrically actuated microfluidic system for biosensor applications. BioNanoScience **6**, 139–145 (2016)
18. Fernández-Carballo, B.L., McGuiness, I., McBeth, C., Kalashnikov, M., Borrós, S., Sharon, A., Sauer-Budge, A.F.: Low-cost, real-time, continuous flow PCR system for pathogen detection. Biomed. Microdevices 18(2), 34 (2016)
19. Hsieh, K., Ferguson, S.B., Eisenstein, M., Plaxco, K.W., Soh, H.T.: Integrated electrochemical microsystems for genetic detection of pathogens at the point of care. Acc. Chem. Res. **48**, 911–920 (2015)
20. Rijal, K., Mutharasan, R.: A method for DNA-based detection of E. coli O157:H7 in proteinous background using piezoelectric-excited cantilever sensors. Analyst 138, 2943–2950 (2013)

21. Patel, M.K., Solanki, P.R., Kumar, A., Khare, S., Gupta, S.: Electrochemical DNA sensor for Neisseria meningitidis detection. Biosens. Bioelectron. **25**, 2586–2591 (2010)
22. Petralia, S., Sciuto, E.L., Di Pietro, M.L., Zimbone, M., Grimaldi, M.G., Conoci, S.: Innovative chemical strategy for PCR-free genetic detection of pathogens by an integrated electrochemical biosensor. Analyst **42**, 2090–2093 (2017)
23. Petralia, S., Castagna, M.E., Cappello, E., Puntoriero, F., Trovato, E., Gagliano, A., Conoci, S.: A miniaturized silicon based device for nucleic acids electrochemical detection, 2015. Sens. Bio-Sens. Res. **6**, 90–94 (2015)
24. Pang, S., Gao, Y., Li, Y., Liu, S., Su, X.: A novel sensing strategy for the detection of Staphylococcus aureus DNA by using a graphene oxide-based fluorescent probe. Analyst **138**, 2749–2754 (2013)
25. Hsu, S.H., Lin, Y.Y., Lu, S.H., Tsai, I.F., Lu, Y.T., Ho, H.S.: Mycobacterium tuberculosis DNA detection using surface plasmon resonance modulated by telecommunication wavelength. Sensors **14**, 458–467 (2013)
26. Foglieni, B., Brisci, A., San Biagio, F., Di Pietro, P., Petralia, S., Conoci, S., Ferrari, M., Cremonesi, L.: Integrated PCR amplification and detection processes on a Lab-on-Chip platform: A new advanced solution for molecular diagnostics. Clin. Chem. Lab. Med. 48, 329–336 (2010)
27. Spata, M.O., Castagna, M.E., Conoci, S.: Image data analysis in qPCR: A method for smart analysis of DNA amplification. Sens. Bio-Sens. Res. **6**, 79–84 (2015)
28. Guarnaccia, M., Iemmolo, R., Petralia, S., Conoci, S., Cavallaro, S.: Miniaturized real-time PCR on a Q3 system for rapid KRAS genotyping. Sensors **17**, 831 (2017)

Diamond-Based Multi Electrode Arrays for Monitoring Neurotransmitter Release

Giulia Tomagra, Alfio Battiato, Ettore Bernardi, Alberto Pasquarelli, Emilio Carbone, Paolo Olivero, Valentina Carabelli and Federico Picollo

Abstract In the present work, we report on the fabrication of a diamond-based device targeted to the detection of quantal neurotransmitter release. We have developed Multi-electrode Arrays with 16 independent graphitic channels fabricated by means of Deep Ion Beam Lithography (DIBL). These devices are capable of detecting the in vitro exocytotic event from neurosecretory cells, while overcoming several critical limitations of standard amperometric techniques.

Keywords Diamond-based sensor · Electrochemical detection
Neuronal network · Ion beam lithography

G. Tomagra (✉)
Drug Science and Technology Department, University of Torino, Corso Raffaello 30,
10125 Torino, Italy
e-mail: gtomagra@unito.it

A. Battiato · P. Olivero · F. Picollo
Section of Torino, Istituto Nazionale di Fisica Nucleare (INFN), Via Pietro Giuria 1,
10125 Torino, Italy

E. Bernardi
Physics Department, University of Torino, Via Pietro Giuria 1, 10125 Torino, Italy

A. Pasquarelli
Institute of Electron Devices and Circuits, Ulm University, Albert Einstein Allee 45,
89069 Ulm, Germany

E. Carbone · V. Carabelli
Drug Science and Technology Department, Inter-departmental Center (NIS),
University of Torino, Corso Raffaello 30, 10125 Torino, Italy

P. Olivero · F. Picollo
Physics Department, Inter-departmental Center(NIS), University of Torino,
Via Pietro Giuria 1, 10125 Torino, Italy

© Springer Nature Switzerland AG 2019
B. Andò et al. (Eds.) *Sensors*, Lecture Notes in Electrical Engineering 539,
https://doi.org/10.1007/978-3-030-04324-7_17

1 Introduction

Exocytosis is a key process of synaptic transmission that occurs when the presynaptic terminal of a neuron is depolarized by an upcoming action potential. Presynaptic Ca^{2+} channels open and the Ca^{2+} flowing into the nerve terminal triggers the exocytosis of presynaptic vesicles. The released neurotransmitters diffuse into the synaptic cleft from the presynaptic terminal to the inter-synaptic space and activate the post-synaptic receptors of neighbouring neurons [1]. Charged oxidizable molecules released from single excitable cells are commonly detected using amperometry, an electrochemical technique that allows resolving the kinetics of fusion and opening of single secretory vesicles with high temporal resolution [2, 3].

Currently, carbon fiber electrodes are conventionally employed for amperometric measurements of neurotransmitters release, but this technique has some limitations: (i) its complexity requires trained operators; (ii) it needs long acquisition times in (single-cell measurements) and (iii) detects only oxidizable molecules. Recently, planar multielectrode devices have been developed to overcome these limitations by exploiting several materials: indium, tin oxide (ITO) diamond-like carbon (DLC), boron-doped nanocrystalline diamond, noble metals (Au, Pt) and silicon-based chips [4].

A new promising substrate for cellular biosensors realization is diamond, that offers a wide spectrum of properties such as an excellent optical transparency, from infrared to near-ultraviolet [5], chemical inertness [6], biocompatibility [7, 8] and the possibility of tuning its electrical properties by directly writing sub-superficial electrodes by MeV ionic lithography [9–12].

In this paper, we describe the microfabrication of monocrystalline diamond substrates for the realization of microelectrode array cellular sensors based on graphitic micro channels (μG-SCD MEA).

2 μG-SCD MEA Microfabrication

Ion beam lithography in diamond is a widely explored technique, which was employed to realize several structures [13–15] and devices, such as waveguides [16, 17] micromechanical resonator [18, 19], photonic structure [20, 21] particle detectors [22–25] and microfluidic channels [26, 27]. This technique exploits the metastable nature of the diamond by giving access to properties of various carbon allotropes, which are completely different (i.e. diamond is an optically transparent electrical insulator, while the graphite is an opaque conductor).

MeV collimated ions are employed to introduce structural damage within the diamond lattice, by inducing the formation of vacancies that promote the progressive creation of a network of sp3- and sp2-bonded carbon atoms. Above a critical level which is usually called graphitization threshold ($1 \times 10^{22} \div 9 \times 10^{22}$ cm^{-3} [28–30]), the complete amorphization of the irradiated material is reached and therefore a high

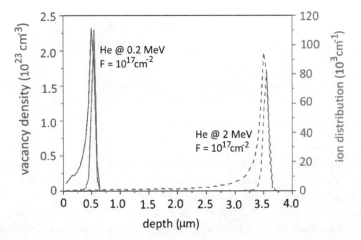

Fig. 1 SRIM Monte Carlo simulation: graphitization threshold is reported in dashed line, the zone of interest is in correspondence of the intersection of the Bragg peak, with the correspondence threshold

temperature thermal annealing ($>900\,°C$) results in the conversion to graphite is obtained. This phenomenon allows the formation of electrically conductive graphitic paths embedded in an insulating diamond matrix.

The employed diamond substrates are typically single crystals with dimensions of $4.5 \times 4.5 \times 0.5\,mm^3$, cut along the (100) crystalline direction and optically polished on the two opposite large faces. The crystals are classified as type IIa, with nitrogen and boron concentrations lower than 1 ppm and 0.05 ppm, respectively. Ion irradiation was performed at room temperature with a broad beam of light ions (i.e. He) with energy comprised between 0.5 MeV and 2 MeV at the INFN National Laboratories of Legnaro, facility equipped with linear accelerator employed for multidisciplinary experiments [31, 32]. The implantation fluence must be defined in order to overcome the critical damage density in correspondence of the Bragg peak [33].

Monte Carlo simulations performed with SRIM code [34] allow evaluating the ion-induced to structural damage on the basis of specific irradiation parameters. For example, Fig. 1 shows that an implantation fluence of $1 \times 10^{17}\,cm^{-2}$ is sufficient to amorphize the diamond with a 2 MeV He beam.

The implantations are performed employing two high-resolution masks that define the 3D geometry of the graphitic structures. The first mask is made up of a metal sheet on which apertures are created through high power laser ablation and it is employed to block the ion beam with the exception of the aperture thus defines the length and width of the graphitic channels. The second mask consists of a metal film deposited directly on the diamond surface by means of thermal evaporation allowing the beam energy modulation that affects the penetration depth. This is functional to connect the buried structures with the sample surface (Fig. 2).

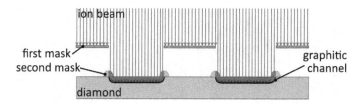

Fig. 2 Deep ion beam lithography of synthetic diamond

With these masks, it is possible to implant simultaneously an array of 16 graphitic channels, whose end-points are exposed to the surface acting as multiple bio-sensing electrodes for in vitro cellular recordings [35, 36].

After implantation, high temperatures treatment is performed ($>950°$) in high vacuum ($\sim 10^{-6}$ mbar) for 2 h in order to induce the permanent conversion of the amorphized regions to a graphite-like phase. 16 graphitic micro-channels (width: $\sim 20\,\mu$m, length: 1.4–1.9 mm, thickness: ~ 250 nm) were obtained at a depth of few μm [37].

The graphitic channels are soldered by flip-chip technique to a polymer-ceramic composite chip carrier (Roger 4003) to allow the connection with the custom front-end electronics. The chip carrier is equipped with an incubation chamber that contains the culture medium necessary for in vitro experiments [38].

The front-end electronics consists of 16 low-noise transimpedance amplifiers with an input bias current of ~ 2 pA and a gain of 100, followed by Bessel low pass filters of the 6th-order with a cut-off frequency of 1 kHz. The filtered signals are acquired with an ADC module (National Instrument USB-6229) having 16-bit resolution over an input range of ± 1 V at a sampling rate up to 25 kHz per channel. The data-acquisition module is interfaced with a PC through a Hi-speed USB link and controlled by a program developed in LabView environment [39].

Fig. 3 Current-voltage characteristics of 16 graphitic electrodes

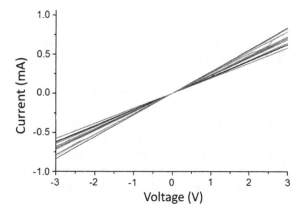

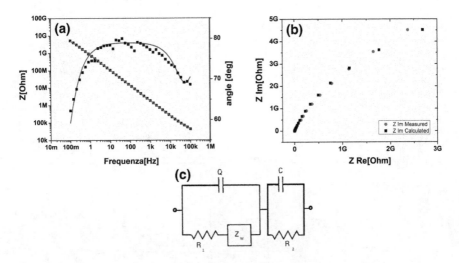

Fig. 4 Figure 4 **a** Bode plot of a representative channel. The black squares represent the measured values, while the solid lines represent the fit considering the equivalent circuit reported in panel (**c**). **b** Nyquist plot. The red circles represent the measured values, while the black squares represent the values calculated using the fit results. **c** Equivalent circuit used to fit the experimental data. The first circuit mesh corresponds to a double layer imperfect capacitor Q, placed in parallel with the resistance due to charge transfer and the impedance due to the diffusion (the Warburg element Z_W). In addition to these elements, a second RC circuit (to the right) is considered to account for the bulk of the electrode

3 Electrical Characterizations

I-V characteristics of the graphitic electrodes were measured to identify their conduction properties. Figure 3 shows linear trends indicating that the electrodes have an ohmic conduction with resistances comprised between $5\,k\Omega$ and $9\,k\Omega$, which correspond to a resistivity of $\rho \sim 1.3\,m\Omega\,cm$, once the geometrical parameters are suitably taken into account. This value is in very satisfactory agreement with that of nanocrystalline graphite ($\rho \sim 1.3\,m\Omega\,cm$) [11, 40].

In Electrochemical Impedance Spectroscopy measurements (EIS), a sinusoidal voltage of $10\,mV$ and variable frequency is applied, with the help of a potentiostat, between the working and the counter electrode and monitored by means of a reference electrode (3-electrode cell), thus measuring the total complex impedance of the circuit. EIS measurements were performed in a grounded Faraday cage to avoid external electric interferences, using a water-based electrolyte (PBS, phosphate buffer saline). The AC signal frequency varied between $0.1\,Hz$ and $100\,kHz$ with seven discrete points per decade, while the DC potential was kept constant at $0\,mV$.

Figure 4a, b reports the data of a representative channel. The module and phase of the impedance $|Z|$ as a function of the modulation frequent (*Bode plot*) were fitted using the equivalent circuit reported in Fig. 4c.

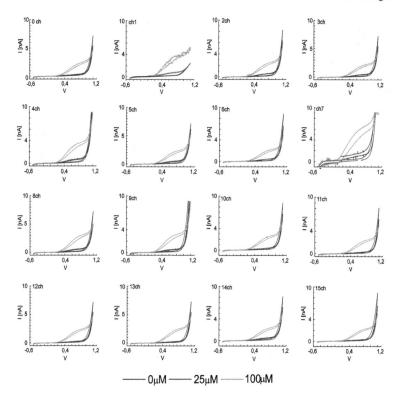

Fig. 5 Steady-state ciclic voltammograms of tyrode in control (0 μm) and with dopamine solution at different concentration (25, 100 μm)

Best-fit parameters were used to calculate the imaginary part of the impedance as function of the real part of the impedance and were compared with the experimental data, as shown in Fig. 4b, indicating good agreement between model and experimental data. The capacitance of the double layer (Q) resulting from the fit is consistent for all the channels (data not reported here) and his impedance is evaluated as $Z_Q = (3.1 \pm 1.2)\,\Omega$ assuming the pulse $\omega = 1$.

Cyclic voltammetry (see Fig. 5) was performed to evaluate the sensitivity of μG-SCD MEA electrodes to detect dopamine. A physiological saline solutions, Tyrode solution, containing (in mM) 128 NaCl, 2 $MgCl_2$, 10 glucose, 10 HEPES, 10 $CaCl_2$ and 4 KCl (pH 7.4), and a Tyrode solution containing dopamine at different concentrations (25 and 100 μm), were employed in these tests. A triangular voltage waveform, ranging from −0.5 and +1.1 V, and with 20 mVs^{-1} scan rate was applied to the graphitic electrodes.

The solution was grounded with a quasi-reference Ag/AgCl electrode. No redox activity was observed using the Tyrode solution in the anodic interval of the hydrolysis window, i.e. up to a polarization voltage of +0.9 V. Under these conditions, a leakage current of less than 10 pA was measured at +0.6 V.

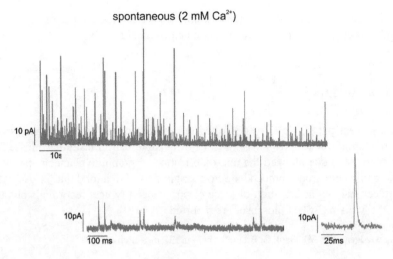

Fig. 6 An example of amperometric recording from PC12 cells

The electrochemical window of the graphitic microelectrodes allowed the detection of the dopamine oxidation peak, between +0.6 and +0.8 V [41], as shown in Fig. 5.

4   Measurements of Quantal Dopamine Release

Ca^{2+}-dependent neurotransmitter release from presynaptic terminals is one of the main mechanisms regulating signal transmission between neurons. Its dysfunction is at the basis of various neurodegenerative diseases. Thus, any technical improvement that facilitates the study of synaptic activity is essential for better understanding the molecular basis of neurosecretion.

Currently, the technique conventionally used to study catecholamine secretion is the amperometric recording with carbon fiber electrodes (CFE). This approach allows the study of quantal release associated to each single quantal event using a redox reaction mechanism between the secreted substance and the carbon fiber. This technique is limited to the measurement of a single cell at a time.

The μG-SCD MEA overcome these limitations recording by simultaneously recording amperometric spikes from more cells, up to 16, maintaining the same high sensitivity and submillisecond temporal resolution of CFEs [42–44]. Figure 6 shows an amperometric recording from an exemplifying electrode. The amperometric signals are collected for 120 s at a sampling rate of 25 kHz. Each spike represents the amperometric current (oxidation) associated with the catecholamine (dopamine) content of a PC12 cells, plated and cultured on the μG-SCD MEA.

PC12 cells are a cell-line derived from rat pheochromocytoma of the adrenal medulla, used as a model of dopaminergic neurosecretory cell.

5   Conclusion

In the present paper, we have described the fabrication of diamond-based multi-electrode array for in vitro measurements of quantal neurotransmitter release. The presented technique allowed the microfabrication of graphitic channels that act as sensing microelectrodes embedded into a single-crystal diamond matrix. An extensive electrical characterization of these electrodes and typical recordings obtained with the MEA μG-SCD biosensor are reported.

Acknowledgements   We thank G. Bruno for help in EIS measurement.

References

1. Südhof, T.C., Rizo, J., Su, T.C., Südhof, T.C., Rizo, J.: Synaptic vesicle exocytosis, cold Spring Harb. Perspect. Biol. **3**(12), 114 (2011)
2. Mellander, L.J., Trouillon, R., Svensson, M.I., Ewing, A.G.: Amperometric post spike feet reveal most exocytosis is via extended kiss-and-run fusion. Sci. Rep. **2** (2012)
3. Simonsson, L., Kurczy, M.E., Trouillon, R.L., Hook, F., Cans, A.S.: A functioning artificial secretory cell. Sci. Rep. **2** (2012)
4. Carabelli, V., et al.: Planar diamond-based multiarrays to monitor neurotransmitter release and action potential firing: new perspectives in cellular neuroscience. ACS Chem. Neurosci. **8**(2), 252264 (2017)
5. Granado, T.C., et al.: Progress in transparent diamond microelectrode arrays. Phys. Status Solidi **212**(11), 2445–2453 (2015)
6. Nemanich, R.J., Carlisle, J.A., Hirata, A., Haenen, K.: CVD diamondResearch, applications, and challenges. MRS Bull. **39**(06), 490–494 (2014)
7. Ariano, P., et al.: Cellular adhesion and neuronal excitability on functionalised diamond surfaces. Diam. Relat. Mater. **14**(37), 669–674 (2005)
8. Ariano, P., Lo Giudice, A., Marcantoni, A., Vittone, E., Carbone, E., Lovisolo, D.: A diamond-based biosensor for the recording of neuronal activity. Biosens. Bioelectron. **24**(7), 2046–2050 (2009)
9. Olivero, P., et al.: Direct fabrication of three-dimensional buried conductive channels in single crystal diamond with ion microbeam induced graphitization. Diam. Relat. Mater. **18**(58), 870–876 (2009)
10. Picollo, F., et al.: Formation of buried conductive micro-channels in single crystal diamond with MeV C and He implantation. Diam. Relat. Mater. **19**(56), 466469 (2010)
11. Picollo, F., et al.: Fabrication and electrical characterization of three-dimensional graphitic microchannels in single crystal diamond. New J. Phys. **14** (2012)
12. Prawer, S., Kalish, R.: Ion-beam-induced transformation of diamond. Phys. Rev. B **51**(22), 1571115722 (1995)
13. Bosia, F., et al.: Finite element analysis of ion-implanted diamond surface swelling. Nucl. Instrum. Methods Phys. Res. Sect. B Beam Interact. Mater. Atoms **268**(19), 2991–2995 (2010)
14. Bosia, F., et al.: Modification of the structure of diamond with MeV ion implantation. Diam. Relat. Mater. **20**(56), 774778 (2011)

15. Bosia, F., et al.: Direct measurement and modelling of internal strains in ion-implanted diamond. J. Phys. Condens. Matter **25**(38), 385–403 (2013)
16. Lagomarsino, S., et al.: Evidence of light guiding in ion-implanted diamond. Phys. Rev. Lett. **105**(23), 233903 (2010)
17. Castelletto, S., et al.: Diamond-based structures to collect and guide light. New. J. Phys. **13**(2), 025020 (2011)
18. Mohr, M., et al.: Characterization of the recovery of mechanical properties of ion-implanted diamond after thermal annealing. Diam. Relat. Mater. **63**, 7579 (2016)
19. Fu, J., et al.: Single crystal diamond cantilever for micro-electromechanical systems. Diam. Relat. Mater. **73**, 267272 (2017)
20. Drumm, V.S., et al.: Surface damage on diamond membranes fabricated by ion implantation and lift-off. Appl. Phys. Lett. **98**(23), 231904 (2011)
21. Lee, J.C., Magyar, A.P., Bracher, D.O., Aharonovich, I., Hu, E.L.: Fabrication of thin diamond membranes for photonic applications. Diam. Relat. Mater. **33**, 4548 (2013)
22. Forneris, J., et al.: A 3-dimensional interdigitated electrode geometry for the enhancement of charge collection efficiency in diamond detectors. EPL (Europhysics Letter) **108**(1), 18001 (2014)
23. Forneris, J., et al.: IBIC characterization of an ion-beam-micromachined multi-electrode diamond detector. Nucl. Instrum. Methods Phys. Res. Sect. B Beam Interact Mater. Atoms **306**, 181–185 (2013)
24. Olivero, P., et al.: Focused ion beam fabrication and IBIC characterization of a diamond detector with buried electrodes. Nucl. Instrum. Methods Phys. Res. Sect. B Beam Interact. Mater. Atoms **269**(20), 2340–2344 (2011)
25. Lo Giudice, A., et al.: Lateral IBIC characterization of single crystal synthetic diamond detectors. Phys. Status Solidi—Rapid Res. Lett. **5**(2), 80–82 (2011)
26. Picollo, F., et al.: Fabrication of monolithic microfluidic channels in diamond with ion beam lithography. Nucl. Instrum. Methods Phys. Res. Sect. B Beam Interact Mater. Atoms **404**, 193–197 (2017)
27. Strack, M.A., et al.: Buried picolitre fluidic channels in single-crystal diamond. Proceedings of SPIE **8923**, 89232X (2013)
28. Hickey, D.P., Jones, K.S., Elliman, R.G.: Amorphization and graphitization of single-crystal diamond—a transmission electron microscopy study. Diam. Relat. Mater. **18**(11), 13531359 (2009)
29. Battiato, A., et al.: Softening the ultra-stiff: controlled variation of Youngs modulus in single-crystal diamond by ion implantation. Acta Mater. **116**, 95103 (2016)
30. Uzan-Saguy, C., Cytermann, C., Brener, R., Richter, V., Shaanan, M., Kalish, R.: Damage threshold for ion-beam induced graphitization of diamond. Appl. Phys. Lett. **67**, 1194 (1995)
31. Rigato, V.: Interdisciplinary Physics with Small Accelerators at LNL: Status and Perspectives, pp. 29–34 (2013)
32. Re, A., et al.: Ion Beam Analysis for the provenance attribution of lapis lazuli used in glyptic art: the case of the Collezione Medicea. Nucl. Instrum. Methods Phys. Res. Sect. B Beam Interact. Mater. Atoms **348**, 278–284 (2015)
33. Olivero, P., et al.: Direct fabrication and IV characterization of sub-surface conductive channels in diamond with MeV ion implantation. Eur. Phys. J. B **75**(2), 127132 (2010)
34. Ziegler, J.F., Ziegler, M.D., Biersack, J.P.: SRIM—The stopping and range of ions in matter. Nucl. Instrum. Methods Phys. Res. Sect. B Beam Interact Mater. Atoms **268**(1112), 1818–1823 (2010)
35. Picollo, F., et al.: All-carbon multi-electrode array for real-time in vitro measurements of oxidizable neurotransmitters. Sci. Rep. **6** (2016)
36. Picollo, F., et al.: Realization of a diamond based high density multi electrode array by means of Deep Ion Beam Lithography. Nucl. Instrum. Methods Phys. Res. Sect. B Beam Interact Mater. Atoms **348**, 199–202 (2015)
37. Bernardi, E., Battiato, A., Olivero, P., Picollo, F., Vittone, E.: Kelvin probe characterization of buried graphitic microchannels in single-crystal diamond. J. Appl. Phys. **117**(2) (2015)

38. Colombo, E., et al.: Fabrication of a NCD microelectrode array for amperometric detection with micrometer spatial resolution. Diam. Relat. Mater. **20**(56), 793797 (2011)
39. Picollo, F., et al.: Development and characterization of a diamond-insulated graphitic multi electrode array realized with ion beam lithography. Sensors **15**(1), 515528 (2015)
40. Ditalia Tchernij, S., et al.: Electrical characterization of a graphite-diamond-graphite junction fabricated by MeV carbon implantation. Diam. Relat. Mater. **74**, 125–131 (2017)
41. Gosso, S., et al.: Heterogeneous distribution of exocytotic microdomains in adrenal chromaffin cells resolved by high-density diamond ultra-microelectrode arrays. J. Physiol. **592**(15), 32153230 (2014)
42. Picollo, F., et al.: Microelectrode arrays of diamond-insulated graphitic channels for real-time detection of exocytotic events from cultured chromaffin cells and slices of adrenal glands. Anal. Chem. **88**(15), 7493–7499 (2016)
43. Picollo, F., et al.: A new diamond biosensor with integrated graphitic microchannels for detecting quantal exocytic events from chromaffin cells. Adv. Mater. **25**(34), 46964700 (2013)
44. Carabelli, V., et al.: Nanocrystalline diamond microelectrode arrays fabricated on sapphire techology for high-time resolution of quantal catecholamine secretion from chromaffin cells. Biosens. Bioelectron. **26**(1), 9298 (2010)

Ultrasensitive Non-enzymatic Electrochemical Glucose Sensor Based on NiO/CNT Composite

K. Movlaee, H. Raza, N. Pinna, S. G. Leonardi and G. Neri

Abstract Quantification of glucose is critical in healthcare applications. Herein, in order to take advantage of both catalytic activity of NiO for glucose oxidation and high conductivity of CNTs, NiO films with different thicknesses were deposited on the surface of stacked-cup carbon nano tube (SCCNT) using atomic layer deposition (ALD) technique. Using the NiO/SCCNT composites, we demonstrated the fabrication of an electrochemical sensor with high sensitivity of 1252.3 μA cm^{-2} mM^{-1} and an ultrafast response (<2 s) for glucose determination in alkaline solution (0.1 M KOH). Additionally, exploiting ALD technique provided us with an opportunity to deposit NiO on the SCCNT surface with different thicknesses, which in turn, enabled us to evaluate thoroughly the effect of different thicknesses of NiO on glucose measurement.

Keywords Glucose · Nickel oxide · Electrochemical sensor

1 Introduction

Glucose is a vital biochemical substance whose detection, quantitatively or qualitatively, is of great importance not only in clinical diagnostic tests but also in food, textile and other industries [1]. Diabetes, as a widespread metabolic disorder relevant to glucose, is being changed to a prevalent challenge for human healthcare due to change in lifestyle and environmental pollutants [2]. Evaluation of glucose can be useful as a diagnostic test for diabetes mellitus, which is a global health problem

K. Movlaee (✉)
School of Chemistry, Center of Excellence in Electrochemistry,
University of Tehran, Tehran, Iran
e-mail: k.movlaee@ut.ac.ir

K. Movlaee · S. G. Leonardi · G. Neri
Department of Engineering, University of Messina, Messina, Italy

H. Raza · N. Pinna
Department of Chemistry, Humboldt-Universitat zu Berlin, Berlin, Germany

© Springer Nature Switzerland AG 2019
B. Andò et al. (eds.), *Sensors*, Lecture Notes in Electrical Engineering 539,
https://doi.org/10.1007/978-3-030-04324-7_18

with devastating social and economic impact, as well as for cancer treatment. There-fore, many efforts have been directed to evaluate glucose using different kinds of organic and inorganic materials. Among inorganic substances, nanomaterials, such as Au, Ni, Pd, Pt, Cu, RuO_2, Co_3O_4, WO_3 and NiO as well as a vast variety of their compositions have been exploited in order to enhance the capabilities of sensors [3]. Herein, we took advantage of carbon nanotubes decorated nickel oxide, which had been prepared by a very precise method named atomic layer deposition (ALD). Using ALD technology gave us the opportunity to not only prepare a sensitive and selective sensor for glucose but also to investigate effects of thickness of nickel oxide layer on catalytic activity and performance of the proposed sensor for glucose determination.

2 Experimental

2.1 Preparation of NiO/SCCNT Composites

First, 120 mg of SCCNT were dispersed in 110 mL of HNO_3 in a round bottom flask and refluxed at 105 °C in an oil bath for 6 h with continuous stirring. The mixture was filtrated and repeatedly washed with distilled water until the neutral point, around pH 7. Oxidized SCCNT were collected and kept in an oven at 80 °C for 16 h. Afterwards, samples for ALD were prepared by dispersing 1.5 mg of Oxidized SCCNT in 2 mL of absolute ethanol by ultrasonic agitation for half an hour. Dispersed SCCNT were drop casted on acetone cleaned aluminium foil (10 cm × 10 cm). Then, ALD was performed in a hot walled, closed chamber type GEMSTAR-6 system from ARRADIANCE, inc. using argon (Air Liquide 99.99% purity) as carrier and purging gas for metal precursor. The temperature of the $Ni(Cp)_2$ container was maintained at 90 °C, while ozone (using pure oxygen in a BMT803 N ozone delivery system, ozone was delivered at a concentration of 100 to 250 g/Nm3) was introduced as generated. The temperatures of the metal precursor and oxidant manifolds and the reaction chamber were maintained at 120 °C, 115 °C and 200 °C, respectively. ALD cycles were performed by sequential pulsing $Ni(Cp)_2$ and ozone for 2 s and 0.5 s, with an exposure/purging time as 30 s/20 s and 20 s/10 s, respectively. After ALD, samples were collected by meticulous scratching off the NiO ALD coated SCCNT from the aluminium foil. Prepared composites were denoted as S25, S50, S100, S200 and S400 where the numbers indicate the number of deposition cycle in ALD.

2.2 Electrode Preparation

In order to modify bare SPE, 1.0 mg of each SCCNT/NiO composites, i.e. S25, S50, S100, S200 and S400, were ultrasonically dispersed in closed polyethylene vials containing distilled water (1 mL). After obtaining homogeneous dispersion of all CNT/NiO composite, desirable amount of CNT/NiO dispersions were directly dropped cast onto the surface of carbon working electrode and left at room temperature to dry until further use.

3 Result and Discussion

3.1 TEM

The detailed structure and morphology of the as synthesized NiO/SCCNT samples was analysed by transmission electron microscopy (TEM). Figure 1 shows TEM and high-resolution TEM (HRTEM) images for some samples, i.e. coated by 25, 50 and 100 layers of NiO. The diameter of the SCCNT as found by TEM images is around 75–110 nm. To differentiate between the samples, a thorough investigation was made by taking the TEM images of individual SCCNT/NiO composites. It can be clearly seen that, SCCNT are equally and conformably coated from the inner and the outer surfaces with NiO nanoparticles. The thickness of the NiO coating increases with the no. of ALD cycles, the average thickness calculated for samples coated with 25, 50, 100, 200 and 400 cycles is 0.9 nm, 1.8 nm, 4.1 nm, 6.8 nm and 13.9 nm, respectively. The layer thickness increase proportionately with the no. of ALD cycles and the growth per cycle (GPC) is calculated as 0.034 nm), proving the self-limiting growth behaviour of the ALD process.

3.2 Electrochemical Behavior of Glucose at CNT/NiO Modified Electrode

The effect of the different thickness of deposited NiO onto CNT toward monitoring of glucose was explored by preparing different modified electrode using S25, S50, S100, S200 and S400. Here, we put our attentions on two important figure of merits i.e. sensitivity and linear dynamic range (LDR) for glucose measurement using these modified electrodes. For achieving this, calibration curves for measurement of glucose with different modified electrodes were obtained using cyclic voltammetry and sensitivity (slope of calibration curve) and LDR were extracted from them. As can be seen in Fig. 2, sensitivity of glucose determination increases with increasing the thickness of NiO up to S100, however, for higher thickness, i.e. S200 and S400, sensitivity decreases to some extent. This behavior can be explained as follows: for S25 sensor, the response to glucose is lowest due to the lack of sufficient active species for glucose oxidation. Increasing the NiO thickness endows higher sensitivity, and finally, for S100 the response reaches the highest value because available active sites for glucose oxidation reached the maximum. For higher thickness, e.g., S200 and S400, the available active sites in the proximity of surface cannot increase further,

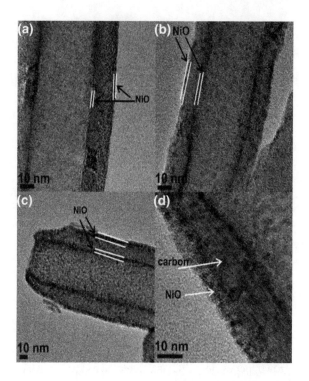

Fig. 1 TEM images of NiO coated SCCNT after deposition of 25 ALD cycles (**a**), 50 ALD cycles (**b**) and 100 ALD cycles (**c, d**)

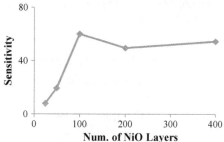

Fig. 2 Effect of different amount of deposited NiO on sensitivity of glucose determination

therefore, sensitivity curve reaches a plateau. On this basis, modified electrode with S100 composite was chosen for further sensing measurements.

The performances of S100 electrode in amperometric mode were assessed by successive addition of glucose aliquots to stirring solution of 0.1 M KOH. Best results regarding sensitivity and linearity were obtained when applied potential was held at 0.65 V. Accordingly, 0.65 V was selected for subsequent amperometric analyses.

Figure 3 exhibits that amperometric measurements of glucose result in the well-defined steady-state shape responses using the proposed sensor.

An attractive aspect of this sensor is ultra-fast response time toward glucose oxidation. The sensor respond almost immediately after addition of glucose and reaches 95% of steady-state current in less than 2 s. The obtained calibration curve

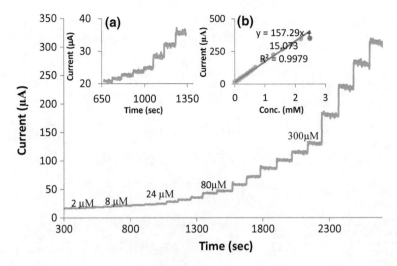

Fig. 3 Current-time responses of NiO/CNT modified electrode acquired in 0.1 M KOH and increasing concentrations of the glucose at Eapp = 0.650 V. Insets illustrate **a** the current-time responses in the low glucose concentration; **b** calibration curve

from amperometric data is shown in the inset of Fig. 3. A good linearity is obtained for this modified electrode over a large range (from micro to millimolar) of glucose concentration. The calibration curve is linear over the concentration range of 2 μM to 2.2 mM with sensitivity of 1310.75 μA cm^{-2} mM^{-1} and determination coefficient (R^2) of 0.9979. Form calibration tests, limit of detection was calculated to be 0.10 μM considering signal-to-noise ratio of 3. These figures of merit compare favorably with literature data, as demonstrated in Table 1.

4 Conclusion

Herein, we took advantage of carbon nanotubes decorated with nanometer nickel oxide layer, which had been prepared by a very precise ALD method. Using ALD gave us this opportunity to not only prepare a sensitive and selective sensor for glucose but also to investigate effects of thickness of nickel oxide layer on catalytic activity and performance of the proposed sensor for glucose determination. These results will be very critical due to the basic knowledge that can be extracted from them in order to design more efficient sensors for glucose.

Table 1 Comparative study of different modified electrode for glucose sensing

Electrode	LDR	LOD (μM)	Sensitivity (μA cm^{-2} mM^{-1})	Response time (s)	References
NiO/GNS[a]	5.0 μM–4.2 mM	5.0	666.71	5	[4]
NiO QDs[b]	1 μM–10 mM and 10–50 mM	26	13.14 and 7.31	–	[5]
NiNPs/ SMWCNTs[c]	1 μM–1 mM	0.5	1438	3	[6]
f-Ni(OH)$_2$- CNT[d]	100 μM–1.1 mM	0.5	238.5	3	[7]
Macro- mesoporous NiO	10 μM–8.3 mM	1.0	243	<5	[8]
NiO/CuO/PANI	10 μM–2.5 mM	2	–	5	[9]
ALD NiO/CNT[f]	2 μM–2.2 mM	0.1	1252.3	<2	This work

[a]Nickel oxide-decorated graphene nanosheet
[b]Nickel oxide quantum dots
[c]Nickel nanoparticles/straight multi-walled carbon nanotubes
[d]Nickel hydroxide nanoflowers/carbon nano tube
[e]Novel nickel and copper oxide nanoparticle modified polyaniline
[f]Atomic layer deposited NiO/Carbon nanotubes

References

1. Galant, A.L., Kaufman, R.C., Wilson, J.D.: Glucose: detection and analysis. Food Chem. **188**, 149–160 (2015)
2. Ocvirk, G., Buck, H., DuVall, S.H.: Electrochemical glucose biosensors for diabetes care. Bioanal. Rev. **6** (2017)
3. Dhara, K., Mahapatra, D.R.: Electrochemical nonenzymatic sensing of glucose using advanced nanomaterials. Microchim. Acta 185 (2018)
4. Zeng, G., Li, W., Ci, S., Jia, J., Wen, Z.: Highly dispersed NiO nanoparticles decorating graphene nanosheets for non-enzymatic glucose sensor and biofuel cell. Sci. Rep. **6** (2016)
5. Jung, D.-U.-J., Ahmad, R., Hahn, Y.-B.: Nonenzymatic flexible field-effect transistor based glucose sensor fabricated using NiO quantum dots modified ZnO nanorods. J. Colloid Interface Sci. **512**, 21–28 (2018)
6. Nie, H., Yao, Z., Zhou, X., Yang, Z., Huang, S.: Nonenzymatic electrochemical detection of glucose using well-distributed nickel nanoparticles on straight multi-walled carbon nanotubes. Biosens. Bioelectron. **30**, 28–34 (2011)
7. Yang, H., Gao, G., Teng, F., Liu, W., Chen, S., Ge, Z.: Nickel hydroxide nanoflowers for a nonenzymatic electrochemical glucose sensor. J. Electrochem. Soc. **161**, B216–B219 (2014)
8. Fan, Y., Yang, Z., Cao, X., Liu, P., Chen, S., Cao, Z.: Hierarchical macro-mesoporous Ni(OH)2 for nonenzymatic electrochemical sensing of glucose. J. Electrochem. Soc. **161**, B201–B206 (2014)
9. Ghanbari, K., Babaei, Z.: Fabrication and characterization of non-enzymatic glucose sensor based on ternary NiO/CuO/polyaniline nanocomposite. Anal. Biochem. **498**, 37–46 (2016)

A Silicon-Based Biosensor for Bacterial Pathogens Detection

Roberto Verardo, Salvatore Petralia◉, Claudio Schneider, Enio Klaric,
Maria Grazia Amore, Giuseppe Tosto and Sabrina Conoci◉

Abstract The miniaturization of integrated nucleic acid testing devices represents a critical step toward the development of portable systems able to offer sample-in-answer-out diagnostic analysis. The conventional molecular testing workflow involves laboratory infrastructures and well-trained staff. Here we present a versatile, user-friendly miniaturized Lab-on-Chip device for the molecular diagnostics of infectious diseases. It is composed of a polycarbonate ring and a silicon chip; a customized reader integrating electronic and optical modules was developed for driving the thermal and optical processes. We report the results obtained using our device for the sample processing and detection of gram-negative opportunistic pathogenic bacteria.

Keywords Point-of-care diagnostics · Nucleic acid testing · Infectious diseases

1 Introduction

The detection of nucleic acids directly from biological samples and its integration into miniaturized biosensors represents a technical challenge in Point-of-Care molecular diagnostics [1]. Its clinical utility is particularly relevant in developing countries where infectious disease diagnosis is still a challenge due to poor laboratory infrastructures [2]. Great efforts have been made for the effective integration of all the main steps of nucleic acid testing (NAT) in miniaturized devices, including DNA extraction [3], PCR amplification and detection [4]. Many PCR-based NAT systems have been developed and reported in the literature; some of these were developed using plastic

R. Verardo · C. Schneider · E. Klaric
LNCIB, Laboratorio Nazionale del Consorzio Interuniversitario
per le Biotecnologie, Padriciano 99, 34149 Trieste, Italy
e-mail: roberto.verardo@lncib.it

S. Petralia (✉) · M. G. Amore · G. Tosto · S. Conoci
STMicroelectronics, Stradale Primosole 50, 95121 Catania, Italy
e-mail: salvatore.petralia@st.com

© Springer Nature Switzerland AG 2019   141
B. Andò et al. (eds.), *Sensors*, Lecture Notes in Electrical Engineering 539,
https://doi.org/10.1007/978-3-030-04324-7_19

materials and include complex microfluidic modules [5], while a few PCR-free systems have been developed [6]. Plastic materials for integrated microfluidic devices
have the advantage of low cost but the miniaturization, integration and automation
required by the biological applications remain the main limitations towards the development of fully integrated point-of-care devices. Various miniaturized systems have
also been reported for the detection of pathogenic bacteria such as Chlamydia trachomatis and Escherichia coli [7], Salmonella enterica [8, 9] and Mycobacterium
tuberculosis (using surface plasmon resonance) [10].

In this study we have developed a fully integrated sample-in-answer-out system
based on a silicon device able to perform sample processing and real-time PCR
amplification in a single microchamber, without the need for nucleic acid purification or movement between different reaction chambers. The system is composed of
a miniaturized silicon chip mounted on a polycarbonate ring to form microchambers
[11]. A portable and easy-to-use reader integrates an electronic board to drive the
silicon device and the optical components to capture the q-PCR fluorescence images.
This work reports the experimental results for the detection of gram-negative (Pseudomonas aeruginosa) bacteria from cultured samples. The DNA extraction was performed by standard Lysozyme and Proteinase treatments, while the real-time PCR
amplification was performed using species-specific PCR primers. For comparison,
the same samples were analyzed using commercial q-PCR equipment (StepOnePlus,
Applied Biosystems).

2 Materials and Methods

2.1 Chemicals and Biological Reagents

All reagents were purchased from Sigma-Aldrich and used as received. The proteinase reagent was purchased from ZyGEM (MicroGEM, United Kingdom). The
PCR reagents were purchased from Kapa Biosystems and used as received. The biological target: Gram-negative (Pseudomonas aeruginosa) bacteria come from cultured samples internally prepared.

2.2 Biosensor Description

The biosensor system (developed by STMicroelectronics) is composed of a miniaturized hybrid silicon-polycarbonate chip with six microchambers (each microchamber
with a total volume of 22 μl) (Fig. 1). The device, based on VLSI-technology, integrating heaters and temperature control sensors in its silicon part allows to perform
PCR thermal cycles with an accuracy of 0.1 °C. A customized reader integrates an

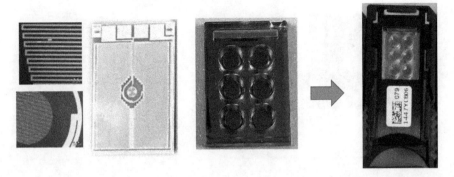

Fig. 1 Miniaturized silicon device composed of heaters and temperature sensors and 6 round microreactors

electronic board to drive the thermal and optical module for PCR amplification and detection, while a smart software was developed for data analysis.

3 Results and Discussion

3.1 Sample Processing and Detection

Gram-negative bacteria were lysed in a microchamber in the presence of lysozyme and proteinase using the following thermal protocol: 15 min at 37 °C (lysozyme activation), 15 min at 75 °C (proteinase digestion) and 15 min at 95 °C (proteinase inactivation). Next, 5 μl of lysate were transferred to a second microchamber and subjected to q-PCR with gram-negative specific primers using the following thermal protocol: 3 min at 95 °C (initial denaturation), 30 s at 60 °C (annealing-extension), for 45 cycles. For comparison, the same samples were run on a StepOnePlus Real Time PCR System (Applied Biosystems).

3.2 Real-Time PCR Experiments

Crude lysates were obtained from bacterial cultures and employed as described in materials and methods section. Five replicas were run for each sample, together with the negative control, NTC (No Template Control). The q-PCR experiments were performed on a miniaturized chip, which was thermally and optically driven by a miniaturized electronic board. The fluorescence signals were collected by the CCD detector inside the board and analyzed by a smart detection software. Figure 2 shows an example of the q-PCR curves while quantitative results as reported in Table 1.

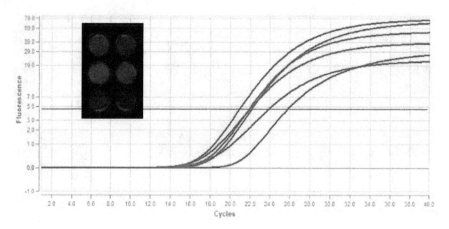

Fig. 2 q-PCR curves

Table 1 q-PCR quantitative results. The experimental results demonstrate the applicability of the new system based on the silicon biosensor compared to standard methods

Sample	Ct from device	Ct from standard instrumentation
PA1	21.1	20.8
PA2	20.8	20.0
PA3	22.2	21.5
PA4	23.2	22.8
PA4	19.8	20.8
Negative control (NTC)	33.3	32.6

4 Conclusion

Here we present a versatile, miniaturized Lab-on-Chip device useful for the molecular diagnostics of infectious diseases. The whole system is composed of a miniaturized silicon-based chip, a reader and a software for data analysis. We report the results obtained for the sample processing and detection of gram-negative opportunistic pathogenic bacteria. The q-PCR experiments revealed that the proposed system is able to perform the species-specific detection on the silicon-based chip. These results, together with the miniaturization and integration are important features towards the identification of an easy-to-use, Point-of-Care molecular diagnostic platform for rapid and sensitive detection of bacterial pathogens without requiring major hands-on time and complex laboratory instrumentation.

References

1. Petralia, S., Verardo, R., et al.: Sens. Actuators B Chem. **187**, 99 (2013)
2. Yager, P., Domingo, G.J., Gerdes, J.: Annu. Rev. Biomed. Eng. **10**, 107–144 (2008)
3. Petralia, S., Sciuto, E., Conoci, S.: A novel miniaturized biofilter based on silicon micropillars for nucleic acid extraction. Analyst **142**, 140–146 (2017)
4. Foglieni, B., Brisci, A., San Biagio, F., Di Pietro, P., Petralia, S., Conoci, S., Ferrari, M., Cremonesi, L.: Clin. Chem. Lab. Med. **48**, 329–336 (2010)
5. Petralia, S., Conoci, S.: PCR technologies for point of care testing: progress and perspectives. ACS Sens. **2**, 876–891 (2017)
6. Petralia, S., Sciuto, E.L., Di Pietro, M.L., Zimbone, M., Grimaldi, M.G., Conoci, S.: Innovative chemical strategy for PCR-free genetic detection of pathogens by an integrated electrochemical biosensor. Analyst **42**, 2090–2093 (2017)
7. Fernández-Carballo, B.L., McGuiness, I., McBeth, C., Kalashnikov, M., Borrós, S., Sharon, A., Sauer-Budge, A.F.: Low-cost, real-time, continuous flow PCR system for pathogen detection. Biomed. Microdevices **18**(2), 34 (2016)
8. Hsieh, K., Ferguson, S.B., Eisenstein, M., Plaxco, K.W., Soh, H.T.: Integrated electrochemical microsystems for genetic detection of pathogens at the point of care. Acc. Chem. Res. **48**, 911–920 (2015)
9. Rijal, K., Mutharasan, R.: A method for DNA-based detection of E. coli O157:H7 in proteinous background using piezoelectric-excited cantilever sensors. Analyst **138**, 2943–2950 (2013)
10. Hsu, S.H., Lin, Y.Y., Lu, S.H., Tsai, I.F., Lu, Y.T., Ho, H.S.: Mycobacterium tuberculosis DNA detection using surface plasmon resonance modulated by telecommunication wavelength. Sensors **14**, 458–467 (2013)
11. Guarnaccia, M., Iemmolo, R., Petralia, S., Conoci, S., Cavallaro, S.: Miniaturized real-time PCR on a Q3 system for rapid KRAS genotyping. Sensors **17**, 831 (2017)

M13 Bacteriophages as Bioreceptors in Biosensor Device

Laura M. De Plano, Domenico Franco, Maria Giovanna Rizzo, Sara Crea, Grazia M. L. Messina, Giovanni Marletta and Salvatore P. P. Guglielmino

Abstract New recognition probes sensible, specific and robust is one of the major problems of biosensor assay. Detection biosensors has utilized antibodies or enzymes as bioreceptors; however, these have numerous disadvantages of limited binding sites and physico-chemical instabilities, can negatively affect capture and detection of target in diagnostic device. In this contest, Phage-Display provides a valuable technique for obtaining large amounts of specific and robustness bio-probes in a relatively short time. This technique relies on the ability of M13 bacteriophages (or phages) to display specific and selective target-binding peptides on major coat protein pVIII of their surface. In this work, we used P9b phage clone, displaying a foreign peptide QRKLAAKLT to selectively recognize *Pseudomonas aeruginosa* like bioreceptor. We describe different methods of functionalization to realize a selective bacteria biosensor surfaces. Several surfaces, such as latex and magnetic beads and polymeric surfaces such as mica, APTES and PEI, were functionalized by covalent bonds or physisorption with P9b. The efficiency of the surface functionalization procedures was evaluated by ELISA and AFM, while capture efficiency of the anchored phages has been assessed by plate count and Fluorescence microscopy. The results of this work pave the way to the use of phages as bioreceptor.

Keywords Phage display · Selective probes · Functionalization · Biosensor Pathogen detection

L. M. De Plano (✉) · M. G. Rizzo · S. Crea · S. P. P. Guglielmino (✉)
Department of Chemical Sciences, Biological, Pharmaceutical and Environmental, University of Messina, Viale F. Stagno d'Alcontres 31, 98166 Messina, Italy
e-mail: ldeplano@unime.it

S. P. P. Guglielmino
e-mail: sguglielm@unime.it

D. Franco
Department of Mathematical and Computational Sciences, Physical Science and Earth Science, University of Messina, Messina, Italy

G. M. L. Messina · G. Marletta
LAMSUN (Laboratory for Molecular Surfaces and Nanotechnology), Department of Chemical Sciences, University of Catania and CSGI, Viale A. Doria 6, 95125 Catania, Italy

© Springer Nature Switzerland AG 2019   147
B. Andò et al. (eds.), *Sensors*, Lecture Notes in Electrical Engineering 539,
https://doi.org/10.1007/978-3-030-04324-7_20

1 Introduction

Advanced bio-selective sensors may meet the requests for isolation, concentration of the agents and their immediate real-time detection. Several biosensors have been described utilized, antibodies, as well as some antibiotics, proteins, DNA/RNA and aptamers, as bioreceptors [1, 2]. However many of these are numerous disadvantages including high cost of production, low availability, great susceptibility to environmental conditions and the need for laborious immobilization methods to sensor substrates. In this contest, Phage-Display provides a valuable technique for obtaining large amounts of specific bio-probes in a relatively short time [3, 4]. Phage-display allows to incorporate random peptide sequences into the major coat protein pVIII of filamentous bacteriophages. The phage able to bind a cell target have been successfully used as molecular recognition agents [5, 6]. The M13 phage is approximately 6.6 nm wide and 880 nm long with circular single-stranded DNA encapsulated in 2,700 copies of the major coat protein pVIII and capped with 5 copies of different minor coat proteins [7].The phages are structurally robust, and have a strong resistance to pH and denaturing agents [8] and its polyelectrolyte nature permit to change the spatial distributions of charges on its surfaces using different pH medium. However, the phages can be used as bioreceptors in sensor devices, as far as it necessarily exposes the binding-peptides on their surfaces. This implies that a crucial problem involves the capability to find the optimal conditions of immobilization on the different surfaces, as it rules their sensing efficiency [9].

In this work, we describe different methods of functionalization to realize a selective bacteria biosensor device, based on immobilization of affinity-selected phage. We used P9b phage clone, displaying a foreign peptide QRKLAAKLT to selectively recognize *Pseudomonas aeruginosa* [10], to tie it on biosensor surfaces. We have studied the use of functional groups present of the phage capsid surface to make the covalent bonds on latex and magnetic beads. Furthermore the analyse of isoelectric point (pI) has permit to optimize the physisorption on the polymeric surfaces such as mica, APTES and PEI using different buffer and pH values. The efficiency of the surface functionalization procedures was evaluated by means of AFM for the morphology of the anchored phages and by means of ELISA assay for the immobilized amount, while capture efficiency of the anchored phages has been assessed by Fluorescence microscopy and plate count. The results of this work pave the way to the use of immobilized phages as bioreceptor.

2 Materials and Methods

2.1 Bacteriophages

P9b phage clone, and St.au 9IVS5 were derived from M13-pVIII-9aa phage peptide library, constructed in the vector pC89, and displayed the foreign peptide

QRKLAAKLT and RVRSAPSSS, which represents a specific and selective probes for *P. aeruginosa* [10] and *S. aureus* [11], respectively. The isoelectric points (pI) of the predicted peptide sequences were calculated by 'compute MW/pI,' also present on the ExPASy proteomics server.

2.2 Functionalization of Magnetic and Latex Beads

ScreenMAG-Amine superparamagnetic beads (1 μm diameter, Chemicell GmbH (Berlin, Germany) and Carboxyl-polystyrene latex beads (0.8 μm diameter, SERVA Electrophoresis GmbH (Heidelberg, Germany) were functionalized using protocol described previous [12, 13]. Phage suspensions in ultrapure water were functionalized with the ratio about 360 phage clones/magnetic beads or with the ratio w/v of 320 μl of phage (title of $1.3 \cdot 10^{12}$ PFU/ml) and 1 ml of beads (10% w/v in ultrapure water). In order to verify phage coating of the beads, we performed an ELISA test with M13-pVIII antibody.

In order to determine maximum capture efficiency, 10^3 *Pseudomonas aeruginosa* ATCC 27853 cells/ml were incubated for 30 min against scalar concentration of phage-coated beads (for latex beads: 5, 7.5, 10 and 12.5% w/v; for magnetic beads: 10^2, 10^4, 10^6 and 10^8 beads). Colony Forming units per millilitre (CFU/ml) counts were determined before and after beads incubation with bacteria and the capture efficiency percentage was calculated. The same capture test was performed with unfunctionalized beads (Blank). Tests were performed in triplicate and results were reported as percentage average of capture.

2.3 Binding of Phage to Polymeric Surface and Capture of Bacteria Target

Surfaces such as quartz and polymeric surface mica, APTES and PEI were functionalized using physisorption protocol described previous work [10, 11]. Phage suspensions in ultrapure water at pH7 and Tris-buffered saline (TBS) (Tris hydrochloride (7.88 g/L) and sodium chloride 140 mM (8.77 g/L) at pI of the bacteriophage, 4.2, 6.3 and 5.4 for pC89 vector, P9b and St.au9IVS5 respectively. The organization of the physisorbed phage layer on the surface was analyzed by AFM analysis. AFM measurements were carried out in tapping mode by using a Nanoscope IIIA-MultiMode AFM (Digital Instruments, Santa Barbara, CA, USA). A negative control, consisting of surface functionalized with pC89 vector was also analyzed. An culture of *P. aeruginosa* or *S. aureus* in PBS at $\approx 4 \times 10^6$ CFU/ml were pre-labeled with DAPI fluorochrome, then was incubated for 15 min at RT with the functionalized surface; washed for three times and observed by fluorescence microscopy (Leica DMRE). Sequential digital images of cell binding were acquired using a CCD camera (Leica

DC300F) and cells number estimated by Scion Image Software (Windows version of NIH Image Software), in terms of integrated density (I.D.).

3 Results and Discussion

In this work, we describe different methods of functionalization to realize a selective bacteria biosensor device, based on immobilization of affinity-selected phage. The surface anchoring has been based on the following two approaches, covalent bonds and physisorption tuned by optimizing pH, ionic force and species. It's possible obtained an optimal functionalization of phage clone using both amino or carboxyl groups present on the its surface. The activation agent EDC in MES reacts with the carboxyl groups, forming highly reactive O-acylisourea derivatives that react readily with the amino groups. In particular to performed phage-coated latex beads the activated carboxyl-groups, exposed on bead surface, covalently bond specific NH_2 phage-chemical groups. On the contrary to obtain phage-coated magnetic beads the P9b phage clone bond covalently the amino groups exposed on magnetic beads, due to intermedia of reaction formed on the COO^- phage-chemical groups. In both cases the data show an efficient coupling of phage onto beads compared to the negative control (Fig. 1a).

The results of capture efficiency were reported as percentage average of captured bacteria (Fig. 1b, c). The phages, coated on the beads using both NH_2 or the COOH present on their surface, maintained the recognition capacity on both type of functionalization despite the covalent bind. In particular, the percentage of cells captured by the phage-coated latex beads increased from 13 to 39.63% as phage-coated beads increased from 5 to 10% w/v. On the contrary, the capture efficiency decreases from 39.63 to 30% when phage-coated beads further increase, suggesting that the ratio between phage-coated beads and *P. aeruginosa* cells was a major factor influencing

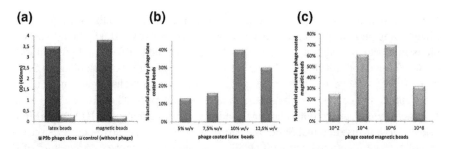

Fig. 1 a ELISA assay on phage-coated beads. Beads functionalized with P9b phage clone and respective negative controls (not coated beads); **b** Dependence of capture efficiency on the amount of used phage-coated latex beads; **c** Dependence of capture efficiency on the amount of used phage-coated magnetic beads. All tests were performed in triplicate and results were reported as percentage average of capture

the capture efficiency. Similarly, phage-coated magnetic beads, at same bacteria concentration, the capture efficiency were clearly increasing until 67% only until at one optimal concentration of 10^6 phage-coated magnetic beads. This is possible because the probability meeting between two elements is low when was presented both small and high quantity of phage-coated beads in the solution with same bacteria concentration. The data show that 10% w/v of P9b-coated latex beads and 106 P9b-coated magnetic beads permit the better capture efficiency of bacteria target.

The wild-type phage coat protein pVIII of pC89 vector (a phage control without the exogenous peptides), influences the isoelectric point (pI) calculated and experimentally confirmed as pH 4.2 [14]. However, the phage clones, present the exogenous amino-acid sequence "in frame" the wild-type shifts the pI at a different value. To evaluate the effect of physisorption for the different pI due to peptides exposed, we used pC89 vector, P9b and St.au9IVS5 specific and selective for *S. aureus* in process. The pI of the peptide sequences, exposed on the clones surfaces, were value at 6.3 and 5.4 for P9b and St.au9IVS5 by protein calculator, respectively. The protein surface charge is positive at pH below pI, and negative at pH above pI, so in the case of pH 7 resulted -3, -1 and -0.1 charge negative for pC89, St.au9IVS5 and P9b, respectively. Our preliminary data showed as the peptide exposed on the pVIII protein of the phages influenced the functionalization according to the phage used, the pH, the buffer solution and the characteristics of surface (Fig. 2).

Only sporadic adhesion for pC89 vector on the mica surface in H_2O pH 7, unlike of St.au9IVS5 which showed a bubbles disposition only when the functionalization was conducted a pH 9. Instead, P9b phage clone showed an adhesion in insula disposition at pH 7, condition in which his charge is near at neutrality. The repulsion between phages clones and mica surface is attributable to electrostatic effects, mica has a surface with negative potential and all the phage clones, in the pH medium used (pH 7), have negative charges exposed on their surfaces. When the phages clone were in TBS pI buffer the ions present influenced the pC89 vector which covers whole the surfaces; likewise St.au9IVS5 phage clone was a similar behavior, however, an better arrangement was observed along the surface. Probably the peptide plays an important role in the stabilization of salt–bridge, which are highly dependent upon factors such as the cost of desolvating the charged groups and the relative flexibility of the side chains involved in the ion pair.

On the contrary the peptide exposed on pVIII proteins of the P9b, influence negatively the density of disposition of the phage on the mica surface. In this surface P9b shows sporadic adhesion in elongated form and in insula disposition in both buffer used.

On surfaces with positive charge such as PEI and APTES the functionalization was conditioned with these charged exposed on the surfaces. Pc89 vector in H_2O pH 7, is strongly attracted of the positive surfaces permit the multi-stratification of the phage on the surfaces. However, in TBS pI, the pC89 vector the present Na^+ ions cover phage surface, reduced the interaction with the positive charges exposed on surface. On the contrary St.au9IVS5 clone has a better distribution in salts presence (TBS pI) compared to H_2O pH7 where the phage clone was conditioned in the self-assembling only for its negative charges exposed. The P9b clone, which is neutral in all the

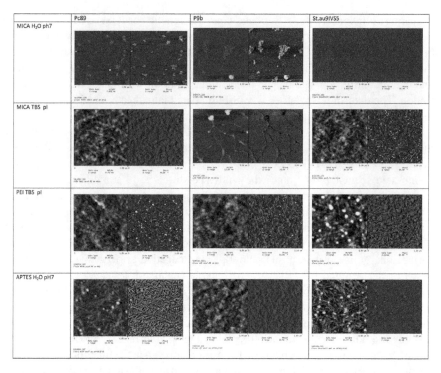

Fig. 2 AFM images of Pc89 vector, P9b and St.au9IVS5 on mica in H_2O at pH 7 (the first line); mica in Tris-buffered saline (TBS) buffer at pI (the second line), PEI in TBS buffer at pI (the third line); APTES in H_2O at pH 7 (the last line)

conditions tested, immobilization on positive surfaces produces the formation of a highly reticulated phage-networks roughly covering the whole substrates; however in TBS pI permits an organized stratification on the surface compared to H_2O pH7.

Also in this case is possible notice that the peptides have an important role in the formation of a shell on the phage structure obtaining the stabilization of salt–bridge and consequently complete stratification compared to pC89 vector. Then the optimal condition of immobilization were on PEI in TBS pI for P9b clone, on mica in TBS pI for St.au9IVS5 and on APTES in H_2O solution for pC89 vector. In these surfaces phage networks show a two-levels organization, respectively corresponding to a phage self-assembly in short fibers (less than 200 nm wide) and large bundles of 5.68 ± 0.68 nm thick and 34.42 ± 3.29 nm wide in the first-level; second-level organization mode occurring on the phage bundles with circular platelets of 3.13 ± 0.24 nm thick and with an average diameter of 40.04 ± 2.56 nm (second-level). Thus, the recognition capability of the targeted bacteria was performed on the optimal functionalized surface. Statistical analysis in terms of I.D. shows a significant coverage of the surface, in particular *P. aeruginosa* and *S. aureus* display the same pattern of the

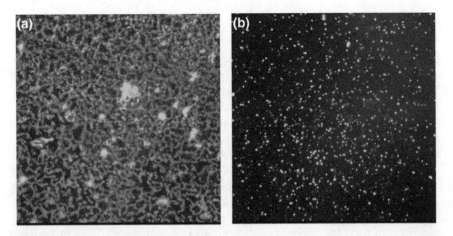

Fig. 3 *P. aeruginosa* captured on PEI physisorbed P9b at PEI; *S. aureus* captured on mica physisorbed St.au9IVS5 at TBS pI

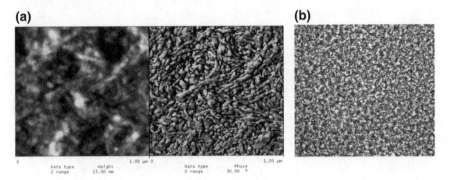

Fig. 4 **a** P9b immobilized on quartz; **b** Phase contrast microscopy images of *P. aeruginosa* captured on P9b-immobilized quartz 1000 × magnification

P9b-coated and St.au9IVS5-coated surface confirmed that the P9b and St.au9IVS5 maintained the capacity to bind the targets (Fig. 3a, b).

In the quartz, which does not present litching characteristic, the phage suspension in TBS pH7 permit an optimal functionalization to have a surface covered (Fig. 4a). The phage functionalized maintains the ability to recognize and capture the target as observed by high-power optical phase contrast microscopy (Fig. 4b). Also in this case the salts play a fundamental role in the multilayer functionalization of phages on the surface.

4 Conclusions

The results of this work pave the way to the use of immobilized phages as bioreceptor. In fact the phage probe could be used to build a micro-biosensor system in which biological sensing element is the selected phage-displayed peptide. Through the use of functional groups on the capsid surface or the analysis of the isoelectric point of the phage clone it was possible to find the optimal conditions of functionalization of different surfaces maintaining the ability to bind and capture target. Then phage-based biosensor offer many unique advantages in the context of product development and commercialization, low time and cost, sensibility and specific. Furthermore, the nature of the bioreceptor layer holds potential utilization for detection of others pathogens agents as bacteria, cancer cells, virus or toxin to which a corresponding phage was selected for. Sensors prepared with phage as probes could be an effective analytical method for detecting and monitoring quantitative changes of bacterial agents under any conditions, including clinical based diagnostics, food monitoring and industrial use.

References

1. Sposito, A.J., Kurdekar, A., Zhao, J., Hewlett, I.: Application of nanotechnology in biosensors for enhancing pathogen detection. WIREs Nanomed. Nanobiotechnol. (2018). https://doi.org/10.1002/wnan.1512
2. Alhadrami, H.A.: Biosensors: Classifications, medical applications, and future prospective. Biotechnol. Appl. Biochem. 2017
3. Chen, J., Duncan, B., Wang, Z., Wang, L.-S., Rotello, V.M., Nugen, S.R.: Bacteriophage-based nanoprobes for rapid bacteria separation. https://doi.org/10.1039/c5nr03779d
4. Qi, H., Lu, H., Qiu, H.J., Petrenko, V.A., Liu, A.: Phagemid vectors for phage display: properties, characteristics and construction. J. Mol. Biol. **417**(3), 129–143 (2012). https://doi.org/10.1016/j.jmb.2012.01.038
5. Carnazza, S., Gioffrè, G., Felici, F., Guglielmino, S.: Recombinant phage probes for Listeria monocytogenes. J. Phys. Condens. Matter **19**, 395011 (13 pp) (2007). http://dx.doi.org/10.1088/0953-8984/19/39/395011
6. Liu, P., Han, L., Wanga, F., Petrenko, V.A., Liu, A.: Gold nanoprobe functionalized with specific fusion protein selection from phage display and its application in rapid, selective and sensitive colorimetric biosensing of Staphylococcus aureus. Biosens. Bioelectron. **82**, 195–203 (2016). https://doi.org/10.1016/j.bios.2016.03.075
7. Butler, J.C., Angelini, T., Tang, J.X., Wong, G.C.L.: Ion multivalence and like-charge polyelectrolyte attraction. Phys. Rev. Lett. **91**, 028301 (2003)
8. Petrenko, V.A., Vodyanoy, V.J.: Phage display for detection of biological threat agents. J. Microbiol. Methods **53**, 253–262 (2003) (PMID: 12654496)
9. Huang, S., Yang, H., Lakshmanan, R.S., Johnson, M.L., Chen, I., Wan, J., Wikle, H.C., Petrenko, V.A, Barbaree, J.M., Cheng, Z.Y.: The effect of salt and phage concentrations on the binding sensitivity of magnetoelastic biosensors for Bacillus anthracis detection. In: Chin, B.A. (ed.) Biotechnol. Bioeng. **101**, 1014–1021 (2008)
10. Carnazza, S., Foti, C., Gioffrè, G., Felici, F., Guglielmino, S.: Specific and selective probes for Pseudomonas aeruginosa from phage-displayed random peptide libraries. Biosens. Bioelectron. **23**, 1137–1144 (2008)

11. De Plano, L.M., Carnazza, S., Messina, G.M.L., Rizzo, M.G., Marletta, G., Guglielmino, S.P.P.: Specific and selective probes for Staphylococcus aureus from phage-displayed random peptide libraries. Colloids Surf. B Biointerfaces **157**, 473–480 (2017)

12. Calabrese, F., Carnazza, S., De Plano, L.M., Lentini, G., Franco, D., Guglielmino, S.P.P.: Phage-coated paramagnetic beads as selective and specific capture system for biosensor applications. In: XVIII AISEM Annual Conference (2015)

13. Lentini, G., Franco, D., Fazio, E., De Plano, L.M., Trusso, S., Carnazza, S., Neri, F., Guglielmino, S.P.P.: Rapid detection of Pseudomonas aeruginosa by phage-capture system coupled with micro-Raman spectroscopy. Vib. Spectrosc. **86**, 1–7 (2016). http://dx.doi.org/10.1016/j.vibspec.2016.05.003

14. Zimmermann, K., Hagedorn, H., Heucks, C.Chr., Hinrichsen, M., Ludwig, H.: The ionic properties of the filamentous bacteriophages Pfl and fd. J. Biol. Chem. **261**(4), 1653–1655 (1986)

One-Step Functionalization of Silicon Nanoparticles with Phage Probes to Identify Pathogenic Bacteria

Maria Giovanna Rizzo, Laura M. De Plano, Sara Crea, Domenico Franco, Santi Scibilia, Angela M. Mezzasalma and Salvatore P. P. Guglielmino

Abstract Optical biosensors are powerful alternatives to the conventional analytical techniques, due to their particular high specificity, sensitivity, small size, and cost effectiveness. Although promising developments of optical biosensors are reported, new bioprobes of cheap and easy synthesis are required, for detection of eukaryotic cells or dangerous infectious agents. In this regard, silicon nanoparticles (SiNPs) can be used as nanoplatform owing to their high specific surface area, optical properties and biocompatibility. They can also be functionalized with bio-probes and used in diagnostic applications. Different methods are described to obtain a stable bond between SiNPs and probes such as nucleotides, antibodies or peptides; however, the latter show many disadvantages about folding instability and sensitivity during the functionalization. Phage Display is a technique for the screening and selection of peptide ligands, that uses an engineered filamentous bacteriophage, mostly made up of 2700 copies of a major coat protein (pVIII) displaying a foreign peptide specific for a target. The bacteriophage or its coat proteins alone can be used as probes to functionalize nanomaterials such as SiNPs. In this work, we propose a new approach to obtain fluorescent bio-probes that can be used for the realization of an optical biosensor. By pulsed laser ablation in liquid (PLAL), SiNPs are functionalized in a "one step" process with phages or isolated pVIII-engineered proteins, selective for *Pseudomonas aeruginosa*. This process led to complexation of SiNPs with both bioprobes proposed. The PLAL did not alter the biological function of phage probes, maintaining their binding capacity to the bacterial target.

M. G. Rizzo (✉) · L. M. De Plano · S. Crea · S. P. P. Guglielmino
Department of Chemical Sciences Biological, Pharmaceutical and Environmental,
University of Messina, Viale F. Stagno d'Alcontres 31, 98166 Messina, Italy
e-mail: mgrizzo@unime.it

S. P. P. Guglielmino
e-mail: sguglielm@unime.it

D. Franco · S. Scibilia · A. M. Mezzasalma
Department of Mathematics, Informatics, Physics and Earth Sciences,
University of Messina, Viale F. Stagno d'Alcontres 31, 98166 Messina, Italy
e-mail: mezzasalma@unime.it

© Springer Nature Switzerland AG 2019
B. Andò et al. (eds.), *Sensors*, Lecture Notes in Electrical Engineering 539,
https://doi.org/10.1007/978-3-030-04324-7_21

Keywords Biosensor · Phage display · M13 pVIII engineered proteins
Silicon nanoparticles · Pulsed laser ablation in liquid

1 Introduction

Nowadays, nanotechnologies are applied in medical field to design new probes for obtaining innovative diagnostic and theranostic systems [1].

Furthermore, the combination of nanotechnology on microfluidic devices has allowed the execution of different analyses with time saving and cost reduction. In addition, the integration of nanoparticles arising from semiconductor materials into different microfluidic based nanosystems may offer a viable alternative to fluorescently labeled particles. In particular, silicon nanoparticles (SiNPs) can be employed in easy-to-use and cheap sensing systems, offer a lot of advantages for their physical properties, their surface state can be easily activated, have stability against photobleaching and a distinctive photoluminescence, exhibit no toxicity and present biocompatibility [2, 3].

Phage Display is a powerful tool used for the screening and selection of peptide ligands to a wide variety of targets, therefore it has been used as a valid substitute for the research of antibodies or peptides. This technique uses M13 filamentous bacteriophage (phage), which consists in a cylindrical shell, mostly made up of 2700 copies of a major coat protein (pVIII) and other four minor coat proteins (pIII; pVI; pVII; pIX), that enclose a circular single-stranded DNA molecule. The "in-frame" insertion of exogenous DNA fragments in the gene encoding the major capsid protein pVIII allows the formation of large molecular libraries, which can be used to discover new bioprobes [4–7].

Bacteriophage or its protein alone can both be used as probes to functionalize several metal (gold or silver [8]) and semiconductor nanoparticles. Furthermore, the whole phage, the isolated pVIII protein or the exogenous peptide alone can be isolated without loss of activity, maintaining their selectivity, specificity and biological activity [9] at different conditions of temperature and pH, or in the presence of acid and organic solvents.

In this work, we propose a new approach for biofunctionalization of SiNPs with M13-engineered bacteriophage or isolated pVIII-engineered proteins, displaying specific peptides that selectively recognize *Pseudomonas aeruginosa* [10]. The "one-step" functionalization is conducted during the pulsed laser ablation in liquid (PLAL) of a silicon plate in a solution containing the bio-probes. This proposed strategy demonstrates its potential use for in vitro applications and could be exploited to realize an optical biosensor to detect a specific target.

2 Results and Discussion

The phage clone used in this work has been selected from landscape M13-pVIII -9aa peptide library. This clone (P9b) displays the foreign peptide QRKLAAKLT which recognizes and specifically binds the 42 KDa outer membrane protein (OMP) of *P. aeruginosa* [10], the most common agent of nosocomial infections.

The two probes (the whole bacteriophage or the isolated pVIII protein alone) have been used separately during PLAL.

The isolation and purification of the major coat protein pVIII of P9b phage clone was performed according to the protocol of Pei Liu et al. [9].

SiNPs were generated and simultaneously (one-step) functionalised by PLAL as follows. High purity (99.99%) monocrystalline silicon plate was immersed in a glass vessel filled with 2.5 mL of an aqueous solution of pVIII protein (25 μg mL^{-1}; pVIII-SiNPs) or a phage suspension in TBS buffer (8 × 10^{11} PFU mL^{-1}; phage-SiNPs). The ablation process was performed using the second harmonic (532 nm) of a neodymium-doped yttrium aluminum garnet (Nd:YAG) laser (model New Wave Mod. Tempest 300), operating at 10 Hz repetition rate with a pulse width of 5 ns [8, 11]. The silicon target was irradiated at the laser fluence of 7.5 J cm^{-2} and for an ablation time of 30 min.

To separate the phage–SiNPs from the unbounded phages and free SiNPs, networks were purified by centrifugation at 20,800 × g for 30 min, while to isolate the complex pVIII-SiNPs ultracentrifugation at 44.700 × g was performed, according to procedure described by Bagga et al. [12].

Preliminarily, we used the above mentioned parameters to test phage and protein stabilities. Despite the high temperatures, a sufficient amount of the phage population complexed with SiNPs and kept its structure intact, as demonstrated in our previous work [13]. In particular, it has been noticed that the phage (800 nm length and 5 nm diameter) was decorated with numerous SiNPs of different sizes as showed in SEM image (Fig. 1).

The binding occurred due to the electrostatic interaction between the charges of SiNPs surface and the phage surface [14], was mediated by ions present in buffer, forming salt bridges in phage–SiNPs network.

EDX analysis showed the presence of N (nitrogen) atomic species typical of proteins exposed on the surface of the bacteriophage; moreover, the presence of Si (silicon), O (oxygen), Na (sodium) and Cl (chloride) confirmed the functionalization of the nanoparticles with the bacteriophage.

Since the isolated pVIII protein can be easily isolated without loss of its activity, we verified whether it could be used as functionalizing agent for SiNPs during PLAL. Although parameters have still to be optimized, the STEM image (Fig. 2) results show the formation of SiNPs complexed with proteins. Furthermore, the interactions with pVIII proteins caused changes in size, shape, and aggregation state of SiNPs. The mean size of the SiNPs and complexes were estimated. The SiNPs had an average diameter of ~17.5 nm, while the pVIII-SiNPs of ~20 nm. Although a portion of these proteins was altered during the laser ablation, a large portion of pVIII proteins

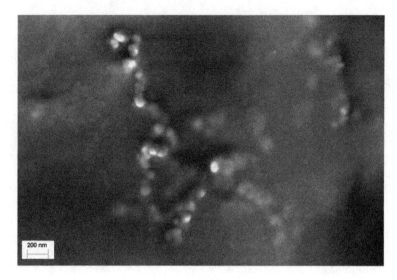

Fig. 1 SEM image shows phage-SiNPs complexes (phages decorated by SiNPs) produced by PLAL in a phage suspension

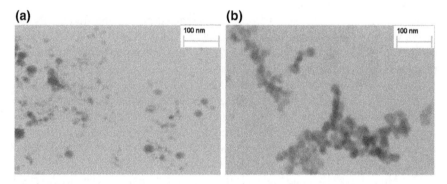

Fig. 2 STEM images: **a** the SiNPs laser-ablated in H_2O; **b** pVIII-SiNPs complexes produced by PLAL in a pVIII protein solution

assembled with SiNPs and interacted with the target. The phage proteins are polyelectrolytes and then tend to aggregate, but during the ablation process, the thermal and electrostatic variations near the plume may initially determine the disaggregation of the proteins. As consequence, the proteins adsorbed on the SiNPs surface will be arranged as monolayer due to the negative charges on the SiNPs (Zeta potential −31 mV), creating a protein corona on every single SiNP [15].

(a)   **(b)**

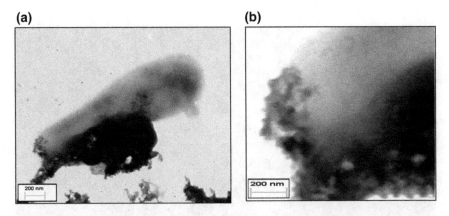

Fig. 3 STEM images: **a** binding of pVIII-SiNPs on *P. aeruginosa* cell; **b** detail of image (**a**)

On the other hand, STEM images showed that a fraction of pVIII-SiNPs could generate clumps due to electrostatic attractions among the exposed groups of the pVIII amino acids, leading to the formation of SiNPs-pVIII-pVIII-SiNPs complexes. Despite this, both single pVIII-SiNPs and larger aggregates are able to recognize the bacterial target. In fact, when the pVIII-SiNPs solution was tested against *P. aeruginosa*, STEM images (Fig. 3) showed the binding of the nanoparticles on the surface of the bacterium, whereas no binding was observed in the control (data not shown). In addition, the image highlights the lytic effect of the peptide on the membrane, confirming the maintenance of the structural and functional characteristics of the exposed peptide.

Consequently, the PLAL did not alter the structure and the properties of the functionalized bioprobes, so the specific peptide maintained its ability to recognize and interact with bacterial target.

To evaluate the possibility of using pVIII-SiNPs complexes as fluorescent probes in the identification of *P. aeruginosa*, a solution of pVIII-SiNPs was incubated with *P. aeruginosa* for 30 min at RT in rotator mixer and then washed in PBS. Finally, the samples were observed by epifluorescence microscope Leica DMRE (Excitation filter BP450-490, Suppression filter LP 515).

Figure 4a shows the presence of *P. aeruginosa* cells covered by the yellow-green complexes of pVIII-SiNPs, while in Fig. 4b, no fluorescence was observed when *P. aeruginosa* cells were treated with SiNPs alone.

These results confirm that pVIII-SiNPs were able to provide a fluorescence response through the luminescent signal to their bacterial target, then they may be used as bio-functional nanoprobes.

(a) (b)

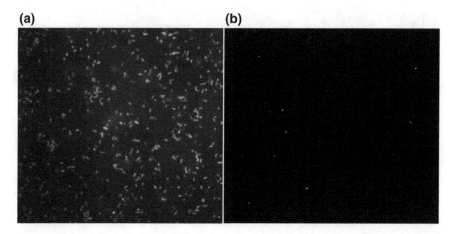

Fig. 4 Epifluorescence images at 63X: **a** Bright, yellow-green *P. aeruginosa* cells stained by pVIII-SiNPs complexes, **b** (Control) *P. aeruginosa* cells with free SiNPs (absence of pVIII engineered protein)

3 Conclusions

Fluorescent silicon-based nanoparticles can be functionalized with M13 engineered bacteriophages or their isolated pVII proteins by a "one step" process without altering their ability to bind the target.

Phages are cheap and easy to produce, and SiNPs luminescence could be a safer and valid alternative to fluorochrome labeling. Moreover, these phage-SiNPs complexes have demonstrated their potential use for in vitro applications and could be exploited as an optical biosensor to detect prokaryotic or eukaryotic targets.

The physical and biological features of these complexes offer convenient multifunctional integration within a single entity with potential for nanotechnology-based biomedical applications.

Therefore, this strategy allows to obtain low-cost and highly-specific luminescent complexes, which may be employed in LOC system for diagnostic applications.

References

1. Solano-Umaña, V., Vega-Baudrit, J.R., González-Paz, R.: The new field of the nanomedicine. Int. J. Appl. Sci. Technol. **5**(1) (2015)
2. Huan, C., Shu-Qing, S.: Silicon nanoparticles: preparation, properties, and applications. In: Invited Review—International Conference on Nanoscience & Technology, China 2013, Chin. Phys. B **23**(8), 088102 (2014)
3. Wang, G., Yau, S.-T., Mantey, K., Nayfeh, M.H.: Fluorescent Si nanoparticle-based electrode for sensing biomedical substances. Opt. Commun. **281**, 1765–1770 (2008)

4. Petrenko, V.A., Vodyanoy, V.J.: Phage display for detection of biological threat agents. J. Microbiol. Methods **53**, 253–262 (2003)
5. De Plano, L.M., Carnazza, S., Messina, G.M.L., Rizzo, M.G., Marletta, G., Guglielmino, S.P.P.: Specific and selective probes for *Staphylococcus aureus* from phage-displayed random peptide libraries. Colloids Surf. B Biointerfaces **157**, 473–480 (2017)
6. Barbas III, C.F., Burton, D.R., Scott, J.K., Silverman, G.J.: Phage Display, A Laboratory Manual. Cold Spring Harbor Laboratory Press, Woodbury, NY (2001)
7. Petrenko, V.A., Smith, G.P.: Phage from landscape libraries as substitute antibodies. Protein Eng. **13**, 101–104 (2000)
8. Scibilia, S., Lentini, G., Fazio, E., Franco, D., Neri, F., Mezzasalma, A.M., Guglielmino, S.P.P.: Self-assembly of silver nanoparticles and bacteriophage. Sens. Bio-Sens. Res. **7**, 146–152 (2016)
9. Liu, P., Han, L., Wang, F., Petrenko, V.A., Liu, A.: Gold nanoprobe functionalized with specific fusion protein selection from phage display and its application in rapid, selective and sensitive colorimetric biosensing of *Staphylococcus aureus*. Biosens. Bioelectron. **82**, 195–203 (2016). https://doi.org/10.1016/j.bios.2016.03.075
10. Carnazza, S., Foti, C., Gioffrè, G., Felici, F., Guglielmino, S.P.P.: Specific and selective probes for *Pseudomonas aeruginosa* from phage-displayed random peptide libraries. Biosens. Bioelectron. **23**, 1137–1144 (2008)
11. Fazio, E., Cacciola, A., Mezzasalma, A.M., Mondio, G., Neri, F., Saija, R.: Modelling of the optical absorption spectra of PLAL prepared ZnO colloids. J. Quant. Spectrosc. Radiat. Trans. **124**, 86–93 (2013)
12. Bagga, K., Barchanski, A., Intartaglia, R., Dante, S., Marotta, R., Diaspro, A., Sajti, C.L., Brandi, F.: Laser-assisted synthesis of *Staphylococcus aureus* protein-capped silicon quantum dots as bio-functional nanoprobes. Laser Phys. Lett. **10**, 065603 (8 pp) (2013)
13. De Plano, L.M., Scibilia, S., Rizzo, M.G., Crea, S., Franco, D., Mezzasalma, A.M., Guglielmino, S.P.P.: One-step production of phage–silicon nanoparticles by PLAL as fluorescent nanoprobes for cell identification. Appl. Phys. A **124**, 222 (2018)
14. Coen, M., Lehmann, R., Groning, P., Bielmann, M., Galli, C., Schlapbach, L.: Adsorption and bioactivity of protein A on silicon surfaces studied by AFM and XPS. J. Colloid Interface Sci. **233**, 180–189 (2001)
15. Shemetov, A.A., Nabiev, I., Sukhanova, A.: Molecular Interaction of Proteins and Peptides with Nanoparticles, vol. 6. American Chemical Society (2012)

FITC-Labelled Clone from Phage Display for Direct Detection of Leukemia Cells in Blood

Domenico Franco, Laura M. De Plano, Maria Giovanna Rizzo, Sara Crea, Enza Fazio, Martina Bonsignore, Fortunato Neri, Alessandro Allegra, Caterina Musolino, Guido Ferlazzo, Sebastiano Trusso and Salvatore P. P. Guglielmino

Abstract Discovery of new markers for the identification and discrimination of cell types is one of the principal objectives in cancer diagnostics. In the last years, many researchers used phage-display technology in vitro and in vivo to obtain random peptide probes able to bind towards cancer targets to be used in diagnostic systems and new targeted drug. In this work, we proposed a Single Drop Biosensor based on phage-labelled probes to detect leukaemia cells in blood from patients affected by chronic lymphocytic leukaemia (CLL). Results show that phage-labelled probes were able to recognize lymphocytes and lymphoblastic cells both in leukemic peripheral blood mononuclear cells and in whole blood from patients affected by CLL. The "proof of concept" proposed, using the phage labelled as bio-probe, could be an alternative way to produce new biosensor for monitoring of chronic pathology. Furthermore the results may have translational relevance for identification and exploring of new ligands directed against cancer hematological cells.

D. Franco (✉) · E. Fazio · M. Bonsignore · F. Neri
Department of Mathematical and Computational Sciences, Physical Science and Earth Science, University of Messina, 98166 Messina, Italy
e-mail: dfranco@unime.it

L. M. De Plano · M. G. Rizzo · S. Crea · S. P. P. Guglielmino
Department of Chemical Sciences, Biological, Pharmaceutical and Environmental, University of Messina, 98166 Messina, Italy
e-mail: sguglielm@unime.it

A. Allegra · C. Musolino
Division of Hematology, Department of Human Pathology in Adulthood and Childhood "Gaetano Barresi", University of Messina, 98125 Messina, Italy

G. Ferlazzo
Laboratory of Immunology and Biotherapy, Department of Human Pathology, University of Messina, 98125 Messina, Italy

S. Trusso
Institute of Chemical-Physical Processes (IPCF)-CNR, 98158 Messina, Italy

© Springer Nature Switzerland AG 2019
B. Andò et al. (eds.), *Sensors*, Lecture Notes in Electrical Engineering 539,
https://doi.org/10.1007/978-3-030-04324-7_22

Keywords Phage-display · Leukemia · Fluorescence imaging

1 Introduction

In recent years, the increasing availability of fluorescent dyes has directed researchers to the development of non-invasive methodologies in diagnostic imaging field. In this context, extracorporeal (ex vivo) diagnostics provides a non-invasive, early, and accurate detection of biological disease markers in the routine screening, thus enabling the appropriate treatment regimen to be chosen. Various nanotechnology platforms have been developed to allow the simultaneous real-time evaluation of a broad range of disease markers [1]. Fluorescence imaging could have many advantages in diagnostic providing valuable information at the structural, functional and/or molecular level of biological markers. Consequently, fluorescence remains one of most diffused diagnostic tools, because extremely versatile and continuously improved by new emerging techniques. The strength of this technique is the ability to identify the target through the fluorescence visualization of labelled probes. Usually, antibodies were used for immunostaining and identification of targets in vitro. However, they expose only two recognition sites on their Fab and are sensitive to temperature, pH and organic solvents that could denature their structure and their biological activity. Other methods, instead, utilize purified peptides selected with phage display technique as imaging probes, which are able to recognize different type of cancer cells, such as breast, colon, prostate cancer [2], or particular targets that are expressed in neoplastic microenvironment [3]. Furthermore, peptides synthesized can be conjugated with radioisotopes [4], fluorochromes or paramagnetic iron particles [5] to bind their target and allow their identification thanks to the tracer to which they are linked. Peptides may be conjugated with drugs, nucleotide sequences [6] and other molecules for drug and gene targeting or penetrate in the cells by exerting cytotoxic activity [7]. Despite the advantages and versatility for many applications, the peptides have the same susceptibility to chemical and physical conditions of antibodies.

Using whole phages rather than synthetic peptides allows overcoming all the limitations discussed above. In fact phages are long term stable and resistant to temperature, pH, organic solvents, denaturing agents and are easier to produce with shorter and cheaper purification steps than peptides or antibodies. Furthermore, phages maintain their viability after conjugation and consequently the possibility to recognize its target compared to synthetic peptides which could lose their tridimensional structure invalidating target recognition. Particularly, M13 phage, genetically engineered on the pVIII protein, not only expose thousands of insert copies, but preserve the folding of exposing peptides on its capsid proteins. In some studies, labelling has been performed using entire M13 bacteriophages, which displayed peptides on PIII minor capsid protein, able to specifically recognize cancer cells [8], osteosarcoma [9], epithelial cell tight junctions [10], cells that early response to anti-angiogenic treatment [11]. Labelling didn't invalidate the recognition of the specific target on PIII protein, since fluorochrome reacts mainly with PVIII capsid protein. Li et al.

have used M13 bacteriophage for dual-modification: fluorescent molecules FL-NHS for bioimaging and folate-azide for giving specificity towards cancer cells overexpressing folate receptors [12].

Avoiding further engineering or chemical modifications of phage capsid proteins, a labelling technique, simple, reliable, cheap and rapid, has been developed using whole phagemid, without interfering on binding affinity and specificity toward cell targets. In this way, an efficient bioactive and bio functionalized probe for ex vivo imaging is obtained.

In this work, we propose a Single Drop Biosensor based on phage-labeled probes to detect leukaemia cells in blood from patients affected by chronic lymphocytic leukemia (CLL). First, we screened phage clone displaying peptides able to specific and strong bind leukemia cells from CLL patients. A labelling technique, with fluorescein isothiocyanate (FITC) fluorochrome, cheap and rapid, has been developed using whole phage clone, without interfering on binding affinity and specificity toward cell targets. Then, phage clone FITC-labelled were tested against leukemic peripheral blood mononuclear cells (PBMC), separated by Ficoll gradient. Finally, in order to realize a rapid and easy diagnostic system, label probe was assayed with whole blood from patients affected by CLL.

2   Materials and Methods

2.1   Bacteriophage

A random M13-pVIII-9aa peptide library (kind gift of Prof. F. Felici) was used for the selection of phage clones specific against leukaemic targets. Firstly, the library was subtracted by pretreatment with PBMCs from healthy subjects then it was used to the selection protocol-binding phage clones for four rounds of affinity selection against PBMCs from two patients affected by chronic lymphocytic leukaemia (CLL), according to our previous work [13]. The most reactive phage clones showing a specific leukemic PBMCs-binding were amplified and their DNA was sequenced to determine the amino acid sequences of the displayed peptides. Sequence similarity searching was carried out by BLAST, an algorithm for comparing primary biological sequence information, then sequences were treated with Epitope-Mediated Antigen Prediction (E-MAP) tools to predict epitopes.

2.2   Phage Labelling with FITC

Phage labelling with FITC was carried out according to the procedure described by Herman et al., and adapted to experimental condition of the present study [9]. 1×10^{13} PFU (Plaque Forming Units) were resuspended in 200 μl Buffer $Na_2CO_3/NaHCO_3$ (pH 9.2) with 5 μl of fluorescein isothiocyanate (FITC, 5 mg/ml). Clones were incubated for 2 h in the dark on rotator (8 rpm) at Room Temperature (RT) to allow reaction with fluorochrome. Sample was incubated at 4 °C over night with 200 µl of PEG/NaCl and then centrifuged at 15300 × g at 4 °C for 1 h. The supernatant was discarded and the pellet resuspended in 100 µL of Tris Buffered Saline [TBS, (7.88 gr/L Tris-HCl, 8.77 gr/L NaCl)]. Labelled clones were stored in the dark at 4 °C until utilization.

2.3 Sample Preparation for Fluorescence Imaging

100 µl of peripheral blood mononuclear cells (PBMCs) (2×10^5 cells) were incubated with 25 µl FITC-labelled phage clone stocks solution (titer 10^{11} PFU/ml, cell/phage ratio 1:5000) for different incubation times (30′, 1 and 2 h) at 37 °C. After incubation, cells were washed three times with Phosphate-buffered saline (PBS) to remove excess unbinding phage and then put on Poly-L-Lysine (PLL) coated glass slides. Wild-type vector pC89 served as negative control for evaluation of background from nonspecific binding. The samples were analyzed by a fluorescence microscope (Leica DMRE) at different magnifications. About imaging test in whole blood, a preliminary step was added to recover blood elements. Blood cells were washed for two times in PBS and then tested as described above.

3 Results and Discussions

Among all the selected clones, we found that sequence similarity searching by BLAST show a statistically significant similarity (about 90% identity in the amino acid sequence) between the foreign peptide displayed by a phage clone, named CLL-1, and a motif of 182 KDa tankyrase 1-binding protein. Tankyrase is a protein, belonging to the ankyrin protein family, with several functions in both the nucleus and the cytoplasm where the majority of tankyrase molecules can be found [14, 15]. Since the tankyrase 1 interacts with several Bcl-2 family proteins in the modulation of apoptosis pathway of the leukemic and other neoplastic cells [16, 17], the subsequent experimental tests were carried out using CLL-1.

First, we evaluated by ELISA test that the CLL-1 bind leukemic PMBCs, using the insert-less vector pC89 as negative control (data not shown). Then, phage clone FITC-labelled were tested against leukemic PBMCs, separated by Ficoll gradient, and the fluorescent detection was compared to insert-less vector pC89 (see Fig. 1). FITC-labelled CLL1 was found to specifically bind leukemic PBMC as microscopy observations revealed bright green fluorescent stained cells. Labelling procedure involves a direct link between phage capsid protein and FITC amine-reactive fluorochrome. On the contrary, only background fluorescence was observed in leukemic PBMC by using FITC-labelled pC89 (insert-less phagemid), confirming leukemic PBMC specificity by ELISA test of the peptide expressed by CLL-1. Although a

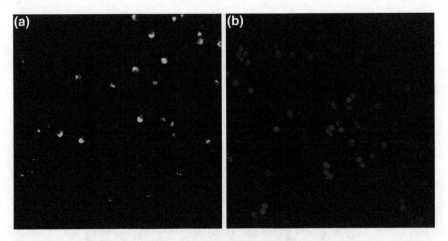

Fig. 1 PBMC targeting by FITC-labelled CLL1 (**a**), selected by a 9-mer random M13 phage display libraries compared to insert-less vector pC89 (**b**) (magnification 40×)

loss of the bond selectivity was expected, CLL-1 showed in vitro imaging a high binding efficiency against their specific cellular targets, probably belonging to lymphocytes/lymphoblasts classes. Results are in agreement to our previous work for neoplastic cells identification by fluorochrome conjugated phage clones [13]. Therefore, the whole phage structure can be used as a robust and versatile probe, in which the recognition sites are contained in each copy of the major coat PVIII proteins and, consequently, homogeneously distributed on the whole phage surface. At a higher magnification, we observed that in some cells green fluorescence was specifically localized in subcellular sites (see Fig. 2) according to previous data about CLL-1 capacity to penetrate inside the cells, although informatics tool does not predict a penetrating ability for foreign peptide (data not shown).

These findings are in agreement with Chang et al. about the distribution of tankirase 1 in multiple subcellular sites, including telomeres and mitotic centrosomes [18]. However further studies should be carried out to confirm the hypothesis of interaction between the tankyrase 1 and CLL-1.

In order to realize a rapid and easy diagnostic system, FITC-labelled CLL-1 clone was tested in whole blood from patients affected by chronic lymphocytic leukaemia. The kinetic of reaction, evaluated at different incubation times (30′, 1 h and 2 h), are showed in Fig. 3.

As shown in Fig. 3, activity of FITC-labelled CLL-1 clone is not influenced by the presence of other blood elements, showing a recognition capacity as soon as after 30 min. Moreover FITC-labelled CLL1 clone was found to specifically bind/recognize lymphocytic and lymphoblastic cells, as microscopy observations revealed bright green fluorescent stained cells as shown in Fig. 4.

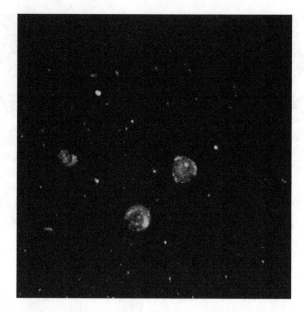

Fig. 2 Specific localization of FITC-labelled CLL-1 in leukemic PBMC subcellular sites (magnification 63×)

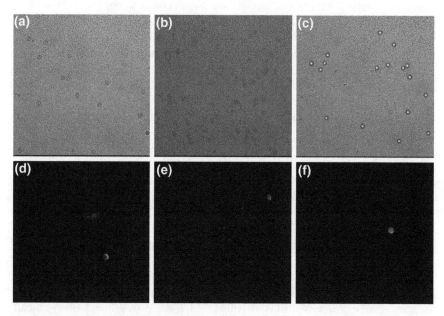

Fig. 3 Light phase contrast (**a**, **b** and **c**) and fluorescence (**d**, **e** and **f**) of leukemic PBMC by FITC-labelled phage clone in blood from patient affected by CLL at 30′ (**a** and **d**), 1 h (**b** and **e**) and 2 h (**c** and **f**) incubation times (magnification 20×)

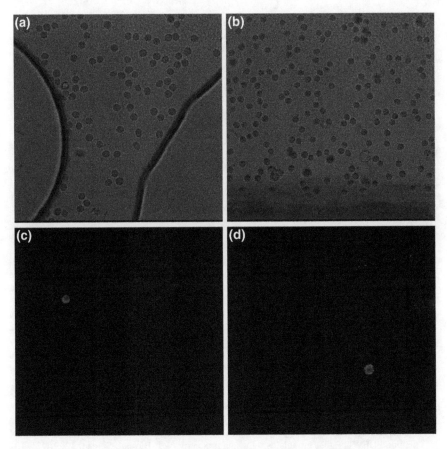

Fig. 4 Light phase contrast (**a** and **b**) and fluorescence (**c** and **d**) of lymphocytic (**a** and **c**) and lymphoblastic (**b** and **d**) cells targeting by FITC-labelled CLL-1 (magnification 40×)

4 Conclusions

Detection of circulating cancer cells in the blood is very useful in order not only to characterize leukaemia typologies but also for the monitoring of therapy, the control of treatment efficacy and, after the therapy, the detection of the minimal residual disease. The phage clone CLL-1, selected for binding CLL cells, exemplifies the ability to use phages in biosensor technology for selective and specific recognition of cancer cells. The "proof of concept" proposed, using the phage labelled as bioprobe, demonstrates a wide detection range, good selectivity, and high reliability for measurements of real blood samples, which may find potential applications in chronic lymphocytic leukaemia control. Furthermore the results may have translational relevance for identification and exploring of new ligands directed against cancer hematological cells.

Acknowledgements This work was supported by Associazione pro Bambini e Adulti Leucemici (A.B.A.L.) onlus Messina (Italy) (http://www.abalmessina.it). Phage-display libraries were a kind gift of Prof. Franco Felici.

References

1. Riehemann, K., Schneider, S.W., Luger, T.A., Godin, B., Ferrari, M., Fuchs, H.: Nanomedicine–challenge and perspectives. Angew. Chem. Int. Ed. **48**, 872–897 (2009)
2. Schally, A.V., Comaru-Schally, A.M., Plonowski, A., Nagy, A., Halmos, G., Rekasi, Z.: Peptide analogues in the therapy of prostate cancer. Prostate **45**, 158–166 (2000)
3. Zhao, Y., Ji, T., Wangm, H., Li, S., Zhao, Y., Nie, G.: Self-assembled peptide nanoparticles as tumor microenvironment activatable probes for tumor targeting and imaging. J. Control. Release **177**, 11–19 (2014)
4. Sosabowski, J.K., Mather, S.J.: Conjugation of DOTA-like chelating agents to peptides and radiolabeling with trivalent metallic isotopes. Nat. Protoc. **1**(2), 972–976 (2006)
5. Yang, L., Mao, H., Cao, Z., Wang, A., Peng, X., Wang, X., Karna, P., Adams, G., Yuan, Q., Staley, C., Wood, W.C., Nie, S., Gao, X.: Development of peptide conjugated superparamagnetic iron oxide (SPIO) nanoparticles for targeted MR imaging and therapy of pancreatic cancer. NSTI-Nanotechnol. **2**, 17–20 (2006)
6. Flierl, A., Jackson, C., Cottrell, B., Murdock, D., Seibel, P., Wallace, D.C.: Targeted delivery of DNA to the mitochondrial compartment via import sequence-conjugated peptide nucleic acid. Mol. Ther. **7**, 50–57 (2003)
7. Akrami, M., Balalaie, S., Hosscinkhani, S., Alipour, M., Salehi, F., Bahador, A., Haririan, I.: Tuning the anticancer activity of a novel pro-apoptotic peptide using gold nanoparticle platforms. Sci. Rep. **6**, 31030 (2016)
8. Ghosh, D., Kohli, A.G., Moser, F., Endy, D., Belcher, A.M.: Refactored M13 bacteriophage as a platform for tumor cell imaging and drug deliver. ACS Synth. Biol. **1**, 576–582 (2012)
9. Herman, R.E., Makienko, E.G., Prieve, M.G., Fuller, M., Houston Jr., M.E., Johnson, P.H.: Phage display screening of epithelial cell monolayers treated with EGTA: identification of peptide FDFWITP that modulates tight junction activity. J. Biomol. Screen. **12**, 1092–1101 (2007)
10. Sun, X., Niu, G., Yan, Y., Yang, M., Chen, K., Ma, Y., Chan, N., Shen, B., Chen, X.: Phage display-derived Peptides for osteosarcoma imaging. Clin. Cancer Res. **16**, 4268–4277 (2010)
11. Cao, Q., Liu, S., Niu, G., Chen, K., Yan, Y., Liu, Z., Chen, X.: Phage display peptide probes for imaging early response to bevacizumab treatment. Amino Acids **41**, 1103–1111 (2011)
12. Li, K., Chen, Y., Li, S., Nguyen, H.G., Niu, Z., You, S., Mello, C.M., Lu, X., Wang, Q.: Chemical modification of M13 bacteriophage and its application in cancer cell imaging. Bioconjug. Chem. **21**, 1369–1376 (2010)
13. Lentini, G., Fazio, E., Calabrese, F., De Plano, L.M., Puliafico, M., Franco, D., Nicolò, M.S., Carnazza, S., Trusso, S., Allegra, A., Neri, F., Musolino, C., Guglielmino, S.P.P.: Phage-AgNPs complex as SERS probe for U937 cell identification. Biosens. Bioelectron. **74**, 398–405 (2015)
14. Smith, S., Giriat, I., Schmitt, A., de Lange, T.: Tankyrase, a poly(ADP-ribose) polymerase at human telomeres. Science **282**, 1484–1487 (1998)
15. Chi, N.W., Lodish, H.F.: Tankyrase is a golgi-associated mitogen-activated protein kinase substrate that interacts with IRAP in GLUT4 vesicles. J. Biol. Chem. **275**, 38437–38444 (2000)
16. Bae, J., Donigian, J.R., Hsueh, A.J.: Tankyrase 1 interacts with Mcl-1 proteins and inhibits their regulation of apoptosis. J. Biol. Chem. **278**, 5195–5204 (2003)
17. Gao, J., Zhang, J., Long, Y., Tian, Y., Lu, X.: Expression of tankyrase 1 in gastric cancer and its correlation with telomerase activity. Pathol. Oncol. Res. **17**(3), 685–690 (2011)
18. Chang, W., Dynek, J.N., Smith, S.: NuMA is a major acceptor of poly(ADP-ribosyl)ation by tankyrase 1 in mitosis. Biochem. J. **391**(Pt 2), 177–184 (2005)

Organised Colloidal Metal Nanoparticles for LSPR Refractive Index Transducers

S. Rella⊙, M. G. Manera, A. Colombelli, A. G. Monteduro, G. Maruccio and C. Malitesta⊙

Abstract This work is focused on optimizing adhesion and distribution of colloidal gold nanoparticles on silanized glass substrates intended as nanostructured plasmonic transducer for sensing applications. This system will be used as platform for subsequent functionalization and/or enzyme immobilization. All preparation steps have been monitored by UV-Vis absorption spectroscopy and X-ray photoelectron spectroscopy (XPS).

Keywords Gold nanoparticles · Silanized substrate · Surface plasmon resonance

1 Introduction

In recent years gold nanoparticles (AuNPs) have been studied intensively due to unique physical, chemical, electrical and optical properties and for this reason they have been investigated as platform for many applications in various fields such as nanotechnology, materials science and chemical sensors. The development and characterization of gold nanoparticles is very interesting from a scientific point of view. In fact, the ease by which the size and the shape of AuNPs can be modified by

S. Rella (✉) · C. Malitesta
Dipartimento di Scienze e Tecnologie Biologiche e Ambientali, Di.S.Te.B.A, Università del Salento, Via Monteroni, 73100 Lecce, Italy
e-mail: simona.rella@unisalento.it

M. G. Manera · A. Colombelli
Istituto per la Microelettronica e i Microsistemi CNR IMM - Lecce, Campus Ecotekne, Via Monteroni, 73100 Lecce, Italy

A. G. Monteduro · G. Maruccio
CNR NANOTEC - Institute of Nanotechnology c/o Campus Ecotekne, Via Monteroni, 73100 Lecce, Italy

A. G. Monteduro · G. Maruccio
Dipartimento di Matematica e Fisica, Università del Salento, Via per Arnesano, 73100 Lecce, Italy

© Springer Nature Switzerland AG 2019
B. Andò et al. (eds.), *Sensors*, Lecture Notes in Electrical Engineering 539,
https://doi.org/10.1007/978-3-030-04324-7_23

tuning the synthetic protocols makes them attractive for sensing applications [1]. For example, the binding event between the recognition element and the analyte can alter physicochemical properties of transducer AuNPs, such as plasmon resonance absorption, conductivity, etc., that in turn can generate a detectable response signal [2]. Surface plasmon resonance (SPR)-based biosensors are increasingly employed, not only in gas sensing, but also in many other important applications in food safety, biology, medical diagnostics and environmental analysis [3]. SPR sensors exploit mono- or polychromatic polarized light, which excites the metal (typically gold—Au or silver—Ag)/dielectric (e.g. target sample—usually a liquid) interface to generate propagating plasmonic waves, highly sensitive to the refractive index (RI) changes in the sample [4]. Different techniques have been used to assemble AuNPs on various surface trying to control nanoparticles dispersion and surface coverage. The physical and chemical properties of these devices will depend not only on the size and shape of the gold nanoparticles but also on their spatial arrangement and on the nature of their interaction with the substrate surface. Usually, coupling agents such as thiol-therminated, amine-therminated silanes are used to anchor gold nanoparticles on solid substrates through terminal functional groups that interact electrostatically and chemically with the nanoparticles [5]. Aminopropyltriethoxysilane (APTES) is one of the most used organosilane agents for the preparation of amine-terminated surfaces. The presence of three hydrolysable ethoxy groups ensure a robust anchoring of the silane to the surface (silanization step), whereas—NH_2 end groups from the aminopropyl groups remain available to immobilize AuNPs through electrostatic interaction. Aim of this work was to find the optimal experimental conditions for fabricating gold nanoparticles layer on a glass substrate which can then be used for LSPR sensors and real applications in bio-chemical analysis. To this purpose we focused on substrates based on gold colloidal nanoparticles immobilized onto silanized glass plates. Stable and uniform substrates have been prepared by using aminopropyltriethoxysilane (APTES) and citrate reduced gold nanoparticles prepared according to the Turkevich method [6].

2 Experimental

2.1 Preparation of Gold Nanoparticles

AuNPs were prepared by sodium citrate (Na_3-citrate, Sigma Aldrich) reduction of the chloroauric acid ($HAuCl_4$, Sigma Aldrich) solution according to known literature procedure [6]. 1 mM HAuCl4 and 38.8 mM trisodium citrate were prepared in pure water (Milli-Q Element, Millipore) as stock solutions. 50 mL of $HAuCl_4$ solution was heated to boiling and 5 mL of Na_3-citrate solution were added. The color of the solution upon addition of citrate becomes black and then turns to light red color, indicating the formation of AuNPs. The absorption spectrum of colloidal AuNPs solution exhibit an extinction band located at approximately 520 nm, typical of nanoparticles having a diameter of about 20 nm.

2.2 Deposition of Colloidal Gold Particles on Silanised Glass Substrate

The glass plates (1×1 cm^2) were immersed in 1:1 MeOH:HCl for 30 min followed by additional incubation (30 min) in concentrated H$_2$SO$_4$ and subsequent rinsing with DI water and drying under N$_2$. The plates were then immersed in ethanol solution of 3-aminopropyltrimethoxysilane (APTES; Sigma Aldrich) in different concentrations (0.1, 0.5, 1, 5% v/v) for 1 h at room temperature followed by rising in ethanol and two times in water with sonication for ten minutes each and drying with N$_2$. It is very important to remove the excess of APTES from glass to prevent gold particle aggregation during the deposition phase. Silanised glass plates were dipped in a tube containing 1 mL of the gold colloidal suspension for 1 h. Afterward, the plates were rinsed thoroughly with DI water and dried with N$_2$. The glass plates were finally thermally annealed for 2 h in ambient air at 550 °C to improve the adhesion of AuNPs to the substrate and to make them more stable.

2.3 Characterization Techniques

Absorbance measurements of synthesized AuNPs solutions and of their planar arrangements on glass substrates have been performed by using a Cary500 UV-visible spectrophotometer. X-ray photoelectron spectroscopy (XPS) was used to characterize the formed layer. XPS provided quantitative information about the elements content following each functionalization step. Moreover, from the XPS high resolution spectra information about the chemical bonds formed between the different species involved in the functionalization was also inferred. XPS measurements have been performed by an Axis ULTRA DLD Spectrometer (Kratos Analytical, UK) with a monochromatic Al Kα source operating at 150 W. The morphology and distribution of NPs deposited onto the glass surface were characterized by Scanning Electron Microscopy (SEM). A compact optical fiber system equipped with a deuterium halogen light source, a portable spectrometer (Thorlabs CCS100/M, wavelength ranging between 350 and 700 nm), and a system of optical fibers in transmission configuration, was used to characterize the plasmonic transducer in liquid phase. White light emerging from the first optical fiber was perpendicularly shed onto the NPs samples. The transmitted light was coupled into a detection fiber and analyzed by using the compact UV–vis spectrometer. All spectra were taken from 450 to 700 nm at room temperature. In order to avoid possible multiple reflections of light on fluid surface, the fiber tip was entirely immersed into the liquid phase. The experimental set-up allows acquiring the sensing response of gold nanoparticles by monitoring the spectral shift in the typical LSPR absorption peak. The sensitivity of the plasmonic transducers was investigated by recording LSPR spectra after their immersion in solutions with different refractive indices: air (n = 1.00), water (n = 1.33) and different water/glycerol solutions at glycerol increasing concentrations (1.333, 1.338, 1.350, 1.365, 1.379 and 1.4).

3   Results

Gold nanoparticles are commonly immobilized on amine-terminated layers through electrostatic interaction. The absorption of AuNPs on the surface of glass substrates was monitored by UV-Vis absorption spectroscopy. In Fig. 1a the absorption spectra of the AuNP-modified surface with different concentrations of APTES are reported.

The absorption spectra of the deposited samples, are strictly related to their morphology. In particular, the presence of large skewed LSPR absorption peak in the UV-Vis spectral range, reveals the possible presence of NPs aggregates. By applying a thermal annealing for 2 h at 550 °C, a more symmetric distribution of nanoparticles, characterized by separation distances of few nanometers can be obtained from the coalescence of closely spaced and aggregated NPs. The morphological evolution of AuNPs is confirmed by a significant variation of their optical properties. As can be noticed in Fig. 1b, after the thermal treatment, the deposited samples exhibit narrow LSPR absorption bands characteristic of well-separated and symmetric nanostructures, as previously reported in literature [7]. In particular, the annealed samples present LSPR absorption peaks centered around 540 nm, which is consistent with the absorption spectra of a homogenous distribution of gold NPs nanoparticles with diameters of few tens of nanometers. This can be confirmed by the scanning electron electron microscopy (SEM) image as shown in Fig. 2. The sample prepared with 0.1% of APTES shows a homogeneous surface with well separated and symmetric nanostructures. The size distribution indicates that the mean size of AuNPs ($\approx$30 nm) is larger than that of sample pre-annealing (20 nm). This happens because annealing leads to the formation of larger spherical particles by fusing of agglomerated ones and to a partial embedding in the glass substrate. At contrary, the sample prepared with a higher concentration of APTES (5% v/v) shows a substrate with non-spherical NPs. This phenomenon has been already reported [7]. Also the optical images of pre-annealed and post-annealed samples indicate a change of the LSPR band through a change of color from dark red to pink (Fig. 1a, b inset).

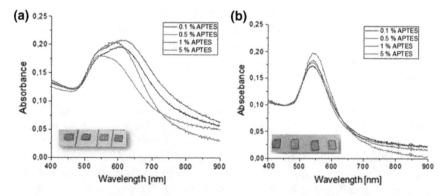

Fig. 1 UV-vis spectra of AuNPs patterned on sinalized glass substrate: **a** first and **b** after annealing; the inset shows the optical images of AuNPs deposited on APTES-functionalized glass surfaces

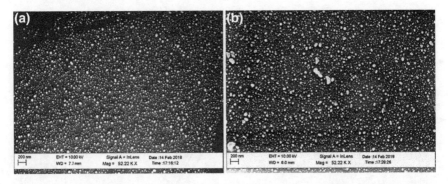

Fig. 2 SEM images of silanized glass surfaces (**a** 0.1% v/v APTES; **b** 5% v/v APTES) after nanoparticles deposition

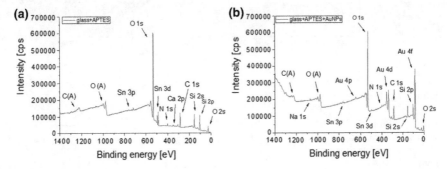

Fig. 3 XPS survey spectra of **a** APTES (0.1%) treated glass substrate and **b** APTES (0.1%) treated glass substrate decorated by AuNPs

The efficiency of silane binding and AuNPs deposition was investigated through XPS analysis. The technique was used to characterize the chemical structure and composition of the silane-treated glass surfaces before and after gold depositions. Figure 3a, b shows XPS survey spectra of APTES-treated glass and gold-coated silanised, respectively. On both survey spectra, are present oxygen, carbon, silicon and tin. In Fig. 3a, a nitrogen peak is clearly present confirming the presence of APTES on the glass substrates. Following gold deposition, Fig. 3b shows the presence of Au signal due to the immobilization of gold nanoparticles on the silanized substrate. In particular, the Au 4f peak is centered at 84 eV confirming the presence of metal gold [8].

For investigating plasmonic properties, APTES-treated glass substrates decorated with AuNPs were contacted with water/glycerol solutions of different refractive index (RI).

When exposed to solutions characterized by increasing RI (from 1.333 to 1.4), spectral shift of the LSPR absorption peak from 540 to 544 nm, was detected (Fig. 4). The optical response (Fig. 5) of the sample exhibits a linear dependence on the RI of the external environment, confirming the ability of this system to detect even small

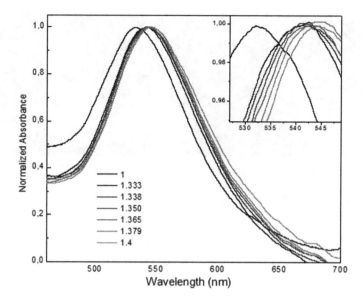

Fig. 4 Absorption spectra of sample after the contact with glycerol-water mixtures of different refractive indices

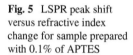

Fig. 5 LSPR peak shift versus refractive index change for sample prepared with 0.1% of APTES

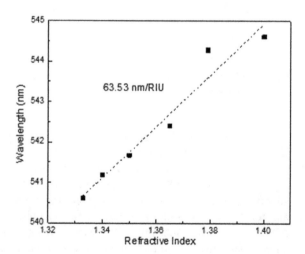

refractive index changes. From these curves, a quantitative information about the Refractive-Index Sensitivity (RIS) can be obtained, considering the wavelength shift of the signal per RI unit (nm RIU-1). In particular, a RIS of 64–77 nm/RIU was estimated for the fabricated transducer, in line with literature results [9].

4 Conclusions

We have demonstrated a simple nanostructured plasmonic transducer for sensing application. Uniform substrates have been prepared by electrostatic binding of citrate stabilized NPs on aminosilane terminated glass. The efficiency of silane binding and AuNPs deposition was investigated by UV absorption spectroscopy, SEM and XPS analysis. The refractive index sensitivity (RIS) of obtained transducer is 64–77 nm/RIU. The result indicates the ability of the system to detect even small changes in the refractive index of the environment confirming their perspective applications as functional transducer in plasmonic sensors.

Acknowledgements S. R. greatly thanks Regione Puglia for financing the project "Biosensori per il monitoraggio della qualità delle acque a base di nanoparticelle di oro funzionalizzate mediante enzimi specifici" (FutureInResearch No. YFN6JP8). Authors thank Michele Lanzillotta for experimental work during his Master's degree thesis.

References

1. Penn, S.G., He, L., Natan, M.J.: Nanoparticles for bioanalysis. Curr. Opin. Chem. Biol. **7**, 609–615 (2003)
2. Shipway, A.N., Katz, E., Willner, I.: Nanoparticle arrays on surfaces for electronic, optical, and sensor applications. Chem. Phys. Chem. **1**, 18–52 (2000)
3. Shankaran, D.R., Gobi, K.V., Miura, N.: Recent advancements in surface plasmon resonance immunosensors for detection of small molecules of biomedical, food and environmental interest. Sens. Actuators B Chem. **121**, 158–177 (2007)
4. Piliarik, M., Vaisocherová, H., Homola, J.: Surface plasmon resonance biosensing (2009)
5. Ben Haddada, M., Blanchard, J., Casale, S., Krafft, J.-M., Vallée, A., Méthivier, C., Boujday, S.: Optimizing the immobilization of gold nanoparticles on functionalized silicon surfaces: amine-vs thiol-terminated silane. Gold Bull. **46**, 335–341 (2013)
6. Morel, A.-L., Boujday, S., Méthivier, C., Krafft, J.-M., Pradier, C.-M.: Biosensors elaborated on gold nanoparticles, a PM-IRRAS characterisation of the IgG binding efficiency. Talanta **85**, 35–42 (2011)
7. Karakouz, T., Maoz, B.M., Lando, G., Vaskevich, A., Rubinstein, I.: Stabilization of gold nanoparticle films on glass by thermal embedding. ACS Appl. Mater. Interfaces **3**, 978–987 (2011)
8. Seitz, O., Chehimi, M.M., Cabet-Deliry, E., Truong, S., Felidj, N., Perruchot, C., Greaves, S.J., Watts, J.F.: Preparation and characterisation of gold nanoparticle assemblies on silanised glass plates. Coll. Surf. A Physicochem. Eng. Asp. **218**, 225–239 (2003)
9. Nath, N., Chilkoti, A.: A colorimetric gold nanoparticles sensors to interrogate biomolecular interactions in real time on a surface. Anal. Chem. **74**, 504 (2002)

Human Organ-on-a-Chip: Around the Intestine Bends

Lucia Giampetruzzi⊙, Amilcare Barca⊙, Chiara De Pascali⊙,
Simonetta Capone⊙, Tiziano Verri⊙, Pietro Siciliano⊙, Flavio Casino
and Luca Francioso⊙

Abstract The small intestine is the central component of the gastrointestinal (GI) tract (gut) where nutrients are absorbed into the body. Its functional structure is mainly based on its extremely extended surface area, further increased by a specific carpet of villi, responsible for the translocation of nutrients from the GI lumen into the bloodstream. Also, in the small intestine, the absorption processes of the orally administered drugs are basically related to the pharmacokinetics [1]. The deficit of cell culture methods to maintain in vivo–like functions forces researchers to optimize and apply methods in which cells are seeded and cultured under controlled and dynamic fluid flow [2]. Moreover, the lack of predictive human organ models has increased the necessity of approaches for proper mimicking of organ function in vitro, studying physiological parameters that regard mechanical, chemical and physical stimuli crucial for differentiation, morphology and function of the epithelia [3]. In this work we present a Gut-On-Chip (GOC) device, equipped with ITO (Indium tin Oxide) electrodes patterned by wet etching techniques, as a multifunctional microsystem for monitoring epithelial parameters. The potential to support cells adhesion, growth and polarization of a functional monolayer is also investigated in the Caco-2 epithelial-like cell line by in-device seeding and culture. In a perspective, this first prototype has established the basis for several technology integrations to study complex cellular phenomena targeted in key physiological topics (e.g. the tight interplay of different physical effects during mechanotransduction processes) and in pharmacological open issues such as drug absorption and metabolism.

Keywords Gut-On-Chip (GOC) · Organ-On-Chip (OOC) · TEER
Mechanical stimuli · Epithelial-like behavior · Embedded sensors

L. Giampetruzzi (✉) · C. De Pascali · S. Capone · P. Siciliano · F. Casino · L. Francioso
Institute for Microelectronics and Microsystems IMM-CNR, Via per Monteroni "Campus
Ecotekne", 73100 Lecce, Italy
e-mail: lucia.giampetruzzi@le.imm.cnr.it

A. Barca · T. Verri
Department of Biological and Environmental Sciences and Technologies (DiSTeBA), University
of Salento, Via per Monteroni "Campus Ecotekne", 73100 Lecce, Italy

© Springer Nature Switzerland AG 2019   181
B. Andò et al. (eds.), *Sensors*, Lecture Notes in Electrical Engineering 539,
https://doi.org/10.1007/978-3-030-04324-7_24

1 Introduction

The gastrointestinal (GI) tract shows extreme peculiarities at each (from micro to macro) scale of structural-functional organization, due to its complex interplay of residing cells, tissues and organs. The small intestine is one of the most important component of the digestive system. It is a considerably extended tube combined with the occurrence of peristaltic movements that are essential for solute and water movements and/or elimination of waste products. The surface area of this "small" tract is significantly enhanced, 30–600 fold, respectively, by the carpet of finger-like projections called intestinal villi, as well as the microvilli, cellular membrane protrusions present on enterocytes (i.e. differentiated epithelial cells).

Villi and microvilli are structural features specialized both for nutrient/ion/water absorption and secretion, and for mechanotransduction. In this scenario, epithelial tight junctions (TJs) are key elements in maintaining the epithelial barrier function and in regulating the permeability of nano-/micro-/macromolecules. They are present in between the epithelial cells keeping the luminal content from leaking between cells, so almost all nutrients go through the gut epithelium to the circulatory system in a strictly controlled manner.

The importance of the small intestine epithelium along the GI tract also concerns the absorption of orally administered drugs into blood across the highly polarized epithelial cell layer and the intestinal mucosa. In this respect, also the peristaltic movements and the intestinal microbiota are primary modulators of the bioavailability of the orally administered drugs [4]. This suggests that there are many intertwined mechanisms that affect the therapeutic efficacy of drugs, especially in certain pathological conditions of the GI, which comprehensively contribute to increase the difficulty in reproducing accurately the cyto/histological features of the simple columnar epithelium and the principal physiological processes occurring in small intestine [5].

To date the static planar culture models (in vitro, or ex vivo) and ethics issues related to in vivo assays do not allow to take into account more than a few crucial functions of the intestinal tracts. Moreover, due the peculiar geometry of intestinal epithelium (i.e. corrugation, convolution), the fluid dynamics are missed in the static two-dimensional (2D) in vitro models, which often fail to reproduce the physiology of e.g. drug metabolism, drug absorption, drug–drug interaction, etc. [6].

Over time, this failure of reproducing several living organ aspects has fueled the development of many organs on-a-chip approaches (Fig. 1).

Recently, microfabrication techniques have been used to replicate key functional units of small intestine as: unique topography of intestinal epithelium and the 3D cell physiology of intestinal crypts.

These micro-devices, based on micro-engineered biomimetic systems containing microfluidic channels, human cell-lined in order to simulate intestinal epithelia, in its properties, eventually combined with microbial symbionts.

The accurate reproduction of the in vivo environments, with controlled microfluidics parameters helps to reduce the delays and costs of research and to replace animal tests.

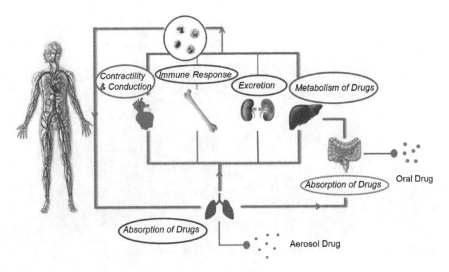

Fig. 1 The Human on a chip concept inspired form Wyss Institute (https://cdn.thenewstack.io/media/2015/04/organs-on-a-chipwyss-institute-3.jpg)

The technology is mainly based on a microfluidic technique, which enables manipulating small amount of liquids, controlling fluid flows in order to deliver nutrients to the biological matrices. The microfluidics is supported by microfabrication that creates microstructures able to control cell shape and function [3].

Considering recent studies of an intestine on a chip, some models were adapted from e.g. lung and heart to mimic the mechanical/structural properties, better reproducing the in vivo environments, but presenting some disadvantages and open issues, yet [7].

The principle limits are the inaccurate simulation of the intestine's peristalsis movements (inducing membrane curving, stretching, bending), the little standardization among companies and research groups, the lack of extensive analyses and considerations on cell culture procedures (cell seeding and attachment, growth, effects of nutrients and stimuli) and the in situ sensing network still lacking [7, 8].

In the perspective of developing a multifunctional Gut-On-Chip (GOC) device, we hereby propose a GOC with an integrated detection of monitoring in real-time several parameters. The transcellular and paracellular transport processes are important issues to investigate the tightness and the mechanisms of absorption and diffusion; similarly, the role of membrane-trafficking/exchange and nutrient/drug fluxes remain underappreciated and unexplored. The device is equipped with micromechanical stimulation sensors that can apply physiologically relevant mechanical stimuli and with electrodes able to control the pH gradients, the plasma membrane potentials and solute gradients as metal ion uptake or translocation. This approach presents many advantages, mainly related to the integrated detection possibilities and multi-parameter investigation (Fig. 2).

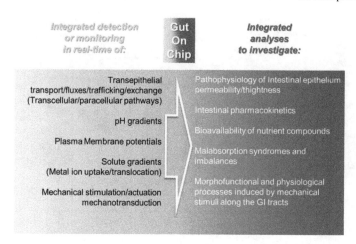

Fig. 2 Potential integrated analyses enabled by Gut-On-Chip devices

Although this study cannot cover all parameters of the gut physiology, by implementing a scheme of an integrated gut on-a-chip it hints the intention of elaborating some microtools equipment to monitor important cell-interaction device parameters, less investigated in microfludic platforms; where they appear more critical and unpredictable than in traditional in vitro cell culture. It also aims to overcome the limits of the novel technology designing a device able to achieve the recent pharmaceutical and medical challenge.

2   Materials and Methods

The GOC platform is composed of a customized commercial Topas polymer-made base with two (lumen vs. blood) fluidic chambers separated by a PET (Polyethylene terephthalate) porous membrane for cell culture (23 μm thickness, 200 nm pore dimension, $5 \cdot 10^8$ pores/cm^2 pore density) (Fig. 3). Membrane size is 11×8.5 mm^2. The fluidic platform is equipped with transparent ITO (Indium tin Oxide) electrodes for TEER monitoring during the cell culture; in fact, that TEER (Transepithelial Electrical Resistance) values are worth to measure integrity of cellular barriers and tight junction dynamics of cultured cells monolayers.

Our re-sealable assembly allows start cell culture with the porous membrane exposed to medium with no cover lid; once cell culture is stabilized, the chip chamber is sealed on top and bottom sides with Corning or Topas slides equipped with 200 nm thick ITO electrodes patterned by wet etching technique. So, 2 independent fluidic cham-bers, apical and basolateral, with a capacity of 50 μl per chamber were created. Major advantage of our device is that it allows software-controlled mechanical deflection of the membrane suspended between the two fluidic chambers, both

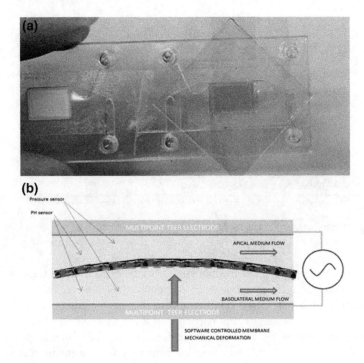

Fig. 3 Modified commercial multifunctional platform for Gut-On-Chip Caco-2 cultures with embedded ITO electrodes (**a**) and schematic of chip concept for multisensorial capabilities (**b**)

above and below the membrane, and simultaneous embedded sensors measurements; which helps in reproducing local mechanical stretches and bends, comparable to an in vivo situation, and monitoring effects in continuum.

We selected Caco-2 cells (human colorectal adenarcinoma-derived) as cellular model [9], suitable to preliminarily investigate cell proliferation and monolayer-to-chip interactions on the PET porous membrane.

The adhesion of Caco-2 cells on chip membrane was evaluated seeding at a density of about 2.5×10^4 cells/cm^2. Then, a flux to completely refresh the medium in the chamber every 10 min was activated and sustained for 24 h post seeding at 37 °C. After 24 h culture, the presence of viable cells was identified by metabolization of the MTT [3-(4,5-dimethylthiazol-2-yl)-2,5-diphenyltetrazolium bromide] compound.

3   Results and Discussions

Considering the importance of creating a biomimetic chip device to expose cells to a microfluidic flow, we basically evaluated the cell interaction with a PET translucent porous membrane (a non-biological matrix), investigating the cell behavior on an

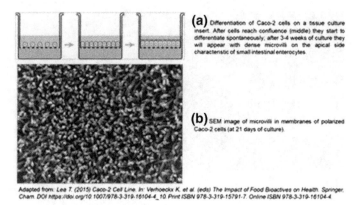

(a) Differentiation of Caco-2 cells on a tissue culture insert. After cells reach confluence (middle) they start to differentiate spontaneously; after 3-4 weeks of culture they will appear with dense microvilli on the apical side characteristic of small intestinal enterocytes.

(b) SEM image of microvilli in membranes of polarized Caco-2 cells (at 21 days of culture).

Adapted from: Lee T. (2015) Caco-2 Cell Line. In: Verhoeckx K. et al. (eds) The Impact of Food Bioactives on Health. Springer, Cham. DOI https://doi.org/10.1007/978-3-319-16104-4_10 Print ISBN 978-3-319-15791-7. Online ISBN 978-3-319-16104-4

Fig. 4 The Caco-2 enterocyte-like model

early GOC prototype in order to improve the weaknesses and the critical aspects de-riving from biological components. For example, the maintenance of sterility in a dynamic biological system, the viability of cells in a small medium volume for long periods in order to form epithelial barriers and the presence of air bubbles effects on culture are challenging issues deriving from the devices micrometric features and peculiarities. Indeed, the multifunctional microsystem for monitoring epithelia parameters has been set up with proper material and microfluidics tubes and it has primary investigated in its ability to support cell adhesion and growth, for several days, before electrical and mechanical test on the epithelium. Notably, the epithelial-like behavior by spontaneous differentiation of Caco-2 into absorptive enterocytes has been previously demonstrated (Fig. 4).

As Caco-2 cells are able to form a functional monolayer, our cell-to-device combination may represent a useful tool to study: (a) the epithelial tightness and the mecha-notransduction as key tools of the regulatory pathways at cellular and epithelial level in the GI tissue districts; (b) solutes gradients and pH variations across the apical and basolateral chambers for intestinal pharmacokinetics; (c) GI drug toxicity and drug screening.

In our experimental setup, viable Caco-2 cells have been detected by metabolization of the MTT compound producing the dark blue-to-black staining (i.e. formazan crystals in viable cells) (see Fig. 5). Figure 5 shows the centripetal migration front of the growing monolayer of cells, highlighting a growth surface coverage of 80–85%. The MTT assay is a preliminary evaluation to demonstrate a considerable efficiency of adhesion, viability and migration of human-derived intestinal cells on microfluidic devices.

The idea of in vivo mimicking conditions in a more reliable way, provides the optimization of the environmental constraints of on-chip cell cultures and the implementation of some enabling technologies to achieve a real time monitoring of parameters that playing an important role in several physiological pathways.

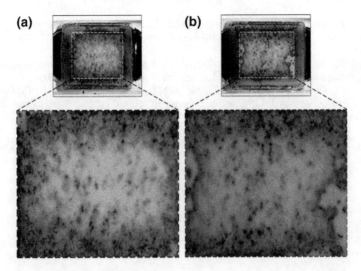

Fig. 5 MTT assay on Caco-2 cells loaded in-device, after cell adhesion and proliferation. Violet dark-blue crystals indicate the presence of viable, proliferating cells in two sample replicates (**a** and **b**)

To measure pH, Zn and Cu passage through the apical and basolateral chamber, miniaturized Ion Selective Electrodes (ISE) will be implemented. Impedance measurements will conduct to allow the detection and evaluation of the metal ions concentrations in situ. To this purpose, the mechanotransduction and sensitivity could represent the keystone to understand important mechanisms to proper reproduce GI tissue and to conduct pharmaceutical tests based on nutrient and drugs intestinal interactions.

References

1. Wilson, I.D., Nicholson, J.K.: Gut microbiome interactions with drug metabolism, efficacy and toxicity. Transl. Res. J. Lab. Clin. Med. **179**, 204–222 (2017)
2. Whitesides, G.M.: The origins and the future of microfluidics. Nature **442**, 368–373 (2006)
3. Huh, D., Hamilton, G.A., Ingber, D.E.: From three-dimensional cell culture to organs-on-chips. Trends Cell Biol. **21**(12), 745–754 (2011)
4. Vermeiren, J., Possemiers, S., Marzorati, M., Van de Wiele, T.: The Gut microbiota as target for innovative drug development: perspectives and a case study of inflammatory bowel diseases. In: Drug Development—A Case Study Based Insight into Modern Strategies. InTech. (2011)
5. Choi, M.S., Yu, J.S., Yoo, H.H., Kim, D.H.: The role of gut microbiota in the pharmacokinetics of antihypertensive drugs. Pharmacol. Res. (2018)
6. Thuenauer, R., Rodriguez-Boulan, E., Römer, W.: Microfluidic approaches for epithelial cell layer culture and characterisation. Analyst **139**(13), 3206–3218 (2014)

7. Kim, H.J., Huh, D., Hamilton, G., Ingber, D.E.: Human gut-on-a-chip inhabited by microbial flora that experiences intestinal peristalsis-like motions and flow. Lab Chip **12** (12), 2165–2174 (2012)
8. Tsui, J.H., Lee, W., Pun, S.H., Kim, J., Kim, D.H.: Microfluidics-assisted in vitro drug screening and carrier production. Adv. Drug Deliv. Rev. **65**(11–12), 1575–1588 (2013)
9. Sambuy, Y., De Angelis, I., Ranaldi, G., Scarino, M.L., Stammati, A., Zucco, F.: The Caco-2 cell line as a model of the intestinal barrier: influence of cell and culture-related factors on Caco-2 cell functional characteristics. Cell Biol. Toxicol. **21**(1), 1–26 (2005)

Portable Optoelectronic System for Monitoring Enzymatic Chemiluminescent Reaction

F. Costantini, R. M. Tiggelaar, R. Salvio, M. Nardecchia, S. Schlautmann,
C. Manetti, H. J. G. E. Gardeniers, D. Caputo, A. Nascetti and G. de Cesare

Abstract This work presents a portable lab-on-chip system, based on thin film electronic devices and an all-glass microfluidic network, for the real-time monitoring of enzymatic chemiluminescent reactions. The microfluidic network is patterned, through wet etching, in a 1.1 mm-thick glass substrate that is subsequently bonded to a 0.5 mm-thick glass substrate. The electronic devices are amorphous silicon p-i-n photosensors, deposited on the outer side of the thinner glass substrate. The photosensors, the microfluidic network and the electronic boards reading out the photodiodes' current are enclosed in a small metallic box ($10 \times 8 \times 15 \, cm^3$) in order to ensure shielding from electromagnetic interferences. Preliminary tests have been performed immobilizing horseradish peroxidase on the inner wall of the microchannel as model enzyme for detecting hydrogen peroxide. Limits of detection and quantification equal to 18 and 60 μM, respectively, have been found. These values are comparable to the best performances reported in literature for chemiluminescent-based optofluidic sensors.

Keywords Photosensors · Amorphous silicon · Microfluidics
Enzymatic reactions · Horseradish peroxidase · Anodic bonding

F. Costantini · M. Nardecchia · A. Nascetti
School of Aerospace Engineering, Sapienza University of Rome, Rome, Italy

F. Costantini · R. Salvio
Department of Chemistry, Sapienza University of Rome, Rome, Italy

R. M. Tiggelaar · S. Schlautmann · H. J. G. E. Gardeniers
MESA+ Institute for Nanotechnology, University of Twente, Enschede, The Netherlands

C. Manetti
Department of Environmental Biology, Sapienza University of Rome, Rome, Italy

D. Caputo (✉) · G. de Cesare
Department of Information Engineering, Electronics and Telecommunications,
Sapienza University of Rome, Rome, Italy
e-mail: domenico.caputo@uniroma1.it

© Springer Nature Switzerland AG 2019
B. Andò et al. (Eds.) *Sensors*, Lecture Notes in Electrical Engineering 539,
https://doi.org/10.1007/978-3-030-04324-7_25

189

1 Introduction

The combination of biosensing systems with microfluidic circuits improves the overall performance of the sensing analysis. The reduced dimensions and volumes in microfluidic channels allow first of all to work with much less sample than using standard equipment, making analysis on drops of blood or even the contents of single cells possible. More importantly, the short distances between analyte molecules and biorecognition elements reduce the diffusion times, which immediately yields a great gain in response time, significantly improving conditions for diffusion-limited processes [1]. Expanding from pure microfluidics to more fully developed lab-on-chip (LoC) solutions, entire sample preparation procedures, biorecognition elements and detection can be integrated in portable systems for health-care and diagnostics. Antibodies, aptamers and enzymes were immobilized by us into lab-on-chip devices for a variety of biosensing applications [2–4]. However, the integration of the sensing element has to be adapted to the technological steps necessary to combine the microfluidic network with the detection systems. For example, the integration of the detection system with the microfluidic network, can be an issue in terms of feasibility and sensitivity. The most used techniques are electrochemical and optical methods. In particular, optical methods rely on the use of fluorescence, absorbance and chemiluminescence. The last one is largely coupled with microfluidics since it does not require an excitation source and does not give background signals [5]. Lately, some groups started to work on the implementation of optical sensing elements directly on the microfluidic platform for chemiluminescence detection. Photosensors where integrated on silicon/PDMS chip for chemiluminescence detection [6, 7].

In this work, we present a novel lab-on-chip system based on a glass microfluidic channel with on-chip amorphous silicon photosensors (a-Si:H) [8] for the monitoring of chemiluminescent based enzymatic sensing assays. The photosensors are positioned underneath the microfluidic channel and allow monitoring the enzymatic assay. A layer of polymer brushes was grown on the inner wall of the microchannel and subsequently an enzyme was immobilized to it by peptide bond. As a proof of principle, horseradish peroxidase (HRP) was used as model enzyme to be anchored for detecting hydrogen peroxide (H_2O_2) in presence of luminol and 4-iodophenol.

2 System Structure and Operation

Figure 1 reports a cross section of the proposed lab-on-chip. It is constituted by two bonded glass substrates, which include both a microfluidic network and a-Si:H photosensors. Substrate 1 is a 1.1 mm-thick glass plate, that has been patterned by wet etching, in order to define the microfluidic network. Substrate 2 is a 0.5 mm-thick glass that has been anodically bonded to substrate 1. After the glass bonding, the a-Si:H photosensors have been deposited on the outer side of the thinner substrate and aligned with the microfluidic network. When a chemiluminescent reaction occurs inside the channels, as a result of a biomolecular recognition, the emitted radiation

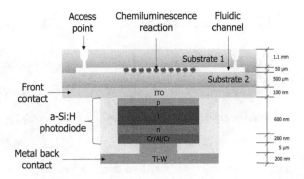

Fig. 1 Cross section of the all-glass lab-on-chip (not in scale): the bottom glass substrate hosts the the a-Si:H sensors on the bottom side, while the top substrate hosts the microfluidic network

is absorbed by the a-Si:H photodiodes, which produce photocurrents proportional to the emitted light. The a-Si:H are p-type doped/intrinsic/n-type doped stacked layers, whose thickness and energy gap has been optimized to detect the spectrum of the chemiluninescent spectrum.

The proposed lab-on-chip structure features therefore on-chip chemiluminescence detection which offers several advantages with respect to off-chip methods and in particular:

- reduction of the distance between the radiation source and the photosensors and as a consequence limited light diffusion and optical losses;
- absence of focusing optics which implies a higher degree of compactness of the system.

3 System Fabrication

The fabrication steps of the lab-on-chip have been optimized in order to keep the compatibility between microfluidics, microelectronics and biochemical requirements. In particular, the deposition and patterning of the a-Si:H photosensors should not affect the chemical surface composition of the glass channels, while the chemical procedures needed to implement the biomolecular recognition and to produce the chemiluminescent signals should not degrade the optoelectronic performance of the a-Si:H photodiodes.

Taking into account these specifications the fabrication of the lab-on-chip has been performed through the following technological steps (see Fig. 2):

- realization of fluidic networks on the bondside of a 1.1 mm-thick Borofloat 33 substrate by wet etching (25% hydrofluoric acid (HF)) yielding a 50 μm-deep and 110 μm-wide channel;
- definition of trenches, through the powderblast technique, in the 1.1 mm-thick glass (at the non-bond side). These trenches define the position of the inlets and outlets but at this phase of fabrication they are not in contact with the microfuidic channels (see Fig. 2b);

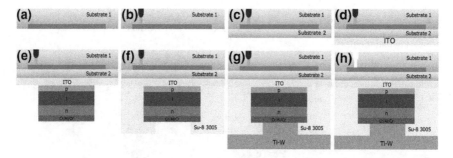

Fig. 2 Fabrication steps of the lab-on-chip: **a** Definition by wet etching of the microfluidic network in substrate 1. **b** Definition of the trenches for the microfluidic access. The purple region is the part removed by powderblasting. The green region is still glass. **c** Anodic bonding of the two glasses. **d** Deposition of the Indium Tin Oxide bottom contact of the photosensors. **e** Deposition by PECVD and patterning by reactive ion etching of the a-Si:H p-i-n junction. **f** Deposition and patterning of the SU-8 3005 insulation layer. **g** Deposition and patterning of the TiW top contact of the sensors. **h** Complete opening of the inlets and outlets of the microfluidic network by laser drilling

- anodic bonding of this substrates to a 0.5 mm-thick BF33 substrates [9];
- realization of the a-Si:H photodiodes by standard microelectronic techniques, including Plasma Enhanced Chemical Vapor Deposition for the deposition of the amorphous silicon layers [8] on the not-bonded side of the 0.5 mm-thick substrate;
- complete opening of the inlets and outlets by laser drilling.

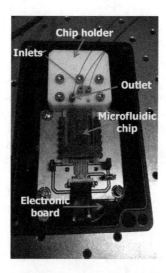

Fig. 3 Picture of the portable optoelectronic system: the metallic box includes the lab-on-chip, the chip holder for the inlet and outlet connections and the electronic boards for the photosensor read-out. A cover lid (not shown in the figure) with two small holes (for microfluidic access) shields the a-Si:H photodiodes from the external radiation

The lab-on-chip and the read-out electronics have been connected together through a custom made connector and enclosed in a metallic box ($10 \times 8 \times 15 \, cm^3$) which ensures shielding from external radiation and electromagnetic interferences. The free-side of the lab-on-chip is also inserted in a Teflon-holder to ensure a sealed connection to the external tubes providing the reagents for the chemical treatments of the inner channels. A picture of the whole system is reported in Fig. 3.

4 Test of the System

The functionality of the entire lab-on-chip has been tested by using horseradish peroxidase (HRP) as model enzyme for detecting H_2O_2. Recognition of this molecule is of great interest because it plays an important role in atmospheric and biochemical processes [10]. The test has been performed by functionalizing, at first, the inner walls of the microfluidic channels. The functionalization procedure envisages the following steps:

- growth of PHEMA on the channel wall by atom transfer radical polymerization;
- flowing of succinic anhydride (SA) solution in order to obtain carboxylic functional moieties PHEMA-SA;
- flowing of a solution of n-hydroxysuccinimide (NHS) to achieve NHS esters functional groups (PHEMA-NHS);
- insertion of a solution of horseradish peroxide (HRP) and incubation over-night at 4 °C to form the PHEMA-HRP brush layer;
- rinsing with blocking buffer for 30 min at 1 μ L/min.

Once the functionalization was accomplished, the enzymatic reaction was conducted by mixing into the microfluidic chip, through two separate inlets, a solution of luminol 1 mM and 4-iodophenol 0.1 mM and a solution of H_2O_2 at different concentrations. The HRP, immobilized on the microfluidic channels, acts as a biosensors for the H_2O_2, catalyzing its reaction with luminol and yielding a chemiluminescent signal proportional to the concentration of hydrogen peroxide. The reaction was monitored by reading-out the sensor photocurrents connected to the low noise electronics. Results demonstrate a limit of detection (LoD) down to 18 μM and an excellent linearity of photosensor response up to 250 μM. These values are within the range of practical interest for this molecule and are comparable to the best performances reported in the literature for chemiluminescent-based optofluidic sensors [11]. We have also verified that, after washing with piranha (solution of sulfuric acid and hydrogen perodixe) and rinsing with deionized water, the same microfluidic chip can be re-used for further analysis.

5 Conclusions

This work has presented a miniaturized lab-on-chip system for monitoring chemiluminescent enzymatic reaction. It is the combination of an all-glass microfluidic chip
and a-Si:H photosensors. The microfluidic chip is achieved by anodic bonding of two
glass substrates. The a-Si:H diodes are deposited directly on one side of microfluidic chip for the on-chip detection of the chemiluminescent signals generated inside
the channels. The whole fabrication process has been designed in order to satisfy
the compatibility between microfluidics, electronics devices and surface chemical
treatments.

The developed system has been tested in the detection of H_2O_2, achieving a LoD
of 18 μM. This result demonstrates the successful integration of the different technologies for detection and quantification of molecule whose biochemical recognition
exploits chemiluminescent signals.

References

1. Luka, G., Ahmadi, A., Najjaran, H., Alocilja, E., de Rosa, M., Wolthers, K., Malki, A., Aziz, H., Althani, A., Hoorfar, M.: Microfluidics integrated biosensors: a leading technology towards lab-on-a-chip and sensing applications. Sensors **15**(12), 30011–30031 (2015)
2. Costantini, F., Sberna, C., Petrucci, G., Manetti, C., de Cesare, G., Nascetti, A., Caputo, D.: Lab-on-chip system combining a microfluidic-ELISA with an array of amorphous silicon pho-tosensors for the detection of celiac disease epitopes. Sens. Bio-Sens. Res. **6**, 51–58 (2015)
3. Costantini, F., Sberna, C., Petrucci, G., Reverberi, M., Domenici, F., Fanelli, C., Manetti, C., de Cesare, G., DeRosa, M., Nascetti, A., Caputo, D.: Aptamer-based sandwich assay for on chip detection of Ochratoxin a by an array of amorphous silicon photosensors. Sens. Actuators B: Chem. **230**, 31–39 (2016)
4. Costantini, F., Tiggelaar, R., Sennato, S., Mura, F., Schlautmann, S., Bordi, F., Gardeniers, H., Manetti, C.: Glucose level determination with a multi-enzymatic cascade reaction in a functionalized glass chip. Analyst **138**(17), 5019–5024 (2013)
5. Zhang, Z., Zhang, S., Zhang, X.: Recent developments and applications of chemiluminescence sensors. Anal. Chim. Acta **541**(1–2), 37–47 (2005)
6. Hofmann, O., Miller, P., Sullivan, P., Jones, T.S., Bradley, D.D.: Thin-film organic photodiodes as integrated detectors for microscale chemiluminescence assays. Sens. Actuators B: Chem. **106**(2), 878–884 (2005)
7. Jorgensen, A.M., Mogensen, K.B., Kutter, J.P., Geschke, O.: A biochemical microdevice with an integrated chemiluminescence detector. Sens. Actuators B: Chem. **90**(1–3), 15–21 (2003)
8. Caputo, D., de Cesare, G., Scipinotti, R., Stasio, N., Costantini, F., Manetti, C., Nascetti, A.: On-chip diagnosis of celiac disease by an amorphous silicon chemiluminescence detector. In: Sensors and Microsystems, pp. 183–187. Springer, Cham. (2014)
9. Costantini, F., Benetti, E.M., Tiggelaar, R.M., Gardeniers, H.J., Reinhoudt, D.N., Huskens, J., Vancso, G.J., Verboom, W.: A Brush?Gel/Metal?Nanoparticle hybrid film as an efficient supported catalyst in glass microreactors. Chem.-A Eur. J. **16**(41), 12406–12411 (2010)
10. Tahirovic, A., Copra, A., Omanovic-Miklicanin, E., Kalcher, K.: A chemiluminescence sensor for the determination of hydrogen peroxide. Talanta **72**(4), 1378–1385 (2007)
11. Kovarik, M.L., Gach, P.C., Ornoff, D.M., Wang, Y., Balowski, J., Farrag, L., Allbritton, N.L.: Micro total analysis systems for cell biology and biochemical assays. Anal. Chem. **84**(2), 516–540 (2011)

A Novel Paper-Based Biosensor for Urinary Phenylalanine Measurement for PKU Therapy Monitoring

Maria Anna Messina, **Federica Raudino, Agata Fiumara, Sabrina Conoci and Salvatore Petralia**

Abstract A novel paper-based biosensors for the measurement of urinary phenylalanine (Phe) was developed and the experimental results here reported. The proposed biosensor is featured by a silicon part integrating temperature sensors and heaters and a polycarbonate ring to form a microchamber. The reagent-on-board format allows a fast and easy self-testing directly from patients. The biosensors is thermally driven by a customized instrument and software. The detection strategy employed is based on the specific reaction of Phenylalanine Ammonia Lyase enzyme to produce ammonia and trans-cinnamic acid from Phenylalanine. The increase of pH value is proportional to the Phe amount and can be monitored by the color changes of a dye solution. The proposed system is suitable to detect the phenylalanine levels in a linear dynamic range concentration from 20 to 3000 μM.

Keywords Biosensors · Phenylalanine · Phenylketonuria

1 Introduction

Phenylketonuria (PKU) is a rare inherited inborn disorder of metabolism mainly caused by a deficiency due a defect in the gene that produce the hepatic enzyme phenylalanine hydroxylase (PAH) [1]. This deficiency provokes the increase of Phenylalanine level in blood causing injurious effects on brain and consequently severe and irreparable intellectual disability [2]. The main treatment for PKU patients is based on a dietary restriction of Phe. The patient with well-controlled Phe levels have normal development. Therefore the control of Phe level in blood or urine is the primary marker for guiding available treatments [3].

M. A. Messina (✉) · F. Raudino · A. Fiumara
Azienda Ospedaliero Universitaria Policlinico Vittorio Emanuele, Via S. Sofia 78, Catania, Italy
e-mail: mmessina@unict.it

S. Conoci · S. Petralia (✉)
STMicroelectronics, Stradale Primosole 50, 95121 Catania, Italy
e-mail: salvatore.petralia@st.com

© Springer Nature Switzerland AG 2019
B. Andò et al. (eds.), *Sensors*, Lecture Notes in Electrical Engineering 539,
https://doi.org/10.1007/978-3-030-04324-7_26

In this context we developed a novel integrated paper-based biosensor for the rapid measurement of urinary Phenylalanine level for the monitoring of dietary therapy efficiency for patients affected by Phenylketonuria disorder. The biosensor is composed by a silicon parts integrating temperature sensors and heaters for temperature control (accuracy of about 0.1 °C) and a polycarbonate ring to form the microwell with a total volume of 200 μl [4]. A paper support for enzyme immobilization was glued upon the silicon part. The device is thermally driven by a customized instrument and software [5]. The PAL enzyme and reagents were properly immobilized at paper surface substrate and the reaction performed at temperature of 35 °C and pH 8.3.

The system integrates a strategy based on the fast deamination reaction of Phenylalanine catalyzed by Phenylalanine Ammonia Lyase (PAL) enzyme to produce trans-cinnamic acid and ammonia. The amount of ammonia produced increase the pH value of sample. This is easily monitored through by a pH indicator (Phenolphthalein) which induces a color change of sample solution. Thus the concentration of Phe can be estimated via colorimetric comparison with a chromatic scale. The proposed system permits the monitoring of Phe in a dynamic range concentration of 20–3000 μM, with a Limit of Detection of about 20 μM. The method was validate by comparison with the standard MS-MS technique [6].

2 Materials and Methods

2.1 Chemicals

All chemicals used on experiments enzyme, salts and additives were purchased by Sigma-Aldrich and used as received.

2.2 Instrumentation

All the analyses were carried out using the following equipments: Electrospray tandem mass spectrometer "Quattro Micro", equipped with 1525 μ Binary HPLC pump and 2777 C auto-sample manager (Waters); Automatic "DBS Puncher" (PerkinElmer); Thermo-shaker "NCS Incubator" (Wallac); pHmeter "FiveEasyTM FE20-KIT" (Mettler Toledo); Analytical balance "ML204" (Mettler-Toledo) and Single ray Spectrophotometer Nanodrop ($V_{max} = 2$ μL) (SpectraMax).

3 Results and Discussion

3.1 Biosensor

The biosensors employed was developed by STMicroelectronics for genetic analysis [7–9]. It is composed by a silicon part containing resistors for the heating and temperature control, a paper substrate glued to the first and finally a polycarbonate ring mounted upon the second layer to form the microchambers. A black plastic holder guarantee the easily device handling and a polycarbonate slide lid guarantee the microchamber sealing during the reaction. A first prototypes was produced used glass slide as solid support and the reaction performed on standard oven. Figure 1 report the paper-glass biosensor (Fig. 1a), the components of paper-silicon based biosensor (from Fig. 1b-1 to b-5) and the whole biosensor (Fig. 1c).

3.2 Phenylalanine Detection Strategy

The detection strategy is based on the deamination reaction of Phenylalanine, to produce trans-cinnamic acid, catalyzed by PAL enzyme. This reaction produces a quantitative amount of ammonia, (which involves a solution pH increasing) with an

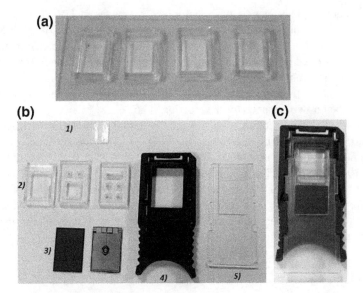

Fig. 1 Paper-based biosensor: **a** paper-glass biosensor, **b** component of paper-silicon based biosensor (1 paper membrane, 2 polycarbonate ring for microrector, 3 silicon part, front and rear view, 4 plastic holder and 5 transparent lid) and **c** whole biosensor

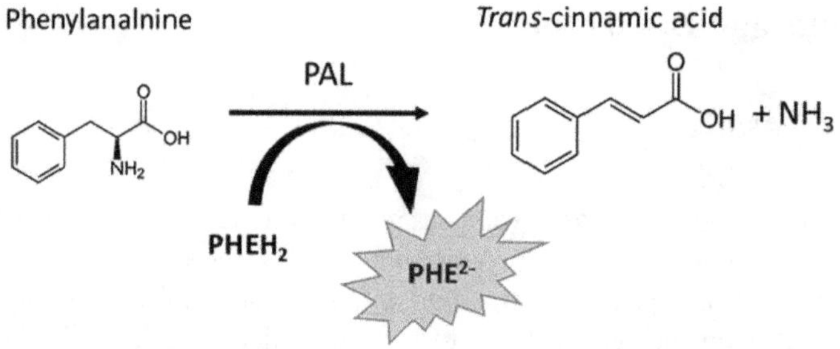

Phenylalanine   *Trans*-cinnamic acid

Fig. 2 Phenylalanine detection strategy

increasing of the pH value of the sample. This increment is monitored through the addition of phenolphthalein as pH indicator that induces color change of solution (Fig. 2). Therefore, the Phe amount was estimated by colorimetric comparison with the chromatic scale. Phenolphthalein was selected as the pH-indicator because the colors of its protonated and un-protonated molecular species change at a pH value (8.3) which correspond to the optimum PAL activity.

3.3   Detection Strategy Optimization and Chromatic-Scale

The detection strategy was properly optimized by means of UV-Vis spectroscopy using standard solution of Phe at a concentration of 86 μM, dissolved in Sodium Posphate 10 mM at pH 8.3, at 35 °C and room temperature. The enzymatic reaction was monitored by optical absorption spectroscopy through the formation of the absorption band centered at 273 nm diagnostic for the tran-cinnamic acid, Fig. 3c report the kinetics trend for the reaction performed at 35 °C. The results as expected indicate a better performance for the reaction at 35 °C.

Additionally the detection strategy was also tested via colorimetric method on standard 96 microplate and on biosensor (see Fig. 3a, c) employing 10 μL of solution containing Phe at concentration 20, 100, 300, 800, 900, 2500 μM, 75 μl of fosphate buffer 10 mM, 5 μL of NaOH 0.1 M and finally 10 μL of PAL (2U/0.28 μl). The solutions were mixed and heated at 35 °C for 20 min. Than a volume of 3 μL of phenolphthalein (1% $^P/_v$) was added at each well and its color compared with the chromatic scale previously prepared using standard solutions of Phe in the range from 20 to 3000 μM (Fig. 3d).

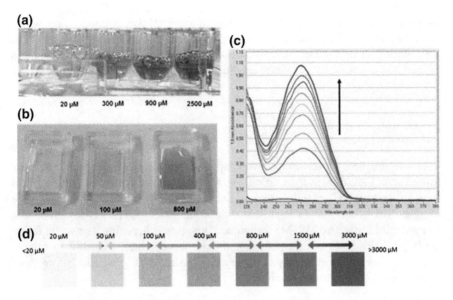

Fig. 3 Colorimetric assay: **a** on standard microplates, **b** on paper-glass biosensor, **c** optical absorption band of trans-cinnamic **a** during the enzymatic reaction at different time **d** chromatic scale

3.4 Phenylalanine Detection on Human Sample

In order to evaluate the assay's performance, mock samples (prepared adding known Phe amount to urine sample) were tested by our colorimetric method performed on paper-based device and the results compared with standard MS-MS technique. The finding data reported in Table 1 indicate a good correlation between the two method.

In details the urine sample tested on paper-based device report a Phe amount in the range of 50–100 μM, data confirmed by the MS-MS measurement (74.98 ± 2.50 μM). Similarly, for the three mock samples containing an amount of nominal Phe of about 302.5, 499.5 and 895.5 μM the follow concentration range were found 100–400, 400–800 and 800–1500 μM. Again the MS-MS data shown a Phe concentration within these range, indicating a good performance of colorimetric method. Although the proposed method give a semi-quantitative response (range Phe

Table 1 Phe amount by proposed method and MS-MS technique on mock sample

Sample	Phe (μM) nominal	Phe (μM) paper-based device	Phe (μM) by MS-MS
Urine sample	*endogenous*	50–100	74.98 ± 2.50
Mock sample 1	302.5	100–400	375.66 ± 11.50
Mock sample 2	499.5	400–800	584.66 ± 14.50
Mock sample 3	895.5	800–1500	828.33 ± 11.51

of concentration, see Fig. 3), it is appropriate for the PKU therapy monitoring, where the measurement of Phe concentration with high resolution is not required. Against that, the low cost of analysis, the easy-to-use procedure, and the reduced analysis time makes this method a potential candidate for a further Point-of-care system.

4 Conclusion

In conclusion, here we reported a novel paper-based biosensor for urinary Phe measurement for PKU therapy monitoring. The biosensor integrates the reagent-on-board for a fast and easy to use self-testing for the detection of Phe concentration on urine samples. The assay exhibits a dynamic range from 20 to 3000 μM. The method was compared with standard MS–MS technique. This approach, paving the way for future development of Point-of-care platform, able to improve the PKU patient's quality of life monitoring the efficiency of dietary therapy in real time without the needs of specialized laboratory and personnel.

References

1. Kure, S., Hou, D.C., Ohura, T., et al.: Tetrahydrobiopterin-responsive phenylalanine hydroxylase deficiency. J. Pediatr. **135**, 375–378 (1999)
2. Scriver, C.R., Kaufman, S.: Hyperphenylalaninemia: phenylalanine hydroxylase deficiency, 8th edn. Mc Graw-Hill (2001)
3. Fonnesbeck, C.J., McPheeters, M.I., et al.: Estimating the probability of IQ impairment from blood phenylalanine for phenylketonuria patients: a hierarchical meta-analysis. J. Inherited Metab. Dis. **36**, 757–766 (2013)
4. Petralia, S., Conoci, S.: PCR technologies for point of care testing: progress and perspectives. ACS Sens. (2), 876–891 (2017)
5. Guarnaccia, M., Iemmolo, R., Petralia, S., Conoci, S., Cavallaro, S.: Miniaturized real-time PCR on a Q3 system for rapid KRAS genotyping. Sensors **17**(831), 1–9 (2017)
6. Messina, M.A., Meli, C., Conoci, S., Petralia, S.: A facile method for urinary phenylalanine measurement on paper-based lab-on-chip for PKU therapy monitoring. Analyst **142**, 4629–4632 (2017)
7. Petralia, S., Sciuto, E.L., Di Pietro, M.L., Zimbone, M., Grimaldi, M.G., Conoci, S.: Innovative chemical strategy for PCR-free genetic detection of pathogens by an integrated electrochemical biosensor. Analyst **142**, 2090–2093 (2017)
8. Santangelo, M.F., Sciuto, E.L., Busacca, A.C., Petralia, S., Conoci, S., Libertino, S.: Si photo-multipliers for bio-sensing applications. IEEE J. Sel. Top. Quantum Electron. 22(3) (2016)
9. Santangelo, M.F., Sciuto, E.L., Busacca, A.C., Petralia, S., Conoci, S., Libertino, S.: Si PM as miniaturised optical biosensor for DNA-microarray applications. Sens. BioSens. Res. (6) 95–98 (2015)

Part III
Physical Sensors

Magnetoencephalography System Based on Quantum Magnetic Sensors for Clinical Applications

Carmine Granata, Antonio Vettoliere, Oliviero Talamo, Paolo Silvestrini, Rosaria Rucco, Pier Paolo Sorrentino, Francesca Jacini, Fabio Baselice, Marianna Liparoti, Anna Lardone and Giuseppe Sorrentino

Abstract In this paper, we present the magnetoencephalography system developed by the Institute of Applied Sciences and Intelligent Systems of the National Research Council and recently installed in a clinical environment. The system employ ultra high sensitive magnetic sensors based on superconducting quantum interference devices (SQUIDs). SQUID sensors have been realized using a standard trilayer technology that ensures good performances over time and a good signal-to-noise ratio, even at low frequencies. They exhibit a spectral density of magnetic field noise as low as 2 fT/Hz$^{1/2}$. Our system consists of 163 fully-integrated SQUID magnetometers, 154 channels and 9 references, and all of the operations are performed inside a magnetically-shielded room having a shielding factor of 56 dB at 1 Hz. Preliminary measurement have demonstrated the effectiveness of the MEG system to perform useful measurements for clinical and neuroscience investigations. Such a magnetoencephalography is the first system working in a clinical environment in Italy.

Keywords SQUID · Magnetometer · Magnetoencephalography

C. Granata (✉) · A. Vettoliere · O. Talamo · G. Sorrentino
Institute of Applied Science and Intelligent Systems "E. Caianiello" of National Research Council, 80078 Pozzuoli (Naples), Italy
e-mail: carmine.granata@cnr.it

P. Silvestrini
Mathematics and Physics Department, University of Campania "L. Vanvitelli", Caserta, Italy

R. Rucco · F. Jacini · M. Liparoti · A. Lardone · G. Sorrentino
Department of Motor Sciences and Wellness, University of Naples Parthenope, Naples, Italy

P. P. Sorrentino · F. Baselice
Department of Engineering, University of Naples Parthenope, Naples, Italy

© Springer Nature Switzerland AG 2019
B. Andò et al. (eds.), *Sensors*, Lecture Notes in Electrical Engineering 539,
https://doi.org/10.1007/978-3-030-04324-7_27

1 Introduction

Magnetometers based on the superconducting quantum interference device (SQUID) are very sensitive low-frequency magnetic field sensors, reaching a spectral density noise of a few fT/Hz$^{1/2}$ [1, 2].

Due to their ultra-high sensitivity, SQUID devices are widely used in several applications [1, 2], such as from biomedicine, non-destructive tests, geophysics, magnetic microscopy and fundamental science. One of the most important application is the biomedical imaging. In particular, the interest is mainly focused on multichannel system for magnetoencephalography (MEG), which provide useful information on brain functionality [2, 3]. MEG systems measure magnetic fields produced by neuronal activity. Reflecting the intracellular electric current flowing in the brain, the MEG measurements provide direct information about the dynamics of evoked and spontaneous neural activity. The magnetic fields generated by brain activity are minimally distorted by the layers surround the brain, allowing for a temporally and spatially accurate reconstruction of the neural signals within the brain (source space). Furthermore, the phase of such signals can be exploited in order to evaluate the amount of information exchanged between brain areas. Among the available brain functional imaging methods, MEG uniquely features both a good spatial and an excellent temporal resolution, allowing useful investigation in neuroscience and neurophysiology. In fact, by using suitable algorithms [4] it is possible to estimates synchronization [5] between areas, thereby providing complementary information to the fMRI.m. Some of the properties of the interactions among brain areas can be analyze through graph theory [6]. In fact, the human brain can be modelled as a network, where the brain areas are the nodes and their interaction are the edges. However, such metrics are influenced by network size or thresholding, making difficult to give a topological interpretation of the results, especially when they come to brain signals [7]. In this paper we will present a multichannel system for Magnetoencephalography operating in a clinical environment.

2 Magnetic Sensors

A SQUID is magnetic flux to voltage transducer. It consist in a superconducting loop interrupted by two Josephson junction [1, 2]. At least the magnetic field sensitivity is proportional to the loop area. However, it is not possible to increase the area of loop to increase the magnetic field sensitivity, because the flux noise increases with the ring inductance. An efficient way to increase the field sensitivity consists of using a proper superconducting detection circuit (flux transformer) consisting of a series of pickup coil having a flux capture area much higher than the SQUID one, and an input coil inductively coupled to the SQUID loop [1]. Typically, the

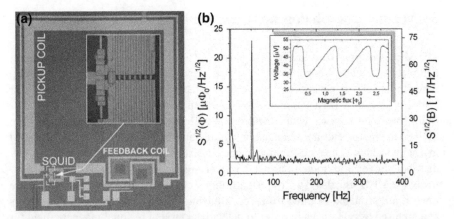

Fig. 1 **a** Fully integrated miniaturized dc SQUID magnetometer. **b** Magnetic flux anf firld spectral density measured at T = 4.2 K in flux locked loop configuration. Theoretical prediction for the white noise is indicated by the straight blue line. The inset report the voltage-magnetic flux characteristic

pickup coil consists of a single square shaped coil including, in one of its sides, a planar multiturn input coil which is located upper to the SQUID loop acting as a secondary coil of flux transformer. A suitable SQUID loop design consists of a single coil having a square "washer" shape. In such a configuration, the SQUID inductance does not depend on external dimension of the washer but only on the hole dimension. The coupling between the washer and the input coil is very good and the input coil inductance is proportional to turn numbers and hole inductance. Hence, the input coil inductance can be adjusted to match a particular load by varying the outer dimension of the washer to accommodate the required number of turns in the input coil. In the Fig. 1a, a fully integrated SQUID magnetometer employed in the MEG system is reported. The above design has been optimized to keep a suitable sensitivity [8]. It has an area less than 10 mm^2 and includes a superconducting flux transformer, an additional positive feedback (APF) circuit and a bipolar feedback coil for low crosstalk operations [9]. The sensing pickup circuit consisting of a superconducting square coil is connected in series with a 8-turn input coil, which is coupled to SQUID loop in a washer configuration. Apart from a better spatial resolution, a small pickup coil minimizes its antenna gain, reducing the effects of radio frequency interference. The spectral density of both flux and magnetic field noise of the miniaturized SQUID magnetometer is reported in Fig. 1b. The sensor exhibits a magnetic flux noise level of 2.2 $\mu\Phi$ 0/$\sqrt{\text{Hz}}$ in the white region corresponding to 5.8 fT/$\sqrt{\text{Hz}}$ [8]. In the inset of Fig. 1b, the magnetic flux-voltage characteristic is reported. It is evident that there are that there are not resonances ensuring a good stability during operation.

3 Magnetoencephalography System

The MEG system shown in Fig. 3, consists of 163 fully integrated dc SQUID magnetometers, featuring adequate field sensitivity and bandwidth for brain imaging. Since these sensors are placed close each to other, the integrated feedback coil for Flux-Locked-Loop (FLL) operation have been properly designed in a bipolar multiturn shape, in order to reduce the crosstalk effect between neighbor sensors. The SQUID magnetometer are arranged on a multisensorial array designed and realized in a helmet shape. The measurement plane consists of 154 SQUID-channels suitably distributed over a fiberglass surface to cover the whole scalp and to record effectively the MEG signals (Fig. 2b) [10]. Further 9 channels, installed on three bakelite towers, are arranged in three triplet each having three orthogonal SQUID sensors and are used as references in order to detect background residual magnetic field far from the scalp (about 9 cm) and to subtract its contribution, via software properly implemented to this aim, from the brain signal detected on the measurement plane.

The sensor array, as shown by Fig. 3a, is located in a fiberglass Dewar with a helmet shaped bottom, at a distance of 18 mm from the outside, where the head of patient is housed. The SQUID sensors are connected to the room temperature

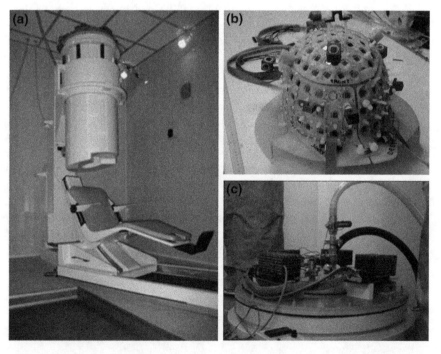

Fig. 2 **a** General view of the Magnetoencephalographic (MEG) system in the shielded room; **b** helmet-shaped array consisting of 163 fully-integrated SQUID magnetometers; **c** top view of MEG system showing the read-out electronics

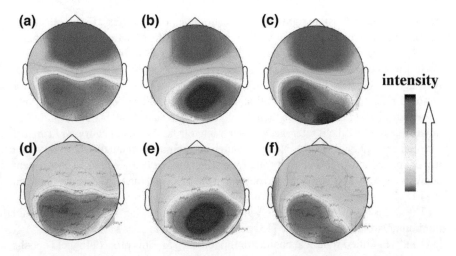

Fig. 3 Preliminary measurement performed by the MEG system. The imaging refers to the activated brain areas detected by the MEG (top row) and EEG (bottom row) during a spontaneous activity (alpha rhythm) (**a, d**) and evoked activity (tapping of the left (**b, e**) forefinger and the right (**c, f**) forefinger

read-out electronics by means of more than 800 copper wires having a diameter of 0.08 mm and twisted in pairs to avoid a parasitic area. The SQUID readout is a direct coupled electronics and is integrated in low-power, miniaturized boards (Fig. 3c). A single card drives six channels and is plugged in a shared motherboard located on the top of the dewar. In turn, the motherboard containing also the control logic unit and the filtering stage before the A/D conversion up to 10 kHz. A remote console allows to manage the electronics parameters setting digital filtering, amplification, under-sampling, on-line average, and on-line software gradiometer composition based on selectable configuration files.

The system is equipped also with a 32 channels system for EEG allowing to record simultaneous both magnetic and electrical signals. To eliminate any ambient disturbance the system operates in a suitable magnetic shielding room (MSR). The MSR consists of an external aluminum layer 12 mm thick and two inner layers of soft magnetic material having a thickness of 1.57 mm. designed on the basis of the environmental magnetic field measured on site before the installation. The resulting shielding factor is about 35 dB at 0.01 Hz, that increases up to 107 dB at 20 Hz.

4   MEG Acquisition and Test Measurements

Before to acuire MEG data, it is necessary to perform the following procedure. Using a suitable tool (Polhemus FASTRAK®) is detected the position of four coils, placed on the forehead and behind the ears of the participants, and of four reference points on the head (nasion, right and left preauricular points, vertex). Hence, electrical currents passing through the four coils are recorded by the MEG sensors. Participants to a study are seated inside a magnetically shielded room to reduce background noise Electrocardiographic (ECG) and Electrooculographic (EOG) signals are co-recorded, for subsequent artefacts removal. Spontaneous brain activity is recorded for almost 5 min, in resting-state condition, with closed eyes. The system has a sampling frequency of 1024 Hz. Data are then band-pass filtered at 0.5–49 Hz. Environmental noise, recorded by the 9 references, is subtracted from the signals through Principal Component Analysis (PCA). Noisy channels are removed manually through visual inspection of the whole dataset by an experienced rater [11]. Independent component analysis (ICA) [12] is performed to eliminate the ECG and the EOG component from the MEG signals.

The data is reconstructed in ninety areas of interest, applying a linearly constrained minimum variance beamformer [13] based on a template atlas or on the native MRI [14] and a modified spherical conductor model [14, 15]. This procedure yield epochs made of 90 time series, one per area of interest (just the cortical regions and the basal ganglia).

In the Fig. 3, preliminary measurements are reported. The first one (Fig. 3a) concerns a spontaneous activity. The so-called alpha rhythm, which appears in a human's brain awake but with eyes closed, has been recorded. The frequency range involved is 9–11 Hz. The second one (Fig. 3b, c), concerns an evoked activity: the brain imaging during the forefinger tapping of a volunteer, has been recorded. Figure 3b refers to the left finger tapping and, as expected, the activate motor cortex area is in the right hemisphere. In the Fig. 3c, instead, the imaging related to the movement of right forefinger is reported; as in the previous case, the contralateral area is involved. In this case the frequency signal are located around 7–9 Hz. The good agreement between EEG and MEG imaging indicates that the system operates properly.

5   Conclusions

A muitichannel system based on high sensitive quantum magnetic sensors for neurological applications has been described. The ultra-low noise of the magnetometers allow to measure the weak magnetic fields associated with the neurological activity. Preliminary test measurements have shows the effectiveness of the magnetoencephalographer. Very interesting measurements concerning Alzheimer disease (AD) and amyotrophic later sclerosis (ALS) on a large cohort of patients are in progress. The connectivity will be obtained by measuring synchronization between brain areas and a graph theoretical approach will be used to study disease staging in ALS. We will compare ALS patients and controls in order to verify that the topology of the brain networks brain show appreciable differences according to the disease stage.

References

1. Granata, C., Vettoliere, A.: Nano superconducting quantum interference device: a powerful tool for nanoscale investigations. Phys. Rep. **614**, 1 (2016)
2. Clarke, J., Braginski, A.I. (eds.): The SQUID Handbook Vol II: Fundamentals and Technology of SQUIDs and SQUID Systems. Wiley-VCH Verlag GmbH & Co. KgaA, Weinheim (2006)
3. Del Gratta, C., Pizzella, V., Tecchio, F., Romani, G.L.: Magnetoencephalography—a noninvasive brain imaging method with 1 ms time resolution. Rep. Progr. Phys. **64**, 1759–1814 (2001)
4. Stam, C.J., Nolte, G., Daffertshofer, A.: Phase lag index: assessment of functional connectivity from multi channel EEG and MEG with diminished bias from common sources. Hum. Brain Mapp. **28**, 1178–1193 (2007)
5. Tass, P., Rosenblum, M.G, Weule, J. Kurths, J., Pikovsky, A., Volkmann, J., Schnitzler, A., Freund, H.-J.: Detection of n:m phase locking from noisy data: application to magnetoencephalography. Phy. Rev. Lett. **81**, 3291–3294 (1998)
6. Bullmore, E., Sporns, O.: Complex brain networks: graph theoretical analysis of structural and functional systems. Nat. Rev. Neurosci. **10**, 186–198 (2009). https://doi.org/10.1038/nrn2575
7. Van Wijk, B.C.M., Stam, C.J., Daffertshofer, A.: Comparing brain networks of different size and connectivity density using graph theory. PLoS One **5**(10), e13701 (2010). https://doi.org/10.1371/journal.pone.0013701
8. Granata, C., Vettoliere, A., Rombetto, S., Nappi, C., Russo, M.: Performances of compact integrated superconducting magnetometers for biomagnetic imaging. J. Appl. Phys. **104**(073905), 1–5 (2008)
9. Granata, C., Vettoliere, A., Russo, M.: Miniaturized superconducting quantum interference magnetometers for high sensitivity applications. Appl. Phys. Lett. **91**, 122509 (2007)
10. Rombetto, S., Granata, C., Vettoliere, A., Russo, M.: Multichannel system based on a high sensitivity superconductive sensor for magnetoencephalography. Sensors **14**, 12114–12126 (2014)
11. Gross, J., Baillet, S., Barnes, G.R., Henson, R.N., Hillebrand, A., Jensen, O., Jerbi, K., Litvak, V., Maess, B., Oostenveld, R., Parkkonen, L., Taylor, J.R., van Wassenhove, V., Wibral, M., Schoffelen, J.-M.: Good practice for conducting and reporting MEG research. Neuroimage **65**, 349–63 (2013). https://doi.org/10.1016/j.neuroimage.2012.10.001
12. Barbati, G., Porcaro, C., Zappasodi, F., Rossini, P.M., Tecchio, F.: Optimization of an independent component analysis approach for artifact identification and removal in magnetoencephalographic signals. Clin. Neurophysiol. **115**, 1220–1232 (2004). https://doi.org/10.1016/j.clinph.2003.12.015
13. Van Veen, B.D., Van Drongelen, W., Yuchtman, M., Suzuki, A.: Localization of brain electrical activity via linearly constrained minimum variance spatial filtering. IEEE Trans. Biomed. Eng. **44** (1997)
14. Tzourio-Mazoyer, N., Landeau, B., Papathanassiou, D., Crivello, F., Etard, O., Delcroix, N., Mazoyer, B., Joliot, M.: Automated anatomical labeling of activations in SPM using a macroscopic anatomical parcellation of the MNI MRI single-subject brain. Neuroimage. **15**, 273–89 (2002). https://doi.org/10.1006/nimg.2001.0978
15. Nolte, G.: The magnetic lead field theorem in the quasi-static approximation and its use for magnetoencephalography forward calculation in realistic volume conductors. Phys. Med. Biol. **48**, 3637–3652 (2003). https://doi.org/10.1088/0031-9155/48/22/002

Calibration System for Multi-sensor Acoustical Systems

Orsola Petrella, Giovanni Cerasuolo, Salvatore Ameduri,
Vincenzo Quaranta and Marco Laracca

Abstract In recent years, the multi acoustic sensors systems have had a major development thanks to their versatility in different fields. These systems, also called acoustic antennas, consist of a set of microphones distributed according to linear, planar or three-dimensional geometries. The acoustic signals detected by the microphones are processed in order to define the location of an acoustic source. The acoustic antennas find large applications in different fields. In automotive they are used to highlight the noise propagation path; in the multimedia, these sensors allow localizing a speaker without portable microphones. Also the civil safety and military fields benefit from these systems: gunshots detection in city areas, fire prevention in wooded zones (Blaabjerg et al. in IV International Conference on Forest Fire Research, 2010), soldiers protection from enemy attacks are just some possible applications (ShotSpotter Gunshot Location System® (GLS). http://www.shotspotter.com). Even in the aerospace field, there are interesting applications such as the monitoring of the air traffic zones (ATZ), locating a plane and tracing its trajectory (Quaranta et al. in ESAV 2011-Tyrrhenian International Workshop on Digital Communications-Enhanced Surveillance of Aircraft and Vehicles, 2011 and Petrella et al. in ICSV19, 2012). The identification of the position of the source requires the knowledge of right acoustic locations of the microphones in the array, generally different from the geometric locations, to this scope a suitable calibration procedure. The proposed method was tuned in a simulation environment to predict signal produced by each microphone. An optimization process was adopted to identify layout configuration guaranteeing the right calibration. The proposed solution was experimentally validated on a two-dimensional acoustic antenna.

Keywords Acoustic array · Accuracy · Calibration

O. Petrella (✉) · G. Cerasuolo · S. Ameduri · V. Quaranta
CIRA-Italian Aerospace Research Center, Via Maiorise, 81043 Capua (CE), Italy
e-mail: o.petrella@cira.it

M. Laracca
Università di Cassino e del Lazio Meridionale, Via G. de Biasio, 43, Cassino (FR), Italy

© Springer Nature Switzerland AG 2019
B. Andò et al. (eds.), *Sensors*, Lecture Notes in Electrical Engineering 539,
https://doi.org/10.1007/978-3-030-04324-7_28

1 Introduction

The development of acoustic arrays of microphones today is of great interest for many applications in different fields and, as a result, under completion in many research projects. The acoustic antenna, through an advanced processing technique called "beamforming" define the position of the source based on the location of the microphones in the antenna and the time delay with which the acoustic signal reaches each microphone with respect to a reference microphone. One of the limitations of these systems is caused by the difference between the geometric and acoustic positions of the microphones due to their position inside the array and the internal features (electronics, inertia of the components). This difference has a dramatic impact in terms of uncertainty in determining the position of the acoustic source. For this reason the acoustic antennas need a dedicated calibration procedure to determine, with appropriate uncertainty, the acoustic position of each microphone used [5, 6]. Comparing with conventional radar, an acoustic system provides some advantages like lower costs, low environmental impact from the point of view of electromagnetic pollution, capacity to detect objects with small radar signature. In military applications, it offers the advantage of being passive, thus non-detectable. In ATZ control applications, and in all those situations where it is required, as well as acoustic source detection, also its spatial localization, a significant aspect is the accuracy that can be achieved in the localization. Logically, the accuracy is connected to acoustic antenna design and realization and it could be improved through a suitable calibration procedure. Various calibration methods and setup are proposed in the literature [7–11] but they do not fully solve the problem of large size acoustic antenna. In this work a methodology and measurement setup is described, tailored for calibration acoustic antennas. The work starts with a brief description of the principles at the basis of the acoustic antennas, to allow a complete understanding of the problems to be solved by the calibration procedure; then a parametric analysis of the proposed methodology was carried out to face its design and realization. Then an optimization procedure was addressed, to spare special size anechoic room, demanded by the large size of the array. Finally, preliminary experimental results on a simplified linear acoustic antenna are reported.

2 Problem Statement

An acoustic antenna is constituted by a set of N microphones allocated in the space, an example is shown in Fig. 1. The microphones allocation could be linear, planar or three dimensional, and with different geometric distributions (random or following specific laws). The acoustic antenna detects the position of the acoustic source by means of a suitable processing of the signals detected by microphones. The signal generated by an acoustic source get to array with a delay that is function of the sound speed and on the distance between each sensor and source. The detection of the

Fig. 1 Microphone array

source location can be done by different way like Acoustic intensimetry, Acoustic Holography, Time Delay of Arrival or Beamforming depending on the number of sources and on the distance. This work focuses on the beamforming technique.

2.1 Acoustic Antenna Calibration

Several critical aspects must be taken into account during the calibration of acoustic antennas. The first one is the right evaluation of microphones positions, strictly related to signal delay detected at each microphone. The delay is in fact a primary parameter due to its greater contribution to the measurement uncertainty in the acoustic source measure. It depends on the electrical and mechanical characteristic of the microphones, on their reception pattern and on the direction the signal coming. Figure 2 shows the effects of these parameters through the distinction between the so called geometric and acoustic locations of the sensors. For the microphone on the left, the source signal crosses the array pattern on a local minimum, differently from what happens for the microphone on the right. This has an impact on the effective time of reception of the signal, thus not depending only by the geometric location of the sensor, but also by its specific array pattern, function of its internal features.

Fig. 2 Difference between acoustic and geometric position

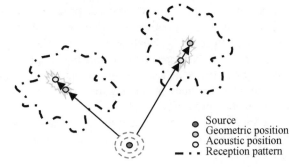

The result of this behavior is an error on the signals delay estimation and, hence, of the locations. Another problem is due to time shift delay.

In fact the triangulation method connotes the cross correlation algorithm based on the assumption that the space shift delay, L, to be estimated is lower than the ratio between the sound speed c and frequency f:

$$L < \frac{c}{f} \tag{1}$$

The signal generated by the ith source is received with a certain delay from the reference microphone r and the jth array microphone. We'll call this delay Wi,j, and is estimated through a cross correlation algorithm [8]. After collecting all delays, we can compute the array vector positions $\bar{x}_j^a$ by minimizing the relation

$$\sum_i^N \left| \left\| \bar{x}_i^s - \bar{x}_j^a \right\| - Wi, j \right| \tag{2}$$

With $\bar{x}_i^s$ the vector position of the ith source. If the time delay is larger than the period T of the acoustic signal, the cross correlation algorithm gives incorrect results: a time delay of $T + \phi$ is erroneously computed as ϕ, this is due to the periodic nature of evaluated signal [12, 13].

In addition, the arrays considered in this work are also too big and integrated with many microphones; as a consequence, the satisfaction of condition (1) must be carefully verified with respect the anechoic room available space ($15 \times 20 \times 15$ m) and operational frequency bandwidth of the array (1–2 kHz).

2.2 Experiment Setup

To solve the above problems, a dedicated strategy was developed. The first necessity is to identify geometric positions that emphasize the array performance. Due to the high number of parameters involved (number and positions of the sources, number and positions of reference microphones) a heuristic strategy based on genetic logic was adopted [14]. The geometric positions were used in the numeric model to obtain the standard deviation and then the acoustic positions. The experimental setup shown in Fig. 3. It is made-up by the acoustic array under calibration, a source array generating the calibration acoustic signals and a reference microphone. A Personal Computer (PC) manages both the generation of the source signals through a signal generator and a power amplifier, and the acquisition of the signals received by the acoustic antenna and the reference microphone. Figure 4 in illustrates a block diagram adopted for the calibration procedure.

Sliding from microphones and sources positions measured through a laser sensor, the pressure signal, p, received by each microphone is estimated as a solution of the spherical wave Eq. (3).

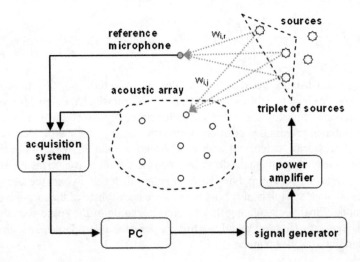

Fig. 3 Experiment setup

Fig. 4 Calibration strategy

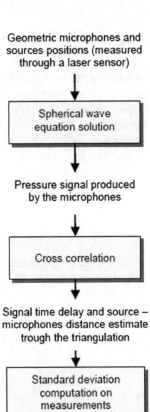

$$p\left(\overline{x_j^a}, t\right) = \frac{1}{4\pi c} \int\limits_0^t \iiint \frac{1}{\rho}\delta[\rho - c(t - t_0)]\delta(\overline{x_0} - \overline{x_j^s})\sin(\omega t_0)d^3x_0 dt_0 \qquad (3)$$

Being ρ the distance between the jth microphone and the position $\overline{x_0}$ of the distributed source element, t the current time, t_0 the time at which the signal is produced, ω the angular frequency and δ the Dirac function. As showed in Fig. 4, moving from an initial approximated position of the array sensors (geometric locations), the acoustical pressure is calculated with wave equation solution; the cross-correlation algorithm is used to estimate the delay between the signals of each array microphone and the reference one. By applying the triangulation algorithm onto the signals produced by all source triplets, it is possible to have an estimate of the location of each microphone. Finally, calculating the standard deviation of the space shift delay on all measurements achieved by all the triplets per each microphone, an estimate of the uncertainty can be obtained.

3 Experiment Result

To validate the methodology, a number of experimental tests were carried out. The experiment setup reported in Fig. 3 was developed considering an Agilent 33220A signal generator connected to a power amplifier 2716 Bruel & Kieran, an LMS International SCADAS SC 310 data acquisition system, an Omnisource 4295 Bruel & kier acoustic source.

The previously described calibration method had already been tested for linear arrays [6], and it had already been established that the source localization improved considerably by using the acoustic positions as input of the realized numerical model (Figs. 5 and 6).

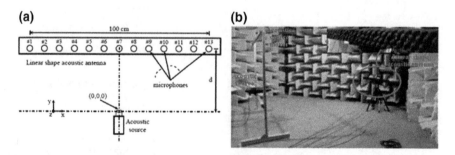

Fig. 5 Schematic (**a**) and picture (**b**) of experiment setup for linear array

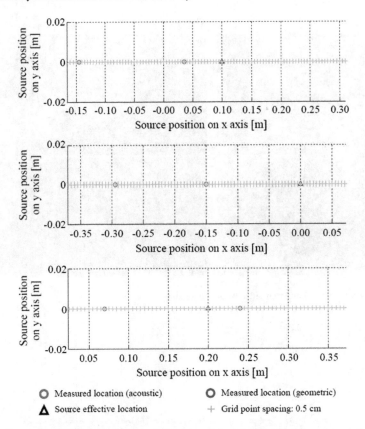

Fig. 6 Comparison between source location measured from both geometric and acoustic microphone positions and the effective source location

At present work a square shape acoustic antenna (two-dimensional array) was used as device under calibration (Fig. 7a). It was constituted by 30 microphones, MPA466 ¼" by BSWA TECH, equally distributed on a linear shape of 1 m length.

The two-dimensional array consists of a wood panel of 1.20 × 1.20 m supported by a wood scaffolding. The two-dimensional array after the realization of the holes for the microphones was covered of sound-absorbing material. All the microphones (Fig. 7b) were numbered and connected to the Data Acquisition System.

To carry out the measurements, the sources distribution used is showed in Fig. 8. The geometric positions were used in the numeric model (Fig. 4) for the determination of acoustic positions. The reference microphone is in a central position.

Fig. 7 Front side (**a**) and rear side (**b**) of square shape acoustic array

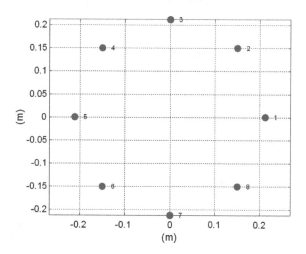

Fig. 8 Sources distribution

The difference between the geometric and acoustic positions obtained with our technique is shown in Fig. 9.

In particular, as we can see from the results of the location of the source, shown in the Figs. 10 and 11 in which the black triangle represents the real position of the source. The blue circle represents the position of the source identified considering the geometric positions of the microphones as input of the localization technique. The

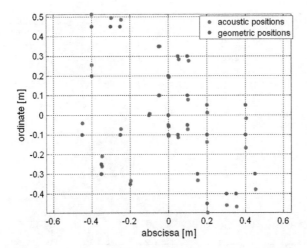

Fig. 9 Geometric (blue) and acoustic positions (red) of the microphones in the two-dimensional array

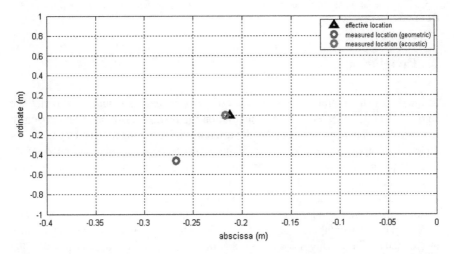

Fig. 10 Localization of the source for the two-dimensional array when the source is positioned at 21 cm from the axis origin

red circle represents the position of the source identified considering the geometric positions of the microphones as input of the localization technique. Using the acoustic positions, the absolute measurement error is reduced by up to 30% compared to that committed with geometric input.

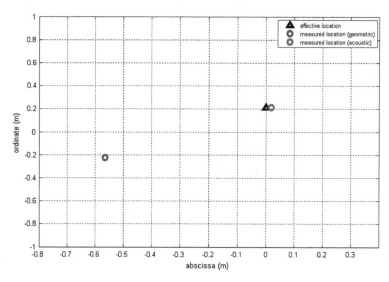

Fig. 11 Localization of the source for the two-dimensional array when the source is positioned in the origin (x = 0, y = 0.2)

4 Conclusions

In this paper, a calibration strategy for arrays of acoustic antennas is shown. The suggested strategy, based on the triangulation method, relies on the comparison of the time delay between array sensor and reference microphone received signals, produced by an assured sum of sources. The accuracy depends on some parameters, like sampling frequency, number of sources, distribution and the distance. A dedicated model was developed, capable of forecasting the pressure signal versus time at each sensor. Two measurement set up were correctly have been made to validate the aforementioned model. The first with a linear array, the second with a square array. An experimental campaign made with both array has confirmed the validity of the method, improving the accuracy and the acoustic source localization after the calibration. In fact, using the acoustic positions instead of the geometric ones, the measurement error decreases by 30% compared to the use of the latter. Additional steps will be focused on testing the whole calibration procedure on large size antennas and on the formulation of the calibration uncertainty.

References

1. Blaabjerg, C., Haddad, K., Song, W., Dimino, I., Quaranta, V., Gemelli, A., Corsi, N., Viegas, D.X., Pita, L.P.: Detecting and localizing forest fires from emitted noise In: VI International Conference on Forest Fire Research, Coimbra, Portugal (2010)

2. ShotSpotter Gunshot Location System® (GLS). http://www.shotspotter.com
3. Quaranta,V., Ameduri, S., Donisi, D., Bonamente, M.: An in-air passive acoustic surveillance system for air traffic control: GUARDIAN Project. In: ESAV 2011-Tyrrhenian International Workshop on Digital Communications-Enhanced Surveillance of Aircraft and Vehicles, Capri, Italy (2011)
4. Petrella, O., Quaranta, V., Ameduri, S., Betta, G.: Preliminary operations for calibrating a phased microphone arrays for air traffic monitoring. In: ICSV19, Vilnius, Lithuani (2012)
5. Petrella, O., Quaranta, V., Ameduri, S., Betta, G.: Acoustic antenna calibration accuracy: a parametric investigation. Inter-noise, New York (2012)
6. Ameduri, S., Betta, G., Laracca, M., Petrella, O., Quaranta, V.: Modelling, parametric analysis, and optimization of an experimental set-up for acoustic antenna calibration. In: Proceeding of IEEE I^{2}MTC 2013, Minneapolis, MN, USA, pp. 1670–1675 (2013)
7. Kroeber, S., Ehrenfried, K., Koop, L.: Design and testing of sound sources for phased microphone array calibration. In: Berlin Beamforming Conference, February, Berlin, Germany (2010)
8. Lauterbach, A., Ehrenfried, K., Koop, L., Loose, S.: Procedure for the accurate phase calibration of microphone array. In: AIAA 2009-3122, 15th AIAA/CEAS Aeroacoustics Conference (30th AIAA Aeroacoustics Conference), Miami, Florida (2009)
9. Gupta, J., Baker, J.R., Ellingson, S.W., Park, H.: An experimental study of antenna array calibration. IEEE Trans. Antennas Propag. 51(3) (2003)
10. Dandekar, K.R., Ling, H., Xu, G.: Smart antenna array calibration procedure including amplitude and phase mismatch and mutual coupling effects. In: IEEE International Conference on Personal Wireless Communications, pp. 293–297 (2000)
11. Son, S., Jeon, S., Hwang, W.: Automatic calibration method for phased arrays with arbitrary aperture shape. Microw. Opt. Technol. Lett. **50**(6), 1590–1592 (2008)
12. Humphreys, W.M., Brooks, T.F., Hunter, W., Meadows, K.R.: Design and use of microphone arrays for aeroacoustics measurements. In: AIAA 98-0471, 36st Aerospace Sciences Meeting & Exhibit, Reno, NV (1998)
13. Goussios, C.A., Papanikolaou1, G.V., Kalliris, G.M.: New design concepts for the construction of an omnidirectional sound source. J. Acoust. Soc. Am. **105**(2), 1053–1053 (1999)
14. David, E.: Goldberg, Genetic Algorithms in Search, Optimization and Machine Learning. Addison-Wesley Longman Publishing Co. Inc., Boston, MA (1989)

Pyroelectric Sensor for Characterization of Biological Cells

S. A. Pullano⦿, M. Greco⦿, D. M. Corigliano⦿, D. P. Foti⦿, A. Brunetti⦿ and A. S. Fiorillo⦿

Abstract Nowadays, cell characterization represents a fundamental task widely diffused in different fields. The main purpose of this paper is the design and development of a device for cellular characterization based on a pyroelectric sensor. After a brief introduction dedicated to the methods actually employed for cell counting, the pyroelectric sensor, the electronic readout unit, and the prototype will be introduced. Different cell concentration samples have been analyzed in order to highlight how the induced pyroelectric response is related to the cell properties. Experimental results shown that sensor output is strongly affected by cell type, concentration and viability. Particularly, it has been observed an increase in output signal related to an increase in cell concentration, mainly due to a lower thermal conductivity. No significant variation has been observed by the drastic reduction of cell viability also by varying cell concentration. The aforementioned results are probably due to the induction of a decrease in mitochondrial activity. Obtained results are very promising for the realization of a low cost laboratory device with all the characteristics listed above.

Keywords Pyroelectric sensor · Biological cell characterization
Thin film devices · Cellular biophysics · Thermal factors
Sensor phenomena and characterization · Counting circuits

1 Introduction

Characterization of cell culture is a routine task accomplished every day on biological fluid for the monitoring of cell viability, concentration and other parameters in order to assess quality control during in vitro experiments [1]. Historically, cell counting is performed through calibrated counting chambers, the aid of a micro-

S. A. Pullano (✉) · M. Greco · D. M. Corigliano · D. P. Foti · A. Brunetti · A. S. Fiorillo
Department of Health Sciences, University Magna Græcia of Catanzaro,
Viale Europa, 88100 Catanzaro, Italy
e-mail: pullano@unicz.it

© Springer Nature Switzerland AG 2019
B. Andò et al. (eds.), *Sensors*, Lecture Notes in Electrical Engineering 539,
https://doi.org/10.1007/978-3-030-04324-7_29

scope, and a trained operator, resulting in a time-consuming task prone to large errors (around 20–30%). Actually, different automated tools are trying to replace manual counters, even though most of the laboratories still used the most dated calibrated chambers. This is mainly due to the large and costly equipment, the need of trained personnel and the necessary maintenances [2]. Nevertheless, a growing interest in the development of novel devices lead to alternative cost-effective techniques with reduced sample-volume, reagents and maintenance [3–5]. In this context, the paper places its attention on an alternative approach for cell counting based on light-induced pyroelectric effect. The sensor is composed by a Thin ferroelectric film made of Polyvinylide-Fluoride (PVDF)-aluminum metalized and coated with graphite. Infrared light illumination is used to induce a rapid rise in temperature within PVDF and thus a charge generation at both surfaces. Current mode front-end amplifier collect and condition the output signal while the dependences of the thermal properties of the sample creates differences in the induced pyroelectric response.

2 Matherials and Methods

A 20 μm PVDF film with a circular geometry (r = 0.25 cm) is placed at the bottom of a sample well. The cylindric well is designed to be filled with 25 μL of sample. On the opposite side, an infrared source centered at a wavelength of 850 nm is fixed at a distance of 2 mm from the PVDF thin film (see Fig. 1). Infrared beam is used to impose a rapid rise in temperature within the upper and lower surface of the pyroelectric element inducing a pyroelectric response.

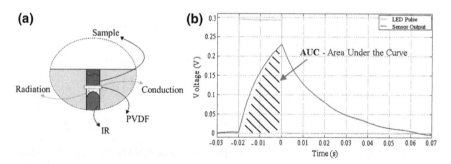

Fig. 1 Schematic representation of the pyroelectric sensor geometry in which part of the thermal energy provided by IR source is dissipated through the series of PVDF and sample by radiation and conduction mechanisms (**a**). Pyroelectric response and infrared stimulation electric signal and the parameter evaluated (**b**)

The sample and the PVDF film are thermally coupled, and can be modeled by a heat capacity C_{th}, and thermal conductance G_{th} (that accounts for heat dissipation through the PVDF and liquid sample). Electrical charge generated by the PVDF, is acquired and processed by a front-end amplifier designed as a charge sensitive amplifier topology. The pyroelectric sensor has been designed to be as minimal as possible avoiding the use of microfluidic components, reducing the fluid handling in just two simple operations: the pipetting of sample into the well and its washing.

Considering the equivalent thermal model and a temporally constant radiant flux, the difference in temperature across the transducer is related to the radiant flux itself and to the thermal conductance of the sample, and thus to its cellular content. Figure 1b reports the pyroelectric response (red line), which is characterized by a fast rise time, due to thermal properties (thermal time constant τ_{th}), and a slower decay which is due to the charge amplifier (electrical time constant τ_e) [6, 7]. The parameter evaluated during the experimental analysis was the area under the curve (AUC), that is, the integral of the sensor output, that was evaluated during infrared stimulation as highlighted in the shaded area (see Fig. 1b).

3 Results

Experimental tests were conducted on human prostate carcinoma cell line, LNCap, widely used in oncology as an androgen responsive prostate cancer model [8]. Cells were cultured in 10 mL flask with commercial medium (RPMI 1640), in a humidified incubator at 37 °C, with 5% CO_2 [9]. Observation on light induced pyroelectric response was obtained at different cell concentrations (Table 1) and by varying cell viability. A 25 µL sample was pipetted into the well and a step-shaped IR beam with a repetition rate of 1 Hz and a duty cycle of 2 and 5% was transmitted through the sample obtaining the results shown in Fig. 2.

Experimental results showed an increase of pyroelectric response influenced by cell concentration (see Fig. 2), while no significant variations were observed in samples with low cell viability (see Fig. 3). As previously stated, as expected, induced pyroelectric effect is mainly modulated by changes of thermal conductance due to the higher concentration of cellular bodies. Consequently, higher cell concentration results in an increase in heat capacity, due to a higher cellular mass per unit volume.

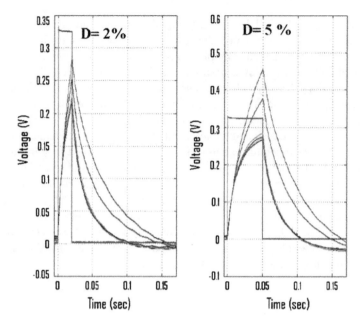

Fig. 2 Pyroelectric sensor response obtained at different concentration by stimulating the PVDF thin film with infrared pulse with a frequency of 1 Hz and a duty cycle (D) of 2 and 5%

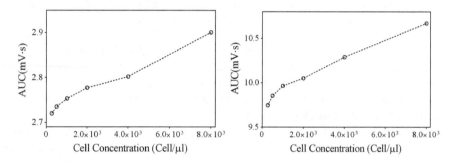

Fig. 3 AUC analysis at few representative concentrations of living cells on raw at a duty cycle of 2 (left) and 5% (right)

Drastic reduction of cell viability due to necrosis (assessed through Tripan blue dye exclusion test) results in a lower/absent mitochondrial activity, and thus in the impairment of cellular membrane with a leakage of cellular content [10]. In this condition, as expected, sensor output remains quite constant without significant variation with cell concentration (see Fig. 4) [11].

Table 1 Characteristics of analyzed sample

Cell Line	Volume	Cell concentration
LNCap	25 μL	0.25×10^3
		0.5×10^3
		1×10^3
		2×10^3
		4×10^3
		8×10^3

Fig. 4 AUC analysis at a duty cycle of 2% on dead cells

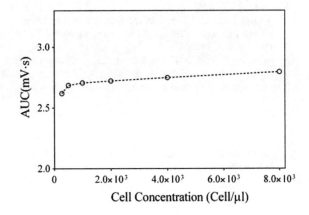

4 Conclusion

The paper presents the design and development of a pyroelectric sensor for the characterization of cell culture. A low-level infrared light stimulation allows the evaluation of a response which showed a dependence from the cell culture parameters (concentration and viability), thus highlighting the possibility of detecting variations of the samples analyzed. Experimental results conducted on human prostate carcinoma cell line showed good linearity of the sensor output vs. cell concentration. Moreover, non-significant variations of the sensor output has been observed at low cell viability. These preliminary findings suggest that pyroelectric sensor can be effectively employed for the development of cell characterization device.

References

1. Herculano-Houzel, S., von Bartheld, C.S., Millerm, D.J., Kaas, J.H.: How to count cells: the advantages and disadvantages of the isotropic fractionator compared with stereology. Cell Tissue Res. **360**(1), 29–42 (2015)
2. Green, R., Wachsmann-Hogiu, S.: Development, history, and future of automated cell. Clin. Lab. Med. **35**(1), 1–10 (2015)

3. Takahashi, T.: Applicability of automated cell counter with a chlorophyll detector in routine management of microalgae. Sci. Rep. **8**, 49–67 (2018)
4. Xu, X., Xie, X., Duan, Y., Wang, L., Cheng, Z., Cheng, J.: A review of impedance measurements of whole cells. Biosens. Bioelectron. **77**(9), 824–836 (2016)
5. Yu, C., Vykoukal, J., Vykoukal, D.M., Schwartz, J.A., Shi, L., Gascoyne, P.R.C.: A three-dimensional dielectrophoretic particle focusing channel for microcytometry applications. IEEE J. Microelectromechanical Syst. **14**(3), 480–487 (2005)
6. Pullano, S.A., Islam, S.K., Fiorillo, A.S.: Pyroelectric sensor for temperature monitoring of biological fluids in microchannel devices. IEEE Sens. J. **14**(8), 2725–2730 (2014)
7. Pullano, S.A., Mahbub, I., Islam, S.K., Fiorillo, A.S.: PVDF Sensor stimulated by infrared radiation for temperature monitoring in microfluidic devices. Sensors **17** (2017)
8. Costa, V., Foti, D., Paonessa, F., Chiefari, E., Palaia, L., Brunetti, G., Gulletta, E., Fusco, A., Brunetti, A.: The insulin receptor: a new anticancer target for peroxisome proliferator-activated receptor-γ (PPARγ) and thiazolidinedione-PPARγ agonists. Endocr.-Relat. Cancer **15**(1), 325–335 (2008)
9. Paonessa, F., Foti, D., Costa, V., Chiefari, E., Brunetti, G., Leone, F., Luciano, F., Wu, F., Lee, A.S., Gulletta, E., Fusco, A., Brunetti, A.: Activator protein-2 overexpression accounts for increased insulin receptor expression in human breast cancer. Cancer Res. **66**(10), 5085–5093 (2006)
10. Strober, W.: Trypan blue exclusion test of cell viability. Curr. Protoc. Immunol. **111**, A3.B.1–A3.B.3 (2015)
11. Pullano, S.A., Greco, M., Messineo, S., Brunetti, A., Fiorillo, A.S.: IR-Light induced pyroelectric effect for cell cultures characterization. In: IEEE Sensors Conference. IEEE, Glasgow (2017)

Characterization of a TMR Sensor for EC-NDT Applications

Andrea Bernieri, Giovanni Betta, Luigi Ferrigno, Marco Laracca and Antonio Rasile

Abstract Non-destructive tests based on eddy currents (EC-NDT) are one of the inspection techniques used to detect and characterize defects in conductive structures. The EC-NDT technique is based on the induction of eddy currents in the material under test and on the analysis of the reaction magnetic field that is generated. In this way, it is possible to detect the presence of a defect and evaluate its geometric characteristics. Generally, magnetic sensors such as AMR or GMR can be used to detect the reaction magnetic field. Recently, magnetic field sensors based on the Tunnel effect (TMR) have been introduced, which seem to have better performances than previous solutions. In this context, the article illustrates the metrological characterization of a TMR sensor for EC-NDT applications, as the information provided by the manufacturer is not complete and sufficient for this type of use. The results obtained show that the TMR sensor is able to provide a higher sensitivity than the AMR and GMR sensors, with a limited measurement uncertainty. This makes it possible to assume that the TMR sensors can be usefully used in EC-NDT applications.

Keywords TMR sensor · Tunneling Magneto-Resistance · Magnetic sensor Eddy current test · Non destructive test

A. Bernieri (✉) · G. Betta · L. Ferrigno · M. Laracca · A. Rasile
Department of Electrical and Information Engineering, University of Cassino
and Southern Lazio, Via G. Di Biasio, 43, 03043 Cassino, Italy
e-mail: bernieri@unicas.it

G. Betta
e-mail: betta@unicas.it

L. Ferrigno
e-mail: ferrigno@unicas.it

M. Laracca
e-mail: m.laracca@unicas.it

A. Rasile
e-mail: a.rasile@unicas.it

© Springer Nature Switzerland AG 2019
B. Andò et al. (eds.), *Sensors*, Lecture Notes in Electrical Engineering 539,
https://doi.org/10.1007/978-3-030-04324-7_30

1 Introduction

The Eddy Current Non-Destructive Testing (EC-NDT) technique is actually widely used in the industrial applications in order to detect the presence of defects in conductive structures. As it is known, the EC-NDT technique is based on the induction of currents in the material under test through a suitable excitation system; these currents generate a reaction magnetic field which changes in presence of defects in the material; by means of an appropriate magnetic field sensor and suitable processing procedures, it is possible to detect the presence of a defect and evaluate its geometric characteristics (length, width, depth) [1–9].

With reference to magnetic field detection systems, many solutions are provided in the literature, based on both magnetic pickups [1] and magnetic field sensors [2–5, 10]. Among the latter, magnetoresistive sensors are among the most used, thanks to their characteristics of limited size, good spatial resolution and adequate sensitivity for EC-NDT applications.

To this category belong the Giant MagnetoResistive sensors (GMR), formed by a multilayer metallic structure with alternating ferromagnetic and non-magnetic layers. In presence of an external magnetic field, the GMR sensors provide changes in electrical resistance that ranges from 10 to 20% up to 70% [11]. However, for optimal use as magnetic field sensors, the GMR sensors require an external control circuitry able to constantly ensure a proper reference magnetization axis and an output signal offset compensation.

Another class of magnetic field sensors based on the magnetoresistive effect is the AMR (Anisotropic MagnetoResistive) sensors. The sensor structure is composed of a thin film of a ferromagnetic material (nickel-iron, permalloy) deposited on a silicon wafer. In presence of an external magnetic field, the AMR sensor resistance changes up to 2–3% [12–15]. Although they exhibit less sensitivity than GMR sensors, AMR sensors are built with internal circuits to compensate for magnetization and offset effects, allowing much easier development of magnetic field measurement applications [12–15].

Recently, magnetic field sensors based on the Tunnel effect (TMR—Tunneling Magneto-Resistance) have been introduced. The TMR sensors are based on a particular multi-layer junction (TMJ—Tunneling Magnetic Junction) composed of two ferromagnetic layers separated by a non-magnetic tunnel barrier. The first ferromagnetic layer is characterized by a "free" magnetic direction, in the sense that it can assume a magnetic polarization with direction depending on the external magnetic field. The second ferromagnetic layer, instead, is "pinned", i.e. the polarization direction is fixed and does not depend, within certain limits, on the external magnetic field. Applying an external magnetic field, the polarization of the free layer is modified according the external magnetic field direction and intensity, determining a variation of overall junction resistance. Figure 1 shows the TMJ resistance against the polarizations of the ferromagnetic layers.

In the presence of a magnetic field, the TMR sensors provide a greater resistance variation than that provided by the previously described AMR and GMR sensors

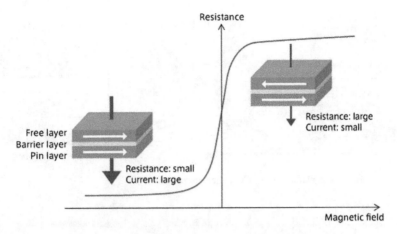

Fig. 1 TMJ resistance against the polarizations of the ferromagnetic layers

and do not require compensating circuits. Compared to a AMR sensitive element, a TMJ element has a higher sensitivity and a wider linear range. Compared to a GMR sensitive element, a TMJ element has a higher sensitivity, less energy consumption, and a wider linear range.

Because of its recent production, the TMR sensor is not well known in terms of metrology performance, and even manufacturers do not provide exhaustive information. For this reason, in order to efficiently use TMR sensors in an EC-NDT probe, the authors have made a preliminary metrological characterization of a TMR sensor by Multi Dimension Technology (model TMR2905D), following the same approach used for the previously characterization of a GMR sensor [10].

2   TMR Sensor Performance Evaluation

Figure 2 shows the measurement station developed to characterize the considered TMR sensor. It is composed by a signal generator coupled with a Kepco bipolar amplifier to feed a calibrated Helmholtz coil in order to generate a controlled excitation magnetic field; the TMR signal output is then amplified and measured by means of a digital multimeter. The TMR sensor is placed in the middle of the Helmholtz coil in order to assure a uniform excitation magnetic field.

The tests are performed applying both DC and AC magnetic fields. Figure 3 shows the obtained DC TMR transfer function in the magnetic field range of ± 30 G and for a TMR supply voltage of 1 V. The analysis of the DC characteristic shows a saturation effect for magnetic field values of ± 10 G, together with a noticeable linearity. Figure 4 shows the TMR sensitivity characteristic, with a mean value of 53.58 mV/G in the magnetic field range of ± 4 G.

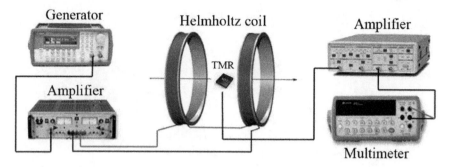

Fig. 2 Measurement station for the TMR characterization

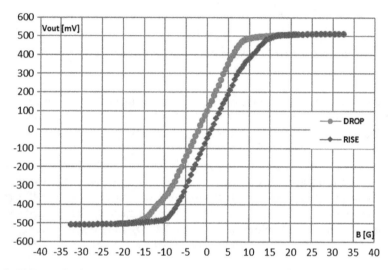

Fig. 3 TMR transfer function for DC magnetic fields in the range of ±30 G

Figure 5 shows the AC TMR transfer function obtained for magnetic field values from 0 to 4 G, in the frequency range from 1 kHz to 50 kHz and for a TMR supply voltage of 1 V. Figure 6 shows the corresponding AC TMR sensitivity characteristic. A mean sensitivity of about 39 mV/G was obtained, together with a good linearity with a maximum variability of 1.32 mV/G for the considered frequency range.

It should be pointed out that only some of these characteristics correspond to those supplied by the manufacturer (when available).

3 Uncertainty Evaluation

For the correct use of the TMR sensor in EC-NDT applications, the main uncertainty contributions that affect sensor performance were evaluated [16, 17], as the manufacturer did not provide any information about it. In particular, for the evaluation of

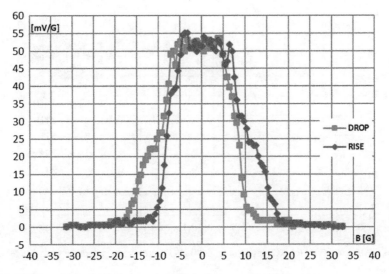

Fig. 4 TMR sensitivity for DC magnetic fields in the range of ±30 G

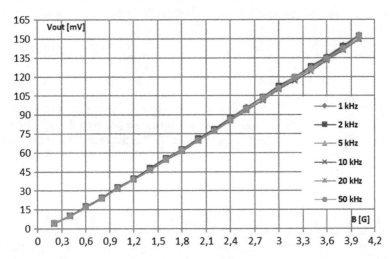

Fig. 5 TMR transfer function for AC magnetic fields (0–4 G) in the frequency range from 1 to 50 kHz

the uncertainty contributions, repeated tests were performed in the magnetic field range of ±4 G (where the considered sensor showed the best performances).

In detail, the uncertainty contributions due to repeatability (σ_{TMR}), sensitivity (μ_{SEN}), hysteresis (μ_{HYS}), non-linearity (μ_{NL}) and frequency variability (μ_{FREQ}) of the sensor response were examined.

The uncertainty due to the repeatability of the sensor response was calculated using the (1), that is, by means of the standard deviation of the sensor response

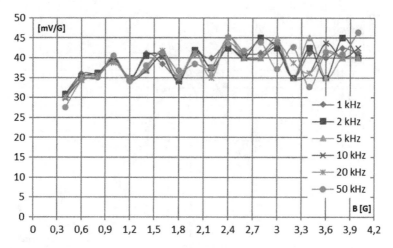

Fig. 6 TMR sensitivity for AC magnetic fields (0–4 G) in the frequency range from 1 to 50 kHz

(Vout) with respect to the applied magnetic field (G), evaluated on N repeated tests (N > 20):

$$\sigma_{TMR} = \frac{\sum_{i=0}^{N-1} Vout_i^2}{\sqrt{N}} = 0.02\text{mV} \tag{1}$$

The uncertainty due to the variability of the sensor sensitivity has been calculated by means of the (2), where $\Delta_{MAX} SEN$ is the difference between the maximum and the minimum value of the sensor output obtained to the same variation of the imposed magnetic field ΔG:

$$\dot{\mu}_{SEN} = \frac{\Delta_{MAX} SEN}{\sqrt{3}} = 0.20\,\text{mV/G} \tag{2}$$

The uncertainty due to the hysteresis of the sensor response was calculated using the (3), where ΔV_{MAX} and ΔV_{MIN} are the maximum and minimum output value of the sensor at the same imposed magnetic field value, with respect to the overall range of magnetic field ΔG in which the hysteresis cycle has been analyzed ($\pm 4G$):

$$\mu_{HYS} = \frac{(\Delta V_{MAX} - \Delta V_{MIN})}{\sqrt{3}} = 0.43\,\text{mV} \tag{3}$$

The uncertainty due to the non-linearity of the sensor response was calculated using the (4), where $\Delta_{MAX} NL$ is the maximum deviation of the sensor response with respect to the ideal output characteristic, depending on the applied magnetic field ΔG:

$$\mu_{NL} = \frac{\Delta_{MAX} NL}{\sqrt{3}} = 1.01 \, \text{mV} \tag{4}$$

The uncertainty due to the frequency variability of the sensor response has been calculated using the (5), where $MAX\_VAR\_FREQ(V_{OUT})$ is the maximum variation of the sensor response with the same applied magnetic field G, in the considered frequency range:

$$\mu_{FREQ} = \frac{MAX\_VAR\_FREQ(V_{OUT})}{\sqrt{3}} = 0.76 \, \text{mV} \tag{5}$$

Finally, the overall uncertainty of the TMR sensor due to all the aforementioned contributions was assessed by (6):

$$\dot{\mu}_{TMR} = \sqrt{\dot{\mu}_{SEN}^2 + \dot{\mu}_{tot}^2} \tag{6}$$

where

$$\mu_{TOT}^2 = \sqrt{\sigma_{TMR}^2 + \mu_{HYS}^2 + \mu_{NL}^2 + \mu_{FREQ}^2} \tag{7}$$

4 Conclusions

The work reports a first characterization of a TMR sensor for EC-NDT applications. The experimental results obtained show that the considered TMR sensor can be usefully used for the measurement of continuous and variable magnetic fields. The characteristics of good sensitivity and linearity and ease of use with respect to the other types of magnetic field sensors (GMR and AMR) make it possible to develop methodologies for analyzing defects in conductive materials by means of eddy current which are more efficient and less critical from the implementation point of view. With this in mind, future research developments will concern the integration of the TMR sensor into an EC-NDT probe for the identification of defects on conductive elements in real operating conditions.

References

1. García-Martín, J., Gómez-Gil, J., Vázquez-Sánchez, E.: Non-destructive techniques based on eddy current testing. Sensors 11(3), 2525–2565 (2011)
2. Bernieri, A., Betta, G., Ferrigno, L., Laracca, M.: Multi-frequency Eddy Current Testing using a GMR based instrument. Int. J. Appl. Electromagn. Mech. 39(1), 355–362 (2012)
3. Bernieri, A., Betta, G., Ferrigno, L., Laracca, M.: Crack depth estimation by using a multi-frequency ECT method. IEEE Trans. Instrum. Meas. 62(3), 544–552 (2013)

4. Bernieri, A., Betta, G., Ferrigno, L., Laracca, M., Mastrostefano, S.: Multifrequency excitation and support vector machine regressor for ECT defect characterization. IEEE Trans. Instrum. Meas. **63**(5), 1272–1280 (2014)

5. Betta, G., Ferrigno, L., Laracca, M., Burrascano, P., Ricci, M., Silipigni, G.: An experimental comparison of multi-frequency and chirp excitations for eddy current testing on thin defects. Measurement **63**, 207–220 (2015)

6. D'Angelo, G., Rampone, S.: Shape-based defect classification for Non Destructive Testing. In: Proceedings of the 2nd IEEE International Workshop on Metrology for Aerospace, Benevento, Italy, 4–5 June 2015

7. D'Angelo, G., Rampone, S.: Feature extraction and soft computing methods for aerospace structure defect classification. Measurement **85**, 192–209 (2016)

8. Ricci, M., Silipigni, G., Ferrigno, L., Laracca, M., Adewale, I.D., Tian, G.Y.: Evaluation of the lift-off robustness of eddy current imaging techniques. NDT and E Int. **85**, 43–52 (2017)

9. D'Angelo, G., Laracca, M., Rampone, S., Betta, G.: Fast Eddy current testing defect classification using Lissajous figures. IEEE Trans. Instrum. Meas. **67**(4), 821–830 (2018)

10. Bernieri, A., Betta, G., Ferrigno, L., Laracca, M.: Improving performance of GMR sensors. IEEE Sens. J. **13**(11), 4513–4521 (2013)

11. Datasheet: AA and AB—series analog sensors. http://www.nve.com/

12. Application note: AN 209 magnetic current sensing. https://aerospace.honeywell.com/

13. Bernieri, A., Ferrigno, L., Laracca, M., Rasile, A.: An AMR-based three phase current sensor for smart grid applications. Sens. J. (2017)

14. Betta, G., Ferrigno, L., Capriglione, D., Rasile, A.: Improving the performance of an AMR-based current transducer for metering applications. In: Lecture Notes in Electrical Engineering Sensors—Springer editor (2016)

15. Betta, G., Ferrigno, L., Capriglione, D., Rasile, A.: An industrial current transducer based on an AMR sensor. In: Proceedings of the Third National Conference on Sensors, Rome, Italy, 23–25 Feb 2016

16. Angrisani, L., Atteo, E., Capriglione, D., Ferrigno, L., Miele, G.: An efficient experimental approach for the uncertainty estimation of QoS parameters in communication networks. In: Proceedings of 2010 IEEE Instrumentation and Measurement Technology Conference (I2MTC), pp. 1186–1191, Austin (TX, USA), 3–6 maggio 2010

17. Angrisani, L., Capriglione, D., Ferrigno, L., Miele, G.: Type A uncertainty in jitter measurements in communication networks. In: Proceedings of Instrumentation and Measurement Technology Conference (I2MTC), pp. 1–6, Hangzhou, China, 10–12 maggio 2011

Thermal, Mechanical and Electrical Investigation of Elastomer-Carbon Black Nanocomposite Piezoresistivity

Giovanna Di Pasquale, Salvatore Graziani, Guido La Rosa, Fabio Lo Savio and Antonino Pollicino

Abstract Polymeric composites, where polymeric matrices are purposefully added with suitable fillers, have raised the interest of the scientific community, since materials with characteristics that depend on the nature of both the polymeric matrix and the filler can be obtained. The paper deals with the investigation of composites based on an insulating polymeric matrix, realized by using polydimethylsiloxane (PDMS) and carbon black (CB), as the filler, for realizing nanocomposites. The PDMS is an insulating matter, while the CB has conducting properties. If a suitable concentration of the CB is used, it is possible, therefore, changing the electrical properties of the composite from insulating to conducting. Such a possibility is, e.g. described in the framework of the percolation theory. Since a deformation of the composite causes a corresponding change in the concentration of the filler, it is possible using the described nanocomposites as piezoresistive elements. Based on the considerations reported above, composites were realized by using different concentrations of the filler, in order to obtain a reasonable value of the composite resistivity. The corresponding thermal, mechanical and electrical properties where, therefore, investigated in typical laboratory conditions.

Keywords Polydimethylsiloxane (PDMS) · Carbon black (CB) · Viscoelasticity Piezoresistivity

G. Di Pasquale
Dipartimento di Scienze Chimiche, University of Catania, Catania, Italy

S. Graziani (✉)
Dipartimento di Ingegneria Elettrica, Elettronica e Informatica,
University of Catania, Catania, Italy
e-mail: salvatore.graziani@dieei.unict.it

G. La Rosa · F. L. Savio · A. Pollicino
Dipartimento di Ingegneria Civile e Architettura, University of Catania, Catania, Italy

© Springer Nature Switzerland AG 2019
B. Andò et al. (eds.), *Sensors*, Lecture Notes in Electrical Engineering 539,
https://doi.org/10.1007/978-3-030-04324-7_31

1 Introduction

In the past years, sensing systems, based on polymers, have raised a significant interest in the scientific community [1]. More specifically, composite materials, consisting of insulating polymeric matrix and a conductive filler have been used for realizing pressure, tactile sensors [2] and gas sensors [3]. In [4], a conductive mixture has been proposed for realizing a glove capable of posture and gesture monitoring. Further applications can be envisaged in the monitoring and preservation of cultural heritage. In [5], a device has been proposed by some of the authors to realize flexible tailored sensors for large deformation monitoring. Results obtained in the characterization of natural rubber, loaded with carbon black, as a piezoresistive material have been reported in [6].

Here, the possibility of using a silicone matrix, loaded by using carbon black nanoparticles, to realize piezoresistive sensing elements is investigated. More specifically nanocomposite based on polydimethylsiloxane (PDMS), loaded with acetylene carbon black (CB), have been considered and the characterization of such composites will be described in the following.

2 The Composite Synthesis and the Thermal Characterization

Samples analyzed in the following, were produced using PDMS, as the matrix, and various amount of CB as filler. The experimental conditions have been verified in order to obtain the most precise control of the degree of crosslinking, which has a high importance, since it influences the mechanical properties of the final polymer. The samples were fabricated using a linear poly(dimethylsiloxane)-hydroxy terminated (PDMS-OH, viscosity 18,000–22,000 cSt from Aldrich). The crosslinking reaction was carried out at room temperature from a mixture of PDMS-OH in presence of 4% (by mass) of TetraEthOxySilane (TEOS) and 1% (by mass) of Tin (II) 2-ethylhexanoate (SNB), as catalyst. A scheme of the crosslinking reactions is shown in Fig. 1.

CB (acetylene 100% compressed, bulk density 170–230 g/L from Alpha Aesar) has been utilized as filler (see Fig. 2). This type of CB is obtained by thermally decomposing acetylene gas. This method provides CB with higher structures and higher crystallinity, good for electric conductive agents.

The PDMS/CB composites have been obtained by solution blending. A small volume (about 2 ml) of dichloromethane was added to a weighted amount of PDMS-OH (4 g) in a Teflon mold, in order to decrease the viscosity of the starting matrix, and then to get better dispersion of the filler. Then, CB was added and the mixture was sonicated at 15 W for 10 min. Samples have been prepared with different percentage by weight of CB (6, 8, 10% of CB).

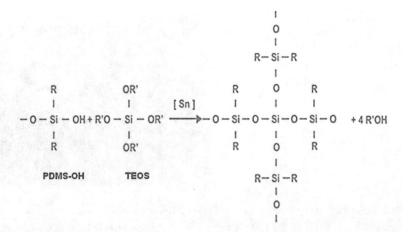

Fig. 1 A scheme of the crosslinking

Fig. 2 A view of CB, used
as filler for the composite
fabrication

Table 1 Weight percentage of CB used for composite realization

Samples	
Neat crosslinked PDMS-OH	PDMS-OH (%)
PDMS-OH + 6% by weight of CB	CB 6
PDMS-OH + 8% by weight of CB	CB 8
PDMS-OH + 10% by weight of CB	CB 10

Details about the composites investigated in the following of the paper are reported in Table 1.

Finally, 4% by weight of crosslinking agent (TEOS) and 1% by weight of SNB catalyst were added and the mixture was mixed for 10 min. The composites were then transferred into metal molds in order to obtain a better control of the thickness of the samples to be subjected to characterization. The crosslinking time was fixed in 4 h. A view of the molding system, along with a produced sample, is shown in Fig. 3.

Fig. 3 The molding system
and a sample as obtained
after the crosslinking phase

Table 2 Degradation temperatures (T_{d10}) of PDMS-OH/Carbon Black composites

Samples	T_{d10} (°C)	Gain in terms of Celsius degrees, compared to the neat matrix
PDMS-OH	389	–
CB 6% (A4106C CB)	405	+16 °C
CB 8% (A4108C CB)	405	+16 °C
CB 10% (A4110C CB)	421	+32 °C

2.1 Thermogravimetric Analysis (TGA) of the Obtained Composites

The Thermogravimetric analysis was performed in order to obtain information on the thermal stability of the nanocomposites. TGAs were carried out in static air in the range 25–700 °C, with a heating rate of 10 °C min^{-1}. The combined action of an oxidizing atmosphere and of high temperatures causes the initiation of degradative processes in the polymers which involve the breaking of the chains and/or variations of the structure and of the terminal groups. From these degradative processes, volatile gaseous products are obtained, which result in a decrease in the residual weight of the sample under test.

A measure of thermal stability can be obtained from the comparison of the degradation temperature determined on the basis of the weight loss. Considering as degradation temperature the one at which a weight loss of 10% (Td_{10}) is recorded, some information can be drawn on the effect of the filler presence.

T_{d10} of neat PDMS-OH is 389 °C (see Table 2), while TGA data for the filler containing samples testify to the good dispersion of the filler as evidenced by the increase of the thermal stability of the composites with respect to the virgin matrix. In fact, T_{d10} of CB 6% and CB 8% samples are 405, 16 °C higher than PDMS-OH, while the one of CB 10% sample is 421, 32 °C higher than neat matrix.

The comparison, between the thermogravimetric curves obtained for PDMS-OH and for the CB 10% samples, is shown in Fig. 4.

Fig. 4 Comparison between
the thermogravimetric curves
of PDMS-OH and composite
with 10% by weight of CB,
respectively

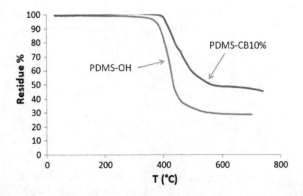

The higher thermal stability found in these samples is due to the fact that dispersed carbon black hinders the formation of radicals so that the temperatures at which there is 10% loss in weight are higher.

2.2   Mechanical Dynamic Analysis (DMA) of the Obtained Composites

Further insight on the chemical structure of the nanocomposites was investigated by performing the Mechanical Dynamic Analysis (DMA) of the samples. Tests were carried out in compression mode at a frequency of 1 Hz in the temperature range $-135 \div 30$ °C (scan rate 10 °C min^{-1}). Tritech2000, from Triton Technology Ltd, was used for the experiments. The presence of the CB determines a shift at higher temperatures of the melting of the crystalline domains of the PDMS (fall of the module ~ -40 °C in the PDMS-OH, and ~ -38, 36 and 29 °C in the CB 6%, CB 8% and CB 10% composites, respectively) and an increase in the storage modulus of the composite at room temperature. A comparison between the storage modulus as a function of temperature in PDMS and CB 10% nanocomposite is shown in Fig. 5.

3   Investigation of the Composite Piezoresistivity

The electric properties of samples based on PDMS and different concentrations of CB, as the filler, were investigated in order study their piezoresistive properties. More specifically, while in a first production phase, the CB varied in the interval 0.1–10% in weight, it was found that relevant changes in the composite resistivity occurred in a smaller range of concentration values. For such a reason, the investigations were further focused for CB concentrations in the range 6–10%, as reported in Table 1.

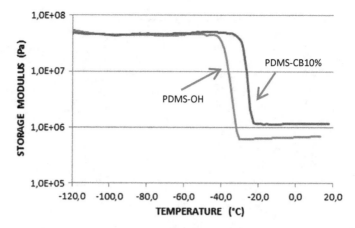

Fig. 5 Comparison between the storage modulus, as a function of temperature, in PDMS and CB 10% composite, respectively

Fig. 6 A CB based sample, extracted from the molded composite

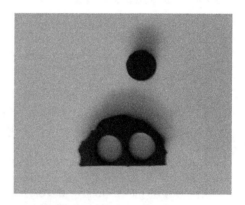

3.1   The Measuring System for Piezoresistivity Investigation

A measurement campaign has been executed in order to investigate the piezoresistive behaviors of the samples. More specifically, all samples have been cut from the molded composites by using a punch and samples thickness has been determined by using a gauge. A view of a sample extracted from a produced composite, as obtained by the molding phase, is reported in Fig. 6.

A system was realized for imposing the deformation to the devices under test. Both a CAD representation and a picture of the set up are reported in Fig. 7. A micrometer (Mitutoyo, model n. 148–316) was used for applying the deformation to the device. The system was, also, equipped with two copper electrodes for measuring the corresponding values of the device resistance.

Loading cycles have been, therefore, applied to the samples with a constant compressive step equal to 2.5%, up to a maximum compressive value equal to 25%.

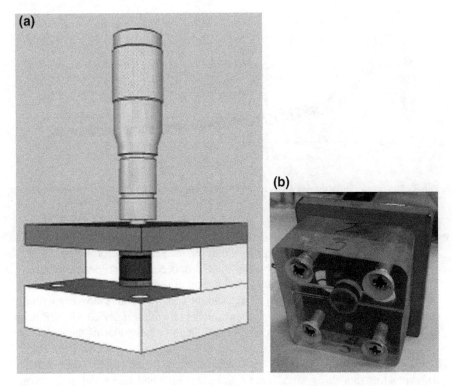

Fig. 7 A scheme of the set up (**a**) and a picture of the realized set up (**b**)

Experiments were performed in hydrostatic conditions. The corresponding values of the sample resistance were measured both by using a linearized Wheatstone bridge and with Agilent 33301A digital multimeters.

3.2 The Experimental Results

It has been reported in the literature that resistive devices, based on polymeric composites, even in the presence of a constant load, show a change in the value of the resistance with time, which is known as resistance creep [2, 6]. For such a reason, a preliminary investigation was performed in order to determine the time scale of such a phenomenon. In Fig. 8, the time plot of the changes in the resistivity values for a sample with 10% of CB, following the application of a 5% compressive load are reported.

More specifically, the results of three experiments are reported in the figure. From the results reported in Fig. 8, it can be concluded that the changes in the resistivity value can be considered occurring in a time interval smaller than about 10 min.

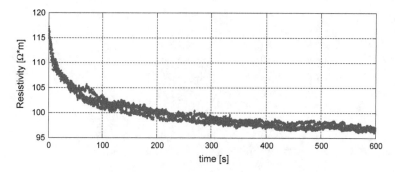

Fig. 8 Time evolution of the resistivity of the CB 10% sample, after applying 5% compressive load

According to the considerations reported above, data used in the following have been acquired 10 min after the application of the mechanical deformation.

Devices have been realized by using the CB concentrations reported in Table 1. The changes in the resistance value as a function of the applied deformation d have been recorded. The corresponding changes in the resistivity value have been, finally, estimated. As an example, in Fig. 9, the results obtained for a device with CB equal to 8% are reported. More specifically, Fig. 9a reports the values of the resistance, while in Fig. 9b, the corresponding resistivity values are shown. In Figs. 9a and 9b, the results obtained by using a polynomial interpolation are also reported. The corresponding modelling RMSEs are equal to 61,7 kΩ and $1.23 * 10^3$, respectively. Similar results have been obtained, also, for samples CB 6% and CB 10%.

From the analysis of the results shown in Fig. 9, it can be observed that for small values of the applied compressive deformation, a piezoresistive effect has been observed, with a reduction of both resistance and resistivity. Nevertheless, for larger values of the deformation, both parameters increase their values.

A further elaboration has been performed for investigating the piezoresistivity the devices, as a function of the CB volume concentration. More specifically, using the density values of the main components (0.97 g/cm^3 and 0.56 g/cm^3, for PDMS and CB, respectively), the values of the volume concentrations for samples CB 6%, CB 8% and CB 10% in absence of any deformation, become 10%, 13%, and 16%, respectively.

Finally, the applied deformations have been transformed into corresponding values of the CB volume concentration [7]. Obtained results are reported in Fig. 10.

The analysis of results reported in Fig. 10 show that though the devices show a piezoresistive behavior, the phenomenon cannot be described as a function of the applied volume filler concentration and further investigations are required for better understanding the involved phenomena.

4 The Viscoelastic Characterization of the Composites

The analysis was carried out in order to characterize the composites from a mechanical point of view, with particular attention to the viscoelastic aspect. The work performed, therefore, should be inserted into a broader context that concerns the study and complete characterization of new types of CB filled polymers.

As previously stated, nanocomposite devices with CB content ranging from 6% to 10% have been investigated and have shown a change in the resistance with the applied deformations. Among the mentioned composites, the results obtained in the viscoelastic characterization of the CB 10% will be reported. More specifically, the investigation is intended to evaluate the influence of the kinematic viscosity of the filled PDMS on the relaxation of the system.

4.1 Description of the Testing Machine

The testing machine used to carry out stress relaxation tests on polymeric materials consists of:

- Aluminum frame on which the various components are fixed;
- Electrostatic actuator Physik Instrumente M-230.25, with relative software, to impose the desired deformations on the specimens;
- Controller Mercury II C-862, to maneuver and control the actuator via PC;
- Load cell Tekkal L2320/50LBS, to measure the force applied to the specimen;
- Board Transducer Techniques TM0-1, to amplify the load cell signal;
- Laser Baumer OADM 12U6460/S35A, to measure the displacement imposed by the actuator;
- Acquisition board NI-DAQ USB-6009 for detecting, processing and transmitting signals from the load cell and the laser to a PC;
- Steel plate, fixed to the base of the machine, to host the specimens to be compressed;
- Calibrated screw, fixed to the load cell and having the task of transmitting the compression force to the specimens, prepared ad hoc in order to have the same diameter of the screw itself;
- A dedicated NI-LabVIEW 2017 software to capture, filter, save and instantly display the output signals (from the load cell and the laser) has been implemented.

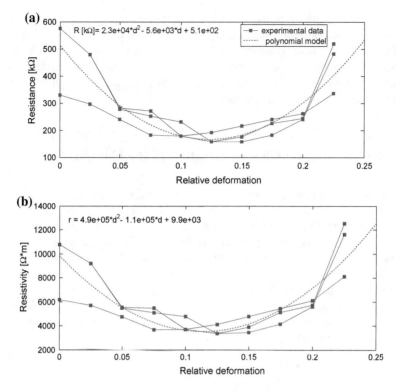

Fig. 9 Dependence of the electrical properties of a CB 8% nanocomposite based device as a function of the applied deformation d. Resistance values (**a**), and resistivity values (**b**)

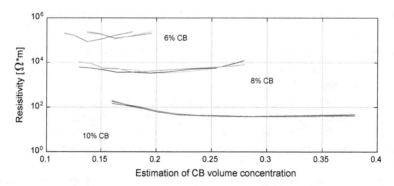

Fig. 10 Dependence of the electrical resistivity on the CB volume concentration

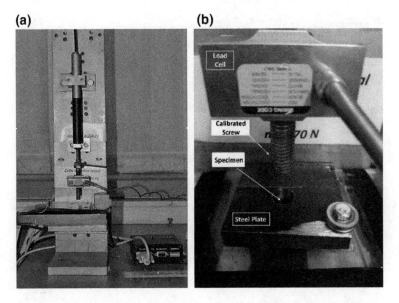

Fig. 11 The set up for the mechanical investigation of the nanocomposite. The compressing system (**a**), and the sample casing (**b**)

The tests were performed under compressive load in confined way inside a steel constraint; in this way, the specimen was compressed under conditions of hydrostatic pressure. Figure 11a, b show the mechanical set up and the casing system used for performing the mechanical investigation.

4.2 The Testing Procedure Acquisition

Relaxation tests were performed with duration of just over an hour each. The tests were carried out under displacement control. Each test requires the actuator to perform a series of movements (loading and unloading cycles, 0.1 mm pre-loading step, compression and unloading steps), as shown in Fig. 12. These phases are commonly used for the initial conditioning of the viscoelastic material, mainly for biological tissues and elastomers. In particular, the following phases were observed: five cycles of loading and unloading at 30% of the specimen thickness, 0.1 mm preloading, compressive loading (start of relaxation test). Between the end of the last loading-unloading cycle and the beginning of the pre-compression, as well as between the end of the latter and the beginning of the final relaxation compression, a waiting time of 10 s has been set, to allow the redistribution of internal stresses created in the polymer by the imposed deformations.

Since deformations of different entities (20, 40 and 50% of the thickness) were imposed on each specimen, it would not have been possible to carry out the unloading

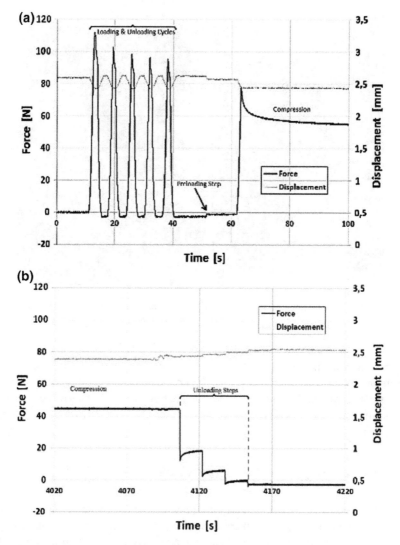

Fig. 12 Initial (**a**) and final (**b**) phases of the relaxation test on CB 10%

phase in the same way for the different types of tests. In all cases, the last step has the same amount (0.1 mm), so as to cancel the effect of the initial preload and return the actuator to the total rest position. Among all the steps, a waiting time of 15 s has been set, to check the elastic response of the material.

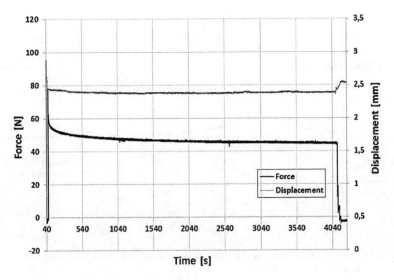

Fig. 13 Relaxation curve of the CB 10% sample

4.3 Results of the Relaxation Phase

The results shown in Fig. 13, obtained by the direct measures (displacement imposed on the specimen, force by the load cell) allow to define the stress-strain behavior.

From the experimental data obtained, assuming an isotropic sample subjected to hydrostatic pressure, it was possible to derive the value of the Young Modulus (E) and Bulk Modulus (K) of the CB 10% sample [8]:

$$E = 5.1 \pm 0.2\,\text{MPa}$$
$$K = 2,820 \pm 5\,\text{MPa}$$

From the equation $= 3\,(1-2v)$ the value of the Poisson coefficient for the material has been obtained (v):

$$v = 0.499699 \pm 1 \cdot 10^{-6}$$

Once the experimental data were acquired, some types of parametric models were studied, in order to choose the model that best suited the real behavior of the material. The experimental data was first filtered using MATLAB. Then, the fitting has been completed using both polynomial and exponential fitting, by spring and damper models in series (Maxwell), in parallel (Kelvin-Voigt, Zener) or more complex. The model with the highest value of R^2, as well as the lowest Root Mean Square Error (RMSE) was the generalized model of Maxwell with three elements in parallel (Fig. 14). The results obtained find excellent concordance in the literature [9].

(a)   **(b)**

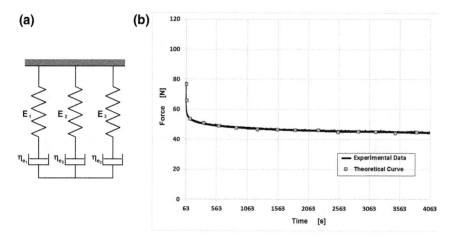

Fig. 14 Scheme of a three-elements of Maxwell generalized model (**a**) and comparison between experimental data and this model (**b**)

Acknowledgements Thi paper is partially supported by the University of Catania Project FIR2014 "Realizzazione di sensori piezoresitivi mediante nano compositi a matrice elastometica".

References

1. De Luca, V., Digiamberardino, P., Di Pasquale, G., Graziani, S., Pollicino, A., Umana, E., Xibilia, M.G.: Ionic electroactive polymer metal composites: Fabricating, modeling, and applications of postsilicon smart devices. J. Polym. Sci., Part B: Polym. Phys. **51**(9), 699–734 (2013)
2. Wang, P., Ding, T.: Creep of electrical resistance under uniaxial pressures for carbon black–silicone rubber composite. J. Mater. Sci. **45**(13), 3595–3601 (2010)
3. Pandey, S., Goswami, G.K., Nanda, K.K.: Nanocomposite based flexible ultrasensitive resistive gas sensor for chemical reactions studies. Sci. Rep. **3**, art. no. 2082 (2013)
4. Lorussi, F., Scilingo, E.P., Tesconi, M., Tognetti, A., De Rossi, D.: Strain sensing fabric for hand posture and gesture monitoring. IEEE Trans. Inf. Technol. Biomed. **9**(3), 372–381 (2005)
5. Giannone, P., Graziani, S.: Flexible tailored sensors for large deformation monitoring. In: Proceedings of I²MTC 2009—International Instrumentation and Measurement Technology Conference, Singapore, Singapore, pp. 593–596 (2009)
6. Giannone, P., Graziani, S., Umana, E.: Investigation of carbon black loaded natural rubber piezoresistivity. In: Proceedings of I²MTC 2015—International Instrumentation and Measurement Technology Conference, Pisa, Italy, pp. 1477–1481 (2015)
7. Ausanio, G., Barone, A.C., Campana, C., Iannotti, V., Luponio, C., Pepe, G.P., Lanotte, L.: Giant resistivity change induced by strain in a composite of conducting particles in an elastomer matrix. Sens. Actutaors A: Phys. **127**(1), 56–62 (2006)
8. Xu, H., Wu, D., Liu, Y., Zhang, Y.: Research on measurement method for polymer melt bulk modulus of elasticity. Procedia Eng. **16**, 72–78 (2011)
9. Yang, L.M., Shim, V.P.W., Lim, C.T.: A visco-hyperelastic approach to modelling the constitutive behaviour of rubber. Int. J. Impact Eng **24**, 545–560 (2000)

Part IV
Optical Sensors

Polishing Process Analysis for Surface Plasmon Resonance Sensors in D-Shaped Plastic Optical Fibers

Nunzio Cennamo, Maria Pesavento, Simone Marchetti, Letizia De Maria, Paola Zuppella and Luigi Zeni

Abstract In this work we want to compare performances of different plasmonic sensors, obtained by different polishing processes of D-shaped Plastic optical fibers (POFs). Three D-shaped POF plasmonic sensor configurations, obtained by three different polishing processes, have been created and experimentally tested. The proposed devices are based on the excitation of surface plasmons at the interface between the under test medium and a thin gold layer directly deposited on plastic fiber core (D-shaped area). The experimental results have shown that the performances are influenced by surface roughness variations in the D-shaped POF region.

Keywords Plasmonic sensors · Plastic optical fiber (POF) · D-shaped POF

1 Introduction

Optical fiber sensors based on surface plasmon resonance (SPR) are today widely proposed for applications in different areas of bio-chemical and chemical sensing [1–4]. Among them, SPR sensors based on plastic optical fibers (POFs) can offer advantages due to their low cost, flexibility, robustness and simplicity of fabrication. Authors reported several bio-chemical applications [5–7] based on an SPR sensor platform in D-shaped POF [8].

N. Cennamo (✉) · L. Zeni
Department of Engineering, University of Campania Luigi Vanvitelli,
Via Roma, 29, Aversa, Italy
e-mail: nunzio.cennamo@unicampania.it

M. Pesavento · S. Marchetti
Department of Chemistry, University of Pavia, Via Taramelli 12, Pavia, Italy

L. De Maria
Department of Transmission and Distribution Technologies, RSE S.p.A,
Via Rubattino 54, Milan, Italy

P. Zuppella
CNR–IFN UOS Padova, Via Trasea 7, Padua, Italy

© Springer Nature Switzerland AG 2019
B. Andò et al. (eds.), *Sensors*, Lecture Notes in Electrical Engineering 539,
https://doi.org/10.1007/978-3-030-04324-7_32

In this work, we compare the optical performances of different hand polished D-shaped POF platforms SPR sensors, in order to establish a mechanical polishing procedure. In particular, three D-shaped POF-SPR sensor configurations, obtained by three different polishing processes, have been created and experimentally tested. In the first case we used only 1 μm polishing paper to obtain the sensor, in the second one only 5 μm polishing paper and, finally, in the last case, we used both 5 μm and 1 μm polishing papers (first 5 μm, then 1 μm).

2 Optical Sensor Configurations

The adopted D-shaped fabrication procedure is based on the insertion of a portion of the POF fiber (about 10 mm long) in a resin support and on a successive hand grinding of the fiber surface. This procedure guarantees an easy and low-cost effective strategy for the removal of the cladding layer in the POF sensing region and for the reduction of the exposed core. In the end, a thin gold film (60 nm thick) can be deposited on the flat hand ground D-shaped POF core, by sputtering process, for exciting SPR resonance at the metal/external medium interface. Figure 1 shows the outline of the D-shaped SPR sensor.

For all three considered platforms, mentioned above, the total depth of the D-shaped POF region and the thickness of the gold film are the same. This analysis is interesting because the performances are influenced by small variations in the morphology of the D-shaped region (i.e. roughness and total depth) resulting from the manual process used for the preparation of the sensor [9, 10].

The proposed devices are characterized by exploiting a halogen lamp (HL–2000–LL, Ocean Optics) to illuminate the optical fiber sensor, observing the transmitted spectra by a spectrometer (FLAME-S-VIS-NIR-ES, Ocean Optics) and normalized to the spectrum transmitted when the outer medium is air [8].

Fig. 1 Outline of the SPR sensor based on a D-shaped POF

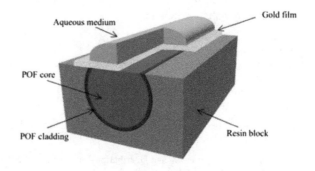

3   Experimental Results

Figures 2, 3 and 4 report, for the three different D-shaped sensors, the experimental SPR transmission spectra (normalized to the spectrum recorded with air as the surrounding medium) referring to five different water-glycerine solutions, with refractive index ranging from 1.332 to 1.380.

In the first case (Fig. 2) we used only 1 μm polishing paper to obtain the sensor, in the second one (Fig. 3) only 5 μm polishing paper and, finally, in the last case (Fig. 4), we used both 5 μm and 1 μm polishing papers.

The experimental results indicate that the configuration created by both polishing papers (Fig. 4) shows better performances.

Table 1 reports the sensors' sensitivity and the detectable refractive index range, obtained by the slopes of the linear fittings of the data (resonance wavelength shift versus refractive index) obtained by the three different polishing processes. Only in the last case (5 μm and 1 μm polishing papers) the linear fitting is a representative fitting of the data ($R^2 = 0.99$) in all the refractive index range.

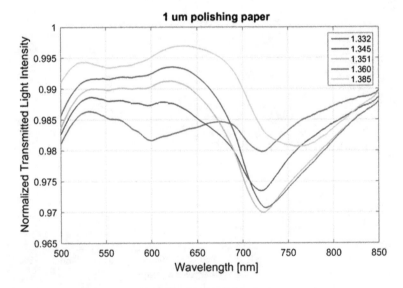

Fig. 2 SPR spectra of the sensor created by 1 μm polishing paper

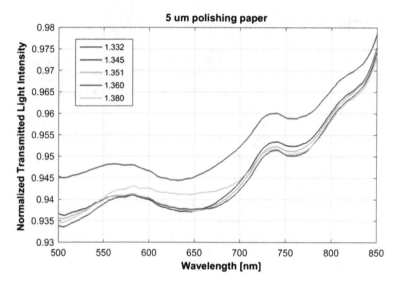

Fig. 3 SPR spectra of the sensor created by 5 μm polishing paper

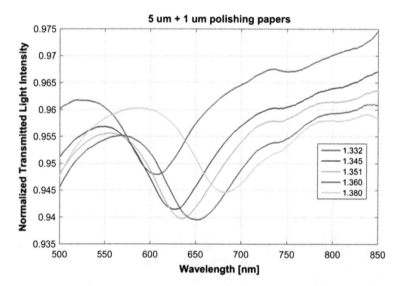

Fig. 4 SPR spectra of the sensor created by 5 μm and 1 μm polishing papers

Table 1 Parameters of the linear fittings of data

Polishing paper	(Slope of linear fitting) $S = \delta\lambda/\delta n_s$ (R^2)	Detectable refractive index range
1 μm	822 [nm/RIU], ($R^2 = 0.78$)	1.35–1.38
5 μm	205 [nm/RIU], ($R^2 = 0.55$)	1.33–1.35
1 μm + 5 μm	1437 [nm/RIU], ($R^2 = 0.99$)	1.33–1.38

4 Conclusions

The experimental results indicate that the configuration created by exploiting both the polishing papers shows better performances. In the future, a mechanical polishing procedure will be set-up starting from these experimental results.

References

1. Homola, J.: Present and future of surface plasmon resonance biosensors. Anal. Bioanal. Chem. **377**, 528–539 (2003)
2. Sharma, K.A., Jha, R., Gupta, B.D.: Fiber-optic sensors based on surface plasmon resonance: a comprehensive review. IEEE Sens. J. **7**, 1118–1129 (2007)
3. Homola, J., Yee, S.S., Gauglitz, G.: Surface plasmon resonance sensors: review. Sens. and Actuators B Chem. **54**, 3–15 (1999)
4. Wang, X.D., Wolfbeis, O.S.: Fiber-optic chemical sensors and biosensors (2013–2015). Anal. Chem. **88**, 203–227 (2016)
5. Cennamo, N., Varriale, A., Pennacchio, A., Staiano, M., Massarotti, D., Zeni, L., D'Auria, S.: An innovative plastic optical fiber-based biosensor for new bio/applications. The case of celiac disease. Sens. Actuators B Chem. **176**, 1008–1014 (2013)
6. Aray, A., Chiavaioli, F., Arjmand, M., Trono, C., Tombelli, S., Giannetti, A., Cennamo, N., Soltanolkotabi, M., Zeni, L., Baldini, F.: SPR-based plastic optical fibre biosensor for the detection of C-reactive protein in serum. J. Biophotonics **9**(10), 1077–1084 (2016)
7. Cennamo, N., Alberti, G., Pesavento, M., D'Agostino, G., Quattrini, F., Biesuz, R., Zeni, L.: A simple small size and low cost sensor based on surface plasmon resonance for selective detection of Fe(III). Sensors **14**, 4657–4671 (2014)
8. Cennamo, N., Massarotti, D., Conte, L., Zeni, L.: Low cost sensors based on SPR in a plastic optical fiber for biosensor implementation. Sensors **11**(12), 11752–11760 (2011)
9. Gasior, K., Martynkien, T., Urbanczyk, W.: Effect of constructional parameters on the performance of a surface plasmon resonance sensor based on a multimode polymer optical fiber. Appl. Opt. **53**(35), 8167–8174 (2014)
10. Gasior, K., Martynkien, T., Wojcik, G., Mergo, P., Urbanczyk, W.: D-shape polymer optical fibres for surface plasmon resonance sensing. Opto-Electron. Rev. **25**, 1–5 (2017)

A Molecularly Imprinted Polymer on a Novel Surface Plasmon Resonance Sensor

Maria Pesavento, Simone Marchetti, Luigi Zeni and Nunzio Cennamo

Abstract As a proof of principle, we have developed a novel surface plasmon resonance (SPR) sensor used to monitor the interaction between a molecularly imprinted polymer (MIP) and a small molecule as the substrate. This plasmonic platform is based on a removable polymethyl methacrylate (PMMA) chip with a thin gold film on the top, two PMMA plastic optical fibers (POFs) and a special holder, designed to obtain the plasmonic phenomenon. We have experimentally tested whether the optical sensitivity is sufficient for monitoring an MIP receptor. The advantage of MIPs is that they can be directly deposited on a flat gold surface by a spin coater machine, without modifying the surface, as needed for the bio-receptors. With this new sensor, it is possible to achieve remote sensing capabilities, by POFs, and also to realize an engineered platform with a removable chip sensor.

Keywords Surface plasmon resonance sensors · Molecularly imprinted polymers
Slab waveguides

1 Introduction

Optical chemical sensors and biosensors have been shown to be able to play an important role in numerous fields, especially the biosensors based on optical fibers and surface plasmon resonance (SPR) phenomenon [1–3]. They selectively recognize and capture the analyte present in a liquid sample so producing a local increase of the refractive index of the dielectric layer in contact with the metal film. The use of SPR sensor devices often requires replaceable parts and disposable chips for easy, fast and on site detection analysis.

M. Pesavento (✉) · S. Marchetti
Department of Chemistry, University of Pavia, Pavia, Italy
e-mail: maria.pesavento@unipv.it

L. Zeni · N. Cennamo
Department of Engineering, University of Campania L. Vanvitelli, Aversa, Italy

© Springer Nature Switzerland AG 2019   259
B. Andò et al. (eds.), *Sensors*, Lecture Notes in Electrical Engineering 539,
https://doi.org/10.1007/978-3-030-04324-7_33

In this framework, we propose a novel low-cost SPR sensor platform for selective detection of analytes in aqueous solutions. It is based on a PMMA slab waveguide covered with a thin gold film and inserted in a special holder, designed to produce the plasmonic resonance at the gold-dielectric interface [4]. On the gold film an MIP receptor has been deposited by a simple procedure, only based on a spin coater machine and an oven [5].

MIPs are synthetic receptors with many favorable aspects with respect to bio-receptors, such as an easier and faster preparation, the possibility of application outside the laboratory and, for example, a longer durability under environmental conditions [6]. The advantage of MIPs is that they can be directly deposited on a flat gold surface without modifying the surface (functionalization and passivation), as needed for the bio-receptors.

In this work, as proof of principle, a selective MIP was considered as the receptor for furfural (furan-2-carbaldehyde, 2-FAL) and the possibility of using the device obtained for detection of 2-FAL in aqueous media was investigated.

The determination of 2-FAL in wine or beer, aqueous matrices of interest in food industry, is becoming a very crucial task in relation to its toxic and carcinogenic effect on human beings. Moreover, 2-FAL is also relevant for food flavor and aroma, which are important factors in terms of quality control and quality assurance. In wines for example, this factor is among the most important ones [7]. During the ageing, the concentration of some compounds, formed during the alcoholic fermentation, decreases and new compounds appear, some deriving from the evolution of the wine components themselves and some extracted from wood. Furanic compounds, in particular 2-FAL, which can be formed in wine by sugar dehydration, and can be extracted from wood as well, has a high impact on the aroma. Moreover, they can be used as ageing markers as well.

2   Plasmonic Platform

Figure 1 shows the novel SPR sensor configuration. It is based on a slab waveguide (a layer of PMMA of 1 cm × 1 cm × 0.5 mm in size) with a thin gold film on the top surface (60 nm thick deposited by the sputter coater Baltec SCD 500) inserted in a special holder, designed to produce the plasmonic resonance at the gold-dielectric interface [4]. As shown in Fig. 1, the light is launched in the slab waveguide through a trench, realized into the holder, and illuminated with a POF, to excite the surface plasmon waves. The output light is then collected by another POF, positioned at the end of the slab waveguide, at an angle of 90° with respect to the trench, and carried to a spectrometer. In this configuration, the trench has been used because a large incident angle is required for SPR excitation [4].

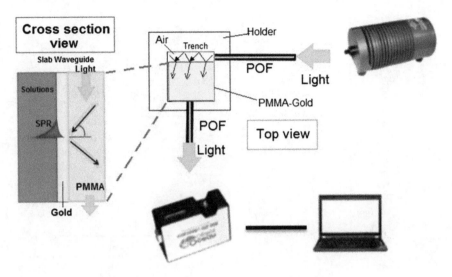

Fig. 1 Sensor system outline and top and cross section view of the chip

On the gold film of the removable chip (slab waveguide) an MIP layer is deposited. The experimental setup is based on a halogen lamp (HL–2000–LL, Ocean Optics), used to illuminate the first POF, and a spectrometer (FLAME-S-VIS-NIR-ES, Ocean Optics) connected to the second POF, used to observe the transmitted spectra (see Fig. 1).

3 Experimental Results

About 50 μL of solutions with different concentrations of 2-FAL were dropped over the MIP layer and the spectrum recorded after ten minutes incubation. The SPR transmission spectra were normalized to the reference spectrum by Matlab software. In particular, the transmission spectra have been normalized to the spectra obtained in air before MIP deposition, in which no plasmon resonance is excited. The dose-response curves were obtained by plotting the resonance wavelengths versus the 2-FAL concentrations.

In a semi-log scale, Fig. 2 shows the resonance wavelengths versus the 2-FAL concentration (ppm) together with the Hill fitting. The obtained results have shown the good performance in terms of sensitivity and limit of detection (LOD) of this novel approach.

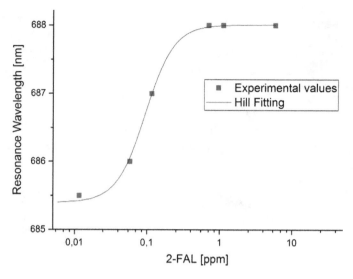

Fig. 2 Resonance wavelengths (nm) versus the concentrations of 2-FAL (ppm), in semi-logarithmic axes, with Hill fitting of data

4　Conclusions

The removable PMMA chip, with gold film on the top, is suitable for chemical applications after the deposition of a specific receptor (MIP). This chip can be changed and optimized for each test. The easy replacement of the chip allows for the production of an engineered platform by simplifying the measurement procedures.

References

1. Sharma, A.K., Jha, R., Gupta, B.D.: Fiber-optic sensors based on surface plasmon resonance: a comprehensive review. IEEE Sens. J. **7**, 1118–1129 (2007)
2. Homola, J.: Present and future of surface plasmon resonance biosensors. Anal. Bioanal. Chem. **377**, 528–539 (2003)
3. Caucheteur, C., Guo, T., Albert, J.: Review of plasmonic fiber optic biochemical sensors: improving the limit of detection. Anal. Bioanal. Chem. **407**(14), 3883–3897 (2015)
4. Cennamo, N., Mattiello, F., Zeni, L.: Slab waveguide and optical fibers for novel plasmonic sensor configurations. Sensors **17**, 1488 (2017)
5. Cennamo, N., De Maria, L., D'Agostino, G., Zeni, L., Pesavento, M.: Monitoring of low levels of furfural in power transformer oil with a sensor system based on a POF-MIP platform. Sensors **15**, 8499–8511 (2015)
6. Uzun, L., Turner, A.P.F.: Molecularly-imprinted polymers sensors: realising their potential. Biosens. Bioelectron. **76**, 131–144 (2016)
7. Camara, J.S., Alves, M.A., Marques, J.C.: Changes in volatile composition of madeira wines during their oxidative ageing. Anal. Chim. Acta **563**, 188–197 (2006)

Design of a Label-Free Multiplexed Biosensing Platform Based on an Ultracompact Plasmonic Resonant Cavity

Francesco Dell'Olio, Donato Conteduca, Maripina De Palo, Nicola Sasanelli and Caterina Ciminelli

Abstract A multi-analyte biosensing platform for the selective label-free detection of protein biomarkers has been designed through a three-dimensional model based on the finite element method. The sensing element of the platform is a planar plasmonic nanocavity, consisting of a one-dimensional periodic structure (Bragg grating) in gold, with a defect, placed on the buried oxide of a silicon-on-insulator substrate. The footprint of this sensing element, which has a good chemical stability, is only 1.57 μm^2. The sensor has a detection limit of 128 pg/mm^2 and a surface sensitivity of 1.8 nm/nm.

Keywords Biosensor · Plasmonics · Bragg grating

1 Introduction

Several diseases, including neoplastic and cardiovascular ones, require early medical diagnosis, in order to detect the pathologies at their first stages, so increasing the chances of survival for patients. Large-scale screening of at-risk patients by biologic fluid analysis through miniaturized lab-on-chip microsystems is a very effective approach to achieve this early medical diagnosis in a noninvasive way [1]. Biologic fluid analysis typically aims at detecting any traces of diagnostic biomarkers [2], but the detection of a single marker at abnormal concentration usually has poor diagnostic value while the identification of abnormalities relevant to the concentration of a set of properly selected biomarkers (biomarker panel) can be strongly correlated to the actual existence of the disease [2]. This is the reason why multiplexed biosensing platforms able to detect abnormal levels of biomarker panels are strongly demanded [3].

F. Dell'Olio · D. Conteduca · M. De Palo · N. Sasanelli · C. Ciminelli (✉)
Optoelectronic Laboratory, Politecnico di Bari, Via Orabona 4, 70125 Bari, Italy
e-mail: caterina.ciminelli@poliba.it

© Springer Nature Switzerland AG 2019　　　　　　　　　　　　　　　　263
B. Andò et al. (eds.), *Sensors*, Lecture Notes in Electrical Engineering 539,
https://doi.org/10.1007/978-3-030-04324-7_34

Label-free optically-based biosensors have demonstrated several advantages with respect to the competing technologies, i.e. higher sensitivity and accuracy, very low detection limit and wide dynamic range, so allowing the detection of low biomarkers concentration (<100 ng/mL) [4, 5]. Furthermore, the strong device compactness and their high suitability for on-chip integration allow realizing multiplexed analysis of several biomarkers with high specificity [6].

The plasmonic biosensing configurations and, in particular, surface plasmon resonance (SPR) biosensors, have been largely used in the field of ultra-sensitive detection of bio-molecules because of their ultra-low resolution, down to 1 ng/mL, which is suitable for almost all emerging applications in medicine [7]. However, SPR biosensors are typically bulky and their miniaturization is challenging. Therefore, different approaches have been proposed in order to enable the miniaturization and the on-chip integration of the plasmonic biosensors. Among these approaches, planar resonant nanocavities, where light is confined in surface plasmon polariton (SPP) waveguides [8–10], are very advantageous because a biosensor with such configuration can be easily arrayed due to a footprint of the order of 1 μm^2, and has a resolution that is compliant to the requirements of many applications in the field of medicine. In addition, a biosensor in which the light beam is confined in planar SPP waveguides exhibits the most attractive advantages of the integrated optical devices, including the possibility to be fabricated by the standard fabrication processes and techniques widely utilized in micro- and nano-electronics.

Here, we report on the design of a new metal-insulator-metal (MIM) ultracompact integrated plasmonic platform, based on multiple label-free plasmonic biosensors for multiplexed biomarker analysis. The platform can be integrated on a silicon-on-insulator (SOI) chip.

2 Design

The configuration of the biosensing platform for detecting and monitoring a number of biomarkers is shown in Fig. 1 [11].

The system includes an array of nanoplasmonic biosensors, assuming to be functionalized with different receptors for the multiplexed biosensor analysis. A mode converter is placed at both the input and output of each biosensor to convert the mode in the plasmonic cavity into a guided mode propagating in a SOI waveguide. At the input of the system there is a $1 \times M$ SOI multimode interferometer (MMI), with M the number of output branches, used as beam splitter to enable the light propagation in each cavity and the biomarkers multi-detection.

In the selection of the platform operating wavelength, two possible values, 1550 and 1300 nm, have been compared. We have verified that tunable laser diodes with low-cost, narrow-linewidth, stable and sufficiently high output power are available at both wavelengths. In addition, at 1550 and 1300 nm the absorption loss of the silica substrate is negligible. Since the cover medium of each nanoplasmonic biosensor is

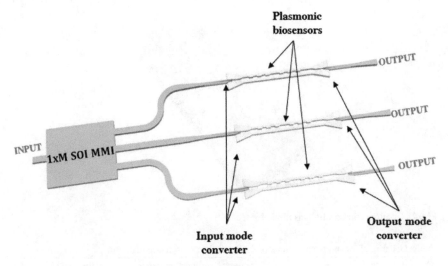

Fig. 1 Schematic of the label-free multiplexed biosensing platform based on the plasmonic biosensor realized by a defective 1D Bragg grating for $M = 3$

the plasma where the biomarkers are dispersed, we have preferred the latter option because the plasma absorption at 1300 nm is significantly lower than at 1500 nm.

The key element of the system is the plasmonic biosensor, realized by 1D Bragg grating with two closely-spaced gold (Au) rails, whose distance is periodically modulated with a defect in the middle to obtain the resonant behavior (see Fig. 2) [12]. The cavity has been designed to obtain the best compromise between the Q-factor and the resonance transmission. A Q-factor of about 20 and a transmission $T = 19\%$ (optical loss $\alpha_{cav} = 7.2$ dB at $\lambda = 1330$ nm) were calculated by three-dimensional finite element method with a number of periods $N = 8$. The total on-chip optical losses for each biosensor is 18 dB. The value of T is appropriate for an accurate measurement of the resonance condition. The archived Q-factor is quite typical for the plasmonic cavities ($Q < 10^2$). With an N value of 6, a T improvement up to 32% can be achieved at the expenses of a Q decrease down to approximately 10.

The Lorentz-Drude model [13] has been selected to derive the wavelength-dependent gold dielectric constant. The main reason of this choice is that the Lorentz-Drude model, unlike other models (e.g., Drude or Johnson and Christy), takes into account the interband transitions and, consequently, allows obtaining numerical results matching very well the experimental ones [14].

The performance of the label-free biosensing platform for several protein biomarkers has been calculated. In particular, we have considered the early diagnosis and monitoring of coronary artery diseases as specific application, by simultaneously sensing three protein biomarkers: C-reactive protein, β_2-microglobulin, and adiponectin. The surface sensing calculated by assuming the selective binding between the target proteins and the bioreceptor molecules attached to the cavity

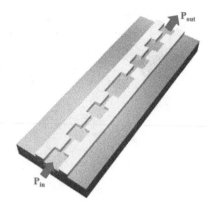

Fig. 2 Configuration of the plasmonic biosensor

Table 1 Parameters and performance of the label-free biosensing platform

Optical device	Parameters and performance	Value
Plasmonic cavity	Cavity length (L_{cav})	3.3 μm
	Q-factor (Q)	20
	Transmission (T)	19%
	Resonance wavelength (λ_R)	1320 nm
	Cavity optical loss (α_{cav})	7.2 dB
1×3 MMI	Length of the MMI (L_{MMI})	37 μm
	Footprint (A_{MMI})	250 μm^2
	MMI optical loss (α_{MMI})	0.28 dB
Mode converter	Length (L_{MC})	10 μm
	Mode converter loss (α_{MC})	3 dB
Label-free biosensing platform	On-chip optical loss (each branch) (α_{TOT})	18 dB
	Total footprint (A_{TOT})	0.011 mm^2
	Surface sensitivity (S_s)	1.8 nm/nm
	Limit of detection (LOD)	0.42 μg/mL

surface is $S_S = 1.8$ nm/nm, while the limit of detection (LOD) of the biosensor is 0.42 μg/mL, much lower than the risky value of the investigated biomarkers.

Table 1 summarize the geometrical parameters and the performance of the designed biosensing platform.

3 Conclusions

The design of a label-free multiplexed biosensing platform based on an ultracompact nanoplasmonic planar cavity has been reported. The achieved result in terms of LOD (= 0.42 µg/mL) confirms that this platform is very suitable for the detection of the specific biomarkers for the diagnosis of cardiovascular diseases at early stages, obtained with a very compact footprint of the system ($A_{TOT} = 0.011$ mm^2). In fact, the selected specific application demands a LOD of 0.5 µg/mL. The simultaneous detection of several biomarkers and the suitability of on-chip integration improve the efficiency of the medical analysis, reducing false-positive cases, which can be obtained more frequently with a single biomarker detection. This improves the prevention and reduces the waiting time for the beginning of the therapy, also minimizing the hospital stay of false-positive patients, who do not require any treatment.

References

1. Lafleur, J.P., Jönsson, A., Senkbeil, S., Kutter, J.P.: Recent advances in lab-on-a-chip for biosensing applications. Biosens. Bioelectron. **76**, 213–233 (2016)
2. Jain, K.: The Handbook of Biomarkers. Springer, New York (2010)
3. Dincer, C., Bruch, R., Kling, A., Dittrich, P.S., Urban, G.A.: Multiplexed point-of-care testing—xPOCT. Trends Biotechnol. **35**, 728–742 (2017)
4. Estevez, M., Alvarez, M., Lechuga, L.: Integrated optical devices for lab-on-a-chip biosensing applications. Laser Photonics Rev. **6**, 463–487 (2011)
5. Ciminelli, C., Campanella, C., Dell'Olio, F., Campanella, C., Armenise, M.: Label-free optical resonant sensors for biochemical applications. Prog. Quantum Electron. **37**, 51–107 (2013)
6. Dell'Olio, F., Conteduca, D., Ciminelli, C., Armenise, M.: New ultrasensitive resonant photonic platform for label-free biosensing. Opt. Express **23**, 28593 (2015)
7. Brolo, A.: Plasmonics for future biosensors. Nat. Photonics **6**, 709–713 (2012)
8. Gao, Y., Xin, Z., Zeng, B., Gan, Q., Cheng, X., Bartoli, F.: Plasmonic interferometric sensor arrays for high-performance label-free biomolecular detection. Lab on a Chip **13**, 4755 (2013)
9. Binfeng, Y., Guohua, H., Ruohu, Z., Yiping, C.: Design of a compact and high sensitive refractive index sensor base on metal-insulator-metal plasmonic Bragg grating. Opt. Express **22**, 28662 (2014)
10. Gazzaz, K., Berini, P.: Theoretical biosensing performance of surface plasmon polariton Bragg gratings. Appl. Opt. **54**, 1673 (2015)
11. Dell'Olio, F., Conteduca, D., De Palo, M., Ciminelli, C.: Design of a new ultracompact resonant plasmonic multi-analyte label-free biosensing platform. Sensors **17**, 1810 (2017)
12. Dell'Olio, F., De Palo, M., Conteduca, D., Ciminelli, C.: Resonant nanoplasmonic platform for fast and early diagnosis of cardiovascular diseases. In: Proceedings of the IEEE 2nd International Forum on Research and Technologies for Society and Industry Leveraging a better tomorrow (RTSI), pp. 1–4. IEEE, New York (2016)
13. Rakic, A.D., Djurisic, A.B., Elazar, J.M., Majewski, M.L.: Optical properties of metallic films for vertical-cavity optoelectronic devices. Appl. Opt. **37**, 5271 (1998)
14. Maier, S.A.: Plasmonics: fundamentals and applications. Springer, New York (2006)

A Novel Intensity-Based Sensor Platform for Refractive Index Sensing

Nunzio Cennamo, Francesco Mattiello and Luigi Zeni

Abstract In the present investigation a new intensity-based sensor platform for refractive index sensing is presented. It is based on a special holder, a slab waveguide and two Plastic optical fibers (POFs). In particular, we present a comparison between two different configurations: the slab waveguide with and without a buffer layer. Advantages of this new approach are the possibility of sensing with a removable chip, the easy production of an engineered platform and the use of a special holder, which is also suitable for thermo-stabilized flow cells implementation.

Keywords Optical sensors · Plastic optical fibers · Slab waveguides

1 Introduction

Refractive index sensors based on optical fibers have more advantages than those based on different approaches, for example, the possibility of remote sensing. Optical fiber sensors are today widely proposed for applications in different areas of bio-chemical and chemical sensing [1–5]. Several optical sensors based on plastic optical fibers (POFs) have been recently proposed by the Authors [6–8]. For example, a plasmonic sensor based on a D-shaped POF is realized with a buffer layer between the exposed POF core and a thin gold film [6].

In this work, the POF is used only to launch the light into the slab waveguide and to collect the light emerging from the waveguide, conveying it to a spectrometer. The sensing region is realized on the PMMA slab waveguide inserted in a special holder. A photoresist (Microposit S1813) buffer layer is deposited over a PMMA chip (slab waveguide) by a spin coater. This photoresist buffer layer is required in order to increase the performances of the sensor. The experimental results indicated that the photoresist layer improves the performances and this configuration could be used for chemical sensing applications.

N. Cennamo (✉) · F. Mattiello · L. Zeni
Department of Engineering, University of Campania L. Vanvitelli, Via Roma, 29, Aversa, Italy
e-mail: nunzio.cennamo@unicampania.it

© Springer Nature Switzerland AG 2019
B. Andò et al. (eds.), *Sensors*, Lecture Notes in Electrical Engineering 539,
https://doi.org/10.1007/978-3-030-04324-7_35

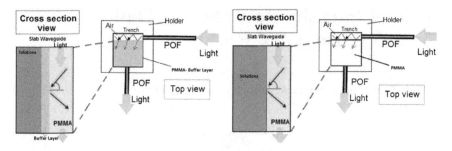

Fig. 1 Top and cross section view of two sensor configurations with and without the photoresist buffer layer

2 Optical Sensor System

Figure 1 shows two novel sensor configurations. They are based on a slab waveguide (a layer of PMMA 1 cm × 1 cm × 0.5 mm in size and with 1.49 RIU) with a photoresist buffer layer (Microposit S1813 Photoresist, 1.61 RIU, 1.5 μm thick deposited on its surface by spin coating) in the first one and without the buffer layer in the second one.

The slab waveguide (with or without the buffer layer) is inserted in a special holder. As indicated in Fig. 1, the exciting light (halogen lamp, HL–2000–LL, Ocean Optics) is introduced in the slab waveguide by a 10 mm long trench (size: 1 mm × 1 mm), realized in the holder, illuminated by a POF (1 mm in outer diameter). On the other hand, another PMMA POF, kept at the end of the slab waveguide at a 90° angle to the trench, is exploited to carry the output light to a spectrometer (FLAME-S-VIS-NIR-ES, Ocean Optics). The trench "air waveguide" has been designed because a large incident angle is required to improve the evanescent field excitation.

A similar approach has been already exploited to realize an innovative, low-cost and simple plasmonic sensor [7].

3 Experimental Results

Figure 2 reports the experimental transmission spectra (normalized to the spectrum recorded with air as the surrounding medium) referring to different water-glycerin solutions, with refractive index ranging from 1.332 to 1.401, for the sensor configuration without the buffer layer. Figure 2 clearly shows that the amplitude of the output decreases when the refractive index increases.

Figure 3 shows the output responses for the same refractive indices when on the PMMA waveguide is present the photoresist buffer layer.

Figure 4 shows the normalized output at 567.5 nm wavelength versus the refractive index for both the sensor configurations. In the same figure is also presented the

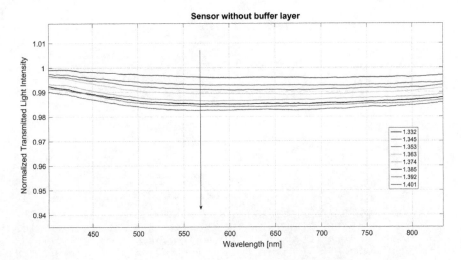

Fig. 2 Sensor configuration without buffer layer. Normalized output for different refractive indices of the aqueous medium

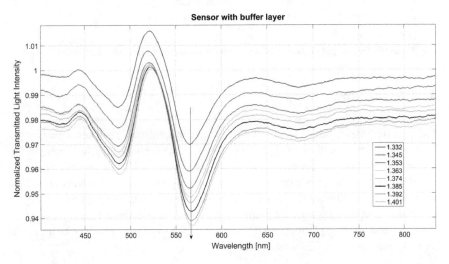

Fig. 3 Sensor configuration with buffer layer. Normalized output for different refractive indices of the aqueous medium

linear fitting to the experimental data. The sensitivity (the slope of the linear fitting) changes when the buffer layer is present. In particular, the buffer layer improves the performances of the sensor system, because the high value of the buffer's refractive index (larger than the PMMA one) improves the interaction between the evanescent field and the aqueous medium.

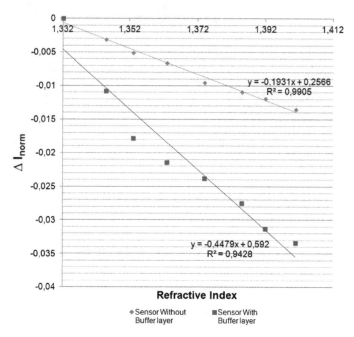

Fig. 4 Normalized output (at 567.5 nm wavelength) versus refractive index and linear fitting

4 Conclusions

The experimental results showed that the sensor's performances improve when the PMMA slab waveguide is coated with a high refractive index buffer layer. In the future, for selective detection of an analyte, we want to deposit by spin coating a chemical receptor layer (Molecularly imprinted polymer) on the slab waveguide. So, the selective detection of the analyte could be possible exploiting an MIP receptor combined with this new optical platform, through intensity-based configuration. In fact, this new sensing platform could be used with a very easy and low-cost experimental setup based on LEDs and photodetectors.

References

1. Wang, X.D., Wolfbeis, O.S.: Fiber-optic chemical sensors and biosensors (2008–2012). Anal. Chem. **85**(2), 487–508 (2013)
2. Monk, D.J., Walt, D.R.: Optical fibers-based biosensors. Anal. Bioanal. Chem. **379**(7–8), 931–945 (2004)
3. Trouillet, A., Ronot-Trioli, C., Veillas, C., Gagnaire, H.: Chemical sensing by surface plasmon resonance in a multimode optical fibre. Pure Appl. Opt. **5**(2), 227–237 (1996)
4. Sharma, K.A., Jha, R., Gupta, B.D.: Fiber-optic sensors based on surface plasmon resonance: a comprehensive review. IEEE Sens. J. **7**, 1118–1129 (2007)

5. Homola, J., Yee, S.S., Gauglitz, G.: Surface plasmon resonance sensors: review. Sens. Actuators B Chem. **54**, 3–15 (1999)
6. Cennamo, N., Massarotti, D., Conte, L., Zeni, L.: Low cost sensors based on SPR in a plastic optical fiber for biosensor implementation. Sensors **11**(12), 11752–11760 (2011)
7. Cennamo, N., Mattiello, F., Zeni, L.: Slab waveguide and optical fibers for novel plasmonic sensor configurations. Sensors **17**(7), 1488 (2017)
8. Sequeira, F., Duarte, D., Bilro, L., Rudnitskaya, A., Pesavento, M., Zeni, L., Cennamo, N.: Refractive index sensing with D-shaped plastic optical fibers for chemical and biochemical applications. Sensors **16**, 2119 (2016)

An Optical Sensing System for Atmospheric Particulate Matter

Luca Lombardo, Marco Parvis, Emma Angelini, Nicola Donato
and Sabrina Grassini

Abstract Atmospheric particulate is one of the main responsible for the environmental pollution of highly populated cities. The particulate matter can be considered an aerosol of many particles, with different shapes, morphologies and sizes down to few micrometers. These small particles are responsible for climate changes and, when inhaled, can reach the lungs causing respiratory and cardiovascular diseases. Usually commercial sensors are based on passive filters and are designed to measure only particles whose size is below a certain value. Therefore, a measuring system capable of assessing the quality of air and evaluating not only the amount of particles, but also the size distribution of the particulate matter in urban atmospheres could be extremely helpful. This paper describes a simple and cheap optical solution, which is able to detect the total amount of solid pollution particles and classify them according to their average size, giving a fast response to the users.

Keywords Optical system · Atmospheric particle distribution
Particulate matter · PMxx · Air pollution monitoring

1 Introduction

Nowadays, atmospheric particulate matter is one of the most dominant factors in air pollution, especially in large cities and industrial sites. Particulate matter (PM) consists in a complex mixture of small solid particles dispersed in air, which have

L. Lombardo (✉) · M. Parvis
Department of Electronics and Telecommunications, Politecnico di Torino, Torino, Italy
e-mail: luca.lombardo@polito.it

E. Angelini · S. Grassini
Department of Applied Science and Material Science, Politecnico di Torino, Torino, Italy

N. Donato
Department of Engineering, University of Messina, Messina, Italy

© Springer Nature Switzerland AG 2019
B. Andò et al. (Eds.) *Sensors*, Lecture Notes in Electrical Engineering 539,
https://doi.org/10.1007/978-3-030-04324-7_36

heterogeneous physical-chemical characteristics and whose size largely vary from hundreds of micrometers down to the sub-micrometer range.

Particulate matter is released in the atmosphere both by natural and anthropic sources. Anthropic sources are typically associated to the human industrial activities and to the combustion of fossil fuels and biomasses. Instead, the principal natural sources of particulate in air are erosion of the Hearts crust, sea-salt sprays, volcanic activities and pollen produced by vegetation [1].

Main problems related to the increase of particulate matter in the atmosphere involve its negative effects on the human health and on the climate change. In particular, a great correlation between the level of atmospheric particulate matter and the onset of respiratory system diseases, such as asthma, emphysema and lung cancer, has been demonstrated. Especially ultra-fine particulates (with average size lower than $2.5\,\mu m$) can deeply penetrate in tissue and also enter in the blood stream causing cardiovascular problems and intoxication [2]. The World Health Organization (WHO) has estimated that about two millions of people die every year for particulate matter pollution [3, 4]. The European Union establishes health based standards for the different pollutants present in the air and since 2008 introduced PM2.5 exposure limits; in 2010 the average exposure indicator (AEI) for PM2.5 has been assessed to be over $20\,\mu g/m^3$, but actually appropriate measures have to be taken for achieving $18\,\mu g/m^3$ by 2020.

Measuring particulate matter, taking into account the particle size distribution in a easy and cheap way, is therefore a big challenge.

In principle, measuring the particulate amount is easy and the gravimetric technique is usually employed. A sampler aspirates a constant air flow through a sampling head; the particulate fraction is collected by a filter. The filter is weighted in fixed temperature and relative humidity conditions ($20\,^\circ C$; 50% RH), before and after the air exposure, and the particulate concentration, expressed in $\mu g/m^3$, is calculated by dividing the total mass of the collected particles for the volume of the sampled air.

The desired dimensional particulate fraction (PMxx) is selected by using a passive filter, which is positioned in front of the weighted one. The passive filter has a mesh suitable to stop particles whose sizes are greater than those required. This way, only particles having dimension smaller than the passive filter mesh are effectively collected and measured, and for example the so-called PM10 value can be easily obtained.

The described solution is currently employed in most of the commercially-available measuring systems, but suffers from some drawbacks: the measurements can be obtained only after some time, the weighting sensitivity must be at micro-gram level, the weighting procedure is complex and requires a long conditioning procedure to prevent temperature and humidity from affecting the measurements. In addition, if the particulate amount at different sizes is requested several measuring units are required thus increasing the overall cost. Several other simple and low-cost solutions [5, 6], based on solid-state lasers and photo-diodes, have been proposed. However, generally, these devices are capable only to count the particles, without determining their size distribution. A solution suitable to discriminate among different sizes is therefore still required.

2   Proposed Architecture

The particulate matter measuring system realized at Politecnico di Torino can be considered a hybrid between an optical particle detector and a classic gravimetric meter. This device is able to detect in quasi-real time the atmospheric particulate matter and to classify it according to the average particle size. Figure 1 shows, as an example, one of the realized prototypes, which contains all the components described in Fig. 2.

Figure 2 shows the block diagram of the measuring system. A small air pump forces a known and constant flow of air through a glass-fiber filter (Grade GF 10 Glass Filter by *GE Healthcare's Life Sciences*) and the particulate matter is captured on its surface. The filter is able to trap particles down to 2.5 μm.

Fig. 1 Prototype of the particulate matter measuring system

Fig. 2 Block diagram of the proposed measuring architecture

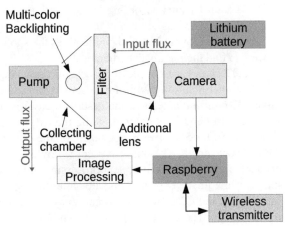

A collecting chamber, realized in ABS by a 3D printer, acts as filter holder and is connected to the pump outlet in order to force the air flow onto a defined area of the filter surface. The collecting chamber provides also a back-lighting system designed to illuminate the filter with a set of seven LEDs at different wavelengths from IR to UV.

The results reported in the following are obtained by using a uEye$^{(TM)}$ video camera with a resolution of 3840×2748 pixels, to take picture of the filter surface. The camera is equipped with a dedicated low-cost optics to focus a small filter area ($500 \times 500 \ \mu m^2$). This setup allows reaching a resolution of about $1 \ \mu m/pixel$, but other cheaper devices can be easily employed.

The images are taken by changing the lighting wavelength. A RaspberryPI is used to control both the camera and the lighting system, and to perform the imaging processing in order to identify the particulate particles present on the filter surface, which appears as dark spots.

The figure also shows the block implementing the wireless connection, which is still under development, and takes advantage of a cloud architecture [7, 8] and of an specific architecture [9–11] already developed by the authors. Eventually, a lithium battery supplies the energy for the entire system so that a completely portable device can be arranged.

The image processing software is written in the Python language and takes advantage of the free OpenCV library [12]. The OpenCV library simplifies the image processing making the identification of the particulate particles and the estimation of their size easy. The low-cost lens employed to focus the image and the simple lighting system let to capture a focused and uniformly lighted image only in the center of the acquired area. Even though this limits the number of captured spots, such a selection let to both open the system to a much cheaper approach and to greatly limit the processing time. By using an image portion of 250 kpixels (i.e. 500×500 pixels corresponding to about $500 \times 500 \ \mu m$) it is possible to limit the processing time of each image to less than 1 s.

The processing selects an image area, converts it to gray-scale using the most solicited color channel according to the lighting wavelength, performs a Gaussian blur to reduce the visual impact of the mesh composing the glass filter, increases the details using a moving kernel then identifies the spots and employs a minimization process to find their equivalent diameter.

Figure 3 shows an example of the described imaging processing. Picture A shows the glass-fiber filter after 24 h of exposure to contaminated air. Only the center area (about 37 mm in diameter in the described case) is interested by the air flow and becomes dirty and gray colored, while the external area remains white. Image B shows the 500×500 pixel image taken, on the filter center, by illuminating it with the UV wavelength. The image is recorded by the camera as a blue image and this channel is used for the subsequent processing. Image C is the gray-scale converted image, while image D is the same area after the Gaussian blurring. On the bottom, pictures C1 and C2 shown the corner areas of the previous two pictures at higher magnification. These last two pictures highlight how the blurring reduces the visual effect of the glass-fiber filter, without changing the aspect of the particulate spots.

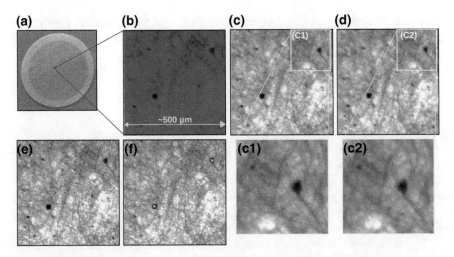

Fig. 3 Example of the image processing steps. The example refers to an image of the glass-fiber filter, after 24 h air exposure, taken by using a wavelength of about 370 nm corresponding to the UV lighting

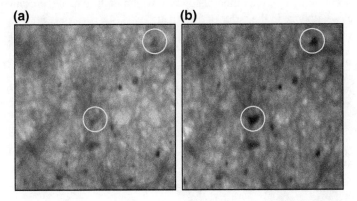

Fig. 4 Example of two images of the same area of a glass-fiber filter after 12 h air exposure, taken with different wavelengths

Finally, image E is the sharpened version of image D and image F shows the spot identification performed by the software.

As already pointed out the entire imaging processing is performed by the Raspberry embedded PC in less than 1 s, so that the entire processing for the seven wavelengths requires less than 7 s. By illuminating the filter with different wavelengths is in principle possible to try to identify different particulate particles taking advantage of the different level of transparency at different wavelengths.

Figure 4 shows as an example two images of the same area of the glass-fiber filter, taken by using a red lighting (625 nm) and UV lighting (350 nm) after 12 h of air exposure. The circles put in evidence the different light transmittance of some particulate particles at the two wavelengths. The difference in contrast of the particulate particles is clearly visible.

3   Results

The proposed system has been tested during some measuring campaigns performed in the metropolitan area close to Politecnico University in Turin, a city in the North of Italy. In the case described here, the measuring system has been positioned at a height of about 1 m very closed to a high traffic road.

Figure 5 shows as an example, the images collected on the glass-fiber filter exposed for 12 h to the urban atmosphere.

Figure 5a shows the image taken on the glass-fiber filter; Fig. 5b shows the result of the imaging processing described in the previous section; eventually Fig. 5c show the particle size distribution. The total air volume fluxed through the filter is of about 0.3 m^3. The total particle counts in the measured area is of about 80 and the obtained size distribution is reported in the histogram, which highlights the presence of a lot of particles with size in the order of 5–10 μm.

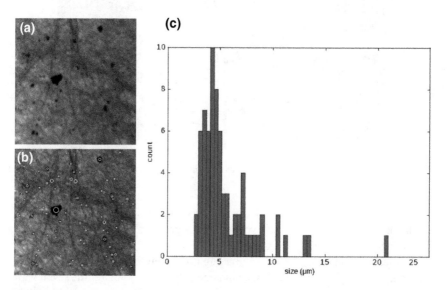

Fig. 5 Example of particulate size distribution obtained by sampling the air for 12 h close to a high traffic road near Politecnico University, in the center of Turin (Italy)

4 Conclusions

The proposed optical sensing system represents a simple and easy way to estimate the particulate particle size distribution in urban atmosphere. The designed and developed prototype is capable of determining particle size down to about 3 μm. Actually, this limitation is due to the glass-fiber filter mesh.

In the prototype implementation a high resolution camera is used, but the measuring campaigns carried out till now highlighted that a resolution of 512×512 pixels is sufficient for providing reliable results, without increasing the cost of the entire measuring system.

The preliminary results obtained by illuminating the glass-fiber filter with the IR-UV multi-lighting system, allows concluding that this low cost spectrometer is promising to provide information on the different nature and chemical composition of the particulate matter, even though more experiments are required to define these capabilities.

The authors are actually working in order to improve the system performance, both reducing the measuring system size and increasing the capability to send the collected data via bluetooth and LoRa protocols.

References

1. Viana, M., et al.: Source apportionment of particulate matter in Europe: a review of methods and results. J. Aerosol Sci. **39**(10), 827–849 (2008)
2. Pope, C.A., et al.: Cardiovascular mortality and exposure to airborne fine particulate matter and Cardiovascular mortality and exposure to airborne fine particulate matter and cigarette smoke shape. Circulation **120**, 941–948 (2009)
3. World Health Organization, Health Effects of Particulate Matter. http://www.euro.who.int/data/assets/pdf_file/0006/189051/Health-effects-of-particulate-matter-final-Eng.pdf (2013). Last checked 14 Nov 2017
4. World Health Organization, Quantifying environmental health impacts. http://www.who.int/quantifying_ehimpacts/news_events/en/. Last checked 14 Nov 2017
5. Dixit, R., Zheng, W., Hatfield, A., Gerardi, J., Hickman, G., Doak, W., Chiarot, P., Klotzkin, D.: Optical backscatter measurement of cloud particulates. In: IEEE Sensors, pp. 1–4 (2013)
6. Li, X., Iervolino, E., Santagata, F., Wei, J., Yuan, C.A., Sarro, P.M., Zhang, G.Q.: Miniaturized particulate matter sensor for portable air quality monitoring devices. In: IEEE Sensors, pp. 2151–2154 (2014)
7. Sisinni, E., Depari, A., Flammini, A.: Design and implementation of a wireless sensor network for temperature sensing in hostile environments. Sens. Actuators A Phys. **237**, 47–55 (2016). ISSN 0924-4247, https://doi.org/10.1016/j.sna.2015.11.012
8. Ascorti, L., Savazzi, S., Soatti, G., Nicoli, M., Sisinni, E., Galimberti, S.: A wireless cloud network platform for industrial process automation: critical data publishing and distributed sensing. IEEE Trans. Instrum. Meas. **66**(4), 592–603 (2017)
9. Lombardo, L., Corbellini, S., Parvis, M., Elsayed, A., Angelini, E., Grassini, S.: Wireless sensor network for distributed environmental monitoring. In: IEEE Transactions on Instrumentation and Measurement, Early access article, December (2017)

10. Corbellini, S., Di Francia, E., Grassini, S., Iannucci, L., Lombardo, L., Parvis, M.: Cloud based sensor network for environmental monitoring. In: Measurement: Journal of the International Measurement Confederation, Early access article (2017)
11. Corbellini, S., Parvis, M.: Wireless sensor network architecture for remote non-invasive museum monitoring. In: Proceedings of the 2016 International Symposium on System Engineering ISSE 2016, Edinburgh, UK, October 3–5, pp. 34–40 (2016)
12. Open Source Computer Vision Library. https://opencv.org. Last checked 1 April 2018

Performances Evaluation of the Optical Techniques Developed and Used to Map the Velocities Vectors of Radioactive Dust

Andrea Malizia and **Riccardo Rossi**

Abstract Radioactive dust mobilization is a risk that can occur in many nuclear plants and, in order to reduce the risk related to this event, it is necessary map the velocity vectors of dust during its mobilization. The authors have designed and used a chain of measurements for air pressure and velocity, temperature, and dust velocity used on the experimental facility STARDUST-Upgrade that can replicate the thermos-fluidodynamic conditions of the loss of vacuum accidents with a pressurization rate in a range of 10–1000 Pa/s and a temperature in a range of 20–140 °C. In this work, the authors present the optical experimental setups and software used to track dust velocities. These techniques are based on the particle tracking velocimetry and flow motion algorithms. Two different experimental setups are used to take into account the different optical properties of dust, each image obtained during the experiments has been analysed with customized software. Three different of algorithms are analysed and criticaly compared in this work: Lucas-Kanade, feature matching and Horn-Schunck. The authors will evaluate the performances of these optical techniques developed and used to map the velocities vectors of radioactive dust.

Keywords Velocity measurements · Flow motion · Dust hazards

1 Introduction

Mobilization of dust plays a crucial role in determining the security risk factors of the nuclear plants and particular industrial plants. The Loss Of Vacuum Accidents (LOVAs) in nuclear plants are events that can provoke the mobilization of radioactive

A. Malizia (✉)
Department of Biomedicine and Prevention, University of Rome Tor Vergata,
Via del Politecnico 1, 00133 Rome, Italy
e-mail: malizia@ing.uniroma2.it

R. Rossi
Department of Industrial Engineering, University of Rome Tor Vergata,
Via del Politecnico 1, 00133 Rome, Italy

© Springer Nature Switzerland AG 2019
B. Andò et al. (eds.), *Sensors*, Lecture Notes in Electrical Engineering 539,
https://doi.org/10.1007/978-3-030-04324-7_37

dust [1]. This dust is combustible, radioactive and toxic. Therefore, dust explosions and dispersions must be avoided to guarantee the safety of the place [2, 3]. Dust explosions is a security concern also for many industrial plants [4, 5].

These accidents are strongly influenced by the local properties of the multiphase flow. Dust concentration, size and velocities are the most important variables together with the analysis of turbulence. Therefore, it is fundamental map the evolution of dust mobilization to predict the creation of critical situations [5, 6].

In the laboratory of the University of Rome Tor Vergata, there is an experimental facility (STARDUST-Upgrade) able to replicate different LOVAs conditions [7]. The fluid-dynamics of STARDUST-Upgrade has been deeply investigated by experimental and numerical studies presented in our previous works [2, 7, 8]. The authors implemented optical techniques on STARDUST-U in order to measure dust velocities and concentrations [8]. The measurements consist of frame recording and analysis. Three different algorithms have been implemented in the analysis software:

1. The Horn-Schunck flow motion algorithm [9];
2. The Lucas-Kanade feature point tracking algorithm [10];
3. Feature matching algorithm [11].

In this work, the authors analyse and discuss the performances of these algorithm to perform measurement of dust velocities.

2 Materials and Methods

2.1 STARDUST-Upgrade

STARDUST-Upgrade is a scaled facility able to replicate Loss Of Vacuum Accidents (LOVAs). It is a stainless steel cylindrical tank, with a radius of 24.5 cm and a length of 92 cm. The vessel can reach a maximum vacuum level of 100 Pa, by means of a vacuum pump. The vacuum leakage is simulated by one of the six inlet valves, that allows air to flow inside [8]. It is possible set the pressurization rates (in a range of 10–1000 Pa/s) by setting the air flow rates in the STARDUST-U facility. The pressurizations rate is the main parameter of a LOVA severity. The pressure of the vessel is monitored by two pressure gauges, an "Edwards" and a "Pirani", which work at different pressure ranges [8]. The temperature is measured by four thermocouples J-type, distributed in different regions of the vessel. STARDUST-Upgrade is provided by quartz windows that allow to interface optical instruments inside the vessel. Now, STARDUST-Upgrade is used to study the mobilization of dust in case of LOVA. Four types of dust are analysed by the authors: flour, stainless steel, carbon dust and tungsten. These dusts are dangerous for nuclear plants (carbon, tungsten, and steel) or industrial plants (flour, carbon, and steel).

The functional scheme of STARDUST-Upgrade can be resumed in three main steps:

1. A weighted quantity of dust is placed on a tray and located inside vacuum chamber;
2. The vessel is hermetically closed to achieve the desired value of vacuum;
3. Once that the desired vacuum is reached, the vessel is isolated from the vacuum line; the iar start to flow in order to reproduce the LOVA conditions and all the sensors and instruments that make the chain of measurement start the acquisition.

All the devices (pressure gauges, thermocouples, valves, etc.) are connected by a COMPAQ DAQ system of the National Instrument. The software that controls the experiment is written in LabVIEW language.

2.2 Optical Measurement of Velocity

Measurements of dust velocity are performed by a dust tracking system. The images of dust are recorded at high frame rates (5000 frame per second-fps) during the accident replication. The acquired images are post-processed by dedicated algorithms to extract information.

Experimental setup for the optical measurements. Two different approaches are used to keep into account the optical properties of the different dusts used. The scattering method is used for the flour and stainless-steel dusts because of their larger scattering cross sections at laser wavelengths in the blue region. The shadow set-up method is used for the carbon and tungsten dusts that have low scattering cross sections.

Scattering setup. The scattering setup use the same principle of Particle Image Velocimetry (PIV) or Particle Tracking Velocimetry (PTV). A diode laser (S3 Artic Series, Wicked Lasers) working at 445 nm and a system of lenses are used to create a light sheet. The system of lenses uses a piano-convex lens and cylindrical lens. Changing the relative distance between the two lenses, the plane aperture can be changed so as the intensity of light. A CCD fast camera (CamRecord 1000, Optronis, Kehl, Germany) is placed orthogonally to the light sheet. When the light interacts with a particle, it is scattered in all direction and then also to the fast camera. The images acquired are represented by a dark background with white particles. Figure 1 shows the scattering setup implemented on STARDUST-U.

Shadow setup. The diode laser spot is converted to a cylinder of light that crosses the vessel, and then it is sent to the CCD fast camera (CamRecord 1000, Optronis, Kehl, Germany). The diode laser (S3 Artic Series, Wicked Lasers) working at 445 nm, the light cylinder, and the fast camera are on the same axis (Fig. 2). When the light interacts with a dust particle, it is absorbed and scattered. The result is an image with a white background and dark dust particles.

Image analysis algorithms. Three different algorithms have been used to track dust. Each of them uses a different approach to the problem that is more suitable

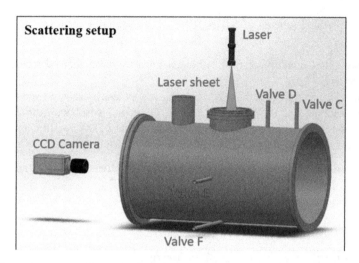

Fig. 1 Scattering setup applied to STARDUST-Upgrade

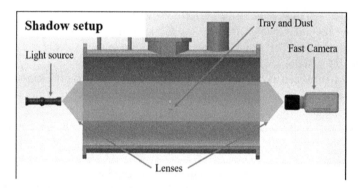

Fig. 2 Shadow setup applied to STARDUST-Upgrade

in certain conditions. These software tools are based on Feature Matching (FM) algorithm, a feature point Lucas-Kanade (LK) algorithm, and a Horn-Schunck (HS) algorithm.

Feature matching algorithm. The feature matching algorithm is based on the principle of tracking the dust particles studying its features. There are a lot of different features, such as greyscale value pyramid and geometrical features, and most of them have been tried and analysed in a previous work [6]. At first, the algorithm detects all the particles in the frame i. It studies and extracts the features of each frame and then it searches for particles with similar features in the following frame $(i+1)$. Known the initial (x_i and y_i) and final (x_f and y_f) positions of each particle, the pixel-meter conversion constant ($C_{m,p}$) and the frame rate of the fast camera (f), velocity components (u and v) can be calculated as follows:

$$\mathbf{u} = \begin{bmatrix} u \\ v \end{bmatrix} = \begin{bmatrix} (x_f - x_i)C_{m,p}f \\ (y_f - y_i)C_{m,p}f \end{bmatrix} \tag{1}$$

Feature point Lucas-Kanade [12–14]. This algorithm is based on the brightness constancy hypothesis, so it supposes that the intensity of a particle does not change from two consecutive frames. Note that this hypothesis is acceptable when large frame rates are used, since brightness changes (due to particle rotation, agglomerated disruption, etc.) are small. This hypothesis can be written as follows:

$$\frac{dI(x,y,t)}{dt} = \frac{\partial I}{\partial x}\frac{\partial x}{\partial t} + \frac{\partial I}{\partial y}\frac{\partial y}{\partial t} + \frac{\partial I}{\partial t} = 0 \rightarrow \frac{\partial I}{\partial x}u + \frac{\partial I}{\partial y}v + \frac{\partial I}{\partial t} = 0 \tag{2}$$

where $I(x, y, t)$ is the pixel intensity at the x and y position and at the time t. Since this equation is an opened problem (two unknowns and one equation), a closure technique is requested. The Lucas-Kanade approach solves the closure problem assuming that pixels in a sub-matrix (called integration window) have the same velocity, and, by a least square minimization method, the velocity equation becomes:

$$\begin{pmatrix} u \\ v \end{pmatrix} = \begin{bmatrix} \sum \frac{\partial I}{\partial x}(p_i)^2 & \sum \frac{\partial I}{\partial x}(p_i)\frac{\partial I}{\partial y}(p_i) \\ \sum \frac{\partial I}{\partial x}(p_i)\frac{\partial I}{\partial y}(p_i) & \sum \frac{\partial I}{\partial y}(p_i)^2 \end{bmatrix}^{-1} \begin{bmatrix} -\sum \frac{\partial I}{\partial x}(p_i)\frac{\partial I}{\partial t}(p_i) \\ -\sum \frac{\partial I}{\partial y}(p_i)\frac{\partial I}{\partial t}(p_i) \end{bmatrix} \tag{3}$$

where p_i indicates the i-th pixel of the integration window. The Lucas-Kanade equation is applied to each detected particle. So, velocity is calculated again by Eq. (1).

Horn-Schunck. This algorithm does not perform particle measurements, but it calculates field variables. It is based on the brightness constancy hypothesis, but it solved the closure problem differently from Lucas-Kanade. It uses an iterative approach to minimize the integral of the square of Eq. (2) over the entire image:

$$\xi^2 = \int\int \left(\frac{\partial E}{\partial x}u + \frac{\partial E}{\partial y}v + \frac{\partial E}{\partial t} \right)^2 dxdy \tag{4}$$

At this equation is usually added a smoothing function to avoid meaningless vectors. The output of this algorithm is two velocity fields, where the velocity is expressed in pixels per frame. The conversion to meters per second is achieved multiply the pixel velocity by the pixel-meter conversion constant ($C_{m,p}$) and the frame rate of the fast camera (f), analogously to Eq. (1).

3 Results and Discussion

3.1 *Fluid-Dynamics Characterisation of the Experiments*

The authors provided information about the experiments performed in this work to analyse the algorithm performances. The experiment has been performed without flow meter, so there is a free and uncontrolled intake of air. Detailed analysis of this accident can be found in a recent work in the literature [10].

Figure 3 shows the mean pressure and temperature with their error bars. These values are obtained as the average of 20 measurements. Note that, excluding the first 0.02 s of the experiment, where the pressure goes from 100 to 3000 Pa, the pressurisation rate is constant at about 3000 Pa/s. That value is constant even without the flow meter because of the mass flow chocking, due to the sonic condition reached inside the nozzle. The dispersion of the pressure values is not high (the maximum relative uncertainty is 4%). The temperature is more dispersed since the temperature boundary condition is not controlled in STARDUST-Upgrade in these experiments (relative uncertainty around 7%).

3.2 *Algorithm Performance Analysis*

The analysis of dust mobilization properties is fundamental to understand which software works better. Dust mobilization is a complex phenomenon, where the dust

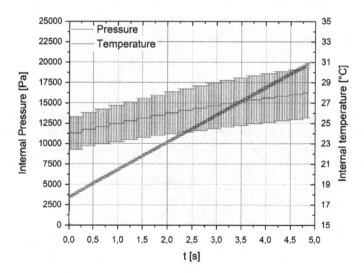

Fig. 3 Mean pressure and temperature versus time during the first 5 s of accident replication

Fig. 4 A frame of mobilized carbon dust with concentration region highlighted

Fig. 5 Mean velocity field of carbon dust in case of loss of vacuum accidents inside STARDUST-Upgrade

is deposited on a wall and, when it mobilize, there is a non-homogeneous concentration distribution. So, in case of dust mobilization, there are regions where the dust concentration is very large (deposited dust), a concentration gradient region and a very low concentration area (see Fig. 4).

The distribution of dust concentration is usually correlated to the velocity field. In fact, at high concentration values, the velocity of dust is very small because of two reasons:

1. High concentration occurs near the walls, where the dust is deposited, and the velocity is zero;
2. A large concentration implies an increase of the particle-particle interactions, which can be seen as an increase in the viscosity of the multiphase flow.

Figure 5 shows the averaged velocity field of a carbon dust during an experiment. In the image can be observed how the small velocities (in blue) are in the high concentration region, the medium velocities (green) in the medium concentration area, and the higher velocities (red) in the highest section of the image, where the concentration tends to zero. Note that only few particles reach very higher velocities.

Uncertainty analysis has been performed to understand which algorithm works better in each area. The relative error has been calculated as the uncertainty of the measurement divided by the velocity magnitude. The uncertainty has been calculated performing the image computation several times and adding at each computation random noise.

The results of uncertainty analysis are shown in Fig. 6. The uncertainty of feature matching is always small (<2%) when a large matching score is required. In this analysis, the matching score is 95%. Lucas-Kanade algorithm has a medium uncertainty that goes from 12 to 16%, so it seems to be not much influenced by the concentration value. Horn-Schunck uncertainty decreases when concentration increases, and

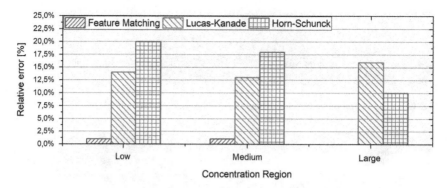

Fig. 6 Error analysis of each algorithm versus the concentration region

Table 1 Number of detected and matched per frame by LK and FM algorithms

Concentrations value	Particle detected per frame [a.u.]	Particle matched per frame (%)		Particle tracked [a.u.]	
	FM and LK	FM (%)	LK (%)	FM	LK
Very Low	2	90.45	100	0.95	1.81
Low	5	9.73	100	0.49	5
Medium	115	1.00	100	1.15	115
High	20	0.00	100	0.00	20

it reaches the best performances in the high concentration area. Furthermore, the number of detected particles plays an important role. In fact, the uncertainty of LK and FM algorithms are calculated on the detected and tracked particles.

Table 1 shows the number of detected and matched particles per frame by each algorithm (LK or FM). The number of detected particles in the low concentration region is very low, according to the concentration value. At medium concentration, the number of detected particles strongly arises (115 particles detected per frame). At high concentration, the number of matched particles falls because they are indistinguishable. The number of matched particles strongly differs between LK and FM. The Lucas-Kanade algorithm tracks all the particles, since there is not a matching score to reach. In the other hand, many of the tracked particles may be wrong, as showed in the error analysis in Fig. 6. The percentage of matched particles of FM is very large at very low concentration, then it drops to zero for high concentration areas.

3.3 Result Discussion

It is not possible have high resolution images of dust particles when dust concentration is very large, but there is a dust cloud. In this case, particles are not detected and then both feature matching and Lucas-Kanade algorithms fail but the Horn-Schunck algorithm has the best performances in this case because of two reasons:

1. It does not follow the particles, but the intensity of each pixel;
2. Smaller velocities values are expected at large concentration, since the particle-particle interactions increase the friction of the disperse phase and Horn-Schunck has good performances with very small velocities.

In case of dust velocities increase (at medium and low concentrations), particles became distinguishable. In this case, the Horn-Schunck algorithm can works only if the velocity is enough small, otherwise it underestimates the measurements. The Lucas-Kanade and feature matching algorithms work since the particles are detected. However, in this case, the Lucas-Kanade is the best option because it is able to track both small and large velocities (pyramidal implementation has been used [10]) and the presence of other particles (that can provoke overlays) does not affect in significant ways the measurements, because the algorithm works on the intensity tracking. Contrariwise, the feature matching algorithm is strongly influenced by the velocity and the relative error usually decrease with the increase of the tracked objects velocities. Then, at these concentrations, Lucas-Kanade is preferable. Moreover, the presence of overlapped particles usually implies a fail of the matching test, since their movements change the features used by the algorithms.

At very low concentration, the velocity of particles may reach very large values, unacceptable for Lucas-Kanade and Horn-Schunck algorithm. Furthermore, the low density of particles makes impossible to perform statistical analysis of the velocities and then high accuracy and certainty of measured values is required. At these concentrations, the best method is the feature matching, because it allows to set the desired matching score. When a large matching score is used (>90–95%), if the matching test succeeds, the accuracy of that computation is very large, and the probability to have a wrong matching is negligible.

The results clearly show that using only one algorithm to accurately measure dust velocity is practically impossible, since each of them works better in different area of the mobilization phenomenon. Therefore, the main idea is to develop a software able to switch the algorithm used in function of the local particle concentration, in order to give an accurate measurement of all the phenomenon.

4 Conclusions

Dust plays a crucial role in several accidents where radioactive, combustible and toxic particles are produced, especially in the nuclear and industrial fields. STARDUST-Upgrade, a facility developed at the laboratory of the University of Rome Tor Vergata, is used to replicate LOVAs. The main scope of the actual and future experimental campaigns of STARDUST-U is to improve the optical diagnostics for dust detection, monitoring, and tracking.

Two experimental setups have been implemented on STARDUST-U: the scattering setup and the shadow setup, the first works with "bright" dusts, while the second with "dark" dusts. The output of each setup is a sequence of images representing the particles of dust at different instants, which must be analysed to extract information about the concentration and the velocities of dust. The classical approaches used for Particle Image Velocimetry or Particle Tracking Velocimetry have led to unacceptable results, since they are optimised to work in specific condition of particle density, size and concentration.

In this work, the authors analysed three different algorithms to track dust velocity: feature matching, Lucas-Kanade, and Horn-Schunck algorithms. Analysis of uncertainty and number of tracked particle have been performed and they showed how each algorithm works better in specific mobilization regions. In fact, the image representing the mobilization phenomenon can be divided into three different areas:

1. Low concentration and large velocity region;
2. Medium concentration and velocity region;
3. High concentration and small velocity region.

The analysis performed led to the following conclusions:

1. In low concentration regions, the Horn-Schunck algorithm can not be used because of the large velocities. The Lucas-Kanade may be used, but the small number of particle per frame detected in this region makes poor the statistical analysis and so it is hard to exclude wrong vectors. The best option it is the feature matching, since the percentage of tracked particles is very high, and the matching score guarantees a good accuracy of the match.
2. At medium concentration, the best option is Lucas-Kanade, since there is a large number of detected particles. So, statistical analysis is possible while feature matching is usually confused by overlapped particles. Velocities are still large for Horn-Schunck.
3. In large concentration areas, Horn-Schunck is the only working algorithm, since the number of detected particles is too small.

Since there is not an algorithm able to work well in each mobilization area, the next step is to develop a software which switch each algorithm in function of the local concentration in the image.

References

1. Honda, T., Bartels, H.-W., Merrill, B., Inabe, T., Petti, D., Moore, R., Okazaki, T.: Analyses of loss of vacuum accident (LOVA) in ITER. Fusion Eng. Des. **45**(4), 361–375 (2000)
2. Malizia, A., Poggi, L.A., Ciparisse, J.-F., Rossi, R., Bellecci, C., Gaudio, P.: A review of dangerous dust in fusion reactors: from its creation to its resuspension in case of LOCA and LOVA. Energies **9**(8) (2016)
3. Denkevits, A., Dorofeev, S.: Dust explosion hazard in ITER: explosion indices of fine graphite and tungsten dusts and their mixtures. Fusion Eng. Des. **75–79**, 1135–1139 (2005)
4. Eckhoff, R.K.: Dust Explosions in the Process Industries. Gulf Professional Publishing, Burlington (2003)
5. Amyotte, P.R., Eckhoff, R.K.: Dust explosion causation, prevention and mitigation: an overview. J. Chem. Health Saf. **17**(1), 15–28 (2010)
6. Eckhoff, R.: Understanding dust explosions. The role of powder science and technology. J. Loss Prevent. Process Ind. **22**(1), 105–116 (2009)
7. Poggi, L.A., Gaudio, P., Rossi, R., Ciparisse, J.-F., Malizia, A.: Non-invasive assessment of dust concentration and relative dustiness in a dust cloud mobilized by a controlled air inlet inside STARDUST-U facility. Reliab. Eng. Syst. Saf. **167**, 527–535 (2017)
8. Poggi, L., Malizia, A., Ciparisse, J., Gelfusa, M., Murari, A., Pierdiluca, S., Re, E.L., Gaudio, P.: Experimental campaign to test the capability of STARDUST-Upgrade diagnostics to investigate LOVA and LOCA conditions. In: 42nd EPS Conference on Plasma Physics, Lisbon, Portugal (2016)
9. Rossi, R., Malizia, A., Poggi, L.A., Peluso, E., Gaudio, P.: Flow Motion and Dust Tracking Software for PIV and Dust PTV. J. Fail. Anal. Prev. **2016**, 1–12 (2016)
10. Schunck, B.G., Horn, B.K.P.: Determining optical flow. Artif. Intell. 185–203 (1980)
11. Lucas, B.D., Kanade, T.: An iterative image registration technique with an application to stereo vision. In: Proceedings of Imaging Understanding Workshop, pp. 121–130 (1981)
12. Shuprajhaa, T., Subasree, S., Vaitheeshwari, M., Sivakumar, S.: A review on image processing techniques using pattern matching in LabVIEW. Int. J. Adv. Eng. Res. Appl. **1**(11), 441–445 (2016)
13. Rossi, R., Gaudio, P., Ciparisse, J.-F., Poggi, L.A., Malizia, A.: Imaging of dust re-suspension in case of LOVA. Fusion Eng. Des. **126**, 156–169 (2018)
14. Bouguet, J.: Pyramidal implementation of the Lucas Kanade feature tracker description of the algorithm. Intel Corporation—Microprocessor Research Labs

Part V
Printed and Flexible Sensors

Low Cost Inkjet Printed Sensors: From Physical to Chemical Sensors

Bruno Andò, Salvatore Baglio, V. Marletta, R. Crispino, S. Castorina, A. Pistorio, Giovanna Di Pasquale and Antonino Pollicino

Abstract Compared to traditional silicon electronics, printed sensors are cheap and suitable for many low cost and disposable devices. Main printing techniques used are screen printing and inkjet printing. In particular we focus on inkjet printing for the rapid prototyping of sensors. Inkjet is a direct, contactless, printing process, with high spatial resolution and compatibility with many substrates. Successful examples of sensors developed by low cost inkjet printers and metal-based inks are reported by authors. In this paper two examples of low cost inkjet printed sensors are given. The first device is an accelerometer aimed to address typical applications in the field of human and seismic monitoring. Main outcomes of the proposed solution are the low frequency operation and the high sensitivity. The realization of a CO_2 gas sensor is also presented. The device makes use of a PEDOT/PSS and Graphene stack and exploits resistive readout.

Keywords Direct printing · Printing techniques · Inkjet printing · Transducers
Low cost · Rapid prototyping · Sensors · Strain measurement · Gas sensors
PEDOT/PSS · Graphene · Interdigitated electrodes · PET · Accelerometer
Seismic monitoring

1 Introduction

Inkjet printing has been gaining considerable interest in recent years as an innovative methodology for the production of low-cost electronic devices [1], in particular sensors [2–6]. This is mainly due to a series of potential advantages offered by this technique, such as rapidity of execution, low costs, low processing temperatures, ease of use and low-cost equipment. Unlike other printing techniques, such as screen

B. Andò (✉) · S. Baglio · V. Marletta · R. Crispino · S. Castorina · A. Pistorio
DIEEI University of Catania, v.le A. Doria, 6, 95127 Catania, Italy
e-mail: bruno.ando@dieei.unict.it

G. Di Pasquale · A. Pollicino
DII University of Catania, v.le A. Doria, 6, 95127 Catania, Italy

© Springer Nature Switzerland AG 2019
B. Andò et al. (eds.), *Sensors*, Lecture Notes in Electrical Engineering 539,
https://doi.org/10.1007/978-3-030-04324-7_38

printing, inkjet printing can also be performed on flexible, or pressure sensitive substrates, as it is a non-contact process. Moreover, inkjet printing does not require the pre-fabrication of a mask. Instead, the pattern is designed using CAD software and sent directly to the printer. Furthermore, due to the nature of the deposition technique, inkjet printing allows for greater control over the pattern. Inkjet printing technology exhibits direct printing feature, high spatial resolution and compatibility with many substrates.

Compared to traditional silicon electronics, printed devices are really cheap and can addresses a wide set of applications requiring low cost and disposable devices [1].

The printing process requires three main components: the inkjet printer, the ink and the substrate.

With reference to the first component, the inkjet printer, in order to implement cheap inkjet printing facility, common office printers can be modified to be used with custom ink. On the other hand, when greater precision and control are required, sophisticated material printers are commercially available that allow for precise adjustment of ink droplet volume, ejection speed, and spacing, allowing for incredible control over the printing process, and are widely used for research purposes.

Commercial printers for inkjet process are typically based on piezoelectric heads which allow for the deposition of functional and structural materials with resolutions in the order of the tens of micrometers [7, 8].

As regards inks, they clearly represents the most important element of the printing process, since the choice of a given material impacts on both the printing process and the resulting device characteristics. In particular conductive inks are needed for electrode fabrication. Common used conductive materials include metallic nanoparticles and nanowires, carbon based materials such as carbon nanotubes and graphene. Several conductive inks alternatives are available on the market, including nanoparticle (NP)-based, organometallic and conductive polymer inks. The choice of the most suitable material depends on its characteristics and the requirements of the specific application.

NP inks commonly consist of a suspension of gold or silver particles, in water or in an organic solvent. The presence of water in the suspension requires the addition of an ionic surfactant to disperse the conductive materials and a heat treatment, known as sintering, to be carried out after the printing process. On the other hand, water based inks are safer and easier to handle compared to solvent-based inks, which can be corrosive and potentially harmful, but exhibits faster dry times with respect to water-based ones. NP inks are characterized by high chemical stability and high electrical conductivity [9].

Organometallic inks are solutions of some kind of metallic compounds dissolved in organic solvents. Since they are in form of a solutions rather than particle suspension, the risk of clogging of the printing head nozzles is eliminated. Moreover, higher conductivity values can be achieved and lower sintering temperatures are required compared to NP inks. Silver-based organometallic inks are quite common, but there is an increasing interest in graphene-based inks [9].

Among materials compatible with inkjet printing technology, polymers like PEDOT-PSS (3, 4-ethylenedioxythiophene) and PANI (Polyaniline) are widely used [6].

The combination of conducting and functional polymers has been proposed as a convenient solution to realize inkjet printed sensors [2]. Metal inks in combination with low cost printers have been successfully used for the rapid prototyping of sensors and conductive patterns [3, 8, 10].

Depending on the printing equipment adopted, inkjet printing can be performed on rigid or flexible substrates. Low-cost office type inkjet printers can handle only flexible materials, while high-end printers can accept rigid substrates too. The latter includes materials like plastic, glass, ceramics and silicon, while commonly used flexible substrates include several type of polymers, such as polyether imide, polycarbonate, polyarylate, polyamide, polyethylene and terephthalate and even paper [9, 11, 12].

In the next section of this paper an analysis of the state of the art in inkjet printed devices, with main focus on sensors, is presented. Then the research activity on inkjet printed sensor of SensorLab@DIEEI, University of Catania, is illustrated by means of two application examples: a fully printed CO_2 gas sensor and a printed accelerometer with resistive readout strategy.

2 State of the Art

The analysis of the state of the art on inkjet printed device presented here is focused on sensors and sensing applications. A wide research activity on inkjet printed chemical and electrochemical sensors exploiting different set of materials and sensing principles is documented in literature.

In [13] a commercial printer and silver- and gold-based inks have been employed to fabricate the electrodes of an electrochemical sensor on paper substrate. Different types of electrode modifications were carried out and the operation of such modified devices as low cost PH sensor and glucose biosensor have been demonstrated.

A low cost, disposable, fully inkjet printed electrochemical sensor on paper which consists of multiple layers of carbon nanotube-printed electrodes is presented in [14]. The operation of the device has been demonstrated in determining the concentrations of analytes with potential step voltammetry method.

Another inkjet printed electrochemical sensor for the measurement of hydrogen peroxide concentration, which is based on carbon nanotube ink and Ag nanoparticles on paper substrate is given in [15].

Carbon based materials, such as graphene, exhibit measurable changes in their electrical conductivity due to absorption of given analytes through their surface. Such effect has been exploited to realize inkjet printed gas sensors based on functionalized carbon-based electrodes (chemiresistive sensing), as reported in [16–18].

Another example of chemiresistive, inkjet printed gas sensor, which makes use of SnO_2 as gas sensitive material, has been demonstrated by Rieu et al. [19].

Fig. 1 **a** Layout and **b** a real
view of one strain sensor
developed (reprinted from
[3] with permission of IEEE)

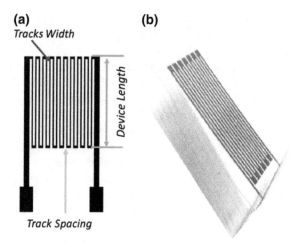

The ability of inkjet printing techniques to accurately place very small volumes of liquids onto a substrate, makes it an ideal technology that could be used within the medical sensor field. In [20] Wang et al. have reported on the fabrication of a glucose biosensor by inkjet printing Glucose oxidase solution on screen printed carbon black electrodes.

In [21] is described a flexible and lightweight chemiresistor made of a thin film, inkjet printed, reduced graphene oxide, which can selectively detect chemically aggressive vapors such as NO_2, Cl_2, etc.

A further example of inkjet printed carbon-based material, in particular single walled carbon nanotubes, sensor is given in [22], where a pH sensor has been demonstrated.

Inkjet printing of conductive inks has also been exploited in order to demonstrate the feasibility of radio frequency (RF) operated passive devices, particularly planar antennas which finds potential application in the field of radio frequency identification (RFID), as reported in [23–25]. An inkjet printed humidity sensor for passive RFID system has also been demonstrated in [26].

Another family of sensors that have received the attention of the scientific community is that of strain and stress sensors. In particular, authors have successfully demonstrated the feasibility of low cost strain sensors through inkjet printing water-based silver ink, by making use of a low cost, commercial, piezoelectric inkjet printer to realize conductive patterns on flexible PET substrates. The fabrication process and the experimental characterization of the devices are reported in [3, 4]. Strain and stress sensors can be printed in several different patterns, including single devices and uni- or bi-dimensional arrays.

In Fig. 1 a typical layout and a real view of strain sensors developed are shown. Printing resolution has been estimated as 200 μm for both track width and spacing, while an average track thickness of 1.90 μm has been evaluated through SEM analyses.

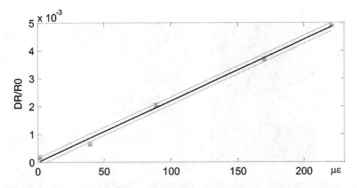

Fig. 2 Response of the low-cost inkjet-printed strain sensors developed at the SensorLab@DIEEI of the University of Catania, Italy (reprinted from [4] under CC BY 4.0)

The sensor response in terms of the relative variation of the sensor resistance as a function of the imposed strain is reported in Fig. 2.

Experiments conducted in order to characterize the sensors behavior have shown considerably large gauge factor values (from 15 to 21, depending on the sensor dimensions), resolution values ranging from 1.2 to 1.9 $\mu\varepsilon$ and repeatability between 2.67×10^{-3} % and 3.04×10^{-3} %, which are coherent with the expected trend.

The next section of this paper will focus on two specific applications of inkjet printing techniques to the fabrication of sensors, which have been developed by the authors at SensorLab@DIEEI, University of Catania: a CO_2 gas sensor and an accelerometer.

3 Inkjet Printed Sensors Application Examples

3.1 CO_2 Gas Sensors

In [6] we presented a CO_2 gas sensor realized by ink-jet printing an interdigitated electrodes pattern with silver-nanoparticle conductive solution through a commercial printer, followed by a double layer of PEDOT/PSS conductive polymer and Graphene onto a PET substrate. The sensor exploits the change in electrical conductivity of graphene due to gas molecules adsorption (chemiresistive sensing).

The development of low cost sensors for Carbon dioxide (CO_2) is becoming mandatory for applications in several fields from greenhouse and agricultural to exhalations monitoring also for food and air quality monitoring, and all those application fields where the use of disposable devices is highly recommended.

PEDOT/PSS is one of the most promising conductive polymer, thanks to its water dispersibility, good conductivity, high transparency in the visible range, excellent solution processability and mechanical flexibility [27, 28]. It can serve as host and

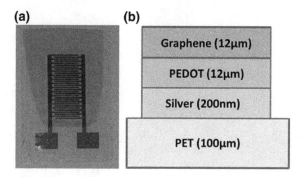

Fig. 3 **a** The real view of the sensor. Device dimensions: 1.8 cm by 1 cm with a thickness of about 140 μm. IDT fingers: width (200 μm), length (6 mm), thickness (200 nm), track spacing (350 μm); **b** Schematization of the sensor structure. (Reprinted from [6] with permission of Elsevier)

guest materials in hybrid systems, in particular with carbon-based materials, as carbon nanotubes and graphene [29].

Graphene also exhibits unique electrical and optical properties that make it an attractive candidate for use as a chemiresistor for chemical and biological detection [21, 30, 31]. Examples of graphene based CO_2 gas sensors are available in the literature [32, 33].

Here we focus on a CO_2 gas sensor device which exploits a stack structure made by a sensitive layer of PEDOT/graphene deposited on inkjet printed electrodes, as described in [6]. The device consists of a PET substrate, where an IDT structure has been realized by a low cost inkjet printer and a silver nanoparticle based ink (Metalon® JS-015 by Novacentrix). The PET substrate is the Novele™ IJ-220 Printed Electronics Substrate for low-cost and low-temperature applications.

Onto the silver electrode, a double active layer has been deposited. As first a layer of PEDOT/PSS, commercially available as an aqueous dispersion, CLEVIOS™ PHCV4 by H. C. Starck, has been deposited by a calibrated spreader (12 μm). The PEDOT/PSS ink has been prepared diluting the starting solution with distilled water (1:1, v/v). After heating at 80 °C for 50 min, the remaining dark blue PEDOT/PSS film is conductive, transparent and durable. Successively, a layer (12 μm) of a solution of pristine graphene powder (N002-PDR purchased from Angstron Materials, USA), concentration 0.1 mg/mL, containing Triton-X100 non-ionic surfactant (0.4 wt%), has been deposited over the PEDOT/PSS coat by the a calibrated spreader. Before use, it has been sonicated to obtain a fine dispersion. After deposition, the structure has been annealed at 80 °C for 50 min.

The main role of the PEDOT layer is to optimize the nominal resistance of the graphene layer in order to improve the sensor response. The sensitive layer of graphene acts also as a shield for the PEDOT layer against the effect of exogenous quantities, such as the humidity.

A real view of the sensor developed is shown in Fig. 3a, while the stack structure of the device is given in Fig. 3b.

The main outcome of the proposed solution is related to its low cost, conferred by the technologies adopted to realize both the electrodes and the functional layer. Such technology allows for the rapid prototyping of cheap devices on flexible substrates which can be properly designed with different shapes.

The working principle of the sensor exploits the changes on the electrical conductance of the graphene layer due to the adsorption of gas molecules on its sensing surface. The resistive output of the sensor is then conditioned by a traditional bridge configuration, followed by gain and demodulation blocks.

Since the printed sensor operation requires consecutive heating cycles, a heater was placed under the substrate and a Pt100 sensor was used to measure the substrate temperature. Each heating cycle allows for heating the device up to 60 °C which is compatible with the PET substrate adopted.

The experimental set-up developed for the sake of the device characterization consists of a CO_2 tank and an insulated chamber with a controlled gas injection system, equipped with flowmeters and electrically controlled valves, and managed by a LabView based tool. The injection protocol consists of three steps; as first the valve of the CO_2 tank is open and the gas flows outside the chamber until the gas flux reaches a steady state regime; in the second step the CO_2 flux is switched into the chamber to obtain the desired CO_2 concentration. The flowmeter PFM710 by SMC is used for monitoring the CO_2 flux injection; in the third step all valves are closed and the flux is stopped.

A reference sensor, with an operating range of (400–4200) ppm, was used to perform an independent measurement of the CO_2 concentration inside the chamber.

A data acquisition system is used to acquire signals provided by the reference sensors and the printed sensor under test. The sampling frequency has been fixed to 40 kHz. The data acquisition system is managed by the same LabVIEW environment implementing the flow control.

Before each experiment, clean air is forced inside the chamber in order to reduce the CO_2 concentration to typical environmental values (around 400 ppm). After ten heating cycles at room gas concentration, a controlled quantity of CO_2 is forced inside the chamber, followed by 10 heating cycles.

The gas sensor response in terms of the relative variation of the output resistance as respect to its value in clean air at 20 °C, dR/R, for increasing values of the CO_2 concentration has been characterized for different values of the operating temperature. The calibration diagrams for two operating temperatures (50 and 60 °C) are shown in Fig. 4. The uncertainty band has been estimated in the 2-σ level. Linear models, fitting the experimental data, for the two operating temperature have been taken into account.

The device responsivity, in terms of the variation of the output resistance for a variation of the gas concentration, in case of the two operating temperatures 50 °C and 60 °C, is 4.0 $\mu\Omega/\Omega$/ppm and 4.7 $\mu\Omega/\Omega$/ppm, respectively. The corresponding values of the resolution are 400 and 420 ppm. The resolution has been estimated by considering the standard deviation of the output resistance, observed in case of clean air, divided by the device responsivity. Of course this term represents the theoretical

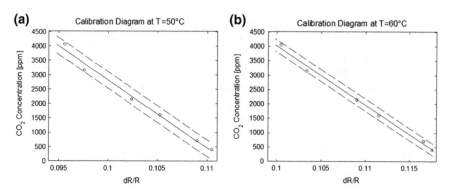

Fig. 4 The calibration diagram for two operating temperatures: **a** 50 °C and **b** 60 °C. The uncertainty bands are represented by the dotted lines

limit of the sensor, which is mainly constrained by the experimental set-up and the conditioning electronics.

The experimental results obtained demonstrate the functionality of the proposed sensor and encourage the development of CO_2 sensors exploiting the suggested sensing strategy.

Although the sensor has not been investigated to reveal cross-sensitivity to other species, it must be observed that the basic idea of the proposed sensing methodology is the development of a cheap and disposable device to be used for the Early Warning of gas presence and for specific applications in controlled environments. Anyway, future efforts will be dedicated to investigate the device cross-sensitivity and to eventually implement further technology steps to reduce such effect.

Moreover, as demonstrated by the experimental results the device is able to work at low temperature (in the order of 50 °C), which is a novelty as respect to few examples of printed gas sensors already available in the literature. Actually, although room temperature CO_2 sensors have been already proposed [32], to our knowledge the sensor developed is the first example of a low cost printed sensor which can be operated at low temperatures.

3.2 Accelerometer

Inkjet printing technique has been used in [5] to demonstrate the feasibility of a low cost printed accelerometer. The device consists of PET membrane clamped to a fixed support by means of four crab-leg springs. Membrane suspension has been achieved by cutting away portions of the PET substrate used for printing. Four strain gauges, one for each crab-leg spring, connected in series, have been patterned on the PET substrate by means of a low cost commercial inkjet printer and a silver nanoparticles based conductive inks. Thus the four strain gauges allows implementing a resistive

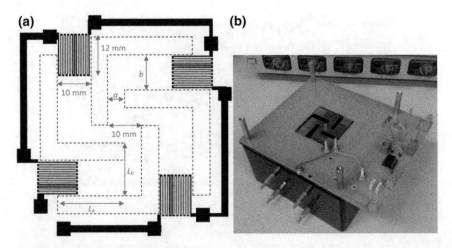

Fig. 5 **a** Layout of the developed inkjet printed accelerometer. The dashed lines represent the empty areas. **b** Real view of the assembled final prototype of the inkjet printed accelerometer developed (reprinted from [5] with permission of IEEE)

readout strategy for the device. The accelerometer layout has been designed following a rigorous approach, with given specifications in terms of frequency and device responsivity values, fitting the needs of human and seismic monitoring application fields.

The device layout is shown in Fig. 5a, together with the dimensions of each part. The PET substrate used is 140 μm thick. The dashed lines represent the empty areas. While Fig. 5b shows a real view of the assembled final device prototype. As it can be noticed in Fig. 5b, a proof-mass has been attached to the center of the membrane in order to fit the frequency response and the responsivity of the device to the application need. A 0.550×10^{-3} kg mass has been used. Moreover, Fig. 5b shows also that the printed device has been attached to a conventional PCB substrate carrying the signal conditional electronics.

The experimental setup used to characterize the sensor response is shown in Fig. 6.

It consists of a mechanical vibration exciter with a moveable platform, driven by means of a dedicated power amplifier, both produced by TYRA GmbH. Test signal for the vibration exciter have been generated by a function generator. As a reference accelerometer, a 333B40 Modal array, ceramic shear ICP®, has been used. Output signals have been acquired by means of a digital oscilloscope.

Experiments on the device impulse response have shown an observed natural frequency of 37 Hz. The frequency response of the accelerometer is reported in Fig. 7 and shows a flat-band behavior up to 20 Hz, in line with the application requirements.

The printed accelerometer responsivity is 9.4 mV/g at 10 Hz and 41 mV/g at 35 Hz. The resolution in the frequency band of interest, 20 Hz, is compatible with the applications addressed by the developed device, 0.126 g at 10 Hz.

306   B. Andò et al.

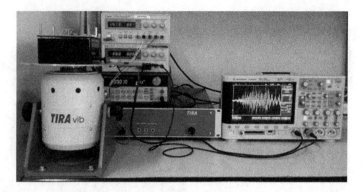

Fig. 6 Experimental setup adopted to characterize the behavior of the inkjet printed accelerometer (reprinted from [5] with permission of IEEE)

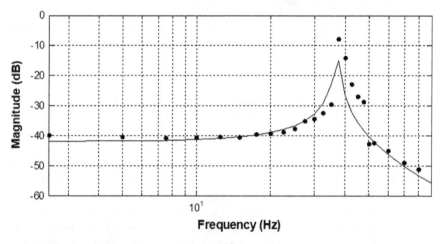

Fig. 7 Frequency response of the device: experimental (black dots) and predicted (solid line). Reprinted from [5] with permission of IEEE

Specifications obtained during the device characterization are in line with devices realized by more complex and expensive printed technology [34].

4 Conclusions

Whenever short production time, low cost, material saving, disposable devices, as well as high spatial resolution and good reproducibility become essential application requirements, inkjet printing represents the best solution for the rapid prototyping of electronic devices, especially sensors.

In this paper we provided a review a inkjet printing techniques, providing a survey of available conductive ink types, substrate materials and printing equipment.

The state of the art of the technology has been analysed with particular focus on the fabrication of sensors. Several examples of sensor devices, both fully inkjet printed and hybrid fabrication processes, have been reported here.

Finally, the state of progress of the research activity of the SensorLab@DIEEI, at the University of Catania, was presented here by means of two concrete application examples: a CO_2 gas sensor and an accelerometer realized by means of jet printing of ink of conductive elements on a flexible PET substrate. For the CO_2 sensor, inks based on silver nanoparticles and PEDOT/PSS conductive polymers were used. The sensitive element was made with an ink based on graphene. The device exploits the graphene conductivity variation due to the absorption of gas molecules.

The accelerometer was made in the form of a PET membrane suspended through four springs on each of which was printed a strain sensor with ink based on silver nanoparticles. The device exploits the variation of resistance induced by the deformation of the suspension elements due to accelerations.

Both devices have been characterized from the experimental point of view and have shown characteristics and performances in line with the current state of the art or with devices manufactured with more complex and expensive technologies.

These results encourage further research efforts aimed at a more in-depth knowledge of materials and processes and the development of new sensors and devices with inkjet printing.

References

1. Mäntysalo, M., et al.: Capability of inkjet technology in electronics manufacturing. In: Proceedings of the IEEE 59th Electronic Components and Technology Conference, San Diego, CA, USA, pp. 1330–1336 (2009)
2. Andò, B., Baglio, S.: Inkjet-printed sensors: A useful approach for low cost, rapid prototyping. IEEE Instrum. Meas. Mag. **14**(5), 36–40 (2011)
3. Ando, B., et al.: All-inkjet printed strain sensors. IEEE Sens. J. **13**(12), 4874–4879 (2013)
4. Andò, B., et al.: Low-cost inkjet printing technology for the rapid prototyping of transducers. Sensors **17**, 748 (2017)
5. Andò, B., et al.: A Low-Cost Accelerometer Developed by Inkjet Printing Technology. IEEE Trans. Instrum. Meas. **65**(5), 1242–1248 (2016)
6. Andò, B., et al.: An inkjet printed CO_2 gas sensor. Procedia Eng. **120**, 628–631 (2015)
7. FUJIFILM Dimatix, Inc. http://www.dimatix.com
8. microdrop Technologies GmbH. http://www.microdrop.de
9. Al-Halhouli, A., et al.: Inkjet printing for the fabrication of flexible/stretchable wearable electronic devices and sensors. Sens. Rev. **38**(4), 438–452 (2018)
10. Andò, B., et al.: A nonlinear energy harvester by direct printing technology. Procedia Eng. **47**, 933–936 (2012)
11. Yang, L., et al.: Integration of sensors and inkjet-printed RFID tags on paper-based substrates for UHF "cognitive intelligence" applications. In: 2007 IEEE Antennas and Propagation Society International Symposium, Honolulu, HI, pp. 1193–1196 (2007)
12. Kim, S., et al.: Inkjet-printed RF energy harvesting and wireless power transmission devices on paper substrate. In: 2013 European Microwave Conference, Nuremberg, pp. 983–986 (2013)

13. Määttänen, A., et al.: A low-cost paper-based inkjet-printed platform for electrochemical analyses. Sens. Actuators B Chem. **177**, 153–162 (2013)
14. da Costa, T.H., et al.: A paper-based electrochemical sensor using inkjet-printed carbon nanotube electrodes. ECS J. Solid State Sci. Technol. **4**(10), S3044–S3047 (2015)
15. Shamkhalichenar, H., et al.: An inkjet-printed non-enzymatic hydrogen peroxide sensor on paper. J. Electrochem. Soc. **164**(5), B3101–B3106 (2017)
16. Mirica, K.A., et al.: Rapid prototyping of carbon-based chemiresistive gas sensors on paper. Proc. Natl. Acad. Sci. **110**(35), E3265–E3270 (2013)
17. Arena, A., et al.: Flexible ethanol sensors on glossy paper substrates operating at room temperature. Sens. Actuators B Chem. **145**(1), 488–494 (2010)
18. Steffens, C., et al.: Low-cost sensors developed on paper by line patterning with graphite and polyaniline coating with supercritical CO_2. Synth. Met. **159**(21–22), 2329–2332
19. Rieu, M., et al.: Inkjet printed SnO_2 gas sensor on plastic substrate. Procedia Eng. **120**, 75–78 (2015)
20. Wang, T., et al.: Fabrication of a glucose biosensor by piezoelectric inkjet printing. In: Third International Conference on Sensor Technologies and Applications, Athens, Glyfada, pp. 82–85 (2009)
21. Dua, V., et al.: All-organic vapor sensor using inkjet-printed reduced graphene oxide. Angew. Chemie Int. Ed. **49**, 2154 – 2157
22. Qin, Y., et al.: Inkjet-printed bifunctional carbon nanotubes for pH sensing. Mater. Lett. **176**, 68–70
23. Shao, B., et al.: Process-dependence of inkjet printed folded dipole antenna for 2.45 GHz RFID tags. In: 3rd European Conference on Antennas and Propagation, Berlin, pp. 2336–2339 (2009)
24. Zheng, L., et al.: Design and implementation of a fully reconfigurable chipless RFID tag using inkjet printing technology. In: 2008 IEEE International Symposium on Circuits and Systems, Seattle, WA, pp. 1524–1527 (2008)
25. Amin, Y., et al.: Inkjet printed paper based quadrate bowtie antennas For UHF RFID tags. In: 2009 11th International Conference on Advanced Communication Technology, Phoenix Park, pp. 109–112 (2009)
26. Virtanen, J., et al.: Inkjet-printed humidity sensor for passive UHF RFID systems. IEEE Trans. Instrum. Meas. **60**(8), 2768–2777 (2011)
27. Jonas, F., Heywang, G.: Technical applications for conductive polymers. Electrochim. Acta **39**(8–9), 1345–1347 (1994)
28. Groenendaal, L., et al.: Poly(3,4-ethylenedioxythiophene) and its derivatives: past, present, and future. Adv. Mater. **12**(7), 481–494 (2000)
29. Kim, G.H., et al.: Thermoelectric properties of nanocomposite thin films prepared with poly(3,4-ethylenedioxythiophene) poly(styrenesulfonate) and graphene. Phys. Chem. Chem. Phys. **14**(10), 3530–3536 (2012)
30. Hill, E.W., et al.: Graphene sensors. IEEE Sens. J. **11**(12), 3161–3170 (2011)
31. Fowler, J.D., et al.: Practical chemical sensors from chemically derived graphene. ACS Nano **3**(2), 301–306 (2009)
32. Yoon, H.J., et al.: Carbon dioxide gas sensor using a graphene sheet. Sens. Actuators B Chem. **157**(1), 310–313 (2011)
33. Muhammad Hafiz, S., et al.: A practical carbon dioxide gas sensor using room-temperature hydrogen plasma reduced graphene oxide. Sens. Actuators B Chem. **193**, 692–700 (2014)
34. Qiao, D., et al.: A single-axis low-cost accelerometer fabricated using printed-circuit-board techniques. IEEE Electron Device Lett. **30**(12), 1293–1295 (2009)

DNA-Based Biosensor on Flexible Nylon Substrate by Dip-Pen Lithography for Topoisomerase Detection

V. Ferrara, A. Ottaviani, F. Cavaleri, G. Arrabito, P. Cancemi, Y.-P. Ho, B. R. Knudsen, M. S. Hede, C. Pellerito, A. Desideri, S. Feo, Giovanni Marletta and B. Pignataro

Abstract Dip-pen lithography (DPL) technique has been employed to develop a new flexible biosensor realized on nylon with the aim to detect the activity of human topoisomerase. The sensor is constituted by an ordered array of a DNA substrate on flexible nylon supports that can be exploited as a drug screening platform for anti-cancer molecules. Here, we demonstrate a rapid protocol that permits to immobilize minute quantities of DNA oligonucleotides by DPL on nylon surfaces. Theoretical and experimental aspects have been investigated to successfully print DNA oligonucleotides by DPL on such a porous and irregular substrate.

V. Ferrara · G. Marletta
Department of Chemical Science, University of Catania,
v.le A. Doria 6, 95125 Catania, Italy

A. Ottaviani · A. Desideri
Department of Biology, University of Rome Tor Vergata,
via della Ricerca Scientifica, 00133 Rome, Italy
e-mail: desideri@uniroma2.it

F. Cavaleri · G. Arrabito · C. Pellerito · B. Pignataro (✉)
Dipartimento di Fisica e Chimica, Università di Palermo,
v.le delle Scienze, 90128 Palermo, Italy
e-mail: bruno.pignataro@unipa.it

G. Arrabito
e-mail: giuseppedomenico.arrabito@unipa.it

P. Cancemi · S. Feo
Dipartimento di Scienze e Tecnologie Biologiche, Chimiche e Farmaceutiche (STEBICEF),
Università di Palermo,
v.le delle Scienze, 90128 Palermo, Italy

Y.-P. Ho
Department of Biomedical Engineering, The Chinese University
of Hong Kong, Hong Kong SAR, China

B. R. Knudsen · M. S. Hede
Department of Molecular Biology and Genetics and INANO,
Aarhus University, Møllers Allé 3, Aarhus, Denmark

© Springer Nature Switzerland AG 2019
B. Andò et al. (eds.), *Sensors*, Lecture Notes in Electrical Engineering 539,
https://doi.org/10.1007/978-3-030-04324-7_39

Keywords Flexible device · Molecular printing · Biosensor · Topoisomerase

1 Introduction

1.1 Flexible Devices

Flexible sensors can be envisioned as promising components for smart sensing applications, including consumer electronics, robotics, prosthetics, health care, safety equipment, environmental monitoring, homeland security and space flight [1]. Proofs-of-concept have been demonstrated from hazardous gas sensing to diagnostic devices, from e-skin sensitivity systems to mechanical stress chips [2–5]. Flexible sensors are typically realized on "paper-like" substrates, which are generally bendable and stretchable, often soft, cheap, not fragile, easy-to-transport, green and biocompatible. These properties make them suitable to be integrated into daily clothes, allowing the user to receive valuable and timely information without compromising the functionality or comfort of their garments [6]. Textiles represent the main class of substrates for realizing such a kind of devices and others easy-to-handle sensors. Among them, nylon has shown interesting properties at the biointerface and a surface chemistry easy to manipulate and control in order to immobilize biomolecules on its surface [7].

1.2 Printed Biosensor for Topoisomerase Detection

Printing techniques permit to easily immobilize biomolecules on substrate surface in order to obtain ordered patterns or printed circuits at the micro-scale [8]. These technologies provide cost-effective routes for processing diverse materials at temperatures that are compatible with plastic and fabrics, simplified processing steps, reduced materials wastage [9]. In this regard, dip-pen lithography (DPL) is an emerging printing approach based on the atomic force microscope (AFM) permitting to print molecular inks on solid supports [10]. It is particular suitable for printing fragile biomolecules because of the possibility to keep them in native hydrated conditions, avoiding phenomena such as protein unfolding and nanostructure destabilization, e.g. DNA-nanocages collapse after drying [11]. In addition, DPL offers a strongly reduction of material consumption, a particularly important aspect to limit the costs of sensor fabrication, especially if based on expensive biomolecules, such as biomolecules.

Here, we focus on the fabrication of a flexible nylon-based biosensor by DPL, with the aim to evaluate the human topoisomerase IB (hTopo IB) activity at the interface. Topoisomerases are enzymes that alter the topological state of DNA by catalysing the breaking and re-joining of the double helix strands [12]. Their activity can be exploited to develop point-of-care systems for pathology diagnosis, e.g. malaria, with the *Plasmodium* topoisomerase as biomarker, or for drug screening, e.g. anticancer molecules targeting topoisomerases. In particular, in the latter case, the strategy is based on the selective interaction of anticancer drugs against hTopo IB. In fact, small molecules, as camptothecin and its derivatives, act by stabilizing the hTopo IB-DNA complex. This is done by binding to the transient covalent complex in a way that slows the religation step and therefore increases the lifetime of the complex [13]. The drug-complex interaction strongly reduces the probability for subsequent hTopo IB catalytic steps, eventually leading to DNA damage and cell death [14]. A surface bound DNA substrate can be utilized to quantify the hTopo IB activity after interaction with an organic compound, to obtain information about the ability of the ligand to inhibit or poison the enzyme acting as anticancer drug.

In the following, we present a strategy based on dip-pen lithography to fabricate our platform. Taking advantage of the selective interaction between DNA and topoisomerases, a proof-of-concept system was developed by assembling a DNA substrate on nylon that can be selectively recognized and bound by hTopo IB. We use fluorescence to confirm the correct assembly of the DNA substrate.

2 Experimental Aspects

2.1 Materials

Positively charged nylon membranes (Amersham Hybond-N+) were purchased from Life Science. Oligonucleotides (CL35linker: 5'–NH$_2$—ACCCCTATTTGTCGCTCACAGGAAAAAAGACTTAG–3', CP25Cy5:5'-TAAAAATTTTTCTAAGTCTTTTTTC–3', R11ATTO488:5'-AGAAAAATTTT-ATTO488-3') were provided by IBA Lifesciences. Glycerol, urea, sodium dodecyl sulphate, sodium hydroxyde and all the reagents of buffers MESTBS (20 mM TRIS, 150 mM NaCl, 4,5% milk powder, 5 mM EDTA, 0.2% NaN$_3$, 1 mg/mL herring DNA, pH 7.35), TETBS (20 mM TRIS, 150 mM NaCl, 5 mM EDTA, 0.05% Tween-20, pH 7.5), Hybridization buffer (10 mM TRIS, 300 mM NaCl, 1 mM EDTA), TE (300 mM NaCl, pH 7.5) were from Sigma Aldrich. The ink droplets were deposited by DPL (BioForce Nanoenabler) through a surface patterning tool characterized by an open microfluidic system connected to a micro-channel that goes through the tip.

2.2 Fabrication Protocol

The CL35linker sequence was spotted by DPL (Nano eNabler™ Molecular Printing System—BioForce Nanosciences) on the as-received nylon (GE Healthcare Amersham Hybond-N+) membranes in form of ordered arrays at different concentration of oligonucleotide (1, 10, 100 μM) and 30% v/v of glycerol as co-solvent in water. The

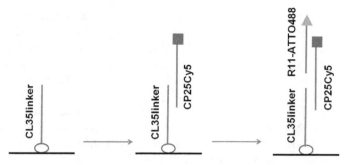

Fig. 1 Schematic representation of the biosensor fabrication steps

relative humidity and the dwell-time were fixed at 70% and 10 s respectively. Then, membranes were exposed to UV crosslinking (Hoefer Ultraviolet Crosslinker) twice and soaked in MESTBS (30 min) and TETBS (15 min) buffers under stirring. The buffers are needed to block the unreacted sites at the photo-activated nylon surface and wash unbound molecules, respectively. Then, the covalently bound CL35linker was hybridized with CP25Cy5 (200 nM in the TE buffer). Hybridization was carried out as follows: 10 min at 65 °C, cooling at 35 °C, then 1 h at room temperature, and 4 °C overnight. Finally, the sequence R11ATTO488 was annealed in 10 min at 37 °C. Stripping tests were carried out by soaking the membranes in aqueous denaturants solutions for 15 min. Dot blots spots were imaged by Transilluminator tool (Bio-Rad Gel Doc). Spots fabricated by DPL were imaged by the laser confocal fluorescent microscope Olympus FV1000 at 10X magnification, and fluorescence signal was evaluated with IMARIS software.

2.3 Biosensor Assembly

DNA can be considered an ideal molecule for nanoscale engineering, since it possesses many unique properties, including its programming capacity between complementary strands, its biological function, biocompatibility and excellent stability [15]. The biosensor was realized by exploiting the high specificity of the DNA single strand (ssDNA) complementarity. The first step is the printing and immobilization of the CL35linker oligonucleotide on nylon by DPL. Subsequently, the CL35linker is hybridized with CP25-Cy5 and R11-Atto488 to form the final DNA substrate (Fig. 1).

The process of droplet deposition on nylon surfaces by DPL can be described according to an imbibition process of a liquid droplet on a deformable porous substrate [16]. When the tip approaches the surface, a liquid meniscus composed of the liquid ink itself and water from the atmosphere is formed between the tip and the surface [17]. In particular, the tip of the DPL deforms the receiving surface of the nylon in order to favour liquid droplet imbibition. Accordingly, a gradient in pressure in the liquid across the wet substrate induces a gradient in the solid matrix which leads

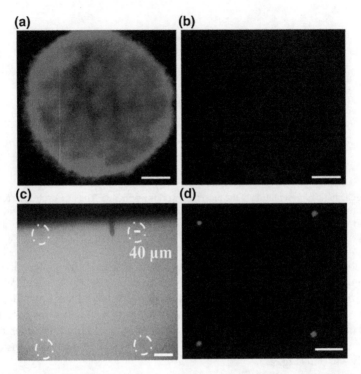

Fig. 2 Confocal microscopy pictures of 0.5 μL dot blot experiments after CP25-Cy5 (**a**) and R11-Atto488 (**b**) hybridization. DPL deposited spots of CL35linker before (**c**) and after CP25-Cy5 hybridization (**d**). Scale bars are equal to 150 μm

to deformation of the nylon substrate. Droplets can be optically visualized as fainted stains on the nylon surface (Fig. 2). The relative humidity (RH) is a fundamental parameter to control in order to obtain a reproducible printing process. We found optimal deposition at RH as high as 70%, whereas lower humidities did not permit an efficient deposition of the molecular ink. Moreover, the ink formulation is also important for a proper ink deposition. First, the presence of high boiling point co-solvent allows to avoid evaporation phenomena during the printing. Here, we chose glycerol as co-solvent (30% v/v) for our aqueous inks, because of its biocompatibility, high miscibility with water, and its ability to stabilize biomolecular structures limiting intermolecular interactions [18]. In addition, it is important to highlight that DNA oligonucleotides deposition can be accomplished by DPL in diffusive or liquid writing modes. The present fabrication protocol uses the latter because the large size and the low molecular diffusion rates of DNA molecules through the water meniscus (i.e., 10^{-12}–10^{-13} m^2/s) make the deposition in diffusive regime less efficient than writing DNA via liquid inks, where the molecules are transferred within the liquid meniscus at higher diffusion rates (i.e., 10^{-5}–10^{-6} m^2/s) [17]. To print the CL35 linker on nylon N+, which is positively charged to enhance DNA binding, the ink was formulated in ultra-pure water. The analogous formulation in TE buffer showed a low deposition yield, likely due to the charge shielding effect of the ions between

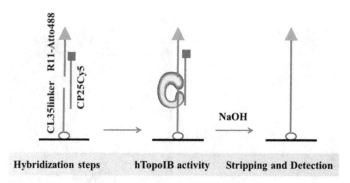

Fig. 3 Schematic representation of the biosensor working principle. The DNA substrate assembled by the three oligos (CL35linker, R11-Atto488 and CP25Cy5) is an optimal binding sequence for hTopo IB (here depicted as a yellow cartoon) which can covalently bind CL35linker and R11-Atto488

the positively charged substrate and the negatively charged oligonucleotides which, accordingly, make droplet imbibition in nylon significantly less efficient. We found that a 10 μM oligonucleotide concentration permitted to obtain spots deposition with good reproducibility. Finally, the dwell-time is also an important parameter to control the spots size and, for porous substrates as nylon, to limit the indentation of the tip into the fibres of the fabric. For a dwell-time 10 s, the spot dimensions are typically ~40 μm (Fig. 2c).

After deposition, CL35linker was hybridized with CP25-Cy5, the cyanine-5 fluorescence-labelled complementary probe (red fluorescent dye), and subsequently the free CP25-Cy5 end was hybridized with R11-Atto488, an Atto-488 fluorescence-labelled sequence (green fluorescent dye). The hybridization experiments were carried out before as preliminary dot blot experiments (Fig. 2a, b) by deposition of 0.5 μL of CL35linker ink on nylon and after on DPL functionalized substrates (Fig. 2c, d).

Cyanine5 and Atto488 form a well-known common pair of fluorophores used for FRET. For this reason, the green fluorescence for Atto488 is weak when both CP25-Cy5 and R11-Atto488 are bound to the substrate, and difficult to be detected for DPL spots (Fig. 2b). The assembled biosensor is designed to be selectively recognized and bound by hTopo IB in serum sample (Fig. 3).

The biosensor working principle is based on the selective sealing of the assembled DNA substrate by hTopo IB. After printing of the CL35linker on nylon, the oligonucleotide is hybridized with CP25-Cy5 and R11-Atto488 to form the final DNA substrate. The choice of DNA sequences was based on the well-known tetrahymena hexadecameric sequence, which represent a highly preferred binding sequence for hTopo IB [19]. In addition, the DNA substrates were chosen to have a nick a few bases downstream to the preferred cleavage site for hTopo IB. Thus, the enzyme can cleave and religate the nick by covalently binding CL35linker and R11-Atto488. Finally, the stripping process removes the non-covalent bound CP25-Cy5 and green fluorescence can be detected and directly related to the enzyme activity when the

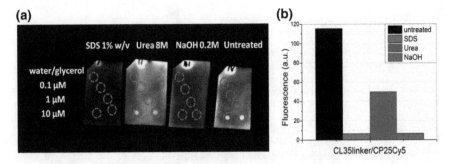

Fig. 4 Transilluminator pictures of CL35linker/CP25-Cy5 dot blots of 0.5 μL on nylon (**a**) and fluorescence signal quantification (**b**) after stripping. Red circles indicate dot blots prepared on nylon

reaction is done in a Mg^{2+} depleted buffer, where only hTopo IB is known to seal the nick in the substrate. Therefore, in a sample from a human patient, the green fluorescence after stripping can be exploited to evaluate the hTopo IB activity on the DNA substrate, and potentially predict the chemo-response of the patient. To achieve the CP25-Cy5 removal from the complementary sequences, we tested four different solutions of denaturant agents, such as sodium dodecyl sulfate (SDS) 1% w/v, urea 8 M and NaOH 0.2 M for 15 min (Fig. 4).

The quantification of fluorescence signal decrease for the CL35linker/CP25-Cy5 0.5 μL dot blots showed that the best result was obtained after membranes treatments with NaOH, that interferes with H-bonds between the DNA base pairs [16]. Good results were also obtained with SDS, while urea led to a lower fluorescence decrease.

3 Conclusions and Future Perspectives

In conclusion, we defined a fabrication approach to obtain an inexpensive, flexible and easy to integrate biosensor, based on DNA substrates assembled on porous nylon substrates by DPL. These DNA substrates can measure the hTopo IB activity in biological fluids. The present biosensor is an example that shows as DPL can play a fundamental role in fabrication of biochips and more in general of lab-on-chip systems, offering a suitable solution for point-of-care diagnostic on chemically modified platforms. In fact, DPL allows to strongly reduce material consumption and obtain biomolecules functionalized interface with a high density of spots per unit of area, that can be examined for a more robust statistical evaluation during pathologies diagnosis and drug screening. Then, DPL can be considered a powerful tool for the realization of a novel class of printed diagnostic devices for biomedical and pharmacological applications. Finally, to our best knowledge, the biosensor is the first example of DPL patterns on fibrous substrate like nylon, a "paper-like" support.

References

1. Segev-Bar, M., Haick, H.: Flexible sensors based on nanoparticles. ACS Nano **7**, 8366–8378 (2013). https://doi.org/10.1021/nn402728g
2. Sun, Y., Wang, HH.: Electrodeposition of Pd nanoparticles on single-walled carbon nanotubes for flexible hydrogen sensors. Appl. Phys. Lett. **90**, (2007).https://doi.org/10.1063/1.2742596
3. Farcau, C., Moreira, H., Viallet, B., Grisolia, J., Ciuculescu-pradines, D., Amiens, C., Ressier, L.: Monolayered wires of gold colloidal nanoparticles for high-sensitivity strain sensing. J. Phys. Chem. C **115**, 14494–14499 (2011). https://doi.org/10.1021/jp202166s
4. Segev-Bar, M., Landman, A., Nir-Shapira, M., Shuster, G., Haick, H.: Tunable touch sensor and combined sensing platform: toward nanoparticle-based electronic skin. ACS Appl. Mater. Interfaces. **5**, 5531–5541 (2013). https://doi.org/10.1021/am400757q
5. Martinez, A.W., Phillips, S.T., Whitesides, G.M., Carrilho, E.: Diagnostics for the developing world: micro fluidic paper-based analytical devices **82**, 3–10 (2010). https://doi.org/10.1007/s10337-013-2413-y
6. Windmiller, J.R., Wang, J.: Wearable electrochemical sensors and biosensors: a review. Electroanalysis **25**, 29–46 (2013). https://doi.org/10.1002/elan.201200349
7. Farahmand, E., Ibrahim, F., Hosseini, S., Rothan, H.A., Yusof, R., Koole, L.H., Djordjevic, I.: A novel approach for application of nylon membranes in the biosensing domain. Appl. Surf. Sci. **353**, 1310–1319 (2015). https://doi.org/10.1016/j.apsusc.2015.07.004
8. Arrabito, G., Pignataro, B.: Solution processed micro- and nano-bioarrays for multiplexed biosensing. Anal. Chem. **84**, 5450–5462 (2012). https://doi.org/10.1021/ac300621z
9. Khan, S., Lorenzelli, L., Dahiya, R.S.: Technologies for printing sensors and electronics over large flexible substrates: a review. IEEE Sens. J. **15**, 3164–3185 (2015). https://doi.org/10.1109/JSEN.2014.2375203
10. Piner, R.D., Zhu, J., Xu, F., Hong, S., Mirkin, C.A.: "Dip-Pen" Nanolithography. Science **283**, 661 LP-663 (1999)
11. He, Y., Ye, T., Su, M., Zhang, C., Ribbe, A.E., Jiang, W., Mao, C.: Hierarchical self-assembly of DNA into symmetric supramolecular polyhedra. Nature **452**, 198–201 (2008). https://doi.org/10.1038/nature06597
12. Wang, J.C.: DNA topoisomerases. Nat. Rev. Mol. Cell Biol. **582**, 209–219 (2009). https://doi.org/10.1007/978-1-60761-340-4
13. Leppard, J.B., Champoux, J.J.: Human DNA topoisomerase I: relaxation, roles, and damage control. Chromosoma **114**, 75–85 (2005). https://doi.org/10.1007/s00412-005-0345-5
14. Zuccaro, L., Tesauro, C., Kurkina, T., Fiorani, P., Yu, H.K., Knudsen, B.R., Kern, K., Desideri, A., Balasubramanian, K.: Real-time label-free direct electronic monitoring of topoisomerase enzyme binding kinetics on graphene. ACS Nano **9**, 11166–11176 (2015). https://doi.org/10.1021/acsnano.5b05709
15. Wang, L., Arrabito, G.: Hybrid, multiplexed, functional DNA nanotechnology for bioanalysis. Analyst **140**, 5821–5848 (2015). https://doi.org/10.1039/C5AN00861A
16. Anderson, D.M.: Imbibition of a liquid droplet on a deformable porous substrate. Phys. Fluids **17**, 87104 (2005). https://doi.org/10.1063/1.2000247
17. Arrabito, G., Reisewitz, S., Dehmelt, L., Bastiaens, P.I., Pignataro, B., Schroeder, H., Niemeyer, C.M.: Biochips for cell biology by combined dip-pen nanolithography and DNA-directed protein immobilization. Small **9**, 4243–4249 (2013). https://doi.org/10.1002/smll.201300941
18. Na, G.C.: Interaction of calf skin collagen with glycerol: linked function analysis. Biochemistry **25**, 967–973 (1986). https://doi.org/10.1021/bi00353a004
19. Andersen, A.H., Gocke, E., Bonven, B.J., Nielsen, O.F., Westergaard, O.: Topoisomerase I has a strong binding preference for a conserved hexadecameric sequence in the promotor region of the rRNA gene from Tetrahymena pyriformis. Nucleic Acids Res. **13**, 1543–1557 (1985). https://doi.org/10.1093/nar/13.5.1543

Aerosol Jet Printed Sensors for Protein Detection: A Preliminary Study

Edoardo Cantù, Sarah Tonello, Mauro Serpelloni and Emilio Sardini

Abstract The possibility to implement engineered devices to obtain feedbacks from biological samples is taking diagnostics and biotechnological research to a new level. Aerosol Jet Printing (AJP) represents a promising technique for this kind of applications. The benefits are related to manufacturing of customized, complex geometries with a high resolution even on irregular surfaces without masks. In this paper, we present the realization of an Aerosol Jet Printed microdevice addressed to protein detection and quantification in biological samples. The sensor was realized with particular attention to materials and geometry choices. The thickness of printed layers was measured thanks to a profilometer, while an electrical evaluation was performed thanks to a multimeter, in order to measure the electrical resistance offered by the printed elements. Finally, the possibility to immobilize antibodies on sensor surface for electrochemical protein quantification was assessed with fluorescence imaging. Final results obtained stated the possibility to print with AJP high resolution lines, with proper values of resistivity. Imaging findings showed a good adhesion of antibodies on the electrodes, Ag-based Anodic stripping voltammetry confirmed sensors' capability to quantify proteins, proposing the designed sensors as promising for the analysis of real biological samples.

Keywords Microsensor · Protein detection · 3-D printing · Aerosol jet printing

1 Introduction

Printed electronics is emerging as promising candidate in fields like diagnostics or tissue engineering with technological platforms giving feedbacks on biological samples or physiological processes. Moreover, the recent attention for disposable, low-cost and reliable biomolecule-to-chip interface systems for high-throughput in vitro tox-

E. Cantù · S. Tonello (✉) · M. Serpelloni · E. Sardini
Department of Information Engineering, University of Brescia, Via Branze, 38, 25123 Brescia, Italy
e-mail: s.tonello@unibs.it

© Springer Nature Switzerland AG 2019
B. Andò et al. (eds.), *Sensors*, Lecture Notes in Electrical Engineering 539,
https://doi.org/10.1007/978-3-030-04324-7_40

icity assays and pharmacology, is becoming a urgent need due to novel international regulatory guidelines [1].

Methods such as screen printing or ink-jet printing are most frequently selected for these applications: electrochemical sensors for biotechnological applications produced with these techniques range from chemicals detection to DNA or protein recognition [2, 3]. These two techniques have been also combined, in order to detect ascorbic acid with a disposable paper-based electrochemical sensor fabricated using screen-printing for base material and inkjet-printing for modifying functional material [4]. Inkjet printing was also employed to develop a method to fabricate hundreds of nano-porous gold electrode arrays on cellulose membranes for electrochemical oxygen sensing, using ionic liquid (IL) electrolytes [5].

Despite the cost and time effectiveness of these technique, they present some issues in term of reproducibility, resolution and difficulty to realize 3D structures useful for a proper management of liquid samples.

In this picture, the possibility to improve resolution, customization, standardization, and to realize complex customized 3D structures makes additive manufacturing a promising potential candidate to bring the production of biosensors to a next level [6, 7]. AM represents a process of joining materials to make objects from 3D model data, usually layer upon layer, following the definition given by the American Society for Testing Materials (ASTM) International Committee on Additive Manufacturing Technologies [8]. AM follows the popularisation of 3D solid modelling Computer Aided Design (CAD) [9].

Aerospace, automotive, biomedical and many other engineering fields benefit from the ability of AM to shape complex geometries with a high resolution and to customize the design for any needs (speeding up product development in automotive field [9], production of rough, engineered surface for more effective bone integration in medical field [10]).

In light of this, the present work proposes the use of a relatively new AM technique, Aerosol Jet Printing (AJP), to realize a customized measuring device with electrochemical sensors, addressable for the analysis of biological samples.

After an overview on the AJP method, the detailed materials and methods followed for device fabrication and testing will be described. Finally, preliminary results related to sensors geometrical and electrical features, together with an evaluation of the device compatibility with diagnostic assay routine for protein detection will be presented and discussed.

1.1   AJP: Introduction and Functioning Principle

AJP is a quite novel additive manufacturing technique belonging to the family of Aerosol-based direct-write (A-DW), subset of droplet-based direct-write. It is also known as M^3D (maskless mesoscale materials deposition) and it was developed by Optomec under the Defense Advanced Research Projects Agency (DARPA) Mesoscopic Integrated Conformal Electronics (MICE) program. AJP functioning principle presents an aerosol beam directed toward a substrate to realize specific surface features (e.g., dots or lines) without using masks or post-patterning (i.e., laser trimming) [11].

The overall process is characterized by four steps: (i) atomization of the ink, (ii) densification of the generated aerosol, (iii) focusing of the aerosol, (iv) deposition of the droplets on the substrate [12].

Using a carrier gas (nitrogen or compressed dry air), a pneumatic Collison atomizer produces the aerosol starting from inks or dispersions with viscosities in a range between 1 and 1000 mPa.s. Following, a virtual impactor allows the carrier gas to be removed and the aerosol droplet size distribution to be adjusted. The typical setup allows the production of droplets smaller than 5 μm in diameter. The aerosol is focused in the print head, thanks to a sheath gas surrounding cylindrically the flow. The complex is accelerated when entering the print nozzle, thus to reach at the nozzle exit an inner radius inferior than 50 mm. Increasing the sheath gas flow additionally, the aerosol diameter can be reduced. The droplets, thanks to their high momentum, do not change their own direction, following the gas flow at the sample surface, and impact the substrate (2–5 mm below the nozzle exit, with the surface perpendicular to it). The substrate could be heated to evaporate deposited solvents present inside the starting ink.

The performances obtained with commercially available system, able to print traces from 10 μm to 5 mm in width at translation speeds up to 200 mm/s with a wide range of inks allowed to successfully use this technique for advanced applications, ranging from energy harvesting and flexible electronics to devices for bio-electronics applications [13].

2 Materials and Methods

2.1 Sensors Design and Fabrication

After evaluating commercially available and literature designs [14], sensors' final layout was defined, using the 3-electrodes system (working (WE), counter (CE) and reference (RE) electrodes) commonly adopted for electrochemical measurements. The geometry was realized using AutoCAD, complete with a custom gadget by Optomec to ensure a proper ink filling. Figure 1 shows all the layers corresponding to the different employed inks: silver (in yellow) for the conductive tracks, carbon (in black) for WE and CE, silver chloride (in white) for RE and polyimide (in orange) for creating a sort of delimiting wall to help containing water-based samples. The selected geometry was reproduced on the substrate four times scaled in different dimensions, with a diameter of WE of 4 mm, 3 mm, 2 mm and 1 mm respectively, with three sensors for each geometry to obtain a good repeatability (Fig. 1). Considerable attention was put in selecting the proper thickness of the lines and the distances between them, so that, tuning printer parameters according to ink viscosity, a homogeneous filling

Fig. 1 AutoCAD drawing
of the final prototype

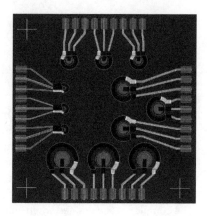

of the electrodes and pads could be ensured. The general design proposed might be customized through a proper functionalization with biomolecules that will ensure the specificity for the target analyte.

Alumina substrates were selected due to its mechanical properties and porosity, which ensures a proper ink adhesion. Regarding inks, the employed materials are: silver (UTDots Inc., cured at 220 °C per 10 min) for conductive tracks, silver chloride (Fujikura Kasei. Co. Ltd., cured at 120 °C per 30 min) for RE, carbon (Creative Materials Inc., cured at 115 °C per 5 min) for WE and CE and polyimide (HD MicroSystems Inc., cured at 120 °C per 20 min) for the final insulating layer.

Each ink was properly tuned with its own thinner to reach the proper viscosity and these two parts were placed in our Optomec's AJ300 printing machine, considering its own specific process parameters (Table 1), in order to realize the final sensor presented in Fig. 2a.

Further, in order to optimize liquid management. 3D wells were realized on each sensor by gluing on them glass-fiber washers.

Table 1 Printing process parameters

Process parameters	Ag	AgCl	C	PI
Sheath gas flow (SCCM)	50	130	40	150
Exhaust flow (SCCM)	830	600	1000	1300
Atomizer flow (SCCM)	810	550	900	1270
Process speed (mms^{-1})	2	2	2	3
Plate temperature (°C)	60	65	75	70

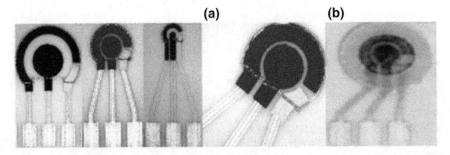

Fig. 2 **a** Examples of the printed sensors on the alumina substrate, from left to right: 4, 3, 1 and 2 mm geometries; **b** printed circuits with glued glass-fiber washers

2.2 Sensors Testing

In order to ensure the possibility to use the fabricated AJP sensors in experiments for proteins detection in biological samples, different features of the sensors were investigated.

First of all, a geometrical analysis was performed, in order to determine the thickness of printed lines for each type of ink and, for the overall circuit, to know the total area covered by the paths. The total area was evaluated as a sum of consecutive trapezoid areas.

A diamond stylus-based profilometer was used for step height measurements, (Alpha-Step IQ Kla Tencor) with a range 8 nm–2 mm and an uncertainty of 0.1%. The system is provided of an integrated optical microscope which allows an automatized selection of the region of interest. The stylus speed can be tuned between 2 and 200 μm/s. The bending radius (nominal) of the diamond tip is 5 μm. Data are recorded and can be analyzed in real time by a dedicated software.

For sensors analysis, a testing force of 62×10^{-6} N was considered, together with the following process parameters:

- Scan speed: 50 μms^{-1} for PI, 20 μms^{-1} for Ag, AgCl and C;
- Scan length: 1000 μm;
- Sampling rate: 50 Hz.

Electrical resistance of the printed paths was measured using a digital bench-top multimeter (Hewlett-Packard 34401a), applying a testing probe to each extremity of each path, thus measuring the resistance offered by all its length. Both the resistance of sample geometries of pure inks and of each electrode of the complete sensors were measured. This allowed to compare resistivity values of each ink's datasheet with the ones obtained during tests and to evaluate the contribute of each material in the complete sensor. For each element, ten measures were performed, calculating then mean values and standard deviations.

Once these values were obtained, it was possible to get resistivity considering the definition of resistance:

$$R = \rho \cdot l \cdot S^{-1} \tag{1}$$

R is resistance, ρ is resistivity, l is the length of the considered path and S its section.

After completing the geometrical and electrical analysis, the optimal volumes (measured as amount of liquid, in μl), for each geometry, were defined depositing on each sensor using a micropipette different volumes of Phosphate Buffered Saline (PBS) (Sigma Aldrich) to optimize the values for future tests with biological samples. This test involved previously only the WE, to optimize the amount of volume for each step of the biofunctionalization, and then all three electrodes, to optimize the optimal volume for the final measurement.

After optimizing working volumes, the effectiveness of primary antibodies coating on the carbon WE was evaluated using a near infra-red imaging system (Odyssey® Fc Dual-Mode Imaging System from LI-COR Biosciences). More specifically, biomolecules adhesion onto carbon WE were qualitatively detected by labelling each antibody with a fluorescent tag, and then recording the emitted light in the near infrared region using the Odyssey intensity quantifier.

A coating with an 8 μl/ml solution of anti-interleukin 8 primary antibody (Duo Set kit) was performed, keeping two electrodes as control (blank samples), covered with pure PBS solution. After that, sensors were placed in a humid chamber prepared through a box and wet paper to avoid solution evaporation and incubated for a night at 4 °C.

The next day, after washing with a solution of PBS with 0.05% Tween, sensors were incubated with a solution of secondary antibodies, labelled with a fluorescent tag functional for the final imaging. After 2 h, sensors were washed and excited with a 685 nm light source, thus measuring the emitted light in the near infrared region with the Odyssey LI-COR system, in order to reveal the coating deposition.

Finally, a voltammetry based protein quantification was performed, using standard solutions of Interleukin 8 (IL-8), a member of the CXC chemokine subfamily, considered as universal biomarkers—from cancer and inflammation to neurodegeneration [15]—thus generalizing the use of the present PoC platform to various clinical fields. IL-8 strong interaction with its capture antibody, allows to reduce the variability to the functionalization phase. The protocol adopted for IL-8 quantification in this calibration phase of the circuit was characterized by a specific functionalization of the sensor using immunocomplexes formed by a capture and a detection antibody, using a dedicated kit (DuoSet development system for ELISA, Human CXCL8/IL-8), and by an Anodic Stripping Voltammetry (ASV) based measurement technique, optimized in [16, 17].

All electrodes were exposed to the same bio-functionalization steps as following: (i) overnight immobilization of IL-8 antibody to sensor surfaces via drop-casting; (ii) 2-hours incubation with IL-8 sample (10 ng/ml); (iii) 2 h incubation with biotin-labelled detection antibody; (iv) addition of streptavidine-tagged Alkaline Phosphatase (AP) enzyme that catalyzes the oxidation of ionic Ag ($AgNO_3$) to metallic Ag, thanks to the reaction happening in presence of Ascorbic acid (AA-p), as described in [18].

Table 2 Thickness and sections of deposited inks

Material	Thickness (μm)	Standard deviation	Section (μm^2)
Ag	6.8	±1	854.2
AgCl	4	±1	392.3
C	6.5	±0.2	365.3
PI	2.7	±2	/

Once completing the bio-functionalization, sensors were covered with optimized measurement volumes of PBS, a constant potential of −0.12 V was applied for 10 s and then a linear sweep voltammetry was performed at a scan rate of 40 mV/s up to +0.4 V, measuring Ag oxidation current. All measurements were performed using a potentiostat (Palmsens, Compact Electrochemical Interfaces) controlled using the devoted software.

3 Results

3.1 Geometrical Analysis

Results obtained from the geometrical analysis (Table 2) showed congruent thicknesses of all the path with the nominal values stated in the datasheet of the Optomec Printing System (single layer thickness in the range 100 nm–10 μm). The differences between the results obtained for each ink can be interpreted and compared taking into account different process parameters, number of printed layers and further considering viscosity: higher thicknesses were obtained for inks with higher viscosity.

3.2 Electrical Analysis

Data presented in Table 3 appeared in agreement with the nominal values of the manufacturers, taking into account the specific process parameters.

More in detail, Ag experimental resistivity (12.2×10^8 Ωm) can be compared with the nominal one (3×10^{-8} Ωm), considering as most relevant affecting variables the substrate (ceramic is not included among the optimal indicated by UTDots). AgCl experimental resistivity value (71.3×10^{-8} Ωm) can be compared with the nominal one reported by Fujikura Kasei. Co. Ltd (56×10^{-8} Ωm), considering the use of the thinner to achieve the viscosity required for printing. Finally, C experimental resistivity (159.6×10^{-8} Ωm) appear increased compared to the theorical one given by Creative Materials (15.5×10^{-8} Ωm), probably due to the usage of

Table 3 Resistance and resistivity of deposited inks

Material	Resistance (Ω)	Path length (mm)	Resistivity (Ωm)
Ag	4.9	13.5	30.8×10^{-8}
	1.6	11	12.2×10^{-8}
	7.3	9.5	65.8×10^{-8}
AgCl	21	12	71.3×10^{-8}
	19.1	10	78.2×10^{-8}
	25.3	11.5	89.8×10^{-8}
C	67.5	13	122.9×10^{-8}
	39.9	13	72.6×10^{-8}
	84.3	12.5	159.6×10^{-8}

a non-optimal substrate and due to different post-processing parameters. Measurements performed on the complete sensors suggested values of resistance coherent with the different dimensions of each geometry, with standard deviation suggesting a good reproducibility of the purposed technique for printing the same geometry.

3.3 Fluorescence Imaging

Results shown in Fig. 3 highlight a significant difference between blank electrodes (black in the image, not emitting any light) and antibodies coated ones (appearing as a clearer emitting spot in the image). The red arrow indicates the zone successfully covered by antibodies.

These findings confirmed the possibility to use these materials and AJP technique to produce electrodes, which allow a homogeneous adhesion of antibodies, essential to perform a complete functionalization to perform immune-sensing of proteins. The homogeneous deposition of primary antibodies would allow a more effective quantification of proteins, thus reducing the interference of currents due to un-specific adsorption.

Fig. 3 Three alumina-substrate sensors covered by protein

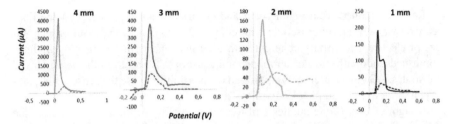

Fig. 4 Plots obtained during protein quantification test for the alumina substrate sensor; each plot measures current (expressed in μA) as a function of potential (expressed in V). Dotted lines represent "blank samples"

3.4 Protein Quantification

Significant differences in term of peak height could be appreciated comparing results from sensors tested with protein solution and blank ones. This suggests the possibility to successfully apply the described ASV protocol for protein quantification on all the AJP sensors (Fig. 4).

The higher difference could be recorded for sensors with 4 mm WE, reproducing the geometry of products already available on the market. However, very interesting results could be obtained with the smaller geometries. The results obtained for the other two intermediate geometries on alumina can be considered comparable. Even if the normalized current peak is significantly smaller than the one obtained with the other geometries, they can be considered as a of good compromise considering all the electronic and the biological specific requirements.

Regarding 1 mm geometry, the current peak obtained (~200 μA) is very high, considering the significantly smaller area, and it is probably due to the closeness of the electrodes that allow virtually all current to be "captured" and little to be lost. The negative aspect related to this geometry is connected to the biological aspect, considering future applications for the detection of disease-related biomarkers of present in a very low concentration (<1 ng/ml) for which in 2 μl of sample (the one required for this smallest geometry) there is a high risk not to deposit any protein on sensor WE.

4 Conclusions

3D printing technologies are increasingly emerging as potential tool to bring significant improvements in the medical and biotechnological fields. Further the possibility to print with high resolution conductive materials open a wide variety of possibilities for producing all the sensors that are involved in bio-electronics applications.

In this picture, this work reports the realization of a microsensor platform for protein detection, using a new 3D printing technique: Aerosol Jet Printing.

Results obtained from geometrical and electrical tests suggested that the lines realized via AJP are microscopic with controllable thickness and with proper values of electrical resistivity coherent with each ink. Furthermore, imaging findings confirmed good adhesion of antibodies on sensors WE and to both the employed substrates. ASV measurements confirmed the possibility to use the sensors for quantifying protein in a sample (10 ng/ml).

In light of this, future studies will test the capability of the sensors to trace the presence of proteins in different biological fluids. To achieve this goal, future developments will address complete calibrations of the platform for protein detection, taking into account all the interferences that might be introduced from the chemistry of real samples (e.g. blood, plasma).

Furthermore, future development might address techniques for improving the confinement of liquid samples both for WE functionalization and for the final measurement, using biocompatible soft materials, (e.g. PDMS). Through a proper modification of the sensor geometry presented and the usage of proper inks, AJP technology would allow the possibility to integrate the electrodes on a customized substrate with 3D wells. Overall, the sensors might be properly integrated with dedicated conditioning electronics and microfluidic, in order to optimize both the electronic performances and liquid samples managing.

References

1. Spanu, A., Lai, S., Cosseddu, P., Tedesco, M., Martinoia, S., Bonfiglio, A.: An organic transistor-based system for reference-less electrophysiological monitoring of excitable cells. Sci. Rep. **5**(1), 8807 (2015)
2. Couto, R.A.S., Lima, J.L.F.C., Quinaz, M.B.: Recent developments, characteristics and potential applications of screen-printed electrodes in pharmaceutical and biological analysis. Talanta **146**(228), 801–814 (2016)
3. Li, J., Rossignol, F., Macdonald, J.: Inkjet printing for biosensor fabrication: combining chemistry and technology for advanced manufacturing. Lab Chip **15**(12), 2538–2558 (2015)
4. Kit-Anan, W., et al.: Disposable paper-based electrochemical sensor utilizing inkjet-printed Polyaniline modified screen-printed carbon electrode for Ascorbic acid detection. J. Electroanal. Chem. **685**, 72–78 (2012)
5. Hu, C., Bai, X., Wang, Y., Jin, W., Zhang, X., Hu, S.: Inkjet printing of nanoporous gold electrode arrays on cellulose membranes for high-sensitive paper-like electrochemical oxygen sensors using ionic liquid electrolytes. Anal. Chem. **84**(8), 3745–3750 (2012)
6. Ragones, H., et al.: Disposable electrochemical sensor prepared using 3D printing for cell and tissue diagnostics. Sensors Actuators B Chem. **216**, 434–442 (2015)
7. Yang, H., Rahman, T., Du, D., Panat, R., Lin, Y.: 3-D printed adjustable microelectrode arrays for electrochemical sensing and biosensing. Sens. Actuators. B. Chem. **230**, 600–606 (2016)
8. Wohlers Report 2015: 3D Printing and Additive Manufacturing State of the Industry, Annual Worldwide Progress Report (2015)
9. Gibson, I.: The changing face of additive manufacturing. J. Manuf. Technol. Manag. **28**(1), 10–17 (2017)
10. Guo, N., Leu, M.C.: Additive manufacturing: technology, applications and research needs. Front. Mech. Eng. **8**(3), 215–243 (2013)
11. Hoey, J.M., Lutfurakhmanov, A., Schulz, D.L., Akhatov, I.S.: A review on aerosol-based direct-write and its applications for microelectronics. J. Nanotechnol. (2012)

12. Binder, S., Glatthaar, M., Rädlein, E.: Analytical investigation of aerosol jet printing. Aerosol Sci. Technol. **48**(9), 924–929 (2014)
13. OPTOMEC: Aerosol Jet ® Printed Electronics Overview, p. 6
14. Alex Gong, C.S., Syu, W.J., Lei, K.F., Hwang, Y.S.: Development of a flexible non-metal electrode for cell stimulation and recording. Sensors (Switzerland) **16**(10) (2016)
15. Kim, H.J., Li, H., Collins, J.J., Ingber, D.E., Beebe, D.J., Ismagilov, R.F.: Contributions of microbiome and mechanical deformation to intestinal bacterial overgrowth and inflammation in a human gut-on-a-chip. PNAS **13**, E7–E15 (2015)
16. Tonello, S., et al.: Wireless point-of-care platform with screen-printed sensors for biomarkers detection. IEEE Trans. Instrum. Meas. **66**(9) (2017)
17. Tonello, S., et al.: Enhanced sensing of interleukin 8 by stripping voltammetry: carbon nanotubes versus fullerene. In: IFMBE Proceedings, vol. 65 (2017)
18. Chen, Z.P., Peng, Z.F., Jiang, J.H., Zhang, X.B., Shen, G.L., Yu, R.Q.: An electrochemical amplification immunoassay using biocatalytic metal deposition coupled with anodic stripping voltammetric detection. Sensors Actuators B Chem. **129**(1), 146–151 (2008)

Novel Coplanar Capacitive Force Sensor for Biomedical Applications: A Preliminary Study

Andrea Bodini⬤, Emilio Sardini⬤, Mauro Serpelloni⬤
and Stefano Pandini⬤

Abstract Nowadays the world of sensors is gaining a primary importance in the electronics field, thanks to the boom of the smartphones and IoT market. Capacitive sensors are widely involved in the most of applications, from the biomedical to the gaming industry, and they can sense a wide variety of physical quantities. In this work we focused on force sensing capacitive sensors, trying to mix two of the most in-fashion markets, sensors and polymers. Polymers industry is constantly growing and, by the constant synthesis of new bio-based molecules, it will quickly enter the most of the technology markets. We made a soft capacitive force sensor, by inserting a coplanar capacitor in a polymeric wafer structure. The involved technology is derived from the touch-sensing technique, usually involved in appliances' user interfaces. This sensor is easy to make and cheap, and it was tested with a force of 1 N. It is waterproof and non-sensitive to moisture variation in the outer environment. It shown a sensitivity of 172 fF/N, with a resolution of 80 mN. As it is only a preliminary study, more investigations are needed in order to obtain a deeper characterization versus different environment conditions and with higher force stimuli. It will be also relevant to evaluate the behavior of the sensor by using different polymers.

Keywords Force sensor · Coplanar capacitive sensor · Biomedical

A. Bodini (✉) · E. Sardini · M. Serpelloni
Department of Information Engineering, University of Brescia, 25123 Brescia, Italy
e-mail: a.bodini004@unibs.it

S. Pandini
Department of Industrial and Mechanical Engineering, University of Brescia, 25123 Brescia, Italy

© Springer Nature Switzerland AG 2019   329
B. Andò et al. (eds.), *Sensors*, Lecture Notes in Electrical Engineering 539,
https://doi.org/10.1007/978-3-030-04324-7_41

1 Introduction

1.1 State of the Art

Knowing force or pressure level is a key feature in many applications, from the manufacturing to the biomedical, through the gaming to the smartphones. A well knowing of this physical quantity has a fundamental relevance in many fields, it can help the success of a surgery as well as improving the gaming experience of a player. The ways to obtain this information are many. Sensors employ different physical principles for this purpose, capacitance, piezo resistance, diffraction and many others. Piezoresistive sensors were the most used but, for their temperature-dependent response and a lower resolution, their getting supersede by capacitive sensors. Furthermore, capacitive touch-sensors are making their fortune because of their wide application in consumer electronics. Capacitive sensors technology is growing also in accordance to the great progress of the polymer industry. The continuous introduction of new bio-based polymers is pushing the sensors' industry to another level, adding lot of new possibilities, starting from biology. For example, for intraocular pressure measuring (IOP) [1], graspers for surgery [2], robot assisted single cell microinjection [3], as human tactile receptors simulators [4] and many others. They are also used in the manufacturing field as next generation contact detection for prevent operators' injuries, for example, when working with co-op robots.

They are also more involved in consumer fields, as gaming, gesture recognition and augmented reality controls [5].

Standard capacitive sensors inherit their working principle from their structure, force sensing capability is directly related to their 3D assembly. Nowadays the making of 3D structures is not that difficult as it was 10 years ago, thanks to the rapid improvement of the addictive manufacturing techniques [6–8]. This however requires a 3D printer capable to deposit layers with high dimensional precision on all the 3 axes, that requires a high-level printer which is very expensive.

Addictive manufacturing is not the only technique for making non-planar structures. Also screen printing can be involved for this kind of purpose, but it requires expensive machines and a complicated process [9].

1.2 Coplanar Capacitors

This research aims at moving towards less difficult processes, reducing the method complexity and then, the most important, the overall costs. How to obtain the same capacitive sensors performance, without using their peculiar 3D structure? Again, the consumer electronics market drives the choice. Since around 2007 we are facing a proper boom of the touch-sensing technologies, in all the fields in which a human-machine interface (HMI) is needed. The touch sensing technique is based on different

Fig. 1 Basic coplanar capacitor principle

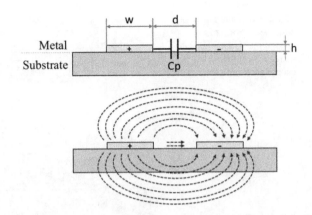

physical principles which could consider, variations of resistance, capacitance, SAW (surface acoustic waves) path perturbations and many more [10]. In the early designs these interfaces were made with resistive touch panels. Unfortunately, they were sensitive to temperature variations and not very accurate. There was also a problem in the dimension of the minimum feature size. Later, capacitive touch screens replaced resistive because of their lower cost, independence from temperature variations and accuracy/sensitivity. It was also easier to make the sensing matrix thicker to improve the spatial resolution. Nowadays this technology reached its maturity thanks to the wide employment in the consumer and manufacturing markets, with also the addition of different features as force touch or 3D touch. These additional features permit to obtain a value of the applied force but are very complex to implement.

For resolving this fabrication difficulty, we tried to mix the coplanar capacitor solution with polymers.

First of all is necessary to explain how a coplanar capacitor work.

Think about standard parallel plates capacitor, the electric field is confined between the two plates and a minimum part of it will leak out because of the fringing effect. Putting the plates on the same plane, we deform the area in which the electric field is confined, and we shape it as a "U". A basic schematic representation is depicted in Fig. 1.

Because it is very difficult to quantify the fringing effects, we should impose that the standard path of the field lines will be curved. There is a better and deeper explanation of this principle in [11]. These capacitors are then characterized by a dual side field. This particular feature will be an advantage or not based on the application. In this case this kind of behavior is un-wanted. The way to remove this effect is to add a ground plane on the opposite side of the plates. This method also helps to increase the low intrinsic capacitance value adding to the original Cp the capacitance of the ground plane (Fig. 2).

A coplanar capacitor, for its nature, senses the variation in the surrounding environment. Approaching it with the fingers, it will measure a sensible variation in the capacitance value. What happens if the capacitance variation will be linked to the

Fig. 2 Coplanar capacitor
with the addition of the
ground plane. This helps to
increase the intrinsic
capacitance value

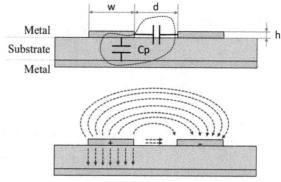

Fig. 3 Coplanar capacitor
measure when approaching
with a finger

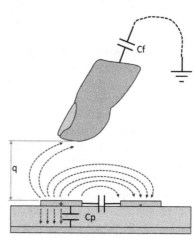

distance from the capacitor by a well-known amount? How can we relate the variation
of the distance q in Fig. 3 to the total capacitance seen at the capacitor heads?

The solution will be the insertion of something with a fixed elastic coefficient,
with a low shape memory. Polymer industry comes to help us. With the insertion of
a polymeric layer between the finger and the capacitor, we can link the capacitance
variation to the applied force.

2 Sensor Fabrication and Preliminary Tests

2.1 Sensor Fabrication Process

The sensor is fully biocompatible and suitable for biomedical environments in a wide
variety of fields of application. It is shaped as an interdigitated spiral, as shown in
Fig. 4, with a maximum radius of 8.5 mm in order to increase its sensitivity to the

Fig. 4 Coplanar capacitor spiral design. The outer diameter is 8.5 mm, the path width is 0.203 mm separated by a 0.1778 mm gap to each other

Fig. 5 A section of the sensor. The thicknesses are, copper 35 μm, polyimide 25 μm, PVS 2 mm and EVA 4.25 mm. EVA's starting thickness is 1.5 mm, but after the thermoforming process it lowers its width of the 25%

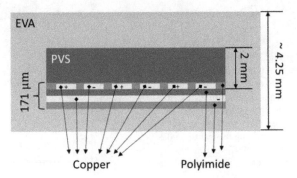

applied force. The sensor is made as a wafer of four different layers, two externals made by ethylene vinyl acetate (EVA) for the enclosure, a central made by poly vinyl siloxane (PVS) and the polyimide flexible layer with the copper plates.

The structure is fabricated through a thermoforming process. A sensor representation in shown in Fig. 5. The sensing flexible area is 130 μm thick and the total height of the sensor is more or less 4.25 mm.

The PCB design of the sensing part is made to prevent the interaction of the environments to both sides, making it sensitive only to a one-side force application.

2.2 Sensor Preliminary Test

Once the first prototypes have been built, shown in Fig. 6, they have been tested with an INSTRON® 3366 machine attached to an HP4194A impedance analyzer. In Fig. 7 are then reported trials results over time and a first capacitance over force characterization in Fig. 8. The sensor reported a sensitivity of 172 fF/N, a resolution of 80 mN and low zero drift, in the order of the units of fF. There is also a hysteresis in the order of 20 fF. These parameters are obtained through a first order linearization with a $R^2 = 0.9798$. In Fig. 7 are reported three different runs of the test, Test1 and Test2 are obtained keeping the INSTRON tip in the center of the sensor, Test3 is

Fig. 6 Flex PCB of the first
prototype of the sensor

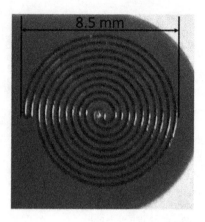

Fig. 7 The INSTRON
reading versus time on the
top, on the bottom the sensor
reading versus time. There
are three different tests,
Test1 and Test2 are made
positioning the test tip in the
center, the Test3 is made
moving the tip to the outside

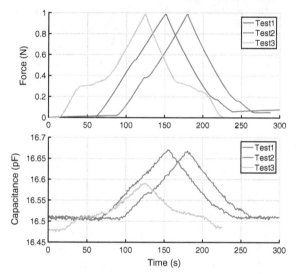

indeed obtained moving it on the edge of the sensing area. There is a lowering in the
sensitivity of the sensors, making it also sensitive to the position of the force application. This could be employed also for tracking the position of the force application
point. Further investigations are now needed in order to achieve a more accurate
sensor characterization.

Fig. 8 Capacitance over Force variation. Test3 curve is different by the others because the test tip was moved on the border of the spiral, instead of the center as in Test1 and Test2. There is a decrease in the sensitivity moving from the center

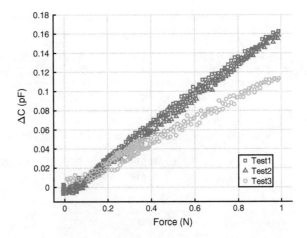

3 Conclusions

In this work a coplanar capacitive force sensor had been developed and preliminary tested. The overall system is fabricated by the mixing of two of the most relevant fields in the actual market, capacitive sensors and polymers. It reported a good response in terms of capacitance over force variation and further developments are needed to improve this value. The sensing element is made with standard flexible PCB techniques in order to reduce the production costs. The polymers involved in the process are all bio-compatible and low cost. Finally, the sensor reported a sensitivity of 172 fF/N and a resolution of 80 mN.

References

1. Ha, D., De Vries, W.N., John, S.W.M., Irazoqui, P.P., Chappell, W.J.: Polymer-based miniature flexible capacitive pressure sensor for intraocular pressure (IOP) monitoring inside a mouse eye. Biomed. Microdevices **14**, 207–215 (2012)
2. Dai, Y., Abiri, A., Liu, S., Paydar, O., Sohn, H., Dutson, E.P., Grundfest, W.S., Candler, R.N.: Grasper integrated tri-axial force sensor system for robotic minimally invasive surgery. In: Proceedings of the 19th Annual International Conference of the IEEE Engineering in Medicine and Biology Society EMBS, pp. 3936–3939 (2017)
3. Wei, Y., Xu, Q.: Design of a PVDF-MFC force sensor for robot-assisted single cell microinjection. IEEE Sens. J. **17**, 3975–3982 (2017)
4. Joo, Y., Yoon, J., Ha, J., Kim, T., Lee, S., Lee, B., Pang, C., Hong, Y.: Highly sensitive and bendable capacitive pressure sensor and its application to 1 V operation pressure-sensitive transistor. Adv. Electron. Mater. **3**, 1–10 (2017)
5. Kim, J., Kwak, Y.H., Kim, W., Park, K., Pak, J.J., Kim, K.: Flexible force sensor based input device for gesture recognition applicable to augmented and virtual realities. In: 2017 14th International Conference on Ubiquitous Robots and Ambient Intelligence, pp. 271–273 (2017)
6. Zhuo, B., Chen, S., Guo, X.: Micro structuring polydimethylsiloxane elastomer film with 3D printed mold for low cost and high sensitivity flexible capacitive pressure sensor. In: 2017 IEEE

Electron Devices Technology and Manufacturing Conference EDTM 2017—Proceedings, vol. 8, pp. 148–149 (2017)

7. Qin, H., Cai, Y., Dong, J., Lee, Y.-S.: Direct printing of capacitive touch sensors on flexible substrates by additive e-jet printing with silver nanoinks. J. Manuf. Sci. Eng. **139**, 31011 (2016)

8. Zhuo, B., Chen, S., Zhao, M., Guo, X.: High sensitivity flexible capacitive pressure sensor using polydimethylsiloxane elastomer dielectric layer micro-structured by 3-D printed mold. IEEE J. Electron Devices Soc. **5**, 219–223 (2017)

9. Fernandes, J., Jiang, H.: Three-axis capacitive touch-force sensor for clinical breast examination simulators. IEEE Sens. J. 7231–7238 (2017)

10. Walker, G.: Touch sensing. In: Interactive Displays: Natural Human-Interface Technologies, pp. 27–105 (2014)

11. Mamishev, A.V., Sundara-Rajan, K., Yang, F., Du, Y., Zahn, M.: Interdigital sensors and transducers. Proc. IEEE **92**, 808–844 (2004)

Graphene-Like Based-Chemiresistors Inkjet-Printed onto Paper Substrate

F. Villani, F. Loffredo, Brigida Alfano, Maria Lucia Miglietta, L. Verdoliva, M. Alfè, V. Gargiulo and Tiziana Polichetti

Abstract In this work, the possibility of manufacturing chemiresistive volatile organic compounds sensing devices by inkjet printing an aqueous suspension of graphene-like layers on paper substrates has been studied. The electrical responses of the devices have been analyzed upon exposure to ethanol at room temperature in dry ambient and analyzed in terms of the conductance variation. Additionally, a reproducibility analysis on the device performances has been investigated.

Keywords Graphene-like layers · Inkjet printing · Aqueous dispersion
Chemiresistive sensors · Volatile organic compounds · Ethanol

1 Introduction

The detection of volatile organic compounds (VOCs) is a current topic that attracts the scientific community for being crucial in several applications mainly related to environment, health and industrial processes monitoring. As a consequence, the demand for low-cost, low-power and portable VOCs detectors is increasing. Such request can be addressed by merging material sensing properties with sustainable fabrication processes and lightweight and flexible substrates.

In this framework, the inkjet printing (IJP) technology, a deposition method from liquid phase, well addresses this demand for its patterning capability, which permits an efficient use of different functional inks so reducing the amount of waste products, and the employable (nonflexible and flexible) substrates. These peculiarities of the IJP technique are the main advantages exploited in several electronic applications [1–5], including the sensor devices' field [6–9].

F. Villani (✉) · F. Loffredo · B. Alfano · M. L. Miglietta · L. Verdoliva · T. Polichetti
Italian National Agency for New Technologies, Energy and Sustainable Economic Development (ENEA), C.R. Portici - P.le E. Fermi 1, 80055 Portici, Naples, Italy
e-mail: fulvia.villani@enea.it

M. Alfè · V. Gargiulo
Combustion Research Institute (IRC) - CNR, P.le V. Tecchio, 80, 80125 Naples, Italy

© Springer Nature Switzerland AG 2019
B. Andò et al. (eds.), *Sensors*, Lecture Notes in Electrical Engineering 539,
https://doi.org/10.1007/978-3-030-04324-7_42

Concerning the sensing materials, in the last years the graphene related materials have focused great interest, since they combine excellent detection sensitivity with remarkable transduction properties for detecting gas/vapors besides a wide range of chemicals and bio-molecules [10–15]. Among these nanomaterials, graphene-like (GL) layers have been recently proposed as new chemiresistive material for alcohol detection in vapor phase [16–18]. GL layers consist in water-stable small graphenic fragments composed by three/four stacked graphene layers decorated at the layer edge with oxygen functional groups (mainly carboxylic/carbonylic). Differently from graphite oxide (GO), in GL layers the basal planes are quite defect-free, that is a clear advantage for the electron conductivity and the electronic properties [17, 19]. Additionally, GL nanostructures exhibit a typical self-assembling behavior after drying on a flat surface.

In the present work, we exploited the sensing capabilities of GL layers to ethanol in order to investigate their potentiality to be processed by means of IJP technique. We fabricated chemiresistors by printing an aqueous suspension of GL layers onto glossy paper with interdigitated gold electrodes. The electrical responses of the sensor devices have been analyzed in terms of the conductance variation ($\Delta G/G_0$) when exposed to ethanol as analyte.

2 Experimental

All the chemicals (analytical grade) have been purchased from Sigma Aldrich and used as received. Carbon Black N110 type (furnace CB, H/C 0.058, BET surface area 139 cm^2/g) has been provided by Sid Richardson Carbon Co.

GL layers in aqueous suspension has been prepared through a double-step oxidation/reduction method starting from nanostructured carbon black [16, 17]: CB powder has been oxidized with concentrated nitric acid (67 wt%) at 100 °C under stirring (90 h). The solid dark brown hydrophilic precipitate obtained, named GLox, has been recovered by centrifugation, washed twice with distilled water and dried at 100 °C. Then GLox has been treated for 24 h at 100 °C under reflux and stirring with hydrazine hydrate (for 20 mg of GLox 450 μl of $N_2H_4xH_2O$ have been used). At the end of the reaction the excess of N_2H_4 has been neutralized with diluted HNO_3 and the resulting black solid (GL layers) has been recovered by centrifugation, washed with distilled water in order to remove all traces of unreacted reagents and then stored as aqueous suspension (1 g/L mass concentration, pH 3.5–3.7). A scheme of the double-step oxidation/reduction process for the GL synthesis is displayed in Fig. 1.

Commercial glossy paper has been employed as substrate. Before depositing the GL layers-based ink, interdigitated Cr (30 nm)/Au (120 nm) electrodes have been e-beam evaporated on top of the substrate. The final system substrate-electrodes has been employed as transducer.

The inkjet equipment was a Dimatix Materials Printer 2831 (DMP2831) of FUJIFILM—USA suitable for the print of functional inks onto flexible and rigid

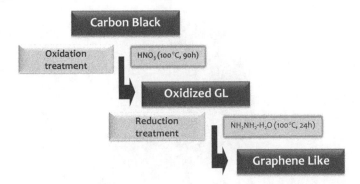

Fig. 1 Scheme of the double-step oxidation/reduction process for the synthesis of GL layers

substrates. This system uses a piezoelectric drop-on-demand technology to eject droplets through a multi-nozzle printhead. The pattern of the printed sensing material was a rectangular surface. The printing parameters have been optimized in order to obtain uniform, conductive and reproducible sensing films. For all the chemiresistor devices the sensing GL-based film has been realized by printing multiple overlapped layers.

In order to analyze the surface morphology of the printed films, atomic force microscopy (AFM) measurements have been carried out in tapping mode by the Veeco Dimension Digital Instruments Nanoscope IV system.

The electrical characterizations have been carried out in a Cascade Summit 12 000 probe station equipment, connected to a Keithley4200 semiconductor characterization system. All the measurements have been performed in air.

Tests for sensing measurements upon the analyte have been performed in a stainless steel chamber placed in a thermostatic box, at room temperature (keeping constant the temperature, $T = 22\ °C$) in N_2 in dry ambient. The sensing analysis has been carried out by biasing the single device at the voltage of 1 V upon exposure of ethanol 50 ppm. For the calibration analysis the measurement protocol has consisted of several sequential exposures at different analyte concentrations, where each exposure step has been preceded and followed by baseline and recovery phases, respectively.

3 Results and Discussion

In our previous works [6–8] we demonstrated that IJP technology is able to deposit graphene-based ink in a controlled manner so to produce sensor devices with reproducible performances in terms of electrical response upon gas exposures. In that case, the ink was based on Liquid Phase Exfoliation (LPE) graphene with marked sensing properties towards NO_2 and less pronounced towards NH_3.

Fig. 2 Picture of the typical GL-based device printed onto glossy paper

Here we aim to investigate new graphene-based materials having sensing character towards organic vapors in order to enlarge the selectivity window.

In this perspective, we have focused our study on a new nanomaterial based on graphene-like layers dispersed in an aqueous solution. The GL-based ink has been printed onto paper/electrodes transducers and the typical sensor device is shown in Fig. 2. The picture highlights the good uniformity of the printed surface pattern with well defined film edges. This macroscopic feature suggests the right control of the adopted deposition method that is basic for a reproducible process.

Behind the macroscopic quality it is crucial to evaluate the microscopic distribution of the nanomaterial that can affect the absorption process of the analyte molecules in the sensing phase. Therefore, we have investigated the surface morphology of the printed film by atomic force microscopy and the section analysis along a 1 μm size is displayed in Fig. 3. The AFM section profile points out a quite regular distribution of small aggregates (half-height width ~50 nm) resulting in a root mean square roughness estimated equal to 12 nm. To a complete evaluation of the GL layers distribution induced by IJP deposition we have to take into account also the surface morphology of the substrate characterized, as shown in Fig. 3, by a grain network having a root mean square roughness of 9 nm.

The material electrical property has been investigated through the volt-amperometric characterization and the measurement results are reported in Fig. 4, where the data I–V detected for two samples, chosen as example to analyze the repeatability, are shown. These samples have been prepared in the same conditions according the same printing parameters: the clearly linear I–V curves underline the evident ohmic behavior of the material in the investigated voltage range. Moreover, the close behavior of the electrical responses of the examined samples, as well as for other samples prepared in the same manner and here not reported, suggests once again that the deposition method assures a process repeatability. The base resistance calculated for the samples of Fig. 4 was 7 kΩ in both cases.

The sensing properties of the printed aqueous GL-layers chemiresistors have been investigated by exposing them to organic vapors of ethanol. This analysis has been performed through tests upon 30 min long exposure to the analyte and evaluating the

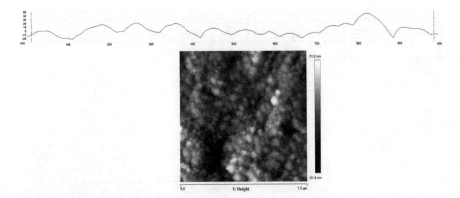

Fig. 3 AFM section profile along a 1 μm size of the GL film printed onto paper substrate (up); AFM topography of paper substrate (scan size 1 × 1 μm²) (down)

Fig. 4 I–V curves of two GL films printed in the same conditions

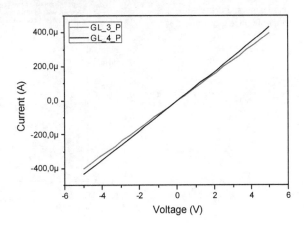

conductance variation ($\Delta G/G_0$), namely the percentage variation of the conductance with respect to its initial value detected at the beginning of the exposure (G_0). In Fig. 5 the electrical responses of two twin chemiresistors upon a single pulse of ethanol 50 ppm are shown for comparison. The devices' performances point out the sensitivity of GL towards ethanol that might be attributed to the presence of residual oxygen functional groups, mainly carboxylic groups. Additionally, the electrical responses of the sensors highlight that, combining the self-assembling property of the GL layers with the controlled deposition method IJP, it is possible to assure a good process repeatability.

To complete the sensing characterization, the GL-based device has been calibrated, with a resolution of few ppm, in the range 0–100 ppm of ethanol in dry ambient. The graph shows the max reactivity versus increasing and decreasing ethanol concentration: over 60 ppm concentration a linearity distortion is observed probably due to a contribution of glossy paper on the sensing response (Fig. 6).

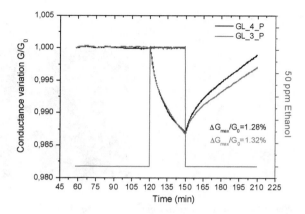

Fig. 5 Conductance variation ($\Delta G/G_0$) of two GL sensor devices printed in the same conditions (twin samples) upon exposure of ethanol 50 ppm

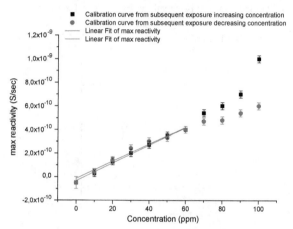

Fig. 6 Max reactivity versus increasing and decreasing ethanol concentration

4   Conclusion

In this study we have demonstrated the feasibility of manufacturing printed organic vapors sensors onto flexible substrates, by merging low-cost and sustainable process and substrate with a new nanomaterial having the suitable sensing properties. In detail, we have realized chemiresistive VOC sensing devices by inkjet printing an aqueous suspension of graphene-like (GL) layers onto glossy paper. The device performances have been investigated upon exposure to ethanol 50 ppm in N_2 in dry ambient.

Furthermore, combining the self-assembling property of the GL layers with the capability of IJP to deposit in controlled manner, we have also demonstrated the process repeatability in terms of surface morphology (macroscopic and microscopic distribution of the printed nanomaterial) and electrical responses (base resistance and conductance variation).

References

1. Sowade, E., Ramon, E., Mitra, K.Y., Martínez-Domingo, C., Pedró, M., Pallarès, J., Loffredo, F., Villani, F., Gomes, H.L., Terés, L., Baumann, R.R.: Sci. Report **6**, 33490 (2016)
2. Gomes, H.L., Medeiros, M.C.R., Villani, F., Canudo, J., Loffredo, F., Miscioscia, R., Martinez-Domingo, C., Ramon, E., Sowade, E., Mitra, K.Y., Baumann, R.R., McCulloch, I., Carrabina, J.: Microelectron. Reliab. **55**, 1192 (2015)
3. Borriello, C., Miscioscia, R., Mansour, S.A., Di Luccio, T., Bruno, A., Loffredo, F., Villani, F., Minarini, C.: Physica Status Solidi A Appl. Mater. Science **212**, 2677 (2015)
4. Bruno, A., Villani, F., Grimaldi, I.A., Loffredo, F., Morvillo, P., Diana, R., Haque, S., Minarini, C.: Thin Solid Films **560**, 14 (2014)
5. Grimaldi, I.A., Barra, M., Carella, A., Di Girolamo, F., Loffredo, F., Minarini, C., Villani, F., Cassinese, A.: Synth. Met. **176**, 121 (2013)
6. Villani, F., Schiattarella, C., Polichetti, T., Di Capua, R., Loffredo, F., Alfano, B., Miglietta, M.L., Massera, E., Verdoliva, L., Di Francia, G., Beilstein, J.: Nanotechnology **8**, 1023 (2017)
7. Schiattarella, C., Polichetti, T., Villani, F., Loffredo, F., Alfano, B., Massera, E., Miglietta, M.L., Di Francia, G.: Lect. Notes Electr. Eng. **431**, 111 (2018)
8. Ricciardella, F., Alfano, B., Loffredo, F., Villani, F., Polichetti, T., Miglietta, M.L., Massera, E., Di Francia, G.: Proceedings of the 2015 18th AISEM Annual Conference, AISEM 2015, vol. 2015, p. 7066858
9. De Girolamo Del Mauro, A., Grimaldi, I.A., Loffredo, F., Massera, E., Polichetti, T., Villani, F., Di Francia, G.: J. Appl. Polymer Sci. **122**, 3644 (2011)
10. Llobet, E.: Sens. Actuators B **179**, 32 (2013)
11. Baptista, F.R., Belhout, S.A., Giordani, S., Quinn, S.J.: Chem. Soc. Rev. **44**, 4433 (2015)
12. Varghese, S.S., Lonkar, S., Singh, K.K., Swaminathan, S., Abdala, A.: Sens. Actuators B **218**, 160 (2015)
13. Yuan, W., Shi, G.: J. Mater. Chem. A **1**, 10078 (2013)
14. Wang, T., Huang, D., Yang, Z., Xu, S., He, G., Li, X., Zhang, L.: Nano-Micro Lett. **8**, 95 (2016)
15. Liu, Y., Dong, X., Chen, P.: Chem. Soc. Rev. **41**, 2283 (2012)
16. Alfè, M., Gargiulo, V., Di Capua, R.: Appl. Surf. Sci. **353**, 628 (2015)
17. Alfè, M., Gargiulo, V., Di Capua, R., Chiarella, F., Rouzaud, J.N., Vergara, A., Ciajolo, A.: ACS Appl. Mater. Interfaces **4**(9), 4491 (2012)
18. Gargiulo, V., Alfano, B., Di Capua, R., Alfé, M., Vorokhta, M., Polichetti, T., Massera, E., Miglietta, M.L., Schiattarella, C., Di Francia, C.: J. Appl. Phys. **123**(2), 024503 (2018)
19. Papari, G.P., Gargiulo, V., Alfè, M., Di Capua, R., Pezzella, A., Andreone, A.: J. Appl. Phys. **121**, 145107 (2017)

Carbon Black as Electrode Modifier in Prussian Blue Electrodeposition for H$_2$O$_2$ Sensing

Daniel Rojas⊙, Flavio Della Pelle⊙, Michele Del Carlo
and Dario Compagnone⊙

Abstract Carbon Black Nanoparticles (CBNPs) onto Screen-Printed electrodes (SPE) are proposed as an electrode modifier for assisting electrodeposition of PB for non-enzymatic hydrogen peroxide electrochemical sensing. CBNPs allows an effective PB electrodeposition on SPE modified electrodes and enhance the electrochemical rate constant for the reduction of hydrogen peroxide (H$_2$O$_2$) compared to bare SPE.

Keywords Carbon Black · Prussian Blue · Screen-Printed electrodes
Hydrogen peroxide

1 Introduction

Nowadays nanomaterials used as sensitivity, stability and selectivity enhancers are widespread in electroanalytical chemistry [1]. In the last years Carbon Black (CB) has been used in Screen-Printed Electrodes (SPE) to modify the surface for analytical purposes. CB is a product of different industrial processes [2], which is formed by a primary structure made of spherical carbon nanoparticles with a diameter ranging from 30 to 100 nm and a secondary structure of aggregates of 100–600 nm. It has several applications in the industry as reinforcement material for rubber and plastic, as black dye and as a capacitor due to its electrical properties. The low cost, compared to other carbonaceous material like graphene and carbon nanotubes, and its versatility makes CB ideal in the electrochemistry field. Good examples are reported in the literature on the use of CB electrodes: electrochemical sensing of antioxidants [3, 4], dopamine [5], pesticides [6] and relevant biological molecules [7]. On the other hand, Prussian Blue (PB) is one of the most know electrocatalyst for H$_2$O$_2$ reduction and has been widely used for non-enzymatic sensing of H$_2$O$_2$. PB allows low potential

D. Rojas (✉) · F. D. Pelle · M. Del Carlo · D. Compagnone
Faculty of Bioscience and Technology for Food, Agriculture
and Environment, University of Teramo, 64023 Teramo, Italy
e-mail: jdrojastizon@unite.it

© Springer Nature Switzerland AG 2019
B. Andò et al. (eds.), *Sensors*, Lecture Notes in Electrical Engineering 539,
https://doi.org/10.1007/978-3-030-04324-7_43

and interference-free detection of H_2O_2 in oxygenated environment, nonetheless it has some disadvantages such as stability at physiological pH and high crystallization rate. The latter hinder potential nanostructuring of the surface and application in biological media [8]. To overcome these shortcomings modification of the electrode with soft or hard templates, polymers, carbonaceous materials or different metals are used in different combinations to design specific analytical platforms for each application. In this work we report the use of Carbon Black (CB) as electrode modifier with electrosynthesis of PB on SPE's surfaces; this resulted in improved hydrogen peroxide electrocatalysis.

2 Materials and Methods

2.1 Materials

Experiments were carried out with MilliQ water from a Millipore MilliQ (Millipore, Bedford, MA, USA), system. All inorganic salts, organic solvents and hydrogen peroxide (30% solution) were obtained at the highest purity from Sigma-Aldrich. SHSY5Y cells were obtained from Sigma (Sigma–Aldrich). H_2O_2 concentration was periodically standardized by titration with $KMnO_4$. Screen-Printed electrodes (SPE) were purchased from Dropsens S.L. (ref. SPE).

2.2 Instrumentation

All electrochemical measurements were carried out in Autolab PGSTAT 12 potentiostat from Metrohm (Utrecht, The Netherlands) connected to a personal computer. The software used was Nova 2.1 (EcoChemie B.V.). The flow injection (FIA) system consisted on a Minipuls 3 (Gilson Inc., Middleton, WI, USA) peristaltic pump, wall-jet cell (ref. FLWCL) (Dropsens, Spain). Sample volume was 50 μL, the working electrode potential chosen was −50 mV (vs. internal reference). The running buffer solution in FIA experiments was 0.05 M phosphate buffer pH 7.4 containing 0.1 M KCl.

2.3 Preparation of SPE-PB and SPE-CB-PB Electrodes

A CBNPs dispersion of 1 mg/mL in water and dimethylformamide (DMF) (1:1 ratio) was prepared. The dispersions were obtained using a bath sonicator for 30 min. SPE-CB electrodes were prepared by drop-casting 10 μL of CBNPs. Prussian Blue electro-deposition was carried out cycling the potentials between +400 and +800 mV (vs. int. ref) for different number of cycles in a solution containing 0.1 M KCl, 0.1 M HCl and 5 mM concentration of Fe^{3+} and $[Fe(CN)_6]^{3-}$. When the electrodeposition is carried out on bare SPE we obtain SPE-PB, in the case of using SPE-CB we obtain SPE-CB-PB. The potential of the internal reference electrode against an Ag|AgCl|KClsat reference electrode was measured as -120 mV. Electrodes were further modified with 2 μL of a Nafion ethanolic solution (0.5% v/v). Current density was calculated considering the geometrical area of SPE.

3   Results and Discussion

PB is usually synthetized using Fe^{3+} and $[Fe(CN)_6]^{3-}$ anions as precursors. Usually, ferric ions are selectively reduced to form Prussian Blue on common electrode's surfaces such as platinum, gold or glassy carbon. The selection of the potential at which PB is electrodeposited is crucial to allow the selective reduction of the precursors since the reduction of both anions at the same time leads to an irregular structure of PB, hindering its potential for the electroreduction of H_2O_2. As shown in the voltammogram of Fig. 1, the reduction peaks of Fe^{3+} and $[Fe(CN)_6]^{3-}$ are not resolved using SPE. Interestingly, in the case of CB-SPE the peak potential of Fe^{3+} is anodically shifted of 280 mV, and peak current is increased by a 1.5 factor confirming an electrocatalytic behavior of CB toward Fe^{3+} reduction. On the other hand, for the reduction of $[Fe(CN)_6]^{3-}$, only a slight increase of peak intensity was observed, keeping the peak potential at the same value of unmodified SPE. A study on the effect of electrode surface coverage on the electrocatalytic rate was then run cycling the potential between +400 and +800 mV for different times to obtain different surface coverage of PB.

The bare (SPE) and CB-modified SPE (CB-SPE) were modified using 5 to 20 electrodeposition cycles. The electrocatalytic properties of these electrodes were evaluated considering the electrochemical rate constant (k) calculated using a wall-jet electrode configuration in continuous flow. This experimental approach uses a semi-empirical model which allows to separate the kinetic and mass transport terms contribution to total current density [9]. Equation 1 shows the relationship between the kinetic term (expressed as the kinetic constant (k)) and the mass transport term (expressed as the effective mass transfer coefficient (k_D)). As shown in Eq. 2, k_D is proportional to the ¾ power of the flow rate and to a b parameter dependent on the flow cell geometry (for further information, lector is referred to Ref. [9]).

$$\frac{1}{j} = \frac{1}{nFC_0}\left(\frac{1}{k_D} + \frac{1}{k}\right) \tag{1}$$

$$k_D = bV^{3/4} \tag{2}$$

Thus, one can obtain k by the intercept of the regression of j^{-1} versus $V^{-3/4}$, Fig. 2 shows the linear regression for different electrodes. As seen in Table 1, the

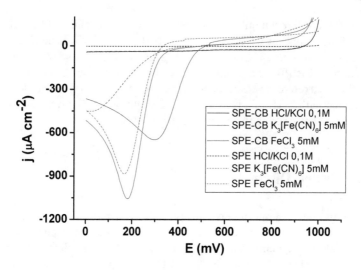

Fig. 1 LSV of $K_3[Fe(CN_6]$ and $FeCl_3$ on SPE and SPE-CB in HCl/KCl 0.1 M solutions

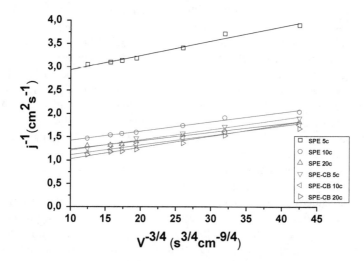

Fig. 2 Dependence of the current density on the flow rate for different electrodes

electrochemical rate constant is higher for SPE-CB for a given cycle number. Even, SPE with 20 cycles has a k comparable to SPE-CB with only 5c.

This highlights the advantage of using CB as an effective SPE modifier for PB electrodeposition and further electrocatalysis of H_2O_2. Calibration curves obtained with SPE-PB 20c and SPE-CB-PB 20c gave a linear range of 0.5–500 μM, detection limits of 0.11 and 0.09 μM and sensitivities of 0.27 and 0.47 μA μM^{-1} cm^{-2} respectively.

Table 1 Electrochemical rate constant of each electrode

Electrode	k (cm s^{-1}) 10^3
SPE-PB 5c	0.39
SPE-PB 10c	0.84
SPE-PB 20c	0.98
SPE-CB-PB 5c	1.04
SPE-CB-PB 10c	1.14
SPE-CB-PB-20c	1.42

4 Conclusions

CBNPs were successfully employed as SPE modifier for electrodeposition of PB from the precursors Fe^{3+} and $[Fe(CN)_6]^{3-}$. The CB-SPE were able to shift the reduction potential of ferricyanide 280 mV, allowing the selective electroreduction of the two precursors. As previously reported in literature, the reduction of both precursors at same time leads to an irregular crystalline structure of PB, lowering it electrocatalytic performance towards H_2O_2 reduction. In fact, a higher electrochemical rate constant was found when PB was electrodeposited on CB-SPE-PB in contrast to SPE-PB. Thus, a higher electrocatalytic activity in CB-SPE-PB was demonstrated. These results are the basis for further developing of these electrodes for non-enzymatic electrochemical sensing of H_2O_2 and biosensing of analytes of interest linking peroxidase enzymes to electrode's surface.

References

1. Pelle Della, F., Compagnone, D.: Nanomaterial-based sensing and biosensing of phenolic compounds and related antioxidant capacity in food. Sensors **18**, 462 (2018)
2. McCreery, R.L.: Advanced carbon electrode materials for molecular electrochemistry. Chem. Rev. **108**, 2646–2687 (2008)
3. Della Pelle, F., Di Battista, R., Vázquez, L., Palomares, F.J., Del Carlo, M., Sergi, M., et al.: Press-transferred carbon black nanoparticles for class-selective antioxidant electrochemical detection. Appl. Mater. Today. Elsevier **9**, 29–36 (2017)
4. Carlo Del, M., Amine, A., Haddam, M., Pelle, F., Fusella, G.C., Compagnone, D.: Selective voltammetric analysis of o-diphenols from olive oil using Na_2MoO_4 as electrochemical mediator. Electroanalysis **24**, 889–896 (2012)
5. Della Pelle, F., Vázquez, L., Del Carlo, M., Sergi, M., Compagnone, D., Escarpa, A.: Press-printed conductive carbon black nanoparticle films for molecular detection at the microscale. Chem.—A Eur. J. **22**, 12761–12766 (2016)
6. Della Pelle, F., Del Carlo, M., Sergi, M., Compagnone, D., Escarpa, A.: Press-transferred carbon black nanoparticles on board of microfluidic chips for rapid and sensitive amperometric determination of phenyl carbamate pesticides in environmental samples. Microchim. Acta. **183**, 3143–3149 (2016)
7. Della Pelle, F., Va, M.V., Blando, L., Santamarı, A.: Electrochemical behaviour of microwave-assisted oxidized MWCNTs based disposable electrodes: proposal of a NADH electrochemical sensor. Electroanalysis **30**, 1–9 (2018)

8. Chu, Z., Liu, Y., Jin, W.: Recent progress in Prussian blue films: Methods used to control regular nanostructures for electrochemical biosensing applications. Biosens. Bioelectron. **96**, 17–25 (2017)
9. Karyakin, A.A., Karyakina, E.E., Gorton, L.: The electrocatalytic activity of Prussian blue in hydrogen peroxide reduction studied using a wall-jet electrode with continuous flow. J. Electroanal. Chem. **456**, 97–104 (1998)

Part VI
Sensing Systems

PPG/ECG Multisite Combo System Based on SiPM Technology

Vincenzo Vinciguerra, Emilio Ambra, Lidia Maddiona, Mario Romeo, Massimo Mazzillo, Francesco Rundo, Giorgio Fallica, Francesco di Pompeo, Antonio Maria Chiarelli, Filippo Zappasodi, Arcangelo Merla, Alessandro Busacca, Saverio Guarino, Antonino Parisi and Riccardo Pernice

Abstract Two versions of a PPG/ECG combined system have been realized and tested. In a first version a multisite system has been equipped by integrating 3 PPG optodes and 3 ECG leads, whereas in another setup a portable version has been carried out. Both versions have been realized by equipping the optical probes with SiPM detectors. SiPM technology is expected to bring relevant advantages in PPG systems and overcome the limitations of physiological information extracted by state of the art PPG, such as poor sensitivity of detectors used for backscattered light detection and motion artifacts seriously affecting the measurements repeatability and pulse waveform stability. This contribution presents the intermediate results of development in the frame of the European H2020-ECSEL Project ASTONISH (n. 692470), including SiPM based PPG optodes, and the acquisition electronic components used for simultaneous recording of both PPG/ECG signals. The accurate monitoring of dynamic changes of physiological data through a non-invasive integrated system, including hemodynamic parameters (e.g. heart rate, tissue perfusion etc.) and heart electrical activity can play an important role in a wide variety of applications (e.g. healthcare, fitness and cardiovascular disease). In this work we describe also a method to process PPG waveform according to a PPG process pipeline for pattern recognition. Some examples of PPG waveform signal analysis and the preliminary results of acquisitions obtained through the intermediate demonstrator systems have been reported.

Keywords PPG · SiPMs · Pattern recognition

V. Vinciguerra (✉) · E. Ambra · L. Maddiona · M. Romeo · M. Mazzillo · F. Rundo · G. Fallica
STMicroelectronics, ADG Central R&D, Stradale Primosole 50, 95121 Catania, Italy
e-mail: vincenzo.vinciguerra@st.com

F. di Pompeo · A. M. Chiarelli · F. Zappasodi · A. Merla
G. D'Annunzio University of Chieti-Pescara-Italy, Via dei Vestini, 33, 66100 Chieti, Italy

A. Busacca · S. Guarino · A. Parisi · R. Pernice
Università di Palermo, Viale delle Scienze, Bldg. no. 9, 90128 Palermo, Italy

© Springer Nature Switzerland AG 2019
B. Andò et al. (eds.), *Sensors*, Lecture Notes in Electrical Engineering 539,
https://doi.org/10.1007/978-3-030-04324-7_44

1 Introduction

PhotoPlethysmoGraphy (PPG) is a noninvasive optical technique that measures blood volume changes during the heart pulsation. PPG is widely used in commercial and clinical devices to evaluate cardiovascular indicators such as oxygen saturation, beat to beat pressure and arterial compliance [1]. In a typical PPG system, visible or infrared wavelength photons coming from a light emitting diode (LED) go through the skin layers up to the underlying tissues and are revealed by a photodetector either for the case of backscattered or transmitted beams. PPG signal and its physiological relation have been widely studied [2]. Recorded pulse has a direct relationship with perfusion and the greater the blood volume the more the light source is attenuated. Since the arterial volume changes during each cardiac cycle, because of the propagating pulse pressure wave, the upcoming light is modulated accordingly. The pulsatile (AC) component of the signal, related to the heart beat, is superimposed to a much larger slow oscillating (DC) component related to tissue and baseline blood volume absorption. Although PPG measurements, because of light diffusion, generally integrate the signal coming from blood vessels of different caliber, different experimental setups can be more sensitive to one vessel type with respect to the others [2]. PPG measurements can be performed in backscattering and/or transmission mode. Most of the commercial devices work in transmission mode on the fingertip or earlobe. In these devices, the tissue is irradiated by a LED and the light is measured by a photodiode on the other side of the tissue. These measurements are limited to microvascular assessment on specific body sites. In fact, larger arteries lie deeper in the tissue and are difficult to be inspected. The depth sensitivity, affecting the capability of exploring the larger arteries, depends on both the wavelength [3] and the sensitivity of the detector employed.

In this perspective the use of Silicon Photomultipliers (SiPMs) operating with intrinsic avalanche gain (up to 10^6) and single photon sensitivity (down to 5–10 detected photons) is expected to bring relevant advantages in PPG systems in terms of higher AC-to-DC ratio in PPG pulse waveform, high repeatability and immunity to motion artifacts and reduced power consumption [4]. The enhanced sensitivity of the SiPMs can add new measurements capabilities to existing PPG solo systems (e.g. blood pressure) or in multisite PPG systems (e.g. pulse wave velocity (PWV)), opening new interesting markets for these devices. Besides the expected innovation in PPG systems, the accurate monitoring of dynamic changes of physiological data through a non-invasive integrated system, including hemodynamic parameters (e.g. heart rate, pulse wave velocity etc.) and heart electrical activity, can play an important role in a wide variety of applications (e.g. healthcare, fitness and cardiovascular disease). To this purpose, there is also a great interest to develop integrated, low-power consumption and portable photoplethysmography-electrocardiography (PPG/ECG) combo systems for assessing the above-mentioned physiological parameters and their ubiquitous monitoring over time. The EU H2020-ECSEL ASTONISH project (grant agreement 692470) [5] will deliver an advanced multisite PPG-ECG combo system for the assessment and monitoring of relevant cardiovascular diseases occur-

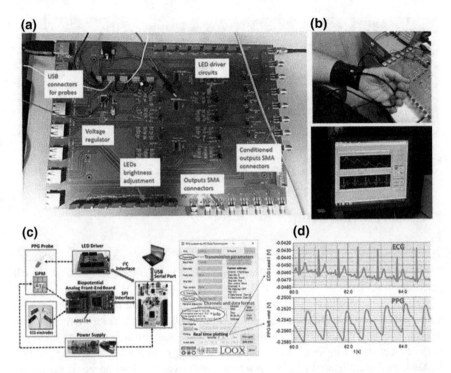

Fig. 1 SiPM based PPG/ECG combo systems demonstrators. **a** Interface board containing suitable circuits for LED and SiPMs operation. **b** IR PPG transmission wristband with USB communication data transfer and comparison of the PPG/ECG signals of the multisite system of ST. **c** Schematic of the portable PPG/ECG demonstrator with two channels realized at UNIPA. **d** Example of synchronized PPG and ECG signals. PPG pulsatile component is characterized by a larger peak of absorption (local minima) during the diastolic phase in which the blood volume change is maximum. A secondary small peak or a bump, due to wave reflection, is also visible. The latency between the ECG R-peak (the largest one) and the PPG onset reflects the propagation of the shock wave in the vascular system

ring as result of ageing, hypertension, and atherosclerosis, providing also information on arterial stiffness. This will be done through a proper combination of advanced technology blocks including high sensitivity SiPM based PPG probes, innovative electronic hardware for fully synchronized multisite measurements, advanced algorithms for data filtering, analysis and pattern recognition and suitable software interface for measurements calibration, data recording and analysis. In this contribution, we present the intermediate results of this development including SiPM based PPG optodes, the acquisition electronics components used for simultaneous PPG/ECG measurements (Fig. 1) and a PPG pattern recognition pipeline [6]. In particular, the preliminary acquisitions obtained through an intermediate demonstrator system composed by 3 PPG optodes and 3 ECG leads along with some examples of PPG waveform signal analysis are reported (Fig. 2).

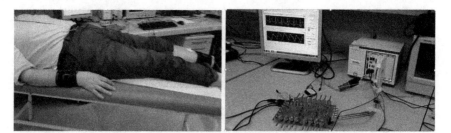

Fig. 2 Experimental setup and first measurements obtained on an intermediate PPG/ECG combo system composed by 3 synchronized PPG probes and 3 ECG leads

2 Experimental Setup

The PPG probes used for the development of the PPG/ECG combo systems introduced in the previous section were equipped with large area n-on-p SiPM [7–10] detectors manufactured at STMicroelectronics in Catania [10].

The devices were fabricated on p-type silicon epitaxial wafers and formed of n^+-p microcells. N-on-p SiPMs were chosen for this application considering their higher photon detection efficiency (PDE) in the visible and near infrared wavelength ranges with respect to the p-on-n version of the technology. The probe consisted of two LEDs 1 cm distant from the SiPM, which was at the center of the probe. Used SiPMs have a geometrical fill factor of 67.4%. Figure 1b shows an IR PPG transmission wristband with USB communication data transfer. The SiPMs array is packaged in a surface mount housing (SMD) with 5.1×5.1 mm^2 total area. Light source was a LED emitting at 940 nm wavelength. The SiPMs were equipped with an optical "long-pass" plastic filter having a cut-off wavelength at $\lambda = 700$ nm in the near infrared [11]. The SiPM was mounted on a small PCB, about 3×2 cm^2, containing only the detector and passive components. To have a probe with a smooth front surface, a sheet of black rubber was applied, with suitable holes in correspondence of sources and detectors and in such a way to assure optical isolation between sources and detectors. This very simple probe was highly flexible and comfortable. Given the average travel speed of the pulse pressure-wave, a time resolution in the range of milliseconds is needed. This has been achieved using a high resolution (24 bit) National Instruments NI-PXIe-4303 ADC for both PPG and ECG. This device has also a very high synchronization time between two different channels which is significantly lower than 1 μs. An interface board (Fig. 1a) has been internally developed including the power supply, the amplification stages and conditioning circuits for output SiPMs signals, the LED driver circuits and suitable connectors for PPG optodes and ECG probes. This system will allow multisite PPG and ECG measurements up to a total of 32 channels. Measurements have been conducted in backscattering mode on the radial and tibial arteries. Simultaneously, the subject ECG was acquired to monitor the heart activity and to assess the PPG Pulse Transit Time. In a typical measurement session, the subject was lying in supine position and instructed to avoid any unnecessary movement. Three probes were placed in correspondence of the right and left wrist radial artery and left ankle as shown in Fig. 2. Following the ECG standard procedure, three electrodes have been placed to acquire LEAD I, LEAD II and LEAD III derivations.

3 Data Analysis

The assessment of quality of PPG waves was conducted by using a fully automatic processing pipeline. A toolkit based on Matlab scripts and functions has been implemented according to the following four main steps:

- QC: overall **quality check** assessment
- AI: **artifact identification**
- PRW: **pattern recognition** of PPG **waveforms**
- FDA: **First derivative analysis** for the assessment of the main PPG parameters.

In the first step (QC), the overall quality of the PPG raw data was considered. The pulsatile (AC) and the slow varying (DC) contents of the signal were considered and compared. Pulsatile component was obtained by filtering the raw data with a band pass zero-lag Butterworth filter in the 1–10 Hz frequency range. Parameters used were the AC/DC ratio and the variation of the AC and DC amplitudes with respect to their initial values. The second parameter can be slightly varying with the pressure of the probe over the skin [2]. Thus, the stability of the AC/DC ratio was used to check for possible decoupling of the probe from the subject skin. Without any external pressure of the probe, the AC/DC ratio is generally larger than 1% and smaller than 10%. The second step (AI) was an automated artifact detection algorithm based on the evaluation of the variance of the signal over time. Artefact affected signal epochs were identified by means of a statistical threshold and discarded. An iterative procedure allowed the identification of all the contaminated periods.

The Pattern Recognition Algorithm (Fig. 3) allowed to identify good PPG waveforms with an accuracy above 90%.

The proposed pattern recognition system supposes to find compliant PPG waveforms between two minimum local points so that firstly performs a simple segmentation of the pre-filtered PPG signals through the detected local minimum.

Afterwards, the bio-inspired core of the proposed pipeline performs ad hoc nonlinear dynamic evolution.

Pattern recognition of PPG waveforms relies on a theoretical waveform derived from a reaction diffusion nonlinear mathematical model properly configured [12, 13]. The nonlinear mathematical model was implemented by means of Cellular Neural Networks (CNNs) as reported in [13]. Due to its analogic implementation, the CNNs are able to perform such operations with high-speed computational capability i.e. near real-time.

Ad hoc normalized sample cross-correlation analysis was performed in order to have a compliance measure for the analyzed PPG waveform. Only the PPG patterns

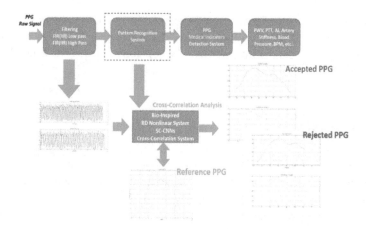

Fig. 3 PPG pattern recognition pipeline

showing high normalized sample cross-correlation (≥ 0.90) were considered as compliant whereas the other ones were discarded [6]. The results of the proposed pipeline applied for PPG signal processing confirmed the robustness and effectiveness of the approach herein described showing very promising sensitivity and specificity. The developed algorithm showed encouraging results in discriminating good PPG signals and artifacts. This novel approach may be extended to the classification of PPG waveforms of different physiological and pathological conditions. Some pulse parameters were obtained by searching for local maxima and minima of the pulse in the proper order [14]. A sample of PPG waveforms together with the related features is reported in Fig. 4. In the last step, a further standard PPG first derivative analysis was performed on the selected waveforms [14]. Good epochs were divided in trials corresponding to single heart beats. The ECG was used to define the trial onset based on the ECG R-peak. In this way, all the parameters found were time-locked to the heart cycle. Waveform foot (foot), primary peak (P1), dicrotic notch (DN) and the secondary peak (P2) were the main pipeline output. Figure 5 shows the first derivative analysis on the average PPG waveform of the radial site as an example.

The parameters were obtained by searching for local maxima and minima of the pulse in the proper order [14]. A sample of PPG waveforms together with the related features is reported in Fig. 5.

Fig. 4 Examples of features identification of a collection of PPG waveforms

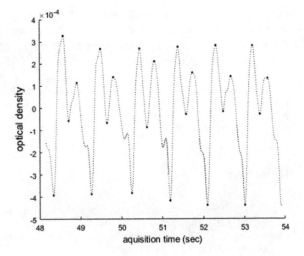

Fig. 5 First derivative analysis of the waveform obtained by averaging over the selected recorded PPG trials

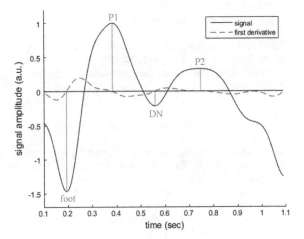

4 Conclusions

The results presented in this work demonstrate the capabilities of the SiPM photodetectors technology in combination with ECG electrodes in order to build a multisite and a portable PPG/ECG system to be employed for PPG measurements in a backscattering mode. Measurements were performed by using infrared light of an LED used as optical light source and a suitable optical longpass filter on SiPM detector for environmental light rejection. For both the systems, to analyze data, we have implemented also an automatic process pipeline able to perform data cleaning, selection of good PPG trials and carry on a first derivative analysis. A novel approach based on pattern recognition was also presented. The analysis showed the feasibility of the pipeline on the acquired data. Future directions may exploit SiPMs high sensitivity and large dynamic range by employing them in the PPG evaluation of deep arteries for which a small pulsatile signal is expected. These measurements could be conducted both in backscattering and transmission mode and in different body sites than just fingers and ear lobes.

Acknowledgements The research activity leading to the results shown in this work was partially funded from the H2020-ECSEL Joint Undertaking under grant agreement n° 692470 (ASTONISH Project).

References

1. Allen, J.: Photoplethysmography and its application in clinical physiological measurement. Physiol. Meas. **28**(3), R1–R39 (2007)
2. Reisner, A., et al.: Utility of the photoplethysmogram in circulatory monitoring. J. Am. Soc. Anesthesiol. **108**(5), 950–958 (2008)
3. Liu, J., Yan, B.P.-Y., Dai, W.-X., Ding, X.-R., Zhang, Y.-T., Zhao, N.: Multi-wavelength photoplethysmography method for skin arterial pulse extraction. Biomed. Opt. Exp. **7**(10), 4313–4326 (2016)
4. Mazzillo, M., et al.: Silicon photomultiplier technology at STMicroelectronics. IEEE Trans. Nucl. Sci. **56**(4), 2434–2442 (2009)
5. ASTONISH project homepage. http://astonish-project.eu. Accessed 20 Mar 2018
6. Vinciguerra, V., et al.: Progresses towards a processing pipeline in photoplethysmogram (PPG) based on SiPMs. In: 2017 European Conference on Circuit Theory and Design (ECCTD), pp. 1–5 (2017)
7. Agrò, D., et al.: PPG embedded system for blood pressure monitoring. In: Proceedings of AEIT International Annual Conference (AEIT), pp. 1–6 (2014)
8. Pagano, R., et al.: Improvement of sensitivity in continuous wave near infra-red spectroscopy systems by using silicon photomultipliers. Biomed. Opt. Express **7**(3), 249662 (2016)
9. Chiarelli, A., et al.: Wearable continuous wave functional near infrared spectroscopy system based on silicon photomultipliers detectors: in-vivo assessment of primary sensorimotor response. Neurophotonics **4**(3), 035002 (2017)
10. Buzhan, P., et al.: Silicon photomultiplier and its possible applications. Nucl. Instrum. Methods Phys. Res. A **504**(1–3), 48–52 (2003). Bland, J.M., Altman, D.G.: Measuring agreement in method comparison studies. Stat. Methods Med. Res. **8**(2), 135–160 (1999)
11. Mazzillo, M., et al.: Noise reduction in silicon photomultipliers for use in functional near-infrared spectroscopy. IEEE Trans. Radiat. Med. Plasma Sci. **1**(3), 212–220 (2017)
12. Rundo, F., Battiato, S., Giuffrida, E.U.: A cellular neural network for zooming digital colour images. In: International Conference on Consumer Electronics, pp. 1–2 (2008)
13. Arena. P., et al.: A CNN-based chip for robot locomotion control. IEEE Trans. Circuits Syst. **52**(9) (2005)
14. Elgendi, M.: On the analysis of fingertip photoplethysmogram signals. Curr. Cardiol. Rev. **8**(1), 14–25 (2012)

A Small Footprint, Low Power, and Light Weight Sensor Node and Dedicated Processing for Modal Analysis

Federica Zonzini, Luca De Marchi and Nicola Testoni

Abstract Structural Health Monitoring functionalities are aimed at constantly assessing the health of a building in order to prevent dramatic consequence of a damage. This work describes a well-defined wireless sensor network system installed over a steel beam capable to perform modal parameters estimation, such as natural vibration frequencies and modal shapes. Signal Processing Techniques were aimed at computing Power Spectral Density of the acceleration signals acquired, dealing with parametric and non parametric approaches. Algorithms in frequency domain, together with the Second Order Blind Identification method were implemented for modal shapes reconstruction. Beside a satisfactory agreement between the theoretical model and the output response of the algorithms implemented, versatility, easiness of reconfiguration, scalability and compatibility with long term installation are among the most powerful advantages of the architecture proposed. Light weight, low power consumption also enhance the capabilities of the system to provide real-time information in a relatively cheap way.

Keywords Modal analysis · Structural health monitoring · Low power

1 Introduction

The actual trends in civil engineering increased the capability to build complex structures, designing elements which are characterized either by the peculiarity of their architectural properties or by composing materials. This evolution in structural development requires continuous information to be extracted and analyzed in order to assess the integrity of buildings, preferably according to automatic techniques which

F. Zonzini (✉) · L. De Marchi · N. Testoni
DEI, University of Bologna, Viale Del Risorgimento 2, 40136 Bologna, Italy
e-mail: federica.zonzini@studio.unibo.it

L. De Marchi
e-mail: l.demarchi@unibo.it

N. Testoni
e-mail: nicola.testoni@unibo.it

© Springer Nature Switzerland AG 2019
B. Andò et al. (Eds.) *Sensors*, Lecture Notes in Electrical Engineering 539,
https://doi.org/10.1007/978-3-030-04324-7_45

simplify and accelerate data acquisition and management. Furthermore, power consumption, costs and weight shall be minimized not only for energetic reasons but also to make the presence of devices as unobtrusive as possible. Historical buildings can also benefit from these studies, since nowadays most of them are used for different purposes. Consequently, evaluating the robustness of infrastructures becomes fundamental for human safety.

In such a scenario, electronic devices have been designed according to Structural Health Monitoring (SHM) requirements, primarily with the intent of providing a constant monitoring of the structure under test and correspondingly identifying damages in case of occurrence. These two aspects together enable to generate periodic reports about the state of the structure over a wide period of time, mainly related to the effects that usual stresses (i.e. wind, weather) can evoke. Traditional systems performance and requirements have primarily demonstrated to produce expensive designs and steady deployment, connected respectively to the high accuracy and sensitivity of sensors and to the fixed positions associated to each of them. Moreover, problems arise when dealing with long-term acquisition phases due to the presence of piezoelectric transducers whose cost compel researchers to limit the duration of experimental campaigns and reuse them. Another important aspects to be underlined is a strongest consciousness of the fact that the signals recorded by piezoelectric devices are not completely suited to investigate the dynamic behavior and the stability of infrastructures with sufficient precision.

Recently, evident improvements have brought about distributed sensor network solutions which provide embedded data processing, mainly achieved by installing a certain number of nodes along the whole structure. The usage of Micro-Electro Mechanic (MEM) sensors provides, among the many benefits, versatility, scalability, non-invasiveness and long-term analysis. Positions can be changed, acquisition process reconfigured digitally without any particular consequence for the structure, hence containing obtrusiveness. Synchronized vibrations recorded by inertial elements within nodes installed in any point of the building can be examined in order to compute modal parameters, which consists of natural frequencies, damping ratio and modal shapes. The basic idea of SHM is to continuously analyze the internal vibration features of a building and compare them to structural properties measured under nominal conditions, thus revealing possible changes in normal modes of oscillation which may differ for number, width and frequency. In particular, when damages do not occur, the behavioral dynamics is unchanged and only minimal drifts can be noticed.

This work focuses on a sensor network with embedded data processing for real-time SHM, specifically developed to cope with low power consumption, light weight, small size requirements. In order to assess the functionality of the system, vibrational measurements were recorded from a steel beam undergoing a mechanical stress. The presented setup is simple but efficient for these purposes, being very common in modal analysis scenarios. In the second section a general overview of the node is presented, followed by Sect. 3 in which the description of the algorithms implemented to evaluate natural frequencies and modal shapes of the experimental setting, proving the reliability of the nodes in a SHM context.

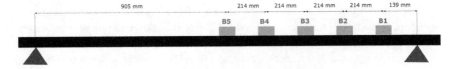

Fig. 1 Experimental setup for modal analysis verification and comparison

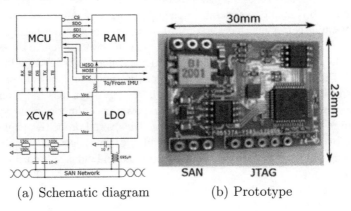

(a) Schematic diagram   (b) Prototype

Fig. 2 Developed sensor node

2 Sensor Node

In this paper, a simple but effective setup for modal analysis was developed to assess the capability of the proposed sensor node network to cope with Real-time Structural Health Monitoring (SHM) functionalities. The importance of this discipline relies on the capability to periodically analyze the vibration features of a structure through a network of smart sensors acquiring information from a plurality of transducers. Whenever a change is detected with respect to the ordinary structural properties, the state of the entire system under test can be inferred. As reported in Fig. 1, five sensor nodes equally spaced were installed in the first half of a $L = 1900$ mm steel beam with cross-section base $b = 60$ mm and height $h = 10$ mm in free vibrations conditions.

Each sensor is roughly 30 mm × 23 mm, drawing power from a Sensor Area Network (SAN) bus based on Data-over-Power (DoP) communication which simultaneously allows for data transmission and power supply. At an architectural level, four main blocks cooperate to achieve data collection and processing as schematically depicted in Fig. 2.

Following the flow of information, real-time acceleration samples are recorded by means of a ST Microelectronics LSM6DSL iNEMO Inertial Measurement Unit (IMU) with a 3D accelerometer characterized by a maximum dynamic range of ±16 g. They are subsequently sent to a ST Microelectronics Microcontroller Unit (MCU) which is the core unit of this device, especially designed for low consumption even in nominal condition. It includes a 40 KiB SRAM memory for temporary storage of the samples, but also an embedded 256 KiB FLASH memory is present giving the opportunity to implement signals preprocessing capabilities. The presence of a Serial Peripheral Interfaces (SPI) and Universal Synchronous Asynchronous Receiver Transmitter (USART) enables to transmit input/output serial data independently of the nominal execution of tasks. Collected data are sent to an external SRAM memory which is mainly inserted to expand the storage capability of the device. A ST XCVR transceiver interfaces the MCU to the bus thanks to a mesh of passive elements. Additionally, recordings flow through a gateway to be sent to a user defined cloud system. A Low-Drop Out regulator provides energy, establishing the voltage to 3.3 V. The sensor node so far described performs an overall consumption lower than 40 mW, with small size and light weight, being it less than 5 g. Considering that five sensors have been installed, the global power in continuous monitoring does not go over 300 mW, weighing less than 50 g. The overall properties explained are enforced by the extreme scalability and versatility of the sensor node implemented, considering that up to 64 elements can be connected and managed sharing the same bus.

3 Modal Estimation

3.1 Natural Frequency Estimation

Signal Processing Techniques have been applied to the raw acquired signals to extract natural frequencies and modal shapes. The results of the implemented algorithms were compared with the theoretical predictions obtained from the physical model of the beam to assess the performances of the designed circuitry. In particular, the extracted three natural frequencies are compared with the nominal values estimated as 5.28, 21.14, and 47.51 Hz through the theoretical formula:

$$f_n = \frac{1}{2\pi} \frac{(n \cdot \pi)^2}{L^2} \sqrt{\frac{E \cdot I}{A \cdot \rho}} \tag{1}$$

where $\rho = 7880 \, \text{kg/m}^3$ is the density of the steel, $A = bh$ and $I = bh^3/12$ are respectively the cross-section area and moment of inertia respectively, and n is the frequency index. Here, for accurately measured beam dimensions and weight, a value of the Youngs modulus $E = 195 \, \text{GPa}$ was assumed to minimize the error on f_1.

The evaluation of the modal parameters in a phase when no damage occurs is of extreme importance because it creates a set of benchmark values that must be as accurate as possible in order to prevent potential errors during the real-time estimation stage.

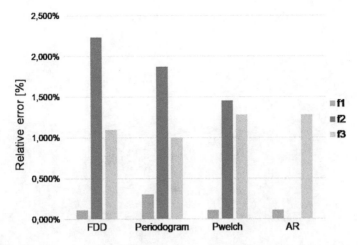

Fig. 3 Error distribution in vibrational modes extractions comparing different techniques of PSD estimation

Several procedures were considered to compute the Probability Density Function (PSD) of the data series, based on non-parametric and parametric approaches. Among the former, we considered the Frequency Domain Decomposition (FDD) [1] periodogram estimation [2] Welch evaluation [2] whilst the latter refer to Autoregressive (AR), and AR+Noise models [2] which are suited for modal analysis in presence of strong acquisition noise.

Results The analysis of the results obtained can be deployed at two different levels. First, according to the relative error depicted in Fig. 3, we may consider that the error is not uniformly distributed over the whole spectrum, since it is mostly relevant at higher frequencies where Signal to Noise Ratio (SNR) is worse and the overall energy of the structure is lower. Second, it is worth noticing that AR models seem to have the best performance, allowing to assert that noise may affect the measurements and it has to be correctly considered. Percentages are always lower than 2.5%, thus assessing the reliability of the architecture proposed. Furthermore, since the second vibrational mode of the structure is detected with the lowest precision, it is possible to argue that it may be related to the solicitation induced over the beam. Generally, the PSD computed with the techniques mentioned, demonstrates that, for the first part of the spectrum, there is an evident vertical alignment between the frequencies computed and those expected, meanwhile the output is less precise when frequency increases. Furthermore, as predicted by the model and clarified by the spectrum reported in Fig. 4, an additional frequency near 97 Hz can be detected. This fourth vibrational mode was not considered for further studies since it is close to Nyquists frequency. For this reason, the errors can creep due to the scarcity of samples to be used in order to recreate the associated modal shapes, notwithstanding the fact that it can be extracted from the spectrum itself.

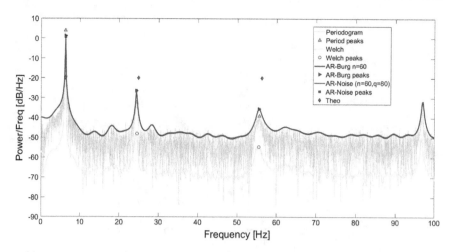

Fig. 4 Example of PSD estimation with parametric and non parametric approaches

3.2 Modal Shapes Reconstruction

Beside data and power communication, the bus connecting the sensor nodes also natively allows for data acquisition time base synchronization and consequently the output-only estimation of modal shapes. For this purpose, algorithms in frequency domain were developed, followed by the application of the Second Order Blind Identification (SOBI) method, a strategy which reveals independent components hidden within a set of measured signal mixtures [3].

Frequency Domain Decomposition Technique Frequency Domain Decomposition (FDD) method identifies modal parameters of a dynamic system by applying the Singular Value Decomposition (SVD) technique to the output spectral density matrix [1]. This algorithm works as an output-only estimation technique, whose computation merely consists of two different steps: given a data set, it estimates the PSD and consequently filters n dominating peaks, where n refers to the degree of freedom of the system under consideration. Formally, its operating principle relies on the key-function named Frequency Response Function (FRF) matrix:

$$[G_{yy}(\omega)] = [H(\omega)][G_{xx}(\omega)][H(\omega)]^H \tag{2}$$

where $[G_{xx}(\omega)]$ and $[G_{yy}(\omega)]$ denote respectively PSD matrix of the inputs and the outputs, $[H(\omega)]$ represents FRF matrix and ^{H} is the conjugate transpose operator. Applying the SVD to the output spectrum $[G_{yy}(\omega)]$, it is obtained

$$[G_{yy}(\omega)] = [U][V][U]^H \tag{3}$$

with [U] being the orthogonal matrix of the singular vectors and [V] is the singular value diagonal matrix organized by column. Concerning the recorded data we have that

$$y(t) = [\Phi]q(t) \tag{4}$$

where the response of the structure $y(t)$ comes from the decomposition of these output signals into participations from the different modes $[\Phi]$ expressed via the modal coordinates $q(t)$. Using the Correlation matrix and applying the SVD to it, it is finally demonstrated that

$$[G_{yy}(\omega)] = [\Phi][G_{qq}(\omega)][\Phi]^H \tag{5}$$

The assumptions are that $[G_{yy}(\omega)]$ is a diagonal matrix, i.e. the modal coordinates are uncorrelated, and that the mode shapes (the columns in $[\Phi]$) are orthogonal. Comparing (3)–(5) and assuming that the decomposition described by (3) is unique, it follows that the singular vectors coincide with the estimation of the mode shapes and the corresponding singular values present the response of each mode (Single Degree of Freedom systems) expressed by the spectrum of each modal coordinate.

Second Order Blind Identification Technique SOBI algorithms found their theoretical formulation on the assumption that the second order momentum, that is the expected covariances, is completely representative of the observed data. Considering a set of acquired signals $y(t)$, (6) defines covariances between the values of two different signals $y_i(t)$ and $y_j(t)$, where τ is the time-lag or delay:

$$[C_y^\tau]_{i,j} = \text{cov}(y_i(t + \tau)y_j(t)) \tag{6}$$

Furthermore, for $i = j$, (6) turns into the auto-covariances $[C_y^\tau]_i$ between the same signal $y_i(t)$ at different time steps. Combining together these quantities, this technique exploits the diagonalization of the time-lagged covariance matrix, resulting as in (7)

$$[C_x^\tau] = \mathbb{E}\left\{x(t + \tau)x^H(t)\right\} \tag{7}$$

where ^{H} denotes the conjugate transpose operator.

The time structure introduced contributes to relax the requirement of non Gaussianity for the Independent Components (ICs), replaced by the condition that all the independent sources have different and nonzero auto-covariances. As described in [3], imposing that the covariances for $\tau = 0$ have zero value, the decorrelation between the observed signals $y(t)$ can be assured. As a consequence, the obtained signals satisfy the following property:

$$[C_x^\tau] = 0, \tau = 0 \tag{8}$$

Nevertheless, the correlation matrix does not contain enough information for the separation. In fact, due to the independence property of the ICs, all their lagged covariances, and not only one, should be zero:

$$[C_x^\tau] = 0, \forall \tau \tag{9}$$

The SOBI estimation is therefore performed starting from the assumption that each observed data $y(t)$ is obtained as a linear combination of unknown sources $x(t)$ through a mixing matrix [A]:

$$y(t) = [A]x(t) \tag{10}$$

enforcing that time-lagged covariances are zero as well. This leads to

$$\mathbb{E}\left\{y_i(t + \tau)y_j(t)\right\} = 0, \forall\, i, j, \tau \tag{11}$$

The problem is solved whenever [A] is computed [3]. Modal shapes are contained in the column of the mixing matrix, thus enabling the reconstruction process.

Results The results depicted in Fig. 5 display the first three mode shapes extracted with two different techniques superimposed to the theoretical mode shapes. The experimental behavior of the structure fits the model almost perfectly at low frequency, whereas the deviation is more remarkable going up in the spectrum. This behavior is justified by the fact that SNR is not sufficient to discriminate among true signal and noise, especially for the higher components which mainly suffer from this effect. According to Fig. 4, the first vibrational mode contains almost the

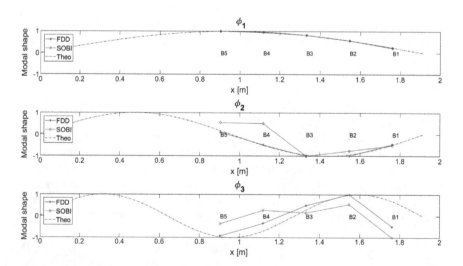

Fig. 5 Modal shapes extracted with different techniques: FDD, SOBI. It is worth noting that the experimental curves fit the model almost perfectly at low frequencies, whereas the deviation is some-how more evident at higher frequencies

overall energy of the acquired acceleration, appearing to be almost 40 dB higher than the other analyzed modes. It is also important to highlight that, although the SOBI method is totally unsupervised, its performances are almost equivalent to FDD, making it suitable for autonomous damage detection system. Being almost equal the outcome of the tested algorithms, it is necessary to underline the advantages that a blind strategy can produce. As a matter of fact, the computation of a Blind Source Separation (BSS) method relies on a restricted a-priori knowledge about the nature of the extracted signal, therefore it is entirely independent from the observed phenomenon. In addition, this technique is a completely unsupervised estimation which does not require to be supplied with users information about the expected frequency, such as in FDD. Moreover, being the SVD an expensive algorithmic process to be computed, especially when the dimension of data is very large, the computational cost associated to frequency methods cannot be forgotten.

Some aspects have to be underlined when dealing with the idea of embedding real-time signal processing capabilities in a low power sensor node, thus making realistic the possibility to offer an immediate estimation of modal parameters and consequently testing the health state of a given structure. The first point to be faced is connected with the data sources every algorithm requires, dividing the techniques between single-sensor driven methods and sensor-array driven estimators. For example, FDD naturally works with matrices, where every column (row) is a collection of samples provided by a single sensor over the global acquisition time. On the contrary, parametric methods for vibrational modes estimation can be applied onto single vectors, therefore they are best suited for single nodes. The second relevant aspect is ruled by the specificity of the processed samples. Indeed, SOBI performances are ideal when data under test refer only to the natural damped decaying of signals, whereas other techniques performance improves when samples cover a wide time-interval. In such a scenario, there are two different possibilities to exploit real-time estimation, simply derived from the intrinsic nature of the algorithms tested. In a master-slave architecture, a specifically crafted sensor node can perform the whole computation starting from samples provided by all the other sensors. A different strategy is based on the hypothesis of a multi-master system in which the processing phase is provided by each node separately and finally the outcomes are merged. The choice of strategy impacts on the algorithmic complexity and the computational requirements needed, which have to reach the best trade-off between power consumption and computational time, especially in terms of synchronization, coherency and storage capability.

4 Conclusions

The presented experiments reveal the potentialities of the implemented sensor network for SHM applications thanks to its versatility and high scalability, becoming a suitable candidate for a relatively cheap and low consumption system capable to provide real-time information.

References

1. Hoa, L.T., Tamura, Y., Yoshida, A., Anh, N.D.: Frequency domain versus time domain modal identifications for ambient excited structures. In: International Conference on Engineering Mechanics and Automation (ICEMA 2010), pp. 1–2 (2010)
2. Stoica, P., Moses, R.L., et al.: Spectral Analysis of Signals, vol. 1. Pearson Prentice Hall, Upper Saddle River (2005)
3. Poncelet, F., Kerschen, G., Golinval, J.-C.: Experimental modal analysis using blind source separation techniques. In: International Conference on Noise and Vibration Engineering, Leuven (2006)

IEEE 21451-001 Signal Treatment Applied to Smart Transducers

F. Abate⊙, M. Carratù⊙, A. Espírito-Santo, V. Huang, G. Monte and V. Paciello⊙

Abstract Control and monitoring systems base their decisions on the information provided by transducers. These signals must be acquired, processed and transmitted. There is an exchange between sensor signal processing and data rate, more signal processing inside the transducer implies lower data rate over the networks. This paper presents the standard IEEE 21451-001 whose main purpose is to extract knowledge of transducer signals that can be shared with others to increase system reliability, infer shapes, normal/abnormal states and to provide a normalized building structure for extracting knowledge. This standard extracts information directly from sampling based on a more complete structure. The standard is described highlighting the purpose and the main algorithms.

Keywords IEEE 1451 · Smart transducers · Oversampling · Sensor networks

1 Introduction

The real world is inferred through sensor signals. The expanded application of embedded systems has direct impact on transducers design. The microcontrollers "all inclusive" and the Application-Specific Integrated Circuit (ASIC)s offer every-

F. Abate · M. Carratù
Department of Industrial Engineering, University of Salerno, Salerno, Italy

A. Espírito-Santo
Department of Electromechanical Engineering, University of Beira Interior, Covilhã, Portugal

V. Huang
Georgia Institute of Technology, 791 Atlantic Drive NW, Atlanta, GA 30332, USA

G. Monte
Facultad Regional del Neuquén, Universidad Tecnológica Nacional, Buenos Aires, Argentina

V. Paciello (✉)
DIEI University of Cassino and Southern Lazio, Cassino, Italy
e-mail: v.paciello@unicas.it

© Springer Nature Switzerland AG 2019
B. Andò et al. (eds.), *Sensors*, Lecture Notes in Electrical Engineering 539,
https://doi.org/10.1007/978-3-030-04324-7_46

thing needed to turn a regular transducer into a smart transducer, with the integration of a sensor element into a single unit. Considering size, power consumption, cost and integration, it is logical to infer that signals will be fully processed at the point of acquisition. Although today this is only applicable to sensors with complex signal processing related to specific objectives, it will occur for most transducers in the future, even for small and low-cost sensors [1–5]. The world is moving toward a fully interconnected state. Even though commercial and industrial networks have different objectives and requirements like determinism, security and safety issues, they share many concepts and the division line is diffuse. These networks are based on three fundamental concepts [6]: Sense, Connect and Process/share.

For each concept, answers are needed for three questions: What? How? And Where? For example, how to share? Or where to process? Or what to transmit? These answers define the whole system and new paradigms arise like in the IoT (Internet of Things) cloud, fog and edge computing [7]. As networks become more complex, there is a need to process data away from the centralized points. The vertiginous technological evolution requires changing paradigms. This is the case of sensing. Almost all acquisition systems are based on uniform sampling to get digital data. The sample value is only a small fraction of the information embedded into a sensor signal and it does not provide a straightforward structure for extracting information in real time. This standard introduces a new sampling technique that facilitates signal features extraction in real time. The signal treatment outcomes should be normalized to allow interchange of knowledge among transducers. There is no universal treatment for sensor signals that allows signal analysis, synthesis and a dialogue among transducers to validate the measurement processes and to provide a platform for data mining. This standard will allow smart transducers from different manufacturers to interchange signal information to infer states of the instrumentation system, which helps to achieve higher reliability and better real world inference. In the next sections, the standard is presented with emphasis in core algorithms.

2 Standard Structure

2.1 Sampling of Sensor Signals

Even though uniform sampling is the principal approach used, it does not provide a straightforward platform for information and knowledge extraction. It gives us only the signal value at a specific instant. In a signal, the information is embedded in the relationship among samples at different time instants.

The main idea for representing a sensor signal is to use a concatenation of known function segments instead of samples. The signal information is related directly to the sequence of segments. In [8] a signal is considered that lives in a union of different subspaces. Since the standard approach is to build a model to extract knowledge in real time, and it must be simple enough to embed it into transducers, a finite set of

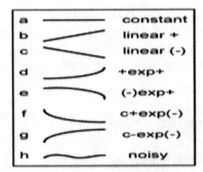

Fig. 1 Class segments are based on the value of the left and right samples and the interpolation error. In classes a, b and c, the segment ends due to it has reaches a maximum length

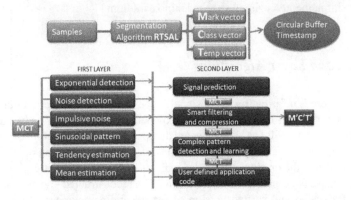

Fig. 2 Block diagram of the proposed standard, divided in two layers. First layer uses MCT algorithm. Second layer uses MCT and outputs from first layer algorithms

subspaces represented by simple trajectories is used, see Fig. 1. The sensor signal is represented by the concatenation of a normalized function that can be dilated, contracted and adapted to follow the real signal in a subspace bounded by an error as shown in Fig. 2. Eight segment classes are enough to approach the real signal. Classes "a", "b" and "c" exist due to a real constrain of the segment length. Class "h" must be avoided by splitting it into different classes.

The sensor signal is described by the sequence of values of α, β and τ. An algorithm is needed for comparing the real signal, suitable for embedding it into transducers, to determine which segment is the best for a particular signal segment. In [6] an algorithm was proposed for sensor signal preprocessing. The central idea is to compare real signal trajectory against linear trajectory. Checking how the real departs from linear allows deciding when the simplified segment must end since the error is greater than a prefixed value. This algorithm is described in [9, 10]. It is called RTSAL "Real Time Segmentation and Labeling". It employs oversampling to check at a rate fast enough, the real signal against simplified trajectories. The algorithm outputs are the values of α, β and τ that represent signal segments. Three vectors, Mark (α), Class (β) end Time (τ), in short MCT characterize the digital sensor signal. MCT vectors are the building blocks for this standard. The algorithm uses linear interpolation to track linear trajectories, therefore it is simple and can be executed in real time.

2.2   Standard Proposed Algorithms

Figure 2 shows the standard structure and algorithms divided into two layers. The first one includes most common and useful information extracted from sensor signals.

The second layer includes advanced feature extraction and employs MCT vectors and outcomes from first layers as inputs. First layer algorithms are described in [10]. They are implemented using a state machine paradigm.

2.3   Second Layer Algorithms

Since that there is information about signal shape is possible to predict future behavior. For example, if it has been detected exponential pattern, the steady state value can be predicted. Or if it has been detected dumped oscillations, predict the next one.

With MCT vectors, it is possible to reconstruct an approximation of the sampled signal. Since that the original sampled signal lies in an interpolation-error controlled subspace, there is no possibility for aliasing. The reconstructed signal is a concatenation of a scaled and shifted version of the generating segments. Since that the signal has been rebuilt using simplified trajectories, a low pass version is obtained. The number of iterations can be controlled by observing if the desire objectives have been reached (Fig. 3).

Figure 4 shows signal reconstruction after eight iterations and Fig. 5 shows that after eight iterations the number of local maximum and minimum remains constant and the period can be computed. If the computed period has low variance, the signal is periodic. Recall that local maximum a minimum are immediate outcomes of MCT vectors, i.e., the union of an ascending with a descending segment is a maximum.

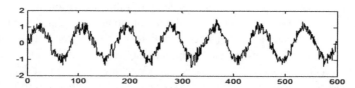

Fig. 3 Test signal for filtering. The objective is to determine if it is periodic or not and to compute period and phase. Number of extreme points = 175

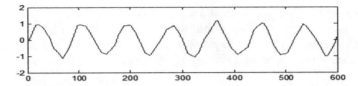

Fig. 4 Signal of Fig. 3 after eight iterations. Number of extreme points $= 14$

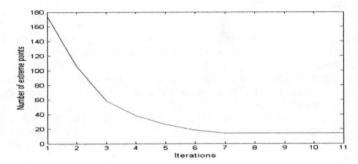

Fig. 5 After eight iterations the number of extreme points remains constant, therefore noise has been removed and period can be computed

The low pass version of MCT vectors could be used as a compressed version of the sampled signal. The resulting $M'C'T'$ vector is the building block for reconstructing the original signal considering the interpolation error and the number of iterations.

Based on MCT vectors, first layer output and operations on MCT structure, algorithms are proposed for detecting and learning specific patterns. A work was done for detecting the type and grade of damage in vibration signals related to bearing defects [11]. Also, a research was carried out for processing ECG signals using these concepts [12–14].

3 Conclusions

A brief description of the IEEE 21451-001 was presented. Strength of this standard is that it is applicable to a wide range of sensors from industry to IoT. The algorithms were simulated with MATLAB and implemented in microcontrollers of 8/32 bits in C programming language.

Thinking of a fully interconnected world, data grow faster than channel capacity, so the traffic over the networks will be around useful information extracted directly from smart signal sampling processes. Therefore, standardization processes will play a lead role in facilitating knowledge sharing.

For example, let's suppose we want to determine if the signal in Fig. 3 is periodic and to estimate the period. Each time a new MCT vector is obtained, a fraction of the high frequency content is lost and the reconstructed signal tends to be smoother than the previous one, which will lead to decreasing number of local maxima and minima. Therefore, a method based on number of extreme values is designed to choose a suitable iteration for filtering.

References

1. Di Lecce, V.: Towards intelligent sensor evolution: a holonic based system architecture. In: SENSORCOMM 2012: The Sixth International Conference on Sensor Technologies and Applications (2012)
2. Carratù, M., Pietrosanto, A., Sommella, P., Paciello, V.: Suspension velocity prediction from acceleration measurement for two wheels vehicle. In: I2MTC 2017 (2017). https://doi.org/10.1109/i2mtc.2017.7969943
3. Angrisani, L., Capriglione, D., Cerro, G., Ferrigno, L., Miele, G.: On employing a Savitzky-Golay filtering stage to improve performance of spectrum sensing in CR applications concerning VDSA approach. Metrol. Meas. Syst. 23(2), 295–308. https://doi.org/10.1515/mms-2016-0019
4. Capriglione, D., Carratù, M., Liguori, C., Paciello, V., Sommella, P.: A Soft stroke sensor for motorcycle rear suspension (2017). https://doi.org/10.1016/j.measurement.2017.04.011
5. Betta, G., Capriglione, D., Cerro, G., Ferrigno, L., Miele, G.: The effectiveness of Savitzky-Golay smoothing method for spectrum sensing in cognitive radios. In: 2015 XVIII AISEM Annual Conference, Trento, 2015, pp. 1–4. https://doi.org/10.1109/aisem.2015.7066819
6. Sivakumar, K.: Internet of Things (IoT). Cisco Systems. http://www.ethernetsummit.com/English/Collaterals/Proceedings/2014/20140430_1D_Sivakumar.pdf (2014)
7. Edge Computing-Ryan LaMothe Research Scientist Pacific Northwest National Laboratory, Jan 2013
8. Lu, Y.M., Do, M.N.: A theory for sampling signals from a union of subspaces. IEEE Trans. Signal Process. 56(6), 2334–2345 (2008)
9. Monte, G.: Sensor signal preprocessing techniques for analysis and prediction. In: 34th Annual Conference of IEEE. Industrial Electronics, 2008. IECON 2008, pp 1788–1793. ISBN 978-114244-1766-7
10. Monte, G., Liu, Z., Abate, F., Paciello, V., Pietrosanto, A., Huang, V.: Normalizing transducer signals: An overview of a proposed standard. In: Proceedings IEEE International Instrumentation Measurement Technology Conference (I2MTC), Montevideo, Uruguay, May 2014, pp. 614–619
11. Monte, G., Abate, F., Huang, V, Paciello, V., Pietrosanto, A.: Real time transducer signal features extraction: a standard approach. In: 2015 IEEE 13th International Conference Industrial Informatics (INDIN). http://dx.doi.org/10.1109/INDIN.2015.7281740
12. Monte, G., et al.: A novel time-domain signal processing algorithm for real time ventricular fibrillation detection. J. Phys.: Conf. Ser. 332, 012015 (2011). https://doi.org/10.1088/1742-6596/332/1/012015
13. Abate, F., Paciello, V., Pietrosanto, A., Guia, S.S., Santo, A.E.: Period measurement with an ARM microcontroller. In: Proceedings of the 2015 18th AISEM Annual Conference, AISEM 2015, pp. 1–4. ISBN 9781479985913. https://doi.org/10.1109/aisem.2015.7066785
14. Abate, F., Paciello, V., Pietrosanto, A., Monte, G.: Preliminary analysis of a real time segmentation and labeling algorithm. In: 2015 IEEE Workshop on Environmental, Energy, and Structural Monitoring Systems, EESMS 2015—Proceedings, pp. 215–219 (2015) ISBN 9781479982141. https://doi.org/10.1109/eesms.2015.7175880

Accuracy and Metrological Characteristics of Wearable Devices: A Systematic Review

Gloria Cosoli and Lorenzo Scalise

Abstract The aim of this paper is to study the state of the art about the metrological characteristics and the accuracy of wearable devices, tested in comparison to a gold standard instrument. A bibliographic research has been made on the main scientific databases (e.g. Scopus and Web of Science). Papers have been included on the basis of established criteria (e.g. the wearable device has to be commercial). At present, neither a standard protocol nor fixed metrological characteristics can be identified in the literature. Among the most discussed wearable devices, there are certainly Fitbit, Jawbone, Garmin and Polar ones. Chest-strap monitors generally result to be more accurate than wrist-worn devices, which, on the other hand, are cheaper and more comfortable. Given the lack of standards in the validation process, the data appear to be very irregular (even among studies conducted on the same device) and consequently barely comparable. It would be extremely important to conduct a pilot study on a few devices, validating them according to an established test protocol and comparing the results to a gold reference instrument (e.g. ECG for Heart Rate assessment). In this way, it would be possible to start building a database of the accuracy and the metrological characteristics of wearable devices.

Keywords Wearable devices · Health monitoring · Physiological parameters
Metrological characteristics · Measurement accuracy

1 Introduction

Substantial evidence leads to state that an active lifestyle is surely a cornerstone of improved health and disease prevention [1, 2]. In order to monitor physical activity, different methods can be used: energy expenditure measurements, questionnaires or activity monitors [3]. The real-time monitoring of human physiologic function

G. Cosoli · L. Scalise (✉)
Department of Industrial Engineering and Mathematical Sciences, Università Politecnica delle Marche, v. Brecce Bianche, 60131 Ancona, Italy
e-mail: l.scalise@univpm.it

© Springer Nature Switzerland AG 2019
B. Andò et al. (eds.), *Sensors*, Lecture Notes in Electrical Engineering 539,
https://doi.org/10.1007/978-3-030-04324-7_47

377

and performance during different tasks in different fields (e.g. sport medicine, free-living conditions, mental exercises and pathologic states) is very important in order to provide a picture of the subject's physical activity pattern. Moreover, it is suitable to increase the subject's awareness about his physical activity and an activity monitor acts as a motivational tool [4, 5] and also pushes towards healthy habits. This is of utmost importance, particularly thinking of a more and more sedentary population not only in the western world [2], because of advances in society and technology [6], but even worldwide. A sedentary lifestyle leads to serious risk factors for death, for what concerns cardiovascular and cardiometabolic diseases, diabetes, obesity, chronic diseases and even some forms of cancer [7–10]. Consumer-based monitors could be a valid means to cope with this matter; in fact, at present their variety is increasing (just as a few well-known examples, let us think to Fitbit, Jawbone and Apple Watch), as well as their popularity [11, 12]. Wearable electronics to track physical activity and physiological parameters includes different devices, such as smartwatch, wristbands and data-logger, joined by some common characteristics. Indeed, they are small, unobtrusive, relatively cheap, user-friendly and wearable at different locations (e.g. hip, wrist and chest). On the other hand, it is right to wonder if such instruments are sufficiently accurate and reliable in order to use them as a solid basis for trustworthy considerations on a subject's performance, especially if an expert analyses them to give the subject some recommendations. However, to this day, to the best of authors' knowledge, only few studies consider the accuracy, the validity and the metrological characteristics of these devices, even if the comparison between different devices clearly requires a good inter-device reliability. Not even the manufacturers take care of providing the consumers with such parameters.

Therefore, the aim of this review is to examine the state of the art about investigations on wearable monitoring devices, focusing the attention on the highlighted metrological properties and the accuracy that should always be taken into consideration before the use of such instruments, tested in comparison to a gold standard instrument. In particular, attention will be focused on cardiac parameters and energy expenditure.

2 Materials and Methods

For the purpose of conducting a review about the validation of wearable devices for the activity monitoring, focused on their metrological characteristics, a bibliographic research was done, taking advantage of numerous academic databases (e.g. Scopus, Web of Science, Google Scholar, PubMed and IEEE Xplore), looking for wearable devices capable to measure the quantities of interests (i.e. cardiac parameters and energy expenditure).

A selection of papers was made on the basis of the following characteristics:

- Studies validating activity monitors comparing them to a gold standard instrument;
- Studies involving commercial devices (e.g. Fitbit or Apple Watch);
- Studies reporting some metrological characteristics of the examined devices.

The used key words were "activity", "monitor", "device", "wearable", "accuracy" and "validation". The abstracts of the found papers were screened in order to verify that they met the chosen inclusion criteria. In affirmative case, the full-text of the paper was obtained. In order to collect all the information about title, authors, journal and publication date, open-source Zotero tool was used.

It is important to underline the fact that the algorithms used by commercial wearable devices are proprietary software (often based on machine learning techniques), so it is not possible to know how the measure results have been effectively obtained. This is a drawback also for the researchers who want to compare the performance of different devices, since the results could not be definitely comparable.

The different measurement techniques applied in such devices have been analysed and their performance has been evaluated, although neither a standard protocol nor fixed metrological characteristics can be identified in the literature.

3   Results

With regard to cardiac activity assessment (i.e. Heart Rate, HR, Heart rate Variability, HRV, and RR intervals), different sensing methods are employed: contact methods (i.e. electrodes) mainly in chest-strap monitors (e.g. Polar), photoplethysmographic (PPG) techniques in wrist-worn devices (e.g. Fitbit and Jawbone, wearable also in different locations, e.g. arm, back, hips, thigh, ankle and waist) and also in smartphone apps. Accuracy results on a few commercial wearable devices are reported in Table 1; it generally decreases together with an increasing activity level.

On the other hand, an energy expenditure estimation can be obtained by combining accelerometer based measurements (e.g. steps or activity estimation) and HR results.

Given the lack of standards in the validation process, the data appear to be very irregular (even among studies conducted on the same device) and consequently barely comparable (e.g. performance is evaluated by means of different quantities, such as error, accuracy or correlation coefficient).

The gathered information about single devices are reported in the following, where they are divided between wrist-worn and chest-strap ones.

3.1   Wrist-Worn Monitors

As previously stated, wrist-worn monitors are those devices wearable on the wrist or at different locations (e.g. arm or ankle), especially when they are clip-shaped. Nowadays, their use is spreading and their popularity is rapidly growing, thanks to the fact that they are relatively inexpensive, small, user-friendly and more and more fashionable. It is very important to take into account the device positioning, adjusting the wristband length in order to have a good contact with the subject's skin. Moreover, it would be significant to do tests both in laboratory and in free-living conditions.

Table 1 Performance evaluation of some commercial wearable devices in the measurement of HR and energy expenditure parameters, compared to reference methods (i.e. ECG and calorimetry, respectively); r is the correlation coefficient

Measured quantity	Sensing principle	Device	Performance
HR (bpm)	PPG	Apple Watch	Accuracy = 99.9%, r = 0.81
		Basis Peak	r = 0.92–0.95
		Fitbit Charge HR	r = 0.83
		Microsoft Band	Error < 10%
		Motorola Moto 360	Accuracy = 92.8%
		Polar V800	r > 0.999
		Samsung Gear S	Accuracy = 95%, precision = 20.6%, r = 0.67
		Samsung Gear Fit	Accuracy = 97.4%
		Samsung Gear II	Accuracy = 97.7%
		Scosche Rhythm	Error = 4.0%, r = 0.93
	ECG electrodes	Polar S810i	r = 0.98–1
		Polar T31	r = 0.997–0.999
		Suunto Memory Belt	Error = 0.11%, r > 0.99
Energy expenditure	Accelerometer and HR results	Actical	Error = 14%
		ActiGraph	Error = 12.6%
		Basis Peak	Error = 8%
		Fitbit Ultra	Error = 6–20%, r = 0.98
		Fitbit Zip	r = 0.81–0.86
		Fitbit Flex	Error = 16.8%
		Jawbone UP	Error = 12–30%, r = 0.65–0.74
		Jawbone UP24	Error = 0.8–14.3%
		Withings Pulse	r = 0.79

Actical Actical is a triaxial accelerometer-based device capable to measure energy expenditure and step count. It is used in research field, also as a reference [13]; it has been tested during different activities, reporting an error of 14%, seeming to underestimate calories during cycling [14].

ActiGraph ActiGraph has a uniaxial accelerometer allowing activity and energy expenditure (during activity) to be measured. With regard to calories, in a study on healthy subjects tested during different activities (e.g. walking and running) an absolute error of 12.6% was reported [15]. In another study, an underestimation of 26% was found [14].

Apple Watch Apple Watch functioning is based on a PhotoPlethysmoGraphic (PPG) sensor, so that it requires an optimal adhesion to the wrist skin and the processing algorithm has to take into account the skin colour, in order to make the proper corrections [16]. It can be disturbed by rapid hand movements, wrist hair and sweat [17]. If compared to chest-strap monitors, the results are worse [18]; its HR monitoring has been judged not accurate [19]. On the contrary, a very high accuracy is reported in [20]: 99.9% for HR, when compared to a professional pulso-oximeter during walking test. Also Wallen et al. (2016) obtained a bias of 1.3 bpm with a correlation coefficient $r = 0.81$ in a cycling test on healthy subjects [16].

Basis Peak Basis Peak functioning is based on a PPG sensor, a triaxial accelerometer, two thermometers and a sensor for the skin galvanic response. In this way, it can be used to measure HR, calories, steps and activity (distinguishing among walking, running and cycling). It is small, lightweight and waterproof (up to 5 atm), with a 2–3-day battery. When compared to ECG during different activities (e.g. walking and cycling), it proved to satisfy the validation criteria for HR monitor, with a bias equal to -2.5 bpm at rest ($r = 0.92$) and -4.9 when HR > 116 bpm ($r = -0.77$) [21]. Its performance is comparable to the Apple Watch one. In the same study, it has been noted a slight decline of the performance at high exercise intensity. In a study based on a treadmill protocol, it is reported an error of 3.6% ($r = 0.95$) [22]. With regard to energy expenditure, it seems to underestimate it, with a mean error of -28.7 kcal (8%) during different activities [23]. In that study, when compared to calorimetry, it is the only wrist-worn sensor not significantly differing from calorimetry, even if it presents high differences on single measurements.

Fitbit Ultra Fitbit Ultra is a clip-on device, wearable on the belt or in a pocket; it tracks steps, climbed stairs, travelled distance, energy expenditure and activity intensity. The performance of Fitbit Ultra (provided with a triaxial accelerometer) has been evaluated to assess steps and energy expenditure (comparing the results with a direct observation of steps and with a portable metabolic system) using a protocol considering walking, running and agility exercises, tested on a population of 20 people [24]. It was noticed that there is an underestimation of energy expenditure (17% in agility tasks, with a correlation coefficient $r = 0.89$). A similar protocol consisting of walking and jogging (but also with slopes) is proposed in [13], where the device (worn in correspondence of the iliac crest) is tested on 23 subjects and compared to Actical and indirect calorimetry for what concerns steps and energy expenditure, respectively. The correlation coefficient is $r = 0.81$–0.87 for energy expenditure, proving that the device is reliable except in case of slopes. The same device was also tested in free-living conditions (for 8 days) by a single subject wearing 10 devices in his trouser pockets, in order to evaluate the inter-device reliability (regarding steps, distance and energy expenditure), which is considered good ($r = 0.90$) [25].

Fitbit Zip With respect to Fitbit One, Fitbit Zip is cheaper but has better performance. It is a clip-on tracker, measuring steps, distance and calories. Based on a piezoelectric triaxial accelerometer, it is small and lightweight; its battery duration is of 5–10 days and it has a memory of 23 days. In a 48-h monitoring (with the device on the waist), it was reported that Fitbit Zip underestimates energy expenditure (bias: -497 kcal,

R = 0.81) [26]. In another study, it is reported a RMSE equal to 40.8 kcal (absolute value of 10.1%) [15]. Energy expenditure was underestimated also in [27], where it is reported an overestimation of moderate-vigorous activity (r = 0.86). Fitbit Zip can be considered quite reliable both in laboratory and in free-living conditions.

Fitbit Flex Fitbit Flex is a small wristband device, with a battery duration of 5–10 days and a memory up to 23 days. It can measure steps, distance, energy expenditure and activity. It is relatively inexpensive and its popularity is growing. In [28], Fitbit Flex is tested for the performance on the assessment of steps and energy expenditure (one device on each wrist) during treadmill. It results that energy expenditure is overestimated (mean error: 0.2–2.6 kcal), even if in Evenson et al. (2015) it is underestimated (unless compared to indirect calorimetry) [27]. With regard to energy expenditure, in a study testing the device at different activity intensities it is reported a bias of −20.4 kcal and a mean error of 16.8%, with an overestimation only during aerobic exercise [29]. Finally, in [30] it is reported an overestimated activity in case of moderate-vigorous intensity.

Fitbit Charge HR Fitbit Charge HR is a user-friendly, cheap and attractive device; it uses an optical sensor (to measure HR) and a triaxial accelerometer. This wristband can measure HR, steps, distance, energy expenditure, climbed height and activity. If compared to other HR monitors, it provides a poor performance [31], in fact it does not satisfy the validation criteria (r = 0.83 at rest, r = 0.58 when HR > 116 bpm), with a performance declining with increasing intensity of activity [21, 32, 33]. This is confirmed when the device is compared to a chest-strap monitor during treadmill exercises [22]. Leth et al. (2017) report an underestimation of HR, with a mean error of −3.42% [32].

Fitbit Tracker Fitbit Tracker functioning is based on an accelerometer. It has been proved to be reliable during walking and jogging for energy expenditure (r = 0.56–0.72), except when used in sloping areas [13]. Dannecker et al. (2013) report an underestimation of energy expenditure (error: 28%) during a test protocol including different activities (e.g. walking and cycling) [14].

Jawbone Jawbone is a relatively inexpensive device allowing the user to track activity mainly thanks to accelerometer technology.

Jawbone UP Jawbone UP uses a triaxial microelectromechanical accelerometer and allows the user to monitor steps, sleep (duration and quality), activity and energy expenditure. It is waterproof (up to 1-m depth) and has a battery lasting up to 10 days. It has to be worn on the wrist and does not have a display. With regard to energy expenditure during tests on a treadmill, in [15] it is reported an absolute error of 12 it is reported a bias of 18.57 kcal (r = 0.72 during walking, r = 0.65 during running). In another study, energy expenditure is overestimated (20%, r = 0.99) during running, but underestimated during agility exercises (−30%, r = 0.88) [24]. A study conducted in 48-h free-living conditions reports an underestimation of energy expenditure (bias: −898 kcal, r = 0.74), and an overestimation of activity (bias: 22.7 min, r = 0.81) [26].

Jawbone UP24 Jawbone UP24 technology is based on a conductive sensor and a triaxial accelerometer. This wristband is able to measure steps, energy expenditure (at rest and during activity, separately), walked distance and sleep (quality and duration).

Bai et al. (2016) obtained an underestimation in energy expenditure (-18.5%) during different activities [29], confirmed in [27].

Microsoft Band Microsoft Band is a wristband whose functioning is based on PPG; it tracks HR, exercise, calorie burn and sleep quality. It was compared to a chest-strap monitor, finding an error of 4.8% (r = 0.96) in a study using a treadmill protocol [22]. When compared to ECG, the error was <5% in 76.29% of the cases, <10% in 86.26% of the cases [34].

Motorola Moto 360 Motorola Moto 360 is an elegant smart watch able to measure steps and HR. With regard to HR, it was proved to have an accuracy equal to 92.8% [20].

Polar V800 Polar V800 is a multisport training computer, wearable on the wrist [35]. It was tested in an orthostatic test, comparing the results with 3-lead ECG (i.e. gold standard instrument) [36]. With regard to RR intervals, high correlation was found (r > 0.999), without significant differences for HRV parameters (r > 0.99).

Samsung Gear S Samsung Gear S is a smartwatch able to measure HR, calories (by means of accelerometer data, without using HR results) and steps. In a study with a protocol involving treadmill and cycling, with regard to HR a bias of -7.1 bpm (r = 0.67) was found [16]. In another study, an accuracy of 95.0% and precision of 20.6% are reported [20], when compared to a professional pulso-oximeter.

Samsung Gear Fit Samsung Gear Fit is a smartwatch measuring HR and steps. Compared to a professional pulso-oximeter, an accuracy of 97.4% for HR was found [20].

Samsung Gear II Comparing Samsung Gear II smartwatch to a professional pulso-oximeter, an accuracy of 97.7% for HR was found [20].

Scosche Rhythm Scosche Rhythm is a PPG-based HR monitor armband (i.e. an infrared LED and a photodetector) wearable on the forearm. With a treadmill protocol, it was proved to measure HR with an error of 4.0% (r = 0.93) [22].

Withings Pulse Withings Pulse functioning is based on a triaxial accelerometer, allowing the user to measure steps, walked distance, climbed height, energy expenditure, activities and sleep (duration and quality). It is small and lightweight; its performance is poorer when worn on the wrist than on the shirt collar or on the waist [23]. In a study conducted on healthy subject in 48-h free-living conditions (with the device on a belt), the device was compared to different accelerometer/multisensory based devices (i.e. BodyMedia SenseWear and ActiGraph GT3X+). An underestimation of calories (bias: -533 kcal, r = 0.79) is reported. With regard to energy expenditure, it seems not suitable, reporting significant differences with calories measured by means of a portable calorimeter [23].

3.2   Chest-Strap Devices

Chest-strap HR monitors are generally considered more reliable than wrist-worn devices, because they are positioned closer to the heart (obviously, the electrode-skin contact has to be very good) [37]. Moreover, they are cheaper, even if less comfortable than a smartwatch. If you rely on accuracy, chest-strap sensors are surely superior.

Polar Chest Strap Polar Chest Strap is based on ECG sensor. The strap has to be worn below the thorax, adjusting the belt with the sensor at the centre of the chest. With respect to Apple Watch, it was demonstrated to be more accurate, with a higher sampling frequency [17].

Polar S810i Polar S810i has an ECG sensor for the measurement of RR interval (with a sampling period equal to 1 ms). Polar S810i performance was compared to ECG, considering both HR and HRV parameters; an optimal correlation was found, with r = 1 at rest, r = 0.98 during exercise [38]. Such an optimal performance is confirmed also in [39] and in [40], where it is reported r = 0.99 for RR in corrected tacograms, r > 0.97 for HRV parameters. It is fundamental to put gel on the belt in order to improve the skin-electrode contact; moreover, before conducting HRV analysis, tacograms should be corrected.

Polar T31 Polar T31 is a codified transmitter capable to send information to a wristband; it is thin, lightweight and completely waterproof. It was tested at rest conditions and compared to ECG [41]; correlation was optimal, with r = 0.999 for HR and r = 0.997 for RR intervals.

Suunto Memory Belt Suunto memory Belt is equipped with ECG sensors. It was compared to ECG, resulting in an error of 0.11% for RR and in a very high correlation for HRV parameters (r > 0.99) [42].

4 Discussion and Conclusion

Being interested in accuracy and metrological characteristics of wearable devices, studying the state of the art immediately highlights the absence of a standard test protocol, as well as the lack of standard parameters (e.g. uncertainty or error [43]) to consider in order to evaluate the device performance. Moreover, in the reviewed papers the devices were tested on populations different both in size and characteristics (e.g. healthy or not). It is important to take into account that the accuracy depends on different factors linked to population and test protocol:

- activity intensity [22];
- skin photosensitivity, melanin concentration and pigmentation, when the sensor is PPG [34, 44];
- sweat, movement artefacts, sensor positioning [45].

Wrist-worn devices (mostly based on accelerometer and PPG sensors) are generally less accurate than chest-strap devices [46] with regard to heartbeat related parameters. On the other hand, the accuracy of several wrist-worn devices is often not adequate when compared to reference instrumentation. At this purpose, it is worth noting that a device, even if not comparable to gold standard, can be suitable for some purposes, such as in sport applications. So, it should be taken into consideration the application field and the variables to monitor, which will not be the same in different areas of interest. Depending on such items, also accuracy requirements could be different; for example, if a healthcare provider wants to take a decision on the basis of a measurement on his patient, the uncertainty will have to be lower than in the case a common user would like to track his activity during the day.

Given the fact that there are neither a standard test protocol nor fixed parameters usually taken into consideration for a device validation, it would be of utmost importance to conduct a pilot study fixing these lacks, validating the devices according to an established test protocol and comparing the results to a gold reference instrument (e.g. ECG for Heart rate assessment). Such an investigation could consider only a few wearable devices (taking into consideration as many measured parameters as possible), identifying some characteristics that can be considered essential for the accuracy evaluation, which should be made following the guidelines for the expression of uncertainty in measurement [43]. If the results from this study were good, the investigation could be extended to more devices (e.g. the most recent ones, like BioRing and the clothing with electronics integrated directly into the textile substrates [47]), so as to start building a database with the metrological characteristics of wearable devices. Such a database would be extremely useful for the experts in the field (e.g. health care professionals and fitness coaches) to obtain information worthwhile for the development of work plans on the patients.

References

1. Haskell, W.L., et al.: Physical Activity and Public Health. Updated Recommendation for Adults from the American College of Sports Medicine and the American Heart Association. Circulation, Ago (2007)
2. Katzmarzyk, P.T.: Physical activity, sedentary behavior, and health: paradigm paralysis or paradigm shift? Diabetes **59**(11), 2717–2725 (2010)
3. Van Remoortel, H., et al.: Validity of activity monitors in health and chronic disease: a systematic review. Int. J. Behav. Nutr. Phys. Act. **9**, 84 (2012)
4. Kurti, A., Dallery, J.: Internet-based contingency management increases walking in sedentary adults. PubMed—NCBI (2013)
5. Naslund, J.A., et al.: Feasibility of popular m-health technologies for activity tracking among individuals with serious mental illness. Telemed. J. E-Health Off. J. Am. Telemed. Assoc. **21**(3), 213–216 (2015)
6. Hallman, D.M., et al.: Prolonged sitting is associated with attenuated heart rate variability during sleep in blue-collar workers. Int. J. Environ. Res. Public. Health **12**(11), 14811–14827 (2015)
7. Colditz, G.A.: Economic costs of obesity and inactivity. Med. Sci. Sports Exerc. **31**(11), S663–S667 (1999)
8. Ford, E.S., Caspersen, C.J.: Sedentary behaviour and cardiovascular disease: a review of prospective studies. Int. J. Epidemiol. **41**(5), 1338–1353 (2012)
9. Owen, N., et al.: Sedentary behavior: emerging evidence for a new health risk. Mayo Clin. Proc. **85**(12), 1138–1141 (2010)
10. Tremblay, M.S., et al.: Physiological and health implications of a sedentary lifestyle. Appl. Physiol. Nutr. Metab. Physiol. Appl. Nutr. Metab. **35**(6), 725–740 (2010)
11. Meyer, J., Hein, A.: Live long and prosper: potentials of low-cost consumer devices for the prevention of cardiovascular diseases. Medicine **20**(2), e7 (2013)

12. Piwek, L., Ellis, D.A., Andrews, S., Joinson, A.: The rise of consumer health wearables: promises and barriers. PLOS Med. **13**(2), e1001953 (2016)
13. Adam Noah, J., et al.: Comparison of steps and energy expenditure assessment in adults of Fitbit Tracker and ultra to the Actical and indirect calorimetry. J. Med. Eng. Technol. **37**(7), 456–462 (2013)
14. Dannecker, K.L., et al.: A comparison of energy expenditure estimation of several physical activity monitors. Med. Sci. Sports Exerc. **45**(11), 2105–2112 (2013)
15. Lee, J.-M., Kim, Y., Welk, G.J.: Validity of consumer-based physical activity monitors. Med. Sci. Sports Exerc. **46**(9), 1840–1848 (2014)
16. Wallen, M.P., et al.: Accuracy of heart rate watches: implications for weight management. PLoS ONE **11**(5), e0154420 (2016)
17. Ge, Z., et al.: Evaluating the accuracy of wearable heart rate monitors. In: 2016 2nd International Conference on Advances in Computing, Communication, Automation (ICACCA), pp. 1–6 (2016)
18. Glenn, K.: Wrist-worn Heart Rate Monitors Less Accurate Than Standard Chest Strap—American College of Cardiology (2017)
19. Lamkin, P.: Mio boss: Fitbit and Apple are getting heart rate monitoring wrong (2015)
20. El-Amrawy, F., Nounou, M.I.: Are currently available wearable devices for activity tracking and heart rate monitoring accurate, precise, and medically beneficial? Healthc. Inform. Res. **21**(4), 315–320 (2015)
21. Jo, E., et al.: Validation of biofeedback wearables for photoplethysmographic heart rate tracking. J. Sports Sci. Med. **15**(3), 540–547 (2016)
22. Stahl, S.E., et al.: How accurate are the wrist-based heart rate monitors during walking and running activities? Are they accurate enough? BMJ Open Sport Exerc. Med. (2016)
23. Woodman, J., et al.: Accuracy of consumer monitors for estimating energy expenditure and activity type. PubMed—NCBI Med. Sci. Sports Exerc. (2016)
24. Stackpool, C.M.: Accuracy of various activity trackers in estimating steps taken and energy expenditure (2013)
25. Dontje, M.L., et al.: Measuring steps with the Fitbit activity tracker: an inter-device reliability study. J. Med. Eng. Technol. **39**(5), 286–290 (2015)
26. Ferguson, T., et al.: The validity of consumer-level, activity monitors in healthy adults worn in free-living conditions: a cross-sectional study. Int. J. Behav. Nutr. Phys. Act. **12**, 42 (2015)
27. Evenson, K.R., et al.: Systematic review of the validity and reliability of consumer-wearable activity trackers. Int. J. Behav. Nutr. Phys. Act. **12**, 159 (2015)
28. Diaz, K.M., et al.: Fitbit®: an accurate and reliable device for wireless physical activity tracking. Int. J. Cardiol. **185**, 138–140 (2015)
29. Bai, Y.: Comparison of consumer and research monitors under semistructured settings. Med Sci Sports Exerc. **48**(1), 151–158 (2016)
30. Reid, R., et al.: Validity and reliability of Fitbit activity monitors compared to ActiGraph GT3X+ with female adults in a free-living environment. J. Sci. Med. Sport (2017)
31. Prospero, M.: Who has the most accurate heart rate monitor?, Tom's Guide, June 2016
32. Leth, S., et al.: Evaluation of commercial self-monitoring devices for clinical purposes: results from the future patient trial, Phase I. Sensors **17**(1) (2017)
33. Montgomery-Downs, H.E., Insana, S.P., Bond, J.A.: Movement toward a novel activity monitoring device. Sleep Breath. **16**(3), 913–917 (2012)
34. Parak, J., Korhonen, I.: Evaluation of wearable consumer heart rate monitors based on photopletysmography. Proc. IEEE Eng. Med. Biol. Soc. **2014**, 3670–3673 (2014)
35. Polar I V800: Polar Italia (2017)
36. Giles, D., Draper, N., Neil, W.: Validity of the Polar V800 heart rate monitor to measure RR intervals at rest. Eur. J. Appl. Physiol. **116**, 563–571 (2016)
37. Stables, J.: Heart rate monitors: chest straps v wrist. Wearable (2017)
38. Vanderlei, L.C.M., et al.: Comparison of the Polar S810i monitor and the ECG for the analysis of heart rate variability in the time and frequency domains. Braz. J. Med. Biol. Res. Rev. Bras. Pesqui. Medicas E Biol. **41**(10), 854–859 (2008)

39. Porto, L.G.G., Junqueira, L.F.: Comparison of time-domain short-term heart interval variability analysis using a wrist-worn heart rate monitor and the conventional electrocardiogram. Pacing Clin. Electrophysiol. PACE **32**(1), 43–51 (2009)
40. Gamelin, F.X., Berthoin, S., Bosquet, L.: Validity of the polar S810 heart rate monitor to measure R-R intervals at rest. Med. Sci. Sports Exerc. **38**(5), 887–893 (2006)
41. Radespiel-Tröger, M., et al.: Agreement of two different methods for measurement of heart rate variability. Clin. Auton. Res. **13**(2), 99–102 (2003)
42. Bouillod, A., et al.: Accuracy of the Suunto system for heart rate variability analysis during a tilt-test. Braz. J. Kinanthropometry Hum. Perform. **17**(4), 409–417 (2015)
43. BIPM—Guide to the Expression of Uncertainty in Measurement (GUM) (2017)
44. Fallow, B.A., Tarumi, T., Tanaka, H.: Influence of skin type and wavelength on light wave reflectance. J. Clin. Monit. Comput. **27**(3), 313–317 (2013)
45. Takacs, J., et al.: Validation of the Fitbit One activity monitor device during treadmill walking. J. Sci. Med. Sport **17**(5), 496–500 (2014)
46. Sawh, M.: BioRing. Wearable (2016)
47. ComfTech—Textile sensors (2017)

Short Range Positioning Using Ultrasound Techniques

Antonella Comuniello, Alessio De Angelis and Antonio Moschitta

Abstract In this paper, a positioning technique based on measuring the Time Difference of Arrival (TDoA) of ultrasonic chirp pulses is proposed. The implementation issues of the proposed method are discussed and the development of two versions of low-complexity prototype is presented. Experimental results from a measurement campaign using the realized prototype are presented, demonstrating sub-centimeter position measurement accuracy in the short range, thus validating the proposed technique.

Keywords Positioning technique · TDoA measurement · Ultrasonic chirp pulses

1 Introduction

Positioning techniques are a strong enabler for various innovative services, often related to the Internet of Things paradigms, that include for instance Domotics, Ambient Assisted Living, and Line Traceability. Ultrasound transmissions enable low cost and accurate solutions, competitive in Line of Sight and short-range scenarios with respect to other technologies, such as those based on image processing or on transmissions of electromagnetic signals. Ranging and positioning are usually achieved using time domain measurements, such as Time of Flight, Round Trip Time, or Time Difference of Arrival (TDoA). The latter solution is interesting and explored in the described activity, since it does not require synchronization of the involved nodes. Moreover, current embedded and sensor technologies consent to easily generate, acquire, and process ultrasound signals, coupling transducers to

A. Comuniello (✉) · A. De Angelis · A. Moschitta
Department of Engineering, University of Perugia, Perugia, Italy
e-mail: antonella.comuniello@studenti.unipg.it

A. De Angelis
e-mail: alessio.deangelis@unipg.it

A. Moschitta
e-mail: antonio.moschitta@unipg.it

© Springer Nature Switzerland AG 2019
B. Andò et al. (eds.), *Sensors*, Lecture Notes in Electrical Engineering 539,
https://doi.org/10.1007/978-3-030-04324-7_48

389

low-cost embedded platforms. In a previous activity, a simple 2D positioning system was proposed, based on TDoA measurement performed by wireless capable nodes [1]. Such approach is extended in this activity, where the performance of a partially distributed planar (i.e. 2D) positioning system was studied, as a function of the data acquisition sampling frequency and resolution. Two system configurations were developed using available embedded platforms. The main design requirements for the prototype were low-cost and low-complexity. Thus, at first, the proposed method in [1] was implemented, using the bandpass sampling technique. Such an approach allows for reducing both the sampling rate and the number of samples acquired for a given chirp pulse duration, at the expense of timing resolution degradation. Then, the system was reconfigured, gradually increasing the sampling frequency and the resolution. Performance improvement was assessed, by comparing the achieved timing and ranging accuracies.

2 The Positioning System

2.1 System Architecture

The considered system, as shown in Fig. 1, includes a mobile node acting as active beacon, a set of wireless anchors, acting as listeners, and a Master node, acting as supervisor. No synchronization is assumed between the mobile node and the anchors, but the anchors are assumed to be time-synchronized with each other. One of the anchors acts as leader (Trigger source), triggering the remaining ones using wired connection. The proposed approach was experimentally validated using embedded platforms, delegating data acquisition and part of the processing tasks to low-cost microcontrollers. The realized prototype is based on a set of Cypress Programmable System on Chip (PSoC) devices [2]. The mobile node, as shown in the left side of Fig. 2, is realized using a PSoC 4 BLE microcontroller equipped with a Bluetooth Low Energy (BLE) radio module [2]. Three anchors are realized each using a PSoC 5 Low-Power (LP) [3], as shown in the right side of Fig. 2. The mobile node, the Master node and the Trigger Source node are equipped with a Bluetooth Low Energy (BLE) radio module. All nodes but the Master are equipped with an ultrasound transceiver, namely the Murata MA40S4R piezoelectric transducer [4].

2.2 System Operation

As anticipated in the previous section, the Master node enables the estimation of the mobile node position, triggering both the mobile node and the anchors' set using the BLE radio interface [1].

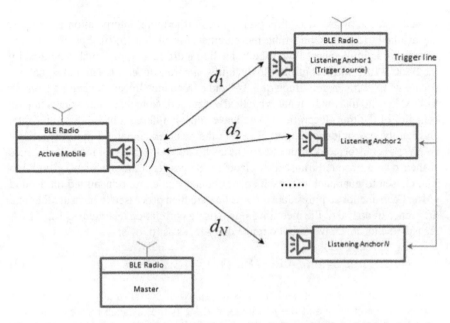

Fig. 1 The considered positioning system. One possible implementation is depicted: the anchor nodes are synchronized by means of a wired connection providing the trigger signal

Fig. 2 The realized prototype. Left: Mobile node, right: Listening anchor

The mobile node begins transmitting a repetitive sequence of ultrasound linear chirp pulses, given by

$$s(t) = s_0\left(T_C\left\langle\frac{t}{T_C}\right\rangle\right)$$

$$s_0(t) = \begin{cases} A\,\sin(2\pi f(t)t), & f(t) = f_0 + \frac{f_1-f_0}{2T_C}t,\ 0 \le t \le T_C \\ 0, & elsewhere \end{cases} \qquad (1)$$

where f_0 is the lowest frequency, f_1 is the highest frequency, T_C is the duration of a single chirp pulse, and $\langle\cdot\rangle$ is the fractional part operator [1]. The chirp signal is used because of its correlation properties. Furthermore, interpolation techniques are available for improving time measurement resolution [5, 6]. Specifically, the transmitter is implemented by directly feeding a digital output signal generated by the PSoC 4 BLE microcontroller to the ultrasonic transducer. This digital signal is a square wave, with an amplitude of 3.3 V and a frequency linearly changing from 38 to 42 kHz in 20 ms. Such signal is cyclically repeated. Note that, due to the bandpass behavior of the transducer, the square wave chirp is filtered, obtaining a sinusoidal chirp at the transmitter's output. Note that the received signal is further filtered by receiver's acoustoelectric transducer. The anchors start listening for incoming signals at their own ultrasound transceiver electrical output for a time window of duration T_W, chosen to guarantee that each anchor collects an entire transmitted ultrasound pulse. Each anchor acquires data records that are then processed to measure the time difference of arrival of the incoming signal using correlation techniques. The TDoA measurements are converted into range differences as follows:

$$d_{ij} = v\tau_{ij} = R_i - R_j, \quad i, j = 1, \ldots, N, \quad i \neq j, \tag{2}$$

where N is the number of anchors, v is the speed of sound in air, τ_{ij} is the TDoA between the i-th node and the j-th node, and R_m is the distance between the m-th anchor and the mobile node, given by

$$R_m = \sqrt{(x_m - x)^2 + (y_m - y)^2 + (z_m - z)^2}, \tag{3}$$

where (x_m, y_m, z_m) are the known coordinates of the m-th beacon, while (x, y, z) are the coordinates of the mobile node, which have to be estimated [1]. Such estimation is performed using a least squares approach. In particular, the following cost function is minimized:

$$J(x, y, z) = \sum \left(d_{ij}(x, y, z) - \tilde{d}_{ij} \right)^2, \tag{4}$$

where $\tilde{d}_{ij}$ is the measured range difference between the i-th node and the j-th node. The minimization of (4) is performed by a Personal Computer connected to the board node via USB.

3 Experimental Results

3.1 Experimental Setup

Two versions of the measurement setup are developed, called setup 1 and setup 2. Their fundamental building blocks are designed and tuned in order to achieve a good

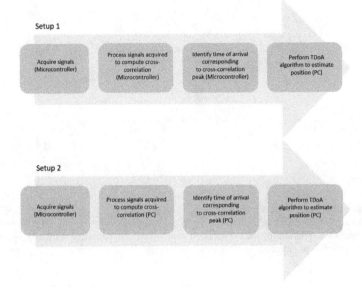

Fig. 3 Flow-chart of the measurement procedure in the two considered setups

tradeoff between complexity, accuracy, and update rate. The operation of the two setups is described by flow-charts shown in Fig. 3.

The setup 1, shown in Fig. 4, applies the bandpass sampling approach, with the aim of demonstrating the feasibility of a positioning system based on simple and low-cost hardware [7]. The system operates as follows: the signal from the ultrasonic transducer at the receiver is first amplified using a Programmable Gain Amplifier (PGA) and digitized by an Analog to Digital Converter (ADC) using bandpass sampling. The ADC resolution is set to 8 bits, the sampling frequency is set to 24 kSa/s and the acquisition record length was 480 samples. Thus, the sampling window duration was 20 ms. As an example, the records acquired by three receivers are shown in Fig. 5. Then, the acquired signal is processed, obtaining the cross-correlation sequence with respect to a locally-stored template. Subsequently, the envelope of the cross-correlation sequence is obtained by low-pass filtering the absolute value of the sequence, obtaining a smooth envelope, as depicted in Fig. 6. Then, the peak of each envelope is identified, and its abscissa provides the required delay estimation. Note that the timing resolution of peak identification is improved by quadratic interpolation of the samples adjacent to the peak [8]. The interpolated time instant corresponding to each cross-correlation peak, is then transferred by each receiving node to a PC via USB connection. In the PC, each time value is first converted into a distance estimate, by multiplication by the known speed of sound. Then, range-difference estimates are obtained by subtraction, and the heuristic presented in Sect. 2-C is applied. Finally, the TDoA positioning algorithm described

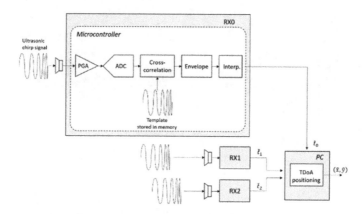

Fig. 4 Block diagram of the partially distributed architecture implemented in setup 1. The individual receiver nodes, RX0, RX1, and RX2, acquire and process the signals. Subsequently, they transmit the value, representing the interpolated time instant corresponding to the cross-correlation peak, to a host PC that performs TDoA based positioning

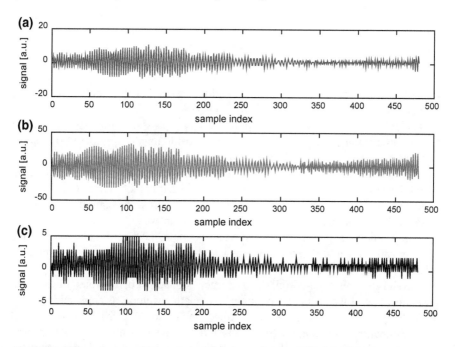

Fig. 5 Signals acquired simultaneously by the three receivers: **a** RX0, **b** RX1, and **c** RX2

in [9] is used to estimate the position of the mobile node. The overall measurement duration is about 0.9 s.

In setup 2, as shown in Fig. 7, the system is reconfigured using the DMA (Direct Memory Access) transfer supported by the PSoC microcontroller. The high DMA

Fig. 6 Envelope of normalized cross-correlation sequences of the implemented setup 1. The blue, red, and black lines represent the cross-correlation between the signal received by beacons 0, 1 and 2, respectively, and the template

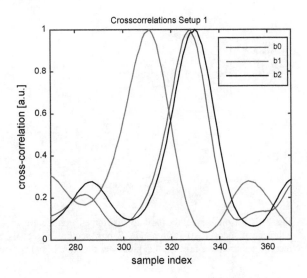

transfer rate permits both oversampling (i.e. using a sampling frequency larger than Nyquist Rate) and increased ADC resolution. Both degrees of freedom were exploited to improve the system. Thus, in setup 2, the sampling frequency is set to 100 kSa/s, while the ADC resolution is set to 12 bits. The same record length of 20 ms was used. Unlike setup 1, in setup 2, individual nodes do not extract time domain measurements, but transmit the received signal records to the PC. In turn, the PC, performs cross-correlation analysis between signals received from couples of beacons, as depicted in Fig. 8, obtaining TDoA estimations. By moving part of the computational burden to the PC, the measurement duration was not increased with respect to setup 1. The performance of the cross correlation is enhanced using interpolation techniques, as in setup 1.

3.2 Experimental Results

The system performance was experimentally validated, realizing both setup 1 and setup 2 and performing measurements. Both setups are respectively shown in Fig. 9a and b. In both setups, the anchors were placed at fixed positions along the x axis. Three beacons were deployed in a co-linear configuration. Instead, the mobile node was placed at a known position in the center of the considered area, chosen according to the directivity specifications of the ultrasound transceivers in [4], with the aim of maximizing the received SNR. In particular, the effect of sensor directivity is dependent on the angle between the direction of maximum directivity of RX, that was set parallel to the y axis, and the line joining TX and RX. After placing the

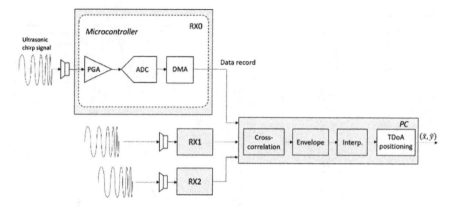

Fig. 7 Block diagram of the partially distributed architecture implemented in setup 2. The individual receiver nodes, RX0, RX1, and RX2, only acquire the signals. Subsequently, they transmit data records acquired to a host PC that processes the signals and performs TDoA based positioning

Fig. 8 Envelope of normalized cross-correlation sequences of the implemented setup 2. The cross-correlation between the signal received by beacons 0 and 1 is denoted by the blue line, that between beacons 0 and 2 by the red line, and that between beacons 1 and 2 by the black line

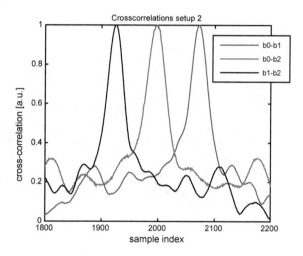

mobile node in the test position 10 repeated acquisitions were performed, followed by 2D position estimation.

In setup 1, as shown in the left of Fig. 10, all nodes are placed on stands and the test position is surveyed by aligning the stands with a grid placed on the floor. Due to nonidealities in the experimental setup, such as misalignment of the stands, the ground-truth accuracy of the test positions can be quantified to be within 1 cm. In setup 2, as shown in Fig. 10b, only the mobile node was placed on a stand.

As a metric for performance evaluation, the positioning error was computed, defined as the Euclidean distance between the true position and the estimated position. Experimental results for the two configurations are shown in Fig. 10, where the mean estimated position is plotted for the test point, together with the corresponding true

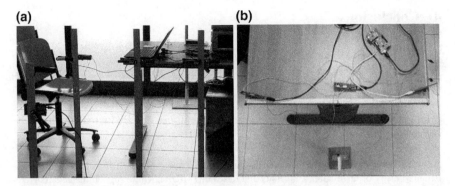

Fig. 9 Experimental setup 1 (**a**) and 2 (**b**)

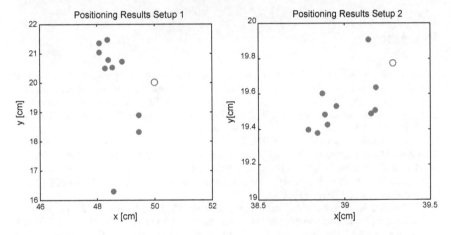

Fig. 10 Plot of the positioning results. The blue circlets indicate the true position, the filled red circlets denote the estimated positions. Left: setup 1, right: setup 2

Table 1 Positioning error for the two considered setups		Error mean (cm)	Error standard deviation (cm)
	Setup 1	2.0	0.8
	Setup 2	0.4	0.1

position. The obtained results show that a low-complexity and low-cost system for TDoA ultrasonic positioning is feasible. The realized system is capable of sub-centimeter accuracy in a range of about 1 m. Table 1 shows the positioning accuracy, expressed by error mean and standard deviation. Notice that the mean positioning error decreased from 2.0 cm, in the case of setup 1, to 0.4 cm in the case of setup 2.

4 Conclusions

A low-cost and low-complexity positioning system based on wirelessly triggered TDoA measurements was developed. Two partially distributed architectures were considered, and the positioning accuracy in a 2D scenario was assessed. It was shown that, under favorable conditions, the proposed low-cost architecture can achieve sub-centimeter accuracy, by using oversampling techniques and by increasing the ADC resolution.

Acknowledgements This research activity was funded through grant PRIN 2015C37B25 by the Italian Ministry of Instruction, University and Research (MIUR), whose support the authors gratefully acknowledge.

References

1. De Angelis, A., Moschitta, A., Comuniello, A.: TDoA based positioning using ultrasound signals and wireless nodes. In: 2017 IEEE International Instrumentation and Measurement Technology Conference (I2MTC), Turin, pp. 1–6 (2017). https://doi.org/10.1109/i2mtc.2017.7969873
2. Cypress Semiconductor Corp., CY8CKIT-042-BLE Bluetooth Low Energy Pioneer Kit Guide (2015)
3. Cypress Semiconductor Corp., CY8CKIT-059 PSoC® 5LP Prototyping Kit Guide (2015)
4. Murata Manufacturing co., ltd., MA40S4R Ultrasonic Sensor Application Manual (2016)
5. Medina, C., Segura, J.C., De la Torre, A.: Ultrasound indoor positioning based on a low-power wireless sensor network providing sub-centimeter accuracy. Sensors **13**, 3501–3526 (2013). https://doi.org/10.3390/s130303501
6. De Angelis, A., Moschitta, A., Carbone, P., Calderini, M., Neri, S., Borgna, R., Peppucci, M.: Design and characterization of a portable ultrasonic indoor 3-D positioning system. IEEE Trans. Instrum. Meas. **64**(10), 2616–2625 (2015). https://doi.org/10.1109/TIM.2015.2427892
7. Vaughan, R.G., Scott, N.L., White, D.R.: The theory of bandpass sampling. IEEE Trans. Signal Process. **39**(9), 1973–1984 (1991). https://doi.org/10.1109/78.134430
8. Marioli, D., Narduzzi, C., Offelli, C., Petri, D., Sardini, E., Taroni, A.: Digital time-of-flight measurement for ultrasonic sensors. IEEE Trans. Instrum. Meas. **41**(1), 93–97 (1992). https://doi.org/10.1109/19.126639
9. Chan, Y.T., Ho, K.C.: A simple and efficient estimator for hyperbolic location. IEEE Trans. Signal Process. **42**(8), 1905–1915 (1994). https://doi.org/10.1109/78.301830

Estimating the Outdoor PM10 Concentration Through Wireless Sensor Network for Smart Metering

D. Capriglione⏺, M. Carratù⏺, M. Ferro⏺, A. Pietrosanto⏺ and P. Sommella⏺

Abstract The original proposal of Advanced Metering Infrastructure based on short-range communication (wM-Bus) is suggested for the continuous monitoring of Particulate Matter within Smart Cities. A prototype of water meter equipped with a low cost off-the shelf PM sensor has been developed as remote node to be adopted in the radio Local Area network. Result of the metrological characterization against the quality requirements of the PM measurement according to European regulations as well as the simulation of a typical Smart Metering scenario in terms of communication performance confirm the feasibility of the proposed AMI for an effective adoption within urban areas.

Keywords Air pollution · Smart City · wM-Bus

1 Introduction

The monitoring of pollution level in urban areas must be considered one of the most important services for the citizens, indeed, the presence of high levels of pollutants are correlated with respiratory illnesses such as bronchitis, asthma, and chronic obstructive pulmonary disease that represent in according to the WHO the cause of death for 12 million of peoples [1].

Typically the air pollution is monitored by expensive (10,000 € to 30,000 € worth) station usually located in fixed point in an urban areas. Moreover, complex and repetitive calibration and maintenance operations are requested in order to obtain a high accurate measurement. Thus, few points of measurements are often available, not ensuring the awareness of the air quality in places not close to the measurement station. In order to obtain a distributed information on air pollution various solutions are adopted such as the use of air-quality mobile stations, passive samplers or models. In the first case, mobile stations are also very expansive as fixed stations and in most

D. Capriglione (✉) · M. Carratù · M. Ferro · A. Pietrosanto · P. Sommella
DIIn - University of Salerno, Via Giovanni Paolo II 132, 84084 Fisciano, SA, Italy
e-mail: dcapriglione@unisa.it

© Springer Nature Switzerland AG 2019
B. Andò et al. (eds.), *Sensors*, Lecture Notes in Electrical Engineering 539,
https://doi.org/10.1007/978-3-030-04324-7_49

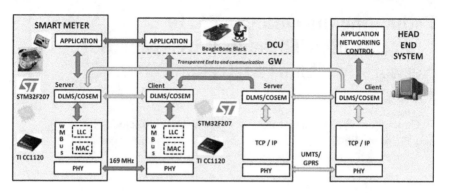

Fig. 1 Block scheme of the proposed Advanced Metering Infrastructure

cases they do not guarantee to reach a sufficient level of information to make a good density of spatial sampling. The passive samplers although they are less expensive than fixed and mobile stations, are not able to appreciate episodes of pollution with a high dynamics. Finally, the adoption of models requires a high level of knowledge and data that are not available in the most cases [2].

Recent technological advances in the field of embedded electronic and IoT communication technologies lead to the development and proposal of low-cost sensors (~15€) for pollution monitoring and wireless/wired interfaces (GPRS, Wi-Fi, LAN) enabling people [3–7] to share information about the urban air quality.

Following the Smart City paradigm, the authors intend to perform the air pollution monitoring through the Advanced Metering Infrastructure (AMI) that is the backbone of pilot projects where measurements of private (gas, water, electricity) consumption or public services are already remotely managed [8].

Twofold advantages are expected: the (zero-cost) sharing of an existing network infrastructure currently used by the smart meter to delivery information to the end users and the high spatial density of the measurement points guaranteed by the pervasive meter installation which may be exploited to achieve a tri-dimensional air pollution map.

2 The System Under Test

The AMI is a network that automatically provides utility companies with real-time data about consumptions that come from smart meters according to the hierarchical topology schemed in Fig. 1.

A set of wireless sensor nodes (leaf nodes) are connected to the master node, the Data Concentrator Unit (DCU/GW), which in turn forwards information to the Central Access System (SAC) where data are processed and stored. The leaf node battery powered hardware is featured only with a 169 MHz wM-Bus [9] radio module,

Fig. 2 The Particulate
Matter sensor
GP2Y1010AU0F

whilst master node hardware is provided also with long-range transmission capacity (GSM/GPRS antenna) to get to the SAC via cellular network. DCUs are either AC powered or solar cell recharged battery powered, it depends on the installation. Both leaf node—master node communications and master node—SAC communications are based on DLMS/COSEM protocol [10]. Each master node is responsible for the concentration and management of data generated from a number of leaf nodes, which are to be sent to the SAC.

The adoption of Advanced Metering Infrastructure (AMI) based on short-range communication (wM-Bus) is proposed for the continuous monitoring of Particulate Matter within urban areas in order to exploit the widespread deployment of gas and water meters: a prototype of smart meter equipped with a low cost off-the shelf PM sensor has been developed as remote node to be adopted in the radio Local Area Network.

In detail, the GP2Y1010AU0F sensor manufactured by Sharp (see Fig. 2) was proposed to measure the outdoor PM level [11] and integrated into the visual water meter described in [12].

In order to estimate the metrological performance of the developed digital sensor, the calibration of 30 PM sensors was performed by adopting the highly accurate Dylos Pro-1100 device (as reference instrument) and obtaining the calibration curve reported in Fig. 3. More in details, the PM sensors and the reference instrument were placed inside a hardboard box (internal volume $= 1$ m^3). A cigarette (kept at the middle of the box) was used as PM source for calibration, whereas the sensors (previously synchronized with time resolution of 1 s) were collecting samples at one minute intervals. After lighting the cigarettes, the box was closed and left for a total of 300 min. After a sharp increasing in the corresponding readings, the PM sensors remained saturated until the 150th minute, and then gradually decreased.

By adopting the cubic polynomial fitting suggested in [13], the calibration curve has been computed for each PM sensor. The results are shown in Fig. 3 in terms of mean value and standard deviation within the output range of interest ($0 \div 500 \, \mu g/m^3$).

Fig. 3 Calibration curve
(blue line) of the PM sensor
and corresponding
uncertainty interval (red
lines)

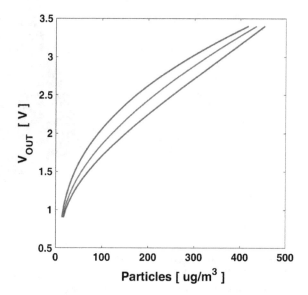

As expected the measurement uncertainty exhibited by the developed PM sensor is quite poor (ranging from 5 to 25% of the corresponding reading) if compared with the data quality assured by fixed measurement stations.

3   The Feasibility Study

Result of the previous metrological characterization has been the basis of the feasibility study about a microscale model to be adopted for PM10 monitoring (as integration of the fixed stations), which takes into account the quality requirements of the PM measurement according to European directive [14] about air quality evaluation. Indeed, the daily PM10 concentration should be estimated on hour basis with instrumentation able to assure data coverage (not lower than 75%, equal to 18 daily measurement results) and data uncertainty (with 95% confidence level) not greater than 25% of the limit daily value for preserving the public health, that is equal to 50 $\mu g/m^3$ (against which the dally mean value has to be compared).

In other words, the width of each (1-h) measurement result (expressed as 95% Confidence Interval) should lower than 25 $\mu g/m^3$. Thus, in order to effectively adopt the proposed Wireless Network, the corresponding metrological performance of the corresponding PM sensors has to be compensated through the averaging operation by exploiting the data availability from spatially distributed sensors within the WSN.

In detail, the comparison between the worst case for the estimated measurement uncertainty (80 $\mu g/m^3$ at PM concentration equal to 230 $\mu g/m^3$) and the prescribed data quality (12.5 $\mu g/m^3$) leads to the requirement for the concentration ratio, which should be assured by the distributed WSN, as detailed in the following scenario, where the wM-Bus protocol limitations are considered.

Indeed, according to the national regulation for natural gas AMI (UNI-TS 11291, [15]) a smart meter with short range radio capability is allowed to send 4 wM-Bus frames (maximum length equal to 255 byte) a day, by adopting the N2-a channel (featured by 12.5 kHz bandwidth centered at 169 MHz, GFSK modulation and transmission rate equal to 4.8 kbit/s). Each one of the four transmission start times should be randomly selected by the autonomous smart meter within the corresponding 6-h window (6:00:00 is considered as the conventional daily starting time, when the software routine for the random selection is executed). Thus, the smart meter equipped with the PM sensor may be programmed to run the PM measurement once an hour in order to include the available 1 byte PM readings (stored in the volatile memory) into the data field of the next wM-Bus frame (whose typical length ranges from 70 to 120 byte for spontaneous upward transmission from gas and water meters compliance with DLMS/COSEM).

Numerical simulation of the micro-scale model has been carried out by considering for each node of the proposed WSN a corresponding measurement equal to the random variable which takes into account the probability distribution of the outputs from a fixed measurement station and the low-cost PM sensors. In detail, the former distribution has been estimated by averaging the 1-h measurement results recorded during 1 month by a fixed air quality station placed in Salerno, Italy whereas, the latter (uniform) distribution of the PM sensor output takes into account the calibration results. Finally, different values (in the range 75%-95%) have been considered for the successful communication rate of the uplink transmission from WSN nodes to DCU/GW.

As main result of the simulation, the necessary 1-h PM measurements about the micro-scale (with the prescribed data quality) are always available at level of the Central Access System (as average of the corresponding readings from the smart meters) when the minimum concentration ratio of 1 DCU/GW to 100 remote nodes is considered. The requirement is typically fulfill by the smart meter planning in urban areas (characterized by the population density greater than 6000 inhabitants/km^2), where each DCU/GW is able to effectively manage the bidirectional short-range communication with hundreds of sensor nodes located until to 300 m.

References

1. Bentayeb, M., Wagner, V., Stempfelet, M., Zins, M., et al.: Association between longterm exposure to air pollution and mortality in France: a 25-year follow-up study. Environ. Int. **85**, 5–14 (2015)
2. Castell, N., Dauge, F.R., Schneider, P., Vogt, M., et al.: Can commercial low-cost sensor platforms contribute to air quality monitoring and exposure estimates. Environ. Int. **99**, 293–302 (2017)
3. Cerro, G., Ferdinandi, M., Ferrigno, L., Molinara, M.: Preliminary realization of a monitoring system of activated carbon filter RLI based on the SENSIPLUS® microsensor platform. In:

2017 IEEE International Workshop on Measurement and Networking (M&N), Naples, pp. 1–5 (2017). https://doi.org/10.1109/iwmn.2017.8078361

4. CitiSense Homepage. http://www.citi-sense.eu. Accessed 03 Mar 2018
5. Everyaware Homepage. http://www.everyaware.eu. Accessed 03 Mar 2018
6. Betta, G., Capriglione, D., Ferrigno, L., Laracca, M.A.: Measurement-driven approach to assess power line telecommunication (PLT) network quality of service (QoS) performance parameters. Meas. Sci. Technol. **20**(10), art. no. 105101 (2009)
7. Bernieri, A., Betta, G., Ferrigno, L., Laracca, M.: Improving performance of GMR sensors. IEEE Sens. J. **13**(11), art. no. 6547705, 4513–4521 (2013)
8. Carratù, M., Ferro, M., Paciello, V., Pietrosanto, A., Sommella, P.: Performance analysis of wM-Bus networks for smart metering. IEEE Sens. J. **17**(23), 7849–7856 (2017)
9. Communication systems for meters and remote reading of meters Part 4: Wireless meter readout (Radio meter reading for operation in SRD band). EN 13757-4:2013 (2013)
10. IEC Electricity metering—Data exchange for meter reading, tariff and load control—Part 53: COSEM application layer. IEC 62056-53 (2006)
11. Carratù, M., Ferro, M., Pietrosanto, A., Sommella, P.: Adopting smart metering RF networks for particulate matter distributed measurements. In: IMEKO TC19 Symposium: Metrology on Environmental Instrumentation and Measurements (EnvIMEKO 17), pp. 50–55, Aguascalientes, Mexico (2017)
12. Di Leo, G., Liguori, C., Paciello, V., Pietrosanto, A., Sommella, P.: Towards visual smart metering exploiting wM-Bus and DLMS/COSEM. In: IEEE International Conference on Computational Intelligence and Virtual Environments for Measurement Systems and Applications (CIVEMSA), pp. 1–6, Shenzhen, China (2015)
13. Rajasegarar, S., Zhang, P., Zhou, Y., et al.: High resolution spatio-temporal monitoring of air pollutants using wireless sensor networks. In: IEEE 9th International Conference on Intelligent Sensors Networks and Information Processing Symposium on Intelligent Sensors, pp. 1–6, Singapore (2014)
14. Directive 2008/50/EC of the European Parliament and of the Council of 21 May 2008 on ambient air quality and cleaner air for Europe (2008)
15. UNI/TS 11291-1:2013: UNI, Italian Committee for Standardization, pp. 1–28 (2013)

Machine Learning Techniques to Select a Reduced and Optimal Set of Sensors for the Design of Ad Hoc Sensory Systems

Luigi Quercia and Domenico Palumbo

Abstract The first step of this research has been to discriminate, by means of a commercial electronic nose (e-nose), the maturity evolution of seven types of fruits stored in refrigerated cells, from the post-harvest period till the beginning of the marcescence. The final aim was to determine a procedure to select a reduced set of sensors that can be efficiently used to monitor the same class of fruits by a low cost system with few, suitable sensors without loss in accuracy and generalization. To define the best subset we have compared the use of a projection technique (the Principal Component Analysis, PCA) with the sequential feature selection technique (Sequential Forward Selection, SFS) and the Genetic Algorithm (GA) technique by using classification schemes like Linear Discriminant Analysis (LDA) and k-Nearest Neighbor (kNN) and applying two data pre-processing methods. We have determined a subset of only three sensors which gives a classification accuracy near 100%. This procedure can be generalized to other experimental situations to select a minimal and optimal set of sensors to be used in consumer applications for the design of ad hoc sensory systems.

Keywords Sensors selection · Classification algorithms · Electronic nose
Fruit monitoring · PCA

1 Introduction

The e-nose is largely used for food applications [1] because it offers many advantages, among which, mainly, the fact that the analysis is inexpensive, not disruptive and that the data processing is done in real time. Moreover, in the last years, the ability to realize portable, reduced and inexpensive e-nose systems [1, 2] has opened the possibility of a pervasive development of sensorial techniques which could be applied not only in the post-harvest period, in the storage cells, but also wherever we want

L. Quercia (✉) · D. Palumbo
ENEA, Casaccia Research Centre, Via Anguillarese 301, 00123 Rome, Italy
e-mail: luigi.quercia@enea.it

© Springer Nature Switzerland AG 2019   405
B. Andò et al. (eds.), *Sensors*, Lecture Notes in Electrical Engineering 539,
https://doi.org/10.1007/978-3-030-04324-7_50

evaluate the fruits freshness "on the go" such as when the fruits are on the shelves for consumer applications. Furthermore portable, reduced and inexpensive e-nose systems are wished not only in food industry [3], but also for quality control, explosive detection, air quality control [1], and many other applications [4–7], requiring new experimental methodologies to design and optimize ad hoc sensory systems for each specific application. In our research we would verify the feasibility of using specific e-nose experimental data to select a minimal and optimal set of sensors to be used in consumer applications for the design of such sensory systems.

In general, feature selection is a critical problem in machine learning applications and in bioinformatics [8, 9]. The main problem is to determine which and how many optimal features we need in our applications. The classical PCA technique can't be appropriate to our aims because the data variance can't be always well correlated to the useful information. Using a commercial e-nose, we will apply the PCA to our data and we will compare it with other techniques of feature selection using different classification algorithms.

2   Materials and Methods

We have used to monitor the fruits a commercial e-nose (PEN3, Airsense Analytics GmbH) equipped with a 10 metal-oxide semiconductor (MOS) gas sensor array. It has been already used with success in many studies [10–14].

Seven different fruits types have been used in this study: bananas, clementines, strawberries, pears, peaches, plums and grapes. After the first experimental campaigns suggesting the critical importance of using a realistic quantities of fruits and storage cell dimensions, we have performed all campaigns filling the storage cell, about $4\,m^3$, with fruits at 80% of the volume, that is to say about 300 kg of fruits. Each type of fruits has been monitored in different campaigns under comparable experimental conditions. The monitoring period is different for each fruit type because this is dependent on the storage life variability of each fruit type to the storage process. The temperature was set to 5 °C. Fruits, when commercially available, were placed in the refrigerated cell as soon as possible. To follow the evolution of our samples headspaces we have utilised the e-nose coupled with an automatic electro valves system designed and realized in order to be synchronized with the e-nose operations. The measurements were taken at up to six different sampling points every 4 h because we have experimentally verified that in this period the headspace reaches always the equilibrium. We can ignore the perturbation introduced by the monitoring procedure because during 15 min of aspiration at the flux of 400 ml/min, we remove only 6 L on a total volume of 4000 L. Every 4 h the e-nose switches in "sample time" and aspires from a test point for 15 min, then the electro valves are closed and the e-nose goes in "flush time" for 25 min, usually enough to recover all sensors. During the sample time the e-nose collect a sample every second. After an initial period during which the signals of the 10 sensors of the e-nose rise, there is a period in which the signals are almost constant. Usually the last points in the flat zone are used for successive analysis but, in our case, because there could be the influence of the refrigerator fan motor, we use the average of all the points in the flat zone. Moreover the measurements used in this study for data analysis were only those taken at the same hour: about 5 a.m. The averages of the 10 sensors for each day, will give us the evolution of the headspace during the monitoring phase until the fruits reach the putrescence phase (which we define as storage life) clearly pointed out by mould development.

2.1 Data Pre-processing

The manipulation of e-nose data is an important aspect of the information extracting process, especially when we have to compare data from different e-nose systems, or different types of samples, i.e. fruits and meat, or different types of fruits. Many data normalization techniques have been implemented [15, 16].

Usually the most used techniques are the sensor *auto scale*, in which each sensor is scaled so that the resulting signal has zero mean and variance 1, and the sensor *vector normalization*, in which each sensor is scaled in the interval [0 1] [15, 17–19]. Both the techniques transform the data, changing the scale between the sensors, so that the inter-information between the sensors is lost. We won't use this method [20] because auto scale solution is usually adopted only when there is a big difference in the scale of the data. Moreover, both auto scale and vector normalization, amplify all the sensors at the same way, also the sensors who are not informative [17, 20]. To overcome these problems we use a *"global" normalization* so that a complete data set of each fruit is normalized in [0 1] using the global min and max of the fruit data set itself, following the equation:

$$k_{i,j} = \frac{x_{i,j} - x_{\min\_fruit}}{x_{\max\_fruit} - x_{\min\_fruit}} \tag{1}$$

where i is referred to the sensor and j to the sample, $x_{\min\_fruit}$ and $x_{\max\_fruit}$ are the minimum and maximum of the fruit data set analysed. We apply the previous equation for each fruit data set. The *global normalization* has the advantage that the relative information between the sensors isn't changed so that the electronic "fingerprints" of each fruit is retained and the sensors with less information aren't amplified.

Other authors didn't apply any data pre-processing especially when they study only one type of fruit [10–12].

Because we are comparing different fruits and there can be some scale differences between them we will use the *global normalization*; we will compare this normalization with the raw data without normalization and we will refer to this data using the label *no normalization*.

2.2 Feature Selection and Classification Algorithms

After the pre-processing step, we have applied different techniques to the entire set of the seven fruits types to select some reduced features which retain the "valuable information". An exhaustive search of the combinations of k features over n total features, is possible but we have to evaluate, for each k, n!/[k!(n − k)!] combinations [21] and, in total:

$$\sum_{k=1}^{n} \frac{n!}{k!\,(n-k)!} \tag{2}$$

which, in the case of n big, is impracticable.

Many techniques can be implemented to reduce the number of combinations to be evaluated [22]. The feature selection techniques compared here are the PCA, the SFS and the GA whereas their evaluation has been performed using the KNN and LDA classification algorithms. These classification algorithms gave us, in a preliminary phase, better classification accuracy than Support Vector Machines (SVMs) and Artificial Neural Networks (ANNs) also according to Tang et al. [2]. In the KNN classification we have used the neighbors equal to 2, the distance metric equal to Euclidean and the distance weight set to Equal without standardize the data. The classification accuracy is quantified, in all cases, by the 5 folds cross-validation loss. These parameters will be used in all the simulations, which have been done using Matlab R2015b [23].

The PCA analysis is largely used in clustering analysis to obtain preliminary information especially because it is an unsupervised technique. The technique is also used in feature selection in many fields [24] because it's very simple: it only discharge the components which have less variance. It should be noted that the use of the PCA without the evaluation of a classification function is inadequate, so, following Pardo et al. [25], we will use the PCA in conjunction with classification algorithms.

The SFS algorithm and the SBS, Sequential Backward Selection, algorithm are techniques which sequentially add, or remove, respectively, to the previous set only one feature not included in the previous features set. At every step, p sets are evaluated. The performance of each of the p sets can be evaluated using a classification procedure. The best set is chosen and a new set is formed. Due to the nature of the processes, the SFS and SBS have memory of the past so that a new set has only 1 feature different from the previous set, i.e. the procedure only investigate a limited number of combination and can be trapped in local minima [26].

The Genetic Algorithm, starting from a random population, transforms the actual population in a new population acting like the natural evolution [27]. All the individuals of a new population are evaluated, the best individuals are chosen and a new generation is formed following the rules of genetics. The procedure stops when we reach the best population (according to a fitness function which can be a classification function) or when a maximum number of iterations is reached. The GA

Table 1 Storage lives of the seven fruit types

Fruit	Bananas	Clementines	Grapes	Peaches	Pears	Plums	Strawberries
Days	20	21	19	46	30	33	12

algorithm doesn't implement an exhaustive search but is able to avoid local minima and find a near optimal solution [28] with the best ratio between the recognition rate and the number of evaluations [29]. In this research we have adapted the Sheffield "Genetic Algorithm Toolbox for MATLAB, v1.2" developed at the Department of Automatic Control and Systems Engineering of The University of Sheffield [30] especially changing the fitness function to determine the classification accuracy.

3   Results

The storage lives of the fruits analyzed in our campaigns have a large variability (see Table 1). To have the same conditions, as needed by GA elaboration, we started analyzing only the first 12 days, correspondent to the strawberries storage life.

Principal Component Analysis
We have to remember that, considering each type of fruits separately, the sensors with the prominent variances are different to that of the other fruits due to the fact that different fruits emitted different types of VOCs. Moreover, considering the global variance of the overall fruits without distinction between the species, whereas the pre-processing phase doesn't change the relative variance, and hence the sensors sequence for a type of fruit, it changes the global variance of overall fruits so that we have two different sets of sensors for the two types of normalizations, see Tables 2 and 3.

Determined the two sequences of sensors of the overall data fruits for the two types of normalizations, we have evaluated the classification accuracy of the 10 sequences of sensors using the KNN and LDA for both no and global normalization, starting from using one sensor up to using ten sensors. For each combination of sensors we have calculated the average of 100 different runs using the built-in Matlab functions.

Sequential Forward Selection algorithm
Also for the SFS procedure with the KNN and LDA classification, we have used both no and global data normalization. The functions and options used are the same used in the PCA analysis. The SFS algorithm has been described in Sect. 2.2. We have to note that, opposite to the sensor set extracted by the PCA analysis, in this case the optimum sensor sequence couldn't be the same at each run due to the random division of the data with the "5 folds" partition, so in the case of LDA and no normalization we have extracted the two most frequent sequences, {2 1 3 4 9 7 5 6 10 8} and {2 1 9 4 3 7 10 5 6 8}, which give the lowest error over 100 runs, see Tables 4 and 5. We also have to note that the more sensors we have to extract, the more different

Table 2 PCA fruit variances. First 12 days. *No normalization*

Sensor	2	6	8	3	9	1	7	5	4	10
Overall variances (%)	63.329	8.292	6.923	5.356	4.729	3.664	3.625	3.417	0.549	0.111

Table 3 PCA fruit variances. First 12 days. *Global normalization*

Sensor	2	6	8	1	9	7	4	3	5	10
Overall variances (%)	62.408	11.692	11.253	5.595	4.597	1.754	1.405	0.773	0.417	0.104

Table 4 Selected sensors by PCA and SFS with LDA. First 12 days. In the SFS sequences there were more than one combination of sensors with the same error

	PCA LDA	SFS LDA	
No norm.	2 6 8 3 9 1 7 5 4 10	**2 1 3 4 9 7 5 6 10 8**	2 1 9 4 3 7 10 5 6 8
Global	2 6 8 1 9 7 4 3 5 10	**5 6 1 2 4 3 8 7 9 10**	5 6 2 1 4 3 7 9 8 10

Table 5 Selected sensors by PCA and SFS with KNN. First 12 days

	PCA KNN	SFS KNN	
No norm.	2 6 8 3 9 1 7 5 4 10	**2 9 3 4 1 5 7 6 10 8**	2 9 4 3 1 5 6 7 8 10
Global	2 6 8 1 9 7 4 3 5 10	**5 6 2 1 9 3 4 8 7 10**	5 9 7 4 1 2 3 6 10 8

sequences we have with the same cross-validation loss. For example, if we have to extract one sensor, in the case of SFS with KNN and global normalization (but similar behavior happens in the other cases), we have a sensor, the number 5, with the 82% of frequency whereas, selecting this sensor number 5, we have three sequences with two sensors {5 2}, {5 6} and {5 9} with a frequency near the 33% each; we apply the same procedure when we would select three or more sensors. The same happens in the case of SFS with LDA and global normalization and in the SFS with KNN correspondent cases and the same procedure has been done to select the sequences and to calculate the cross-validation losses.

Genetic algorithm

We have applied the GA to our data with the same options adopted for the PCA and SFS algorithms. As in the SFS algorithm, the sensor sequences aren't unique and, in addition to the previous case, the sensor sequences at a defined step could be different from the previous sensor sequences for more than one feature: for example in the case of GA KNN with no normalization we pass from the sequence with five sensors {1 2 7 9 10} to the sequence with six sensors {1 2 3 4 5 9} changing three different sensors. So, for each searching sensor set, from 1 to 10 sensors, and for each combination of algorithm and pre-processing, we have done, at least, 100 runs each to determine the most frequent sequences which give the lowest error. We have to note, in this case, that the more the sensors we have to extract the more different sequences we have with the same cross-validation loss but, differently from the SFS case, with 1, 2 or 3 sensors, we have a predominant combination, as shown in Table 6 for the first three sensors, so, for GA, we consider only one sequence of sensors.

4   Discussion

Figure 1 summarizes the results obtained with the three methods for the first 12 days. The figure suggests that the best result is obtained in the case of GA-KNN-no normalization (1% and 0.6% of cross validation loss with 2 and 3 sensors respectively),

Table 6 Selected sensors by Genetic Algorithm. First 12 days. For this Table, we show all the selected sensors at each step because in this case a sequence can differ from the previous one for more than one sensor. The per cents, only showed for simplicity for the first 3 sequences, are the frequencies of the combinations over 100 runs

N° of sensors	GA LDA				GA KNN			
	No norm		Global		No norm		Global	
1	2	100%	5	100%	2	100%	5	100%
2	9 10	82%	5 6	100%	2 9	100%	5 9	71%
3	1 8 9	67%	1 2 4	85%	2 9 5	39%	5 9 7	78%
4	1 5 7 9		4 5 6 9		2 9 5 1		5 9 7 4	
5	1 5 6 7 9		4 5 8 9 10		2 9 1 7 10		5 9 7 4 10	
6	1 3 5 6 7 9		1 2 3 4 5 6		2 9 1 3 4 5		1 2 3 5 8 10	
7	1 3 4 5 6 7 9		1 4 5 7 8 9 10		1 2 3 4 5 8 9		1 2 3 5 7 8 9	
8	1 3 4 5 6 7 9 10		2 3 4 5 6 8 9 10		1 2 3 5 7 8 9 10		1 3 4 5 7 8 9 10	
9	1 3 4 5 6 7 8 9 10		1 2 3 4 5 7 8 9 10		1 2 3 4 5 6 7 8 9		1 2 3 4 5 7 8 9 10	
10	1 2 3 4 5 6 7 8 9 10		1 2 3 4 5 6 7 8 9 10		1 2 3 4 5 6 7 8 9 10		1 2 3 4 5 6 7 8 9 10	

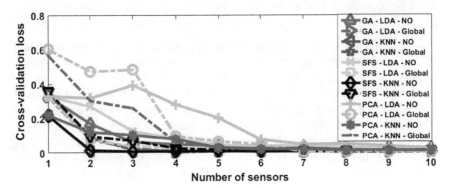

Fig. 1 Cross-validation loss, first 12 days. The cross-validation losses showed for the SFS cases, are related to the sequences in bold in Table 4 and in Table 5. In all the cases, the cross-validation losses showed, are the average of 100 different runs

although from three features onwards, excluding the case of GA-LDA-no normalization, the GA cases are equivalent.

The worst sensor subset is determined by the PCA method. In effect, although the PCA is capable of choosing the sensor 2 as one of the most important sensor in terms of classification accuracy, as done by the GA and the SFS methods, it doesn't take into account the importance of the sensor 9 which, in the case of *no normalization*, has "only" the 4.73% of the overall variance compared to the 8.3% of the sensor 6 but this difference is highly correlated to our problem of correctly classify the seven fruits: the PCA is incapable of discriminate between "useful information", for our scopes, and "noise".

The optimum data pre-processing method depends by the classification algorithm: in the KNN case, excluding the PCA case, the best accuracy is obtained by the cases in which there isn't data pre-processing whereas in the LDA case, the best accuracy is obtained for the case of data with *global normalization*. In the LDA case, the PCA method hasn't a uniform behaviour.

5 Conclusions

In this paper we have used the experimental data to compare and evaluate two data pre-processing methods, two classification algorithms and three feature selection techniques. The data have been obtained during many campaigns on different fruits monitored by a PEN3 e-nose in realistic refrigerated cells simulating commercial storage and transport conditions. We reached three main conclusions:

(1) optimal data pre-processing method depends on the classification algorithm;
(2) KNN classification algorithm gives the best classification result;
(3) about sensors selection we can summarize our result as follows:

    (a) PCA technique gives the worst result: variance comes out to be drastically different from information;
    (b) GA technique is the best but also the SFS technique, for a low number of features to be selected, gives optimal results. GA should be preferred, especially when the number of features is high, for a reduced computational cost and sensitivity to eventual local minima.
    (c) GA technique with KNN algorithm and "no normalization" pre-processing method came out to be the best combination. Using only three sensors, i.e. the sensors{2 9 5}, we have a classification accuracy near the 100%.

These three sensors could be inserted in low cost systems for consumer applications, to automatically detect not only the type and variety of fruit but, primarily, his freshness.

Acknowledgements This work has been partially funded by Italian Ministry of Economic Development thanks to "Ricerca di Sistema elettrico", "ORTOFRULOG", and "Magazzino Viag-giante" projects. The authors are thankful to Paolo di Lorenzo for his support in the experimental campaigns.

References

1. Wilson, A.D., Baietto, M.: Applications and advances in electronic-nose technologies. Sensors **9**, 5099–5148 (2009)
2. Tang, K.T., Chiu, S.W., Pan, C.H., Hsieh, H.Y., Liang, Y.S., Liu, S.C.: Development of a portable electronic nose system for the detection and classification of fruity odors. Sensors **10**, 9179–9193 (2010). https://doi.org/10.3390/s101009179

3. Dutta, R., Hines, E.L., Gardner, J.W., Kashwan, K.R., Bhuyan, M.: Tea quality prediction using a tin oxide-based electronic nose: an artificial intelligence approach. Sens. Actuators B Chem. **94**, 228–237 (2003). https://doi.org/10.1016/s0925-4005(03)00367-8

4. de Vries, R., Brinkman, P., van der Schee, M.P., Fens, N., Dijkers, E., Bootsma, S.K., de Jongh, F.H.C., Sterk, P.J.: Integration of electronic nose technology with spirometry: validation of a new approach for exhaled breath analysis. J. Breath Res. **9**, 046001 (2015)

5. Wang, J., Gao, D., Wang, Z.: Quality-grade evaluation of petroleum waxes using an electronic nose with a TGS gas sensor array. Meas. Sci. Technol. **26**, 085005 (9 pp) (2015)

6. Wongchoosuk, C., Lutz, M., Kerdcharoen, T.: Detection and classification of human body odor using an electronic nose. Sensors **9**, 7234–7249 (2009). https://doi.org/10.3390/s90907234

7. Yin, X., Zhang, L., Tian, F., Zhang, D.: Temperature modulated gas sensing E-nose system for low-cost and fast detection. IEEE Sens. J. **16**(2) (2016)

8. Butler, H.J., Ashton, L., Bird, B., Cinque, G., Curtis, K., Dorney, J., Esmonde-White, K., Fullwood, N.J., Gardner, B., Martin-Hirsch, P.L., Walsh, M.J., McAinsh, M.R., Stone, N., Martin, F.L.: Using Raman spectroscopy to characterize biological materials. Nat. Protoc. **11**, 664–687 (2016). https://doi.org/10.1038/nprot.2016.036

9. Ollesch, J., Drees, S.L., Heise, H.M., Behrens, T., Bruning, T., Gerwert, K.: FTIR spectroscopy of biofluids revisited: an automated approach to spectral biomarker identification. Analyst **138**, 4092 (2013)

10. Benedetti, S., Buratti, S., Spinardi, A., Mannino, S., Mignani, I.: Electronic nose as a non-destructive tool to characterise peach cultivars and to monitor their ripening stage during shelf-life. Postharvest Biol. Technol. **47**, 181–188 (2008). https://doi.org/10.1016/j.postharvbio.2007.06.012

11. Hernández Gómez, A., Wang, J., Hu, G., García Pereira, A.: Electronic nose technique potential monitoring mandarin maturity. Sens. Actuators B Chem. **113**, 347–353 (2006). https://doi.org/10.1016/j.snb.2005.03.090

12. Hernández Gómez, A., Wang, J., Hu, G., García Pereira, A.: Discrimination of storage shelf-life for mandarin by electronic nose technique. LWT-Food. Sci. Technol. **40**, 681–689 (2007). https://doi.org/10.1016/j.lwt.2006.03.010

13. Baldwin, E.A., Bai, J., Plotto, A., Dea, S.: Electronic noses and tongues: applications for the food and pharmaceutical industries. Sensors **11**, 4744–4766 (2011). https://doi.org/10.3390/s110504744

14. Rizzolo, A., Bianchi, G., Vanoli, M., Lurie, S., Spinelli, L., Torricelli, A.: Electronic nose to detect volatile compound profile and quality changes in 'Spring Belle' peach (Prunus persica L.) during cold storage in relation to fruit optical properties measured by time-resolved reflectance spectroscopy. J. Agric. Food Chem. **61**, 1671–1685 (2013). https://doi.org/10.1021/jf302808g

15. Scott, S.M., James, D., Ali, Z.: Data analysis for electronic nose systems. Microchim. Acta **156**, 183–2007 (2007). https://doi.org/10.1007/s00640-006-0623-9

16. Gorunescu, F.: Data Mining, Concepts, Models and Techniques. Intelligent Systems Reference Library, vol. 12. Springer, Berlin, Heidelberg (2011). ISBN: 978-3-642-19720-8 (Print) 978-3-642-19721-5 (Online); https://doi.org/10.1007/978-3-642-19721-5

17. Gutierrez-Osuna, R.: Pattern analysis for machine olfaction: a review. IEEE Sens. J. **2**(3), 189–202 (2002)

18. Boilot, P., Hines, E.L., Gongora, M.A., Folland, R.S.: Electronic noses inter-comparison, data fusion and sensor selection in discrimination of standard fruit solutions. Sens. Actuators B Chem. **88**, 80–88 (2003)

19. Dutta, R., Hines, E.L., Gardner, J.W., Udrea, D.D., Boilot, P.: Non-destructive egg freshness determination: an electronic nose based approach. Meas. Sci. Technol. **14**, 190–198 (2003)

20. Lavine, B.K.: Chemometrics: clustering and classification of analytical data. In: Encyclopedia of Analytical Chemistry, pp. 1–21. Wiley, Chichester (2000)

21. Gardner, J.W., Boilot, P., Hines, E.L.: Enhancing electronic nose performance by sensor selection using a new integer-based genetic algorithm approach. Sens. Actuators B Chem. **106**, 114–121 (2005). https://doi.org/10.1016/j.snb.2004.05.043

22. Ehret, B., Safenreiter, K., Lorenz, F., Biermann, J.: A new feature extraction method for odour classification. Sens. Actuators B Chem. **158**, 75–88 (2011). https://doi.org/10.1016/j.snb.2011.05.042
23. The MathWorks, Inc., Natick, MA, USA
24. Leopold, J.H., Bos, L.D., Sterk, P.J., Schultz, M.J., Fens, N., Horvath, I., Bikov, A., Montuschi, P., Di Natale, C., Yates, D.H., Abu-Hanna, A.: Comparison of classification methods in breath analysis by electronic nose. J. Breath Res. **9**, 046002 (2015)
25. Pardo, M., Kwong, L.G., Sberveglieri, G., Brubaker, K., Schneider, J.F., Penrose, W.R., Stetter, J.R.: Data analysis for a hybrid sensor array. Sens. Actuators B Chem. **106**, 136–143 (2005). https://doi.org/10.1016/j.snb.2004.05.045
26. Guo, D., Zhang, D., Zhang, L.: An LDA based sensor selection approach used in breath analysis system. Sens. Actuators B Chem. **157**, 265–274 (2011). https://doi.org/10.1016/j.snb.2011.03.061
27. Oluleye, B., Leisa, A., Leng, J., Dean, D.: A genetic algorithm-based feature selection. Br. J. Math. Comput. Sci. **4**(21), 3090–3106 (2014)
28. Questier, F., Walczak, B., Massart, D.L., Bouconb, C., de Jong, S.: Feature selection for hierarchical clustering. Anal. Chim. Acta **466**, 311–324 (2002)
29. Pardo, A., Marco, S., Calaza, C., Ortega, A., Perera, A., Sundic, T., Samitier, J.: Methods for sensor selection in pattern recognition. In: Gardner, J.W., Persaud, K.C. (eds.) Electronic Noses and Olfaction, pp. 83–88. IoP Publishing, Bristol (2000)
30. Chipperfield, A., Fleming, P., Pohlheim, H., Fonseca, C.: Genetic algorithm toolbox for use with MATLAB, Free Computer Programs on the Website of the Department of Automatic Control and Systems Engineering, University of Sheffield (1994)

Multi-sensor Platform for Automatic Assessment of Physical Activity of Older Adults

Andrea Caroppo⑩, Alessandro Leone⑩ and Pietro Siciliano⑩

Abstract This work presents a multi-sensor platform integrating one or more commercial low-cost ambient sensors and one wearable device for the automatic assessment of the physical activity and sedentary time of an aged person. Each sensor node could operate in a stand-alone way or in a multi-sensor approach; in the last case, fuzzy logic data fusion techniques are implemented in a gateway in order to improve the robustness of the estimation of a physiological measure characterizing the level of physical activity and specific parameters for the quantification of a sedentary lifestyle. The automatic assessment was conducted through two main algorithmic steps: (1) recognition of well-defined set of human activities, detected by ambient and wearable sensor nodes, and (2) estimation of a physiological measure, that is (MET)-minutes. The overall accuracy for activity recognition, obtained using simultaneously ambient and wearable sensors data, is about 5% higher of single sub-system and about 2% higher of that obtained with more than one ambient sensor. The effectiveness of the platform is demonstrated by the relative error between IPAQ-SF score (used as ground-truth, in which a low score corresponds to a sedentary lifestyle whereas a high score refers to moderate-to-vigorous activity level) and average measured (MET)-minutes obtained by both sensor technologies (after data fusion step), which never exceeds 7%, thus confirming the advantage of data fusion procedure for different aged people used for validation.

Keywords Intelligent environments · Ambient assisted living
Human activity recognition · Ambient sensor · Wearable sensor

1 Introduction

In the world the older population continues to grow at an unprecedented rate. According to the "An Aging World: 2015" report [1], 617 million people, 8.5% of the world's population, are 65 and over, a growth trend that in 2050 will make the world's elderly

A. Caroppo (✉) · A. Leone · P. Siciliano
National Research Council of Italy, Institute for Microelectronics and Microsystems, Lecce, Italy
e-mail: andrea.caroppo@cnr.it

© Springer Nature Switzerland AG 2019
B. Andò et al. (eds.), *Sensors*, Lecture Notes in Electrical Engineering 539,
https://doi.org/10.1007/978-3-030-04324-7_51

417

population reach 1.6 billion, almost 17% of the human beings that will live on our planet. The greatest concern for the health of the growing elderly population will increasingly be non-communicable diseases, with an unsustainable impact on already fragile health systems. To all this must be added other risk factors—such as smoking and alcohol abuse, the inadequate use of fruit and vegetables, the low levels of physical activity (PA) and the increment in sedentary time (ST)—that directly or indirectly contribute to the global increase in diseases.

Therefore, both PA and ST are important factors to consider in the health of older adults. Further work is required in this area with regard to cause of these outcomes and potential confounding factors. At this stage it is important to define the current level of PA and the amount of ST to allow for the size of the issue to be identified.

The interpretation of ST has been under scrutiny already, anyhow general consensus of opinion indicates that it should refers to any waking movement characterized by an energy expenditure (EE) ≤ 1.5 metabolic equivalents (METs), a value generally assumed in sitting or reclining posture. This means that any time a person is sitting or lying down, they are engaging in ST. Typical ST for older adults includes television (TV) viewing, reading and sleep on the couch or in bed.

Smart monitoring of seniors behavioral patterns and more specifically activities of daily living (ADLs), PA and ST, have attracted immense research interest in the scientific community. Development of smart decision systems to support the promotion of health smart homes has also emerged taking advantage of the plethora of smart, inexpensive and unobtrusive monitoring sensors, devices and software tools. Moreover, smart sensors are becoming more and more a technology key in Ambient Assisted Living (AAL) scenarios and, since in AAL context it is important to make people feel comfortable with technologies, acceptability is a fundamental aspect because it conditions the real use of these new devices by older adults.

In the recent years the analysis of human behavior in a domestic environment has been widely investigated, focusing on measurement of PA and ST. For example, commercial devices that integrates pedometers have been used to study activity levels [2] even if this device presents significant limitations in its manage because it results to be not efficient for the discrimination of different walking velocities [3].

SenseWear Armband (SWA) is another interesting solution that has been developed for the approximation of PA. This device is efficient to collect a diversity of physiological signs, but it is used principally for an accurate estimation of Energy Expenditure (EE) and step count during treadmill effort, providing a reasonably accurate measure of step count. Moreover, the results report in [4] showed that SWA permits to quantify the amount of PA, providing a methodology for automatic decision-making system for the increasing of activity in aged people. While armbands technologies have proven to be fine devices for tasks of daily life they have not been appropriate for higher intensity exercise, so its manage for the evaluation of physical and sedentary levels is not ideal.

In a recent paper [5] the authors compares objective and subjective assessment measures of PA and ST in young adults, using SWA for the objective measurement and PPAQ [6] to capture self-reported PA over the observation period. The result obtained show considerable discrepancies in the measures of PA/ST levels, with young adults interested in weight gain prevention that engage in both high levels of moderate-to-vigorous intensity PA (MVPA) and ST, with participants self-reporting fewer MVPA minutes and more ST compared to objective estimates. The limitation of this work is related to the lack of elderly subjects for the validation of the methodology.

The work proposed by Claridge et al. [7] is aimed at estimating objective and subjective quantification of habitual physical activity (HPA) and sedentary time in ambulatory and non-ambulatory adults with cerebral palsy, where objective measures of HPA and ST were obtained by using ActiGraph GT3X. The findings of authors support the use of accelerometers to objectively measure HPA and sedentary behavior in adults, even if they highlight that participants showed little reluctance to wearing accelerometers.

The present work describes a multi-sensor platform able to automatically assess the level of PA and ST of older adults. Both ambient and wearable sensors are used in the integrated version of the platform and this type of hardware architecture is motivated by the fact that in this way we expands the number of end-users, as they may accept only a type of sensor technology. Moreover, a fusion module was implemented for the improvement of activity recognition task. The smart sensors are able to recognize the same set of activities, extensively used for human behavior understanding [8]. The assessment of PA/ST was evaluated saving on a database the recognized activities and their relative duration. Subsequently, every activity was converted in (MET)-minutes which is derived from the time engaged in an activity with consideration to the number of METs of the same (statistically derived from a sample of persons).

The remainder of this paper is structured as follows. Section 2 reports an overview of the platform, technical specifications regarding the used sensors and a description of computational framework for human activities recognition in correspondence of each sensing technology involved. The same section describes also the study design and the methodology used in order to assess the level of PA and ST. The experimental results are reported and discussed in Sect. 3. Finally, conclusions and directions for future research are provided in Sect. 4.

2 Materials and Methods

2.1 Platform Overview

The topology of the platform is reported in Fig. 1. It consists of a number of detector nodes managing several ambient sensor nodes and one detector node that manages the wearable sensor node. Moreover, the platform integrates one coordinator node (an embedded personal computer) that receives high-level reports from each detector node. The actual version of the platform exploits hardware components selected in order to meet typical requirements of AAL applications.

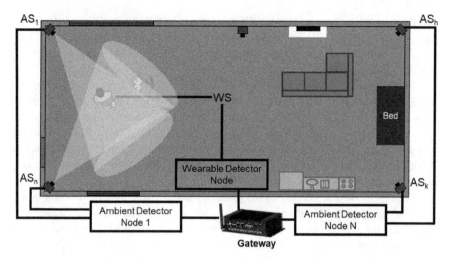

Fig. 1 Conceptual representation of the network topology ($AS_{1...n...k}$ are the ambient sensors, WS is the wearable sensor, the gateway is an embedded PC)

The computational framework is conceived as a modular, distributed and integrated into a larger AAL system through open middleware. Each ambient detector node includes a fuzzy rule-based data fusion logic that aggregates information from the vision sensor in order to provide more accurate information regarding activity detection task rather than each one of the individual source alone.

2.2 Ambient Sensor and Relative Framework for Activity Recognition Task

The ambient sensor used in the present work is Microsoft Kinect V2 (Fig. 2a), a low-cost wired structured light sensor, able to acquire 3D point clouds of the observed environment (even in overlapped views when more than one Kinect sensor is used to monitor the environment, with target in a range up to 10 m from the optical center of the cameras) [9]. Kinect was primarily designed for natural interaction in a computer game environment. However, the characteristics of the captured data have attracted the attention of researchers from other fields including mapping, 3D modeling and human activity recognition. The sensor consists of an infrared laser emitter, an infrared camera and an RGB camera. The laser source emits a single beam which is split into multiple beams by a diffraction grating to create a constant pattern of speckles projected onto the scene. This pattern is captured by the infrared camera and is correlated against a reference pattern. The reference pattern is obtained by capturing a plane at a known distance from the sensor, and is stored in the memory of the sensor. When a speckle is projected on an object whose distance to the sen-

(a)   **(b)**

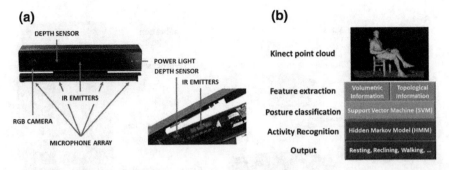

Fig. 2 **a** Microsoft Kinect for Windows V2, **b** main steps of computational framework for activities recognition using Kinect sensor

sor is smaller or larger than that of the reference plane the position of the speckle in the infrared image will be shifted in the direction of the baseline between the laser projector and the perspective center of the infrared camera. These shifts are measured for all speckles by a simple image correlation procedure, which yields a disparity image. For each pixel the distance to the sensor can be retrieved from the corresponding disparity. The Kinect sensor is able to capture RGB, depth, and infrared streams with frame rate of 9–30 Hz based on resolution. The default display resolution of these streams is 640×480 pixels, but it can be increased up to 1280×1024 with a lower frame rate. The RGB stream comes in 8 bit resolution in either color format of VGA or UYVY whereas the depth stream is 11 bit allowing for 2048 different depth sensitivity levels. Moreover, 3D sensors enable a new vision-based monitoring modality by which the subject's privacy problem can be mitigated since only distance data unable to reveal the subject's identity are processed.

The computational framework for the ambient sensor deals with feature extraction, human posture classification and activity recognition steps (Fig. 2b). Features are extracted from Kinect point cloud by using a twofold volumetric-topological descriptor having different levels of detail and computational complexity [10]. Given the coarse-to-fine features extracted, the posture classification step was performed via Support Vector Machine (SVM) classifier. The last step provides the activity recognition process, modelling each activity by means of a discrete Hidden Markov Model (HMM), whose observed symbols are the postures output of the previous step of the framework.

2.3 Wearable Sensor and Relative Framework for Activity Recognition Task

The wearable device used in this version of the platform is the Wearable Wellness System (WWS), produced by Smartex [11]. It is made up of a sensorized garment

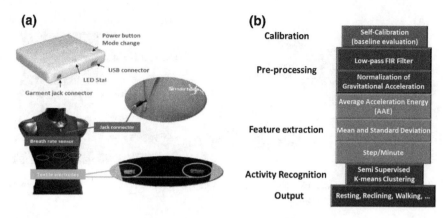

Fig. 3 **a** Wearable Wellness System compound by smart garment and an electronic device for acquisition, storage and wireless transmission of accelerometer data, **b** main steps of computational framework for activities recognition using tri-axial accelerometer data

and an electronic device (SEW) for the acquisition, the storage and the wireless transmission of the data. It has been designed to continuously monitor main vital parameters (Electrocardiography-ECG, Heart rate, Breathe rate) and the movements. The WWS garment is suitable and comfortable, reducing the well-known usability problems of the smart wearable devices. Moreover, it can be washed and it can be in tight contact with the body without any creases (Fig. 3a).

It integrates two textile electrodes, one textile piezoresistive sensor and a pocket, placed on the chest, for the SEW electronic device in which it is integrated a tri-axial MEMS accelerometer for the movement monitoring. The WWS can operate in streaming, via Bluetooth up to 20 meters (in free space), or in off-line modality, storing the data on a micro-SD integrated in the SEW device. In streaming mode the duration of its rechargeable battery life is about 8 h, while in off-line the duration is more than 18 h. For the elaboration, the raw acceleration data have been sent to the embedded PC with a 25 Hz frequency, that it is enough to recognize the physical activity [12].

The data are in the decimal format and represent the acceleration values with full scale in the range of ± 2 g and a 10 bit resolution for an high sensitivity. The MEMS accelerometer is DC coupled, so it measures both static and dynamic acceleration along the 3 axes and allows to get information on the 3D spatial relative position (compared to the Earth gravity vector) of the person who wears it. Indeed, if the accelerometer relative orientation is known, the resulting data can be used to determinate the angle α of the user position with respect to the vertical direction.

The main computational steps of the software architecture for wearable sensor concerns data acquisition, data pre-processing, system calibration, feature extraction and classification steps (Fig. 3b). The samples coming from the device are filtered out by a low pass FIR filter to reduce the noise due to the electronic components, environment and human tremors. A calibration procedure was accomplished by recovering the initial conditions after the device mounting, followed by an algorithmic step focused on extraction of useful features (Average Acceleration Energy-AAE, mean and the standard deviation over three acceleration axes, step/minute). Finally, the classification of the activities is obtained by implementing the effective and robust semi-supervised k-means clustering algorithm.

2.4　Methodology for Automatic Assessment of Physical Activity

The computational frameworks described in the previous sections transform the sensors into "smart sensors", able to recognize a set of well-defined activities. At this stage, the evaluation of the PA and ST are obtained by analyzing the duration of the activities detected and linking this duration to a physiological measure able to objectively assess the level of PA and to evaluate the amount of ST. In this work that common value is (MET)-minutes. MET is a unit for the estimation of the oxygen amount used by the body during PA. It is, in short, a way to compare the amount of exertion required for different activities between people. One MET is the base unit of METs in relation to activity. A single MET is equivalent to the energy, or oxygen, utilized by the body while at rest, or while participating in other idle activities, such as sitting quietly or reading a book, for example. The harder your body works during any given activity, the more oxygen is consumed and the higher the MET level. Activity that burns three to six METs is considered moderate-intensity physical activity. Activity that burns more than six METs is considered vigorous-intensity physical activity. (MET)-minutes are simply the time engaged in an activity with consideration to the number of METs. So if a subject walked at a pace (equivalent to 5 METs) for 30 min it would be calculated as follows:

$$5 \text{ METs} \times 30 \text{ minutes} = 150 \text{ (MET)-minutes} \tag{1}$$

Consequently, the activity of each participant was converted into (MET)-minutes per week (=MET level × minutes of activity/day × 5 times/week) according to IPAQ [13]. It is important to emphasize that MET values are derived from the analysis of the literature in which they are experimentally and statistically derived from a sample of persons as indicative averages, since the level of intensity could deviate from the representative experimental conditions used for the calculation of the standard MET values. In the present work, participants were classified into four classes of physical activity as: class 1, sedentary (MET-minutes × 5 times/week ≤ 450); class 2, some activity (450 < MET-minutes × 5 times/week ≤ 1000); class 3, moderate activity (1000 < MET-minutes × 5 times/week ≤ 2200), or class 4, high activity (MET-minutes × 5 times/week < 2200). The intervals were determined from the analysis of the scientific literature that addresses this issue, evaluating the average weekly level of (MET)-minutes, and proportioning it compared to the temporal monitoring period of the present study.

Table 1 Confusion Matrix for ambient sensor (multiple camera view), RES = resting, RCL = reclining, STD = seated, STRE = standing/relaxed, WLK1 = walking at 1.5–3.0 km/h, STRD = standing for domestic work, WLK2 = walking at 3.0–6.0 km/h. The values in bold represent the correct predictions of the considered activity in terms of overall accuracy

		Predicted activity (%)							Activity description	Met
		RES	RCL	STD	STRE	WLK1	STRD	WLK2		
Actual activity (%)	RES	**94**	6						Resting	0.7
	RCL	5	**92**	3					Reclining	0.8
	STD		4	**96**					Seated (reading or writing)	1.0
	STRE				**98**	1	2		Standing (relaxed)	1.2
	WLK1				1	**94**	1	4	Walking (1.5–3.0 km/h)	1.9
	STRD				2	2	**93**	3	Standing (domestic work)	2.0
	WLK2					2	1	**97**	Walking (3.0–6.0 km/h)	3.4

3　Results and Discussion

The activity detection performance (expressed in terms of average accuracy) has been estimated by using a common experimental setup in which the participants were asked to perform a predefined set of activities. During such experimental session, data were collected simultaneously by ambient-installed Kinects and the smart garment worn by each participant (only 3 participants agreed to wear the sensorized garment). The confusion matrix of classified activities are reported in Table 1 (ambient sensors with multiple view) and Table 2 (wearable sensor). The tables show different kinds of activities (and related MET units) useful for the evaluation of PA and ST.

The results obtained show a satisfactory classification procedure using both the sensing technologies. In particular, it is clear that the use of more ambient sensors provides the highest performance in terms of overall accuracy, with only slight confusion in distinguishing activities involving similar postures. The average accuracy obtained with the wearable sensor is also satisfactory, even if lower performance is obtained in correspondence of activities performed in standing position (relaxed, domestic work or walking at different speeds).

Moreover, the estimation of accuracy level that the platform can achieve in terms of activities recognition using data fusion process was evaluated. In Table 3, In order to facilitate performance comparison, the overall accuracy related to each standalone sub-system and the one related to the integrated system is reported.

The overall accuracy obtained using simultaneously data provided by ambient and wearable sensors is about 5% higher of single sub-system and about 2% higher of the activity recognition rate obtained with more than one ambient sensor, thus confirming the advantage in using a coordinator node that is able to manage and elaborates the high level outputs (activities) provided by each sub-system.

Table 2 Confusion Matrix for wearable sensor (multiple camera view), RES = resting, RCL = reclining, STD = seated, STRE = standing/relaxed, WLK1 = walking at 1.5–3.0 km/h, STRD = standing for domestic work, WLK2 = walking at 3.0–6.0 km/h. The values in bold represent the correct predictions of the considered activity in terms of overall accuracy

		Predicted activity (%)							Activity description	Met
		RES	RCL	STD	STRE	WLK1	STRD	WLK2		
Actual activity (%)	RES	**95**	5						Resting	0.7
	RCL	2	**90**	4	2		2		Reclining	0.8
	STD		4	**96**					Seated (reading or writing)	1.0
	STRE		1		**91**	3	4	1	Standing (relaxed)	1.2
	WLK1				2	**90**	2	6	Walking (1.5–3.0 km/h)	1.9
	STRD		1		6	3	**88**	2	Standing (domestic work)	2.0
	WLK2				1	5	2	**92**	Walking (3.0–6.0 km/h)	3.4

Table 3 Performance evaluation of activity recognition module in terms of overall accuracy (bold value indicates the best result)

Sub-system	Overall accuracy (%)
Ambient sensor (single view)	91.57
Ambient sensor (multi-view)	94.71
Wearable sensor	91.71
Integrated system	**96.68**

Differences between the results of IPAQ-SF score (expressed in MET-minutes x weekdays) and the average objective measurement obtained during the experimental period by ambient and wearable sensors are reported in Table 4. They have been estimated in terms of relative error (RE) defined as follows:

$$RE = \frac{|Q(i) - S(i)|}{S(i)} \times 100\% \tag{2}$$

where $Q(i)$ is the IPAQ-SF score of the ith subject, whereas $S(i)$ is the average (MET)-minutes $\times$ weekdays of the ith subject measured with both sensing technologies. From the results obtained it is evident that the methodology appears to be less reliable when the level of physical activity increases, and this is due to the classification performance obtained by each sensor with respect to activities of the following categories: walking (at different speeds) and standing (for domestic work).

Table 4 Relative error expressed in % between IPAQ-SF score and average measured (MET)-minutes obtained by ambient sensors, wearable sensor and both sensor technologies

User #	IPAQ-SF Score	RE (%)		
		Ambient	Wearable	Integrated system
1	1150	11.36	Not applicable	Not applicable
2	560	3.22	5.41	2.51
3	1245	9.45	11.53	6.63
4	375	3.84	Not applicable	Not applicable
5	750	5.07	7.71	4.16
6	480	3.50	Not applicable	Not applicable
7	760	5.14	Not applicable	Not applicable

4 Conclusion

The objectives of this study were to (1) examine the use of smart sensors (ambient and wearable) for human activity monitoring and (2) to assess the level of PA/ST of the observed subject in objective way, in order to provide a metric to the physician and make it capable to determine interventions that can, for example, reduce sedentary behaviour and increase at same time the physical activity during free-living activities in a domestic environment. The use of MET-minutes as objective measure for the assessment of the PA levels was appropriate analyzing the results obtained and comparing them with the results of IPAQ-SF score. Moreover, further advantages of the present work are to be found in the system architecture topology and in the design of the computational framework characterized by a modular, distributed and open architecture that permits a simple integration of both hardware resources and software modules with the aim to extend the number of services offered and the number of potential end-users, increasing the level of acceptability thanks to the use of different sensors. The suggested computational framework was implemented and tested for embedded processing in order to meet typical in-home application requirements such as low power consumption, noiselessness and compactness.

Future studies will be addressed to the inclusion of a larger sample size in the experimental stage and to the evaluation of sensors/detectors both ambient and wearable alternative to those already used, with the aim of integrating different smart sensors and raise the level of acceptance by end-users.

References

1. He, W., Goodkind, D., Kowal, P.: An aging world: 2015. US Census Bureau, pp. 1–165 (2016)
2. Bassett Jr., D.R., Wyatt, H.R., Thompson, H., Peters, J.C., Hill, J.O.: Pedometer-measured physical activity and health behaviors in United States adults. Med. Sci. Sports Exerc. **42**(10), 1819 (2010)

3. Lee, M., Kim, J., Jee, S.H., Yoo, S.K.: Review of daily physical activity monitoring system based on single triaxial accelerometer and portable data measurement unit. In: Machine Learning and Systems Engineering, pp. 569–580. Springer Netherlands (2010)
4. Dwyer, T.J., Alison, J.A., Mc Keough, Z.J., Elkins, M.R., Bye, P.T.P.: Evaluation of the SenseWear activity monitor during exercise in cystic fibrosis and in health. Respir. Med. 103(10), 1511–1517 (2009)
5. Unick, J.L., Lang, W., Tate, D.F., Bond, D.S., Espeland, M.A., Wing, R.R.: Objective estimates of physical activity and sedentary time among young adults. J. Obes. 2017 (2017)
6. Paffenbarger, R., Wing, A., Hyde, R.: Paffenbarger physical activity questionnaire. Am. J. Epidemiol. 108, 161–175 (1978)
7. Claridge, E.A., McPhee, P.G., Timmons, B.W., Martin, G.K., MacDonald, M.J., Gorter, J.W.: Quantification of physical activity and sedentary time in adults with cerebral palsy. Med. Sci. Sports Exerc. 47(8), 1719–1726 (2015)
8. Kellokumpu, V., Pietikäinen, M., Heikkilä, J.: Human activity recognition using sequences of postures. In: MVA, pp. 570–573 (2005)
9. Microsoft Kinect V2. https://support.xbox.com/en-US/xbox-on-windows/accessories/kinect-for-windows-v2-info. Accessed 14 Mar 2018
10. Diraco, G., Leone, A., Siciliano, P.: Geodesic-based human posture analysis by using a single 3D TOF camera. In: Proceedings of ISIE, pp. 1329–1334 (2011)
11. Smartex Wearable Wellness System (WWS). http://www.smartex.it/en/our-products/232-wearable-wellness-system-wws. Accessed 14 Mar 2018
12. He, Y., Li, Y.: Physical activity recognition utilizing the built-in kinematic sensors of a smartphone. Int. J. Distrib. Sens. Netw. 9(4), 481580 (2013)
13. Craig, C.L., Marshall, A.L., Sjorstrom, M., Bauman, A.E., Booth, M.L., Ainsworth, B.E., Pratt, M., Ekelund, U., Yngve, A., Sallis, J.F., Oja, P.: International physical activity questionnaire: 12-country reliability and validity. Med. Sci. Sports Exerc. 35(8), 1381–1395 (2003)

Failure Modes and Mechanisms of Sensors Used in Oil&Gas Applications

M. Catelani, L. Ciani and M. Venzi

Abstract This paper focuses on failure modes and mechanisms of the main sensors used in Oil&Gas field, especially for Safety application, in the Safety Instrumented System design. Moreover, the role of the on-board diagnostic will be also discuss and how it improves the capability to detect a failure mode in the sensors used in safety critical applications.

Keywords Sensor · Failure mode · Failure mechanism · Diagnostic

1 Introduction

The operation of many industrial processes, especially in chemical and oil&gas fields, involves inherent risk to persons, property, and environment. The goal of functional safety is to design, built, operate and maintain systems in such a way to prevent dangerous failures or, at least, to be able to control them in case of hazardous conditions. In order to reduce the risk arising from industrial plants, it might be necessary to automatically activate safety measures when required to avoid dangerous situations: functional safety of Electrical/Electronic/Programmable Electronic Safety-Related Systems is achieved with Safety Instrumented Systems (SIS). These systems are specifically designed to protect personnel, equipment, and the environment by reducing the likelihood or the impact severity of hazardous events [1–3].

Safety Instrumented Systems (see Figs. 1 and 2) are typically constituted by a combination of three fundamental blocks [4, 5]:

- Sensor(s) detects a physical quantity and provides a corresponding electrical output. Field sensors are used to collect information and determine an incipient danger; these sensors evaluate process parameters (e.g. temperature, pressure, flow, etc.) in order to determine if single equipment or the whole process or plant is working

M. Catelani · L. Ciani (✉) · M. Venzi
Department of Information Engineering, University of Florence, via S. Marta 3, Florence, Italy
e-mail: lorenzo.ciani@unifi.it

© Springer Nature Switzerland AG 2019
B. Andò et al. (eds.), *Sensors*, Lecture Notes in Electrical Engineering 539,
https://doi.org/10.1007/978-3-030-04324-7_52

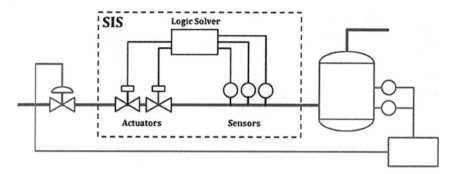

Fig. 1 SIS—Safety Instrumented System

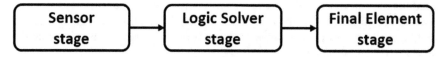

Fig. 2 Safety Instrumented System functional block diagram

properly and it is in a safe state. Such sensors do not monitor the normal process but they are usually dedicated to SIS.

- Logic solver(s) receives the information collected by the sensor and elaborates it to take the best response. It is typically a controller that takes actions according to the defined logic in order to prevent hazardous conditions.
- Final element(s) implements the outcomes of the logic solver. This actuator is the last element of the loop and in many industrial applications is represented by a pneumatic valve.

Two important parameters for safety assessment are Diagnostic Coverage (DC) and Safe Failure Fraction (SFF) [6]. DC is the ratio of the probability of detected failures to the probability of all the dangerous failures and it is a measure of system ability to detect failures; SFF, instead, indicates the probability of the system failing in a safe state so it shows the percentage of possible failures that are self-identified by the device or are safe and have no effect.

$$DC = \frac{\lambda_{DD}}{\lambda_{DD} + \lambda_{DU}} \tag{1}$$

$$SFF = \frac{\lambda_{SU} + \lambda_{SD} + \lambda_{DD}}{\lambda_{SU} + \lambda_{SD} + \lambda_{DU} + \lambda_{DD}} \tag{2}$$

where λ_{SU} is the safe undetected failure rate, λ_{SD} is the safe detected failure rate, λ_{DU} is the dangerous undetected failure rate and λ_{DD} is the dangerous detected failure rate.

2 Sensors in Oil&Gas Application

This paragraph introduces a brief description of the main used sensor classes for Oil&Gas SIS application: temperature, pressure, flow and level. The temperature sensors allow the users to detect any temperature change measuring the amount of heat energy which is produced by the equipment. Temperature sensors are classified in contact devices (i.e. must be in physical contact with the monitored item) and non-contact devices (i.e. that works using convection and radiation to monitor changes).

The main kinds of temperature sensors are listed below [7, 8]:

- Thermocouple (TC): two junctions of different metals welded together. The first one (i.e. the reference junction, also called cold junction) is maintained at constant temperature while the other one is the measuring point. This sensor provides an output voltage proportional to the temperature difference between the two junctions. It could work approximately from -200 to $2000\ ^{\circ}$C.
- Thermostat (TS): a bi-metallic strip formed bonding together two different metals. The mechanical bending of the strip is due to the different linear expansion rate of the two metals produced by a temperature change.
- Thermistor (RTD): passive resistive devices that changes its resistance because of changes in temperature.

Three types of pressure measurements are possible: absolute pressure (i.e. measurement referred to perfect vacuum), gauge pressure (i.e. measurement referred to ambient pressure) and differential pressure (i.e. difference between two points of measurement). Usually, pressure sensors provide all these kinds of measurements. The main pressure sensors are the following [9, 10]:

- Capacitive: a parallel plate capacitor formed by a fixed base plate and a deformable conductive member (diaphragm). Pressure deforms the diaphragm inducing a variation in the capacitance.
- Piezoelectric: uses the piezoelectric effect to measure the strain upon the sensing area and consequently the pressure.
- Piezoresistive: uses the piezoresistive effect to detect strain due to applied pressure.

The rate at which a fluid moves through a cross section is called flow. There are several types of flow meters that derive the flow rate from other physical quantities using fundamental fluid flow principles. The main types of flow meter are listed below [11, 12]:

- Differential pressure: uses Bernoulli's principle in order to evaluate the flow rate, which is one of the most common ways to evaluate a flow measurement. The pressure dropping across the sensor is proportional to the square of the flow rate.
- Direct force: are governed by balancing forces within the system.
- Ultrasonic—Doppler effect: uses frequency shift of an ultrasonic signal which is reflected by discontinuities in motion in order to measure the fluid velocity.
- Ultrasonic—Transit time: uses two different transducer together. It measures the flight time of a signal sent from the up transducer to the down transducer.

Level sensors detect the level of liquids and fluids in two different ways: continuous or discrete. A continuous level meter measures the quantity of liquid in a container; a discrete fluid meter indicates if the liquid is above or below a reference line. There are a lot of different types of level sensors [13, 14]:

- Magnetic and mechanical: direct contact or magnetic system control the mechanical switch operation.
- Ultrasonic: contactless sensor used in case of highly viscous fluids that works through an ultrasonic signal to measure the distance between the sensor and the reference surface.
- Hydrostatic: provides an indirect measure of the fluid column height obtained by the fluid pressure measure.

3 Failure Modes and Failure Mechanism of Sensors

In order to perform the availability and reliability of the systems, it's important to study the failure mechanisms and failure modes that can occur to these four families of sensors. The use of the standard ISO 14224 is mandatory to asses this type of analysis [15]. The international standard provides a base for the collection of data and for the reliability to all the equipment in oil&gas field. The main objectives of the standard are: design and configuration of a system, RAMS requirement for industrial plants, life cycle cost and maintenance optimization.

According to [15] a failure mechanism is the physical, chemical or other process that leads to a failure and a failure mode is the effect by which a failure is observed on the failed item.

Table 1 shows an extract of the failure mechanisms and relative failure modes for temperature sensors used in Oil&Gas safety application.

Figure 3 shows the effect percentage of the most probable failure modes for the thermocouples and thermistors, the two main temperature sensors used in Oil&Gas SIS. The left pie chart is referred to the thermocouples, the right one described the

Table 1 Examples of failure mechanisms and modes for temperature sensors

Failure mechanism	Description	Failure mode
Vibration	Abnormal vibration	Erratic output
Corrosion	All types of corrosion, both wet (electrochemical) and dry (chemical)	External leakage
		Low output
		Erratic output
Breakage	Fracture, breach and crack	No output
Overheating	Material damage due to overheating/burning	High output
		Low output
		External leakage

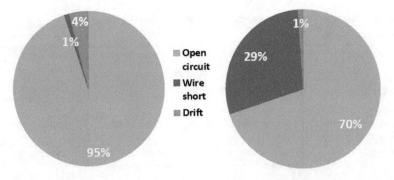

Fig. 3 Pie chart of failure mode percentages for thermocouples and RTDs used in Oil&Gas SIS

Table 2 Examples of failure mechanisms and modes for pressure sensors

Failure mechanism	Description	Failure mode
Deformation	Distortion, bending, buckling, denting, yielding, shrinking, blistering, creeping, etc.	Erratic output
		Spurious operation
Corrosion	All types of corrosion, both wet (electrochemical) and dry (chemical)	External leakage
		Low output
		Erratic output
Breakage	Fracture, breach and crack	No output
Fatigue	In case the cause of breakage can be traced to fatigue	High output
		Low output
		Spurious operation

Table 3 Examples of failure mechanisms and modes for level sensors

Failure mechanism	Description	Failure mode
Vibration	Abnormal vibration	Erratic output
Deformation	Distortion, bending, buckling, denting, yielding, shrinking, blistering, creeping, etc.	Erratic output
		Spurious operation
Looseness	Disconnection, loose items	Erratic output
		Spurious operation

thermistors. For both sensors the most probable failure mode is the Open circuit, therefore it needs more attention during the design phase of the system. Diagnostic and optimization of the maintenance policies are mandatory in order to mitigate the occurrence of this failure mode.

Table 2, Table 3 and Table 4 show an extract of the failure mechanisms and failure modes for respectively pressure sensors, level meters and flow sensors used in Oil&Gas safety application.

Table 4 Examples of failure mechanisms and modes for flow sensors

Failure mechanism	Description	Failure mode
Deformation	Distortion, bending, buckling, denting, yielding, shrinking, blistering, creeping, etc.	Erratic output
		Spurious operation
Looseness	Disconnection, loose items	Erratic output
		Spurious operation
Clearance/alignment failure	Failure caused by faulty clearance or alignment	Erratic output
Sticking	Sticking seizure, jamming due to reasons other than deformation or clearance/alignment failures	Erratic output
Cavitation	Relevant for equipment such as pumps and valves	Erratic output

4 Diagnostic in Oil&Gas Safety Sensors

This study shows how the on-board diagnostic improves the capability to detect a failure mode in the sensors used in safety application. The work focuses on a single flow meter *Electromagnetic Flow (8732E)* by Rosemount Corporation; it is equipped with on-board diagnostic and it could be programmed to internally identify its failure modes and communicates its health state.

In case the on-board diagnostic is set off the logic solver detect the failure modes that lead the output signal out of range. The failure rate results are shown in Table 5.

Diagnostic Coverage and Safe Failure Fraction are the following:

$$DC = \frac{128}{128 + 309} = 29.2\% \tag{3}$$

$$SFF = \frac{284 + 0 + 128}{284 + 0 + 128 + 309} = 57.1\% \tag{4}$$

In case the on-board diagnostic is set on the results are shown in Table 6. Diagnostic Coverage and Safe Failure Fraction are the following:

$$DC = \frac{916}{916 + 309} = 74.7\% \tag{5}$$

Table 5 Failure rates of flow meter without on-board diagnostic

λ_{SU}	λ_{SD}	λ_{DU}	λ_{DD}
284 FIT	0 FIT	309 FIT	128 FIT

Table 6 Failure rates of flow meter with on-board diagnostic

λ_{SU}	λ_{SD}	λ_{DU}	λ_{DD}
284 FIT	0 FIT	309 FIT	916 FIT

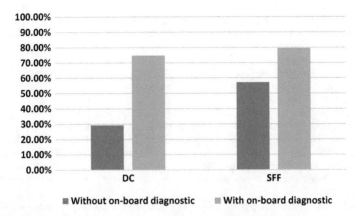

Fig. 4 Comparison of DC and SFF values with and without on board diagnostic

$$SFF = \frac{284 + 0 + 916}{284 + 0 + 916 + 309} = 79.5\% \tag{6}$$

According to Eqs. (3)–(6) Fig. 4 shows the results of DC and SFF compared with and without diagnostic.

5 Conclusions

This paper proposes an overview about the main sensors used in safety application in Oil&Gas field, including temperature, pressure, level and flow sensing units. An extract of the main failure modes and failure mechanism of these sensors are analyzed and described, focusing on the effect percentage of the most probable failure modes for the thermocouples and thermistors.

Finally, the paper focuses on the effects of on-board diagnostic on an industrial flow meter used in a Safety Instrumented System. The on-board diagnostic highly improve the probability of detection of the dangerous failure, with a consequent increase of the Diagnostic Coverage and the Safe Failure Fraction.

References

1. IEC 61508: Functional safety of electrical/electronic/programmable electronic safety related systems, Part 1–6, Technical report, International Electrotechnical Commission (2010)
2. Catelani, M., Ciani, L., Luongo, V.: Safety analysis in oil & gas industry in compliance with standards IEC61508 and IEC61511: methods and applications. In: IEEE International Instrumentation and Measurement Technology Conference (I2MTC), Minneapolis, MN (USA), pp. 686–690 (2013)

3. Catelani, M., Ciani, L., Mugnaini, M., Scarano, V., Singuaroli, R.: Definition of safety levels and performances of safety: applications for an electronic equipment used on rolling stock. In: 2007 IEEE Instrumentation & Measurement Technology Conference (IMTC), pp. 1–4 (2007)
4. Safety Instrumented Systems (SIS)—Safety Integrity Level (SIL) Evaluation Techniques: ISA-TR84.0.02, ISA, Research Triangle Park, NC (1999)
5. Lundteigen, M.A.: The effect of partial stroke testing on the reliability of safety valves. In: Conference: ESREL 2007, Stavanger
6. Rausand, M.: Reliability of Safety-Critical Systems. Wiley Inc. Publication (2014)
7. Analog Device: In: W. Jung (ed.) Op Amp Application Handbook. Elsevier, USA (2005)
8. Zhengbing, H., Jotsov, V., Jun, S., Kochan, O., Mykyichuk, M., Kochan, R., Sasiuk, T.: Data science applications to improve accuracy of thermocouples. In: 2016 IEEE 8th International Conference on Intelligent Systems (IS), pp. 180–188 (2016)
9. Fraden, J.: Handbook of Modern Sensors: Physics, Designs and Applications, 3rd edn. Springer, USA (2004)
10. Nallathambi, A., Shanmuganantham, T., Sindhanaiselvi, D.: Energy scavenging aspects of cantilever based MEMS piezoelectric pressure sensor. In: 2016 International Conference on Emerging Technological Trends (ICETT), pp. 1–6 (2016)
11. Analog Device Inc.: In: Sheingold, D.H. (ed.) Transducer Interface Handbook. USA (1980)
12. Aziz, E., Kanev, Z., Barboucha, M., Maimouni, R., Staroswiecki, M.: An ultrasonic flowmeter designed according to smart sensors concept. In: Proceedings of 8th Mediterranean Electrotechnical Conference on Industrial Applications in Power Systems, Computer Science and Telecommunications (MELECON 96), vol. 3, pp. 1371–1374 (1996)
13. Chhantyal, K., Viumdal, H., Mylvaganam, S., Elseth, G.: Ultrasonic level sensors for flowmetering of non-Newtonian fluids in open Venturi channels: using data fusion based on artificial neural network and support vector machines. In: 2016 IEEE Sensors Applications Symposium (SAS), pp. 1–6 (2016)
14. Jahn, A., Ehrle, F., Roppel, C.: A level sensor for fluids based on hydrostatic deformation with piezoelectric generated sounds in a low frequency range. In: 2014 6th European Embedded Design in Education and Research Conference (EDERC), pp. 245–249 (2014)
15. ISO 14224: Petroleum, petrochemical and natural gas industries—collection and exchange of reliability and maintenance data for equipment (2016)

Lab-on-Disk Platform for *KRAS* Mutation Testing

Iemmolo Rosario⊙, Guarnaccia Maria, Petralia Salvatore⊙,
Cavallaro Sebastiano⊙ and Conoci Sabrina⊙

Abstract Colorectal cancer (CRC) is one of the most common cancers worldwide.
In the United States is currently the third deadliest cancer with more than 1 million patients diagnosed annually of which 50% will develop metastatic disease. In some subtypes of CRC, the *KRAS* mutation status has emerged as an important diagnostic/prognostic marker for the response to treatment with anti-EGFR drugs in patients with metastatic CRC. Currently, the direct sequencing remains the gold standard technique for the diagnosis of DNA mutations, but the low sensitivity, time-consuming and the need for operating in rooms with dedicated instrumentation makes this method disadvantageous in daily practice. In recent years, new technologies characterized by different sensitivities, specificities and complexities are starting to be used in research and clinical studies for the detection of DNA genotyping. In this work, we propose a novel portable Lab-on-Disk platform developed by STMicroelectronics as a competitive device able to perform TaqMan-based real-time PCR for the rapid and simultaneous detection and identification of KRAS gene mutations.

Keywords Optical biosensor · Colorectal cancer · Diagnostics · RtPCR
DNA mutations

1 Introduction

Colorectal cancer (CRC) is one of the most common causes of cancer-related death in both men and women, with an incidence of almost a million cases annually [1, 2]. In addition to lifestyle and environmental risk factors, the development of CRC is

I. Rosario · G. Maria · C. Sebastiano
Institute of Neurological Sciences—Italian National Research Council, Via Paolo Gaifami,
18–95126 Catania, Italy
e-mail: iemmolo.rosario@gmail.com

P. Salvatore · C. Sabrina (✉)
STMicroelectronics, Stradale Primo Sole, 50–95121 Catania, Italy
e-mail: sabrina.conoci@st.com

© Springer Nature Switzerland AG 2019
B. Andò et al. (eds.), *Sensors*, Lecture Notes in Electrical Engineering 539,
https://doi.org/10.1007/978-3-030-04324-7_53

associated with genetic defects, such as chromosomic instability and several genetic alterations involving different genes implicated in proliferation, differentiation, apoptosis and angiogenesis pathways [1, 3, 4]. The genetic heterogeneity of CRC allows to a more precise diagnosis of predisposing familial syndromes [5] and to predict prognosis and direct therapies [1]. In fact, several point mutations in specific RAS genes, involved in the G-protein signal transduction pathway, are known to influence diagnosis and treatment of colorectal cancer [6–8]. Among them, mutations in *KRAS* can be diagnosed in approximately 35–45% of patients, with a high concordance between primary and secondary lesions [9, 10]. The most frequent mutations in *KRAS* are single base missense mutations, 98% of which are found at codons 12 and 13 [6]. In particular, these mutations cause a constitutive activation of KRAS impairing the intrinsic GTPase activity of Ras and confer resistance to GTPase activators, thereby causing Ras accumulation in its active guanosine triphosphate (GTP)-bound state, sustaining the activation of Ras signaling [11]. This leads to the activation of multiple downstream proliferative signaling pathways, such as MAPK, the RAF/MEK/ERK and the PI3 K/AKT signaling cascades that result in a sustained proliferation signal within the cell even in the absence of growth factor stimulation [12]. Despite the increased activity of the signaling pathways, the mutation alone causes the loss of RasGTPase enzymatic function only [13]. To date, several studies conducted to explore the role of *KRAS* mutations in relation to carcinogenesis highlighted that its mutational status is a frequent predictive and prognostic marker of tumor progression and sensitivity or resistance to anti-EGFR therapy [14]. These results suggest that the prediction of response to EGFR with respect to the patients' genotype is one of the main tasks in colon cancer management to ensure maximum efficacy with minimal adverse effects. Different approaches can be utilized to identify the *KRAS* mutational status but the most common require skilled personnel, high costs and specialized equipment [15, 16]. Moreover, these methods need to be standardized regarding their sensitivity, specificity, and cost *per* analysis before they might be considered as standard reliable practice to be employed for diagnostics purposes. There are two main challenges that need to be standardized to achieve more reproducible and consistent results: heterogeneity of the testing materials and differences in the detection limits among methods [17].

The real time Polymerase Chain Reaction (RT-PCR) is a standard method for the genetic study. To allow a large use of this method in medical practice, great efforts have been made from scientists to develop easy-to-use and miniaturized devices allowing testing analysis to be conducted by unspecialized personnel at very low cost [18, 19]. These platforms called Genetic "Point-of-Care" (PoC) are designed to be used outside the central laboratory and directly near the patient [20]. The genetic PoC must integrate the three main steps necessary for molecular analysis: sample preparation, PCR amplification and detection. This goal can be achieved by various formats such as Lab-on-chip, Lab-on-tube, Lab-on-Disk systems [21–23]. The new frontiers in the genetic PoC are oriented to assays based on PCR-free technologies. The cooperative hybridization is an innovative method employed to capture and detect a whole genome at device surface [24]. In this paper, we propose the use of a novel and portable Lab-on-Disk platform developed by STMicroelectronics, for the rapid and simultaneously detection of the main *KRAS* mutations based on Real Time PCR methodology. Based on hybrid silicon-plastic technology, this platform is composed of a miniaturized silicon chip, with onboard the temperature sensors, heaters, and the compact tool that drives the real-time PCR process [25].

Compared with commercial platforms, the Lab-on-Disk offers a rapid and sample-in answer-out analysis allowing a new approach to genetic counseling and testing for several diseases.

2 Materials and Methods

LAB-on-Disk system. It is composed by a polycarbonate disk containing 4 disposable silicon chips (Fig. 1). Each chip integrates heaters and temperature control sensors in the silicon part to perform PCR thermal cycles with an accuracy of 0.1 °C. The polycarbonate part is glued to a silicon part to form microchambers with a total volume of 25 μL. A Lab-on-disk reader for the RT-PCR experiments was developed by STMicroelectronics. The system is composed by an electronic board to manage temperature cycling, and by an optical module with two wavelength (FAM and VIC dyes) for the RT-PCR optical detection.

DNA extraction. Genomic DNA from in home Formalin Fixed and Paraffin Embedded Colorectal Cancer (FFPE-CRC) samples was extracted using the automated workstation EZ1 (Qiagen, Germany) according to manufacturer's instruction. Quality and yields of the DNA were evaluated with NanoDrop 1000 spectrophotometer (Thermo Fisher Scientific, USA). Informed consents were obtained from patients for the use of their samples and for the access to medical records for research purposes.

Primers and Probe design. The assay was designed to detect the most common *KRAS* codons 12 and 13 mutations (c.34G > A; c.34G > C; c.34G > T; c.35G >

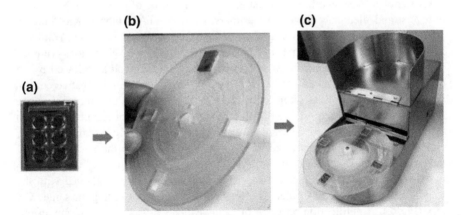

Fig. 1 Lab-on-Disk platform: **a** 6-wells silicon chip; **b** plastic disk where silicon microchips are inserted and **c** lad-disk-reader

Table 1 Primer pairs used to detect *KRAS* gene codon 12 and 13 mutations[a]

Target	Sequence
KRAS exon 2 forward wild type	GACTGAATATAAACTTGTGGTAGTTGGA
KRAS exon 2 forward G12S	AATATAAACTTGTGGTAGTTGGAGCTA
KRAS exon 2 forward G12D	AACTTGTGGTAGTTGGAGCTGA
KRAS exon 2 forward G12A	CTTGTGGTAGTTGGAGCTGC
KRAS exon 2 forward G12 V	ACTTGTGGTAGTTGGAGCTGT
KRAS exon 2 forward G12R	AATATAAACTTGTGGTAGTTGGAGCTC
KRAS exon 2 forward G12C	AATATAAACTTGTGGTAGTTGGAGCTT
KRAS exon 2 forward G13D	GTGGTAGTTGGAGCTGGTGA
KRAS exon 2 reverse	CATATTCGTCCACAAAATGATTCTGA
KRAS exon 2 internal probe	5′-FAM-CTGTATCGTCAAGGCACTCT-MGB-3′

[a]In bold, nucleotides matching the relative mutation

A; c.35G > C; c.35G > T; c.38G > A). Primer-Blast (NCBI, Bethesda, MD, USA) was used to select high-quality primers for each mutation with an optimal length of 24–28 bp and an optimal annealing temperature between 59 and 61 °C (Table 1). The specificity to the target sequence was verified in silico with the UCSC Genome Browser and Primer-BLAST. The 3′ terminus of forward primer was adapted to anneal specifically the mutated DNA template in order to obtain eight different allele specific forward primers. In addition, an oligonucleotide probe, common to the wild type and mutant *KRAS* alleles, was designed and FAM-labeled at the 5′ terminus while a Minor Grove Binding (MGB) site was added at the 3′ terminus.

KRAS genotyping assay. Genotyping of human *KRAS* was performed by Lab-on-Disk platform as previously described [25]. Briefly, TaqMan Genotyping assay was performed in order to evaluate the mutational status of *KRAS* in FFPE-CRC DNA samples. A PCR mix was prepared to contain 10 ng of genomic DNA, 1 μM of reverse and allele-specific forward primers, 100 nM of FAM-labeled *KRAS* probe, 2.5 μL of 2X TaqMan Genotyping Master Mix (Thermo Fisher Scientific, Waltham, MA) in a final volume of 5 μL *per* chamber. Thermocycling conditions were set as follow: an initial denaturation step at 97 °C × 10′ (one cycle), followed by 45 cycles at 97 °C × 15″ and 64 °C × 60″. In every run, positive and negative control samples were co-amplified. Region of Interest (ROI) areas in the microchip were manually selected to acquire the fluorescent signal at the end of each amplification cycle. However, dedicated software allows a correction of the ROI diameters at the end of the analysis for best fitted data. In order to determine the performance of the novel Lab-on-Disk platform, the Cycle Thresholds (Cts) values of the TaqMan-based real-time PCR specificity assay were compared to those obtained with the standard equipment Light-Cycler 1.5 (Roche Diagnostic, USA) using the same PCR mix (Fig. 2). Amplification program of LightCycler 1.5 was set as follow: an initial denaturation step at 95 °C × 10′ (one cycle) followed by 45 cycles of 95 °C × 15″, 62 °C × 10″, and a final extension of 72 °C for 10 min. *KRAS* mutation status

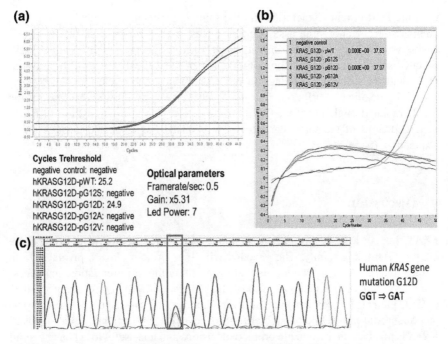

(a)

Cycles Trehreshold
negative control: negative
hKRASG12D-pWT: 25.2
hKRASG12D-pG12S: negative
hKRASG12D-pG12D: 24.9
hKRASG12D-pG12A: negative
hKRASG12D-pG12V: negative

Optical parameters
Framerate/sec: 0.5
Gain: x5.31
Led Power: 7

(b)

(c)

Human *KRAS* gene
mutation G12D
GGT ⇒ GAT

Fig. 2 Genotyping assay of formalin fixed and paraffin embedded colorectal cancer specimen carrying *KRAS* G12D mutation: **a** genotyping assay performed by the Lab-on-Disk platform has a more elevated efficiency if compared to standard real time PCR (LightCycler 1.5) performances (**b**). The obtained cycle thresholds were confirmed by three independent experiments; **c** electropherogram of formalin fixed and paraffin embedded colorectal cancer specimen used to compare performances of the two platforms. Blue square highlights the G > A transition in *KRAS* codon 12 of the same clinical sample

of the samples were previously evaluated through direct sequencing using BigDye terminator v.3.1 kit (Applied Biosystems, Foster City, California) according to manufacturer's instructions, on ABI PRISM 310 genetic analyzer (Applied Biosystems, Foster City, California).

Statistical Analysis. One-way analysis of variance (ANOVA) was used to compare differences among groups and statistical significance was assessed by the Tukey–Kramer post hoc test. The level of significance for the statistical test was $p \leq 0.05$. Data were reported as Mean ± standard deviations.

3 Results

In this work, a TaqMan Genotyping assay was implemented on the Lab-on-Disk platform to detect the mutational status of human *KRAS* in FFPE Colorectal cancer samples. The specificity of the assay was compared to that obtained with a standard rea time PCR (Light-Cycler 1.5) and the same experimental protocol. In Fig. 2, a comparison of genotyping assay performances of the Lab-on-Disk (panel a) and LightCycler 1.5 (panel b) instruments using a *KRAS*-G12D (dbSNP: rs121913529) mutated sample are shown. The optimization of the optical sensor of the Lab-on-Disk platform allowed a distinct decrease of about 12.43 cycle threshold values for wild type allele (***$P < 0.0001$) and about 12.17 cycle threshold values for the G12D mutated allele (***$P < 0.0001$), respectively. The G>A transition in *KRAS* gene codon 12 of the same sample was previously diagnosticated through direct sequencing (panel c).

4 Discussion

KRAS (human homolog of rodent sarcoma) is the most studied gene of the RAS GDP/GTP-binding family. The encoded wild type protein plays a pivotal role in intracellular signal transduction, including proliferation, differentiation and senescence [26]. Activating mutations in codon 12, 13 or 61 of *KRAS* gene occur in 30–40% of CRC patients [27, 28], making *KRAS* mutational analysis one of the most important diagnostic/prognostic parameter for this neoplasm. In fact, mutational status of *KRAS* is involved in therapeutic protocol decisions and it is useful to evaluate overall and progression free survival [14, 29, 30].

In this work, we propose a new method for mutational analysis of *KRAS* in FFPE-CRC samples using Lab-on-Disk platform. This innovative platform is able to detect point mutations with high specificity and sensitivity. The first innovation is represented by a plastic disk in which four different silicon microchips can be inserted. This allows to increase the number of amplification reactions *per* single run. The improved performances and the technological innovation achieved, make the Lab-on-Disk platform a competitive device in the genotyping of human genes with a considerable advantage over traditional platforms, which results in a reduction of costs and time required for the analysis, maintaining high efficiency and reliability.

5 Conclusion

The Lab-on-Disk platform was able to perform highly specific mutational analysis in human genomic DNA proving to be highly competitive in both time-consumption, cost of analysis, and high reliability when compared to standard equipment. Thanks to its features, the Lab-on-Disk platform offers a rapid and sample-in answer-out analysis, and represents a reliable Point-of-Care for rapid and accurate genotyping of human DNA samples.

References

1. Kim, H.S., et al.: The impact of KRAS mutations on prognosis in surgically resected colorectal cancer patients with liver and lung metastases: a retrospective analysis. BMC Cancer **16**, 120 (2016)
2. Siegel, R.L., Miller, K.D., Jemal, A.: Cancer statistics, 2018. CA Cancer J. Clin. **68**(1), 7–30 (2018)
3. Schweiger, M.R., et al.: Genomics and epigenomics of colorectal cancer. Wiley Interdiscip. Rev. Syst. Biol. Med. **5**(2), 205–219 (2013)
4. Chubb, D., et al.: Genetic diagnosis of high-penetrance susceptibility for colorectal cancer (CRC) is achievable for a high proportion of familial CRC by exome sequencing. J. Clin. Oncol. **33**(5), 426–432 (2015)
5. Boland, P.M., Yurgelun, M.B., Boland, C.R.: Recent progress in Lynch syndrome and other familial colorectal cancer syndromes. CA Cancer J. Clin. (2018)
6. Jancik, S., et al.: Clinical relevance of KRAS in human cancers. J. Biomed. Biotechnol. **2010**, 150960 (2010)
7. Dienstmann, R., Vilar, E., Tabernero, J.: Molecular predictors of response to chemotherapy in colorectal cancer. Cancer J. **17**(2), 114–126 (2011)
8. De Roock, W., et al.: Effects of KRAS, BRAF, NRAS, and PIK3CA mutations on the efficacy of cetuximab plus chemotherapy in chemotherapy-refractory metastatic colorectal cancer: a retrospective consortium analysis. Lancet Oncol. **11**(8), 753–762 (2010)
9. Knickelbein, K., Zhang, L.: Mutant KRAS as a critical determinant of the therapeutic response of colorectal cancer. Genes Dis. **2**(1), 4–12 (2015)
10. Mariani, P., et al.: Concordant analysis of KRAS status in primary colon carcinoma and matched metastasis. Anticancer Res. **30**(10), 4229–4235 (2010)
11. Zenonos, K., Kyprianou, K.: RAS signaling pathways, mutations and their role in colorectal cancer. World J. Gastrointest. Oncol. **5**(5), 97–101 (2013)
12. Eser, S., et al.: Oncogenic KRAS signalling in pancreatic cancer. Br. J. Cancer **111**(5), 817–822 (2014)
13. Garassino, M.C., et al.: Different types of K-Ras mutations could affect drug sensitivity and tumour behaviour in non-small-cell lung cancer. Ann. Oncol. **22**(1), 235–237 (2011)
14. Jones, R.P., et al.: Specific mutations in KRAS codon 12 are associated with worse overall survival in patients with advanced and recurrent colorectal cancer. Br. J. Cancer **116**(7), 923–929 (2017)
15. Shackelford, R.E., et al.: KRAS testing: a tool for the implementation of personalized medicine. Genes Cancer **3**(7–8), 459–466 (2012)
16. Giamblanco, N., et al.: Ionic strength-controlled hybridization and stability of hybrids of KRAS DNA single-nucleotides: a surface plasmon resonance study. Coll. Surf. B **158**, 41–46 (2017)
17. Guedes, J.G., et al.: High resolution melting analysis of KRAS, BRAF and PIK3CA in KRAS exon 2 wild-type metastatic colorectal cancer. BMC Cancer **13**, 169 (2013)
18. Foglieni, B., et al.: Integrated PCR amplification and detection processes on a Lab-on-Chip platform: a new advanced solution for molecular diagnostics. Clin. Chem. Lab. Med. **48**(3), 329–336 (2010)
19. Petralia, S., et al.: Silicon nitride surfaces as active substrate for electrical DNA biosensors. Sens. Actuators B Chem. **252**, 492–502 (2017)
20. Petralia, S., Conoci, S.: PCR technologies for point of care testing: progress and perspectives. ACS Sens. **2**(7), 876–891 (2017)
21. Petralia, S., Sciuto, E.L., Conoci, S.: A novel miniaturized biofilter based on silicon micropillars for nucleic acid extraction. Analyst **142**(1), 140–146 (2017)
22. International drug monitoring: the role of national centres. Report of a WHO meeting. World Health Organ Technical Report Series, vol. 498, p. 1–25 (1972)
23. Petralia, S., et al.: A miniaturized silicon based device for nucleic acids electrochemical detection. Sens. Bio-Sens. Res. **6**, 90–94 (2015)

24. Petralia, S., et al.: An innovative chemical strategy for PCR-free genetic detection of pathogens by an integrated electrochemical biosensor. Analyst **142**(12), 2090–2093 (2017)
25. Guarnaccia, M., et al.: Miniaturized real-time PCR on a Q3 system for rapid KRAS genotyping. Sensors (Basel) **17**(4) (2017)
26. Kranenburg, O.: The KRAS oncogene: past, present, and future. Biochimica et Biophysica Acta (BBA)—Rev. Cancer **1756**(2), 81–82 (2005)
27. Amado, R.G., et al.: Wild-type KRAS is required for panitumumab efficacy in patients with metastatic colorectal cancer. J. Clin. Oncol. **26**(10), 1626–1634 (2008)
28. Neumann, J., et al.: Frequency and type of KRAS mutations in routine diagnostic analysis of metastatic colorectal cancer. Pathol. Res. Pract. **205**(12), 858–862 (2009)
29. Sinicrope, F.A., et al.: Association of dna mismatch repair and mutations in braf and KRAS with survival after recurrence in stage III colon cancers: a secondary analysis of 2 randomized clinical trials. JAMA Oncol. **3**(4), 472–480 (2017)
30. Khan, S.A., et al.: EGFR gene amplification and KRAS mutation predict response to combination targeted therapy in metastatic colorectal cancer. Pathol. Oncol. Res. **23**(3), 673–677 (2017)

Study Toward the Integration of a System for Bacterial Growth Monitoring in an Automated Specimen Processing Platform

Paolo Bellitti⬤, Michele Bona⬤, Stefania Fontana⬤, Emilio Sardini⬤
and Mauro Serpelloni⬤

Abstract As bacterial infection diseases represent a relevant threat for human health worldwide, many efforts are spent in accelerating the diagnostic process of biological specimens. The WASPLab automated platform, by COPAN Italia S.p.A., detects bacterial growth by processing the images of the Petri dishes containing a sample to analyze. This work presents a study performed on a developed system that exploits impedance measurement to monitor bacterial growth in Petri dishes in real time. It is part of an activity aiming at system integration in the WASPLab, to enhance its monitoring capabilities and flexibility. Through repeated 24-h tests executed with the system, we successfully detected *S. aureus* growth in Petri dishes that were inside one of the WASPLab incubators, starting from impedance measurements performed at 50–150 Hz. In particular, depending on the parameter being observed, detection time was between four and six hours, for an initial bacterial concentration in the order of $4.5 \cdot 10^7$ CFU/ml. These preliminary results represent the first step for evaluating system integration in the WASPLab.

Keywords Bacterial growth detection · Impedance measurement WASPLab platform

1 Introduction

Bacterial infections count as a major source of disease worldwide, especially when they are not properly treated. As an example, nearly 50 million infection-related sepsis cases are estimated each year, with more than 5 million deaths [1]. The problem is even more serious because of the increasing capability of some bacterial strains to resist to specific antibiotic therapies, as indicated in a report released in January 2018

P. Bellitti · M. Bona (✉) · E. Sardini · M. Serpelloni
Department of Information Engineering, University of Brescia, 25123 Brescia, Italy
e-mail: m.bona002@unibs.it

S. Fontana
Copan Italia S.p.A, 25125 Brescia, Italy

© Springer Nature Switzerland AG 2019
B. Andò et al. (eds.), *Sensors*, Lecture Notes in Electrical Engineering 539,
https://doi.org/10.1007/978-3-030-04324-7_54

by the World Health Organization [2]. For these reasons, a quick diagnosis showing that a biological specimen is infected needs to be achieved, since it allows delivering a proper therapy in a timely manner, preserving subject's health.

The market provides numerous commercial systems able to meet this necessity. In particular, such systems detect bacterial growth in a biological sample much faster than traditional methods based on colony count [3, 4]. In this way, only the specimens that are really contaminated are sent sooner to dedicated laboratories for specific tests, accelerating analysis process and therapy delivery to a patient. They exploit different measurement principles for their operation. For instance, the BacT/ALERT (bioMérieux SA) [5, 6] and the BACTEC™ (BD) [7] detect a fluorescence variation, which is due to CO_2 production from bacterial metabolism. The BacTrac (SY-LAB Geräte Gmbh) [8] and the RABIT (Don Whitley Scientific Ltd.) [9] perform a measurement of electric impedance, which changes because of pathogens activity. Then, the Biacore (GE Healthcare) exploits the principle of Surface Plasmon Resonance to detect the presence of bacteria on sensor surface [10]. Finally, the Thermal Activity Monitor (TA Instruments) evaluates bacterial growth from the heat generated by related chemical reactions, implementing isothermal microcalorimetry method [11].

An approach different from those followed by the previously mentioned systems is exploited by the WASPLab platform (COPAN Italia S.p.A.) [12]. WASP stays for Walk-Away Specimen Processor. Referring to Fig. 1, which shows it for convenience, the WASPLab works in the following way. A Petri dish is inoculated with a possibly infected biological sample in WASP station, by a robotic manipulator. Then, it moves to one of two temperature-controlled incubators, in order to enhance bacterial growth, which is monitored by taking pictures of it at different moments in the image acquisition stations. Pictures are analyzed from a digital imaging interface, and they are stored in WASPLab Central, which shares data about the analysis to other connected systems. Finally, dishes are accumulated in silos at the end of the analysis process. All steps are managed in a completely automated way.

The WASPLab platform is a very advanced solution for bacterial growth detection. However, there is the possibility to even enhance its capabilities and flexibility. For instance, analyzed Petri dishes need to be moved between incubators and image acquisition stations every time a picture has to be taken. This repeated change in the environmental conditions may have a negative influence on bacterial growth. Then, pictures of two Petri dishes at most can be acquired at a time, because there are two image acquisition stations in the WASPLab, each receiving the Petri dishes from one of the incubators. Finally, carrying out a measurement with a sensor system would permit to have at disposal quantitative data about bacterial growth in real time.

In a previous work [13], we described a system that performs this task through repeated impedance measurements related to a Petri dish. Such system has been developed with the final aim of integrating it in the WASPLab once it is optimized, in order to monitor bacterial growth in Petri dishes while they remain always inside the incubators. Furthermore, we illustrated an experimental analysis executed in a laboratory setup, to evaluate system performances regarding measurement accuracy and bacterial growth monitoring. On the other hand, this work presents a study performed with the system operating when analyzed Petri dishes were incubated directly

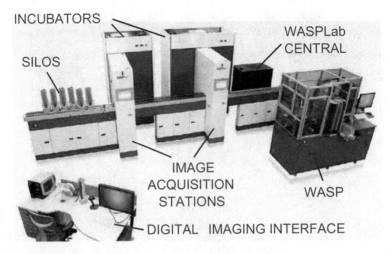

Fig. 1 Picture of the WASPLab automated specimen processing platform

inside the WASPLab. Through the study, we could observe system behavior in the field, which is the first step toward its integration in the WASPLab. The following section includes a brief description of the system and illustrates how the study was carried out. Then, third section presents the obtained preliminary results.

2 Materials and Methods

2.1 The Measuring System

For this work, we realized an improved version of the measuring system, with respect to the one presented in [13], which enhances its portability. Used system is shown in Fig. 2. It presents three different principal parts.

The first part is an instrumented Petri dish, which has been realized by adding two macroelectrodes of specific geometrical characteristics to the same disposable object that is used during the analysis with the WASPLab. It contains an agar-based culture medium for bacterial proper growth. From the frequency response of the instrumented Petri dish, we obtained an equivalent lumped-parameter model representing either electrode/medium interface, through double layer capacitance C_{DL} and charge transfer resistance R_{CT}, or medium conductivity, through resistance R_{M}. These parameters are the quantitative data about bacterial growth that the developed system is able to provide. They are obtained from Petri dish impedance measurements performed at two fixed working frequencies, by implementing specific mathematical formulas [13].

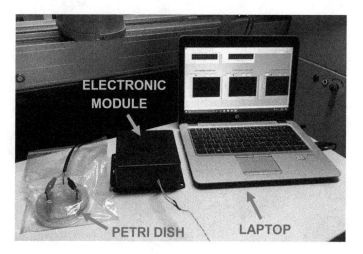

Fig. 2 The measuring system in its improved version

The second part is an electronic module, which is designed for measuring the impedance related to the instrumented Petri dish at the working frequencies [13]. It is entirely inside a box of dimensions $120 \times 120 \times 55$ mm. Then, it is configured according to a particular necessity, by properly choosing its components. The system can work with more electronic modules, if the number of the Petri dishes to analyze simultaneously augments. This increases its modularity.

Then, the third part is a laptop, executing a LabVIEW Virtual Instrument (VI) that drives the electronic module and elaborates the impedance measurements to find C_{DL}, R_{CT}, and R_M. This VI allows system user changing measurement parameters when necessary and monitoring bacterial growth by looking at C_{DL}, R_{CT}, and R_M real time trend. Furthermore, it presents additional features, with respect to the program described in [13]. First, since the system is designed to implement an automatic bacterial growth monitoring, it generates an alarm as it detects that C_{DL}, R_{CT}, or R_M variations have overcome a threshold. Second, when the alarm occurs, an e-mail is sent automatically to a defined address, reporting that bacterial growth has been detected and at what time. Third, obtained data are stored not only in laptop's own memory, but also in a folder that is shared with connected devices through the cloud. In this way, information about bacterial growth is available from remote locations too.

2.2 Study Setup and Protocol

The study presented in this paper was entirely carried out in COPAN Italia S.p.A. (Brescia, Italy), with the developed system working with the WASPLab platform, as shown in Fig. 3. During the study, we executed repeated bacterial growth monitoring

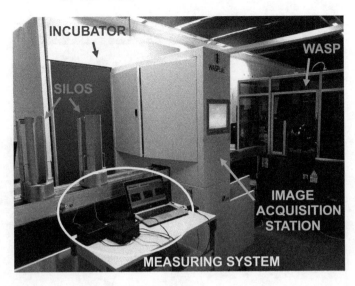

Fig. 3 System working with the WASPLab platform

tests with the same setup, whose block scheme is represented in Fig. 4. Furthermore, every test was conducted following a defined protocol.

Three instrumented Petri dishes were filled with Tryptone Soy Agar medium. Two of them were also inoculated with *S. aureus* ATCC 6538, considering an initial concentration in the order of $4.5 \cdot 10^7$ CFU/ml. Such value is high, but we kept it to verify that the system worked properly, in a favorable condition. Inoculation took place in a safe environment, at ambient temperature. Then, remaining Petri dish was not inoculated. In this way, we had a reference to evaluate the trend related to the others.

The next step after inoculation was system preparation. We configured three electronic modules, in a way they excited one Petri dish each with waveforms of amplitude equal to 1 V_{pp} and measured Petri dish impedance at frequencies $f_1 = 50$ Hz and $f_2 = 150$ Hz. In addition, their internal gain factor was obtained by attaching a 100 Ω commercial resistor to the system and acquiring the corresponding impedance.

When the system was ready, we put the Petri dishes in one of the WASPLab incubators, which had already turned on, in order to make its internal temperature uniform. In particular, once incubator was closed after Petri dishes positioning, its control system drove the temperature to stay in a range between 34.8 and 35.2 °C, given a set point equal to 35.0 °C. Every Petri dish was attached to the terminals of an electronic module. As all modules are equal to each other, such operation was performed without respecting a precise order.

Last step was triggering the VI running on the laptop to acquire bacterial growth data. We set its parameters to allow a continuous impedance measurement at f_1 and f_2 and provide a single value of C_{DL}, R_{CT}, and R_M every two minutes, for 24 h.

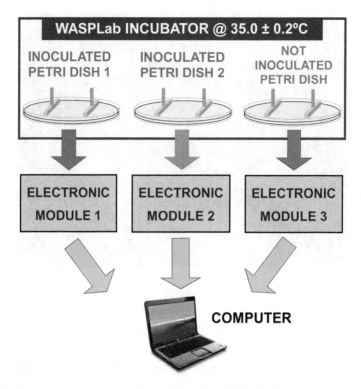

Fig. 4 Block scheme of the setup used for testing the system during the study

3 Preliminary Results

3.1 Double Layer Capacitance C_{DL}

Figure 5a illustrates C_{DL} time behavior related to the Petri dishes analyzed during one of the tests. All represented curves have an initial transient of about one hour, which is due to medium temperature variation from initial state (inoculation at ambient temperature) to the condition in which dishes are inside WASPLab incubator (about 35 °C). Then, once initial transient gets over, the curve related to the Petri dish that was not inoculated presents a gradual decreasing trend, caused by medium partial drying. On the contrary, the curves referring to inoculated Petri dishes reflect a typical growth trend. In fact, after a lag phase in which it is stable, C_{DL} starts to augment (log phase). Such increase derives from bacterial growth, which generates charged molecules accumulating at electrode/medium interface, as reported in the literature [3]. The transition between lag phase and log phase identifies system detection time, which occurs at four hours. Finally, the curves show a brief stationary phase, in which C_{DL} is stable again, and death phase, in which C_{DL} decreases, as bacterial activity is coming to an end. Even though both curves allow recognizing all growth phases, they are characterized by a different dynamics. For instance, log phase lasts about one hour less for inoculated Petri dish 1 than for inoculated Petri dish 2. This is caused by a variance in bacterial distribution along electrode/medium interface between the two inoculated Petri dishes, as inoculation step was carried out by hand and, therefore, it is not an exactly repeatable operation. In fact, a low variability in double layer characteristics may have great consequences on C_{DL} values. In any case, such discrepancy does not lead to a relevant difference in detection time.

3.2   Charge Transfer Resistance R_{CT}

Figure 5b shows the time trend characterizing the other interface parameter, i.e., resistance R_{CT}. This figure highlights that R_{CT} behavior reflects C_{DL} curves shape shown in Fig. 5a. In fact, all curves present an initial transient and those related to inoculated Petri dishes help identifying bacterial growth phases, even if with different dynamics.

However, R_{CT} has an opposite direction with respect to C_{DL}. In fact, for every C_{DL} increasing tract, there is a R_{CT} descending phase, and vice versa. In particular, bacterial growth leads to a decrease of R_{CT}, since it causes an accumulation of charged molecules at electrode/medium interface, which augments the charge transfer capability of double layer. Since R_{CT} and C_{DL} trends are similar, system detection time extracted from the analysis on charge transfer resistance is comparable to the one obtained from observing double layer capacitance.

3.3   Medium Resistance R_M

Finally, Fig. 5c represents R_M time behavior, which is identical for the two curves related to the inoculated Petri dishes (although there is a difference between the corresponding values), unlike what happens for C_{DL} and R_{CT}. Furthermore, Fig. 5c highlights that such behavior does not reflect the one characterizing the other parameters. In fact, after a two-hours initial transient, R_M decrease during log phase for inoculated Petri dishes is slower than R_{CT}'s. In addition, this trend lasts until the end of the test. Anyway, decrease is in agreement with what is stated in the literature, as it is due to an increase of medium conductivity caused by bacterial metabolism [3]. Consequently, system detection time obtained from R_M observation is about six hours, i.e., it is greater than what is found from C_{DL} and R_{CT} study. On the other side, the curve related to not inoculated Petri dish has a progressively increasing trend after the transient, which is caused by medium partial drying.

Even though its analysis leads to the identification of a greater detection time, as compared to those found by studying C_{DL} and R_{CT} trends, R_M has a more stable behavior. In fact, its curves do not present the anomalous peaks that are visible when

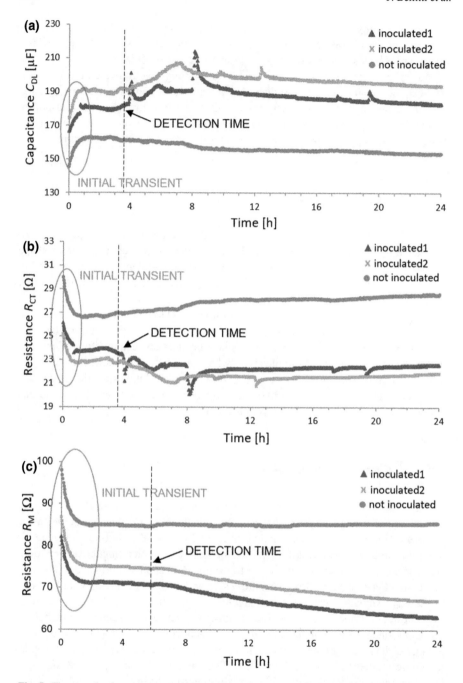

Fig. 5 Time trends of parameters providing information about bacterial growth, obtained from one of the performed tests. **a** Capacitance C_{DL}. **b** Resistance R_{CT}. **c** Resistance R_M

observing the other two parameters. Consequently, since the measuring system has the advantage of monitoring three parameters at the same time, a preliminary alert about bacterial growth detection can be generated from C_{DL} and R_{CT} observation. Then, a full alarm can be produced when the system detects R_M variation too.

4 Conclusions

This paper has presented a study performed on a system designed for bacterial growth monitoring in Petri dishes. System operation relies on the impedance measurement, at two fixed working frequencies, of the analyzed Petri dishes, which are instrumented with electrodes. Firstly, we have given a general description of the system. Then, we have illustrated how the study was conducted, i.e., S. aureus growth was monitored when Petri dishes were inside an incubator of an automated platform, called WASPLab and commercialized by company COPAN Italia S.p.A. Finally, we have reported preliminary achieved results, which highlight system capability to detect bacterial growth in the field.

Preliminary results from the performed study pave the way for system integration in the WASPLab platform or, at least, for a connection between them, once the former is optimized for such purposes. This would lead to several advantages regarding the automated analysis carried out by the WASPLab. First, bacterial growth is monitored in Petri dishes that are inside the incubators for the entire test duration. Second, since the system provides three output parameters, additional real time data about any growth phase is obtained, from all analyzed Petri dishes. Such data could be shared between other connected systems, included the remote ones, allowing all authorized people to be informed anytime and anywhere. Finally, the system is easily reconfigurable, to meet user's needs and particular applications. This contributes to augment WASPLab flexibility.

Future research activity will deal with further growth tests, considering different initial concentrations and different bacteria. In addition, design features for system optimization will be introduced, in order to favor its integration in the WASPLab.

Acknowledgements The authors thank Giorgio Triva and Roberto Paroni, from COPAN Italia S.p.A., for their appreciated support in the realization of the presented study.

The research activity is part of the Adaptive Manufacturing project (CTN01_00163_216730), which is financed by the Italian Ministry for the Instruction, University, and Research, through the Italian "Cluster Tecnologico Nazionale Fabbrica Intelligente".

References

1. Fleischmann, C., Scherag, A., Adhikari, N.K.J., Hartog, C.S., Tsaganos, T., Schlattmann, P., Angus, D.C., Reinhart, K.: Assessment of global incidence and mortality of hospital-treated sepsis current estimates and limitations. Am. J. Respir. Crit. Care Med. **193**(3), 259–272 (2016). https://doi.org/10.1164/rccm.201504-0781OC
2. Global Antimicrobial Resistance Surveillance System (GLASS) report: early implementation 2016–2017. WHO Library Cataloguing-in-Publication Data. ISBN 978-92-4-151344-9 (2018)
3. Yang, L., Bashir, R.: Electrical/electrochemical impedance for rapid detection of food-borne pathogenic bacteria. Biotechnol. Adv. **26**(2), 135–150 (2008). https://doi.org/10.1016/j.biotechadv.2007.10.003
4. Ivnitski, D., Abdel-Hamid, I., Atanasov, P., Wilkins, E.: Biosensors for detection of pathogenic bacteria. Biosens. Bioelectron. **14**(7), 599–624 (1999). https://doi.org/10.1016/s0956-5663(99)00039-1
5. Totty, H., Ullery, M., Spontak, J., Viray, J., Adamik, M., Katzin, B., Dunne, W.M., Deol, P.: A controlled comparison of the BacT/ALERT® 3D and VIRTUO^{TM} microbial detection systems. Eur. J. Clin. Microbiol. Infect. Dis. **36**(10), 1795–1800 (2017). https://doi.org/10.1007/s10096-017-2994-8
6. Jacobs, M.R., Mazzulli, T., Hazen, K.C., Good, C.E., Abdelhamed, A.M., Lo, P., Shum, B., Roman, K.P., Robinson, D.C.: Multicenter clinical evaluation of BacT/Alert Virtuo blood culture system. J. Clin. Microbiol. **55**(8), 2413–2421 (2017). https://doi.org/10.1128/JCM.00307-17
7. Chang, J., Park, J.S., Park, S., Choi, B., Yoon, N.S., Sung, H., Kim, M.N.: Impact of monitoring blood volume in the BD BACTEC^{TM} FX blood culture system: virtual volume versus actual volume. Diagn. Microbiol. Infect. Dis. **81**(2), 89–93 (2015). https://doi.org/10.1016/j.diagmicrobio.2014.11.001
8. Fernández, P., Gabaldón, J.A., Periago, M.J.: Detection and quantification of Alicyclobacillus acidoterrestris by electrical impedance in apple juice. Food Microbiol. **68**, 34–40 (2017). https://doi.org/10.1016/j.fm.2017.06.016
9. Zsivanovits, G., Szigeti, F., Mohacsi-Farkas, C.: Investigation of antimicrobial inhibition effect of quince fruit extract by rapid impedance method. In: Proceedings of the International Scientific-Practical Conference "Food, Technologies and Health", pp. 264–270. Plovdiv, Bulgaria (2013)
10. Fratamico, P.M., Strobaugh, T.P., Medina, M.B., Gehring, A.G.: Detection of Escherichia coli O157:H7 using a surface plasmon resonance biosensor. Biotechnol. Tech. **12**(7), 571–576 (1998). https://doi.org/10.1023/A:1008872002336
11. Braissant, O., Wirz, D., Göpfert, B., Daniels, A.U.: Use of isothermal microcalorimetry to monitor microbial activities. FEMS Microbiol. Lett. **303**, 1–8 (2010). https://doi.org/10.1111/j.1574-6968.2009.01819.x
12. WASPLab webpage, http://products.copangroup.com/index.php/products/lab-automation/wasplab. Accessed on 29 March 2018
13. Bellitti, P., Bona, M., Borghetti, M., Sardini, E., Serpelloni, M.: Flexible monitoring system for automated detection of bacterial growth in a commercial specimen processing platform. In: Proceedings of the IEEE RTSI 2017—IEEE 3rd International Forum on Research and Technologies for Society and Industry, pp. 207–212. IEEE, Modena, Italy. https://doi.org/10.1109/rtsi.2017.8065950

A Virtual ANN-Based Sensor for IFD in Two-Wheeled Vehicle

D. Capriglione⊙, M. Carratù ⊙, A. Pietrosanto⊙ and P. Sommella⊙

Abstract In the context of automotive and two-wheeled vehicles, the comfort and safety of drivers and passengers is even more entrusted to electronic systems which are closed-loop systems generally implementing suitable control strategies on the basis of measurements provided by a set of sensors. Therefore, the development of proper instrument fault detection schemes able to identify faults occurring on the sensors involved in the closed-loop are crucial for warranting the effectiveness and the reliability of such strategies. In this framework, the paper describes a virtual sensor based on a Nonlinear Auto-Regressive with eXogenous inputs (NARX) artificial neural network for instrument fault diagnosis of the linear potentiometer sensor employed in motorcycle semi-active suspension systems. The use of such a model has been suggested by the particular ability of NARX in effectively take into account for the system nonlinearities. The proposed soft sensor has been designed, trained and tuned on the basis of real samples acquired on the field in different operating conditions of a real motorcycle. The achieved results, show that the proposed diagnostic scheme is characterized by very interesting features in terms of promptness and sensitivity in detecting also "small faults".

Keywords Soft sensor · NARX · Fault diagnosis

1 Introduction

Thanks to the progressive diffusion of MEMS [1, 2]. (Micro Electro Mechanical System), today's vehicles are increasingly adopting sensors and electronic instrumentation to support different types of applications like Electronic fuel injection (EFI), Antilock-braking system (ABS), Electronic stability program (ESP), Active or Semi-Active suspension system and so on [3]. As an example, the motorcycle suspension is one of the most critical subsystem, which directly affects the comfort and the vehicle handling: it ensures the contact between tires and road, and at the

D. Capriglione (✉) · M. Carratù · A. Pietrosanto · P. Sommella
Department of Industrial Engineering, University of Salerno, Via Giovanni Paolo II, 132, Fisciano, SA, Italy
e-mail: dcapriglione@unisa.it

© Springer Nature Switzerland AG 2019
B. Andò et al. (eds.), *Sensors*, Lecture Notes in Electrical Engineering 539,
https://doi.org/10.1007/978-3-030-04324-7_55

same time isolates the vehicle frame from the road roughness [4–8]. An active or semi-active suspension system and suitable strategies are able to change the damping coefficient as a function of the suspension stroke velocity, the pitch rate and/or other measurements about the vehicle dynamics from a set of sensors typically including MEMS like accelerometers, gyroscope and magnetic encoders [9, 10]. Driven by the primary goals of cost-saving and safety, the use of the MEMS has led the sensor fusion to become an interesting research topic through the development of new solutions such as virtual sensors: the process of estimating any system or process variable by adopting mathematical models, replacing some physical devices and using data acquired from some other available sensors. They represent a good solution successfully applied to solve various problems such as the backup of measurement systems, the what-if analysis, the prediction for real-time plant control and sensor validation. A virtual sensor may be useful for multipurpose, first of all, the inferential model intended to reduce the measuring hardware requirements may result into a significant source of budget saving and increasing system reliability. Probably the main application of the virtual sensor is the Sensor Validation following the (physical/analytical) redundancy-based approach typically exploited in the automotive safety [10]. Focusing the attention on the semi-active suspension systems, typically their effectiveness can drastically decrease because the aging and/or faults that can mostly occur on the rear suspension stroke sensor. Indeed, the sensor devoted to the measurement of such a quantity is typically a linear potentiometer whose electrical contacts suffer of deterioration due to the continuous attrite and crawling imposed by motorcycle normal operating. As a consequence, a suitable Instrument Fault Detection (IFD) scheme should be developed for identifying faults that could occur on such a sensor with the aim of improving the reliability and effectiveness of the suspension control strategy, and the driver (and passenger) safety [11, 12]. In addition, to develop an on-line IFD scheme it should be also based on simple models requiring computational burden compatible with typical control units adopted in motorcycle context. In this framework, the authors propose a suitable IFD scheme based on the analytical redundancy provided by a virtual sensor specifically designed for the residual generation of rear suspension stroke sensor. In a more detail, the residual function that allows identifying the fault on such a sensor is evaluated as difference between the output of a suitable trained Nonlinear Auto-Regressive with eXogenous inputs (NARX) artificial neural network and the actual output measured by the sensor. Then, the analysis of the residual on a suitable time interval allows identifying the presence of a fault on the rear-stroke sensor.

The obtained results are very encouraging because the proposed scheme shows a very good promptness also in identifying "small faults" which are typically due to the device wear and tear and aging, or to other influence factors as the variation of the sensor power supply which results as changing of the input/output curve of the sensor.

2 The Two-Wheeled Vehicle Under Test

The design and validation of the virtual sensor measuring the rear suspension position as well as the development of the corresponding IFD scheme have been performed according to the data-driven approach introduced in [9] by adopting the SUZUKI GSX-1000 model as the two-wheeled vehicle under test.

As depicted in Fig. 1, the motorcycle has been equipped with (i) semi-active suspension system (including the magneto-rheological fork and rear shock damper); (ii) a set of hard sensors measuring the main quantities of interest for controlling the vertical dynamics according to specifications reported in Table 1.

Moreover, the measurement set-up also includes a suitable data acquisition system designed for sampling and storing the data collected by the sensors. In detail, data

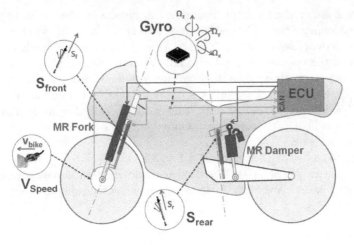

Fig. 1 Scheme of the measurement set-up for the motorcycle under test

Table 1 Specification of the measurement set-up for the test motorcycle

Output	Symbol	Sensor type	Manufacturer	Model	Mounting notes
Front suspension stroke	S_{front}	Linear displacement Sensor	Penny and Giles	SLS130	Fixed to the handlebar and the fork bottom
Rear suspension stroke	S_{rear}				Mounted between the frame and rear wheel
Motorbike speed	V_{bike}	Magnetic encoder	Dorman	970-011	Fixed to the front wheel
Pitch velocity	$Gyro$	Gyroscope	STMicroelectronics	L3GD20	Fixed to the frame

sampling has been carried out at the frequency of 1 kHz with a 12-bit ADC (about the output from the analogical sensors), whereas a CAN-BUS data logger working at 1 Mbps has been adopting for data recording.

A measurement campaign involving the motorcycle under test has been performed with reference to the 8 km long lap, which includes various profiles (cobblestone stretch, urban and extra-urban road, concentrated obstacles) able to introduce the typical excitation modes of the suspension system. As a result, more than 40 test laps have been run (mean lap time equal to 500 s) by achieving 6 h data logging about the sensors output in faulty-free conditions.

3 The Virtual Sensor

The virtual sensor adopted for predicting the rear suspension position has been developed by adopting a NARX Neural Network. This choice has been driven by the very attractive proved features of NARX neural networks in accurately prediction of non-linear dynamic systems the good capability of noise filtering [13]. The proposed architecture takes into account the front suspension position, the pitch rate of the motorcycle body and the longitudinal speed of the vehicle. Indeed, such a quantities are really correlated to the rear suspension as also described by the half car model (this one is valid only in steady state conditions but highlights the relationship among the above quantities) [14].

A sketch of the NARX is reported in Fig. 2.

As for the number of hidden layers and nodes different values have been analyzed by adopting the Matlab Neural Toolbox ranging in the following sub-sets:

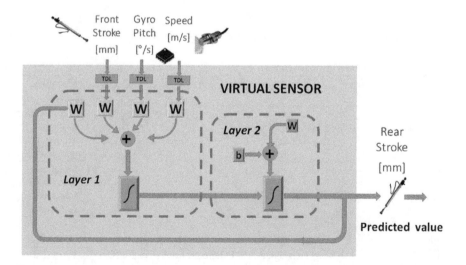

Fig. 2 Block-diagram of the proposed IFD scheme

- the number of neurons in the (unique) hidden layer in the range [5–20];
- the tapped delay of the input signals in the range [50–200] ms, according to the motorcycle dynamics [10];
- the tapped delay of the output signal in the range [50–200] ms.

The learning (training + test) set was constituted by 200,000 experimental samples acquired in different working conditions of the system under test [15]. To train the network, to identify the best architecture and to achieve reliable performance results, a k-fold validation technique was adopted. As a result, the configuration showing the best accuracy in terms of regression capability (i.e. the capability in reproducing the expected sensor outputs) and repeatability was achieved by considering 13 neurons in the hidden layer, and tapped delays (for the input and output) equal to 100 ms.

4 The IFD Scheme

Based on the data acquisition and adoption of the virtual sensor previously introduced, an original IFD scheme is proposed by including two steps (see the simplified block-diagram in Fig. 3):

- Residual Generation: the stage computes the difference between the measured and predicted values of the rear suspension stroke (respectively provided by the linear potentiometer and the virtual sensor), by highlighting the fault symptom;
- Decision Making: the stage implement the rules needed to correct the fault of interest.

Fig. 3 Block-diagram of the proposed IFD scheme

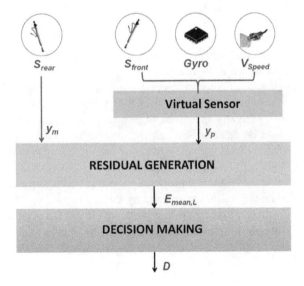

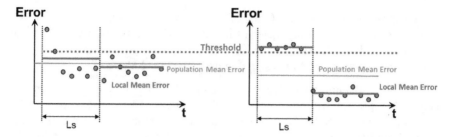

Fig. 4 Examples of local accuracy about the sensor output prediction and threshold selection about the residual

As previously reported, the instantaneous prediction of the soft sensor is satisfying for most of the experimental dataset. However, some conditions remain (see Fig. 4), where the percentage difference between the ground-truth and the predicted position are significant. To limit the effects of the poor prediction, the performance of the virtual sensor has been evaluated in terms of the *Sliding Occurence Error* (*SOE*) curve, that may be interpreted as the survivor function of the error tolerance. Then, a strategy based on the moving average is employed for computing a more accurate residual:

$$E_{mean,L}(i) = \frac{1}{L_S} \sum_{k=0}^{L_S-1} \left| \frac{y_p(i-k) - y_m(i-k)}{y_m(i-k)} \right| \tag{1}$$

where y_p and y_m are the predicted and measured rear suspension stroke, L_s is the number of samples included in the moving window length L.

As an example, about the worst predicted cases by the virtual sensor (10% of the experimental dataset), the minimum value for the residual $E_{mean,L}$ is less than 5%, when L equal to 500 ms is considered.

The IFD scheme for the rear stroke sensor is proposed to reveal the small faults, also known as "un-calibration faults", mainly due to the device wear and tear and aging, or to other influence factors as the variation of the sensor power supply and which results as changing of the input/output curve of the sensor. Such a kind of fault generally appears as slight amplitude deviation from the expected behavior and could be detected through the plausibility checks typically implemented in automotive ECUs only after hours or days from the occurrence, when the performance degradation implies unacceptable risk levels.

According to the proposed Decision Making step (see Fig. 5), a fault is detected when the residual computed by the corresponding block exceeds a fixed threshold $T\%$ longer than the sliding window L. Focus has been devoted to analyze the optimal value for the window length L when the level of the un-calibration $T\%$ is equal to 10%.

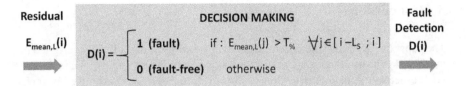

Fig. 5 Detection rule for revealing the un-calibration faults of the rear stroke sensor

Table 2 Performance of the proposed IFD scheme for the rear stroke sensor

L (s)	FA%	MD%	CD%	$t_{d,mean}$
0.1	74.3	0.0	25.7	0.7
0.5	48.0	0.0	42.0	0.9
1.0	0.4	3.6	**96.0**	19.9
2.0	0.0	20.2	79.8	33.4

The instrument fault detection scheme has been verified against $N_{faults} = 1000$ faults randomly introduced in the (measured) rear stroke samples of dataset, by considering the following performance indexes:

- the percentage *FA%* of false alarms, when threshold is exceeded for predicted samples corresponding to faulty-free sensor output;
- the percentage *MD%* of missed detections, when either threshold is not exceeded for predicted samples corresponding to faulty sensor output or threshold is exceeded after a maximum delay $t_{d,max}$ with respect to the fault insertion time;
- the percentage *CD%* of correct fault detections, when threshold is exceeded for predicted samples corresponding to unhealthy sensor output by the maximum observation time $t_{d,max}$.

Experimental results are summarized in Table 2 for L varying from 100 ms to 2 s, when $t_{d,max} = 60$ s is considered. A satisfying performance is exhibited by the proposed fault detection scheme when the virtual sensor is adopted by averaging the prediction every second ($L_s = 100$). Shorter sliding windows allow to achieve poor performance in terms of *FA%* because of the prediction limits by the NARX model about the signal tracking for 10% of the Test set samples whereas, a larger sliding window leads to poor performance in terms of *MD%* because the threshold exceeding is not completely satisfied for all the output samples within the observation time.

Moreover, as shown from the mean delay time $t_{d,mean}$ for the correct fault detection the sliding window length L equal to 1 s allows to achieve a very prompt response of the IFD scheme.

5 Conclusions

The paper has described the development of a virtual sensor specifically designed for the fault diagnosis of the rear stroke suspension sensor in motorcycle. The fault diagnosis procedure is based on the processing of residuals given at each time instant by the difference between the measured value and the one predicted by the virtual sensor. The scheme has been tuned for achieving a good tradeoff between sensitivity and promptness in detecting also "small faults", i.e. ones which typically related to the device wear and tear and aging, or more in general to the variation of the input/output curve of the sensor. Really, this kind of faults are very difficult to identify in practice because they could bring to very small residuals which can be confused with values achievable in normal and fault-free operating conditions if the false alarm percentage has to be kept low as well. In this framework, the use of a Nonlinear Auto-Regressive with eXogenous inputs (NARX) artificial neural network and suitable residuals post-processing methods have revealed particularly effective for these purposes. In particular, after a suitable tuning phase, a Correct Detection percentage approaching 95% and False Alarms percentage less than 1% have been achieved for un-calibration faults. The corresponding mean detect time was less than 20 s which is a very interesting performance in terms of promptness (as matter of fact, these kind of faults do not affect dramatically the suspension system behavior and passenger's safety if isolated within a reasonable time interval). Further developments will concern with the extension of the Instrument Fault Detection scheme to the other sensors involved in the semi-active suspension control strategy and in verifying the on-line implementation of the whole diagnostic procedure for on-board applications [16].

References

1. D'Angelo, G., Laracca, M., Rampone, S., Betta, G.: Fast eddy current testing defect classification using Lissajous figures. IEEE Trans. Instrum. Meas. **67**(4), 821–830 (2018)
2. Bernieri, A., Ferrigno, L., Laracca, M., Rasile, A.: An AMR-based three-phase current sensor for smart grid applications. IEEE Sens. J. **17**(23), art. no. 7974752, 7704–7712 (2017)
3. Marek J.: Automotive MEMS sensors—trends and applications. In: International symposium on VLSI technology, systems and applications (VLSI-TSA). http://doi.org/10.1109/VTSA.2011.5872208
4. Liguori, C., Paciello, V., Paolillo, A., Pietrosanto, A., Sommella, P.: Characterization of motorcycle suspension systems: comfort and handling performance evaluation. In: Proceedings of IEEE Instrumentation and Measurement Technology Conference, pp. 444–449. ISBN: 978-146734622-1. https://doi.org/10.1109/i2mtc.2013.6555457
5. Carratù, M., Pietrosanto, A., Sommella, P., Paciello, V.: Suspension velocity prediction from acceleration measurement for two wheels vehicle. In: Proceedings of I2MTC 2017. https://doi.org/10.1109/i2mtc.2017.7969943
6. Liguori, C., Paciello, V., Paolillo, A., Pietrosanto, A., Sommella, P.: On road testing of control strategies for semi-Active suspensions. In: Proceedings of IEEE Instrumentation and Measurement Technology Conference, art. no. 6860931, pp. 1187–1192. ISBN: 978-146736385-3. https://doi.org/10.1109/i2mtc.2014.6860931

7. Liguori, C., Paciello, V., Paolillo, A., Pietrosanto, A., Sommella, P.: ISO/IEC/IEEE 21451 smart sensor network for the evaluation of motorcycle suspension systems. IEEE Sens. J. **15**(5), 2549–2558. https://doi.org/10.1109/jsen.2014.2363945

8. Paciello, V., Sommella, P.: Smart sensing and smart material for smart automotive damping. IEEE Instrum. Measur. Mag. **16**(5), art. no. 6616288, 24–30. https://doi.org/10.1109/mim.2013.6616288

9. Capriglione, D., Carratu', M., Liguori, C., Paciello, V., Sommella, P.: A soft stroke sensor for motorcycle rear suspension. Measur. J. Int. Measur. Confed. **106**(1), 46–52 (2017). https://doi.org/10.1016/j.measurement.2017.04.011

10. Capriglione, D., Carratu', M., Pietrosanto, A., Sommella, P.: NARX ANN-based instrument fault detection in motorcycle. Measur. J. Int. Measure. Confed. **117**, 304–311. https://doi.org/10.1016/j.measurement.2017.12.026

11. Catelani, M., Ciani, L.: A fault tolerant architecture to avoid the effects of Single Event Upset (SEU) in avionics applications. Measur. J. Int. Measur. Confed. **54**, 256–263 (2014). https://doi.org/10.1109/i2mtc.2017.7969915

12. Leturiondo, U., Salgado, O., Ciani, L., Galar, D.: Architecture for hybrid modelling and its application to diagnosis and prognosis with missing data. Measur. J. Int. Measur. Confed. **108**, 152–162 (2017). https://doi.org/10.1016/j.measurement.2017.02.003

13. Zhang, J., Yin, Z., Wang, R.: Nonlinear dynamic classification of momentary mental workload using physiological features and NARX-model-based least-squares support vector machines. IEEE Trans. Hum-Mach. Syst. **47**(4), 536–549 (2017)

14. Spelta, C., Delvecchio, D., Savaresi, S.M.: A comfort oriented control strategy for semiactive suspensions based on half car model. In: Proceedings of ASME Conference DSCC20102, pp. 835–840 (2010)

15. Angrisani, L., Bonavolontà, F., Liccardo, A., Schiano Lo Moriello, R., Ferrigno,. L., Laracca, M., Miele, G.: Multi-channel simultaneous data acquisition through a compressive sampling-based approach. Measur. J. Int. Measur. Confed. **52**(1), 156–172. https://doi.org/10.1016/j.measurement.2014.02.031

16. Capriglione, D., Liguori, T., Pietrosanto, A.: Real-time implementation of IFDIA scheme in automotive systems. IEEE Trans. Instrum. Meas. **56**(3), 824–830 (2007). https://doi.org/10.1109/tim.2007.894899

A Smart Breath Analyzer for Monitoring Home Mechanical Ventilated Patients

Antonio Vincenzo Radogna⊙, Simonetta Capone⊙,
Giuseppina Anna Di Lauro, Nicola Fiore, Valentina Longo⊙,
Lucia Giampetruzzi⊙, Luca Francioso⊙, Flavio Casino, Pietro Siciliano⊙,
Saverio Sabina⊙, Carlo Giacomo Leo⊙, Pierpaolo Mincarone⊙
and Eugenio Sabato

Abstract In this work we developed a Smart Breath Analyzer device devoted to the tele-monitoring of exhaled air in patients suffering Chronic Obstructive Pulmonary disease (COPD) and home-assisted by mechanical ventilation. The device based on sensors allows remote monitoring of a patient during a ventilotherapy session, and transmit the monitored signals to health service unit by TCP/IP communication through a cloud remote platform. The aim is to check continuously the effectiveness of therapy and/or any state of exacerbation of the disease requiring healthcare. By preliminary experimental tests, the prototype was validated on a volunteer subject.

Keywords COPD · Telemedicine · Breath analyzer

A. V. Radogna (✉) · S. Capone · V. Longo · L. Giampetruzzi · L. Francioso · F. Casino
P. Siciliano
Institute of Microelectronics and Microsystems (CNR-IMM), 73100 Lecce, Italy
e-mail: antonio.radogna@le.imm.cnr.it

S. Capone
e-mail: simonetta.capone@cnr.it

G. A. Di Lauro · N. Fiore
Dedalo Solutions srl, 56037 Peccioli, PI, Italy

S. Sabina · C. G. Leo
Institute of Clinical Physiology (CNR-IFC), 73100 Lecce, Italy

P. Mincarone
Institute for Research on Population and Social Policies, National Research Council
(IRPPS-CNR), 72100 Brindisi, Italy

E. Sabato
"A. Perrino" Hospital, Pulmonology Ward, 72100 Brindisi, Italy

© Springer Nature Switzerland AG 2019   465
B. Andò et al. (eds.), *Sensors*, Lecture Notes in Electrical Engineering 539,
https://doi.org/10.1007/978-3-030-04324-7_56

1 Introduction

Chronic obstructive pulmonary disease (COPD) is a chronic inflammatory lung disease that causes obstructed airflow from the lungs, with emphysema and chronic bronchitis being the two most common conditions that contribute to COPD. Symptoms include breathing difficulty, cough, mucus (sputum) production and wheezing; frequent exacerbations are associated with disease progression. COPD is caused by long-term exposure to irritating gases or particulate matter, most often from cigarette smoke [1]. COPD is one of the major causes of chronic morbidity worldwide and it's currently the fourth leading cause of death in the world, but projections place it in third place among the causes of death by 2020. It represents a significant burden for patients, carers and health services worldwide [2, 3].

Standard pulmonary functions tests (such as spirometry and lung volume tests) are used for COPD diagnosis. Other tests include pulse oximetry and arterial blood gas tests [4]. However many limitations and issues remain in COPD diagnosing and monitoring, due to the disease heterogeneity encompassing a variety of phenotypic expressions and several complications, including respiratory infections (i.e. pneumonia) that make it much more difficult to breathe and causing further damage to lung tissue.

There is an unmet need for simple additional testing approaches that would enable early diagnosis, differentiation from other respiratory diseases (as asthma), prevention of acute exacerbations and home COPD patient management by health service [5, 6]. Home Mechanical Ventilation is applied to patients with different chronic respiratory diseases and, in order to be effective, it requires accurate individual titration. Nowadays this can be obtained only by frequent home interventions by specialized hospital staff and this is associated with waiting time and high costs health service.

The aim of this work was to develop a simple device, a Smart Breath Analyzer (SBA), that allows remote monitoring of a COPD patient by a physician at the hospital during a session of ventilation therapy by sensors and may reduce the number of emergency-room visits. The SBA transmits the monitored signals to the healthcare center by TCP/IP communication through a cloud remote platform in order to check continuously the effectiveness of therapy and/or any state of exacerbation of the disease requiring healthcare.

2 Smart Breath Analyzer

2.1 Rationale

Nowadays, it's recognized by the scientific and medical community that a relatively new concept, the analysis of exhaled air, might prove useful as a new non-invasive, safe and fast tool in medical applications supporting (not substituting) standard tools and methodologies of clinical practice in disease screening, diagnostics/prognostics and patient monitoring [7, 8].

In particular, in this work we introduced breath analysis by gas and Volatile Organic Compounds (VOCs) sensors in home ventilotherapy by developing a Smart Breath Analyzer (SBA), a potential new telemedicine medical device designed as external unit to ventilator, that monitors additional parameters to those registered by the ventilator [9–11]. The aim is to innovate patient monitoring by the acquisition of different signals from gas sensors, exposed to the exhaled air flow, being able to reveal any state of exacerbation of the disease requiring nursing or medical intervention.

2.2 Architecture

The SBA device was designed in order to properly fit the fluidic needs for gas flows and to hold all the electronic circuitry for sensors interfaces. Two printed circuit boards (PCBs) were designed and realized: the first, called *SenseShield*, is the interface board between the second board, called *MultiSense* and the main board. All sensors are physically connected, through a fluidic circuit, to a gas tight chamber mounted on the *MultiSense* board for the sensing of the patient's exhaled breath. The sensors adopted in the SBA device are the following:

- Infrared CO_2 sensor;
- Electrochemical O_2 sensor;
- An array of 3 Metal Oxides-based chemoresistive sensors to fingerprint the exhaled Volatile Organic Compounds (VOCs) pattern;
- Temperature and relative humidity sensor.

The overall hardware architecture was schematized in Fig. 1. The main board plays a key role for coordinating the entire system and, in the early stage of design, a survey was carried out to select the right platform for the application. Despite the large amount of development boards and evaluation kits in today's *embedded* and *Internet-of-Things* market, including performance targeted ARM Cortex-A or Intel x86 based *System-on-Chip*, the choice was done on a simpler core architecture to take advantage of the cost, firmware development time, application flexibility and smarter possibilities for future developments.

In particular, Arduino MEGA 2560 was adopted due to its low cost, low power consumption and small size factor; it also benefits of a simple and open source Integrated Development Environment (IDE) and a huge community support. The target of the main board is to communicate with sensors and device on the system, to properly control the sequence of operations and to communicate with a remote server through TCP/IP protocol for data storage and processing. The SBA has also two LED devices, remotely controlled through the processing application on the server, for direct communication with users. Finally, the TCP/IP communication is performed thanks to a Ethernet shield 2 board. Except for the CO_2 sensor, that is electrically connected directly to the Arduino, all sensors are connected to the *MultiSense* board. A simplified diagram of the *MultiSense* board with emphasis on the gas fluidics in the sensor chamber is reported in Fig. 2.

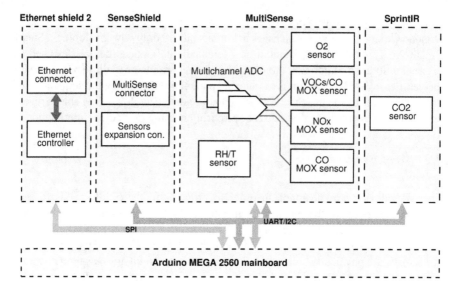

Fig. 1 Hardware architecture of SBA device

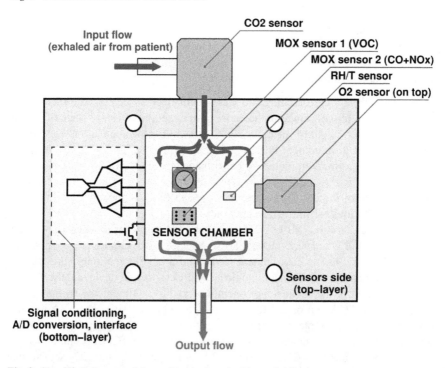

Fig. 2 Simplified diagram of the multisense board with gas fluidics

2.3   Preliminary Test

The SBA device was connected to the patient's breathing circuit through two T-connection; a small fraction of the main exhaled flow, that is the input flow in the above figure, is piped in the sensor chamber and the output flow is returned to the breathing circuit. It was verified that this solution doesn't cause any pressumetric

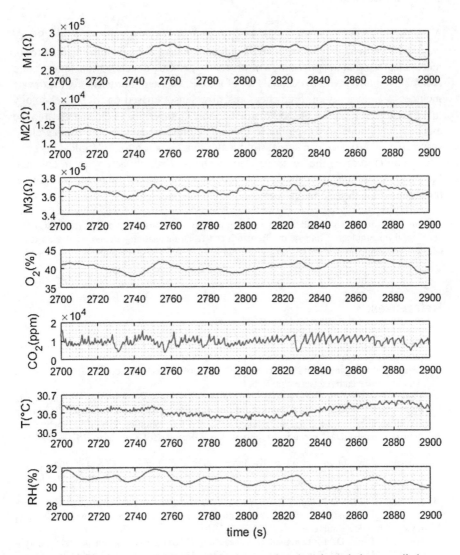

Fig. 3 Sensors signals versus time monitored by the smart breath analyzer during a ventilotherapy session

or volumetric variation in the breathing circuit, allowing the ventilator's normal operation.

The prototype was validated on a volunteer subject; the acquired signals from all the sensors are shown in Fig. 3.

3 Conclusions

A novel low cost device for breath analysis, called Smart Breath Analyzer (SBA), was developed and tested. The device is aimed to enhance the monitoring features of mechanical ventilators, providing sensors for O_2 and CO_2 concentration, temperature and relative humidity in the exhaled air of a COPD patient. Furthermore, the device includes a small array of three chemoresistive sensors based on MOX technology that provide a fingerprint of the Volatile Organic Compounds (VOCs) pattern in the exhaled air. Thanks to the ability to transmit monitored signals to a remote cloud server, the SBA device can successfully enter in a tele-monitoring clinical program whom aim is to make the home-assisted mechanical ventilotherapy more efficient. Future works are concerned the development of smart embedded algorithms to predict the status of the disease in real-time.

Acknowledgements The Smart Breath Analyzer was developed inside the project ReSPIRO (Rete dei Servizi Pneumologici: Integration, Research and Open-innovation—Bando Aiuti a Sostegno Cluster Tecnologici Regionali, project cod. F29R1T8) founded by Apulia Region.

References

1. Postma, D.S., Bush, A., van den Berge, M.: Risk factors and early origins of chronic obstructive pulmonary disease—review. Lancet **385**, 899–909 (2015)
2. World Health Organization 2017: The top 10 causes of death (Fact sheet no 310). http://www.who.int/mediacentre/factsheets/fs310/en/
3. Vestbo, J., Hurd, S.S., Agustí, A.G., Jones, P.W., Vogelmeier, C., Anzueto, A., Barnes, P.J., Fabbri, L.M., Martinez, F.J., Nishimura, M., Stockley, R.A., Sin, D.D., Rodriguez-Roisin, R.: Global strategy for the diagnosis, management, and prevention of chronic obstructive pulmonary disease GOLD executive summary. Am. J. Respir. Crit. Care Med. **187**(4), 347–365 (2013)
4. Global initiative for chronic obstructive lung disease, pocket guide to COPD diagnosis, management, and prevention. A guide for Health Care Professionals. 2017 Report. © 2017 Global Initiative for Chronic Obstructive Lung Disease, Inc
5. Dixon, L.C., Ward, D.J., Smith, J., Holmes, S., Mahadeva, R.: New and emerging technologies for the diagnosis and monitoring of chronic obstructive pulmonary disease: a horizon scanning review. Chronic Respir. Dis. **13**(4), 321–336 (2016)
6. Lange, P., Halpin, D.M., O'Donnell, D.E., MacNee, W.: Diagnosis, assessment, and phenotyping of COPD: beyond FEV1—review. Int. J. COPD **11** (Special Issue 1st World Lung Disease Summit) (2016)
7. Mathew, T.L., Pownraj, P., Abdulla, S., Pullithadathil, B.: Technologies for clinical diagnosis using expired human breath analysis. Diagnostics **5**, 27–60 (2015)

8. Amann, A., Miekisch, W., Schubert, J., Buszewski, B., Ligor, T., Jezierski, T., Pleil, J., Risby, T.: Analysis of exhaled breath for disease detection. Annu. Rev. Anal. Chem. **7**, 455–482 (2014)
9. Bofan, M., Mores, N., Baron, M., Dabrowska, M., Valente, S., Schmid, M., Trové, A., Conforto, S., Zini, G., Cattani, P., Fuso, L., Mautone, A., Mondino, C., Pagliari, G., D'Alessio, T., Montuschi, P.: Within-day and between-day repeatability of measurements with an electronic nose in patients with COPD. J. Breath Res. **7**, 017103 (8 pp.) (2013)
10. Shafiek, H., Fiorentino, F., Merino, J.L., López, C., Oliver, A., Segura, J. de Paul, I., Sibila, O., Agustí, A., Cosío, B. G.: Using the electronic nose to identify airway infection during COPD exacerbations. PLoS ONE **10**(9), e0135199 (16 pp.) (2015)
11. Schnabel, R.M., Boumans, M.L.L., Smolinska, A., Stobberingh, E.E., Kaufmann, R., Roekaerts, P.M.H.J., Bergmans, D.C.J.J.: Electronic nose analysis of exhaled breath to diagnose ventilator associated pneumonia. Respir. Med. **109**, 1454–1459, (2015)

A Nonlinear Pattern Recognition Pipeline for PPG/ECG Medical Assessments

Francesco Rundo, Salvatore Petralia⬭, Giorgio Fallica
and Sabrina Conoci⬭

Abstract In this contribution, an innovative platform for ECG assessment from PPG signals for automotive applications is presented. The platform we propose is based on an optical miniaturized probe coupling a LED emitter with a silicon photomultipliers (SiPM) detector. The optical probe is able to measure PPG signal from the palm of hands. The new nonlinear Pattern Recognition Pipeline we developed associate the "Diastolic phase" of the heart to a "Reaction" physical dynamic while the "Systolic phase" can be mathematically modelled having a "Diffusion" physical proprieties. Results show there is specific cross-correlation between ECG signal and first-derivative of processed PPG waveform for the same person.

Keywords PPG · ECG · Pattern recognition

1 Introduction

The monitoring of physiological parameters such us heart electrical activity (ECG) and heart rate variability through a non-invasive miniaturized systems can be fundamental in a wide variety of medical applications (such us cardiovascular diseases) and sports training.

Electrocardiography (ECG) is a very common cardiologic exam obtained by measuring the electrical activity of the heart over a period of time by means of electrodes placed on the skin of specific body positions. These electrodes detect the electrical changes arising from the heart muscle (electro-physiologic pattern) due to depolar-

F. Rundo (✉) · S. Petralia · G. Fallica · S. Conoci
STMicroelectronics—Automotive and Discretes Group—Central R&D, Catania, Italy
e-mail: francesco.rundo@st.com

© Springer Nature Switzerland AG 2019
B. Andò et al. (eds.), *Sensors*, Lecture Notes in Electrical Engineering 539,
https://doi.org/10.1007/978-3-030-04324-7_57

izing and repolarizing process occurring during each heartbeat [1]. The recording of ECG requires specific medical instrumentation needing also medical offices to be measured.

The PhotoPlethysmoGraphy (PPG) signal is a physiological waveform due to the synchronous changes of the blood volume upon each heart pumps that reaches and distends also the arteries and arterioles in the subcutaneous tissue of the body periphery.

The change of the blood volume caused by the pressure pulse can be monitored by illuminating the skin with specific wavelength light and measuring the amount of light either transmitted or reflected by a photodetector: a peck appears for each cardiac cycle [2, 3].

The shape of the PPG waveform differs from subject to subject and since the blood flow to the skin can be influenced by other physiological parameters, PPG can also be used to monitor breathing, hypovolemia and circulatory conditions.

Figure 1 shows typical ECG and PPG compliant waveforms

On the basis of the above considerations, PPG can be more easy recorded by low-cost optical systems in a non-invasive manner than ECG. For that reasons the use of PPG signal offers significant advantages for Point-of-Care (PoC) medical devices [4].

In this work we describe a robust and efficient bio-inspired pattern recognition pipeline for the reconstruction of ECG signal from PPG.

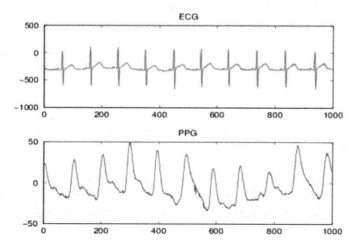

Fig. 1 Typical ECG and PPG waveforms

2 Materials and Methods

The physiological probe used for the measurements of PPG signal was fabricated by STMicroelectronics [5, 6]. It was composed by a LED light emitter coupled with a Silicon Photomultipliers (SiPMs) detector.

The LED was OSRAM LT M673 emitting at 529 nm featured by an area of 2.3×1.5 mm^2, viewing angle of $120°$, spectral bandwidth of 33 nm.

The SiPMs was a microdevice of total area of 4.0×4.5 mm^2 and contain 4871 square microcells with 60 μm pitch. A bandpass filter with a pass band centered at 542 ± 70 nm was glued on the SiPM package (Loctite® 352TM adhesive).

A proper printed circuit board (PCB) was developed and used to interface the PPG probe and NI (National Instruments) acquisition instrumentation. The board was featured by a 4 V portable battery, a power management circuits, a conditioning circuit. Measurements were carried out in reflectance mode on the right radial artery.

A LabVIEW software was developed to manage the acquisition of the PPG signals (sampling frequency 1 kHz).

The overall dataset was stored in a log file that was then handled by MATLAB based algorithm for the PPG signal pattern recognition.

3 Results and Discussion

The herein proposed pipeline is schemed in Fig. 2.

It includes a number of processing modules, such as:

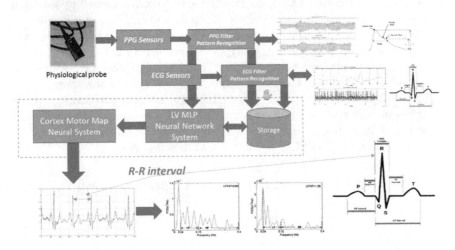

Fig. 2 PPG/ECG pipeline

- a block "PPG Filter Pattern Recognition". This block make an appropriate filtering of the PPG signal by removing signal artefacts that can affect the following process steps. Moreover, this block is suitable to segment each single conformed PPG waveforms discarding the corrupted ones in order to have—at the end—a segmented well conformed PPG time-series;
- a block "ECG Filter Pattern Recognition". As per the above reported block, this block also applies proper filters to the ECG signals to remove possible artefacts. As per the previous step, this block carried out the segmentation of the single conformed ECG waveforms discarding the corrupted ones. The output of this block is a well-conformed ECG timeseries;
- a block "Storage". In this block, both the PPG and the ECG filtered signals are stored before the processing with the bio-inspired algorithm;
- a block "LV MLP Neural Network System". In this block a properly configured LeVenberg-Marquardt Neural Network [7] carried out a first modelling of reconstructed ECG signals from the corresponding PPG ones. The network is trained by the LV modified Error back propagation learning algorithms. The input segmented PPG samples are fed into the network with the target in order to reconstruct the corresponding ECG samples. The PPG and ECG timeseries are resized in order to have same length. The learning is performed by using each resized segmented filtered-PPG waveform as input and corresponding resized pre-filtered ECG waveform as target for back propagation schema. All the data are normalized into [0.2–0.9] range.
- a block "Cortex Motor Map Neural System" used to improve the neural approximation PPG to ECG performed by previous block by means of LV Neural Network [6]. The tests shown that LV neural network does not exhibits the capability to perform careful reconstruction of ECG signal starting from PPG ones. By means of Cortex Neural Motor Map, the function approximation capability of overall system is improved, so that the whole neural system (LV Neural Network plus Cortex Motor Map) is able to perform an ECG time-series reconstruction from corresponding PPG samples. Actually, the reconstructed ECG samples have fed into Cortex Motor Map as input vector. The classical Winner Take All (WTA) algorithm discriminates the winner neuron (i.e. the neuron which minimize the Euclidian distance between normalized reconstructed ECG samples) and weight values of such neurons of Motor Map. The winner neuron performs a random update of the corresponding output weight "w^{out}" which will be used to improve the reconstructed ECG(ECG^{CMM}) as per the following equation:

$$ECG^{CMM}(t+1) = ECG'(t) + a(t) \cdot \sum_{x,y \in N_w} B(x,y,t) \cdot \left(w^{out}(t)\right)$$

$$w^{out}(t+1) = w^{out}(t) + c(t)$$

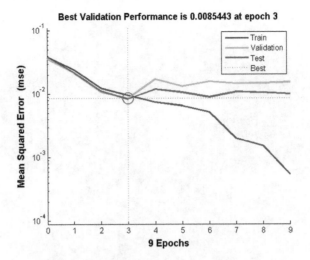

Fig. 3 LV neural network: learning error dynamics

where ECG'(t) is the reconstructed ECG timeseries from previous LV Neural Network block, the a(t) and B(t) are the learning rate and neighborhood function respectively, while c(t) identifies a random variable which will be used to search the right values to improve the ECG reconstruction. The WTA algorithm performs update of input and output weights only if mean square error between above reconstructed ECG^{CMM} timeseries and real ECG ones is minimized under prefixed precision.

Figures 3 and 4 shows the learning error minization of LV Neural Network and the Motor Map input and output weights distribution.

Figure 5 reports the obtained results. It can be noticed that the "reconstructed ECG signal" from PPG is perfectly superimposable with the "real ECG signal" in the part we needs for medical assessment in automotive application.

In automotive application we need to reconstruct the so called HRV (Hear Rate Variability) which is a well-defined indicator suitable to detect the level of drowsiness of driver [8, 9].

The HRV is computed from R-peak of ECG (max value of ECG pattern). As clearly reported in Fig. 3, the proposed algorithm perfectly reconstruct the R peak of each ECG waveform so that we are capable to reconstruct a robust HRV diagram. This allows to be implemented in a car more easily respect to ECG which required complex system (sensors, at least two contact points as per Einthoven triangle, etc.).

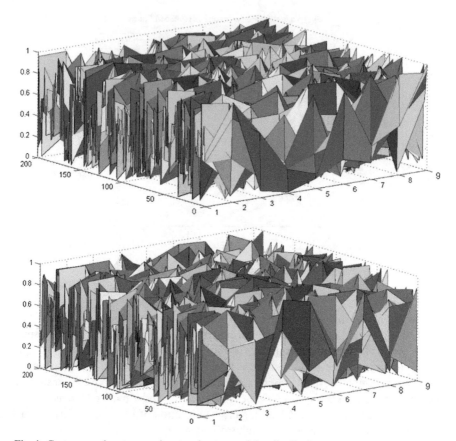

Fig. 4 Cortex neural motor map input and output weights distribution

4 Conclusion

The monitoring of physiological parameters such us heart electrical activity (ECG) and heart rate variability through a non-invasive miniaturized systems can be fundamental in a wide range of PoC applications from medical assessment in automotive application.

In this work we have described a robust and efficient bio-inspired pattern recognition pipeline for the reconstruction of ECG signal from PPG acquired by a miniaturized physiological optical probe.

The results highlight a fully compliance between the reconstructed ECG signal from the PPG pattern with target to have a great precision in the R-peak detection in reconstructed timeseries. This is very promising for the development of PoC device for ECG reliable measurements in medical environments for different applications.

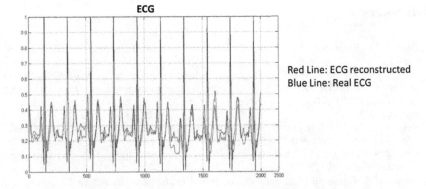

Red Line: ECG reconstructed
Blue Line: Real ECG

Red: ECG reconstructed
Blue: Real ECG

RR interval for HRV!

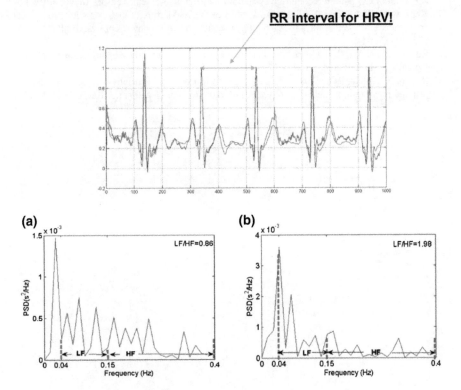

Fig. 5 Typical results HRV reconstructed from PPG and ECG

References

1. Birse, R.M., Knowlden, P.E.: Oxford Dictionary of National Biography. Oxford, Blackwell (2004)
2. Oreggia D., Guarino S., Parisi A., Pernice R., Adamo G., Mistretta L., Di Buono P., Fallica G., Cino C. A. and Busacca A. C.. Physiological parameters measurements in a cardiac cycle via a combo PPG-ECG system. In: Proceedings of the AEIT International Annual Conference, pp. 1–6 (2015)
3. Reisner, A., et al.: Utility of the photoplethysmogram in circulatory monitoring. J. Am. Soc. Anesthesiol. **108**(5), 950–958 (2008)
4. Petralia, S., Conoci, S.: PCR technologies for point of care testing: progress and perspectives. ACS Sens. **2**, 876–891 (2017)
5. Mazzillo, M., Condorelli, G., Sanfilippo, D., Valvo, G., Carbone, B., Fallica, G., Billotta, S., Belluso, M., Bonanno, G., Cosentino, L., Pappalardo, A., Finocchiaro, P.: Silicon photomultiplier technology at STMicroelectronics. IEEE Trans. Nuclear Sci. **56**, 2434–2442 (2015)
6. Rundo, F., Conoci, S., Ortis, A. Battiato S.: An advanced bio-inspired PhotoPlethysmoGraphy (PPG) and ECG pattern recognition system for medical assessment. Sensors **18**, 405 (2018)
7. Kanzow, C., Yamashita, N., Fukushima, M.: Levenberg-Marquardt methods with strong local convergence properties for solving nonlinear equations with convex constraints. JCAM **172**, 375–397 (2004)
8. Rundo, F., Fallica, P.G., Conoci, S., Petralia, S., Mazzillo, M.C.: Processing of electrophysiological signals, IT patent no. 102017000081018
9. Rundo, F., Fallica, P.G., Conoci, S.: A method of processing electrophysiological signals, corresponding system, vehicle and computer program product, IT patent no. 102017000120714

Electronic System for Structural and Environmental Building Monitoring

Leonardo Pantoli, Mirco Muttillo, Giuseppe Ferri, Vincenzo Stornelli, Rocco Alaggio, Daniele Vettori, Luca Chinzari and Ferdinando Chinzari

Abstract In this work, we present an innovative sensing and monitoring system of indoor environment parameters and structural elements of a building. The system has been conceived and optimized for modern wooden structures; it is organized with a control unit and sensing nodes that can be arranged freely. This architecture allows to provide continuous information about inside and outside ambient temperatures, moisture conditions and tri-axis inclination of structural elements. The data, continuously monitored, are collected on a web server able to check the overall status of the building and to generate automatic warnings and alerts for any criticisms. A web-application has been also developed for system monitoring, giving the possibility to create summary report and to analyse the data profiles. The system is currently working on a novel public construction fabricated in XLAM and designated for school activities in Lucca (Italy). All measurement results will be shown in the full paper.

Keywords Monitoring system · Wooden structures · Sensors system

1 Motivation

Wood is being billed as the answer to creating greener cities, lightweight, safe and sustainable. New types of ultra-strong timber are partly driving the trend; in addition, cross-laminated timbers, for instance , create a stronger weave more fire resistant

L. Pantoli (✉) · M. Muttillo · G. Ferri · V. Stornelli · D. Vettori
Department of Industrial and Information Engineering and Economics,
University of L'Aquila, Via G. Gronchi 18, 67100 L'Aquila, Italy
e-mail: leonardo.pantoli@univaq.it

R. Alaggio
Department of Civil, Construction-Architectural and Environmental Engineering,
University of L'Aquila, Via G. Gronchi 18, 67100 L'Aquila, Italy

L. Chinzari · F. Chinzari
L.E.R. S.R.L, Via Ventiquattro Maggio, 46, 00187 Rome, Italy

© Springer Nature Switzerland AG 2019
B. Andò et al. (eds.), *Sensors*, Lecture Notes in Electrical Engineering 539,
https://doi.org/10.1007/978-3-030-04324-7_58

than steel and more elastic of traditional structures. Obviously, a disadvantage of the timbers is that they require a higher maintenance and their properties are sensitive to environmental parameters, in particular moisture. This has led to the typical choice of using external coating to preserve the wood integrity, but adds also a further difficulty making hard the periodic check of the timber quality [1, 2]. Anyway, the benefits related to the use of wooden structure are more relevant of the relative criticisms and this justifies the rapid spread of this technology solution as constitutive element of modern buildings.

2 Proposed Monitoring System

This work has been conceived with the aim to overcome the above described drawbacks on one side and to further increase the human wellness and safety on the other side, by continuous monitoring both the environmental parameters and structure elements. In fact, in the last times an important research effort has been put also in the buildings monitoring [3–11]. The system here proposed is an innovative sensor network managed by microcontrollers; each sensing node allows to check inner and outdoor temperature on the considered wall, its three-axis inclination and moisture conditions at the intersection of each wall with transverse structural elements. All sensors, developed with dedicated interfaces, are usually placed below the internal/external coating, directly connected on the wooden elements for better reliability. Only an inspection box is necessary for each sensing node. In Fig. 1, a block scheme of the node structure and the first prototype are shown. All nodes communicate with the data logger by means of a wired link on a RS485 protocol; this control unit, beyond managing the sensor network, is in charge of transmit the real-time data on a web server by using a GSM on board module. In Fig. 2a the data logger architecture is shown, while in Fig. 2b a test bench of the proposed system is reported. Finally, Fig. 3a illustrates a possible installation scenario. The collected data can be checked and managed by a web application, as shown in Fig. 3b. In this way, the status of each sensing node can be continuously monitored and a dedicated software provides also an alert system by generating automatic warnings and alarms for any criticisms. The limits thresholds can be self-defined by the user.

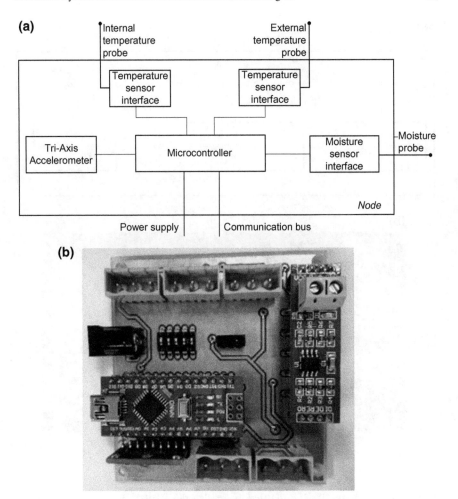

Fig. 1 **a** Block scheme of each sensing node; **b** prototype node (without sensors)

3 Test Case: Example of Application

The described has been installed for the first time about two years ago on a novel public construction fabricated in XLAM and designated for school activities in Lucca (Italy) (Fig. 4). It is still currently working there, without any needs of maintenance or anomalies. In Fig. 5a a simplified building plant is reported together with the position of the sensing nodes (1)–(5) and of the data logger (DL). The system has been installed with wired connections in the interspaces of the wooden walls according to the interconnection scheme of Fig. 5b. Each measurement parameter is monitored with a sampling time of 1 s, and the collected data are sent by the data logger to the web server about in real time by means of a dedicated GSM/GPRS connection.

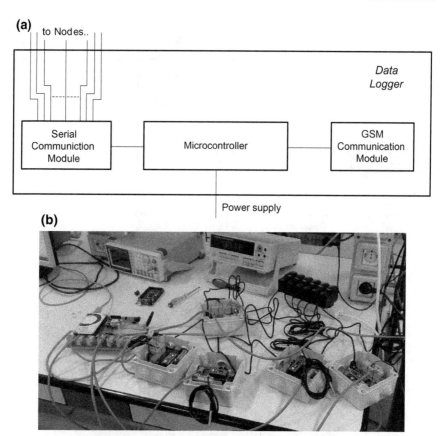

Fig. 2 **a** Block scheme of the data logger; **b** test bench of the proposed system

The relative web-application that has been also developed for system monitoring, gives the possibility to create summary report and to analyze the data profiles. Just an example of measurements performed in the construction period is reported in Fig. 6, in which the data profile of outdoor temperatures (a), inner temperatures (b) and wall inclinations (c) are reported over time for each sensing node. The accuracy and reliability of the proposed system is clearly shown in Fig. 6a, where a couple of sensors plainly show how the outside temperature of two walls significantly increase during the day with respect to the others of the structure. By analyzing this phenomenon, the cause has been traced on the different solar irradiation that interest the building and this is proved, beyond of the orientation of the building itself, also by considering that the measured temperature of all nodes fit well during the dark hours.

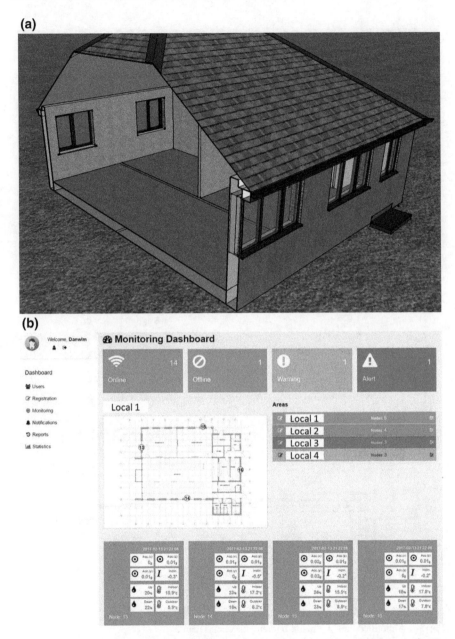

Fig. 3 **a** Example of application: sensing elements with red circles; node with red rectangle. H_{in} states for moisture sensor; T_{in} and T_{out} indicate the inner and outdoor temperature sensors, respectively; M is the motion probe. **b** Graphical web interface with the control panel for monitoring and data analysis of the sensor system

Fig. 4 Test case: Scuola Primaria "Fornaciari"—Lucca

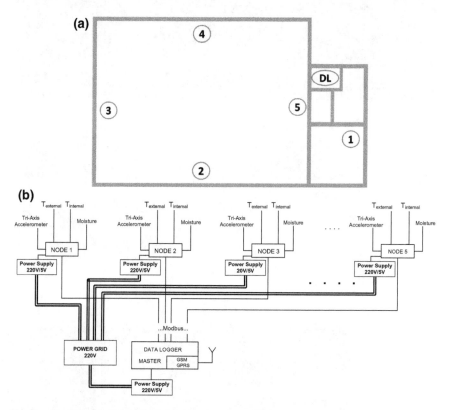

Fig. 5 **a** Simplified building plant; **b** interconnection scheme

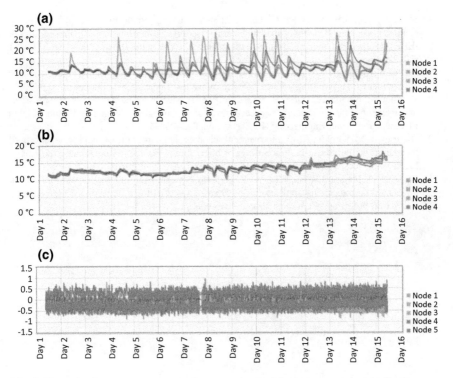

Fig. 6 Example of measurements: **a** outdoor temperature; **b** inner temperature; **c** inclination

References

1. Okamoto, S.: Vibration and scattering monitoring of Japanese roofing tile by accelerometer. In: Proceedings of International Conference on Fluid Power and Mechatronics, Beijing. pp. 30–35 (2016)
2. Cuadra, C.H., Shimoi, N., Nishida, T., Saijo, M.: Estimation of dynamic properties of traditional wooden structures using new bolt sensor. In: Proceedings of 13th International Conference on Control, Automation and Systems (ICCAS), Gwangju, pp. 1593–1598 (2013)
3. Moghavvemi, M., Ng, K.E., Soo, C.Y., Tan, S.Y.: A reliable and economically feasible remote sensing system for temperature and relative humidity measurement. Sens. Actuators A **117**(2), 181–185 (2005)
4. Ausanio, G., Barone, A.C., Hison, C., Iannotti, V., Mannara, G., Lanotte, L.: Magnetoelastic sensor application in civil buildings monitoring. Sens. Actuators A **123–124**, 290–295 (2005)
5. Santana, J., van den Hoven, R., van Liempd, C., Colin, M., Saillen, N., Zonta, D., Trapani, D., Torfs, T., Van Hoof, C.: A 3-axis accelerometer and strain sensor system for building integrity monitoring. Sens. Actuators A **188**, 141–147 (2012)
6. Kifouche, A., Baudoin, G., Hamouche R., Kocik, R.: Generic sensor network for building monitoring: design, issues, and methodology. In: Proceedings of IEEE Conference on Wireless Sensors (ICWiSe), Miri, pp. 1–6 (2017)

7. Shen, W., Xue, H.H., Newsham, G., Dikel, E.: Smart building monitoring and ongoing commissioning: a case study with four canadian federal government office buildings. In: Proceedings of IEEE International Conference on Systems, Man, and Cybernetics (SMC), Banff, pp. 176–181 (2017)
8. Včelák, J., Vodička, A., Maška, M., Mrňa, J.: Smart building monitoring from structure to indoor environment. In: Proceedings of Smart City Symposium Prague (SCSP), Prague, pp. 1–5 (2017)
9. Tanasiev, V., Necula, H., Darie, G., Badea, A.: Web service-based monitoring system for smart management of the buildings. In: Proceedings of International Conference and Exposition on Electrical and Power Engineering (EPE), Iasi, pp. 025–030 (2016)
10. Dmitriev, V.N., Sorokin, A.A.: Improvement schemes elements of measuring instruments used for monitoring buildings. In: Proceedings of International Conference on Actual Problems of Electron Devices Engineering (APEDE), Saratov, pp. 227–234 (2014)
11. Dessales, D., Poussard, A.M., Vauzelle, R., Richard, N.: Case study of a wireless sensor network for a building monitoring application. In: Proceedings of IEEE International Conference on Green Computing and Communications, Besancon, pp. 651–654 (2012)

Closed-Loop Temperature Control CMOS Integrated Circuit for Diagnostics and Self-calibration of Capacitive Humidity Sensors

Moataz Elkhayat, Stefano Mangiarotti, Marco Grassi and Piero Malcovati

Abstract A temperature control loop for capacitive humidity sensors has been developed in 0.35 μm CMOS technology and characterized with experimental lab measurements. The proposed circuit is employed for self-diagnostics of the humidity sensor as well as for auto-calibration, thanks to correlation between relative humidity and temperature. An input Switched-Capacitor (SC) thermometer read-out circuit compares the measured temperature value with a setpoint while a SC PWM PI circuit drives a power n-MOS to deliver heat to the sensor. Measurements results summary is reported. The temperature loop accuracy is ±0.25 °C in a range from room temperature to 50 °C, while consuming 0.9 mA from a single 3.3 V supply for the control and about 43 mW for delivering heating power.

Keywords Auto-Diagnostics · Capacitive humidity sensor
Temperature control loop · Self-Calibration

1 Introduction

Capacitive humidity sensors feature several advantages over other types of humidity sensors, due to their fair linearity response, adequate accuracy, low fabrication cost, and small size with possibility of easy integration in standard CMOS technologies [1–4]. This research work is about the design of an interface circuit for capacitive humidity sensors to generate an equivalent predictable artificial relative humidity

M. Elkhayat · S. Mangiarotti · M. Grassi (✉) · P. Malcovati
Department of Electrical, Computer, and Biomedical Engineering,
University of Pavia, Pavia, Italy
e-mail: marco.grassi@unipv.it

M. Elkhayat
e-mail: moataz.elkhayat01@universitadipavia.it

S. Mangiarotti
e-mail: stefano.mangiarotti01@universitadipavia.it

P. Malcovati
e-mail: piero.malcovati@unipv.it

© Springer Nature Switzerland AG 2019
B. Andò et al. (eds.), *Sensors*, Lecture Notes in Electrical Engineering 539,
https://doi.org/10.1007/978-3-030-04324-7_59

variation, exploiting a controlled temperature step. Correlation between RH and temperature for a given absolute humidity value is the key rule for getting the defined RH step for diagnostics and calibration purpose. A prototype is developed in CMOS 0.35 μm, operating from a single internal 3.3 V supply.

Sensitivity and offset of humidity sensors can move away from the values obtained during initial calibration because of ageing, long term drift, or exposure to relatively high temperature for a long time (typically $T > 50$ °C at $RH > 70\%$). Therefore, self-diagnosis and self-calibration techniques capable of periodically updating the calibration coefficients could strongly improve the performance and reliability of the sensors over time. Since the sensor response is strongly temperature dependent, a good stability over time is easier with the adoption of a temperature control loop also during normal operation mode, leading also to higher RH measurement accuracy, reducing also small ageing drift effects.

2 Temperature Control System Circuit

The temperature control loop shown in Fig. 1 is driven by a PI (Proportional-Integral) controller with parameters $TI = 1.04$ s and $Kp = 10$. These parameters are optimized on the basis of previous analysis on specific capacitive humidity sensors, calibrated according to an appropriate algorithm (Ziegler-Nichols rules).

Voltage gain and integrator are realized using switched-capacitor (SC) techniques, with two non-overlapped clock phases: $\phi1$ and $\phi2$ with a frequency of 100 Hz. A comparator and a sawtooth generator block are designed to work as a PWM modulator at 100 kHz. The output of the PWM circuit is connected to a power n-MOS sized with $W = 1000$ μm and $L = 0.4$ μm ($R_{DS} = 4$ Ω), which delivers the needed current to R-heater. The thermal coupling inside the sensor between the adjacent insulated R-heater and R-thermo platinum resistors has been modeled in VerilogA for transistor level system simulations during preliminary study. The first resistor has a nominal value of 250 Ω and has been connected to the power n-MOS to regulate the output current and therefore the temperature, while the other terminal of R-heater is connected to an external voltage supply varying from 3.3 V (equal to chip supply voltage value) up to 10 V depending on the desired maximum temperature. The temperature signal from the R-thermo (nominal value is 2.5 kΩ) is biased by a programmable current generator and it is converted into a voltage which then is amplified by a factor of 10 by a first SC fixed gain stage. The total DC closed loop gain of the system is 100. The fabricated chip has the option to drive an external nMOS gate to use discrete power devices till up to 10 V voltage supply.

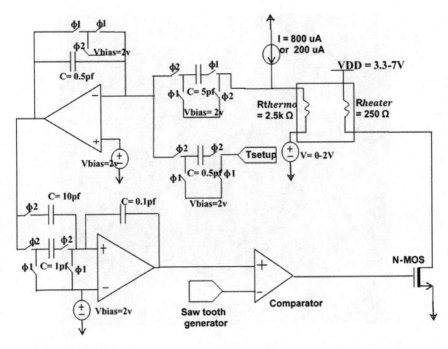

Fig. 1 Schematic of the complete system architecture

3 Experimental Measurements

The read-out circuit has been fabricated in a 0.35 μm CMOS technology with 3.3 V supply voltage. The photograph of the chip with 24 I/O pads occupying an area of 500 μm × 500 μm is shown in Fig. 2, including auxiliary supplies for pad protections and buffers. Preliminary measurements have been carried out by using an equivalent PT2500 resistor due to typical platinum fabrication material of humidity sensors embedded heaters-thermometers. The platinum resistor, with a temperature coefficient $\alpha = 3.925$ m °C^{-1}, featuring 2.5 kΩ at 0 °C, is thermally coupled with a power resistor, featuring 250 Ω at ambient temperature of about 25 °C. The measurements are divided into two groups, depending on the value of the heating supply voltage and thus the maximum temperature. The first group of measurements is performed with 3.3 V supply to the internal nMOS for a temperature range typically between 25 and 30 °C. The second group of measurements have been carried out using an external discrete power device (a high base input resistance NPN BDX53C Darlington BJT) at 10 V supply to drive the sensor heater for a temperature range typically from 30 to 50 °C.

The first actual measurement, exploiting internal nMOS power device, has been performed biasing the thermometer at 766.2 μA by means of the internal chip programmable current generator and setting to 2.15 V the reference voltage bias. A

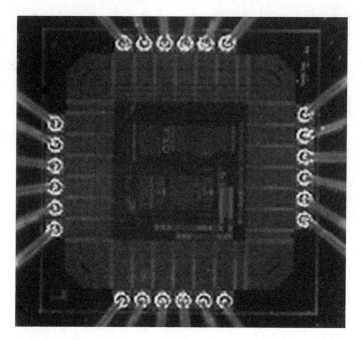

Fig. 2 Chip photograph (500 μm × 500 μm)

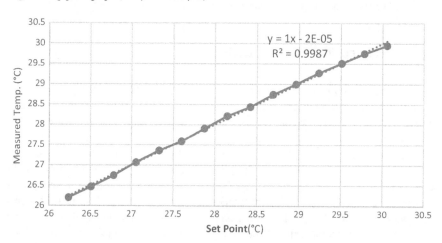

Fig. 3 Measurements for the range [27–30 °C]

PT2500 resistor is used as thermometer and a 1 W power, 250 Ω resistor is connected to the internal nMOS drain pad for heating according to the set-point temperature. The measurements have been carried-out by changing the setpoint value and measuring the PT2500 value which is equivalent to have the information on the actual temperature. Figure 3 shows steady temperature value at different set-points.

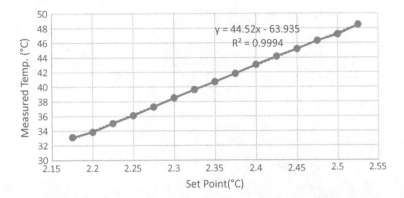

Fig. 4 Measurements for the range [30–50 °C]

In the second group of measurements, performed using the high base input resistance NPN BDX53C Darlington BJT supplied at 10 V, the thermometer current is set to 199μA, while the reference voltage bias to 2.11 V. A PT2500 resistor is used again as thermometer and a 1 W power, 250 Ω resistor is connected to the external Darlington BJT collector for heating the sensor according to the setpoint temperature. The measurements have been carried-out by sweeping the setpoint and measuring the PT2500 value which is equivalent to the desired temperature. Figure 4 shows a graph of the variation of steady temperature sensor in measurement with different set-points.

All performed measurements data have been carried out in an ad hoc MATLAB routine that calculates the temperature accuracy for each range of temperatures. Figures 5 and 6 show the system transient response. The temperature ripple in measurements is lower than 0.2 °C peak, while the closed loop system accuracy in temperature regulation is ±0.11 °C when using 3.5 V supplied internal heater power MOS and ±0.25 °C exploiting the external BJT Darlington driver.

4 Conclusions

The proposed temperature control circuit consists of a programmable current thermometer read-out, a SC voltage amplifier, a SC PI analog controller for PWM modulation to drive the heater power device. The core supply voltage is 3.3 V, while the maximum current consumption for the control part is 0.9 mA, including external thermometer bias current, leading to an internal maximum power dissipation of about 1 mW (the remaining 2 mW are dissipated by the external thermometer). The maximum current drawn instead by the external heater actuator is about 13 mA for maximum duty cycle, but this power contribution is almost all dissipated outside the chip, i.e. into the external connected resistive heater (43 mW), whose temperature

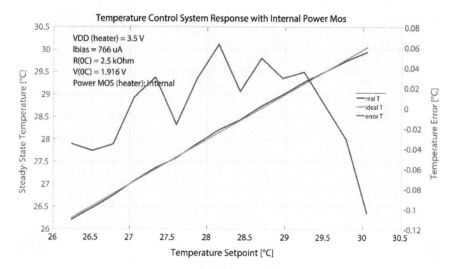

Fig. 5 System response (internal power MOS at 3.3 V supply)

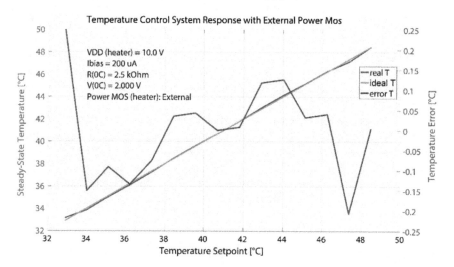

Fig. 6 System response (external BJT Darlington at 10 V supply)

will range between 20 and 45 °C by regulation through a loop set-point. In fact, the resistance of the output power transistor is about 4 Ω against 250 Ω of the heater resistance. The heater will be connected to a 3.3 V supply with common ground. The circuit is driven by two main clocks. The first clock (external) drives the switched-capacitor architecture for the control loop and is set to 100 Hz, while the second clock (generated internally by a relaxation oscillator) is set to 100 kHz to generate a sawtooth waveform for the PWM heating actuator driver.

Finally, it is possible to demonstrate a change in relative humidity controlled by temperature variation. The self-calibration enabled by temperature control strongly improves the performance and reliability of the sensors over time. The temperature control loop implementation in CMOS technology offers high accuracy (better than 0.25 °C) and a strong reduction of drift effects. The temperature control of humidity sensors enables many applications, from the simple heating of the sensor in order to treat critical situations such as condensation or accumulation of water, to more complex sensor self-diagnostic system that allows figuring out the malfunction in humidity reading and sensor self-calibration.

References

1. Kim, J.H., et al.: High sensitivity capacitive humidity sensor with a novel polyimide design fabricated by MEMS technology. In: International Conference on Nano/Micro Engineered and Molecular Systems (2009)
2. Zhao, C.L., et al.: A fully packaged CMOS interdigital capacitive humidity sensor with polysilicon heaters. IEEE Sens. J. **11**(11), 2986–2992 (2011)
3. Lazarus, N., et al.: CMOS-MEMS capacitive humidity sensor. J. Microelectromech. Syst. **19**(1), 183–191 (2010)
4. Elkhayat, M., Mangiarotti, S., Grassi, M., Malcovati, P., Fornasari, A.: Capacitance humidity micro-sensor with temperature controller and heater integrated in CMOS technology. In: CNS 2016 Proceedings, Lecture Notes in Electrical Engineering, pp. 383–387. Springer (2017)

An UAV Mounted Intelligent Monitoring System for Impromptu Air Quality Assessments

M. Carrozzo, S. De Vito, E. Esposito, F. Formisano, M. Salvato,
Ettore Massera, Girolamo Di Francia, P. Delli Veneri, M. Iadaresta
and A. Mennella

Abstract Air quality is a source of increasing concern in several cities, due to the adverse health effects of significant pollution levels. For this reason, the need to assess the concentration of pollutants at high temporal and spatial resolution is perceived as very urgent. As such, there is a growing interest in building pervasive networks integrating different sensing technologies to achieve this capability. In this view, low cost smart sensors data could be fused together with fixed but more accurate multisensor devices and certified analyzers data to build hi-res maps (<100 m) of pollutant concentrations. Low cost UAVs are versatile platforms capable to host playloads integrating multi sensor technologies for short term, mobile monitoring tasks. The recently proposed tethered UAV platforms can be deemed as interesting solutions to build and rapidly deploy impromptu networks of air quality analyzers for mid-term environmental monitoring actions. In this work, we propose the integration of the ENEA MONICA multisensor platform as a measurement payload for the TopView SAV-ES UAV. The prototype has been flow on board of the SAV-ES platform for test flight targeting the measurement of a plume generated with a brushwood controlled fire. The functional proof of concept flight have confirmed the possibility to use MONICA as a measurement payload for the targeted platform.

Keywords Air quality assessments · UAV · Intelligent monitoring systems

M. Carrozzo · S. De Vito (✉) · E. Esposito · F. Formisano · M. Salvato · E. Massera · G. Di Francia
P. Delli Veneri
Laboratory DTE-FSN-DIN, ENEA C.R. Portici, P. le E. Fermi, 80055 Portici, NA, Italy
e-mail: saverio.devito@enea.it

E. Esposito
e-mail: elena.esposito@enea.it

M. Iadaresta · A. Mennella
TOPVIEW s.r.l., Via Alessandro Pertini, 25D, 81020 San Nicola La Strada, CE, Italy

© Springer Nature Switzerland AG 2019   497
B. Andò et al. (eds.), *Sensors*, Lecture Notes in Electrical Engineering 539,
https://doi.org/10.1007/978-3-030-04324-7_60

1   Introduction

New Air Quality (AQ) monitoring systems based on solid state sensors are under development with some of them being already available on the market (for example AQMesh [1]). However, several issues, such as the lack of selectivity of chemical sensors with respect to the interferent gases present in the air, the poor long-term stability and the lack of repeatability of the manufacturing process, limit their use on a large scale.

In the last few years, the interest in the development of mobile implementations of air quality sensors has significantly grown. Regulatory proof, accurate, estimation of atmospheric pollutants concentrations still appears far to be achieved by low cost systems, but some recent developments show that Machine Learning approaches [2] to on field recorded data may allow the achievement of Data Quality Objectives [3] for indicative measurements for some pollutants, in particular inorganic ones.

These results highlight that, the inability to obtaining a dense network of conventional monitoring stations, due to their size and the maintenance required by them, may be relieved by the possible deployment of a dense network of low cost sensors, managed with Internet of Things hardware (IOT).

Unmanned Aerial Vehicle systems (UAVs) represent a good solution for the monitoring of a single measurement point or for an extemporaneous network, for example for the mapping of little areas presenting a complex morphology, such as cultivated areas, industrial plants, ports, etc.

UAV mounted Chemical sensing solutions (MOX sensors based) are currently under investigations of other research institutes [4]. All solutions are subject to common problems related to the weight of Payload, connectivity requirements, quality of the measurements and reliability of the sensors over time. Also noteworthy are the effects due to the aerodynamic perturbations generated by drone propellers.

2   Methodology: The Tethered Air Quality Drone Architecture

The basic design of the system consists of four main pillars (see Fig. 1).

The implementation of the designed HW/SW architecture aimed to compact dimensions and UAVs transport suitability. It consists of the Monica chemical multisensor [5, 6] plus an IoT processing unit for recording, geo-referencing, data processing and data transmission.

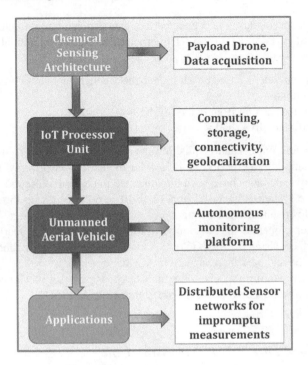

Fig. 1 Basic monitoring system design

2.1 Payload Part 1: MONICA Multisensor Node

Monica (v 2.0), developed by ENEA (Department of Energy Technologies, Photovoltaic and Smart Network Division, Innovative Devices Lab.) is a chemical multisensor device based on Alphasense A4 electrochemical sensing unit and T, RH environmental sensors (sensirion SHT75). The gas sensors are targeted to CO (CO-A4), NO_2 (NO_2-A43F) and O_3 (O_3-A431), that are among the most relevant air pollution in the city. A known cross interference between NO_2 and O_3 affect the respective sensors while all the sensors show a relevant response to temperature. The CO sensor shows a noise level of 20 ppb and a sensitivity of 220–375 nA/ppm at 2 ppm. Conversely, the NO_2 sensor should show a noise level of 15 ppb and a sensitivity of -175 to -450 nA/ppb at 2 ppm. Finally, the O_3 sensor datasheet information shows a noise level of 15 ppb and a sensitivity of -200 to -650 nA/ppm at 1 ppb.

MONICA sensing array is connected to a data acquisition unit, namely STM32 Nucleo board (v. L432 KC) from ST microelectronics. The microcontroller unit is connected to:

- the power subsystem, based on a lithium battery which guarantees more than thirty hours of autonomy;
- the Alphasense analog front end (AFE 810-0020-00);
- EDR 2.0 Bluetooth™ shield.

The firmware controls data acquisition and conversion providing a duty cycle with a variable sampling rate period of 1–75 s for the entire sensor array. Furthermore oversampling techniques have been implemented to perform noise reduction.

The entire system is enclosed in a $12 \times 10 \times 6$ cm box with three 3×3 cm fans that helps maintaining a constant air flow near the sensitive edge of the sensor units. In order to operate, the sensor node is usually connected to a smartphone using the BT interface. Using the MONICA™ Android™ app, the smartphone provide for geolocalization and backend connectivity via 4G networks.

For UAVs, mounting a special IoT system has been purposely designed and implemented to substitute and improve the smartphone capabilities.

2.2 Payload Part 2: IoT Processing Unit

The payload have been implemented by connecting the MONICA device with a Raspberry Pi unit featuring a BT radio, a GPS unit for auxiliary geolocation purposes (Pycom Pytrack) and a radio transmission subsystem (WiFi/LTE) for guaranteeing the backend connectivity under radio coverage (see Fig. 2).

The Raspberry Pi is also responsible for implementing a calibration function based on shallow neural networks trained on lab based multivariate (target gas + temperature) calibration recipe.

Several technologies have been identified to realize a software stack IoT:

Fig. 2 The 3D printed Monica Payload case, including the Monica node, the Raspberry Pi module and battery

- Python, for portability between platforms;
- JS for integration with native Web development APIs;
- PHP and MongoDB for storing data acquired from sensor nodes.

Raw and calibrated data are sent to, and stored in, the MONICA backend, based on a MONGODB NOSQL database, as JSON packets.

The *Monica Environmental Network Analysis* (MENA) interface, based on a Javascript/Php engine including Google API Maps calls, have been hence used to visualize and download the data for the reconstruction of the flight trajectory and measured gas concentration.

2.3 UAV Platform

The TopView SAV-ES UAV system is a customized UAV (Unmanned vehicle system) based on the DJI S900 platform with N3 flight control subsystem (MTOW1 = 8 kg). Precision geolocalization is obtained through an on board GNSS UBLOX M8 N constellation receiver module with RTK2 capabilities.

Maximum flight duration is 18 min if operated on Lipo battery while it is virtually unlimited if operated with Safe-T system as a tethered device.

The Safe-T system is a ground power supply, that provides the electric power to the aircraft through a system of three cables with three-phase alternating voltage. It is a case powered by primary supply voltage equal to 220 Vac/16 A 50 Hz, in which there is an automated rotating axis. Around this axis, the winch (3 mm thick) that contains the power cables connected to the APR, is wound. When the aircraft takes off, it unrolls the winch (maximum length equal to 100 m), which is always kept with a given mechanical voltage from the automated axis. The cables voltage varies in [200, 600] V AC, while the power is limited to a maximum of 5 A, due to the size and the weight.

An AC/DC converter turns the alternating voltage into DC power and supplies the APR. The hexarotor requires a supply voltage equal to 22.2 V DC, with an average power consumption of 2.5 KW.

The system is also equipped with a Wi-Fi module, that allows to access to different data, reported in Table 1.

Table 1 Data achieved by Wi-Fi module

Data type
Internal case temperature
Cable mechanical voltage
Unrolling speed
Cable length
Instantaneous electric power absorbed by the aircraft
Flight time

Table 2 System's application field (SAF)

SAF	Description
Static surveillance	Events, forest fire management, anti-poaching
Pop-up telecommunications	
Traffic management	
Industrial inspections	Oil and Gas, public services, civil work structures. Remote inspections of industrial facilities are often expensive, dangerous and sometimes impossible to achieve. The rapid growth of unmanned systems is offering the ability to overcome the limits and the lower costs; free flight drones are still limited due to their limited flight time and their security level is low
Air quality monitoring	Impromptu monitoring networks, spills source detection, source emission monitoring
TV broadcast	Live air links on sporting events, concerts, shows

The UAV Tethered Platform system uses conveyed wave technology in order to exchange data with the APR in flight. Thus, the information coming from the aircraft, shall be sent via the three conductor cables. Finally, there is an Ethernet port that allows to collect the data. The main system's application fields are collected in Table 2.

This system represents a step forward in the APR framework and further developments are possible in order to improve the communication between the aircraft and the GCS.

3 Experimental Results: First Flight Session

The prototype has been mounted on board of the SAV-ES platform (see Fig. 3) for test flight targeting the measurement of a plume generated with a brushwood controlled fire (see Fig. 4).

Previously, Payload and the PyMonica applications have been used in field tests from September to November, mapping different routes, on foot, by bike and by car, between the cities of Portici and Naples. The test of the payload installed on board of the UAV has been carried out in early November at Castel Campagnano, Caserta (Italy). Payload worked optimally in all field test sessions. The flight session chart is shown in 3D picture (Fig. 5), in which it's clear the passage of the Payload in the controlled fire plume oriented along the mutating wind directions (red color).

Fig. 3 The topview SAES-ES drone with the mounted Monica Payload

Fig. 4 The SAES-EV while in flight and monitoring the controlled fire plume

Field tests are analyzed by the user through the MENA panel. In Fig. 6, the 2D projection of the 3D flight session graph is shown; the recorded data have been imported and normalized in Matlab, clearly showing the correlation between NO_2 and O_3 sensors responses. It is evident the sensor dynamic influence, in particular for CO sensor.

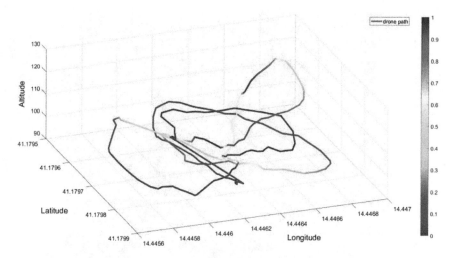

Fig. 5 View of drone test session. Normalized CO (ppm)

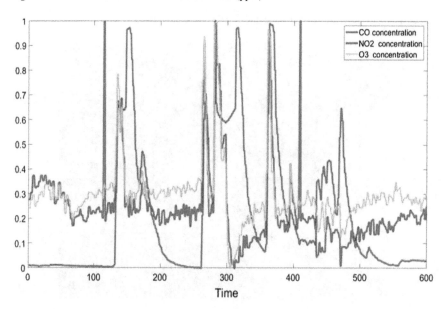

Fig. 6 Session chart of normalized data

The MENA panel show the results (see Fig. 7), in terms of chart, map and raw data, of the test flight targeting the measurement of a plume generated with a brushwood controlled fire. The sensors response peaks that can be observed correspond to the flight over the plume.

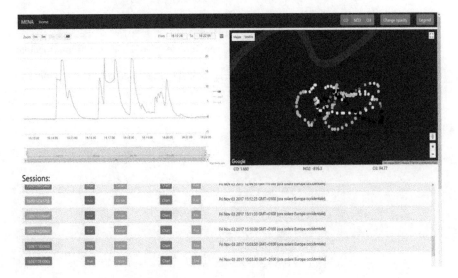

Fig. 7 MENA web panel, showing flight session results

4 Conclusions

The functional proof of concept flight have confirmed the possibility to use MONICA as a measurement payload for the targeted platform. In order to refine the payload shape and intake functionality, the in-flight air flow will be analyzed by dynamic simulations The flying MONICA device calibration will be refined by using data recorded during a 2 weeks co-location period with an ARPA managed certified environmental monitoring station unit. This will allow to take into account, in the training process, the interferences induced by the simultaneous presence of non-target gases, so to obtain a dataset more representative of real world situations.

The expected error reduction on field will make the system more accurate for its application as a UAV payload.

References

1. https://www.aqmesh.com
2. De Vito, S., et al.: Calibrating chemical multisensory devices for real world applications: an in-depth comparison of quantitative machine learning approaches. Sens. Actuators B Chem. **255**(2), 1191–1210 (2018)
3. Directive 2008/50/EC of the European Parliament, the Council of 21 May 2008 on ambient air quality, and cleaner air for Europe

4. Rossi, M., et al.: Gas-Drone: portable gas sensing system on UAVs for gas leakage localization. In: IEEE Sensors Proceedings, pp. 1431–1434 (2014)
5. MONICA Eppela Website: https://www.eppela.com/it/projects/9652-monica-il-tuo-navigatore-personale-antismog. Accessed on April 2018
6. Capezzuto, L., et al.: A maker friendly mobile and social sensing approach to urban air quality monitoring. In: IEEE Sensors Proceedings, pp. 12–16 (2014)

Part VII
Sensors Applications

Fluxgate Magnetometer and Performance for Measuring Iron Compounds

Carlo Trigona, Valentina Sinatra, Bruno Andò, Salvatore Baglio, Giovanni Mostile, Alessandra Nicoletti, Mario Zappia and Adi R. Bulsara

Abstract It has been observed that patients with some neurodegenerative diseases (such as Alzheimer, Parkinson etc.) accumulate metals in their nervous system, in particular the area of interest is the basal ganglia region. Several solutions can be considered such as Magnetic Resonance Imaging (MRI), Positron Emission Tomography (PET), and SQUID magnetometers. It should be noted that the first approach works without radioactive tracers, however in presence of patients with implanted devices the MRI cannot be used. The PET is an invasive technique which requires use of radioactive tracers and the SQUID is a solution that can be pursued but it is expensive and it is a sophisticated method of measurement with a complex set-up and it is an expensive cryogenic instrumentation. In this paper, the possibility to use the Flexible RTD-Fluxgate magnetometer as alternative low cost solution, able to detect weak magnetic fields or field perturbations with low dimensions, high sensitivity, low power consumption and an intrinsic digital form of the output signal as respect the classical second harmonic fluxgate, will be addressed. Experimental results are shown that encourage to pursue this approach in order to obtain simple devices that can measure several quantities of known ferromagnetic compounds accumulated in a localized area in order to be used as diagnosis method.

Keywords Neurodegenerative disease monitoring
Flexible RTD-Fluxgate magnetometer · Measurements of iron compounds

C. Trigona (✉) · V. Sinatra · B. Andò · S. Baglio
Dipartimento di Ingegneria Elettrica, Elettronica e Informatica, University of Catania, Viale Andrea Doria 6, 95125 Catania, Italy
e-mail: carlo.trigona@dieei.unict.it

G. Mostile · A. Nicoletti · M. Zappia
Department "G.F. Ingrassia", Section of Neurosciences, University of Catania, Catania, Italy

A. R. Bulsara
Space and Naval Warfare Systems Center Pacific, Code 7100, 53560 Hull Street, San Diego, CA, USA

© Springer Nature Switzerland AG 2019
B. Andò et al. (eds.), *Sensors*, Lecture Notes in Electrical Engineering 539,
https://doi.org/10.1007/978-3-030-04324-7_61

1   Introduction

Clinical studies have demonstrated that patients with neurodegenerative diseases, e.g. Alzheimer's disease, Parkinson's disease, Huntington's disorders, amyotrophic lateral sclerosis, and prion diseases, accumulate metals and, specifically, iron compounds in specific brain regions, called basal ganglia [1].

In particular literature highlights that it is possible to correlate the quantity of metal accumulated in the brain with a neurodegenerative disease [2].

For this reason, several techniques have been exploited in literature in order to measure brain iron levels in neurodegenerative diseases, including PET, SQUID magnetometers and especially MRI [3–5].

It must be observed that PET and a similar solution named Single Photon Emission Computed Tomography (SPECT) [6] are invasive techniques which require use of radioactive tracers. The approach based on SQUID is less invasive and less dangerous as respect the PET, however this magnetometer is expensive and it is a sophisticated method of measurement which can operate in cryogenic conditions and it is a complex architecture. The MRI is a method able to measure brain iron levels without an invasive approach. It is worth noting that this method cannot be used on patients with implanted devices considering that potential hazards (such as motion, dislocation or torquing of the implanted medical system, heating of the leads, and its damage) may occur. In this work the authors present a novel sensing approach for measuring iron compounds based on a Flexible Core Residence Times Difference (RTD) Fluxgate Magnetometer [7–9]. The proposed sensor presents several advantages compared with other magnetometers, such as: low cost solution, compact device, high performance in terms of sensitivity and resolution, low power consumption and an intrinsic digital form of the output signal, high spatial resolution [10].

2   Measurement Method and Experimental Setup

In this section the employment of a Flexible RTD-Fluxgate magnetometer [7–10] to measure iron concentrations comparable with the ferrous contents in a specific brain region, called basal ganglia, and distinctive of healthy and pathological conditions, is illustrated. It is worth noting that the sensor is a groundbreaking device to sense quasi-static variations of the external magnetic field. The architecture of the sensor includes: a ferromagnetic core having a diameter of 100 μm, a relative permeability of about 80,000 and an hysteretic input-output characteristic; a primary and a secondary coils, identified as excitation and pick-up coils respectively, directly wound around the ferromagnetic core and composed of 900 turns (see Fig. 1).

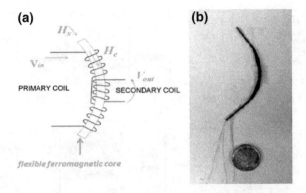

Fig. 1 **a** The architecture; **b** the flexible prototype

The measurement procedure has been assembled considering the working principle of the RTD-Fluxgate, where the information connected to the measured output of the pick-up coil (*Vout*) is exploited. The readout strategy is focused on the measurement of two time intervals, T^+ and T^-, called residence times, which represent, from a physical point of view, the times spent by the hysteretic-core magnetization variable in the two (stable) steady states. Their difference is indicated with the index RTD, acronym for Residence Times Difference. In absence of an external magnetic field, the RTD = $T^+ - T^- = 0$; in the presence of the target signal, the RTD $\neq 0$. In order to obtain an estimation of RTD the following measurement method has been implemented: a waveform generator is used to generate a sinusoidal voltage, *Vin*, which is transformed in a sinusoidal current through a V/I converter in order to drive the excitation coil. The output signal of the pick-up coil, which has the shape of a "spike train response", has been connected (through an instrumentation amplifier) to the Schmitt trigger (correlated with a level shifter) in order to convert the measurand into a square waveform. The data are then acquired by a DAQ board (NI USB-6255) and they are processed in LabVIEW environment through a specific routine to obtain a measure of RTD from the input square waveform. The relation between RTD and the duty cycle (*D*) of the square waveform has been determined as follows:

$$T = T^+ + T^- \rightarrow T^- = T - T^+ \tag{1}$$

$$T^+ = DT \tag{2}$$

$$RTD = T^+ - T^- = DT - \left(T - T^+\right) = DT - (T - DT) = T(2D - 1) \tag{3}$$

The (3) equation is implemented in LabVIEW routine in order to collect a signal for 25 s; the DAQ board is set to acquire a square waveform, for this reason it is synchronized with rise and fall fronts of the input signal. The whole measurement procedure can be summarized in Fig. 2.

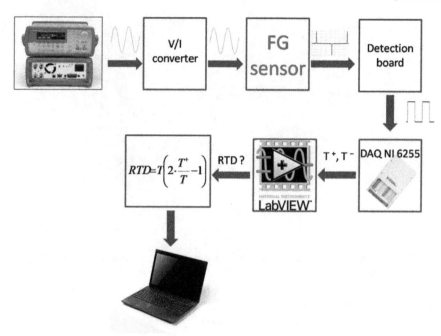

Fig. 2 The measurement procedure

The measurement method has demanded the realization of a suitable experimental setup in order to evaluate the iron compound contents (see Fig. 3). A function generator (Agilent 33220A) is used to drive the primary coil with a sinusoidal current having an amplitude of 7 mApp and a frequency of 150 Hz. The output signal (*Vout*) is sent to a detection board and it is visualized on an oscilloscope (Agilent DSO-X 3024A with a band of 200 MHz, four channels and 4 GSa/s). Two power supplies are employed to provide the V/I converter and the detection board. The sensing element that detects the target compounds is a Flexible core RTD-Fluxgate magnetometer. It is remarkable to underline the presence of a permanent magnet having length 2.5 cm, diameter 0.5 cm, and a static magnetic field (evaluated at 5 cm of separation) of about $40 \, \mu T$. The sensor and the magnet are maintained fixed at an angle of 90°, as shown in Fig. 4. The role of the permanent magnet is invaluable to polarize the ferromagnetic samples used to represent the iron contents in the basal ganglia region in a human brain.

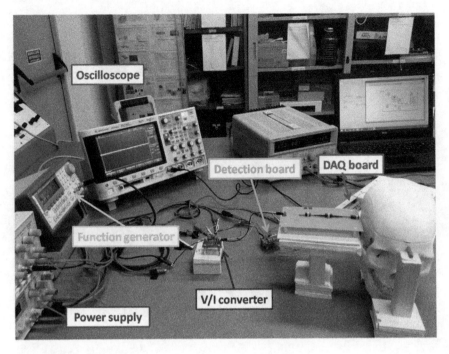

Fig. 3 Experimental setup

3 Experimental Results

The obtained experimental results are presented in this section. It is important to observe that a peculiar PVC head has been employed to emulate a human cranium and it has been filled with 500 mL of a suitably saline solution, keeping in mind that water and NaCl are well represented in a human brain [11]. The sensor capability of detecting and measuring the iron contents has been tested and validated in [9], where specific ferromagnetic samples, whose content can be identified as a hallmark of neurodegenerative diseases, have been prepared; in detail, ferromagnetic material, having a density of about 866 kg/m^3 and known magnetic properties, has been dispersed in oil (to avoid agglomeration [12]) and it has been inserted into a 1.5 mL conical plastic container. Several ferromagnetic samples with different iron compound contents have been realized, from 2.5 to 50 mg, and they have been inserted inside the PVC cranium. In a previous paper [9] a measurement campaign has been carried out in the presence and absence of the iron compounds, in order to estimate the variation of the RTD, confirming the sensor capability to distinguish different iron compound contents. In this work a particular attention has been reserved to the employment of the permanent magnet, as essential element in the experimental

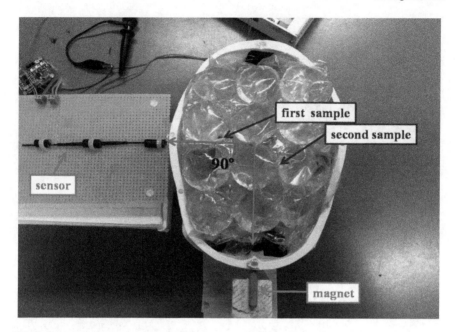

Fig. 4 Focus on placement of sensor and permanent magnet

setup; it is worth to highlight that the magnet generates a constant magnetic field and not variable, therefore it is not dangerous for the possible application in a human context. The presence of the magnet allows to polarize the ferromagnetic material that composes the ferromagnetic samples; therefore, a stronger and less noisy signal on the pick-up coil is expected. The pursuit to obtain more steady output signal is correlated to the necessity to use the magnetometer in an unshielded environment, as it can be a medical laboratory.

For this motive, the authors have decided to analyze the effect and the influence of the permanent magnet on the ferromagnetic samples and as consequence on the output signal, taking account of the same experimental setup. In Figs. 5 and 6 the RTD value as function of the iron compound content with and without magnet is shown.

As can be observed, the Gaussian curves present a decrement in terms of RTD as a function of the reduction of iron contents in both study cases. The zero target condition, without magnet in Fig. 5 and with magnet in Fig. 6, is indicated with the red curve (dotted line). It is intriguing to note that the Gaussian curves represent the distribution of the RTD values for each iron compound; in particular the presence of the magnet (see Fig. 6) allows to achieve a reduction of the Gaussian curves width, and therefore a minor dispersion of RTD values. Eventually, it is important to underline

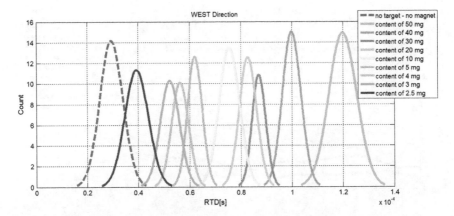

Fig. 5 RTD for several values of ferromagnetic samples (from 2.5 to 50 mg) without magnet

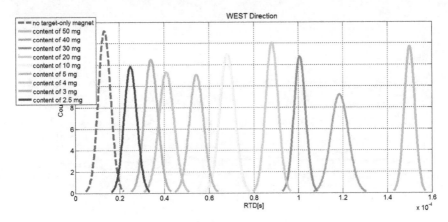

Fig. 6 RTD for several values of ferromagnetic samples (from 2.5 to 50 mg) with magnet [9]

that the Gaussian curves are well separate and discernible exactly where the magnet is used, whereas a remarkable overlapping is visible in Fig. 5. The same trend can be identified taking account of the RTD values as a function of the time. In detail, an acquisition time of 25 s has been established and the RTD values with and without magnet are presented respectively in Figs. 7 and 8. It is interesting to highlight that the signal is noisier with partial overlapping of the RTD values when the magnet is absent (see Fig. 7); opposite trend can be noticed in Fig. 8 where the RTD curves are distinct completely with a small exception between the concentrations of 3 and 4 mg.

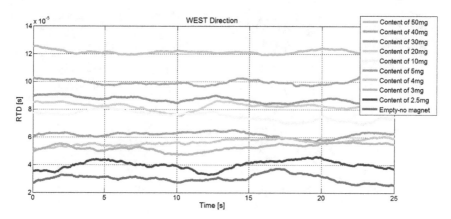

Fig. 7 RTD values as a function of the time without magnet

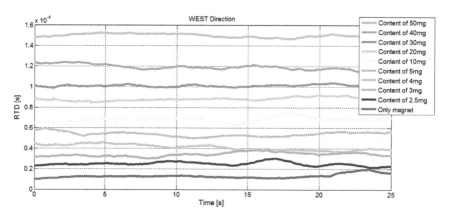

Fig. 8 RTD values as a function of the time with magnet

4 Conclusion

In this paper the innovative opportunity of using a Flexible RTD-Fluxgate magnetometer in monitoring of iron compound contents, that can be correlated to neurodegenerative diseases, has been presented. Furthermore the authors have paid attention to the introduction of a permanent magnet in the experimental setup in order to increase and to stabilize the output signal. The achieved experimental results confirm that the permanent magnet, generating a constant magnetic field of about 40 μT, is able to polarize the ferromagnetic samples and a stronger signal has received in the output in terms of RTD with a weaker fluctuation.

References

1. Langkammer, C., Krebs, N., Goessler, W., Scheurer, E., Ebner, F., Yen, K., Ropele, S.: Quantitative MR imaging of brain iron: a postmortem validation study. Radiology **257**(2), 455–462 (2010)
2. Mostile, G., Cicero, C.E., Giuliano, L., Zappia, M., Nicoletti, A.: Iron and Parkinson's disease: a systematic review and meta-analysis. Mol. Med. Rep. **15**(5), 3383–3389 (2017)
3. Liang, S.H., Holland, J.P., Stephenson, N.A., Kassenbrock, A., Rotstein, B.H., Daignault, C.P., Vasdev, N.: PET neuroimaging studies of [^{18}F]CABS13 in a double transgenic mouse model of Alzheimer's disease and nonhuman primates. ACS Chem. Neurosci. **6**(4), 535–541 (2015)
4. Dobson, J.: Magnetic iron compounds in neurological disorders. Ann. N. Y. Acad. Sci. **1012**(1), 183–192 (2004)
5. Yan, S.Q., Sun, J.Z., Yan, Y.Q., Wang, H., Lou, M.: Evaluation of brain iron content based on magnetic resonance imaging (MRI): comparison among phase value, R2* and magnitude signal intensity. PLoS ONE **7**(2), e31748 (2012)
6. Pimlott, S.L., Sutherland, A.: Molecular tracers for the PET and SPECT imaging of disease. Chem. Soc. Rev. **40**(1), 149–162 (2001)
7. Ando, B., Ascia, A., Baglio, S., Bulsara, A.R., Trigona, C., Pitrone, N.: Residence times difference (RTD) Fluxgate magnetometer for magnetic biosensing. In: AIP Conference Proceedings, vol. 1025, no. 1, pp. 139–149 (2008)
8. Ando, B., Ascia, A., Baglio, S., Bulsara, A.R., Trigona, C., In, V.: RTD fluxgate performance for application in magnetic label-based bioassay: preliminary results. In: 28th Annual International Conference of the IEEE EMBC 2006 Engineering in Medicine and Biology Society EMBS'06, pp. 5060–5063 (2006)
9. Trigona, C., Sinatra, V., Andò, B., Baglio, S., Mostile, G., Nicoletti, A., Zappia, M., Bulsara, A.R.: Measurement of iron compound content using a flexible core Fluxgate magnetometer at room temperature. IEEE Trans. Instrum. Meas. **67**(4), 971–980 (2018)
10. Andò, B., Baglio, S., La Malfa, S., Trigona, C., & Bulsara, A.R.: Experimental investigations on the spatial resolution in RTD-fluxgates. In: Instrumentation and Measurement Technology Conference, 2009, I2MTC'09, pp. 1542–1545 (2009)
11. Goo, K.G.: The normal and pathological physiology of brain water. Advances and Technical Standards in Neurosurgery, vol. 23, pp. 47–142. Springer, Vienna (1997)
12. Popplewell, P., Rosensweig, R.E.: Magnetorheological fluid composites. J. Phys. D Appl. Phys. **29**(9), 2297–2303 (1996)

Micro Doppler Radar and Depth Sensor Fusion for Human Activity Monitoring in AAL

Susanna Spinsante, Matteo Pepa, Stefano Pirani, Ennio Gambi and Francesco Fioranelli

Abstract Among the older adults population, falls represent a serious health problem, and a considerable economic issue for the society as a whole, due to their many consequences. In order to design reliable systems for automatic fall detection, able to distinguish falls from activities of daily living, the sensor fusion approach may be exploited. In this paper, the quite innovative fusion of Micro Doppler Radar and Kinect sensors to achieve acceptable accuracy and sensitivity in fall detection is investigated. The results show that by fusion, it is possible to provide a 100% fall detection sensitivity, over a dataset collected by taking into account ten different actions, with proper configuration of the acquisition setup and algorithmic parameters.

Keywords Fall detection · Sensor fusion · Depth · Micro doppler radar

1 Introduction

Since the proportion of elderly people is increasing worldwide, and many of them prefer living at home rather than in nursing homes, a wide range of technology-based applications have been developed to assist elderly in their own premises, within the Ambient Assisted Living (AAL) domain. When dealing with older adults, one of the most crucial issues to tackle is the risk of fall. As reported by the World Health Organization (WHO) in [1], approximately 37.3 million falls that are severe enough to require medical attention occur each year. Falls represent either a serious health

S. Spinsante (✉) · M. Pepa · S. Pirani · E. Gambi
Dipartimento di Ingegneria dell'Informazione, Universita' Politecnica delle Marche,
60131 Ancona, Italy
e-mail: s.spinsante@staff.univpm.it
URL: https://www.dii.univpm.it

F. Fioranelli
School of Engineering, University of Glasgow, Glasgow G12 8QQ, UK

© Springer Nature Switzerland AG 2019
B. Andò et al. (Eds.) *Sensors*, Lecture Notes in Electrical Engineering 539,
https://doi.org/10.1007/978-3-030-04324-7_62

problem for the person involved, and a considerable economic issue for the society as a whole. Among non-fatal falls, the largest morbidity occurs in people aged 65 years or older, with the reduction in physical activity, muscle atrophy and less social interactions. As such, systems for automatic fall detection and notification (like those exploiting inertial sensors, infrared, vibration, acoustic, and magnetic sensors, video cameras, RGB-Depth (RGB-D), and radar sensors) can provide a valuable support to alleviate the impact of falls on elderly's quality of life. Additionally, reliable monitoring systems can be beneficial to comprehensively evaluate the pattern of life of an individual. This includes, for instance, how active the person is, how often he/she moves in different locations of the house and what activities are performed, in particular the so-called Activities of Daily Living (ADLs). Irregularities in the pattern of life of an individual can be used for early detection of deteriorating health conditions (e.g. initial symptoms of dementia), providing the opportunity for timely and more effective treatment [2, 3] when integrated to classical clinical health information [4], and even enabling fall prevention.

Within the domain of non-wearable approaches to home monitoring, proposals adopting RGB-D sensors have gained interest in recent years, as they enable event detection without infringing the monitored subject's privacy. However, RGB-D sensors suffer from limited detection range. Additionally, depth sensors using the structured light approach are prone to changing light conditions and possible destructive interference, in multiple sensors configuration [5]. For these reasons, RGB-D sensors alone may not suffice in providing enough data to reliably understand human actions, so in this paper we investigate the fusion of RGB-D and Micro Doppler Radar (MDR) sensors, to provide a reliable monitoring solution aimed not only at fall detection, but at ADLs recognition [6, 7] too. The radar part may provide longer detection ranges and insensitivity to light conditions; the MDR sensor, by nature, does not record images of any type so it keeps the privacy-preserving feature of the system. The fusion approach [8] applied to MDR and depth sensors is quite novel, and most of the literature is from the co-authors; fusion is performed at features level, by concatenating the features obtained from both the sensing systems, before applying the fall detection algorithm.

The paper is organized as follows: in Sect. 2, materials and methods are presented, with details about the sensors used, the test protocol and setup for datasets collection, and the fall detection algorithm applied on the data. Sect. 3 describes and discusses the experimental results obtained; finally, Sect. 4 concludes the paper.

2 Materials and Methods

2.1 Sensors and Data Acquisition Setups

As already anticipated in the former section, two sensors have been chosen in this research: a Micro-Doppler Radar (MDR), and an RGB-Depth Kinect version 2,

(a)   **(b)**

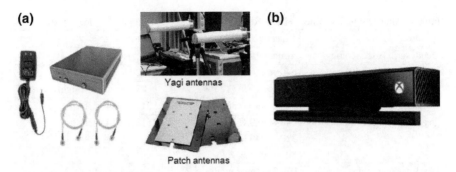

Yagi antennas

Patch antennas

Fig. 1 Sensors used: **a** MDR with Yagi or centered patch (Ancortek 5.8 GHz) antennas, **b** RGB-D Kinect v2

shown in Fig. 1a and b, respectively. RGB-D and radar systems show many important advantages compared with other technologies, especially because they are more easily accepted by the assisted person. First of all, they preserve privacy, since the RGB-D sensor uses only depth data in this context and the radar sensor does not record images of any type by its nature. Moreover, being ambient sensors, they are not cumbersome, and there is no risk that the person forgets to wear them. Radar sensing technology may rise concerns on possible hazards posed by electromagnetic radiation. However, commercial sensors use power levels below 20 dB_m, comparable with the power transmitted by conventional Wi-Fi routers and smartphones. Moreover, perceived risks should be traded off with benefits of continuous monitoring. Nevertheless, these sensors have some intrinsic limitations. For instance, in some conditions, the Doppler information obtained with the radar is not sufficient but, on the other hand, the radar, unlike the RGB-D sensor, has a longer detection range and is insensitive to varying light conditions [6]. In this sense, a sensor fusion-based approach at features level is considered, to exploit the advantages of both the sensors and compensate for their respective limitations.

Micro Doppler Radar Most of the Radar sensors operate in the range of C-band (5.8 GHz), X-band (8 GHz) and K-band (24 GHz). The Doppler shift (DS) is proportional to the carrier frequency of the signal, hence as the frequency increases, the μ-DS becomes more suitable for feature extraction and classification. In this research, an off-the-shelf Frequency Modulated Continuous Wave (FMCW) radar system (Ancortek SDR 5.8 GHz) in the C-band has been adopted [9]. The bandwidth equals 400 MHz and the chirp duration of 1 ms leads to an unambiguous Doppler frequency range of ± 500 Hz, sufficient to capture the whole human μ-DS for indoor activities. Power transmitted by the radar sensor is approximately 19 dB_m. The graphical user interface (SDR-GUI) provided with the hardware has been used for data acquisition, which allows the user to select the operating frequency and parameters, signal waveforms, filter types and data recording mode. It also provides a graphical representation of signals in time domain, frequency domain, combined time and frequency domain in real time. Yagi and patch antennas, both with vertical polarization,

Table 1 Main features of Kinect v1 and Kinect v2 [6]. For uncertainty in (*) refer to [10]

Feature	Kinect v1	Kinect v2
Depth sensing technology	Structured light	Time of flight
RGB image resolution	640 × 480 @15/30 fps	1920 × 1080 @30 fps
	1280 × 960 @12 fps	(15 fps with low light)
IR image resolution	640 × 480 @30 fps	512 × 424 @30 fps
Depth sensing resolution	640 × 480 @30 fps	512 × 424 @30 fps
	320 × 240 @30 fps	
	80 × 60 @30 fps	
Depth sensing range	0.4–3 m (near mode)	0.5–4.5 m (*)
	0.8–4 m (normal mode)	
Field of view	57° ± 5° horizontal	70° ± 5° horizontal
	43° ± 5° vertical	60° ± 5° vertical
Skeleton tracking	Skeleton with 20 joints	Skeleton with 25 joints
	Up to 2 subjects	Up to 6 subjects

are commonly used for data acquisition. The former are expensive and cumbersome, but have a long detection range and narrow beam-width, instead the latter are cost-effective and compact, but have a short detection range and wide beam-width. In particular, Yagi antennas have a 17 dB$_i$ gain and beam-width of 24° ± 2° in azimuth and elevation, while patch antennas have 12 dB$_i$ gain and 40° ± 2° in azimuth and elevation. For this reason, in all recordings the former have been adopted.

RGB-Depth Kinect In 2010 Kinect was officially launched, as a camera-based control method for games. Almost immediately, hackers and independent developers started to create open source drivers for this sensor, to use it in different ways and applications. Two versions of Kinect succeeded over the years: Kinect v1 and Kinect v2. Table 1 reports main characteristics of both. Kinect is actually a multi-sensory device. The depth sensor consists of an IR projector combined with an IR camera: Kinect v1 exploits a structured light approach, whereas Kinect v2 uses the Time of Flight (ToF) principle. In this study, the skeleton joints 3D coordinates provided by a Kinect v2 sensor have been used for processing.

2.2 Test Protocol and Dataset Collection

Three datasets have been recorded in different periods and different environments. From a chronological point of view, the first dataset (named UNIVPM) has been created at Universita' Politecnica delle Marche (Ancona, Italy), in December 2016. The second and third datasets (named respectively UofG I and UofG II) have been created in two different office environments inside the School of Engineering at University of Glasgow (Glasgow, UK), in March and June 2017 respectively. In all datasets, 10

Table 2 The ten activities under investigation

No.	Activity	Description
1	Walking	The act of walking back and forth at a normal speed while moving arms and legs
2	Walking while carrying an object	The act of walking back and forth at a normal speed while holding an object with both hands
3	Sitting down	The act of sitting down on a chair from upright position
4	Standing up	The act of standing up from a chair
5	Bending and picking up an object	The act of performing a step and then bending to pick up an object from the ground with one hand
6	Bending and tying shoelaces	The act of bending on knees and tying laces of one shoe with both hands
7	Drinking	The act of holding a cup on one hand and making one or more sips
8	Talking on phone	The act of extracting the phone from the pocket in the trousers and answering an incoming call
9	Tripping and falling	The act of simulating a trip and subsequent front fall on the ground
10	Checking under the bed	The act of bending on knees and place the head next to the ground to check something under a piece of furniture

actions, described in detail in Table 2, were recorded. Recording time was 5 seconds for actions from 3 to 9, and 10 seconds for actions 1, 2 and 10. The decision to keep the same actions in all tests is motivated by different reasons: it ensures homogeneity and interoperability between the datasets; the scientific community agrees on them as examples of daily activities; they represent a challenge for classification algorithms, being very similar to each other, especially in couples. Activity 9, that corresponds to a fall, is the one we are interested in recognizing with the best accuracy.

In UNIVPM, Kinect and MDR devices were placed in front of the actor performing the action. Transmitting and receiving antennas of the radar are placed in a monostatic manner (Fig. 2a). A total of 7 volunteers, 4 males and 3 females, aged 23–40, were recruited to repeat all the actions 3 times, to increase intraclass variability. One actor repeated the measurements with Kinect and radar in a side configuration. As far as Kinect is concerned, only the skeleton joints 3D coordinates are actually used for processing. However, all the available streams (depth, skeleton, RGB, IR) have been acquired, to have either a ground truth reference of the performed action, and a complete dataset. As far as the MDR is concerned, the operating frequency was

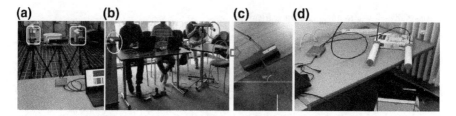

Fig. 2 Laboratory setups for dataset acquisition: **a** UNIVPM dataset. In the yellow boxes (left and right), the bi-static MDR by Ancortek; in the red box (center), the RGB-D sensor; **b** UofG I: the radar system configuration is bistatic, as the transmitter (antenna inside yellow circle) and receiver (antenna inside blue circle) are placed in two distinct positions. The Kinect is inside the red circle; **c** UofG I: the axis of Kinect and the direction of movement of actors are 60° inclined; **d** UofG II: the radar is in the monostatic configuration and the Kinect is in the front view configuration, as in UNIVPM

5.8 GHz, FMCW Sawtooth was selected as operating mode, bandwidth was fixed at 400 MHz and sweep time at 1 ms.

In UofG I, actions were performed only once, but the experiment was enriched with participation of more volunteers: 16 actors (12 males and 4 females), aged 23–58. To further test the robustness of the algorithms in a more realistic manner, Kinect and radar data were acquired with sensors in an unfavourable position: the axis of the Kinect was 60° rotated with respect to the actors (Fig. 2b, c). This emulates a possibly common situation in everyday life, since the probability of having an angle of 0° or 90° is close to zero. As far as the radar is concerned, the bistatic configuration was used. Differently from the monostatic case, in which the antennas are close in space and oriented in the same direction in front of the actor, in the bistatic case an angle is created between the receiver and the transmitter. As a consequence, the Doppler shift is reduced, the signal-to-noise ratio (SNR) is reduced too, and the discrimination between activities becomes more challenging. Additionally, 4 actors repeated the measurements with the radar antennas in a monostatic configuration. This way, a fair comparison with the UNIVPM dataset can be made and the influence of the recording environment can be investigated. In UofG II, 3 repetitions were performed for each activity by 8 actors (5 males and 3 females), aged 24–30; the bistatic configuration of the MDR and front view of Kinect were restored, as shown in Fig. 2d.

2.3   Fall Detection Algorithm

In this research, once the 3D coordinates from 25 joints at each frame are collected, the approach based on temporal pyramid of key poses, developed by Cippitelli et al. and described in [11–13], is applied. The algorithm is structured in four main steps, as shown in Fig. 3 and detailed here:

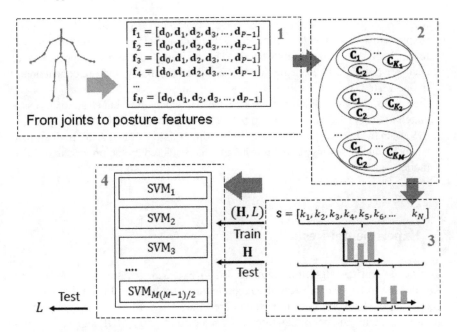

Fig. 3 Main steps of the fall detection algorithm

- Step 1 *Extraction of posture features*: Vectors $\mathbf{p}_1, \mathbf{p}_2, \ldots, \mathbf{p}_N$, containing the 3D coordinates of all joints respectively in frames $1, 2, \ldots, N$ (N = number of frames = FR·T, being FR the frame rate and T the acquisition time), are extracted for each action, being $\mathbf{p}_i = [\mathbf{J}_0, \mathbf{J}_1, \ldots, \mathbf{J}_{P-1}]$ (P = number of joints = 25) and $\mathbf{J}_j = [J_x, J_y, J_z]$ ($\mathbf{J}_i$ = i-th joint). Then, for each individual action, a sequence $\mathbf{F}$ of feature vectors $\mathbf{f}_1, \mathbf{f}_2, \ldots, \mathbf{f}_N$ is extracted. Each $f_i = [d_1, d_2, \ldots, d_{25}]$, being each $\mathbf{d}_j$ the normalized Euclidean distance between each joint ($\mathbf{J}_j$) and the skeleton center of mass ($\mathbf{J}_{cm}$);
- Step 2 *Codebook generation and key poses substitution*: The feature vectors sequences $\mathbf{F}_i$ ($i = 1, 2, \ldots, t_h$, where t_h = number of recordings for action h, with $h = 1, 2, \ldots$, M, and M the number of actions, i.e. the number of classes), representing different recordings of the same kind of action (i.e. different repetitions by one actor or recordings from different actors), are grouped in the same class. The training instances $\mathbf{F}_i$ of the h-th class are then clustered in K_h key poses, represented by the cluster centers $C_1, C_2, \ldots, C_{K_h}$. A codebook is obtained by merging all the key poses obtained for each class. The dimension of this codebook will be therefore $K_1 + K_2 + \cdots + K_M$. At the end of this step, an action, previously represented by a sequence of feature vectors $\mathbf{F} = [\mathbf{f}_1, \mathbf{f}_2, \ldots, \mathbf{f}_N]$, is encoded by a sequence of key poses $\mathbf{S} = [k_1, k_2, \ldots, k_N]$. To reach this goal, for each frame, the specific displacement of joints is associated with the closer key pose;

- Step 3 *Histograms of key poses and temporal pyramid*: A sequence of key poses **S** is split into 2^{l-1} segments, at the l-th level of a temporal pyramid constituted by L levels. Each time a histogram is obtained by counting the number of appearance of each key pose within the segment. At the end, all the histograms are concatenated in one vector **H**;
- Step 4 *Classification*: Each set of histograms **H**, which represent an action, is associated to the corresponding class label (L), using binary Support Vector Machine (SVM) in the one-versus-one modality. During training, **H** and L are both known, whereas, during testing, **H** is known and L is predicted by the previously trained model.

3 Experimental Results

In order to run the algorithm summarized in Sect. 2.3, features from MDR and Kinect have to be selected. About Kinect, features are extracted from the 3D coordinates of 25 skeleton joints at each captured frame, according to step no. 1 of the algorithm; a total number of 22 features are selected from the MDR signals (texture-based ones, spectrogram derived ones, and single-value-decomposition features). Results are given in terms of *accuracy*, defined as the ratio between the number of correctly classified events over the total amount of classified events, and *fall detection (FD) sensitivity*, given as the ratio between the number of correctly classified falls and the total number of falls detected. If we define the classification of an event: TP as true positive, FN as false negative, FP as false positive, and TN as true negative, then $accuracy = (TP + TN)/(TP + FN + FP + TN)$ and $FD\,sensitivity = TP/(TP + FN)$.

Experimental results have been evaluated over the UofG I dataset, being the richest one. First, we obtained the so-called confusion matrix from the Kinect sensor alone, using a Linear Discriminant Classifier, over all the activities (none removed) and 8 clusters in the algorithm. The reported *accuracy* was 85.2% and the *FD sensitivity* was 94%. Figure 4a shows the corresponding confusion matrix. From Table 3 it is possible to check the mean *accuracy*, and the percent values of correctly detected

Table 3 Mean *accuracy*, and percent values of correctly detected falls, over 10 actions, with Kinect sensor

No. actors Training/testing	mean *accuracy*	Correctly detected Falls (%)
11/5	0.52	64.00
12/4	0.73	93.33
13/3	0.80	92.00
14/2	0.84	**100.00**

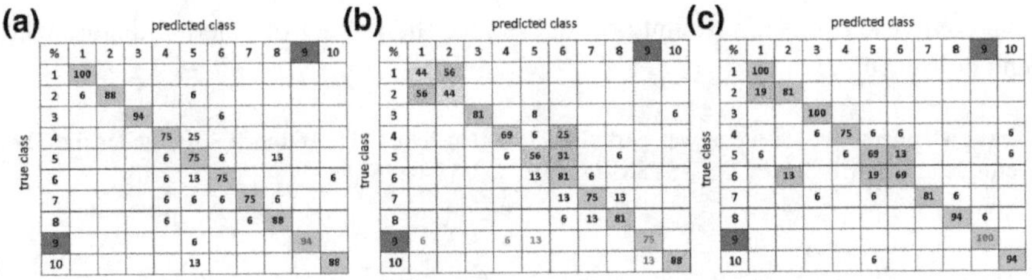

Fig. 4 Confusion matrices obtained by: **a** Kinect only, **b** MDR only, **c** Kinect and MDR fusion

falls, for different combinations of the number of actors used for training and testing the classifier, over all the 10 activities. Combination 14/2 gives a 100% fall detection.

In Fig. 4b, the confusion matrix obtained by applying the same algorithm on features computed over the MDR signals is shown. It is quite evident, from the cells with light red background, that the amount of activities that are not correctly classified by the algorithm increases with respect to the use of the Kinect device. In particular, the algorithm is not good in correctly discriminating activities 1 and 2, i.e. walking and walking while carrying an object, respectively. Similarly, the activity recognition fails for activities 4 and 5, that are both recognized as being the 6 one. As a consequence, the resulting *accuracy* equal to 69.4%, and the *FD sensitivity* equal to 75%, are remarkably smaller than in the Kinect case. Finally, the effects of sensor fusion at features level are visible in Fig. 4c, where the confusion matrix resulting from both Kinect and MDR sensors is provided. Feature vectors from different sensors are concatenated and given as input to the same classifier, to produce a more diverse feature space. Both *accuracy* and *FD sensitivity* feature a sensitive increase, up to 86.3% the former, and up to 100% the latter. This means that given the test dataset and the configuration previously described, the algorithm is able to always classify correctly the fall events related to execution of activity no. 9. Independently from the classifier, results are always better after fusion.

4 Conclusion

Fall detection systems are central in modern healthcare, because of the devastating effects that falls have on patients, the increased burden on family members, and on the social welfare as a whole. Such systems must be able to detect a fall independently on its dynamics, and must have both high sensitivity and specificity: the former means that most falls are detected, and the latter means that false alarms are minimal. In this work, we investigated the possibility to exploit the fusion among Micro Doppler Radar and Kinect sensors to achieve acceptable accuracy and sensitivity in fall detection. This type of approach is quite innovative with respect to the literature, and confirms the opportunity to take advantage from the fusion at features level, to overcome each sensor's limitations and reduce the impact of external disturbances and noise sources.

Acknowledgements M. Pepa was partially supported by COST Action IC1303 under the Short Term Scientific Mission ECOST-STSM-IC1303-200317-084684.

References

1. World Health Organization: falls—fact sheet (2016). http://www.who.int/mediacentre/factsheets/fs344/en/ (last accessed: April 29th, 2018)
2. Koenig, A., et al.: Validation of an automatic video monitoring system for the detection of instrumental activities of daily living in dementia patients. J. Alzheimer's Dis. **44**(2), 675–685 (2015)
3. Daponte, P., De Marco, J., De Vito, L., Pavic, B., Zolli, S.: Electronic measurements in rehabilitation. In: 2011 IEEE International Symposium on Medical Measurements and Applications, pp. 274–279, May 2011
4. Garcia, N.M., Garcia, N.C., Sousa, P., Oliveira, D., Alexandre, C., Felizardo, V.: TICE.Healthy: A perspective on medical information integration. In: 2014 IEEE-EMBS International Conference on Biomedical and Health Informatics, BHI 2014, art. no. 6864403, pp. 464–467 (2014)
5. Martín Martín, R., Lorbach, M., Brock, O.: Deterioration of depth measurements due to interference of multiple RGB-D sensors. In: 2014 IEEE/RSJ International Conference on Intelligent Robots and Systems, Chicago, IL, pp. 4205–4212 (2014)
6. Cippitelli, E., Fioranelli, F., Gambi, E., Spinsante, S.: Radar and rgb-depth sensors for fall detection: a review. IEEE Sens. J. 17, no. 12, 3585–3604 (2017)
7. Li, H., Shrestha, A., Fioranelli, F., Le Kernec, J., Heidari, H., Pepa, M., Cippitelli, E., Gambi, E., Spinsante, S.: Multisensor data fusion for human activities classification and fall detection. In: 2017 IEEE Sensors Proceedings, Glasgow, UK, pp. 1–3 (2017)
8. Luo, R.C., Yih, C.C., Lan Su, K.: Multisensor fusion and integration: approaches, applications, and future research directions. IEEE Sens J **2**(2), 107–119 (2002)
9. Smith, G.E., Woodbridge, K., Baker, C.J.: Radar micro-doppler signature classification using dynamic time warping. IEEE Trans. Aero. Electron. Sys. **46**(3), 1078–1096 (2010)
10. Yang, L., Zhang, L., Dong, H., Alelaiwi, A., Saddik, A.E.: Evaluating and improving the depth accuracy of kinect for windows v2. IEEE Sens. J. **15**(8), 4275–4285 (2015)
11. Cippitelli, E., Gambi, E., Spinsante, S., Florez-Revuelta, F.: Evaluation of a skeleton-based method for human activity recognition on a large-scale RGB-D dataset. In: 2nd IET International Conference on Technologies for Active and Assisted Living (TechAAL 2016), London, UK, pp. 1–6 (2016)
12. Cippitelli, E., Gambi, E., Spinsante, S., Florez-Revuelta, F.: Human action recognition based on temporal pyramid of key poses using RGB-D sensors. In: Blanc-Talon J., Distante C., Philips W., Popescu D., Scheunders P. (eds.) Advanced Concepts for Intelligent Vision Systems. ACIVS 2016. Lecture Notes in Computer Science, vol. 10016. Springer, Cham (2016)
13. Cippitelli, E., Gasparrini, S., Gambi, E., Spinsante, S.: A human activity recognition system using skeleton data from RGBD sensors. Comput. Intell. Neurosci. **2016**, Article ID 4351435, 14 pages (2016). https://doi.org/10.1155/2016/4351435

Characterization of Human Semen by GC-MS and VOC Sensor: An Unexplored Approach to the Study on Infertility

Valentina Longo⑩, Angiola Forleo⑩, Sara Pinto Provenzano,
Daniela Domenica Montagna, Lamberto Coppola, Vincenzo Zara⑩,
Alessandra Ferramosca⑩, Pietro Siciliano⑩ and Simonetta Capone⑩

Abstract Infertility is one of principal health and social problems of this century. Male factors are involved in half of the cases and often the alteration concerns sperm motility. Seminogram is the gold standard technique for semen analysis, but it presents several limits. For this reason, we propose a new method for discriminating asthenozoospermic samples (low sperm motility) from normozoospermic ones (progressive motility > 32%) based on the never explored analysis of the volatile metabolites in the headspace of human semen sample by Gas Chromatograph (GC) equipped with two detectors: a Mass Spectrometer (MS) and a metal oxide based gas sensor sensitive (MOX) to Volatile Organic Compounds (VOCs). VOC sensor signal profiles (resistance vs. time) showed a higher sensitivity to specific organic classes such as aldehydes and ketones. The sensorgrams were preprocessed and analysed by PLS-DA. The results showed that sensorgrams analysis by suitable bioinformatics techniques has a good discrimination power and could support physiological parameters in human semen assessment. The analysis of the human semen Volatilome may be a proof-of-concept for the development of a novel micro-GC device with a sensor array detector, a potential candidate for infertility assessment in clinical practice.

Keywords Human semen · VOC sensor · GC/MOX sensor system

V. Longo (✉) · A. Forleo · P. Siciliano · S. Capone
National Research Council of Italy, Institute for Microelectronics and Microsystems (CNR-IMM),
Lecce, Italy
e-mail: valentina.longo@le.imm.cnr.it

S. P. Provenzano · D. D. Montagna · L. Coppola
Biological Medical Center "Tecnomed", Nardò (Lecce), Italy

V. Zara · A. Ferramosca
Department of Environmental and Biological Sciences and Technologies, University of Salento,
Lecce, Italy

© Springer Nature Switzerland AG 2019   529
B. Andò et al. (eds.), *Sensors*, Lecture Notes in Electrical Engineering 539,
https://doi.org/10.1007/978-3-030-04324-7_63

1   Introduction

Infertility is defined as the inability of a couple to achieve spontaneous pregnancy after 1 year of regular, unprotected sexual intercourse [1]. Almost 8–12% of couples in reproductive age is affected by this global health issue [2]. Male factors contribute to approximately 50% of all cases of infertility and male infertility alone accounts for approximately one-third of all infertility cases.

The diagnostic ability of available male fertility investigative tools is limited.

Semen analysis (SA), also known as "seminogram", is the gold standard technique for determining men's fertility [3]. Unfortunately, SA provides limited information and cannot discriminate fertile from infertile men on an individual basis [4]. Moreover, widely overlapping ranges of seminal parameters have left clinicians in search of better seminal biomarkers. Therefore, new technologies are needed since a high percentage (ranging from 6 to 27%) of infertile men show normal semen parameters.

Advanced research allowed to explore new potential biomarkers for specific alterations in fields of genomics, proteomics and metabolomics, featuring the so-called "omics" era. The researchers are employing these novel biomarkers from blood, urine and breath [5, 6].

Through knowledge of spermatozoa metabolomics, we can reach a comprehensive understanding of what is happening in the cells and find the key of molecular mechanisms that regulate their biology [7].

In particular, untargeted metabolomic profiling is a powerful tool for biomarker discovery by observing the changes in metabolite concentrations of various biofluids and by identifying altered metabolic pathways [8]. An important branch of metabolomics, unexplored for human semen, is volatilomics. Volatilome is the totality of Volatile Organic Compounds (VOCs) derived from cellular metabolism and, recently, it is considered useful to a better understanding of physiological and pathophysiological processes [9].

In this work, we propose a new technique based on the application of a MOX sensor, used as additional detector after gas chromatograph (GC) separation, to evaluate metabolome profile of human semen applied to infertility study. Mass spectrometer (MS), which is the master detector of the system, is here used exclusively to identify the compounds detected by the VOC sensor. The two detectors work in parallel by a two-way splitter, that splits the helium (He) flow, eluting from the chromatographic column, in 1:1 ratio towards the two detectors through two segments of defunctionalized chromatographic column.

For data interpretation, statistical analysis was carried out by PLS-DA, the most known tool to perform classification and regression in metabolomics. One of the main advantages of PLS-DA is that it has the ability to analyze highly collinear and noisy data [10]. This method is particularly appropriate for the analysis of large, highly-complex data sets as are the sensor resistance profiles versus time (labelled "sensorgrams") of all the samples. In addition, PLS-DA applied to sensorgrams may be an useful tool for generating parsimonious models through feature selection and data reduction, as well as providing more predictive results.

2 Materials and Methods

The study was focused on 57 reproductive-age men. First, semen analysis (semino-gram) was carried out on semen samples collected from subjects at the Biological Medical Center "Tecnomed" (Nardò, Lecce—Italy). Next, on the same semen sam-ples, headspace VOC analysis was performed by the GC/[MS + sensor] system.

Initially, semen samples were classified based on sperm motility in two groups: asthenospermic, with a progressive motility (PR) < 32% and normozoospermic (PR > 32%) samples. Subsequently, asthenozoospermic samples have been splitted in severe and slight asthenozoospermic ones.

The seminal VOCs were extracted by Solid-Phase Microextraction (SPME) tech-nique; SPME was carried out by a Carboxen®/Polydimethylsiloxane (CAR/PDMS) fiber (cod. 57318, Supelco) which was exposed to each human semen sample headspace overnight. The GC-MS analysis of the extracted seminal volatiles was performed using a GC (6890 N series, Agilent Technologies) coupled to a MS (5973 series, Agilent Technologies) equipped with a ZB-624 capillary column (Phe-nomenex) with the injector temperature set at 250 °C to allow thermally desorption of VOCs. The carrier gas was high purity helium with a flow rate of 1 ml/min.

GC/MS analysis was carried out in full-scan mode with a scan range 30–500 amu at 3.2 scans/s. Chromatograms were analysed by Enhanced Data Analysis software and the identification of the volatile compounds was achieved by comparing mass spectra with those of the data system library (NIST14, p > 80%).

For GC/sensor analysis the capillary column was inserted, by a splitter (Agilent G3180B Two-Way Splitter), into a tiny chamber hosting a VOC sensor (MiCS-5521, e2v technologies, UK) and it was positioned near sensor surface. The MOX sensor operating temperature was 400 °C. The MOX sensor traces (resistance vs. time, i.e. "sensorgrams") were used for data analysis. The sensorgram data obtained by GC/MOX sensor experiments and physiological data were elaborated by multivariate data analysis techniques using web-based tools available with open access server MetaboAnalyst (version 4). Partial least-squares discriminant analysis (PLS-DA) were applied to the preprocessed sensorgrams. In particular, to ensure that each donor profile is on the same scale, resistance profiles were standardized using range-scaling between 0 and 1 [11].

3 Results

Seminograms allow us to classify the 57 samples in three motility groups. We studied 16 human sperm samples with high motility, 20 samples with slight asthenozoosper-mia and 21 severe asthenozoospermic samples.

In this paper, the use of GC/MS system only aims to support sensor analysis and to identify the compounds inducing resistance variation.

In Fig. 1, overlapping of a chromatogram and the corresponding sensorgram (in relation to run time) for each sample group is shown. General background contamination, such as column bleed (signed by asterisks), is not perceived by VOC sensor.

Forty-nine total VOCs are detected by GC/MOX sensor system. Some of these (43%) are present only in one individual, while the remaining 57% in at least two semen samples.

Despite high variability, some compounds have a different expression in the three sample groups, whereas other compounds are present exclusively in a group. In particular, those VOCs which are closely related with severe asthenozoospermia are shown in Fig. 2.

2D- and 3D-PLS-DA results obtained from sensor data were compared with data obtained by seminograms, considering 2 groups (asthenozoospermic and high motility samples) and 3 groups classification models where severe and slight asthenozoospermic samples are included into two separate groups. 2D-PLS-DA based on semen analysis showed an accuracy of 85% and of 65%, in the discrimination in two and three groups, respectively. In three dimensions, accuracy increase up to 92% (two classes) and 76% (three classes) (Fig. 3, panel A).

Using exclusively sensor resistance profiles, accuracy is of 77% (in 2D) and 78% (in 3D) for discrimination in asthenozoospermic and normozoospermic samples, while when we considered three classes the accuracy decreases to 37% in two dimensions and to 40% in three dimensions (Fig. 3, panel B).

Finally, we have investigated MOX sensor responses to different classes of organic compounds comparing VOCs detected by GC/MS and GC/MOX systems. Our results (Fig. 4) show that VOC sensor has different affinity to organic compound classes, having a higher reactivity with aldehydes (72% of total VOCs), ketones and acetamides (50%).

4 Discussion

Overlay of chromatograms and sensorgrams (Fig. 1) allowed us to highlight that chromatographic contaminants are not a problem for sensor analysis. In fact, while for gas-chromatograms, peaks and ions corresponding to column bleed must be removed, resistance variations of sensor occur only at the elution of those VOCs to which the sensor is sensitive.

Through time-correspondence between peaks of two systems, we are able to determine the VOCs responsible for resistance variation. The percentages of volatile compounds unique of a single patient and occurring in more patients (57% vs. 43%, respectively) reflect the well-known biological variability of human semen. Human semen (such as other biofluids) is a concentrate of compounds derived from cellular metabolism, but also from diet, environment and lifestyle. Result is a "melting pot" of different molecules [5].

Nevertheless, some VOCs have a different expression in patients with asthenozoospermia or with high sperm motility. For example, 1H, 1,2,4-Triazole is present

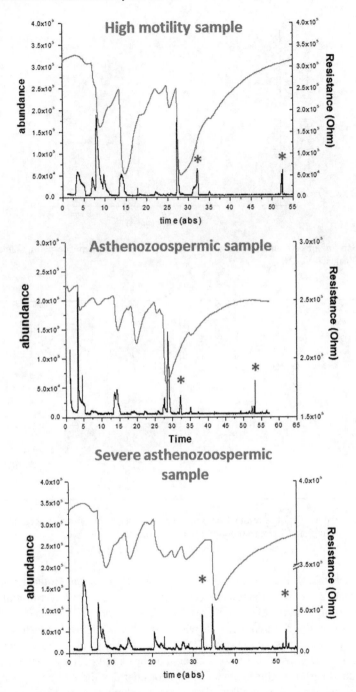

Fig. 1 Overlay of chromatograms and sensorgrams of one sample for each sample groups (high motility, asthenozoospermia and severe asthenozoospermia). Asterisk sign column bleed

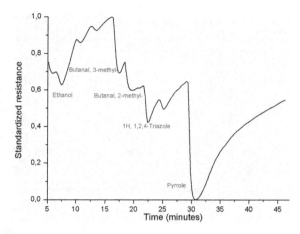

Fig. 2 Sensorgrams (variation of resistance vs. time) of a severe asthenozoospermic sample (PR = 0%) and matched VOCs

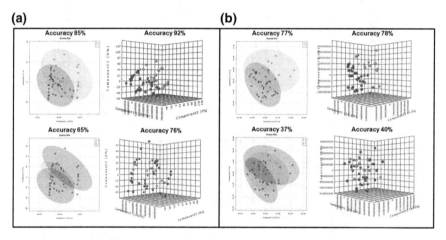

Fig. 3 2D- and 3D-PLS-DA results from seminogram (panel A) and from sensorgram (panel B) data. 2 groups (top) and 3 groups (bottom) discrimination were carried out

exclusively in severe asthenozoospermic samples (Fig. 3) and Butanal, 3-methyl- and Butanal, 2-methyl- are more concentrated in samples with lower motility.

After normalization [11], sample resistance profiles are used to perform PLS-DA and to compare discrimination results with those obtained by PLS-DA of physiological data from seminograms (Fig. 4). This latter present a value of accuracy higher than that PLS-DA based on sensor data both in the classification in two and in three sperm sample groups. This is obvious if we consider that three of parameters in our seminograms is based on motility, that represents the characteristic on the basis of which we have classified our samples.

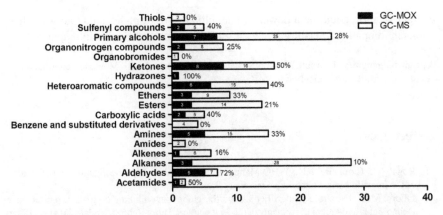

Fig. 4 Covering of VOCs detected by GC/MOS respect those identified by GC/MS system, in relation with organic compound classes

Anyway, the separation in asthenozoospermic and normozoospermic samples with sensorgram data is satisfactory (77–78%). To discriminate asthenozoospermic samples in severe and slight ones with a high accuracy, it is needful to increase the number of samples by a more extended experimental campaign.

VOC sensor showed a different affinity for organic classes, proving a higher response to aldehydes, ketones and acetamides (Fig. 4).

5 Conclusions

Despite very high biological variability of human semen samples, GC/MOX sensor system allowed us to develop a reproducible and novel method to detect human semen headspace VOCs based on which classify semen samples in relation with sperm motility. PLS-DA based on VOCs analysis by sensorgrams provided good classification between asthenozoospermic from normozoospermic samples, compared to PLS-DA based on sperm motility measured by seminogram (SA). This constitutes a first promising result in using semen volatilome detected by a MOX sensor in evaluate semen quality, and hence man fertility, by classification in standard classes.

Setup of a microGC coupled to VOC sensor could represent a new technology that physician and biologists could use to support gold standard method and improve the quality of the medical report on fertility by further additional information parameters.

Of course, further studies are required to extend this study to a larger sample population and to select the pattern of VOCs that can statistically discriminate asthenozoospermic from normozoospermic samples integrating VOC biomarkers analysis to current standard semen analysis. Moreover, large-scale semen volatilome characterization could be a powerful unexplored tool to understand pathway alterations in idiopathic infertility in which semen parameters are physiological.

Acknowledgements The authors would like to thank Dott. Luigi Montano, coordinator of "EcoFood-Fertility", for scientific support.

References

1. Rowe, P.J., Comhaire, F.H.: WHO Manual for the Standardized Investigation and Diagnosis of the Infertile Male. Cambridge University Press (2000)
2. Inhorn, M.C., Patrizio, P.: Infertility around the globe: new thinking on gender, reproductive technologies and global movements in the 21st century. Hum. Reprod. Update **21**(4), 411–426 (2015)
3. WHO: WHO Laboratory Manual for the Examination and Processing of Human Semen. 5th edn. Cambridge University Press (2010)
4. Bieniek, J.M., Drabovich, A.P., Lo, K.C.: Seminal biomarkers for the evaluation of male infertility. Asian J. Androl. **18**(3), 426–433 (2016). https://doi.org/10.4103/1008-682X.175781
5. Amann, A., De Lacy Costello, B., Miekisch, W., Schubert, J., Buszewski, B., Pleil, J., Ratcliffe, N., Risby, T.: The human volatilome: volatile organic compounds (VOCs) in exhaled breath, skin emanations, urine, feces and saliva. J. Breath Res. **8**(3), 034001 (2014). https://doi.org/10.1088/1752-7155/8/3/034001
6. Capone, S., Tufariello, M., Forleo, A., Longo, V., Giampetruzzi, L., Radogna, A.V., Casino, F., Siciliano, P.: Chromatographic analysis of VOC patterns in exhaled breath from smokers and nonsmokers. Biomed. Chromatogr. **32**(4) (2017). https://doi.org/10.1002/bmc.4132
7. Aitken, R.J., Henkel, R.R.: Sperm cell biology: current perspectives and future prospects. Asian J Androl. **13**(1), 3–5 (2011)
8. Kovac, J.R., Pastuszak, A.W., Lamb, D.J.: The use of genomics, proteomics, and metabolomics in identifying biomarkers of male infertility. Fertil. Steril. **99**(4), 998–1007 (2013)
9. Costello, B.D., Amann, A., Al-Kateb, H., Flynn, C., Filipiak, W., Khalid, T., Osborne, D,. Ratcliffe, N.M: A review of the volatiles from the healthy human body. J. Breath Res. **8**(1) (2014)
10. Gromski, P.S., Muhamadali, H., Ellis, D.I., Xu, Y., Correa, E., Turner, M.L., Goodacre, R.: Metabolomics and partial least squares-discriminant analysis—a marriage of convenience or a shotgun wedding. Anal. Chim. Acta **879**, 10–23 (2015)
11. Weber, C.M., Cauchi, M., Patel, M., Bessant, C., Turner, C., Britton, L.E., Willis, C.M.: Evaluation of a gas sensor array and pattern recognition for the identification of bladder cancer from urine headspace. Analyst **136**(2), 359–64 (2011). https://doi.org/10.1039/c0an00382d

A Novel Technique to Characterize Conformational State of the Proteins: p53 Analysis

Saad Abdullah, Mauro Serpelloni, Giulia Abate and Daniela Uberti

Abstract As the technology is advancing, biotechnologist and pharmacologist seems more interested and focused towards the development of innovative sensing solution/technology capable of evaluating proteins without any limitations of time and cost which were encountered/offered by conventional/traditional methods such as ELISA used for protein quantification. To allow continuous monitoring and attain protein sample information in a non-invasive way, spectrophotometry might be considered as an alternate method which analyzes different conformational states of proteins by closely observing the variation in optical properties of the sample. The work presented studies p53 protein conformational dynamics and their involvement in various pathophysiological and neurodegenerative disease/disorders using the spectrophotometer-based method. By utilizing the technique of spectrophotometry, investigations were carried out on three samples containing varied molecular state of p53 (Wild p53, Denatured p53, and Oxidized p53), to detect the difference in light absorption. Overall, this proposes the possibility of a simple, non-invasive and optical based method capable of detecting and identifying different structural states of p53 while overcoming the complexities offered by the conventional procedures.

Keywords Spectrophotometery · p53 · Protein · Absorbance spectrum
Infrared spectra · Protein detection

1 Introduction

The advancements in technology represent an active research area for the study of protein conformational dynamics which in turn can give valuable information regarding certain chronic diseases, like diabetes, cancer, and neurodegenerative disorders.

S. Abdullah (✉) · M. Serpelloni
Department of Information Engineering, University of Brescia, Brescia, Italy
e-mail: s.abdullah@unibs.it

G. Abate · D. Uberti
Department Molecular and Translational Medicine, University of Brescia, Brescia, Italy

© Springer Nature Switzerland AG 2019
B. Andò et al. (eds.), *Sensors*, Lecture Notes in Electrical Engineering 539,
https://doi.org/10.1007/978-3-030-04324-7_64

Therefore, efforts are made to combine newest achievements in the fields of material science, mathematics, engineering, and bioinformatics to design and develop a non-invasive technique for the purpose of investigating various protein conformational states [1]. Since a strong correlation exists between protein conformational states and biological functions [2], biomedical research in the domains of neurology and generiatics focuses at an early stage diagnosis of pathology via reliable identification of biomarkers, i.e., proteins [3].

Among the proteins, which experience specific conformational states to play a role in specific biological functions, the tumor suppressor phosphoprotein p53 is undoubtedly the more studied, due to its high conformational flexibility and the complexity of its biological functions. p53 is usually regarded as the 'guardian of the genome' or the 'cellular gatekeeper', and its significance is highlighted by the discovery of p53 mutations in more than 50% of all human tumors [4]. In fact, p53 plays a vital role in cellular responses to stressors by activating different cellular strategies that involved cell cycle arrest, DNA repairing or apoptosis depending on the intensity of the toxic stimuli. On the other hand, p53 is involved in physiological functions, such as regulation of metabolic pathway, redox homeostasis, and control of immune system [5]. p53 protein represents an interesting redox-sensitive protein involved in different pathophysiological processes, ranging from cancer to neurodegenerative diseases. It is located at the crossroads of complex networks of stress response pathways. Several extracellular or intercellular stresses evoke cellular responses directly or indirectly through activation of p53-redox modulation [6–8]. Many studies demonstrated that the interplay among p53 and Reactive Oxygen or Nitrogen Species (ROS/RNS) is crucial for the cellular fate.

The ability to identify the transitions from wild type to mutated one introduces an extremely promising starting point for developing new therapeutic approaches. Pertinent literature has revealed that the alterations in the conformational states of p53 protein are also associated with the onset of neurodegenerative diseases [9]. Considering all this, the ability to successfully and accurately discriminate different conformations of p53, possibly correlated with specific loss or gain of function, is of significant interest. Currently, biochemical assays (e.g., ELISA) provide information related to quantification of the protein only, but not to the conformational state of this protein.

Aiming to combine sensitivity with non-invasiveness and ease of methodology, spectrophotometry represents one of the most promising, widely used, analytical procedures in biochemistry. This method is based on the two laws of light absorption by solutions, namely Lambert's Law and Beer's Law which states: "the amount of energy absorbed or transmitted by a solution is proportional to the molar absorptivity of solution and the concentration of solute [10]." Beers Lambert law is mathematically expressed as:

$$A = eLc \tag{1}$$

where, A is the absorption, e the molar attenuation coefficient, L the path length and c the concentration of solution.

A literature review conducted shows the application of this concept used for analysis of proteins. The research study conducted by Zhou et al. [11] demonstrates discrimination of different conformational states pH dependent of BSA protein on the basis of difference in absorbance wavelengths. Similarly, infrared spectroscopy was used in [12] to measure different conformational states of protein.

In consideration to the results obtained with other protein, we develop a new methodology that utilizes spectrophotometry to characterize specific and characteristic absorbance spectrums related to different p53 protein conformational state, hence identifying its conformational states which can give new insight in different pathophysiological conditions. Therefore, an experiment was designed to detect the three structural states of p53 proteins namely wild type, oxidative and denatured state, obtained by following a protocol extensively described in the literature [13–15].

2 Methodology

2.1 Sample Preparation

In order to investigate the absorption of different conformational states of p53 protein, p53 wild type recombinant protein was exposed to different oxidant stressors to generate different redox-p53 products: (i) metal chelator agent that distrains Zn atom and induce the opening of the protein [16]; (ii) Fenton reaction, mainly mediated by the OH· derived from the decomposition of H_2O_2 in the presence of redox metals (Fe^{2+} and Cu^+) [17] generates a burst of oxygen radicals involved in protein oxidation. Thus, p53 recombinant protein was incubated for 1 h at 37 °C with the appropriate buffer: (i) 200 μM EDTA and 5 mM DTT (denatured p53); (ii) 10 mM H_2O_2 and 30 μM $FeSO_4$ (oxidized p53). PBS buffer solution alone was used as reference.

2.2 Spectrophotometer Testing

The spectrophotometry measurement was acquired at ambient conditions with Shimadzu UV-2600 system. The equipment has a measuring wavelength ranging from 185 to 1400 nm with a wavelength accuracy of ±0.3 nm. It used deuterium lamp and 50 W halogen lamp, which have a noise level of 0.00003 Abs RMS (500 nm) and adopted UV Probe application for operation of equipment.

The proposed method incorporates the use of spectrophotometer for recognizing the diverse conformations of p53 protein on the basis of distinct and distinguishing absorbance spectrum. The experimental setup employed for measuring the absorption of the three samples, containing the altered form of target protein, is shown in Fig. 1. For the purpose of holding altered form of p53 protein solutions 1 ml capacity quartz

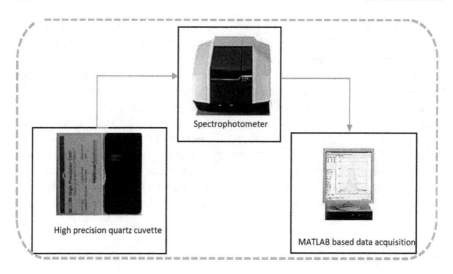

Fig. 1 Experimental setup

cuvette was used. The visual graphical trend of absorbance spectrum of each solution against the obtained data can be appreciated via MATLAB software.

Quartz cuvette containing 1 ml of PBS buffer solution is set in the spectrophotometer to note the absorbance of the reference solution and to set the baseline for the experiment. The procedure is repeated with wide type, denatured type and oxidized states of p53 solution along with PBS as a reference solution to attain and record their absorbance respectively the experiment was repeated thrice to measure the repeatability of the results.

The absorbance data obtained from each solution prepared is then collected on which a MATLAB algorithm is applied to acquire a graphical absorbance spectrum for interpretation and obtain meaningful information related to absorbance of varied p53 protein conformational states.

3   Result

The different conformational states of p53 protein can be identified from their deviant absorbance spectrum obtained through spectrophotometer. The peak absorbance of different p53 protein solution are tabulated in Table 1. Furthermore, the absorbance region for each type of solution is highlighted in Fig. 2. The absorbance range for wild type p53 is identified to lie in the region of 205–206 nm whereas the major absorbance for denatured form of target protein is found within the region of 212–214 nm and shows a maximum peak at 213 nm. Also, a negative peak at around 235–245 nm can be appreciated for the denatured state of p53. On subtracting the reference absorbance spectrum of PBS buffer solution from the solution containing denatured kind of p53

Table 1 Absorbance wavelength of p53 Wildtype and p53 Denatured type and oxidized p53

Solution	Absorbance peak 1 (nm)	Absorbance peak 2 (nm)	Molecular orientation of p53
Wild p53	205	–	
Denatured p53	213	240	
Oxidized p53	230	274	

protein, a positive absorbance with a peak at 240 nm was observed. Moreover, for the oxidized p53 protein the absorbance region detected was found to be expressing in two regions (i) 225–235 nm with a peak absorbance at 230 nm and 270–278 nm with a peak absorbance at 274 nm.

It is evident from the obtained absorbance data shown in Table 1 and absorbance graphs indicated in Fig. 2 that absorbance varies with variation in molecular state of protein. Hence, the result obtained from spectrophotometer, enables to appreciate that each peculiar state of protein shows different absorbance spectrum thus allowing detection of various transitional state of p53 protein.

4 Discussions

For the purpose of investigating various molecular orientations of p53 protein, the technique of spectrophotometer was employed. The graph obtained using spectrophotometry approach shows optimum absorbance for wild form at 205 nm as appreciated from Fig. 2a. For denatured state of p53 protein, positive absorbance height is obtained at 213 nm, and a negative peak was found at 240 nm as indicated from Fig. 2b, c. Similarly, the two peak absorbance at 230 and 274 nm are shown in Fig. 2d were noted for oxidized form of target protein. Hence it is comprehensible from the results obtained that the technique of spectrophotometer can successfully identify various molecular orientations of the protein.

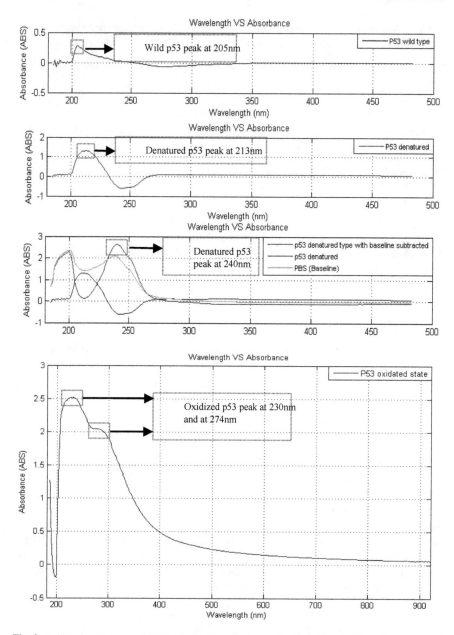

Fig. 2 **a** Absorbance graph of Wild p53. **b** Absorbance graph of Denatured p53. **c** Absorbance of denatured p53 with baseline subtracted. **d** Absorbance graph of oxidized p53

5 Conclusion

The proposed method of spectrophotometer yields distinguishable absorbance peaks for different p53 protein solutions thereby indicating the use of spectrophotometer technique as a novel and independent technology capable of detecting and distinguishing protein conformational states based on variations in absorbance spectrum. For wild p53, denatured p53 and oxidized p53 the absorbance peak wavelength noted are 205 nm, 213 nm and 240 nm, 230 nm and 274 nm respectively.

The positive results obtained from the technique of spectrophotometry has therefore proven to be a novel technological advancement in improving health care by enabling the identification and detection of different molecular orientations and modifications in protein structures hence assisting the clinicians to detect disease in its initial stages by identifying mutations and variations of biomarkers.

References

1. Suprun, E.V., Shumyantseva, V.V., Archakov, A.I.: Protein electrochemistry: application in medicine. A review. Electrochim. Acta **140**, 72–82 (2014). https://doi.org/10.1016/j.electacta.2014.03.089
2. Svobodova, Z., Reza Mohamadi, M., Jankovicova, B., Esselmann, H., Verpillot, R., Otto, M., Taverna, M., Wiltfang, J., Viovy, J.-L., Bilkova, Z.: Development of a magnetic immunosorbent for on-chip preconcentration of amyloid beta isoforms: representatives of Alzheimer's disease biomarkers. Biomicrofluidics **6**(2), 24126–2412612 (2012). https://doi.org/10.1063/1.4722588
3. Humpel, C.: Identifying and validating biomarkers for Alzheimer's disease. Trends Biotechnol. **29**(1), 26–32 (2011). https://doi.org/10.1016/j.tibtech.2010.09.007
4. Levine, A.J.: p53, the cellular gatekeeper for growth and division. Cell **88**, 323–331 (1997)
5. Meplan, C., Richard, M.J., Hainaut, P.: Redox signalling and transition metals in the control of the p53 pathway. Biochem. Pharmacol. **59**(1), 25–33 (2000)
6. Couto, R.A.S., Lima, J.L.F.C., Quinaz, M.B.: Recent developments, characteristics and potential applications of screen-printed electrodes in pharmaceutical and biological analysis. Talanta **146**, 801–814 (2016). https://doi.org/10.1016/j.talanta.2015.06.011
7. Escamilla-Gómez, V., Hernández-Santos, D., González-García, M.B., PingarrónCarrazón, J.M., Costa-García, A.: Simultaneous detection of free and total prostate specific antigen on a screen-printed electrochemical dual sensor. Biosens. Bioelectron. **24**, 2678–2683 (2009). https://doi.org/10.1016/j.bios.2009.01.043
8. Liang, Y.-F., Huang, C.-Y., Liu, B.-D.: A voltammetry potentiostat design for large dynamic range current measurement. In: 2011 International Conference on Intelligent Computation and Bio-Medical Instrumentation (ICBMI), pp. 260–263 (2011). https://doi.org/10.1109/icbmi.2011.44
9. Uberti, D., Lanni, C., Carsana, T., Francisconi, S., Missale, C., Racchi, M., Govoni, S., Memo, M.: Identification of a mutant-like conformation of p53 in fibroblasts from sporadic Alzheimer's disease patients. Neurobiol. Aging **27**(9), 1193–1201 (2006). https://doi.org/10.1016/j.neurobiolaging.2005.06.013
10. Illustrated Glossary of Organic Chemistry. http://www.chem.ucla.edu/~harding/IGOC/B/beers_law.html. Accessed 21 Feb 2018
11. Zhou, J., Rao, X., Tian, J., Wang, J., Li, T.: Study of protein conformation change induced by pH condition using terahertz spectroscopy. In: 2017 10th UK-Europe-China Workshop on Millimetre Waves and Terahertz Technologies (UCMMT), Liverpool (2017). https://doi.org/10.1109/ucmmt.2017.8068501

12. Gupta, S.D., Kelp, G., Arju, N., Emelianov, S., Shvets, G.: Metasurface-enhanced infrared spectroscopy: From protein detection to ceils differentiation. In: 2017 Conference on Lasers and Electro-Optics Europe & European Quantum Electronics Conference (CLEO/EuropeEQEC), Munich (2017). https://doi.org/10.1109/cleoe-eqec.2017.8086872
13. Hainaut, P., Milner, J.: A structural role for metal ions in the "wild-type" conformation of the tumor suppressor protein p53. Cancer Res. **53**, 1739–1742 (1993)
14. Méplan, C., Richard, M.J., Hainaut, P.: Redox signalling and transition metals in the control of the p53 pathway. Biochem Pharmacol. **59**(1), 25–33 (2000). https://doi.org/10.1016/s00062952(99)00297-x
15. Méplan, C., Richard, M.J., Hainaut, P.: Metalloregulation of the tumor suppressor protein p53: zinc mediates the renaturation of p53 after exposure to metal chelators in vitro and in intact cells. Oncogene **19** (2000). https://doi.org/10.1038/sj.onc.1203907
16. Kara, P., de la Escosura-Muñiz, A., Maltez-da Costa, M., Guix, M., Ozsoz, M., Merkoçi, A.: Aptamers based electrochemical biosensor for protein detection using carbon nanotubes platforms. Biosens. Bioelectron. **26**, 1715–1718 (2010)
17. Jeong, B., Akter, R., Han, O.H., Rhee, C.K.: Increased electrocatalyzed performance through dendrimer- encapsulated gold nanoparticles and carbon nanotube-assisted multiple bienzymatic labels: highly sensitive electrochemical immunosensor for protein detection (2013)

Electrical Energy Harvesting from Pot Plants

R. Di Lorenzo, Marco Grassi, S. Assini, M. Granata, M. Barcella
and Piero Malcovati

Abstract In recent years, energy harvesting studies have grown significantly. Energy recovery for low power applications is assuming considerable importance for powering non-essential auxiliary circuits. Among the various sources of energy that can be found in nature, in this paper we consider the Plant-Microbial Fuel Cell (P-MFC), fed by bacteria present in the roots of plants. We show a preliminary study for the optimization of P-MFC energy harvesting. The system considered is a collection of series and parallel connected pot plants. The main principle of operation is the one of a microbial cell. Plants through photosynthesis produce sugars that are subsequently released into the soil through the roots. Bacteria present near the roots consume these sugars and produce ions. Therefore, thanks to an redox process, by introducing two electrodes into the ground (anode and cathode) it is possible to obtain a potential difference that can be exploited as an energy source for indefinitely feeding electronic devices, even where it is not possible to have a direct connection to an outlet.

Keywords Energy harvesting · Plant microbial fuel cell · PMFC · Plant

1 Plant Microbial Fuel Cell

Bioelectrochemical systems (BES) [1–4], are devices in which microorganisms catalyze electrochemical reactions through interactions with the electrodes (anode and cathode). These include MFC and P-MFC. A microbial fuel cell (MFC) is a fuel cell that converts the chemical energy produced by the system into electrical energy, thanks to microbial metabolism and redox phenomena [1–4]. In an electrochemical cell there are always two parts: the anodic compartment and the cathode compartment,

R. Di Lorenzo · M. Grassi (✉) · P. Malcovati
Department of Electrical, Computer and Biomedical Engineering, University of Pavia, Pavia, Italy
e-mail: marco.grassi@unipv.it

S. Assini · M. Granata · M. Barcella
Department of Earth and Environmental Sciences, University of Pavia, Pavia, Italy

© Springer Nature Switzerland AG 2019
B. Andò et al. (eds.), *Sensors*, Lecture Notes in Electrical Engineering 539,
https://doi.org/10.1007/978-3-030-04324-7_65

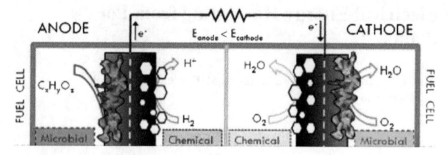

Fig. 1 Operating principle of a plant microbial fuel cell

separated by a selective membrane or PEM (proton exchange membrane). In these systems, the oxidation phenomenon of organic substrates occurs in the anodic compartment, while in the cathodic compartment the reduction of an oxidant (normally oxygen) occurs. Figure 1 shows the operating principle of an MFC. Theoretically, the maximum recoverable electrical power per unit area varies between 1.6 and 3.2 $\frac{W}{m^2}$ (see [2] for more details). Considering an area of natural sediment, with various implementation technologies to recover energy from plants, this value theoretically considering the maximum efficiency, it is practically reduced to 0.24 $\frac{W}{m^2}$. The maximum value is obtainable only if the system is completely drown in water and if the anode is maintained under anaerobic conditions.

The Eqs. (1–2) describe the basic reactions that take place at the anode and at the cathode, while the Eq. (3) illustrates the process that occurs in MFCs, in the case of a glucose-fed system:

$$\text{Anode: } C_6H_{12}O_6 + 6H_2 \rightarrow 6CO_2 + 24H^+ + 24e^- \tag{1}$$

$$\text{Cathode: } 24H^+ + 24e^- + 6O_2 \rightarrow 12H_2O \tag{2}$$

$$C_6H_{12}O_6 + 6O_2 \rightarrow 6CO_2 + 6H_2O + Electricity \tag{3}$$

The P-MFC has a structure that can be divided into layers. Starting from the base of the pot, we find a thread of titanium or gold (anode), the soil with the plant, a selective membrane and a carbon felt inside which another thread of titanium or gold is present. This second wire represents the cathode of the cell. The wire used must have very precise characteristics, since the electrode must be "biocompatible", which means that its properties (charge, structure, composition) must allow microorganisms to rapidly create electrochemically active biofilms and guarantee the "docking" of the conduction mechanisms of electrons to the electrode surface. Furthermore, the material must not oxidize. In the first preliminary experiments a copper wire was used, but after a few days it was completely oxidized, compromising the conduction. Another peculiarity is that the material must have specific electronegativity i.e. the ability to attract electrons. In fact, it must be as electronegative as possible: the more electronegative is the material, the greater is its ability to capture the electrons left free during the chemical process that occurs in the ground. Finally, the electrical conductivity of the electrode must be high, in order to minimize the ohmic losses.

2   Description of the System and Preliminary Measurements

In this work, we considered two families of plants: gramineae and cyperaceae. From the first family, we analyzed Agrostis Capillaris and Sesleria Pichiana. These plants have a root system and a foliage system more developed in the space. These two characteristics can influence positively energy production: if the plant has more leaves, it will have a higher photosynthetic activity. This leads to a greater production of sugars that will be released into the soil, near the root system. Another feature of these plants is that they are suitable for life in marsh habitats.

During the initial experimentation phase the geometry of the P-MFC has been optimized. The internal resistance of the cell was measured using an LCR meter. It has been found that soil resistance, as expected, is considerably lower when the soil is wet, rather than when the soil is dry. To give two reference values, for the P-MFCs developed, when the ground is wet, the resistance is around 8 kΩ, while with dry ground it is equal to 19 kΩ. Another parameter that influences the internal resistance of the P-MFC is the distance between anode and cathode: bringing this distance from 13 to 8 cm and increasing the contact surface, we have achieved a significant decrease in the internal resistance. Considering wet soil, resistance has fallen from 12 to 8 kΩ with also an increase in open circuit voltage (V_{oc}). In the case of a non-wet soil, the resistance has dropped from 24 to 19 kΩ after narrowing electrodes distance. The presence of water [2] in the ground is essential for the operation of P-MFCs. In fact, the system works very well if the soil is wet, while when it dries, the short circuit current drops drastically. In Table 1 the productivity of plants in different environmental conditions is reported.

The values of the open-circuit voltage (V_{oc}) and of the short-circuit current (I_{sc}) measured on four different P-MFC over two weeks with different environmental conditions are reported in Table 1. Each plant produced an open circuit voltage which varied between 550 and 800 mV. During the phase in which the plants were exposed to artificial light and to a period of drought, we can notice a reduction in values of short-circuit currents.

Table 1 Measurements in different environmental conditions

P-MFC 1		P-MFC 2		
V_{OC} (mV)	I_{SC} (μA)	V_{OC} (mV)	I_{SC} (μA)	
425	3	90	8	Artificial light—Dry
678	15	350	6	
545	20	630	6	
667	30	670	7	
672	30	680	100	Natural light—Wet
560	38	660	90	
606	28	680	125	
570	56	690	112	
638	52	450	88	
587	58	223	40	Natural light—Dry
426	98	380	23	
580	56	1074	160	Natural light—Wet
452	21	1040	254	
660	26	870	186	

3 Electronic Harvesting Circuit and Improved Measurements

The boost converter [5], realized using the integrated circuit BQ25504 by Texas Instrument, with an input current of 10 μA allows us to obtain an efficiency of about 80% when the input voltage is around 1 V, as shown in Figs. 2 and 3. Therefore, the optimal configuration for exploiting the P-MFC is obtained by connecting in parallel at least two groups of two cells each connected in series (four P-MFC set). Another option is to connect all plants in parallel. The optimal V_{out} for reducing the losses is 3.1 V. In order to maximize the transferred power Maximum Power Point Tracking must be chosen for operation, which corresponds to an input voltage equal to half of the open circuit voltage value (V_{oc}). When we connect two plants in series it is fundamental that all plants are in perfect health because any ill plant will limit the maximum current for all that branch. Using more plants in parallel allows a redundancy of the system. The power produced for each plant varies from 16 to 20 μW.

Further improved measurements have been performed using a plastic pot without hole on the bottom, a thinner carbon felt, and gold plated wires as terminals connections. The real innovative part of the experimental measurement is the removal of the proton exchange membrane, which means renouncing to have a net distinc-

Fig. 2 Efficiency of the boost converter as a function of the input voltage

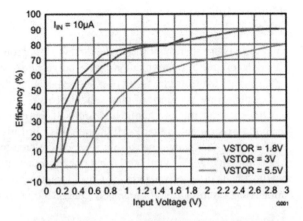

Fig. 3 Efficiency of the boost converter as a function of the input current

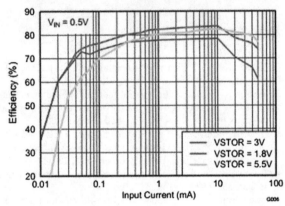

Table 2 Measurements in different environmental conditions

DATE	P-MFC 1		P-MFC 2		P-MFC 3		P-MFC 4	
	Voc (mV)	Isc (mA)	Voc (mV)	Isc (mA)	Voc (mV)	Isc (mA)	Voc (mV)	Isc (mA)
11-08-2017	495	1.79	260	0.56	530	1.84	570	1.85
12-10-2017	540	1.92	270	0.57	540	1.90	575	1.92
01-15-2018	560	1.89	260	0.56	550	1.96	570	1.91

tion between anode and cathode sub-cells. In these updated conditions, significantly higher I_{SC} currents in most of the P-MFC samples observed in the last study have been experienced. Four identical P-MFC cells have been characterized and the results reported in Table 2. Only P-MFC 2 delivered poor energy (i.e. lower Voc and I_{SC}) due to poor health conditions of one of the plants in a series path of the cell.

4 Conclusions

The carried out experiment shows that first P-MFC prototypes have delivered significant data for future employment in energy harvesting systems in isolated environment where access to electricity is not possible, especially where solar energy is not easily accessible for the presence of plants themselves. Better energy delivery is achieved when plants "operate" in humid environment. Furthermore, an important feedback from the experiment is that P-MFC energy is not fully controllable (even if artificially wetting the plants) because it is strictly related to the life cycle of the plants. The optimum P-MFC macro-cell is obtained by connecting together a matrix of NxN healthy plants in series and parallel. Considering an array of few square meters, night ornamental illumination of gardens is guaranteed in terms of energy.

References

1. Wetser, K., et al.: Electricity generation by a plant microbial fuel cell with an integrated oxygen reducing biocathode. Appl. Energy **137**(1), 151–157 (2015)
2. Wetser, K., et al.: Plant microbial fuel cell applied in wetlands: Spatial, temporal and potential electricity generation of Spartina Anglica salt marshes and Phragmites Australis peat soils. Biomass Bioenergy **83**(12), 543–550 (2015)
3. Venkata Mohan, S., et al.: Biochemical evaluation of bioelectricity production process from anaerobic wastewater treatment in a single chambered microbial fuel cell (MFC) employing glass wool membrane. Biosens. Bioelectron. **23**(9), 1326–1332 (2008)
4. Rabaey, K., et al.: Microbial fuel cells: novel biotechnology for energy generation. Trends Biotechnol. **23**(6), 291–298 (2005)
5. Texas Instrument: bq25504 Ultra Low-Power Boost Converter With Battery Management for Energy Harvester Applications, Oct 2011

Preliminary Study on Wearable System for Multiple Finger Tracking

Paolo Bellitti⬛, Michele Bona⬛, Emilio Sardini⬛ and Mauro Serpelloni⬛

Abstract Devices that track the human body movement are heavily used in numerous and various fields like medicine, automation and entertainment. The work proposed is focused on the design of a modular device able track the flection of human hand phalanxes. The overall system composed by two parts: a computer program interface and a modular wearable system applied to the finger whose motion is to be monitored. The wearable device is equipped with an Inertial Motion Unit (IMU) with the purpose to detect the first phalanx orientation and a stretch sensor applied between the first and the second phalanx to recognize the flection angle. The configuration is completed with a microcontroller unit (ATmega328P) and a Bluetooth Low Power Module (RN4871) to ensure a reliable and easy to implement communication channel. We conduct two main set of tests to verify the global functionalities. In the first set the device is used to track the full flexion of a single finger while in the second we test the device capability to recognize different grabbed objects starting from the data retrieved from two fingers. The preliminary results open the possibility of a future development focused on a modular device composed by five elements, one for each hand finger and able to detect complex gesture like pinch, spread or tap.

Keywords Data glove · Finger tracking · Stretch sensor · Inertial motion unit
Human machine interface

1 Introduction

In the last years, new devices and algorithms have been developed to implement innovative and accurate 3D body motion tracking systems [1]. These devices are widespread in fields such as medicine, automation, entertainment and are generically used to implement ergonomic human-machine interfaces. New methods allow

P. Bellitti (✉) · M. Bona · E. Sardini · M. Serpelloni
Department of Information Engineering, University of Brescia,
Via Branze 38, 25123 Brescia, Italy
e-mail: p.bellitti@unibs.it

© Springer Nature Switzerland AG 2019
B. Andò et al. (eds.), *Sensors*, Lecture Notes in Electrical Engineering 539,
https://doi.org/10.1007/978-3-030-04324-7_66

capturing the movement of body parts such as hands, head or eyes, to issue commands to a computer, in order to achieve a more natural interaction [2, 3]. Synergy between technology innovations and research activities led to several commercial devices with a great level of accuracy. Some optical tracking system reach a submillimeter accuracy grade thanks to the use of special markers, which makes them suitable for use in critical fields such as medicine as a support to surgery [4]. The development of new human hand tracking devices is one of the main lines of research. These devices can be categorized based on the motion capturing principle. The two main methods adopted are: optical-based and data-gloves. An optical-based system exploits image processing techniques to retrieve information about motion [5]. This approach leaves the hand free to move since there are not physical link to the measuring system, but the accuracy can be influenced by environmental factors such as illumination or extraneous objects in the camera line of sight. Furthermore, an accurate optical system generally requires expensive dedicate hardware with multiple cameras and illuminators. On the other hand, systems based on data gloves are slightly more invasive but generally produce more reliable data [6]. These devices are usually composed by a fabric glove in which bend or stretch sensors are integrated. The sensors are generally fixed between the interphalangeal joint and used to evaluate the angle formed between the phalanges. The purpose of this work is to develop a wireless wearable device able to track finger movements exploiting stretch sensors and inertial modules embedded in a highly integrated modular device that can be applied on each finger independently. In this way, according to the need, it is possible to reduce the inconvenience of wearing an entire glove when it is not required. For example, movement of a single finger can be used to control a computer cursor or to issue some commands through a simple gesture recognition. In other circumstances, when a virtual object must be manipulated, it can be necessary to track three or more fingers. Thanks to its modularity, the developed system can adapt to these different scenarios. The designed system can measure the flexion of the proximal phalanx analyzing data from an inertial module and exploits a stretch sensor to monitor the medial phalanx. A low power wireless communication method is used to transmit data to a specifically written computer program. The project therefore aims to develop an innovative device that can implement an accurate human-machine interface useful for biomedical, industrial or entertainment applications.

2 Description of the System

The overall system (Fig. 1) is composed by two main parts. The first one is a physical measurement unit that can be applied directly on the finger (on the first and second phalanges) whereas the second one is an ad hoc computer program written in LabVIEW (Virtual Instrument, VI). The measurement unit can recognize finger orientation in the space through an inertial motion unit rigidly coupled to the first phalanx. The angle formed between the first and the second phalanx is retrieved through a stretch sensor. Data are elaborated and prepared for the transmission by a

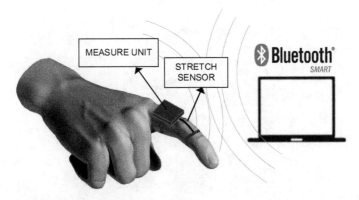

Fig. 1 General representation of the overall system

microcontroller unit and then sent through a Bluetooth low power communication channel to the computer.

The VI presents the data through graphical and numerical indicator. As it can be observed in Fig. 2 program user interface is divided in three main sections (vertically, from left to right). The first one is represented by a real time animation that shows the angles formed between the two proximal phalanges. The animation is driven by the pitch angle retrieved from the accelerometer section of the IMU (first phalanx) and by the resistance value of the stretch sensor. The second part includes two graphical indicators that display pitch angle and stretch sensor resistance respectively. The third part shows the raw acceleration along the three axes and the user selectable path where data file will be saved. The communication (computer side) is obtained through a USB-to-TTL-serial adapter (FTD232, from ftdi) connected to a Bluetooth Low Energy module (RN4020, Microchip). The VI receives raw data from the sensors without any kind of processing, in this way the computational load of the microcontroller on the wearable device is reduced and the elaboration is demanded to the computer. The data between the two Bluetooth modules are exchanged through the Microchip Low-energy Data Profile (MLDP), a proprietary BTLE service. Using this service, after the connection is established between the two modules, virtual UART channel is created and every byte asserted on a module is automatically transferred. The maximum throughput of the MLDP protocol is 50 kbps which is considered enough in this case and even permits to transfer data in ASCII format rather than binary. Therefore debug process is simplified thanks to the readability of the strings sent. In this preliminary phase of the project the ease of implementation is preferred to the energy saving. However, as the wearable device is powered by a small size battery (25 × 25 mm, 44 mAh) the battery life will be a primary aspect in the next prototype. To decrease the power consumption wireless connection speed will be reduced to the minimum that permits the proper system operation. In addition, the specific energy-saving features of the individual components will be enabled to achieve greater efficiency.

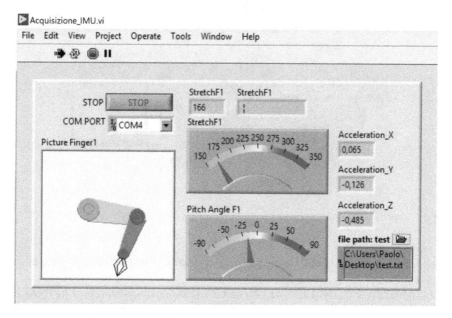

Fig. 2 User interface of the LabVIEW developed program

The measurement unit (Fig. 3) is composed by several parts. A microcontroller unit, ATmega 328P, manages all the peripherals connected, reading the sensors and transmitting the data. To simplify the implementation the microprocessor firmware is written using the Arduino IDE to take advantages of the peripheral libraries. Finger 3-D orientation is captured by an Inertial Motion Unit (LSM9DS1 from STMicroelectronics). This device includes in a single chip an accelerometer, a gyroscope (the inertial section) and a magnetometer. It has dedicated power saving functions that permits to selectively reduce the performance (or turn off) of every section when not in use. The flexion of the proximal phalanx is detected through a stretch sensor (by Images Scientific Instruments Inc.), which is composed by a conductive rubber that increases its resistance when stretched. To adapt to the length of the different fingers every sensor is specifically assembled starting from a defined length of the rubber filament with a pair of crimped conductive clasps at its edge. The rubber filament, according to the datasheet, has a resistance of about 395 Ω/cm. The measurement of stretch sensor resistance is delegated to a specific integrated circuit (MAX31865, Maxim Integrated), a device specifically designed to facilitate the resistance measurement. It has the capability of performing accurate resistance evaluation exploiting 2, 3, or 4 wires measuring technique to compensate the connections resistance influence. In this case, a 2-wire configuration is chose because of the shortness of the links. To ensure a low-power wireless capability, the system is equipped with a Bluetooth Low Energy module (RN4871, Microchip Technology). A two-ring support was developed to firmly apply the stretch sensor between proximal and intermediate phalanges.

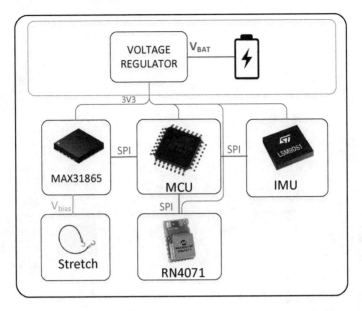

Fig. 3 Block diagram of the measuring unit

3 Experimental Study

Two main tests were performed to examine the system whole functionalities. The first one had the aim to verify system capability to recognize a full flexion and extension movement. As reported in Fig. 4, the system can discriminate between extended and flexed position. The resistance sets associated with the two positions examined are never overlapped. Also considering the variability between successive measurements the sets are always disjointed. In the extended position, according to stretch sensor characteristics [7], resistance value decays slowly, so it is important to fix a threshold below which the finger is considered fully extended. The IMU is used to measure the pitch angle of the proximal phalanx with respect to the horizontal line. Starting from the gravity acceleration detected on the three axes, the angle value is obtained using (1).

$$\theta = \operatorname{atan}\left(\frac{G_y}{\sqrt{G_x^2 + G_z^2}}\right) \tag{1}$$

In Fig. 5, the first test results are reported. In the first track (green) are visible the resistance values from the stretch sensor: lower values for extended position and higher for flexed position. In the second track are reported the pitch angle values of the first phalanx. Comparing this data, it is possible to observe that the system detects both positions correctly. Furthermore, a second set of tests was performed to establish if the system can discriminate between two different grabbed objects. Two

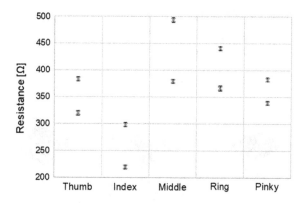

Fig. 4 Stretch sensor resistance for flexed (blue) and extended (orange) position

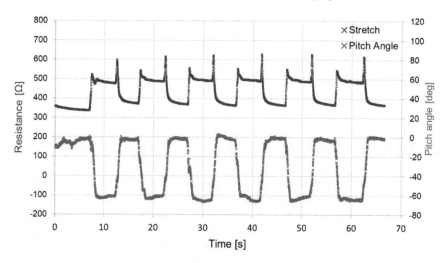

Fig. 5 Full finger extension, signals from IMU (blue) and stretch sensor (green)

stretch sensors were applied to index and middle fingers. The different diameter of each object leads to different flexion angles and then to a different stretch sensor resistance variation. After a first training phase in which the two objects are grabbed and released multiple times an average on the obtained data was performed. The results are arranged in Fig. 6. The left part of the figure shows the resistance value obtained for the index finger. The two sets of data are partially overlapped. On the other hand, the right part reports the resistance values for the middle finger. In this case, it is possible to observe that the overlapping is reduced to very few samples associated to the release position. As it can be observed by crossing the values from the two fingers, the system can detect which object was grabbed.

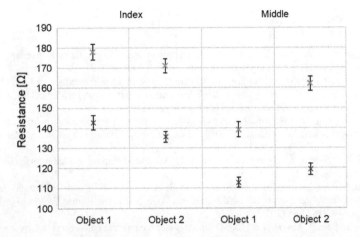

Fig. 6 Object grabbing recognition based on two stretch sensors, blue: object grabbed, green: object released

4 Conclusions

In the presented work, a prototype version of a modular device able to track the movements of a single finger has been developed. The system is composed by a small-size wearable device and a computer program that acts as a readout unit. The communication between the two parts is implemented through a Bluetooth low power connection. The preliminary results show that the device can track simple movements such as full flexion and extension of the first and second phalanges in a normal context where the range movement last about 0.5 s. Further analysis will be conducted to understand the frequency limit of the used stretch sensor used. The inertial unit has a specified bandwidth of 408 Hz, since it is reported [8] that the fastest hand motions including handwriting can be tracked with a sample frequency of 10 Hz. For this reason, the IMU it is certainly suitable. In the second set of experiment, we tested if the system (applied to two fingers) can discriminate between two grabbed objects. Even if these tests were carried out using two objects of very different diameters the authors are confident that crossing the data retrieved from a full 5-fingers implementation can increase recognition accuracy after a proper training phase. The preliminary results open the possibility of a future development focused on a modular device composed by five elements, one for each hand finger and able to detect complex gesture such as pinch, spread or tap and to recognize different predefined object.

Acknowledgements This work was supported by the Italian Ministry of Instruction, University and Research, under Grant PRIN 2015C37B25.

References

1. Dadgostar, F., Sarrafzadeh, A.: Gesture-based human–machine interfaces: a novel approach for robust hand and face tracking. Iran J. Comput. Sci. **1**, 47–64 (2018)
2. Jackowski, A., Gebhard M., Gräser A.: A novel head gesture based interface for hands-free control of a robot. In: Proceedings of the IEEE International Symposium on Medical Measurements and Applications (MeMeA), pp. 262–267, Benevento, Italy (2016)
3. Kumar, P., Rautaray, S., Agrawal, A.: Hand data glove: a new generation real-time mouse for Human-Computer Interaction. In: 2012 1st International Conference on Recent Advances in Information Technology (RAIT) (2012)
4. Brief, J., Hassfeld, S., Eggers, G., Edinger, D.: Accuracy of image-guided implantology. Clin. Oral Implant Res. **16**(4), 495–501 (2005)
5. Sharp, T., Wei, Y., Freedman, D., Kohli, P., et al.: Accurate, robust, and flexible real-time hand tracking. In: Proceedings of the 33rd Annual ACM Conference on Human Factors in Computing Systems—CHI'15As (2015)
6. Fang, B., Sun, F., Liu, H., Liu, C.: 3D human gesture capturing and recognition by the IMMU-based data glove. Neurocomputing **277**, 198–207 (2018)
7. Images Scientific Instruments Inc. Flexible Stretch Sensors. http://www.imagesco.com/sensors/stretch.pdf
8. Durlach, N., Mavor, A.: Virtual reality. National Academy Press, Washington, D.C. ISBN 0309051355 (1995)

Giraff Meets KOaLa to Better Reason on Sensor Networks

Amedeo Cesta, Luca Coraci, Gabriella Cortellessa, Riccardo De Benedictis, Andrea Orlandini, Alessandra Sorrentino and Alessandro Umbrico

Abstract Recent technological advancements in Internet of Things and Cyber-Physical systems are fostering the diffusion of smart environments relying on sensor networks. Indeed, large and heterogeneous amount of data can be provided by sensors deployed in user environments providing valuable *knowledge* to address different user needs and enabling more effective and reliable solutions as well as ensuring personalization and dynamic adaptation. This paper presents a recent research initiative whose aim is to realize autonomous and socially interacting robots by integrating sensor data representation and knowledge reasoning with decision making functionalities within a cognitive control architecture, called Knowledge-based cOntinuous Loop (KOaLa).

Keywords Intelligent environments · Knowledge representation · Ontology
Sensor networks · Artificial intelligence

A. Cesta · L. Coraci · G. Cortellessa · R. De Benedictis · A. Orlandini · A. Sorrentino
A. Umbrico (✉)
National Research Council of Italy, Institute of Cognitive Sciences and Technologies
(ISTC-CNR), Via S. Martino della Battaglia 44, 00185 Rome, Italy
e-mail: alessandra.umbrico@istc.cnr.it

A. Cesta
e-mail: amedeo.cesta@istc.cnr.it

L. Coraci
e-mail: luca.coraci@istc.cnr.it

G. Cortellessa
e-mail: gabriella.cortellessa@istc.cnr.it

R. De Benedictis
e-mail: riccardo.debenedictis@istc.cnr.it

A. Orlandini
e-mail: andrea.orlandini@istc.cnr.it

A. Sorrentino
e-mail: alessandra.sorrentino@istc.cnr.it

© Springer Nature Switzerland AG 2019
B. Andò et al. (Eds.) *Sensors*, Lecture Notes in Electrical Engineering 539,
https://doi.org/10.1007/978-3-030-04324-7_67

559

1 Introduction

Recent technological advancements in Internet of Things (IoT) and Cyber-Physical Systems (CPS) are fostering the diffusion of smart environments relying on sensor networks. Smart environments have created an upward trend in long-term monitoring systems demand for the future. The deployment of such sensor-based research trend is to address the necessities of people in many different environments like e.g., a home, an office or a shop mall, just to mention few examples. These devices usually rely on different types of sensors capable of continuously producing data about, the status of a particular environment or the status of a person. Sensors are capable of *gathering* data that can be processed by *external* services with different purposes. In this context, the continuous improvement of sensing devices in terms of accuracy and reliability as well as the reduction of manufacturing costs and their combination with ICT technologies are paving the way for new research challenges and business opportunities. The management of a continuous flow of data with the need of properly representing and processing such data still represents an open issue and an important research trend. Specifically, there is a lack of integrated solutions for both sensor data representation, data interpretation and knowledge extraction. Also, there are several sensor-based applications that have been successfully developed in different contexts, each of which often leverages its own "internal" representation of sensor data as well as its own processing mechanisms.

This paper presents a recently started research initiative called KOaLa (the Knowledge-based cOntinuous Loop) that tries to propose an innovative approach to deal with sensors, data processing and decision making. KOaLa initiative is a follow up of a previous research project called GiraffPlus [8] whose aim was to realize an integrated system composed by a telepresence robot, a sensor network and personalized services for fostering social interactions and long-term monitoring of senior users living alone in their home. KOaLa aims at enhancing GiraffPlus data analysis services by integrating Artificial Intelligence (AI) techniques to create a continuous robot control loop leveraging automatic reasoning features on sensor data. Specifically, the pursued research objective is to introduce "cognitive" capabilities that allow a system to *proactively* support the daily home living of seniors. The key contribution of KOaLa is to realize a sensor-based reasoning mechanism capable of continuously analyzing sensor data and dynamically make decisions accordingly. KOaLa leverages knowledge representation and reasoning techniques to recognize events/activities within the sensorized living environment and then it leverages automated planning and execution techniques to autonomously decide actions to be executed for either supporting activities or reacting to events.

Sensor data processing and knowledge extraction are performed by leveraging semantic technologies and the Web Ontology Language (OWL) [4]. In particular, KOaLa proposes an holistic approach which relies on an ontology which was defined as an extension of the Semantic Sensor Network (SSN) ontology [7]. The information collected through the sensor network is modeled pursuing an ontological context-based approach to characterize the different types of sensor that can be used, their capabilities as well as the related observations. In addition, the KOaLa ontology introduces concepts that model the environment and the behaviors or situations that may concern the elements composing an application scenario, senior adults and also activities an agent (either a user or a telepresence robot) can perform over time. This paper provides a general description of the envisaged KOaLa cognitive architecture. It focuses on the key role of the ontology and the context-based approach with respect to sensor data management and the knowledge extraction mechanism to realize a flexible and proactive assistive robot.

2 Sensor-Based Applications and Knowledge Extraction

This work covers different fields of AI such as decision making, goal reasoning, autonomous control, knowledge representation and ontological reasoning. There are many works in the literature that are directly or indirectly related to the pursued research objective. Several works deal with the problem of interpreting sensor data to extract useful knowledge about a particular application scenario and leverage such knowledge to realize complex services. The work [2] proposes a knowledge-driven approach based on a ontology defined as extension of the SSN ontology. The aim of the work is to reason about changes in the detected qualities of the pervasive air in a sensorized kitchen. The defined ontology models high-level knowledge about odours, their causes and relations to other phenomenon. Then, an incremental reasoner leverages such knowledge for data processing via Answer Set Programming (ASP). The works [1, 13] propose an ontology-based approach for activity recognition and context-aware reasoning for a home-care service, and a constraint-based approach for proactive human support. Some works deal with the problem of realizing complex architectures for socially-interacting robots capable of "dealing with" humans in robust and flexible way. Some examples are the works [10, 12] that perform human-aware planning by dynamically inferring context and goals by leveraging constraint programming techniques. In particular, [10] proposes an online planning setting capable of dynamically inferring planning goals in order to continuously satisfy human preferences. They rely on the concept of *Interaction Constraints* (IC) which defines a clear model of the interactions between the activities of a human and the activities of other agents. Some other works address the problem of endowing autonomous agents with cognitive capabilities to represent knowledge about contexts and leverage such knowledge to improve the flexibility of control processes in robotics. The work [3] is particularly interesting as it points out the importance of *social-norms* for domestic service robots. It focuses on the concept of *Functional Affordance* and proposes a tight integration of OWL-DL with hierarchical planning to realize robots capable of interacting with the environment in a flexible and robust way. Some other examples are [5, 9] that propose the integration of knowledge processing mechanisms with Hierarchical Task Network (HTN) planning to improve the efficiency and performance of control processes, and the works [11, 14] that propose the integration of knowledge representation with machine learning to improve human-robot interactions as well as flexibility and performance of robot behaviors.

The novelty of our approach consists in the design and development of a cognitive architecture relying on a *holistic* approach to knowledge representation through a well-defined ontology. The key idea is to realize a control architecture capable of leveraging knowledge processing mechanisms to infer a general and abstract model about the application context and, then integrate complex services that can leverage the generated knowledge for different purposes. Specifically, we consider the integration of automated planning and execution techniques to realize a proactive and autonomous robotic service.

2.1 The GiraffPlus Research Project

The GiraffPlus research project [8] represented a successful example of continuous monitoring of older people using sensor networks, intelligent software and a telepresence robot. The objective of GiraffPlus to realize a system capable of supporting elderly people directly at their home through a composition of several services. There was possible to distinguish between services decsigned for primary users (i.e., virtual visits, reminders, messages) and services designed for secondary users (i.e., real-time monitoring visualizations, reports, alarms, warnings). The services at home are provided by the telepresence robot Giraff. The robot allowed older people to communicate to their friends and family, through audio messages and videocalls as well. Seniors could interact with the robot either by means of a touch screen (part of the robot equipment) or by means of vocal commands. Interaction abilities are extremely important to improve the emotional engagement between the user and the robot. For this reason, different interaction abilities were taken into account. GiraffPlus combined the presence of a robot with a network of sensors. The network could be composed by physical sensors (i.e., blood pressure sensors) and environmental sensors (i.e. motion sensor, actuators). Physical (wearable) sensors monitored the health status of users. Environmental sensors monitored both the house and user behaviors according to their deployment.

2.2 Data Needs Semantics

Data coming from sensors must be properly represented and managed in order to produce *useful information* that can be used for different purposes and services. Thus, sensor data must be associated with a *schema* or a *semantics* clearly defining the general rules and properties that must be followed to properly interpret sensor data and extract useful information. Knowledge processing mechanisms are in charge of leveraging such semantics to process data and dynamically generate knowledge about a particular application scenario. There is a lack of standard approaches for representing and managing sensors data. Different sensor manufacturers can use different formats and protocols to represent and communicate data. As a result, sensor-based applications strictly depend on the particular types of sensor and communication protocol used. The data processing logic is highly coupled to the particular physical sensors available and the "syntactic" features of generated data. Such a dependency does not facilitate the deployment of these applications to different contexts with different sensors as well as does not facilitate *interoperability* among different applications.

Standard semantic technologies for knowledge representation and processing like e.g., the Web Ontology Language (OWL) [4], are well-suited to characterize the different properties and features of sensor data. Such technologies can be used to extend sensor-based applications by introducing a *semantic layer* between a *physical layer* which produces data through a particular sensor network and an *application layer* which extracts and processes information coming from the physical layer. In this way, the application layer and the related knowledge processing mechanisms depend only on the semantics of the gathered data. Namely, the data processing logic of a sensor-based application will no longer be coupled to the physical features of the deployed sensors but only to the structure and properties of the extracted information. Semantics plays a key role when dealing with sensors and sensor-based applications. Thus, the ontological approach of KOaLa wants to characterize information, properties and capabilities of physical sensors in order to generate/infer and represent knowledge in a "standard way".

3 KOaLa: Knowledge-Based Continuous Loop

The KOaLa cognitive architecture relies on the integration of two core AI-based modules into a *closed-loop* control cycle. A semantic module deals with data interpretation and knowledge extraction. An acting module deals with plan synthesis and execution by leveraging the timeline-based approach [6] and the PLATINUm planning and execution framework [15]. Figure 1 shows a conceptual view of the architecture by taking into account a typical application scenario of the GiraffPlus project.

The control flow starts with data coming from the sensors deployed inside the house. The *KOaLa Semantic Module* continuously processes sensor data in order to extract knowledge and dynamically build/refine a Knowledge Base (KB) characterizing the status of the home environment. The KB is analyzed to recognize *goals* i.e., high-level activities the system must perform to proactively support the daily home living of the primary user. This module relies on a dedicated ontology (the *KOaLa ontology*) and implements a knowledge processing mechanism to provide sensor data with semantics. Such a process allows the semantic module to dynamically *infer* activities a primary user is performing and/or events occurring inside the house. Then, according to the particular combinations of events and/or activi-

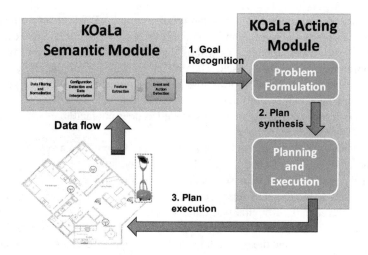

Fig. 1 The KOaLa cognitive architecture with respect to the GiraffPlus case study

ties recognized (i.e., *situations*) a goal recognition process identifies the high-level supporting tasks that must be performed.

Then, the *KOaLa Acting Module* dynamically synthesizes and executes a timeline-based plan according to the received goals. The *Planning and Execution* module "runs" the timelines of the synthesized plan by dispatching commands to the system and by properly managing execution feedbacks. This module implements environment control capabilities in order to actually carry out the high-level activities needed to support a user. A problem formulation process receives *goals* from the semantic module and generates a timeline-based problem specification. The planning and execution module synthesizes a set of timelines that must be executed over time. Each timeline characterizes the sequence of actions/commands a particular feature of the environment must perform to proactively carry out the desired supporting activities.

3.1   A Context-Based Ontological Approach

The semantic module of Fig. 1 is in charge of realizing the cognitive capabilities needed to interpret sensor data and internally represent the *knowledge* about the application scenario. The reasoning mechanism leverages a context-based ontological approach which characterizes the knowledge according to three contexts: (i) the *sensor context*; (ii) the *environment context*; (iii) the *observation context*. The *sensors context* characterizes the knowledge about the sensing devices that compose the environment, their deployment and the properties they may observe. This context directly extends the SSN ontology in order to provide a more detailed representation of the particular types of sensor available and their features/properties.

This context allows the cognitive architecture to dynamically recognize the actual monitoring capabilities of the system but also the set of operations that can be performed according to the detected configuration of the environment. The *environment context* characterizes the knowledge about the structure and the physical elements composing a home environment. It characterizes the properties of the different elements that can be observed and the deployment of sensors on these elements. Thus, this level of abstraction provides a complete representation of the particular configuration of the considered home environment by taking into account the specific deployment of the sensor network. The *observation context* characterizes the knowledge about the *features* of the environment that can actually produce information and the possible *events* and *activities* that can be recognized accordingly. This level of abstraction extends the detected *configuration* of the environment by characterizing the particular types of event and activity that can be actually monitored/recognized, given the particular types of sensor deployed.

A data processing mechanism leverages these contexts to process sensor data and incrementally build/refine the knowledge about the application scenario. Figure 2 shows the main steps fo this data processing mechanism together with their correlations with the sensor, environment and observation contexts. Each step applies a set of dedicated *rules* that elaborate incoming data and refine the KB by *inferring* additional knowledge when possible. The *Configuration Detection and Data Interpretation* step first generates an initial KB by analyzing the "static" information about the configuration of the environment. Then, it refines the KB by interpreting sensor data coming from the environment. The *Feature Extraction* step elaborates the KB by extracting the observable features and properties of the environment according to the configuration. This step processes sensor data in order to properly infer *observations* and refine the KB by taking into account the particular observed values that characterize some property of a features of the home environment. Finally, the *Event and Activity Detection* step further analyzes the inferred observations by taking into account the involved elements, and the related properties. For example, it is possible to infer the event *low temperature* in a particular room of the home environment when the observed temperature of the room is below a known threshold. Different inference rules have been defined to detect different events or activities like e.g., *cooking*, *sleeping* or *watching tv*. Each rule takes into account different "patterns" of observed properties, values and environment features.

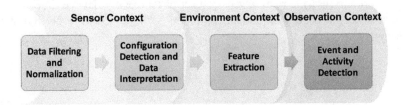

Fig. 2 The KOaLa data processing pipeline

3.2   Linking Knowledge Processing and Planning

Given the KB obtained through the data processing mechanism briefly described in the previous section, a *goal recognition* process completes the analysis of the KB in order to identify *situations* that require the execution of some actions by the system. The goal recognition process is the key architectural element responsible for *linking* knowledge reasoning with automated planning and execution at runtime, and therefore it is responsible for providing the an assistive robot like e.g., the Giraff-Plus robot with *proactivity*. It can be seen as a background process which analyzes the KB and generates *goals* that trigger the *KOaLa Acting Module*. The actual set of operations the system can perform (i.e., the set of goals that can be generated) depends on the particular configuration of the environment and the related *capabilities*. Each goal is associated to a particular situation which represents a particular combination of events and/or activities occurring inside the house. The generated goals and the involved operations may range from general supporting activities like therapy reminders or comfort management to emergency management and health monitoring. For example, if the system recognizes *low temperature* inside a particular room (e.g., the kitchen), and if the system detects the *presence* of the user inside the same room then, the goal recognition can generate a *goal* to *heat* the room in order to improve the comfort of the user. Similarly, if an event like *fallen* or a very *high hearth-rate* is detected then, the goal recognition can generate a *goal* to perform an emergency call and proactively alert user's relatives and/or caregivers.

The *signals* generated by the goal recognition process represent high-level assistive tasks the assistive robot must perform to react to some events or support some activities concerning the user. The acting module is responsible for the synthesis and the execution of the set of actions needed to perform all the desired assistive tasks (i.e., all the *goals* generated by the semantic module). The acting module leverages timeline-based planning and execution technologies [6, 15] to generate and maintain a temporal plan which specifies operations/actions to execute. The key objective of the *KOaLa Acting Module* is to dynamically control the environment through a temporal plan characterizing the behavior of a monitored primary user over time and the corresponding assistive tasks needed. These behaviors are represented in shape of *timelines* that describe the sequences of activities or states a particular feature of the domain (e.g., the primary user or the assistive robot) performs or assumes over time. The behavior of the overall system (i.e., the assistive robot, the sensing devices and the primary user) is represented by a set of temporally flexible timelines that are dynamically executed and *adapted* according to the feedbacks received from the environment.

4 KOaLa and Giraff Working Together

The envisaged cognitive architecture and the resulting tight interaction between a semantic module and an acting module realize an adaptive and flexible assistive system capable of uniformly manage *assistive goals* and *real-time goals*. A system capable of dealing with assistive tasks to support the daily-home living according to the "expected behavior" of a person and simultaneously monitoring the environment to dynamically recognize events/activities that require a proactive support. Real-time goals concern operations the system must perform to proactively react to an event or support an activity triggered by the goal recognition process. Assistive goals instead represent operations that can be planned in advance according to the detected configuration and the expected behavior of a user which can be given as an "external timeline" to the acting module.

Figure 3 shows an example of a generated timeline-based plan containing activities concerning both assistive tasks and real-time goals. The red timeline represents the behavior of the primary user. The green timeline represents the assistive tasks and the real-time goals generated to support the user. Other timelines represent the actions the system perform to achieve the desired goals through the assistive robot (i.e., the GiraffPlus robot). As the plan shows, the system initially detects that the human (i.e., the primary user) is having a meal (see the *cooking* activity followed by the *eating* activity) when the assistive task *HandleCookingDetection* is generated in order to suggest some recipes that comply with the dietary restriction of the user and remind the therapy. Specifically, the robot navigates the home environment to reach

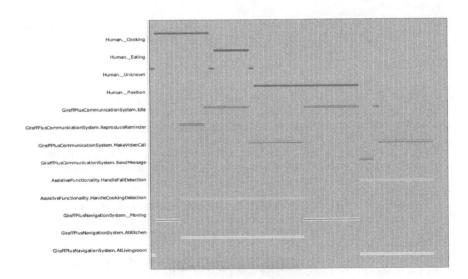

Fig. 3 A timeline-based plan integrating assistive tasks and real-time goals

the kitchen, and then reproduces an audio message to remind dietary restrictions to the user (see the *ReproduceReminder* activity). When the meal is finished (i.e., after the *eating* activity), the robot starts a video call in order to make the user in contact with his/her relatives and check the therapy. After a while, the user falls and a real-time goal is generated to promptly support the "emergency situation" (see the *HandleFallDetection* activity). The robot moves from the kitchen and navigates the home environment to reach the detected position of the user. Then, the location is reached, the robot starts a video call to promptly allow the user to ask for help by calling his/her relatives or associated caregivers.

5 Final Remarks and Future Developments

This paper presented a cognitive architecture integrating sensing, knowledge representation and planning to constitute a cognitive control loop capable of enhancing the proactivity of an assistive robot. It relies on a tight integration of a semantic module and an acting module. A semantic module leverages a dedicated ontology to process sensor data and build a KB. An acting module leverages the timeline-based planning approach to control the overall assistive system. A goal recognition process connects these two modules and provides the key enabling feature to endow the robot with a suitable proactivity level. At this stage, some tests have been performed to show the feasibility of the approach. Further work is ongoing to perform more extensive integrated laboratory tests and better assess the performance and capabilities of the overall system.

References

1. Alirezaie, M., Renoux, J., Kockemann, U., Loutfi, A.: An ontology-based context-aware system for smart homes: E-care@home. Sensors **17**(7) (2017)
2. Alirezaie, M., Loutfi, A.: Reasoning for improved sensor data interpretation in a smart home. CoRR abs/1412.7961 (2014). arXiv.org/abs/1412.7961
3. Awaad, I., Kraetzschmar, G.K., Hertzberg, J.: The role of functional affordances in socializing robots. Int. J. Soc. Robot. **7**(4), 421–438 (2015). https://doi.org/10.1007/s12369-015-0281-3
4. Bechhofer, S.: OWL: Web Ontology Language, pp. 2008–2009. Springer, Boston (2009). https://doi.org/10.1007/978-0-387-39940-9_1073
5. Behnke, G., Ponomaryov, D., Schiller, M., Bercher, P., Nothdurft, F., Glimm, B., Biundo, S.: Coherence across components in cognitive systems–one ontology to rule them all. In: Proceedings of the 25th International Joint Conference on Artificial Intelligence (IJCAI 2015). AAAI Press (2015)
6. Cialdea Mayer, M., Orlandini, A., Umbrico, A.: Planning and execution with flexible timelines: a formal account. Acta Informatica **53**(6–8), 649–680 (2016)
7. Compton, M., et al.: The SSN ontology of the W3C semantic sensor network incubator group. Web Semant Sci. Serv. Agents World Wide Web **17**(Supplement C), 25–32 (2012). http://www.sciencedirect.com/science/article/pii/S1570826812000571

8. Coradeschi, S., Cesta, A., Cortellessa, G., Coraci, L., Gonzalez, J., Karlsson, L., Furfari, F., Loutfi, A., Orlandini, A., Palumbo, F., Pecora, F., von Rump, S., Štimec, A., Ullberg, J., Ötslund, B.: Giraffplus: combining social interaction and long term monitoring for promoting independent living. In: 2013 6th International Conference on Human System Interactions (HSI), pp. 578–585 (2013)
9. Hartanto, R., Hertzberg, J.: Fusing DL Reasoning with HTN Planning. In: Dengel, A., Berns, K., Breuel, T., Bomarius, F., Roth-Berghofer, T. (eds.) KI 2008: Advances in Artificial Intelligence, Lecture Notes in Computer Science, vol. 5243, pp. 62–69. Springer, Heidelberg (2008). https://doi.org/10.1007/978-3-540-85845-4_8
10. Keckemann, U., Pecora, F., Karlsson, L.: Inferring context and goals for online human-aware planning. In: 2015 IEEE 27th International Conference on Tools with Artificial Intelligence (ICTAI), pp. 550–557 (2015)
11. Lemaignan, S., Ros, R., Mosenlechner, L., Alami, R., Beetz, M.: ORO, a knowledge management platform for cognitive architectures in robotics. In: 2010 IEEE/RSJ International Conference on Intelligent Robots and Systems (IROS), pp. 3548–3553 (2010)
12. Pecora, F., Cirillo, M., Dell'Osa, F., Ullberg, J., Saffiotti, A.: A constraint-based approach for proactive, context-aware human support. JAISE **4**(4), 347–367 (2012)
13. Suh, I.H., Lim, G.H., Hwang, W., Suh, H., Choi, J.H., Park, Y.T.: Ontology-based multi-layered robot knowledge framework (OMRKF) for robot intelligence. In: 2007 IEEE/RSJ International Conference on Intelligent Robots and Systems, IROS 2007, pp. 429–436 (2007)
14. Tenorth, M., Beetz, M.: Representations for robot knowledge in the KnowRob framework. Artif. Intell. (2015). https://doi.org/10.1016/j.artint.2015.05.010
15. Umbrico, A., Cesta, A., Cialdea Mayer, M., Orlandini, A.: PLATINUm: a new framework for planning and acting, pp. 508–522. Springer International Publishing, Cham (2017)

Smart Insole for Diabetic Foot Monitoring

Gabriele Rescio⊙, Alessandro Leone⊙, Luca Francioso⊙
and Pietro Siciliano⊙

Abstract Foot ulcer is a severe complication affecting about 25% of diabetes mellitus patients due to a lower blood supply and loss of foot sensitivity (neuropathy). A fast and reliable identification of foot pressure loads and temperature distributions changes on the plantar surface allows to prevent and reduce the consequences of ulceration such as foot or leg amputation. Several wearable technologies have been developed and tested by the scientific community, addressing the "diabetic foot" topic. However, the dimensions of the devices and the combined pressure/temperature monitoring capabilities don't accommodate the requirements from both the end-users and caregivers: normally just one information—pressure loads or temperature map—is acquired, moreover the amount of thermal reading points is lower than 5 and the accuracy of thermal sensors is greater than 0.5 °C. This work presents a smart insole in which both temperature and pressure data in 8 reading points are monitored in remote way for the assessment of the health foot conditions by a caregiver. Minimally invasive and low power temperature and force sensors have been chosen and integrated into two antibacterial polyurethane-based layers architecture, designed in accordance with the typical requirements of diabetic foot. Based on the results, the developed system shows high performance in terms of temperature and pressure detection.

Keywords Foot ulcer · Smart system · Diabetes

1 Introduction

Diabetes represents one of the main problems in health system and a global public health threat that has increased dramatically over the past 2 decades [1, 2]. The dramatic consequences on the quality of life have favored the widely dissemination

G. Rescio (✉) · A. Leone · L. Francioso · P. Siciliano
National Research Council of Italy, Institute for Microelectronics and Microsystems,
Via Monteroni Presso Campus Universitario Palazzina A3, Lecce, Italy
e-mail: gabriele.rescio@cnr.it

© Springer Nature Switzerland AG 2019
B. Andò et al. (eds.), *Sensors*, Lecture Notes in Electrical Engineering 539,
https://doi.org/10.1007/978-3-030-04324-7_68

of several scientific works about the prevention techniques and medical treatments of diabetes. They have shown that with good self-management and health professional support, people with diabetes can live a long and healthy life. One of the common and serious complication in diabetic patients is the foot ulcers that may affect up to 25% of patients in the course of their lifetime [3]. The main causes of the ulcers arising are due to the neuropathy, peripheral artery diseases and vascular alterations. These diseases induce insensitivity conditions that can cause high plantar pressure, deformity and ischemic wounds on the plantar foot up to the ulcers development [1, 3]. When the ulcers appear, the treatment is challenging, expensive and a long term treatment is required. It might need of podiatrist, orthopedic surgeon, endocrinologist, infectious disease physician and nurses. In this scenario the prevention of ulcers assumes a prominent rule. The prevention could forecast appropriate training of patients in order to assimilate the behaviors necessary to reduce the appearance of ulcers (i.e. the glycemic and blood pressure control, lipid management, smoking cessation, feet drying, wearing special and comfortable shoes, ecc). In addition, the use of technologies for the monitoring of relevant vital parameters on the plantar, as temperature and foot load distribution, may be very important to prevent foot ulcers. They may identify the prolonged and excessive pressure at a point of the foot or recognize anomalies in skin conditions and in bloody circulation. Several methods have been developed to measure the pressures and stresses in the plantar tissue by using mainly sensorized mats and force platforms. These strategies are usually realized following pre-defined scheduled medical visits. This approach it could be not very effective because it is desirable to continuously monitor and to detect high pressure values in timely way. Several systems for measuring plantar pressure in the foot are commercially available [4]. These systems are extremely expensive and aim at athletic activities and are not designed for prevention of ulcers.

Another useful parameter for assessing the diabetic foot is the temperature. Progressive degeneration of sensory nerve pathways affect thermoreceptors and mechanoreceptors. High temperatures under the foot coupled with reduced or complete loss of sensation can predispose the patient to foot ulceration. So, the foot thermal monitoring may facilitate detection of diabetic foot problems [5]. The most part of the temperature foot measurements for the diabetic foot regarding with non-invasive, and accurate thermal images analysis or thermography inspections. However, as described for the pressure monitoring, the long-term and continuous measurements of temperature may allow for a more effective ulcers prevention.

Several wearable technologies, for the permanent monitoring of the diabetic foot, have been developed and tested by the scientific community. However, the dimensions of the devices and the combined pressure/temperature monitoring capabilities do not accommodate the requirements from both the end-users and caregivers. Indeed, normally just one information (pressure loads or temperature map) is acquired. Moreover the amount of thermal reading points is lower than 5 and the accuracy of thermal sensors is greater than 0.5 °C. This work presents a smart insole in which both temperature and pressure data in 8 reading points are acquired and then transmitted through a wireless protocol to a gateway in order to monitor foot conditions and inform the caregiver about the health status.

2 Smart Insole System

The architecture of the developed smart insole system consists of three subsystems: (1) sensors system for the temperature and pressure parameters acquisition; (2) wireless module transmission; (3) data aggregation and processing module. The first two subsystem are integrated in the insole, whereas the third subsystem is implemented on a gateway device (Embedded PC) able to send the data to a caregiver through a cloud service (see Fig. 1).

For the evaluation of the pressure and temperature sensors placement, the distribution of load foot pressure was analyzed through the baropodometry P-Walk platform, produced by BTS Bioengineering [6]. As it is possible to see in Fig. 2 for the body weight evaluation on the plantar foot, it can be mainly divided into the toe, metatarsal, midfoot and heel areas.

For the sensor positioning and sensor dimensions have to be find the best trade-off through the covering area and accurately measurements at specific points of the plantar. According to the Ferber et al. [7] and the local load foot analysis, performed by using the BTS G-studio [6], the placing of the pressure sensors was designed as shown in Fig. 3. The temperature sensors were located close the pressure sensor in order to monitor the main pressure point of the foot.

For the temperature analysis, the Maxim MAX30205 sensor was chosen since it presents a greater accuracy (0.1 °C) [8] than works belonging to the state-of-the-art and a low current consumption (600 μA in operative mode and 5 nA of leakage current), useful for the long-term monitoring. For the load pressure acquisition, the IEE CP151 FSR sensor (0.43 mm thick and 6% for the accuracy) [9] has been chosen for the low invasive integration in the flexible thin-film layer (polyamide support with a dielectric and copper thickness of 360 μm and 18 μm respectively, as best trade-off in terms of invasiveness, ergonomics and robustness). Their dynamics are within 100 N on an active area of 113 mm^2, so it is appreciated for dynamic and static foot load pressure measuring. The sensors placement was studied and optimized in order to obtain an accurate monitoring of foot in the 38–43 size range by using only

Fig. 1 Overview of architecture of the smart insole monitoring system

CAREGIVER

Internet

Smart INSOLE

(a) **(b)**

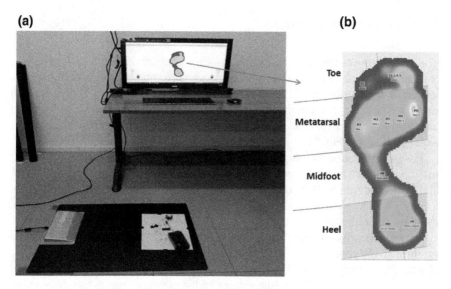

Fig. 2 **a** Setup for evaluation of foot load distribution; **b** higher pressure contact areas of the plantar

Fig. 3 Temperature and
FSR sensors positioning

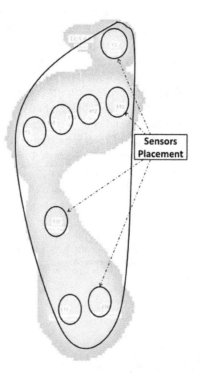

the 39 and 42 foot sizes. The hardware architecture of the electronic insole system consists of two main blocks equivalent to the two subsystems aforementioned. Low-power, low-dimensional, low-profile and high-precision electronic components were identified to enhance the battery autonomy and accuracy level, reducing the size of the system. The acquisition and transmission of data were tested using an external Arduino-based board whereas the interface circuits were integrated on the insole, close to each sensor in order to reduce the noise of the system. Moreover a study of the integration of elaboration/transmission unit was accomplished and the area of the insole board was reserved. A dual-layer circuit board was realized and all components were integrated on the bottom side (apart the flat and very thin FSR sensors) to avoid asperities potentially harmful for the diabetic foot. To measure the foot temperature, a via hole has been realized on the board in correspondence of the sensing pad of each temperature sensor, assuring the thermal transfer between the bottom and the top of the insole, through the application of a silver-based conductive paste. The flexible circuit was incorporated between the first and the second layer of the two antibacterial polyurethane-based layers architecture, designed in accordance with the typical requirements of diabetic foot insoles.

3 Results

The developed smart insole is shown in Fig. 4. The prototype results lightweight and minimally invasive. To evaluate the performance, preliminary laboratory testing were performed in order to compare the temperature values measured by the proposed system and an accurate infrared thermometer on the top of insole.

The mean and the standard deviation of the error calculated for 15 measurements for all temperature points are reported in Table 1. Moreover, to assess the load pressure acquisition, five end-users (47.2 ± 12.3 years old and 26.3 ± 2.1 Body Mass Index)

Fig. 4 **a** Top and **b** bottom view of the flexible smart insole

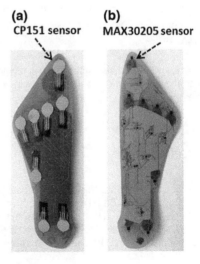

(a) CP151 sensor   **(b)** MAX30205 sensor

Table 1 Main features of the smart insole

Feature	Smart insole system
Temperature accuracy (compared with infrared thermometer)	±0.27 °C (range 25–40 °C)
Max mean peak pressure for all measurements points	235 ± 21 kPa (10 Hz sample rate)
Insole weight	<60 g
Power consumption	Max 86.3 mW
Battery	Lithium-Ion Button cell 600 mAh

performed static and dynamic actions on the BTS baropodometry P-Walk platform, wearing the insoles. The differences between the data acquired with the P-Walk platform and the developed insole are also reported in Table 1 (for the comparison the offset due to the polyurethane layers were evaluated and compensated). Moreover the weight and the power consumption of the system are reported. Based on the results, the developed system shows high performance in temperature/pressure detection.

4 Conclusion

A smart insole system for continuous temperature and load pressure foot monitoring in daily activities was presented for diabetes patients. The data were acquired from eight different points on the foot plantar both for temperature and load pressure parameters useful for the ulcer prevention. The preliminary laboratory tests validate high accuracy level in temperature data acquisition. Moreover good performance was obtained for the foot load distribution evaluation by using minimally and passive pressure sensors. Ongoing studies are focused on the integration of a low-power elaboration and transmission unit to allow for a final coupling with antibacterial socks.

Acknowledgements Authors would like to thank the colleague Mr. Giovanni Montagna for the technical support. This work has been carried out within the Diabesitycare project funded by Apulian Region.

References

1. Yazdanpanah, L., Nasiri, M., Adarvishi, S.: Literature review on the management of diabetic foot ulcer. World J. Diabetes **6**(1), 37–53 (2015)
2. Ogurtsova, K., da Rocha Fernandes, J.D., Huang, Y., Linnenkamp, U., Guariguata, L., Cho, N.H., Cavan, D., Shaw, J.E., Makaroff, L.E.: IDF Diabetes Atlas: global estimates for the prevalence of diabetes for 2015 and 2040. Diabetes Res. Clin. Pract. **128**, 40–50 (2017)

3. Iraj, B.1., Khorvash, F., Ebneshahidi, A., Askari, G.: Prevention of diabetic foot ulcer. Int. J. Prev. Med. **4**(3), 373–376 (2013)
4. DeBerardinis, J., Trabia, M., Dufek, J.S.: Review of foot plantar pressure—focus on the development of foot ulcerations. Open Access J. Sci. Technol. **4**(101158), 7 (2016)
5. Etehadtavakol, M., Ng, E.Y.K.: Assessment of foot complications in diabetic patients using thermography: a review. Appl. Infrared Biomed. Sci. 33–43 (2017)
6. https://www.btsbioengineering.com
7. Ferber, R., Webber, T., Everett, B., Groenland, M.: Validation of plantar pressure measurements for a novel in-shoe plantar sensory replacement unit. J. Diabetes Sci. Technol. **7**(5), 1167–1175 (2013)
8. https://www.maximintegrated.com/en/products/sensors-and-sensor-interface/MAX30205.html
9. https://www.iee.lu/en/technologies/force-sensing-resistor

Identification of Users' Well-Being Related to External Stimuli: A Preliminary Investigation

Filippo Pietroni, Sara Casaccia, Lorenzo Scalise and Gian Marco Revel

Abstract In this paper, the authors have investigated the possibility of evaluating the well-being of the user, in relation to external stimuli, by means of continuous monitoring of physiological quantities. This preliminary analysis has interested the extraction of features from Electro-Dermal Activity (EDA) signal and their correlation with different emotional states (i.e., *Arousal*). A low-cost system for continuous monitoring of EDA has been described, together with the experimental setup and the processing techniques applied. A unique indicator, which combines the features extracted from the raw waveform, has been discussed in the paper and applied in post-processing. The implementation of the processing algorithms and the computation of the novel indicator allow to discriminate, with a statistical significance, the user perception, in case of high emotion events (i.e., from low level Arousal < 3 to high level one, > 6). More investigation is needed to improve the processing technique and validate the preliminary results obtained.

Keywords Electro-Dermal activity · Wearable sensor · Signal processing

1 Introduction

In Europe, the total population older than 65 is expected to increase from 22% by 2015 to 27.5% by 2050, while the population aged over 80 may grow more than 4% in the same period. According to experts, this increasing trend will have a deep impact on several aspects, which affect the quality of life of millions of people around the world [1]. From a technological point of view, emerging assistive technologies together with ubiquitous and pervasive computing could increase the quality of life of aging people who decide to stay at home [2–4]. Moreover, the

F. Pietroni · S. Casaccia · L. Scalise (✉) · G. M. Revel
Università Politecnica delle Marche, Via Brecce Bianche 12, Ancona, Italy
e-mail: l.scalise@univpm.it

F. Pietroni
e-mail: f.pietroni@staff.univpm.it

© Springer Nature Switzerland AG 2019
B. Andò et al. (eds.), *Sensors*, Lecture Notes in Electrical Engineering 539,
https://doi.org/10.1007/978-3-030-04324-7_69

development of dedicated assistive technologies might help in saving resources from medical services. For example, the Italian Smart Cities project "Health@Home: Smart Communities for citizens' wellness" (H@H) has focused on this aspect. One of its main objectives has been the development of a technological framework to collect data from heterogeneous devices, together with the deployment of services (i.e., health care, social care) to assist the user within the home environment [5]. Despite a lot of research is focused on the application of physiological monitoring (e.g., Heart Rate—HR, Heart Rate Variability—HRV, Blood Pressure, etc.), only a few efforts have been made towards monitoring and regulating the aging adult's arousal state, which is usually indicative of stress or mental illness. In fact, these are fundamental aspects in the self-perception of well-being [6]. In addition, the lack of human-machine interfaces in interpreting the patients' emotional states is a severe drawback [7]. Moreover, continuous monitoring of *Arousal* level in the aging adult is essential for understanding and managing personal well-being regarding mental state. Several physiological features have been widely used in the literature, which measure alterations of the central nervous system (e.g., HRV, Nasal Skin Temperature) [8, 9]. One of the most interesting physiological markers is the Electro-Dermal Activity (EDA), as it allows to quantify changes in the sympathetic nervous system. Therefore, EDA has been tested in numerous works to assess the *Arousal* level of the patients. Wearable sensors are most appropriate to continuously measure EDA from the patients, given their performance in providing detailed user-specific information. In the state of the art, there are some examples of EDA sensors, which differ for the data transmission protocol, the sensing elements, or the integration of additional sensor in the measuring process [10]. Differently from other physiological signals, i.e., Electrocardiogram (ECG), the features to be extracted from EDA are not standardized: however, there are several studies that focus on this aspect, together with the identification of algorithms and application in software solution [11–13]. EDA may find application in several fields, from the assessment of cognitive load [14], the quantification of stress [15, 16], the classification of emotional states [17] and the evaluation of the user response after relaxation processes, e.g., music or meditation [18, 19]. The aim of this work is to provide a preliminary investigation of the EDA signal, in relation to different emotional states of the participants involved in the trial. The sensor adopted will be presented, together with the implementation of an existing algorithm to extract useful features from the signals acquired.

2   Materials and Methods

Today, bio-signals are increasingly gaining attention beyond the classical medical domain, into a paradigm, which can be described as "Physiological Computing". The modern uses of bio-signals have become an increasingly important topic of study within the global engineering. So far, physical computing has mostly been characterized by the adoption of hardware, which deals with requirements that are not completely compatible with the needs of bio-signal acquisition, (e.g., high tolerance

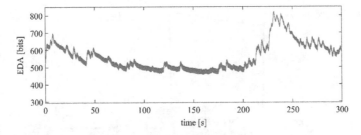

Fig. 1 Raw signal acquired from BITalino EDA sensor module

to noise, low sampling rates, no need for galvanic isolation, etc.). According to this, the BITalino acquisition board has been developed [20]. The platform comes as a single board, with its onboard sensors pre-connected to analog and digital ports on the control block. However, it is designed in such way that each individual block can be physically detached from the main board, allowing people to use it in many different configurations. The hardware is then supported by an open-source software (OpenSignals), which allows the end-user to acquire waveforms through Bluetooth communication protocol and store data. The EDA acquisition module developed for BITalino allows the measurement of skin resistance (0–1 MΩ) through the raw data acquired (pre-conditioned analog output, high SNR, 10 bits). Figure 1 shows an example of such raw signal acquired within a test, which is affected by noise, probably due to the wireless transmission protocol or other devices in the test room.

The application of manipulation processes (i.e., down-sampling and signal filtering, as described in the next section) can reduce such noise and provide a smooth waveform for the extraction of useful features. The complete BITalino board has a cost in the order of 150–200 €, while the EDA module is about 25 €. However, a calibration curve for the EDA signal has been provided for the circuit design of the BITalino board itself, so this should be used with such device.

3 Experimental Setup

The aim of this preliminary investigation is to evaluate the possibility to compute real-time user emotional state from EDA features extracted. Next paragraph describes the experimental setup for the tests conducted. Then, the algorithm to extract information from EDA is discussed, together with the open-source software, which allows a finer analysis and the identification of additional quantities. A novel indicator, which combines the features extracted from the signal, is discussed.

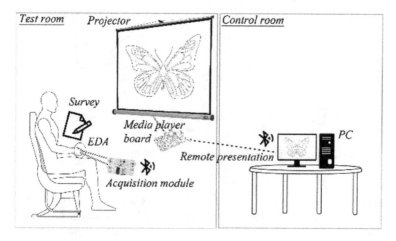

Fig. 2 Concept of the trials conducted

3.1 Participants and Trials

Figure 2 summarizes the concept of the trial conducted. The examiner, which will be in the control room for all the duration of the test, prepares the test room and the participant. The participant is seated in this room, with no external noises during the trial. He is asked to wear the EDA electrodes, one on the left index and the other on the left middle finger and an accelerometer on the left thumb. The BITalino board and the sensor modules have been fixed in the left arm of the subject by means of a prototype container (i.e., a running strap, Fig. 3) to facilitate the installation procedure. The participant will be asked to look at a series of pictures, which will be projected in the room at fixed time steps (i.e., 5 s of blank image with description of what they will see, 6 s for the image projection, 20 s black screen). During the black screen, they will be asked to fill in the survey with their personal reaction to the image they have seen. In the control room, the examiner launches the EDA acquisition software, waits for 5 min (needed to provide a stable EDA measure), then launches the slides presentation. The images have been extracted from a large database provided by the International Affective Picture System (IAPS). In the state of art, this database has been widely adopted to investigate possible correlations between the emotional status and physiological quantities (e.g., EDA [21], EEG [22], Heart Rate Variability [23], or personal sensations [24, 25]).

Figure 3 shows the 9-point evaluation scale used to provide user emotional feedback about the image. The *Arousal* index has been primarily adopted, which indicates if an image induces a passive (1) or active (9) reaction to the user. The image presentation consisted of 36 images, chosen from the IAPS and placed in a random order within the slides, with the following criterion:

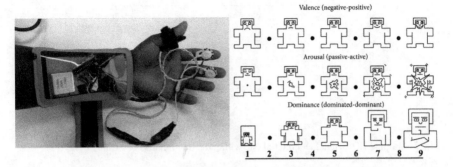

Fig. 3 Left: Prototype of EDA acquisition device. Right: Evaluation scale for IAPS

- *Arousal* index between 1 and 3: 12 (no criterion for *Valence* or *Dominance*);
- *Arousal* index between 4 and 6: 12;
- *Arousal* index between 6 and 9: 12.

The main aspect the authors want to investigate in the paper is if and how EDA features change according to the different *Arousal* level (i.e., the classification provided by literature). However, the personal sensations provided by the user through the surveys have been collected and will be analyzed in future, to find if they agree with the ones derived in the state of the art. This trial has involved 7 healthy and young subjects (5 males, 2 females, age: 30 ± 8 years, weight: 70 ± 12 kg, height: 177 ± 9 cm).

3.2 Data Processing and Features Extraction

The data acquired within the trials have been processed in MATLAB environment. The first elaboration is the implementation of a previous algorithm found in literature [10]. Figure 4 enumerates the different steps that will be described below.

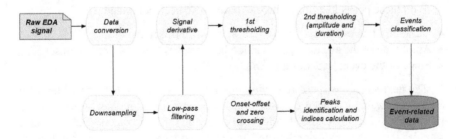

Fig. 4 Processing algorithm for EDA signal manipulation and features extraction

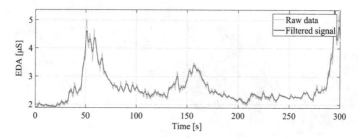

Fig. 5 Original EDA signal and filtered one within the first minutes of a trial

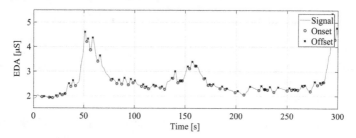

Fig. 6 Identification of all onsets and offsets of EDA signal

First, the raw EDA signal is converted to provide the correct measurement unit (microSiemens, μS). According to the BITalino specifications, this is achieved by converting the values sampled from the channel into a sensor resistance value (MΩ) and then by inverting this value. After this step, a down-sampling is performed, from 1 kHz to 20 Hz. The application of a low-pass 1.5 Hz IIR filter (Butterworth, 1st order) allows to reduce the signal noise, as shown in Fig. 5.

Then, a derivative filter is applied to highlight the differences in the signal slope. The interest is then focused on the positive part of this signal only (i.e., the first threshold applied in this stage). A zero-crossing detection algorithm has been implemented to identify the onset and offset of each EDA event (i.e., they have been identified as the local minimum and maximum of EDA signal and so the zeros of the derivative waveform) (Fig. 6).

With these peaks identified, three features have been computed:

- Duration of an event [s]: time deviation between offset and onset;
- Amplitude of the event [μS]: deviation between EDA at offset and at onset;
- Slope of the event [μS/s]: ratio Amplitude/Duration.

The next step consists of removing possible outliers from the events identified: a threshold value has been identified for Duration feature and has been applied at this stage. According to the algorithm described in [10], the events with number of samples lower than Fs/2 are detected and removed from the list. Next, the last and first sample of two consecutive detected events are evaluated, and the second event is unconsidered if the distance between them is lower than Fs $*$ 5. Finally, these events have been divided into three groups, according to the level of *Arousal* from literature.

Parallel to this evaluation, the open-source Ledalab software has been investigated. This has proven to provide accurate algorithms for the discrimination of EDA related events and is adopted frequently as comparison methodology [26, 27]. According to the results of the previous works in the state of art, the Continuous Decomposition Analysis methodology (CDA) has been adopted to extract EDA features, which are:

- nSCR: number of significant events within the response window;
- Latency [s]: latency of first significant SCR (Skin Conductance Response);
- AmpSum [μS]: sum of SCR-amplitudes of significant SCR;
- SCR [μS]: average phasic driver, representing phasic activity;
- ISCR [μS $*$ s]: area (i.e. time integral) of phasic driver;
- PhasicMax [μS]: maximum value of phasic activity;
- Tonic [μS]: Mean tonic activity of decomposed components.

Detailed information about the decomposition algorithm and features computed can be found in [11]. The second kind of elaboration has consisted in merging the quantities collected (e.g., Duration, Amplitude and Slope computed with the previous algorithm) into a unique indicator. As presented in a different work [28], the indicator collected for the analysis has been the area of a polygon, whose apices are the quantities computed in the previous stage. The equation for the area is the following:

$$IN3 = log_{10}\left(\frac{1}{2} \cdot sin(\alpha) \cdot (Am \cdot Du + Du \cdot Sl + Sl \cdot Am)\right) \qquad (1)$$

where: α is the angle between apices (i.e., 120°); *Am* and *Du* are the computed Amplitude of EDA events [μS] and the computed Duration of the events [s] respectively and *Sl* is the Slope [μS/s]. Similarly, for the data provided from the Ledalab software, the equation to get data from the quantities collected (i.e., Latency, AmpSum, SCR, iSCR, PhasicMax) is:

$$IN5 = log_{10}\left(\frac{1}{2} \cdot sin(\beta) \cdot (La \cdot Am + Am \cdot SCR + SCR \cdot iSCR + iSCR \cdot Ph + Ph \cdot La)\right) \qquad (2)$$

where β is in this case equal to 72°.

4 Analysis of Results

EDA events, identified through the application of the processing algorithm, have been grouped into three categories, as cited above, according to the different *Arousal* level. A possible event, correlated with the projection of a picture, is searched in a time window of about 15 s. Then, for each subject, the investigation has interested the number of events identified and the three features (Duration, Amplitude and

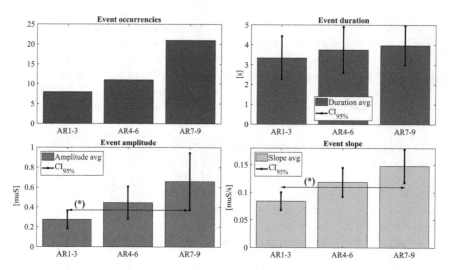

Fig. 7 Example of EDA processing result for Subject 1 (*: statistical difference, p<0.005)

Slope), particularly the average value and the 95% confidence interval. An example is reported in Fig. 7.

As expecting, the number of EDA events increases with the level of *Arousal* perceived by the subject and the same occurs for the numerical values of the features computed (i.e., an increase in the emotional state generates events with higher Amplitude, major Duration, etc.).

In some cases, the differences are statistically relevant, meaning that it could be possible to identify possible thresholds to discriminate these situations (e.g., a threshold value for stress level). This is an interesting preliminary result that may induce the authors to continue the investigation of this methodology. The second analysis has interested the application of the open source software Ledalab and the features extracted with the application of the algorithm embedded. The results obtained (Fig. 8) appears to be in line with the ones obtained from the first processing. In fact, features related to the Amplitude of the EDA signal (i.e., AmpSum, SCR, Phasic) are correlated with the increase of the *Arousal* index. This result suggests that the features extracted by the CDA methodology should be investigated better, i.e., to include them in the algorithm described previously. Then, the novel indicator (IN3 for the first analysis, IN5 for the case of data from Ledalab) has been computed. Figure 9 shows an example of the variations in the area computed, according to an increase in *Arousal* level. It appears that a major effect, so an increase in the area, occurs in the first case, meaning that the algorithm adopted can better discriminate the differences in user's perception.

A statistical analysis, by means of 95% Confidence Interval (CI) for t-Student distribution, has been conducted to verify if the three categories are statistically different one from another.

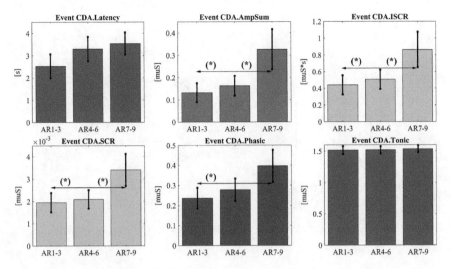

Fig. 8 Results for the application of Ledalab processing tools for all the subjects involved in the trial (*: statistical difference occurred, $p < 0.005$)

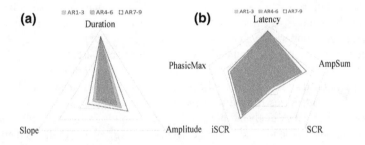

Fig. 9 Polygons computed from the quantities measured. **a** Features from processing algorithm. **b** features from Ledalab software

As shown in Fig. 10, it appears that the index proposed is able to discriminate, with statistical significance, only the extreme conditions (i.e., $Arousal < 3$ and $Arousal > 6$). This result agrees with the one obtained in Fig. 7 for the Slope of the EDA events, suggesting that the proposed indicator is valid. However, a deeper investigation is required to assess if other algorithms or methodologies allow to get more accurate information, or more discrimination power.

5 Conclusions

In this work, the authors have investigated the state of the art and described the application of an EDA measuring device for the assessment of the arousal level of users [10]. Trials have been conducted with the aim of evaluating if the features computed

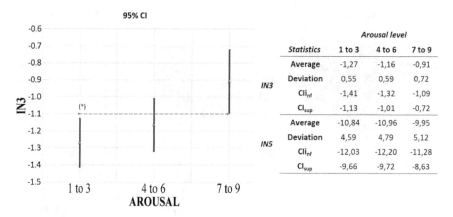

Fig. 10 IN3 95% CI (left) and numerical values for the indicators IN3 and IN5 (right)

from EDA (i.e., Duration, Amplitude, Slope) are correlated with the emotional state of the user (*Arousal* level). The application of the processing algorithm has allowed the identification of such features from the raw EDA signals and results have validated the hypothesis of direct correlation between the increase in the emotional state and higher numerical values of the EDA features. The same results have been obtained for the quantities calculated with the Ledalab software, suggesting that both the processing methodologies are able to discriminate these correlations. The novel indicator proposed and discussed in the paper allows to discriminate not all the cases (i.e., the three categories related to *Arousal*), but is accurate, in case of high user perception. Such EDA analysis methodology can be tuned for different scenarios and applications, to identify additional correlations, e.g. related to thermal comfort, as in [29].

References

1. Mowafey, S., Gardner, S.: A novel adaptive approach for home care ambient intelligent environments with an emotion-aware system. In: Proceedings of the 2012 UKACC International Conference on Control, CONTROL 2012, pp. 771–777 (2012)
2. Martínez Monseny, A., Bobillo Pérez, S., Martínez Planas, A., García García, J.J.: The role of complementary examinations and home monitoring in patient at risk from apparent life threatening event, apneas and sudden infant death syndrome. An. Pediatr. **83**(2), 104–8 (2015)
3. Tsukiyama, T.: In-home health monitoring system for solitary elderly. Procedia Comput. Sci. **63**, 229–235 (2015)
4. Serhani, M.A., El Menshawy, M., Benharref, A.: SME2EM: smart mobile end-to-end monitoring architecture for life-long diseases. Comput. Biol. Med. **68**, 137–154 (2016)
5. Scalise, L., et al.: Implementation of an 'at-home' e-Health system using heterogeneous devices. In: IEEE 2nd International Smart Cities Conference: Improving the Citizens Quality of Life, ISC2 2016—Proceedings (2016)
6. Fernández-caballero, A., et al.: Improvement of the elderly quality of life and care through smart emotion regulation. Ambient Assist. Living Dly. Act. **8868**, 348–355 (2014)

7. Koelstra, S., et al.: DEAP: a database for emotion analysis; using physiological signals. IEEE Trans. Affect. Comput. **3**(1), 18–31 (2012)
8. Healey, J.A., Picard, R.W.: Detecting stress during real-world during tasks using physiological sensors. IEEE Trans. Intell. Transp. Syst. **6**(2), 156–166 (2005)
9. Koji, N., Nozawa, A., Ide, H.: Evaluation of emotions by nasal skin temperature on auditory stimulus and olfactory stimulus. IEEJ Trans. Electron. Inf. Syst. **124**(9), 1914–1915 (2004)
10. Martínez-Rodrigo, A., et al.: Arousal level classification of the aging adult from electro-dermal activity: from hw development to sw architecture. Pervasive Mob. Comput. **34**, 46–59 (2017)
11. Benedek, M., Kaernbach, C.: A continuous measure of phasic electrodermal activity. J. Neurosci. Methods **190**(1), 80–91 (2010)
12. Benedek, M., Kaernbach, C.: Decomposition of skin conductance data by means of nonnegative deconvolution. Psychophysiology **47**(4), 647–658 (2010)
13. Ayata, D., Yaslan, Y., Kamasak, M.: Emotion recognition via random forest and galvanic skin response: Comparison of time based feature sets, window sizes and wavelet approaches. In: 2016 Medical Technologies National Conference, TIPTEKNO 2016 (2017)
14. Nourbakhsh, N., et al.: Using galvanic skin response for cognitive load measurement in arithmetic and reading tasks. In: Proceedings of the 24th Conference on Australian Computer Interaction OzCHI'12, pp. 420–423 (2012)
15. Phitayakorn, R., Minehart, R.D., Pian-Smith, M.C.M., Hemingway, M.W., Petrusa, E.R.: Practicality of using galvanic skin response to measure intraoperative physiologic autonomic activation in operating room team members. Surgery **158**(5), 1415–1420 (2015)
16. Bach, D.R., et al.: Stress detection from speech and galvanic skin response signals. J. Neurosci. Methods **3820**(November), 209–214 (2013)
17. Liu, M., Fan, D., Zhang, X., Gong, X.: Human emotion recognition based on galvanic skin response signal feature selection and SVM. In: 2016 International Conference on Smart City and Systems Engineering (ICSCSE), pp. 157–160 (2016)
18. Larradet, F., Barresi, G., Mattos, L.S.: Effects of galvanic skin response feedback on user experience in gaze-controlled gaming: a pilot study. In: Proceedings of the Annual International Conference of the IEEE Engineering in Medicine and Biology Society, EMBS, pp. 2458–2461 (2017)
19. Lynar, E., Cvejic, E., Schubert, E., Vollmer-Conna, U.: The joy of heartfelt music: an examination of emotional and physiological responses. Int. J. Psychophysiol. **120**, 118–125 (2017)
20. Plácido da Silva, H., Guerreiro, J., Lourenco, A., Fred, A., Martins, R.: BITalino: A Novel Hardware Framework for Physiological Computing (2014)
21. Dunn, B.D., Billotti, D., Murphy, V., Dalgleish, T.: The consequences of effortful emotion regulation when processing distressing material: a comparison of suppression and acceptance. Behav. Res. Ther. **47**(9), 761–773 (2009)
22. Aldhafeeri, F., Mackenzie, I., Kay, T., Alghamdi, J., Sluming, V.: Regional brain responses to pleasant and unpleasant IAPS pictures: different networks. Neurosci. Lett. **512**(2), 94–98 (2012)
23. Choi, K.-H., Kim, J., Kwon, O.S., Kim, M.J., Ryu, Y.H., Park, J.-E.: Is heart rate variability (HRV) an adequate tool for evaluating human emotions?—a focus on the use of the International Affective Picture System (IAPS). Psychiatry Res. **251**, 192–196 (2017)
24. Aluja, A., et al.: Personality effects and sex differences on the international affective picture system (IAPS): a Spanish and Swiss study. Pers. Individ. Dif. **77**(Suppl. C), 143–148 (2015)
25. Thurston, M.D., Cassaday, H.J.: Conditioned inhibition of emotional responses: retardation and summation with cues for IAPS outcomes. Learn. Motiv. **52**, 69–82 (2015)
26. Bach, D.R.: A head-to-head comparison of SCRalyze and Ledalab, two model-based methods for skin conductance analysis. Biol. Psychol. **103**(1), 63–68 (2014)
27. Kelsey, M., et al.: Applications of sparse recovery and dictionary learning to enhance analysis of ambulatory electrodermal activity data. Biomed. Signal Process. Control **40**, 58–70 (2018)

28. Pietroni, F., Casaccia, S., Revel, G.M., Scalise, L.: Methodologies for continuous activity classification of user through wearable devices: feasibility and preliminary investigation. In: 2016 IEEE Sensors Applications Symposium (SAS), pp. 1–6 (2016)
29. Gerrett, N., Redortier, B., Voelcker, T., Havenith, G.: A comparison of galvanic skin conductance and skin wettedness as indicators of thermal discomfort during moderate and high metabolic rates. J. Therm. Biol **38**(8), 530–538 (2013)

Smart Transducers for Energy Scavenging and Sensing in Vibrating Environments

Slim Naifar, Carlo Trigona, Sonia Bradai, Salvatore Baglio and Olfa Kanoun

Abstract The possibility to scavenge energy from vibration and to measure, at the same time, additional information, such as physical characteristics of the incoming source of energy, is of great interest in the modern research. This includes autonomous sensing elements, smart transducers and innovative methods of measurements also in the context of "industry 4.0". The pursued approach concerns an electromagnetic transducer able to harvest energy coming from the environment (kinetic source of energy), as consequence, charges will be accumulated inside a storage capacitor. It is also capable to measure the mechanical power and transmits the information by using an optical method. It is worth noting that the proposed architecture works without conditioning circuits or active elements. The smart transducer for energy scavenging is designed and experiments are performed showing the suitability of the proposed device.

Keywords Smart transducer · Energy harvesting · Mechanical power sensor
Electromagnetic mechanism · Vibrating environments · Industry 4.0

1 Introduction

With the wide advancements in the field of autonomous sensor nodes [1], self-sustained measurement systems [2] and smart WSNs [3], the interest in smart devices

S. Naifar · C. Trigona · S. Bradai · O. Kanoun
Technische Universität Chemnitz, ReichenhainerStraße 70, 09126 Chemnitz, Germany
e-mail: slim.naifar@etit.tu-chemnitz.de

C. Trigona (✉) · S. Baglio
D.I.E.E.I, Dipartimento di Ingegneria Elettrica Elettronica e Informatica, University of Catania, Viale Andrea Doria 6, 95125 Catania, Italy
e-mail: carlo.trigona@dieei.unict.it

S. Naifar · S. Bradai
Laboratory of Electromechanical Systems, National Engineering School of Sfax,
University of Sfax, Route de Soukra km 4, 3038 Sfax, Tunisia

© Springer Nature Switzerland AG 2019
B. Andò et al. (eds.), *Sensors*, Lecture Notes in Electrical Engineering 539,
https://doi.org/10.1007/978-3-030-04324-7_70

has extremely increased also from the point of view of energy harvesters [4, 5]. In this context, significant progresses have been made in several fields including energy scavenging and novel transduction mechanisms [5] used in telecommunications, Internet of Things (IoT), automotive, biomedical and industrial equipments [6–9].

This latter area of application arouses also interest in the perspective to supply or to sustain innovative sensors, MEMS and intelligent transducers [10–12] in the context of the industry 4.0 [13]. Literature presents several methods and architectures in order to scavenge energy from available sources including light, wind, thermal gradient, RF and kinetic vibration of devices or environments [4]. It is worth noting that these sources can be converted into electrical power through the adoption of active materials (PZT, PVDF, etc.), by using electrostatic principles, magnetoelectric principles or through the adoption of electromagnetic transducers [14–17].

In this work, the attention will be focused on kinetic vibration, which represents the most abundant and ubiquitous surrounding energy. The conversion mechanism regards an electromagnetic approach in the perspective to generate greater power density compared with other approaches and longer life time [15].

It must be observed that literature shows several examples of energy harvester, novel designs, mechanisms and innovative principles in order to improve its performance also in terms of capability to generate more power, to increase its efficiency or to modify the spectral response [4, 5, 17]. However, few works have been presented concerning the possibility to use the same transducer as a "smart" device. This enables to scavenge energy from the environment and, at the same time, to sense physical quantities with the prerogative to be passive and to transmit the information of the measurand. In fact, several research groups present devices based on various transduction principles but, very often, focusing the attention on a specific prototype which is only able to store energy or to perform only measurements and to transmit the data by using an active circuit [18].

A preliminary study of a prototype able to perform the mentioned functions is accomplished in [19], where authors propose a piezoelectric transducer able to store energy in a capacitor and simultaneously to measure the external level of acceleration.

The presented work is based on an electromagnetic transducer which is able to scavenge energy from ambient vibration, to measure the level of kinetic power applied to the transducer and to transmit its information through an optical readout and an optical fiber.

It is worth to mention that the proposed architecture is a battery-free electromagnetic oscillator able to work in vibrating environments also in presence of very weak power kinetic sources (in the order of tens of μW). The paper is organized as follows: Sect. 2 presents the working principle of the proposed device; Sect. 3 is devoted to the description of the experimental setup and Sect. 4 to the experimental results and to the characterization of the conceived device.

2 Theory of Operation

This section presents the working principle of the smart device used as energy scavenger and sensor of kinetic power in vibrating environments. The architecture is composed of three sections: (1) a mechanical oscillator (mechanical transducer) able to convert external vibration (input source such as induced movements, shock and impressed kinetic energies) into oscillation of a structure composed of an electromagnetic transducer which is a second order mechanical oscillator (mass-spring-damper system).

The proposed transducer as shown in Fig. 1a is based on the use of a mechanical spring with low stiffness ~60 N/m where a coil is attached, furthermore a proof mass ~3 g is located at the tip of the spring. In order to generate electrical power, magnets are fixed and surrounded by the moving coil. In presence of low-level external kinetic vibration, a variable output voltage (induced Lorentz voltage) will be generated.

(2) A conditioning architecture is used to scavenge the energy coming from the vibrating environment and storing the charges inside a load capacitor (see Fig. 1b) through the adoption of a full bridge rectifier which is composed of three silicon diodes and a Light Emitting Diode (LED); (3) a transmission method based on this latter element (the LED) is used to transmit the information of the input kinetic source by using an optical readout strategy and a fiber to connect the conceived device with an optical receiver. It is worth noting that the LED will intermittently turn ON and OFF when the kinetic vibration level is higher than the threshold value of the diode. The output signal, which is a square waveform, is correlated with the amplitude of the input signal (the mechanical power).

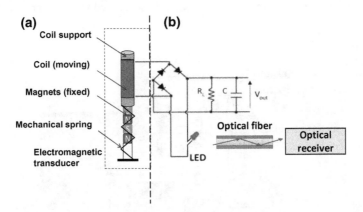

Fig. 1 Schematic of the proposed device: **a** is the proposed transducer and **b** is the conditioning architecture

3   Experimental Setup

The experimental setup is composed of a shaker (VebRobotron Type 11077) driven
by a function generator in order to emulate the external kinetic vibration. A feedback
displacement sensor (Opto NCDT1800) is placed on the shaker close to the electromagnetic transducer in order to monitor the amplitude and the waveform of the
applied vibration. A full bridge rectifier composed of three silicon diodes and a LED
diode is used as an AC/DC converter, in order to harvest the energy in a load capacitor of 22 μF. It is also used to transmit the information (level of kinetic vibration)
by using the LED. The experimental setup includes a photo-receiver (HFBR-2523Z
Series from Avago Technologies) which gives a zero value when the diode is ON,
and a high value when the diode is OFF. As will be presented in the experimental
section, the output signal measured across the optical receiver is correlated with
the amplitude of the kinetic source. Figure 2 shows the experimental setup used to
characterize the device.

4   Results and Discussion

The task to be accomplished in this experiment is the characterization of the harvester
used as a sensor to measure mechanical power. For this purpose, the sensor was
tested with a real vibration profile measured in a lawn mower and reproduced by an
electrodynamic shaker with different scale factors.

Figure 3 represents V_{out}, which refers to the voltage across the capacitor (see
Fig. 1), for 9 different scale factors of the profile. The mechanical powers are calculated by using the Root Mean Square (RMS) of the applied vibration amplitudes and

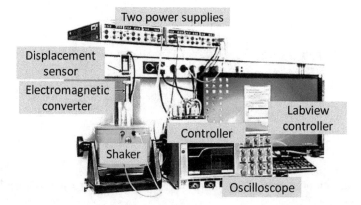

Fig. 2 Experimental setup used for the characterization of the smart device

Fig. 3 Voltage across the capacitor for different applied mechanical powers

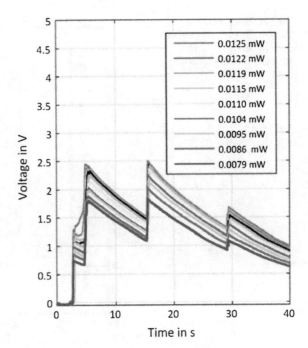

the corresponding accelerations, obtained through the laser displacement sensor for each applied scale factor.

As can be seen in Fig. 3, the maximum output voltages obtained at 7.9 μW and 12.5 μW are around 1.82 V and 2.49 V respectively.

Furthermore, in order to evaluate the performance of the device as sensor, the voltage across the optical receiver (HFBR-2523Z) is analysed. Figure 4 shows the voltage output of the bridge rectifier, V_{out}, and the optical receiver, V_{opt}, for 12.5 μW mechanical power. The aim is to define a relation between the applied mechanical power and the number of pulses in the voltage output of the optical receiver.

The output voltage of the optical receiver is shown in Fig. 5 for four applied scale factors which corresponds to mechanical powers from 10.4 to 12.5 μW. Results indicate that no pulses are obtained for 10.4 μW. It is noted that 1 pulse is observed for 11 μW, 11.9 μW and 5 pulses are detected for 12.5 μW input power respectively (Fig. 5).

Figure 6 shows the calibration diagram of the transducer based on the number of pulses. Five measurements are done for each of the applied mechanical power. The graph shows the linear interpolation of the mean value of the number of pulses (red line) and the two blue dashed lines represent the maximum uncertainty. The number of pulses in the voltage across the optical receiver is evaluated for four excitation amplitudes which correspond to mechanical powers from 10.4 to 12.5 μW. The maximum measured uncertainty is 1.25 pulses. The sensitivity of the sensor in this

Fig. 4 Voltage output of the bridge rectifier (V_{out}) and the optical receiver (V_{opt}) for 12.5 μW mechanical power

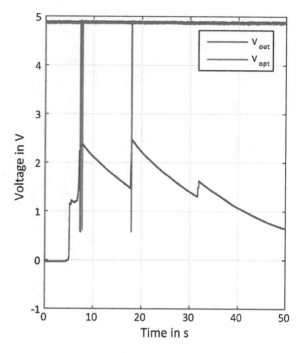

Fig. 5 Output voltages across the optical receiver (V_{opt})

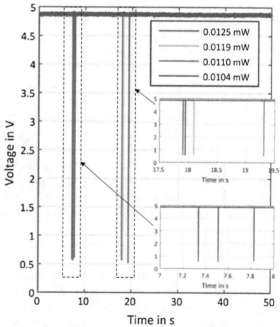

Fig. 6 Calibration diagram
for the sensor based on the
number of pulses

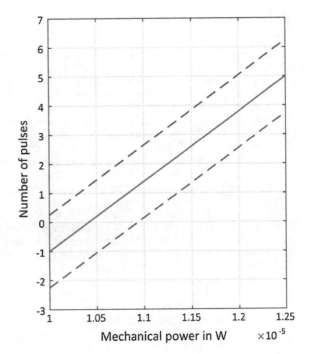

case is about to 2.4 pulses/µW and the resolution estimated during the metrological
characterization corresponds to around 11 µW.

5 Conclusion

In this work an electromagnetic energy transducer, which is capable to harvest kinetic
energy from ambient vibration, to measure the applied mechanical power to the
transducer and to transmit the measurement results through an optical readout, is
conceived, realized and characterized.

The developed system is studied as energy harvester in order to charge a load
capacitor in presence of real vibration applied with different scale factors. In addition,
the system is characterized as a mechanical power sensor. Results indicate that the
sensor has a sensitivity of about 2.4 pulses/µW. The resolution estimated during the
metrological characterization corresponds to 11 µW.

References

1. Toh, W.Y., Tan, Y.K., Koh, W.S., Siek, L.: Autonomous wearable sensor nodes with flexible energy harvesting. IEEE Sens. J. **14**(7), 2299–2306 (2014)
2. Zhang, B., Zhang, L., Deng, W., Jin, L., Chun, F., Pan, H., Wang, Z.L.: Self-powered acceleration sensor based on liquid metal triboelectric nanogenerator for vibration monitoring. ACS Nano **11**(7), 7440–7446 (2017)
3. Andò, B., Baglio, S., La Malfa, S., Pistorio, A., Trigona, C.: A smart wireless sensor network for AAL. In: 2011 IEEE International Workshop on Measurements and Networking Proceedings (M&N), pp. 122–125 (2011)
4. Kazmierski, T.J., Beeby, S.: Energy Harvesting Systems. Springer (2014)
5. Trigona, C., Dumas, N., Latorre, L., Andò, B., Baglio, S., Nouet, P.: Exploiting benefits of a periodically-forced nonlinear oscillator for energy harvesting from ambient vibrations. Procedia Eng. **25**, 819–822 (2011)
6. Kamalinejad, P., Mahapatra, C., Sheng, Z., Mirabbasi, S., Leung, V.C., Guan, Y.L.: Wireless energy harvesting for the Internet of Things. IEEE Commun. Mag. **53**(6), 102–108 (2015)
7. Zhou, F., Joshi, S.N., Dede, E.M.: Thermal energy harvesting with next generation cooling for automotive electronics. In: Thermal Management of Onboard Charger in E-Vehicles Reliability of Nano-sintered Silver Die Attach Materials Thermal Energy Harvesting with, vol. 16 (2017)
8. Chen, Z., Law, M.K., Mak, P.I., Martins, R.P.: A single-chip solar energy harvesting IC using integrated photodiodes for biomedical implant applications. IEEE Trans. Biomed. Circuits Syst. **11**(1), 44–53 (2017)
9. Tuna, G., Gungor, V.C., Gulez, K.: Energy harvesting techniques for industrial wireless sensor networks. In: Hancke, G.P., Gungor, V.C. (eds.) Industrial Wireless Sensor Networks: Applications, Protocols, Standards, and Products, pp. 119–136 (2017)
10. Andò, B., Baglio, S., L'Episcopo, G., Marletta, V., Savalli, N., Trigona, C.: A BE-SOI MEMS for inertial measurement in geophysical applications. IEEE Trans. Instrum. Meas. **60**(5), 1901–1908 (2011)
11. Naifar, S., Bradai, S., Viehweger, C., Kanoun, O.: Survey of electromagnetic and magneto-electric vibration energy harvesters for low frequency excitation **106**, 251–263 (2017)
12. Roundy, S., Rabaey, J.M., Wright, P.K.: Energy Scavenging for Wireless Sensor Networks. Springer, New York, LLC (2004)
13. Bloem, J., Van Doorn, M., Duivestein, S., Excoffier, D., Maas, R., Van Ommeren, E.: The Fourth Industrial Revolution. Things to Tighten the Link Between IT and OT (2014)
14. Spies, P., Pollak, M., Mateu, L.: Handbook of Energy Harvesting Power Supplies and Applications. CRC Press (2015)
15. Shepard Jr., J.F., Chu, F., Kanno, I., Trolier-McKinstry, S.: Characterization and aging response of the d 31 piezoelectric coefficient of lead zirconate titanate thin films. J. Appl. Phys. **85**(9), 6711–6716 (1999)
16. Naifar, S., Bradai, S., Keutel, T., Kanoun, O.: Design of a vibration energy harvester by twin lateral magnetoelectric transducers. In: IEEE International Instrumentation and Measurement Technology Conference I2MTC, pp. 1157–1162 (2014)
17. Bradai, S., Naifar, S., Keutel, T., Kanoun, O.: Electrodynamic resonant energy harvester for low frequencies and amplitudes. In: IEEE International Instrumentation and Measurement Technology Conference I2MTC, pp. 1152–1156 (2014)
18. Lillesand, T., Kiefer, R.W., Chipman, J.: Remote Sensing and Image Interpretation. Wiley (2014)
19. Beninato, A., Trigona, C., Ando, B., Baglio, S.: A PZT-based energy sensor able to store energy and transmit data. In: IEEE Sensors Applications Symposium (SAS), pp. 1–5 (2017)

RMSHI Solutions for Electromagnetic Transducers from Environmental Vibration

Sonia Bradai, Carlo Trigona, Slim Naifar, Salvatore Baglio and Olfa Kanoun

Abstract The demand for harvesting energy from ambient has increased due to the advancement in the field of smart autonomous systems where a self-power source is needed. Kinetic vibration presents one of the main interesting and available source in the environment. However, to store energy from such source, different design requirements should be achieved considering the environmental vibration properties (hundreds of Hz and at low vibration levels, less than few m/s^2). It should be also noted that only hundreds of mV can be generated from vibration converters. In this work, an energy harvester system based on an electromagnetic converter and a passive energy management circuit based on the Random Mechanical Switching Harvester on Inductor (RMSHI) architecture are developed. Results show that also in presence of a generated voltage less than 100 mV, it is possible to store the energy inside a load capacitor. Further, the use of the proposed approach, based on mechanical and passive switch, enables to improve significantly the voltage outcome.

Keywords Electromagnetic energy harvesting · Bistable-RMSHI
Mechanical switches · Wideband vibration · Zero voltage threshold
Random excitation

S. Bradai · C. Trigona (✉) · S. Naifar · O. Kanoun
Technische Universität Chemnitz, Reichenhainer Straße 70, 09126 Chemnitz, Germany
e-mail: carlo.trigona@dieei.unict.it

S. Bradai
e-mail: sonia.bradai@etit.tu-chemnitz.de

S. Bradai · S. Naifar
Laboratory of Electromechanical Systems, National Engineering School of Sfax,
University of Sfax, Route de Soukra km 4, 3038 Sfax, Tunisia

C. Trigona · S. Baglio
D.I.E.E.I., Dipartimento di Ingegneria Elettrica Elettronica e Informatica,
University of Catania, Viale Andrea Doria 6, 95125 Catania, Italy

© Springer Nature Switzerland AG 2019
B. Andò et al. (eds.), *Sensors*, Lecture Notes in Electrical Engineering 539,
https://doi.org/10.1007/978-3-030-04324-7_71

1 Introduction

Self-powered systems present a technological solution for different applications where safety and accessibility are limited [1]. One of the promising solution to power such systems is to harvest energy from environmental sources [2]. During the last decades, different researches have been conducted to design transducers to harvest energy and to power wireless systems or recharge batteries from ambient sources such as thermal gradients [3], kinetic vibration [4], solar [5], etc. Due to its availability in indoor and outdoor environment and its relative high energy density compared to the others [6], vibration presents an interesting available source [7]. Nevertheless, several challenges have to be overcome in this case: first, real vibration profiles have different forms of excitation correlated to the specific application and are more characterized with broadband frequency [8]. Second, real vibration profiles are characterized by a frequency limited to several hundreds of Hz and an amplitude of displacement to several millimeters or centimeters [7–9]. Therefore, harvesting and storing energy from a real vibration profile is very challenging. Different principles can be used, such as: piezoelectric [9, 10], electromagnetic [11, 12] and magnetoelectric [13, 14] and also approaches based on novel active materials [15]. In this paper we will focus the attention on electromagnetic transducer considering its higher performances in terms of efficiency and density of generated power as respect the other already mentioned principles [6]. Based on the state of the art, two main aspects should be taken into account to design the electromagnetic transducer. The first aspect considers its mechanical design which can be based on: (1) the use of a moving magnetic mass and the exploitation of nonlinear behaviors which are typically used [16] to increase the bandwidth of response. However, this concept cannot be used in metallic environments which can influence the dynamic of the moving magnets. (2) the use of a moving coil with fixed magnets. In this case the macroscale system is more linear and the generated voltage at the mechanical resonant frequency is typically limited to hundreds of millivolts [11]. The second aspect is to rectify and store the generated voltage from the transducer. In this case, existing solutions present several limitations. In fact, approaches based on diode bridge rectifiers impose a threshold of about 1.4 V. This represents a limit considering the low-amplitude output generated from the transducer, in this context some solution has been proposed in literature in order to overcome this limit [9]. Other solutions like SSHI (Synchronized Switch Harvesting on Inductor) [17] are useful only in the case of a sinusoidal output voltage which is not the case for real applied excitations where the generated voltage profile is random. It should be noted that this approach is based on active switches (transistor or MOSFET). In this paper, an energy harvester able to scavenge energy from random real profile excitation is presented. The system is based on electromagnetic principle using a moving coil which enables its implementation in different environments. As well an optimized passive management circuit is implemented to store the low-level energy which is based on the RMSHI approach proposed in [18]. It is worth noting that the system here presented is suitable to be used as scavenger or sustainer for sensors [19] and autonomous multi-sensors architectures [20] which can be applied for several fields including telecommunications, automotive, civil engineering, healthcare systems, geophysical [21–24]. The paper is structured as follows: In Sect. 2, the working principle of the proposed device is presented. Section 3 is devoted to present the experimental setup and Sect. 4 for the characterization of the realized system.

2 Working Principle

This section concerns the working principle of the conceived system. The proposed electromagnetic harvester is based on the use of a mechanical spring where a coil is attached and moving in presence of external vibration. 10 ring magnets are placed inside the coil support, as shown in Fig. 1. One of the main limitation of converter based on a moving coil is overcome with the proposed design based on a contactless solution between the coil wires and the mechanical spring which leads to a better robustness for the entire system. Under a specific applied external vibration (i.e. less than 9.81 m/s^2 in presence of wideband noise limited to 50 Hz), an output voltage with the level of few hundreds of mV is generated.

Considering this low-level voltage and the presence of wideband kinetic signals, in order to store the energy, we propose a bistable RMSHI management circuit. Due to its nonlinear (bistable) behavior, the performances of the entire system are improved in the case of real vibration [7, 18], in this specific case we have used lawn mower. The RMSHI bistable circuit is based on the use of a diode bridge, a capacitor, an inductor and a mechanical switch which consists of a cantilever brass-beam and two stoppers. A permanent magnet is attached at the tip of the beam and a fixed magnet is placed in the front of the oscillator in repulsive force condition to obtain the bistable behavior (Fig. 1). The figure also evinces the two mechanical stoppers (S_I, S_S) used as electrical contacts for the circuit. The system switches between the two stable states as consequence of the external applied vibration.

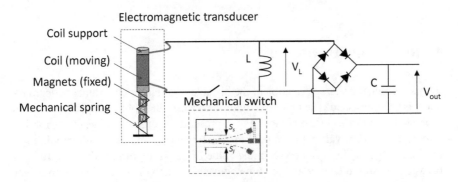

Fig. 1 Schematic of the proposed energy harvesting system

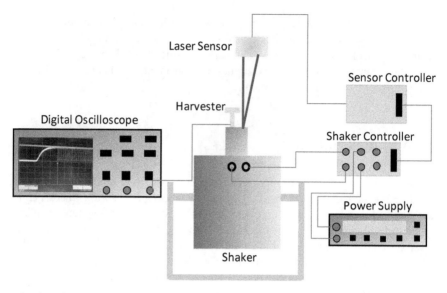

Fig. 2 Schematic for the used experimental setup

3 Experimental Setup

The proposed system is tested under real vibration profiles reproduced by using an electrodynamic shaker (VebRobotron Type 11077) controlled with a closed loop implemented through a LabVIEW routine. The applied profiles are measured using a laser sensor (Opto NCDT1800). The acquisitions are pursued by using an oscilloscope (LeCroyWaveRunner® 6050A). Two power supplies are used to power the laser sensor and the shaker. Figure 2 presents the schematic for the used experimental setup to evaluate the proposed system.

The realized harvester and management circuit parameters are presented in Table 1.

4 Results and Discussion

In this section, the characterization of the bistable RMSHI for electromagnetic transducer is accomplished. The first investigation concerns the optimal capacitance value C, as shown in Fig. 1. In this context, Fig. 3 presents the output voltage (V_{out}) for different capacitances varying from 0.22 to 22 μF when a simulated random vibration profile is applied. As it can be seen, the value of 22 μF represents the best tradeoff between the stored energy and the time constant of the system. A maximum power of about 2.5 nW has been obtained in about 5 s. This value is selected for the rest of the experiments.

Table 1 Parameters of the realized energy harvesting system

Parameter	Value
Electromagnetic converter housing volume	12.3 cm^3
Mechanical spring stiffness	60 N/m
Mass	3 g
Ring magnets	$6 \times 2 \times 2$ mm^3
Coil wire thickness	0.1 mm
Number of turns	600
Coil resistance	35 Ω
Energy management circuit volume	168 cm^3
C	22 μF

Fig. 3 Voltage output using a bridge rectifier with mechanical switch

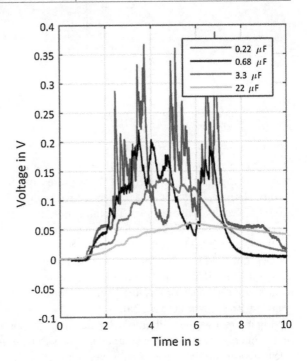

In the following, measurements were conducted using the vibration imposed by a lawn mower and reproduced by using the shaker with different scale factors in order to evaluate the energy output of the converter using the proposed bistable energy management circuit. Figure 4 shows the V_{out} and the corresponding peak to peak voltage output from the converter, V_L, for six different scale factors of the excitation. The output voltages through the converter are 0.06, 0.31, 0.55, 0.75, 0.88 and 0.91 V for the six applied scale factors.

Fig. 4 Voltage output using a bridge rectifier with mechanical switch

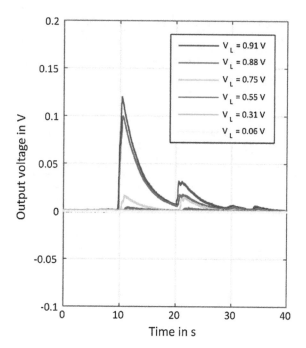

Figure 5 shows the output voltage across the capacitor and the corresponding voltage output from the converter with the presence of the bistable mechanical switch and in absence. The voltage V_L is measured with a differential probe with a scale of 1/10. Results shown in Fig. 5 demonstrate the possibility to harvest the energy in case of low output voltage from the converter which is less than ~100 mV due to the use of RMSHI solution. It must be observed that this value is below the diodes threshold voltage. The use of mechanical switch improves the voltage output from ~0.02 to ~0.5 V. This confirms that, for an applied random excitation, the use of a nonlinear behavior ensures better performance for the system compared to a standard bridge rectifier [25, 26].

Figure 6 shows the calibration diagram of the proposed system. The red line refers to the interpolation of voltage output and the two blue lines represent the maximum measured uncertainty, evaluated at one sigma, for the five measurements done for each vibration profile. The voltage output across the capacitor is measured for the six applied scale factors of the real vibration profile, which corresponds to mechanical powers from about 2 to 12.50 μW. The maximum estimated uncertainty is 0.04 V. The sensitivity of the sensor in this case is about to 0.011 V/μW.

Fig. 5 Comparison of output voltage using a bridge rectifier with and without mechanical switch

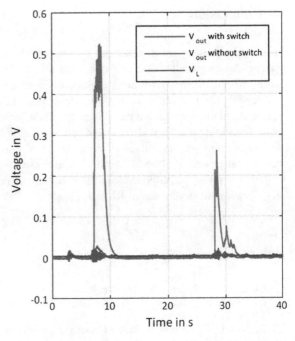

Fig. 6 Calibration diagram of the converter. The maximum uncertainty is evaluated at one sigma

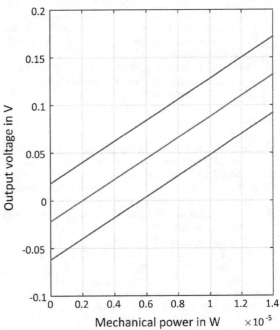

5   Conclusion

This work focuses on the nonlinear bistable RMSHI vibration energy harvesting system. An electromagnetic converter and a bistable based on RMSHI principle energy management circuit are developed and characterized using real vibration profile with different scale factors. It is demonstrated that the system is able to store an amount of energy within a passive energy management circuit also in presence of under threshold voltage, wideband and low amplitude kinetic signals.

A maximum output voltage of about 0.125 V was measured in presence of a mechanical power of about 12.50 μW. The sensitivity of the system corresponds to ~0.011 V/μW. The proposed solution is suitable to be used to collect energy from real vibration due to the use of RMSHI principle.

References

1. Choi, M., Sui, Y., Lee, I.H., Meredith, R., Ma, Y., Kim, G., Li, T.: Autonomous microsystems for downhole applications: design challenges, current state, and initial test results. Sensors **17**(10), 2190 (2017)
2. Kazmierski, T.J., Beeby, S.: Energy Harvesting Systems. Springer (2014)
3. Huesgen, T., Woias, P., Kockmann, N.: Design and fabrication of MEMS thermoelectric generators with high temperature efficiency. Sens. Actuators A **145–146**, 423–429 (2008)
4. Kim, D., Song, H., Khalil, H., Lee, J., Wang, S., Park, K.: 3-D vibration measurement using a single laser scanning vibrometer by moving to three different locations. IEEE Trans. Instrum. Meas. **63**(8) (2014)
5. Viehweger, C., Hartmann, B., Keutel, T., Kanoun, O.: Simulation of shading effects on the power output of solar modules for enhanced efficiency in photovoltaic energy generation. In: 2014 IEEE International Instrumentation and Measurement Technology Conference (I2MTC) Proceedings, pp. 610–613 (2014)
6. Zhu, D.: Sustainable Energy Harvesting Technologies—Past, Present and Future. INTECH, pp. 25 (2011)
7. Trigona, C., Dumas, N., Latorre, L., Andò, B., Baglio, S., Nouet, P.: Exploiting benefits of a periodically-forced nonlinear oscillator for energy harvesting from ambient vibrations. Procedia Eng. **25**, 819–822 (2011)
8. Andò, B., Baglio, S., Trigona, C.: Autonomous sensors: from standard to advanced solutions. IEEE Instrum. Meas. Mag. **13**(3), 33–37 (2010)
9. Maiorca, F., Giusa, F., Trigona, C., Andò, B., Bulsara, A.R., Baglio, S.: Diode-less mechanical H-bridge rectifier for "zero threshold" vibration energy harvesters. Sens. Actuators A **201**, 246–253 (2013)
10. Anton, S.R., Sodano, H.A.: A review of power harvesting using piezoelectric materials. Smart Mater. Struct. **16**, 1–21 (2007)
11. Bradai, S., Naifar, S., Viehweger, C., Kanoun, O.: Electromagnetic vibration energy harvesting for railway applications. In: MATEC Web Conference, International Conference on Engineering Vibration (ICoEV 2017), vol. 148, pp. 12004 (2018)
12. Bradai, S., Naifar, S., Viehweger, C., Kanoun, O.: Survey of electromagnetic and magnetoelectric vibration energy harvesters for low frequency excitation. Meas. J. **106**, 251–263 (2017)
13. Naifar, S., Bradai, S., Keutel, T., Kanoun, O.: Design of a vibration energy harvester by twin lateral magnetoelectric transducers. In: IEEE International Instrumentation and Measurement Technology Conference I2MTC, pp. 1157–1162 (2014)

14. Naifar, S., Bradai, S., Viehweger, C., Kanoun, O., Litak, G.: Response analysis of a nonlinear magnetoelectric energy harvester under harmonic excitation. Eur. Phys. J. Spec. Top. **224**(14), 2897–2907 (2015)
15. Bian, J., Wang, N., Ma, J., Jie, Y., Zou, J., Cao, X.: Stretchable 3D polymer for simultaneously mechanical energy harvesting and biomimetic force sensing. Nano Energy **47**, 442–450 (2018)
16. Bradai, S., Naifar, S., Keutel, T., Kanoun, O.: Electrodynamic resonant energy harvester for low frequencies and amplitudes. In: IEEE International Instrumentation and Measurement Technology Conference I2MTC, pp. 1152–1156 (2014)
17. Wu, L., Do, X.D., Lee, S.G., Ha, D.S.: A self-powered and optimal SSHI circuit integrated with an active rectifier for piezoelectric energy harvesting. IEEE Trans. Circuits Syst. I Regul. Pap. **64**(3), 537–549 (2017)
18. Giusa, F., Giuffrida, A., Trigona, C., Andò, B., Bulsara, A.R., Baglio, S.: Random mechanical switching harvesting on inductor: a novel approach to collect and store energy from weak random vibrations with zero voltage threshold. Sens. Actuators A **198**, 35–45 (2013)
19. Tsai, T.H., Chen, K.: A 3.4 mW photovoltaic energy-harvesting charger with integrated maximum power point tracking and battery management. In: 2013 IEEE International Solid-State Circuits Conference Digest of Technical Papers (ISSCC), pp. 72–73 (2013)
20. Porcarelli, D., Donati, I., Nehani, J., Brunelli, D., Magno, M., Benini, L.: Design and implementation of a multi sensors self sustainable wearable device. In: 2014 6th European Embedded Design in Education and Research Conference (EDERC), pp. 16–20 (2014)
21. Yuan, T., Duraisamy, B., Schwarz, T., Fritzsche, M.: Track fusion with incomplete information for automotive smart sensor systems. In: IEEE Radar Conference, RadarConf, pp. 1–4 (2016)
22. Cho, S., Spencer, B.F.: Sensor attitude correction of wireless sensor network for acceleration-based monitoring of civil structures. Comput.-Aided Civ. Infrastruct. Eng. **30**(11), 859–871 (2015)
23. Catarinucci, L., De Donno, D., Mainetti, L., Palano, L., Patrono, L., Stefanizzi, M.L., Tarricone, L.: An IoT-aware architecture for smart healthcare systems. IEEE Internet Things J. **2**(6), 515–526 (2015)
24. Andò, B., Baglio, S., L'Episcopo, G., Marletta, V., Savalli, N., Trigona, C.: A BE-SOI MEMS for inertial measurement in geophysical applications. Trans. Instrum. Meas. **60**(5), 1901–1908 (2011)
25. Fan, P.M.Y., Wong, O.Y., Chung, M.J., Su, T.Y., Zhang, X., Chen, P.H.: Energy harvesting techniques: energy sources, power management and conversion. In: 2015 European Conference on Circuit Theory and Design (ECCTD), pp. 1–4 (2015)
26. Zhu, D., Beeby, S.: Energy Harvesting Systems: Principles, Modeling and Applications, pp. 1–78. Springer (2011)

Characterization of Sensorized Porous 3D Gelatin/Chitosan Scaffolds Via Bio-impedance Spectroscopy

Muhammad Ahmed Khan, Nicola Francesco Lopomo, Mauro Serpelloni, Emilio Sardini and Luciana Sartore

Abstract Conductive scaffolds are highly used in tissue engineering for bone defect, nerve regeneration, cardiac tissue constructs and many others. Currently, most methods for monitoring cell activities on scaffolds are destructive and invasive such as histological analysis. The research aimed at sensorizing and characterizing a porous gelatin/chitosan scaffold, hence this "Intelligent Scaffold" can behave as a biosensor for evaluating cell behaviour (cell adhesion, proliferation) along with directing cellular growth. Thus, in this research, three-dimensional (3D) gelatin based scaffold has been transformed into conductive scaffold and both the scaffolds are characterized and compared in terms of their electrical conductivity. Carbon black has been used as a doping material to fabricate a Carbon-Gelatin composite conductive scaffold. The scaffolds are prepared by Freeze drying method and carbon black has been homogeneously embedded throughout the gelatin matrix. The scaffold behaviour was characterized by Bio-impedance Spectroscopy method. The preliminary experimental results showed that the conductivity of carbon-gelatin/chitosan scaffold increases around 10 times as compared to simple gelatin scaffold. Thus, these results elucidated the importance of carbon black clustering for development of a conductive network. This shows that carbon black provides conducting path and hence in future, even a small change of cellular activity can be determined by impedance fluctuation within the scaffold.

M. A. Khan · N. F. Lopomo · M. Serpelloni (✉) · E. Sardini
Department of Information Engineering, University of Brescia, Via Branze 38, Brescia, Italy
e-mail: mauro.serpelloni@unibs.it

M. A. Khan
e-mail: m.khan004@unibs.it

N. F. Lopomo
e-mail: nicola.lopomo@unibs.it

E. Sardini
e-mail: emilio.sardini@unibs.it

L. Sartore
Department of Mechanical and Industrial Engineering, University of Brescia, Via Branze 38, Brescia, Italy
e-mail: luciana.sartore@unibs.it

© Springer Nature Switzerland AG 2019
B. Andò et al. (eds.), *Sensors*, Lecture Notes in Electrical Engineering 539,
https://doi.org/10.1007/978-3-030-04324-7_72

Keywords Conductive scaffold · Carbon-Gelatin scaffold
Bio-impedance spectroscopy · Electrical conductivity

1 Introduction

Tissue engineering provides an alternative to conventional and classical transplantation methods, which involves regulation of tissue progression and cell behaviour via designing and development of synthetic extracellular matrix (ECM) that assists tissue regeneration [1, 2]. The fundamental approach used in tissue engineering includes the fabrication of scaffolds, which provide the suitable environment and stimuli for cell growth, differentiation, proliferation and supports functional tissue genesis [3–6].

Many different biopolymers have been engineered and investigated (natural and synthetic) for scaffolds fabrication. Among these, gelatin is widely used because of its unique characteristics including biodegradability, biocompatibility and hydrogel properties [7]. Gelatin is a protein fragment, formed by partial degradation of collagen fiber. It contains free carboxyl groups that can combine with chitosan to form a Gelatin/Chitosan network by means of hydrogen bonding [8, 9]. Gelatin/Chitosan hybrid scaffold has been reported to be useful and effective for skin, nerve, bone, muscles and cartilage tissue engineering [10–14]. Chitosan is a linear polysaccharide and is among the most abundant natural polymers found in nature. It has been widely used in tissue engineering because of its non-toxicity, biodegradability, biocompatibility and anti-bacterial effect [15]. It is produced by alkaline deacetylation of chitin [16–18] and has the capability to interact with adhesion proteins, growth factors and receptors [17, 19].

In addition, along with high strength, high electrical conductivity is also desirable in tissue engineering including neural tissue engineering [20], cardiac tissue engineering [21] and bone engineering [22]. Hence, in order to increase conductivity, carbon-based fillers such as graphene, carbon black, graphene oxide, carbon nanotubes and carbon fibers are widely used [23]. Carbon black fillers are preferred over metal fillers as they don't undergo oxidation, whereas, metal fillers get oxidized and creates an insulation layer on particles surface [24]. Other advantages of hybrid composites made from carbon black fillers include: light weight process capabilities, flexibility, low production costs and absorption of mechanical shock [25].

Along with tissue regeneration, monitoring of cellular activities on scaffold is also very critical in tissue engineering. Currently, most methods for monitoring cell activities are destructive and invasive such as histological analysis [26]. Thus, there is a need of non-invasive, user-friendly, robust and quick sensing technique that can provide continuous real-time monitoring of cellular activities. Hence, to provide a non-invasive solution, "Confocal imaging microscopy" is used for observing cell culture growth and differentiation with higher cellular resolution. However, this approach has its own limitations and needs optimization for tissue engineering application. In this method, the scaffold has to be sliced in thin pieces of 200–300 mm to allow enough light penetration. In addition, the scaffold material should have low auto-fluorescence along with optically transparent feature [27]. In order to overcome this limitation, another technique "electrochemical impedance spectroscopy (EIS)" has been introduced that monitors cell activity by correlating biological cellular phenomena with electrical impedance measurements. EIS has been proved as an efficient method for the real-time analysis of biological systems both in vitro [28–30] and in vivo [31]. According to Giaever and Keese [32, 33], as the cells spread and attach on a conductive surface, they change the area available for current flow and cause a fluctuation in the electrical impedance of system. Therefore, the change in impedance characteristics can be correlated with cell spreading, attachment and other cellular activities. This methodology was then used in [34] to measure the epithelial cells proliferation. Moreover, electrical impedance measurement is also used to monitor attachment, morphology and spreading of fibroblasts cells in culture [35].

Thus, in the presented research, 3D Gelatin/Chitosan (G/Ch) scaffolds are doped with carbon black (C), which transforms the simple G/Ch scaffold into conductive sensorized carbon-based composite scaffold (G/Ch/C). The scaffolds are then characterized by bio-impedance monitoring method. In first part of the paper, methodology and materials has been described; which includes scaffold fabrication and experimental setup used to perform scaffold characterization. Whereas, in the later section, results are discussed in terms of bio-impedance measurement and electrical conductivity analysis.

2 Materials and Methods

2.1 Scaffold Preparation

The simple scaffold is mainly composed of gelatin (G), chitosan (Ch) and poly(ethylene glycol) diglycidyl ether (PEG), whereas in carbon based scaffold, conductive carbon black (C) is added as an additional element. In order to prepare scaffold, 5 g of G was dissolved in 50 ml distilled water with gentle magnetic stirring at maintained temperature of 40 °C. Afterwards, 1.4 g of PEG was added which acts as a cross linker and facilitates cell adhesion. Nextly, chitosan solution (2% wt., 30 g) was added into G/PEG mixture. To obtain G/PEG hydrogel, the reaction mixture was gently stirred at 40 °C for 20 min and poured into the glass plate for gel formation. Then the gel was cut into rectangular bar and frozen by dipping into liquid nitrogen bath maintained at a temperature of −196 °C. The frozen samples were freeze-dried for sublimation of ice crystals. Finally, in order to further increase the degree of grafting, the dried Gelatin/Chitosan scaffolds were placed into an oven under vacuum for 2 h at 45 °C. The preparation procedure for carbon black based scaffold (G/Ch/C) is similar to G/Ch scaffold, only with an addition of carbon black. The final composition of prepared scaffolds are listed in Table 1. The structural, mechanical and biological characterization of prepared scaffold has been reported in [36].

Table 1 Scaffold composition

Scaffolds	Composition				Physical structure
	G (%)	PEG (%)	Ch (%)	C (%)	
G/Ch scaffold	72	20	08	–	
G/Ch/C scaffold	62	14	17	07	

2.2 Experimental Setup and Measurement Protocols for Bio-impedance Spectroscopy

The experimental setup used to analyze impedance response of scaffold is shown in Fig. 1. Both G/Ch and G/Ch/C scaffolds were cut into same rectangular shape and size with length, width and thickness of 13 mm, 1.5 mm and 0.8 mm respectively and were made hydrated with distilled water. The impedance analyzer "hp 4194A" is used with its terminals connected with two ends of scaffold. The analyzer is interfaced with LabVIEW software, which controls the analyzer operation and records the measured impedance values. Impedance Measurements are then analyzed on MATLAB, where its magnitude, phase shift and conductivity is evaluated. The parameters set for impedance analyzer operation includes: Frequency Sweep: 10 kHz–3 MHz, Sweep Type: Linear, Sweep Mode: Repeat, Data Averaging: 4, Integration Time: Medium, Total No. of Points: 401 and Sample Rate: 200 ms.

3 Result and Discussion

3.1 Bio-impedance Measurement

The preliminary experimental results (Fig. 2) show that hydrated scaffold (with distilled water) changes its impedance response with respect to time. Initially, at 7 h, scaffold shows more capacitive behaviour which progressively turns into resistive after 72 h. As scaffold responds differently at different time period in terms of impedance fluctuation, therefore, in future this characteristic would be helpful to monitor the cell activities on scaffold. In addition, the results exhibit that by adding carbon black the impedance value of G/Ch/C (Fig. 2b) scaffold decreases as compared to G/Ch scaffold (Fig. 2a) which will also increase the overall conductivity of carbon based scaffold.

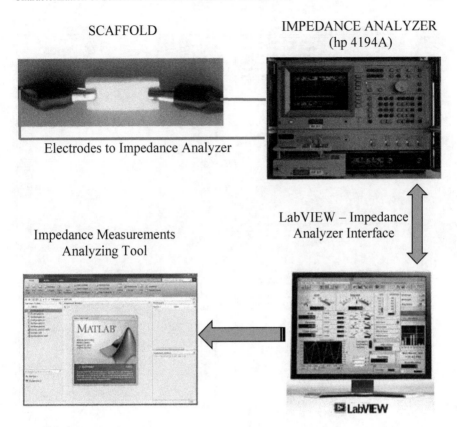

Fig. 1 Schematic representation of experimental setup

3.2 Electrical Conductivity Analysis

From impedance magnitude, resistive values were extracted and it was found that internal resistance within scaffold structure decreases as carbon black was added. This suggests the formation of perfect embedded electric network between carbon black and gelatin matrix. Furthermore, resistive values were used to evaluate the electrical conductivity (σ) of simple Gelatin/Chitosan and carbon-based Gelatin/Chitosan scaffold using "Pouillet's Law Equation" (Eq. 1).

$$\sigma = \frac{L}{R * A} \tag{1}$$

where "σ" is scaffold's electrical conductivity, "L" is the distance between electrodes, "R" is the scaffold's resistance and "A" is cross-sectional area of scaffold. A comparative analysis of the scaffold's conductivity shows that carbon black strongly enhanced the electrical conductivity (around 10 times) as compared to G/Ch scaffold (Table 2).

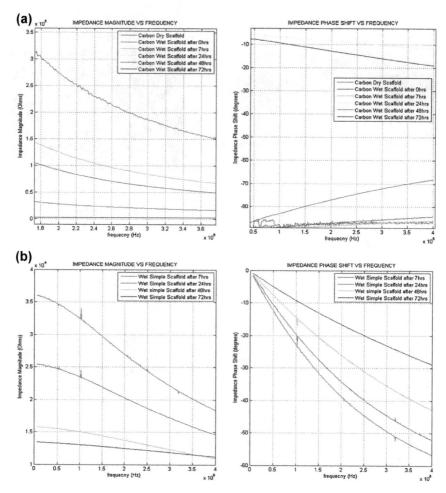

Fig. 2 Impedance Measurement of: **a** wet G/Ch scaffold at different hours **b** wet G/Ch/C scaffold at different hours

4 Conclusion

In this research G/Ch has been sensorized and transformed into a conductive G/Ch/C scaffold. As gelatin based scaffold was not highly conductive, therefore, using a suitable conductive filler was necessary to increase its conductivity. The scaffolds are electrically characterized via bio-impedance spectroscopy method and experimental result shows that carbon black decreases the internal resistance and increases the conductivity of scaffold approximately 10 times which shows the development of conductive network throughout the scaffold. Hence, the obtained results are quite promising and it is expected that in future, cell activities could be monitored

Table 2 Scaffold conductivities at different observation hours

Observation hours (h)	Hydrated G/Ch scaffold		Hydrated G/Ch/C scaffold	
	Conductivity (minimum) (S/m)	Conductivity (maximum) (S/m)	Conductivity (minimum) (S/m)	Conductivity (maximum) (S/m)
07	0.022	0.064	0.24	0.51
24	0.016	0.057	0.13	0.19
48	0.036	0.072	0.31	0.54
72	0.043	0.059	0.70	0.74

non-invasively via impedance measurement technique. Thus, future analyses will include the seeding of different cellular types in order to identify specific impedance variations due to cell activities.

References

1. Yang, F., Murugan, R., Ramakrishna, S., et al.: Fabrication of nano-structured porous PLLA scaffold intended for nerve tissue engineering. Biomaterials **25**, 1891–1900 (2004)
2. Subramanian, A., Krishnan, U.M., Sethuraman, S.: Development of biomaterial scaffold for nerve tissue engineering: biomaterial mediated neural regeneration. J. Biomed. Sci. **16**, 108 (2009)
3. Khan, M.N., Islam, J.M.M., Khan, M.A.: Fabrication and characterization of gelatin-based biocompatible porous composite scaffold for bone tissue engineering. J. Biomed. Mater. Res. Part A **2012**(100A), 3020–3028 (2012)
4. Fowler, B.O., Moreno, E.C., Brown, W.E.: Infra-red spectra of hydroxyl-apatite, octacalcium phosphate and pyrolysed octacalcium phosphate. Arch. Oral Biol. **11**(477), 492 (1966)
5. Rey, C., Shimizu, M., Collins, B., Glimcher, M.J.: Resolution-enhanced Fourier transform infrared spectroscopy study of the environment of phosphate ions in the early deposits of a solid phase of calcium-phosphate in bone and enamel, and their evolution with age. I: investigations in the upsilon 4 PO4 domain. Calcif. Tissue Int. **46**, 384–394 (1990)
6. Walters, M.A., Leung, Y.C., Blumenthal, N.C., LeGeros, R.Z., Konsker, K.A.: A Raman and infrared spectroscopic investigation of biological hydroxyapatite. J. Inorg. Biochem. **39**(19), 3–200 (1990)
7. Haydar, U., Islam, J.M.M.Z., Khan, M.A., Khan, R.A.: Physico-mechanical properties of wound dressing material and its biomedical applica-tion. J. Mech. Beh. Biomed. Mater. **4**, 1369–1375 (2011)
8. Mao, J.S., Liu, H.F., Yin, Y.J., Yao, K.D.: The properties of chitosan–gelatin membranes and scaffolds modified with hyaluronic acid by different methods. Biomaterials **24**, 1621–1629 (2003)
9. Cheng, M., Deng, J., Yang, F., Gong, Y., Zhao, N., Zhang, X.: Study on physical properties and nerve cell affinity of composite films from chitosan and gelatin solutions. Biomaterials **24**, 2871–2880 (2003)
10. Hajiabbas, M., Mashayekhan, S., Nazaripouya, A., Naji, M., Hunkeler, D., RajabiZeleti, S., Sharifiaghdas, F.: Artif. Cells Nanomed. Biotechnol. (2013). http://dx.doi.org/10.3109/21691401.2013.852101

11. Jridi, M., Hajji, S., Ayed, H.B., Lassoued, I., Mbarek, A., Kammoun, M., Souissi, N., Nasri, M.: Int. J. Biol. Macromol. **67**, 373 (2014)

12. Sarem, M., Moztarzadeh, F., Mozafari, M., Prasad Shastri, V.: Mater. Sci. Eng. C **33**, 4777 (2013)

13. Guan, S., Zhang, X.L., Lin, X.M., Liu, T.Q., Ma, X.H., Cui, Z.F.: J. Biomater. Sci. Polym. Ed. **24**, 999 (2013)

14. Martin-Lopez, E., Alonso, F.R., Nieto-Diaz, M., Nieto-Sampedro, M.: J. Biomater. Sci. Polym. Ed. **23**, 207 (2012)

15. Keong, L.C., Halim, A.S.: In vitro models in biocompatibility assessment for biomedical-grade chitosan derivatives in wound management. Int. J. Mol. Sci. **10**, 1300–1313 (2009)

16. Hirano, S., Midorikawa, T.: Novel method for the preparation of N-acylchitosan fiber and N-acylchitosan-cellulose fiber. Biomaterials **19**, 293–297 (1998)

17. Li, Q., Dunn, E.T., Grandmaison, E.W., Goosen, M.F.A.: Applications and proper ties of chitosan. J. Bioact. Compat. Polym. **71**, 370–397 (1992)

18. Majeti, N.V.: A review of chitin and chitosan applications. React. Funct. Polym. **46**, 1–27 (2000)

19. Suh, J.K., Matthew, H.W.: Application of chitosan-based polysaccharide biomaterials in cartilage tissue engineering: a review. Biomaterials **21**, 2589–2598 (2000)

20. Zhang, Z., Rouabhia, M., Wang, Z., et al.: Electrically conductive biodegradable polymer composite for nerve regeneration: electricity-stimulated neurite outgrowth and axon regeneration. Artif. Organs **31**, 13–22 (2007)

21. Martins, A.M., Eng, G., Caridade, S.G., Mano, J.F., Reis, R.L., Vunjak-Novakovic, G.: Electrically conductive chitosan/carbon scaffolds for cardiac tissue engineering. Biomacromolecules **15**(2), 635–643

22. Shahini, A., Yazdimamaghani, M., Walker, K.J., Eastman, M.A., Hatami-Marbini, H., Smith, B.J., Ricci, J.L., Madihally, S.V., Vashaee, D., Tayebi, L.: 3D conductive nanocomposite scaffold for bone tissue engineering. Int. J. Nanomed. **9**, 167–181 (2014)

23. Huang, J.C.: Carbon black filled conducting polymers and polymer blends. Adv. Polym. Technol. **21**(4), 299–313 (2002)

24. Zois, H., Apekis, L., Mamunya, Y.P.: Dielectric properties and morphology of polymer composites filled with dispersed iron. J. Appl. Polym. Sci. **88**(13), 3013–3020 (2003)

25. Tanasa, F., Zanoaga, M., Mamunya, Y.: Conductive thermoplastic polymer nanocomposites with ultralow percolation threshold. Sci. Res. Educ. Air Force-AFASES **2** (2015)

26. Doroski, D.M., Brink, K.S., Temenoff, J.S.: Techniques for biological characterization of tissue-engineered tendon and ligament. Biomaterials **28**, 187 (2007)

27. Smith, L.E., Smallwood, R., Macneil, S.: A comparison of imaging methodologies for 3D tissue engineering. Microsc. Res. Tech. **73**(12), 1123–1133 (2010). https://doi.org/10.1002/jemt.20859

28. Daza, P., Olmo, A., Cañete, D., Yúfera, A.: Monitoring living cell assays with bio-impedance sensors. Sens. Actuators B Chem. **176**, 605–610 (2013)

29. Lei, K.F., Wu, M.H., Liao, P.Y., Chen, Y.M., Pan, T.M.: Development of a micro-scale perfusion 3D cell culture biochip with an incorporated electrical impedance measurement scheme for the quantification of cell number in a 3D cell culture construct. Microfluid. Nanofluid. **12**, 117–125 (2012)

30. Lei, K.F., Wu, M.H., Hsu, C.W., Chen, Y.D.: Real-time and non-invasive impedimetric monitoring of cell proliferation and chemosensitivity in a perfusion 3D cell culture microfluidic chip. Biosens. Bioelectron. **51**, 16–21 (2014)

31. Weijenborg, P.W., Rohof, W.O.A., Akkermans, L.M.A., Verheij, J., Smout, A.J.P.M., Bredenoord, A.J.: Electrical tissue impedance spectroscopy: a novel device to measure esophageal mucosal integrity changes during endoscopy. Neurogastroenterol. Motil. **25**, 574–e458 (2013)

32. Giaever, I., Keese, C.R.: Micromotion of mammalian cells measured electrically. Proc. Natl. Acad. Sci. U.S.A. **88**(17), 7896–7900 (1991)

33. Lind, R., Connolly, P., Wilkinson, C.D.W., Breckenridge, L.J., Dow, J.A.T.: Single cell mobility and adhesion monitoring using extracellular electrodes. Biosens. Bioelectron. **6**, 359–367 (1991)

34. Wegener, J., Keese, C.R., Giaever, I.: Electric cell-substrate impedance sensing (ECIS) as a noninvasive means to monitor the kinetics of cell spreading to artificial surfaces. Exp. Cell Res. **259**(1), 158–166 (2000)
35. Ehret, R., Baumann, W., Brischwein, M., Schwinde, A., Stegbauer, K., Wolf, B.: Monitoring of cellular behaviour by impedance measurements on interdigitated electrode structures. Biosens. Bioelectron. **12**(1), 29–41 (1997)
36. Dey, K., Agnelli, S., Serzanti, M., Ginestra, P., Scarì, G., Dell'Era, P., Sartore, L.: Preparation and properties of high performance gelatin based hydrogels with chitosan or hydroxyethyl cellulose for tissue engineering applications. Int. J. Polym. Mater. Polym. Biomater. (2018). https://doi.org/10.1080/00914037.2018.1429439

Fast Multi-parametric Method for Mechanical Properties Estimation of Clamped—Clamped Perforated Membranes

Luca Francioso⬤, Chiara De Pascali⬤, Alvise Bagolini⬤,
Donatella Duraccio⬤ and Pietro Siciliano⬤

Abstract The knowledge of material properties like Young's modulus and residual stress is crucial for a reliable design of devices with optimized performance. Several works discussed on the determination of the mechanical properties of thin/thick films and microstructures from deflection measurements by a profiler. This work provides an approximate solution for the load-deflection response of perforated membranes clamped on two opposite edges subjected to quasi-point pressure loads applied by a profilometer. Si_xN_y/a-Si/Si_xN_y thin film membranes of different sizes and porosities were fabricated by unconventional 100 °C PECVD process using surface micromachining approach. Tri-layer thin films were mechanically characterized by nanoindentation tests and residual stress measurements based on the wafer curvature method. Load-deflection measurements were done by applying quasi-point loads in the range 4.9–9.8 μN. Finite Element Analysis was used to model the mechanical behavior of the membrane, in agreement with the deflection data measured by profilometer. The elastic modulus measured by nanoindentation was used as reference for the perforated membranes load-deflection analytical function identification. An approximate analytical law was developed, which explicates the maximum deflection amplitude as a function of geometric features (sizes and thickness) and mechanical properties (Young's modulus and residual stress). It was validated numerically and experimentally; it was able to provide an estimation of the residual stress of CCFF perforated membranes, starting from measured data of deflection, for single or multiple loads; also, it can be used in a complementary way to calculate the Young's modulus from deflection data and residual stress information.

L. Francioso · C. De Pascali (✉) · P. Siciliano
Institute for Microelectronics and Microsystems, CNR-IMM, 73100 Lecce, Italy
e-mail: chiara.depascali@le.imm.cnr.it

A. Bagolini
Fondazione Bruno Kessler, Center of Materials and Microsystems, 38132 Povo, Trento, Italy

D. Duraccio
Institute for Agricultural and Earthmoving Machines, CNR-IMAMOTER, 10135 Turin, Italy

© Springer Nature Switzerland AG 2019
B. Andò et al. (eds.), *Sensors*, Lecture Notes in Electrical Engineering 539,
https://doi.org/10.1007/978-3-030-04324-7_73

619

Keywords Load-deflection model · Perforated clamped-clamped membrane
Finite element analysis · Young's modulus · Residual stress

1 Introduction

Performance, reliability and lifetime of microelectromechanical systems (MEMS)
are strongly influenced by the quality of the materials and fabrication processes. The
design of a MEMS device is often supported by prediction models that solve the
electro-mechanical behavior by using proper numerical methods. To obtain more
useful and realistic results, the numerical simulation should be based on measured
materials properties, geometries and applied conditions. The preliminary knowledge
of properties like residual stress and Young's modulus validate the numerical analysis
results, for a better evaluation of the potential performance of the device in specific use
applications and conditions. The mechanical characterization of structural materials
is usually performed on ad hoc thin film test samples, however an in situ evaluation by
direct measurements on the MEMS device could provide more realistic information
on specific properties of interest.

A surface profiler can be used to measure the deflection to quasi-point applied
loads of suspended structures (beams, cantilevers or membranes, for example).
Young's modulus and stress can be indirectly and approximately calculated from
the load-deflection response, by using numerical models of comparison or by applying analytical solutions, exact or approximate, of the load-deflection equation. In
[1–6] the displacement of thin- and thick-films, cantilever, plates and shells, under
uniform loads and different conditions applied on the edges (simple support, clamp-
ing, free-edges) was discussed and correlated to the mechanical properties of the
materials. However, to the best of our knowledge, it is not clearly available an experimentally derived approximate function to solve the relation between mechanical
properties and nonlinear large deflection for specific membranes with two opposite
edges Clamped and the other two Free (CCFF) subjected to quasi-point pressure
loading.

Aimed to solve this lack, this work provides an approximate analytical description,
validated by measurements and simulations, of the nonlinear large deflection of
CCFF perforated membranes subjected to quasi-point pressure loads. Multilayer thin
film membranes of different sizes and porosity were fabricated by unconventional
ultra-low temperature (100 °C) PECVD of Si_xN_y/a-Si/Si_xN_y. Hardness and Young's
modulus of the structural tri-layer were measured by nanoindentation test at different
penetration depths. By using a profilometer and applying the wafer curvature method,
the residual stress of each deposited layer was measured, and that of the tri-layer
was calculated by applying the Stoney's formula [7]. The mechanical behavior of
the membrane, in terms of maximum displacement under load, was analytically
described as a function of different parameters, both geometric and mechanical.
Hence, an estimation of Young's modulus and stress can be indirectly obtained from
the maximum deflection of the membrane under increasing loads measured with a

profilometer. An interesting effect of the micromachining process on the mechanical properties of the investigated thin films was found with respect to the values measured on thin film samples.

2 Fabrication and Mechanical Characterization

2.1 Fabrication of CCFF Perforated Membranes

CCFF membranes were fabricated using standard IC fabrication equipment, 6 inches diameter process substrates and SEMI standard silicon wafers. A 6 µm thick sacrificial photoresist was deposited (AZ 4562), patterned by standard photolithography and baked at 200 °C for 30 min before PECVD. Next, the membrane multilayer was deposited by PECVD of Si_xN_y 190 nm, a-Si 320 nm, Si_xN_y 210 nm, at a temperature of 100 °C. The membrane multilayer was patterned with photolithography and etched using IC standard TEGAL900 plasma etching. Finally, the cap structures were released from the sacrificial layer by oxygen plasma using a Matrix resist stripper with etching temperature of 200 °C, 400 W power and 16 min etch time. CCFF membranes with different porosities and side lengths were realized, with a perforation diameter of 5 µm and a non-perforated frame at the edges, around the central perforated area, about 13–15 µm wide. The porosity, calculated as perforated area per total area of the membrane, was in the range 8–32% for various geometries. A SEM image of suspended CCFF membranes is shown in Fig. 1.

2.2 Mechanical Properties of the Structural Tri-layer

Martens and Vickers micro hardness and Young's modulus of the structural tri-layer were measured by nanoindentation at different penetration depths (100, 400 and 600 nm) using a FISCHERSCOPE HM 2000XYm tool. For each investigated indentation depth, the test was programmed for reaching the final peak within 20 s, with a dwell time of 5 s at the maximum depth to check for residual creep. For each depth, ten indentations were made to account for the scatter caused by foreign particles or porosity. The results pointed out lowest values of hardness and Young's modulus at 100 nm of penetration, while the average values at highest depths are quite similar. Martens Hardness of 6439.8 $\pm$ 108.7 N/mm^2, Vickers Hardness of 1115.4 $\pm$ 22.8 GPa and Young's modulus of 136.6 $\pm$ 1.1 GPa were recorded at 600 nm of penetration depth, which takes into account all structural layers.

The residual stress of the thin films was evaluated by wafer curvature measurements performed before and after deposition using mechanical profilometry (Kla-Tencor P-6). Three profiles were measured along the diameter of each sample, with a separation offset of 5 mm. A standard deviation better than 3% was obtained for

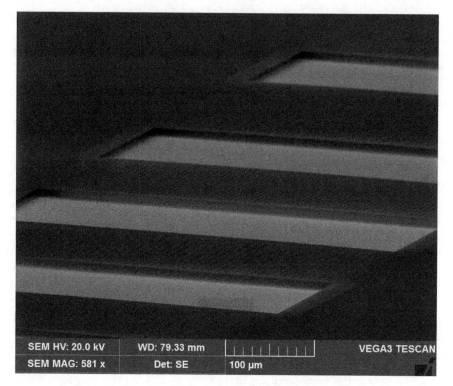

Fig. 1 SEM image of suspended $Si_xN_y/a-Si/Si_xN_y$ perforated membranes

all stress data. A residual stress of 108 MPa (2 MPa std. deviation), compressive, and +261 MPa (1 MPa std. deviation), tensile, was measured for silicon nitride and a-Si, respectively. The residual stress of the tri-layer was calculated by applying the Stoney's equation [7], using the thickness and the residual stress of each layer, and it resulted of about 63 MPa, tensile.

3 Deflection Model of a Perforated Membrane

This work was aimed to counteract the scarce information available in literature up to this date on the nonlinear large deflection of CCFF membranes subjected to a quasi-point loading. First, a mechanical analysis was performed on sub-micrometer thick CCFF perforated membranes, based on a parametric investigation as a function of the mechanical and geometric parameters of major interest.

A given membrane exhibit a higher deflection under load after perforation, the same experienced by an identical unperforated membrane with higher Young's modulus given by [8, 9]:

$$E_{NP} = E_P / (1 - P) \tag{1}$$

with P porosity and E_P elastic modulus of the perforated membrane. However, E is an intrinsic property of the material and invariant with respect to perforation. Conversely, perforation tends to reduce thin film residual stress [10]. The curvature method cannot be applied on perforated thin films, hence more sophisticated techniques (e.g. optical, SEM-assisted) has to be considered for in situ stress measurements.

The deflection law developed in this work is the interpolation function of the maximum deflection experienced by a CCFF membrane under increasing loads. FEA on the load-deflection behavior of CCFF perforated membranes was performed using Comsol Multiphysics software. As indicated by multi-parametric simulations, the maximum deflection under load ω_{max} increases with L linearly, but it decreases with h, W and E with allometric law, given by:

$$\omega_{max}(p) = a p^b \tag{2}$$

with p chosen parameter (h or W or E) and a and b (<0) fit coefficients. Instead, ω_{max} varies with stress and load as

$$\omega_{max}(p) = A + B\,exp(pC) \tag{3}$$

A, B and C are fit coefficients and p is the chosen parameter (σ or F).

From these results, the following deflection law for a CCFF perforated membrane was derived:

$$\omega_{max} = \omega_0 + \omega_{NS} f(\sigma) \tag{4}$$

where ω_0 represents the intrinsic displacement due to residual stress state, in absence of applied load; the second term on the right side of the Eq. (4) describes the displacement due to an external load. If the membrane has negligible stress ($\omega_0 = 0$) and the Eq. (4) reduces to a no-stress form ω_{NS}:

$$\omega_{NS} = 3.4L(F/hWE)^{0.45} \tag{5}$$

Otherwise, $\omega_0 \neq 0$ and ω_{NS} is modulated by the stress though $f(\sigma)$, given by:

$$f(\sigma) = F^{0.1}[2 - exp(-0.004F\sigma)] \tag{6}$$

4 Measurements and Results

Load-deflection measurements were done by a KLA Tencor P-6 profilometer on three perforated membranes of sizes W × L equal to $200 \times 300, 200 \times 400, 300 \times 300$ μm, and porosity ranging between 31.8 and 33.3%. An example of deflection profiles of

Fig. 2 Deflection profiles of a perforated membrane with sizes W × L: 200 μm × 300 μm, under increasing loads, measured by profilometer

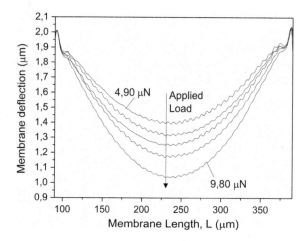

a perforated membrane (200 × 300 μm) under increasing loads, measured by pro-filometer, is reported in Fig. 2. The applied load was in the range from 4.9 to 9.8 μN. Since the perforation doesn't modifies the elastic modulus of the material, the value measured by nanoindentation (136.6 GPa) remains meaningful for the perforated membranes. Instead, the effective residual stress of each membrane was calculated by interpolation of the maximum deflection versus load using the Eqs. (4–6). Figure 3 shows the variation of the maximum deflection versus load, for three investigated CCFF perforated membranes, and the data interpolation by using the Eq. (4). For all membranes, it was found that the stress was smaller than the value measured on the continuous tri-layer (about 63 MPa), as reported in Fig. 4. This result was also confirmed by FEA simulations, in which the best agreement between numerical and measured deflection data (in terms of the maximum deflection under a quasi-point pressure load applied to the center of the membrane) was found for smaller values of stress (Table 1). For those membranes larger than 200 × 300 μm, the deflection law slightly overestimates the residual stress predicted by FEA. Two main effects could explain this reduction: the perforation of the tri-layer and its structuration in form of suspended membrane clamped on two opposite edges.

The results obtained and reported in Table 1 demonstrate that: (i) the numeri-cal model is in full agreement with the experiments; (ii) the approximate analytical function developed to describe the deflection under quasi-point pressure loads of CCFF perforated membranes is in full agreement with the measurements; (iii) per-foration and structuration of the tri-layer in form of CCFF perforated membrane act by reducing the residual stress with respect to the value measured by curvature test on continuous tri-layer.

Fig. 3 Maximum deflection measured on three different perforated membranes under increasing loads: interpolation by using the Eq. 4 and corresponding FEA results. Membranes sizes: W × L: 300 μm × 300 μm—200 μm × 300 μm—200 × 400 μm

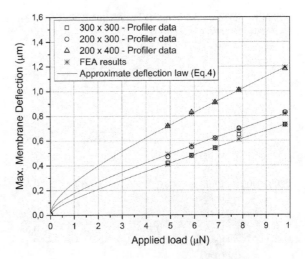

Fig. 4 Residual stress of the investigated membranes, calculated by FEA and by interpolation using the Eq. 4

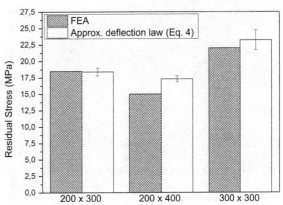

5 Conclusions

An approximate function, experimentally and numerically validated, was developed and used as interpolation function of the maximum deflection of CCFF perforated membranes under increasing quasi-point pressure loads, measured by a surface profiler. The dependence of the load-deflection behavior of the membrane versus different parameters, both geometric and mechanical, was explained and modeled. As main results for the investigated cases, the proposed deflection law provided a deflection estimation in line with experimental data; further, it can also be indirectly used to obtain an estimation of Young's modulus of the investigated system, starting from simple and fast stylus profiler data.

Table 1 Measured and simulated maximum deflections under increasing applied loads, for three perforated membranes

Perforated membrane	Method	Applied load, F				
		4.90 μN	5.88 μN	6.86 μN	7.84 μN	9.80 μN
300 × 300	Profiler deflection (μm)	0.42	0.48	0.54	0.65	0.73
	FEA deflection (μm)	0.40	0.47	0.55	0.62	0.76
200 × 300	Profiler deflection (μm)	0.47	0.55	0.62	0.70	0.83
	FEA deflection (μm)	0.46	0.56	0.63	0.71	0.86
200 × 400	Profiler deflection (μm)	0.72	0.83	0.91	1.01	1.18
	FEA deflection (μm)	0.71	0.84	0.96	1.07	1.28

References

1. Pan, J.Y., Lin, P., Maseeh, F., Senturia, S.D.: Verification of FEM analysis of load-deflection methods for measuring mechanical properties of thin films. In: Technical Digest IEEE Solid State Sensor and Actuators Workshop, Hilton Head Island, SC, 73, 70–73 (1990)
2. Malhaire, C.: Comparison of two experimental methods for the mechanical characterization of thin or thick films from the study of micromachined circular diaphragms. Rev. Sci. Instrum. **83**, 055008 (2012)
3. Tian, Y.B., Zhou, L., Zhong, Z.W., Sato, H., Shimizu, J.: Finite element analysis of deflection and residual stress on machined ultra-thin silicon wafers. Semicond. Sci. Technol. **26**, 105002 (2011)
4. Timoshenko, S., Woinowsky-Krieger, S.: Theory of Plates and Shells, 2nd edn. McGraw-Hill, Inc. (1959)
5. Eaton, W.P., Bitsie, F., Smith, J.H., Plummer, D.W.: A new analytical solution for diaphragm deflection and its application to a surface-micromachined pressure sensor. In: Technical Proceedings of the 1999 International Conference on Modeling and Simulation of Microsystems, Chapter 17, 640–643 (1999)
6. Kimiaeifar, A., Tolou, N., Barari, A., Herder, J.L.: Large deflection analysis of cantilever beam under end point and distributed loads. J. Chin. Inst. Eng. **37**(4), 438–445 (2014)
7. Stoney, G.G.: The tension of metallic films deposited by electrolysis. Proc. R. Soc. Lond. Ser. A **82**(553), 172–175 (1909)

8. Ugural, A.C.: Stress in Plates and Shells. McGraw-Hill, Boston (1999)
9. Kovacs, A., Pogany, M., Mescheder, U.: Mechanical investigation of perforated and porous membranes for micro- and nanofilter applications. Sens. Actuators B **127**(1), 120–125 (2007)
10. Rabinovich, V.L., Gupta, R.K., Senturia, S.D.: The effect of release-etch holes on the electromechanical behavior of MEMS structures. In: International Conference on Solid State Sensors and Actuators, TRANSDUCERS Chicago, IL, 2, 1125–1128 (1997)

Improvement of the Frequency Behavior of an EC-NDT Inspection System

Andrea Bernieri, Giovanni Betta, Luigi Ferrigno, Marco Laracca and Antonio Rasile

Abstract The non-destructive testing based on eddy currents (EC-NDT) is an in-field inspection technique mainly used to detect and characterize defects in conductive materials. This technique is adopted in many industrial, manufacturing and aerospace applications. Today, in aeronautical and industrial applications, new thin carbon fiber materials often substitute metallic materials as aluminum, copper and so on. Therefore, the actual trend of EC-NDT is the in-field inspection of these very thin materials. Considering the skin depth of induced eddy current, the thinness of these materials imposes the use of higher frequencies than those typically adopted in traditional tests. Due to the inductive load of the excitation coil, the voltage required to obtain the desired current values could reach very high values. The use of resonant excitation circuits could mitigate this problem. The article describes the effect of some real resonant circuits on the improvement of feeding high-frequency EC-NDT probes in terms of required voltage and power.

Keywords Eddy current test · Non destructive test · Resonance · High frequency

A. Bernieri (✉) · G. Betta · L. Ferrigno · M. Laracca · A. Rasile
Department of Electrical and Information Engineering, University of Cassino and Southern Lazio,
Via G. Di Biasio 43, 03043 Cassino, Italy
e-mail: bernieri@unicas.it

G. Betta
e-mail: betta@unicas.it

L. Ferrigno
e-mail: ferrigno@unicas.it

M. Laracca
e-mail: m.laracca@unicas.it

A. Rasile
e-mail: a.rasile@unicas.it

© Springer Nature Switzerland AG 2019
B. Andò et al. (eds.), *Sensors*, Lecture Notes in Electrical Engineering 539,
https://doi.org/10.1007/978-3-030-04324-7_74

1 Introduction

Eddy Current based Non Destructive Testing (EC-NDT) is an in-field inspection technique mainly adopted for detecting and characterizing flaws on metallic materials. The basic operating principle of this kind of test is very simple [1]. An excitation coils induces eddy currents in the specimen under test while magnetic sensors, or sensing coils, measure the reaction magnetic fields. The presence of defects in the specimen under test modifies the magnetic reaction fields. This field variation, through suitable processing techniques based on inversion models or experimental analysis, can highlight the presence and size of defect. A key element of an EC-NDT instrument is the EC probe. Typically, it is composed by three elements: (a) the exciting coil, (b) the measurement sensor and, (c), the mechanical facility where both the (a) and (b) elements are located. As for the exciting coil, the maximum allowed current and the factor of merit Q (meant as the ratio between the inductive and resistive impedance of the coil) are very important project parameters. The excitation coil must guarantee hundreds of milliamps of the excitation current and high values of Q factor to induce acceptable eddy current values in the specimen under test.

Today, new carbon fiber materials are involving many industrial and aerospace applications. These materials introduce new defect types and, consequently, they impose new EC-NDT test conditions and probes. In detail, the inspection of these materials is imposing the use of frequency much higher than the previous ones, up to some MHz. In these cases, the higher frequency, the higher the overall impedance of the excitation coil, the higher the voltage excitation required to achieve the desired current values. This leads to problems in developing and realizing the current amplifier that feed the excitation coil increasing the overall cost and reducing the overall bandwidth. To overcome this problem, in this paper the effect of some resonant circuits developed to reduce overall voltage and power required to feed the excitation coils is analyzed. In detail, after some theoretical remarks about the resonating circuits in EC-NDT testing, two set-up have been realized and compared for frequencies up to 1 MHz.

2 Some Theoretical Notes to the Resonance Condition in an EC-NDT Probe

The resonance frequency in a RLC dynamic circuit (see Fig. 1) is the frequency at which the reactive components of the impedance have the same modules and opposite phases. In this condition, the impedance of the circuit is theoretically composed by only a resistive part, and in particular, it reaches the minimum module.

In general, defining the overall impedance as:

$$Z = R + j(X_L - X_C) \tag{1}$$

Fig. 1 RLC series circuit

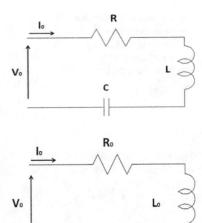

Fig. 2 Simplified electronic circuit for an eddy-current probe in air

where $X_L = \omega L$ and $X_C = 1/\omega C$, under resonance conditions, i.e. when $X_L = X_C$, we have that Z is equal to R.

Resolving with respect to the pulsation, $\omega L = 1/\omega C$, the resonance frequency can be determined by:

$$fr = 1/(2\pi \sqrt{(LC)}) \qquad (2)$$

A coil connected to a power source by a conductor cable represent the typical configuration of an Eddy-Current probe.

The equivalent electrical circuit of this configuration, ignoring the parasitic couplings, consists of a series of a resistor and an inductor (RL) [2, 3]. This is shown in Fig. 2, where the inductor represents the ideal coil and the resistor represents the overall resistance of the coil and of the connection to the electrical source.

Consequently, the equivalent impedance is given by:

$$Z_0 = R_0 + j\omega L_0 \qquad (3)$$

Equation (3) shows that the impedance increases as the frequency increases, so the voltage on the coil required to operate the same current through the system must increase.

This scheme models an ECT coil in the air. When the coil is closer to an electrically conductive material, an electromagnetic coupling occurs between the coil and the surface of the material. The equivalent circuit of the whole system must be extended to model the coupling interaction [4–6].

The coupling with the surface of a conductive material modifies the electrical properties of the probe and the system could be approximated to an equivalent circuit containing the components that modify the electrical characteristics. The coupling interaction can be modeled by the circuit shown in Fig. 3, where the equivalent

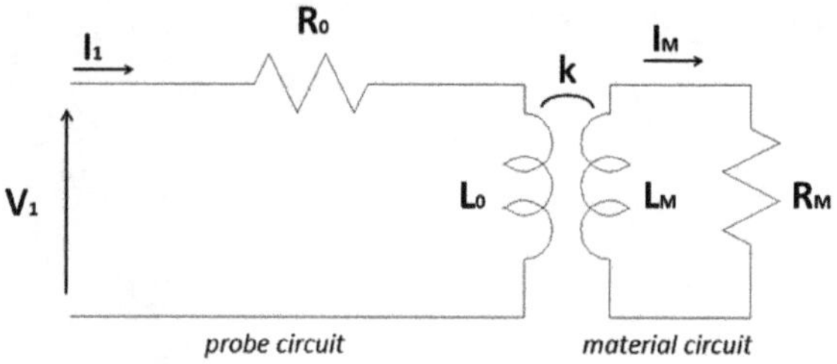

Fig. 3 Simple transformer circuit model for an eddy-current probe coupled to the surface of an electrically conducting material

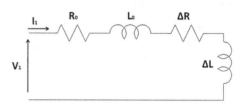

Fig. 4 Coupled ECT system equivalence circuit, adapted from [6]

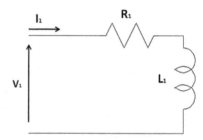

Fig. 5 Coupled eddy-current probe equivalence circuit with effective resistance and inductance

circuit of the system is schematized using the classical model of a transformer with ohmic-inductive elements (RL) [2–4].

The overall impedance, Z_1, of the system can be formulated using Kirchoff's laws [6]:

$$Z_1 = R_0 + j\omega L_0\left(1 - k^2\right) + \frac{\omega L_0 R_M k^2}{R_M^2 + \omega^2 L_M^2}(\omega L_M + j R_M) \qquad (4)$$

where k is the coupling factor.

Using the approximation of the uniformity of the magnetic field interaction, the resistive and inductive components of the secondary circuit are considered to be equal in the module ($\omega L_M = R_M$) as shown by the [6], and (4) can be written as:

$$Z_1 = R_0 + j\omega L_0 + \frac{1}{2}\omega L_0 k^2(1 - j) \qquad (5)$$

$$Z_1 = Z_0 + \frac{1}{2}\omega L_0 k^2(1 - j) \qquad (6)$$

Figures 4 and 5 identifies final results.

where:

$$Z_1 = R_1 + j\omega L_1 \tag{7}$$

$$R_1 = R_0 + \Delta R = R_0 + \frac{1}{2}\omega L_0 k^2 \tag{8}$$

$$L_1 = L_0 + \Delta L_0 = L_0\left(1 - \frac{1}{2}k^2\right) \tag{9}$$

To bring us back to the situation in Fig. 1, a series capacitor must be inserted in the circuit.

The total impedance of the system is:

$$Z = R_0 + \omega L_0 \frac{k^2}{2} + j\frac{\omega^2 C L_0\left(1 - \frac{k^2}{2}\right) - 1}{\omega C} \tag{10}$$

and the expression of the resonance frequency, f_r, is given by:

$$f_r = \frac{1}{2\pi\sqrt{C L_0 - \frac{k^2}{2}C L_0}} \tag{11}$$

3 Some Theoretical Notes on the Double Resonant Circuit

In classic series resonant circuits, the maximum current is limited by the power generator output current capability. When the current needs to be increased a double resonant circuit can further double the coil current. This circuit is typically called double resonant circuit [7] (see Fig. 6). It consists of two equal value capacitors, labeled C, and a magnetic coil (i.e. the EC-NDT probe).

To calculate the resonant frequency, first we need to define the resonant conditions. The requirements for resonant is such that at a given frequency the imaginary

Fig. 6 Double resonant circuit

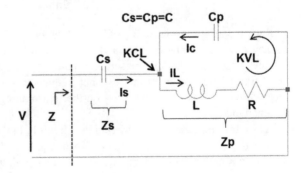

(reactive) portion of the input impedance, Z, is equal zero. The impedance is purely resistive at resonance.

$$Img(Z) = Img(Zs) + Img(Zp) = 0 \tag{12}$$

In Fig. 6 the coil resistance (R) is usually small value and parasitic to the coil. The small resistance has no measurable effect on the resonant frequency circuit. To simplify the resonant frequency calculation, the small resistance is excluded.

The two capacitors forming the resonant circuit are the same value ($C_S = C_P = C$).

Looking at Z_P only, it is a classic parallel resonant. The imaginary portion of Z_P is given by (13). In addition, Z_S is just a capacitor. For a simple capacitor, the imaginary part of Z_S is just its reactance giving by (14).

$$Img\left(Z_p\right) = \frac{1}{\omega C - \frac{1}{\omega L}} \tag{13}$$

$$Img(Z_s) = \frac{1}{\omega C} \tag{14}$$

Now substitute the (13) and (14) into the (12) (resonant condition), it becomes the (15).

$$\frac{1}{\omega C - \frac{1}{\omega L}} + \frac{1}{\omega C} = 0 \Rightarrow 2\omega C = \frac{1}{\omega L} \tag{15}$$

Next is solving for frequency ω. The result is showing in (16) which is the current amplified resonant frequency ω_0.

$$\omega_0 = \frac{1}{\sqrt{2CL}} \tag{16}$$

$$f_0 = \frac{1}{\sqrt{2}} \frac{1}{2\pi \sqrt{LC}} \tag{17}$$

$$C = \frac{1}{2L(2\pi f_0)^2} \tag{18}$$

Remembering that $\omega_0 = 1/\sqrt{LC}$ is the classic (series or parallel) resonant frequency and rearranging the (16) and (17), we can found that the new resonant frequency is related to the classic series or parallel resonant frequency by a factor of $1/\sqrt{2}$ times. The reason for the resonant frequency is lower, is because the double resonant circuit has two capacitors. Further rearranging Eq. 17, the capacitance is obtained in Eq. 18.

As for the coil current, considering the circuit in Fig. 6, applying the Kirchhoff's voltage law (KVL) to parallel resonant loop (C_P, L, and R), ignoring the small coil resistance R and considering that C_P and C_S are equal, we obtain

$$j\omega L I_L + \frac{-j}{\omega C} I_C = 0 \Rightarrow \omega L I_L = \frac{I_C}{\omega C} \tag{19}$$

$$I_L = \frac{I_C}{\omega^2 L C} \tag{20}$$

From (16), the resonance frequency is $\omega = 1/\sqrt{(2LC)}$. Now substituting (16) into (20) and simplifying we obtain:

$$I_L = \frac{I_C}{\left(\frac{1}{\sqrt{2LC}}\right)^2 LC} = 2I_C \tag{21}$$

From Eq. 21 and Fig. 6 we obtain that the magnetic coil current is two times the capacitor C_P or C_S current. From the illustration in Fig. 6, the two capacitor currents are flow into the coil at the same time. Thus, the double resonant circuit is achieving current magnification.

Then, from (21) I_C is one-half of I_L and substitute I_C from (21) into (22) it results that:

$$I_L = I_S + I_C \tag{22}$$

$$I_L = I_S + \frac{1}{2} I_L \tag{23}$$

$$I_L = 2I_S \tag{24}$$

In conclusion, the current coil (I_L) is twice amplified respect to the source current I_S.

4 First Experimental Results

Considering what stated in [8] and the previous authors' experience in EC-NDT and RLC measurements [9–12], a suitable experimental set-up has been realized to test resonant and double resonant circuits.

The use of a variable decade capacitor inside the excitation coil supply circuit allowed the obtaining of the resonance conditions for the different frequencies. Figure 7 show the test set-up for the single resonant circuit.

The considered EC-NDT probe [8, 10] has an excitation coil composed by 180 square coils. It shows a resistance R equal to 1.19 Ω and a inductance L of 87.42 μH. All the experiments have been carried out on aluminum specimen. Initially, an experimental characterization of the frequency behavior of the probe impedance magnitude for frequencies up to 1 MHz was performed. Results are shown in Fig. 8. It is possible to highlight that at the frequency of 1 MHz the impedance magnitude becomes higher than 500 Ω. In this condition an exciting current of 0.3 A, that is typical for EC-NDT, requires more than 150 V across the current amplifier. Then,

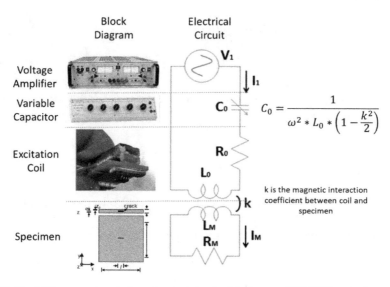

Fig. 7 Block Diagram and equivalent electrical circuit of the improved EC-NDT excitation circuit. The resonant value of C_0 is highlighted

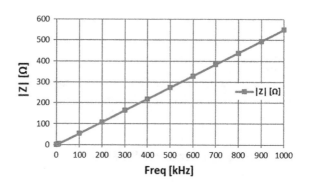

Fig. 8 Experimental Behavior of the impedance magnitude versus the frequency

the performance comparison between single and double resonant circuits have been executed for the frequency range 15 kHz–1 MHz. Results have been also compared with those obtained in absence of resonance. Figures 9 and 10 show the obtained results in the case of resonant excitation (magenta line), double resonance excitation (green line) and absence of resonant excitation (blue line) for the required voltage and power respectively.

It is possible to highlight that the presence of the resonant circuit or double resonant circuit always guarantees a voltage that is much lower than the non-resonant circuit (on average about 12 times lower in the examined situation). In addition, in the case of absence of the resonant circuit, the current amplifier is unable to provide the required current for frequencies greater than 700 kHz while it correctly operates up the 1 MHz in presence of the resonance excitation and double resonance excitation. Moreover the double resonance circuit requires a power lower than the resonance circuit.

Fig. 9 Required excitation voltage for an imposed current of 0.3 A

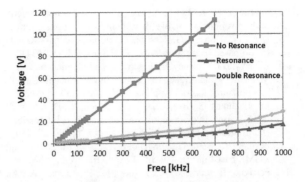

Fig. 10 Required Power for an imposed current of 0.3 A

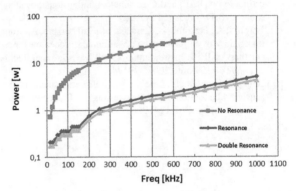

5 Conclusion

The paper shows first experimental results on resonant EC-NDT probes to overcome the problem of feeding high-frequency EC-NDT coils. In particular, realized test highlighted the real benefits of single and double resonant circuits. The use of a double resonance circuit is able to reduce the power required by the generator with respect to the single resonance circuit.

The final aim of the proposed research is the realization of a smart excitation probe with embedded variable electronic capacitors able to impose double resonant conditions automatically for sinusoidal signal from few kHz up to hundreds of MHz.

References

1. Betta, G., Ferrigno, L., Laracca, M., Burrascano, P., Ricci, M., Silipigni, G.: An experimental comparison of multi-frequency and chirp excitations for eddy current testing on thin defects. Measurement **63**, 207–220 (2015)
2. Hughes, R., Fana, Y., Dixon, S.: Near electrical resonance signal enhancement (NERSE) in eddy-current crack detection. In: NDT&E International, pp. 82–89 (2014)
3. Wheeler, H.A.: Formulas for the skin effect. Proc. IRE **30**(9), 412–424 (1942)

4. Owston, C.N.: A high frequency eddy-current, non-destructive testing apparatus with automatic probe positioning suitable for scanning applications. J. Phys. E. Sci. Instrum. **3**(10), 814 (1970)
5. Liu, C.Y., Dong, Y.G.: Resonant coupling of a passive inductance-capacitance- resistor loop in coil-based sensing systems. IEEE Sens. J. **12**(12), 3417–23 (2012)
6. Bleaney, B.I., Bleaney, B.: Electricity and Magnetism. Oxford University Press, Oxford (1989)
7. Magnetic Field Generator Uses New Resonant Circuit. Application note. http://www.accelinstruments.com/Applications/WaveformAmp/Magnetic-Field-Generator.html
8. Lu, J., Huang, S., Pan, K., Qian, Z., Chen, N.: Development of characteristic test system for GMR sensor. In: Proceedings of the 8th IEEE International Symposium on Instrumentation and Control Technology (ISICT), London, pp. 20–23 (2012)
9. Bernieri, A., Betta, G., Ferrigno, L., Laracca, M.: Multi-frequency Eddy Current Testing using a GMR based instrument. Int. J. Appl. Electromagn. Mech., 355–362 (2012)
10. Ferrigno, L., Laracca, M., Liguori, C., Pietrosanto, A.: An FPGA-based instrument for the estimation of R, L, and C parameters under nonsinusoidal conditions. IEEE Trans. Instrum. Meas. **61**(67), 1503–1511 (2012)
11. Bernieri, A., Ferrigno, L., Laracca, M., Tamburrino, A.: Improving GMR magnetometer sensor uncertainty by implementing an automatic procedure for calibration and adjustment. In: Proceedings of IEEE Instrumentation and Measurement Technology Conference (2007). https://doi.org/10.1109/imtc.2007.379174
12. Betta, G., Ferrigno, L., Laracca, M.: Calibration and adjustment of an eddy current based multi-sensor probe for non-destructive testing. In: Proceedings of the ISA/IEEE Sensors for Industry Conference, pp. 120–124 (2002)

Author Index

CPSIA information can be obtained
at www.ICGtesting.com
Printed in the USA
LVHW080102240119
605054LV00003B/21/P

Praise for **Thrive** Books

"Brendan's knowledge is second to none."
—Simon Whitfield, Olympic gold medalist
(triathlon, Sydney 2000)

"*Thrive* is an eye-opening and a life-changing book.
It should replace bibles in hotels."
—Dave Zabriskie, professional cyclist,
Tour de France stage winner, and record holder of the
fastest time trial in Tour de France history

"Brendan Brazier knows how to eat and train for wicked
performance … In his latest book, *Thrive Fitness* … the B.C.-based
athlete goes further, sharing a wealth of fitness information
designed to get you into top shape."
—*Calgary Herald*

"*The Thrive Diet* has revolutionized the way I go about fueling my
body and helped push me to a higher level of performance and
workout recovery. There's no other resource like it out there."
—Mac Danzig, Ultimate Fighter 6 champion

"During my fifteen years in the strength and conditioning industry
I've read just about every nutrition book that's come down the pike.
Not too many of them, however, have had the lasting impact that *Thrive*
has. Because of what I learned from Brendan I have completely revamped
my own nutritional program and in my mid-thirties, I'm feeling and
performing better than ever. More importantly, I've started using his
recommendations with all of my high-level athletes, and they are setting
new personal bests and recovering faster than ever before. I can't say
enough good things about *Thrive* and honestly believe that every
athlete, trainer, or coach owes it to themselves to read it."
—Jason Ferruggia, chief training advisor to
Men's Fitness magazine, author of *Fit to Fight*

PENGUIN CANADA

WHOLE FOODS TO THRIVE

A former professional Ironman triathlete, and a two-time Canadian 50 km Ultra Marathon champion, BRENDAN BRAZIER is the bestselling author of *The Thrive Diet* (Penguin, 2007) and *Thrive Fitness* (Penguin, 2009), as well as the creator of an award-winning line of whole food nutritional products called Vega.

Recognized as one of the world's foremost authorities on plant-based nutrition, Brendan is a guest lecturer at Cornell University and presents an eCornell module entitled "The Plant-Based Diet and Elite Athleticism."

Brendan was chosen as one of the 25 Most Fascinating Vegetarians by *VegNews Magazine* and named one of the Top 40 Under 40 most influential people in the health industry by Natural Food Merchandiser. He has been nominated three times for the prestigious Manning Innovation Award for the creation of the Vega formula.

Also by **Brendan Brazier**

Thrive Fitness

The Thrive Diet

WHOLE FOODS TO
THRIVE

NUTRIENT-DENSE, PLANT-BASED
RECIPES FOR PEAK HEALTH

BRENDAN BRAZIER

PENGUIN
CANADA

PENGUIN CANADA

Published by the Penguin Group

Penguin Group (Canada), 90 Eglinton Avenue East, Suite 700,
Toronto, Ontario, Canada M4P 2Y3 (a division of Pearson Canada Inc.)

Penguin Group (USA) Inc., 375 Hudson Street, New York, New York 10014, U.S.A.
Penguin Books Ltd, 80 Strand, London WC2R 0RL, England
Penguin Ireland, 25 St Stephen's Green, Dublin 2, Ireland
(a division of Penguin Books Ltd)
Penguin Group (Australia), 250 Camberwell Road, Camberwell,
Victoria 3124, Australia (a division of Pearson Australia Group Pty Ltd)
Penguin Books India Pvt Ltd, 11 Community Centre, Panchsheel Park,
New Delhi – 110 017, India
Penguin Group (NZ), 67 Apollo Drive, Rosedale, Auckland 0632, New Zealand
(a division of Pearson New Zealand Ltd)
Penguin Books (South Africa) (Pty) Ltd, 24 Sturdee Avenue, Rosebank,
Johannesburg 2196, South Africa

Penguin Books Ltd, Registered Offices: 80 Strand, London WC2R 0RL, England

First published 2011

1 2 3 4 5 6 7 8 9 10 (WEB)

Copyright © Brendan Brazier, 2011

Graphics created by Tommy Heiden

Manufactured in Canada.

LIBRARY AND ARCHIVES CANADA CATALOGUING IN PUBLICATION

Brazier, Brendan
Whole foods to thrive : nutrient-dense, plant-based recipes for peak health / Brendan Brazier.

Includes bibliographical references and index.
ISBN 978-0-14-317690-9

1. Vegan cooking. 2. Cooking (Natural foods). 3. Food allergy—Diet
therapy—Recipes. 4. Nutrition. I. Title.
TX837.B725 2011   641.5′636   C2011-901486-6

Visit the Penguin Group (Canada) website at **www.penguin.ca**

· Special and corporate bulk purchase rates available;
please see **www.penguin.ca/corporatesales** or
call 1-800-810-3104, ext. 2477 or 2474

CONTENTS

DRINKS

BREAKFASTS

SALADS

SPREADS, DIPS, SAUCES, AND DRESSINGS

INTRODUCTION

Coming from an athletic background, I developed my interest in food simply as a means for enhancing performance. I wanted the best fuel and biological building blocks available. But after several years of being meticulous about what I ate, it dawned on me that I actually knew very little about food itself. While I understood nutrition—the components that make up food—I knew very little about food as a whole: where it came from, who grew it, how much time needed to pass for it to go from seed to ready-to-eat food, etc. And how did my food choices affect all those involved along the way? Then there were the environmental considerations. What was the environmental draw of the laborious process of converting natural resources into edible sustenance?

When I selected what to eat, I understood that choice most certainly had a direct impact on health and performance (which I discuss in Chapter 1), but I was only just beginning to appreciate the broad and significant influence that our individual food choices have on the lives of others and how our food choices impact the environment.

The significance of this began to sink in, and as it did, an appreciation for the scope of influence our food choices had—one that extended far beyond us as individuals—came with it.

In fact, it's that appreciation that led me to write this book. While my interest in food had been sparked by a selfish desire for premium nutrition to fuel athletic performance, as I pried deeper into the world of food, I became fascinated with the system as a whole.

I began asking questions.

Undeniably, unrefined whole food is an essential component to good health, but what attribute determines a food's nutritional quality? Caloric density? Vitamins? Minerals? Phytochemicals? Antioxidants? I wanted to know. And what is the environmental cost of each of these nutritional components? Certainly not all foods are equal in their nutritional makeup, but how do each of their impacts on the environment compare?

I wrote this book in pursuit of those answers. What I found fascinated me. The choices each of us makes every day as to what we'll eat turned out to have a greater impact than I ever could have imagined.

Starting off with a focus on obtaining peak health, I begin by discussing a North American epidemic, one of the leading causes of disease and unrealized potential: stress. The subject of my first book, *The Thrive Diet*, stress has become a ubiquitous part of our modern lives. Unfortunately, its familiar symptoms—difficulty sleeping, inability to lose body fat despite regular exercise, sugar and starch cravings, dependence on stimulants such as coffee to start the day, and general fatigue hitting around two o'clock in the afternoon—have become the rule, not the exception. Depending on its nutritional makeup, food can either contribute to or help alleviate overall stress.

I examine what nutritional characteristics to consider when making food choices to better nourish the body, and therefore to reduce stress through the consumption of higher quality food.

As I found, micronutrient content, known as nutrient density, is the most comprehensive measure of the health-boosting properties of a given food. The greater the nutrient density, the less nutritional stress (the biological strain created when nutritional requirements are not adequately met). I give a detailed account of nutritional stress in Chapter 1.

In addition to advocating that your first consideration in choosing food should be its nutrient density, I propose a set of "guiding principals" to use when selecting what to eat.

In Chapter 2, I look at the environmental toll levied in the food-production process. Our health is, overwhelmingly, tied to the quality of

> Growing and processing food not only consumes land, water, and fossil fuel but also creates carbon emissions.

the food we eat. And food quality is directly tied to the quality of the soil in which it was grown. Therefore, the health of the environment has a direct tie to our health by way of food (not to mention by way of the air we breathe and the water we drink).

Since plants pull minerals from the soils—micronutrients essential for human health—they serve as a conduit, taking the soil—the environment—putting it into a digestible form, and passing it on to us. Each time we take a bite of food, part of the environment literally becomes part of our biological fabric, our bodies. At the risk of sounding like a hippy, the Earth is part of us.

This being the case—if for no other reason than personal self-interest—it's worthwhile taking environmental preservation measures. But there are larger reasons for caring about food. In return for food, we exchange a considerable amount of natural resources. Growing and processing food not only consumes land, water, and fossil fuel but also creates carbon emissions.

In Chapter 3 I examine what others have done to address the vast global environmental strain of food production. The U.K. government leads the way in offering carbon labels on food to give consumers some perspective as to the effect their food choices have on carbon emissions production, and therefore on environmental health. I also explore steps progressive companies, such as Whole Foods Markets, are taking to help consumers make healthy choices by indicating the nutrient density of many of their food items.

Having had the opportunity to work with some of the best minds in the nutrition world, as well as with some leading environmental advocates, I have developed an appreciation of two different perspectives, which haven't often converged.

In being exposed to the connection between personal and environmental health, I dedicate a large part of Chapter 4 to the marriage of the two. I call this connection the "nutrient-to-resource ratio."

I also examine the monetary cost of food. Undoubtedly it's less expensive to gain calories from highly processed food. But what about micronutrients—the true measure of food value—what's the least expensive way to obtain them? I explore this question, displaying my findings by using icons that clearly display the good from the less desirable.

Rounding out the chapter, I visually showcase the environmental strain involved in producing a day of meals following, first, a Standard American Diet, and next following what is commonly perceived as a

"Healthy" American diet. Finally, I look at the environmental strain of producing a day's meal plan using the whole foods recipes in this book. The contrast was sharper than I could have imagined.

In Chapter 4, I go on to consider eight key nutritional components, and their benefits, that are worth seeking when making food choices.

In Chapter 5, I suggest specific foods that have health-boosting and environmental-preserving attributes described earlier in the book. The section "The Whole Foods to Thrive Pantry" provides a list of staples that will help you keep your kitchen well stocked with the essential ingredients for whole food meals.

The book culminates with 200 fabulous recipes, all made with nutrient-dense, plant-based whole foods that are both health-boosting and easy on the environment.

Created with help from some top chef friends, all the recipes in this final chapter adhere to my nutritional philosophy.

Feature Chefs

While I created all the recipes in both *The Thrive Diet* and *Thrive Fitness*, as well as some of the recipes in this book, I grew curious as to how top-tier chefs would approach recipe creation using nutrient-dense, plant-based whole food ingredients.

And, since I'm by no means a chef, I felt there must certainly be flavor profiles and ingredient combinations—unknown to me—that would increase palatability and overall appeal. For this reason, I enlisted the help of a few of my favorite chefs.

The chef creating most of the recipes—about half of the book's total— is Julie Morris. Julie is a Los Angeles–based natural food chef who has the unique ability to draw and balance a wide range of flavors from natural whole foods ingredients.

> All the recipes are nutrient-dense, tread lightly on the environment, and taste amazing.

In keeping with the nutritional philosophy of this book, Julie has taken a truly novel and creative approach in developing delicious, accessible, and easy-to-make recipes out of premium, nutrient-dense, health-boosting ingredients.

The other top chefs I've enlisted have each contributed two to four world-class, plant-based, whole food creations, all of which, I think you'll agree, are truly delicious.

All the recipes here are nutrient-dense, tread lightly on the environment, and taste amazing. One of the most pleasing aspects of these recipes is their diversity. From simple to elaborate, and drawing on a variety of ethnic cuisines, they all showcase the exceptional creativity and scope of these talented chefs.

AMANDA COHEN

Part of the gourmet New York food world for years, in 2008 Amanda opened her own restaurant in New York's East Village and called it Dirt Candy. Known for her ability to cook vegetables in unique and innovative ways, in 2010 Amanda was a contestant on *Iron Chef America*. For more information, visit dirtcandy.com.

MATTHEW KENNEY

Matthew is a chef, a restaurateur, and an author, and is known for his unique brand of organic and vegetarian cuisine. His company, Matthew Kenney Cuisine, is focused on the development of products, books, and businesses that reflect his passion for sustainable living.

He is the founder and operator of The 105degrees Academy, which is a state-licensed educational institution. Matthew created it to share and advance cutting-edge "living cuisine" in an inviting environment.

For more information about Matthew, his many books, the 105degrees Academy, or his work in general, you may visit 105degreesacademy.com.

JULIE MORRIS

Julie Morris is a Los Angeles–based food writer and natural food chef with a talent for creating delicious, health-boosting, plant-based, whole-food recipes. As a chef, she combines complementary ingredients to achieve a unique yet balanced flavor profile in a remarkable way. Julie specializes in the creation of recipes using superfoods. Although a formal definition has not been settled upon, superfoods are most commonly recognized as foods with a high nutrient density.

For more information about Julie, or to read her blog, watch her recipe preparation videos, or learn about her new book, *Superfood Cuisine*, you may visit juliemorris.net.

CHAD SARNO

Now the R&D chef for Whole Foods Markets' "Health Starts Here" program, Chad has been bringing his approach to healthy cuisine to some of the world's premier organic vegan restaurants, spa resorts, film sets, and individuals for over a decade. I was first exposed to Chad's creations when he was executive chef at Saf, a unique upscale raw restaurant in London, England, back in 2008. I've been a fan ever since.

To learn more about Chad, his work, and his company, Vital Creations, you may visit his site rawchef.com.

TAL RONNEN

Chef Tal Ronnen is one of the most celebrated plant-based chefs working today. In the spring of 2008, he became known nationwide as the chef who prepared plant-based meals for Oprah Winfrey's 21-day vegan cleanse. Chef Ronnen also has the honor of being the first to serve a plant-based dinner at the U.S. Senate. Catering the 2010 Physicians Committee for Responsible Medicine (PCRM) gala in Los Angeles, Tal prepared a diverse menu, which was my first exposure to his recipe-creating prowess.

To view Tal's videos and to learn more about him and his book, *The Conscious Cook,* you may visit talronnen.com.

Recipes from My Favorite Restaurants

Throughout the recipe section you'll also find recipes from my favorite restaurants and cafés across North American. Ranging from elaborate five-star formal dining, to casual cafés, and even to basic takeout, these establishments are top-tier and each offers a unique culinary experience, all paralleling my nutritional philosophy.

As someone who crisscrosses North America several times each year, in addition to appreciating the Whole Foods Market and other plant-based, whole food–savvy grocery stores, I have a few favorite restaurants and cafés:

- **Beets Living Foods Café**—Austin, Texas
- **Blossoming Lotus**—Portland, Oregon
- **Candle 79**—New York City, New York
- **Cru**—Los Angeles, California
- **Crudessence**—Montreal, Quebec
- **Fresh at Home**—Toronto, Ontario
- **Gorilla Food**—Vancouver, British Columbia
- **The Green Door**—Ottawa, Ontario
- **Horizons**—Philadelphia, Pennsylvania
- **JivamukTea Café**—New York City, New York
- **Karyn's on Green**—Chicago, Illinois
- **Karyn's Fresh Corner Café**—Chicago, Illinois
- **Life Food Gourmet**—Miami, Florida
- **Live Organic Food Bar**—Toronto, Ontario
- **Millennium**—San Francisco, California
- **Pure Food & Wine**—New York, New York
- **Ravens' Restaurant**—Mendocino, California
- **Thrive Juice Bar**—Waterloo, Ontario
- **Veggie Gril**—Los Angeles, California

You can find more information about each establishment, along with its website URL, address, and phone number, starting on page 302.

1 HEALTH'S DEPENDENCE ON NUTRITION

Stress: just thinking about it can bring it on. When the term was first used in the 1930s, it meant biological trauma, that is, an incident causing physical harm. But in recent decades, we have come to more commonly speak about stress in psychological terms. Its context changed to include the daily events of modern life, not just physical strain. "Modern life stress," as it is sometimes termed, while often originating with worry, or simply a feeling of being overwhelmed, is still displayed through physical symptoms.

In North America, the number of reported incidences of stress-related illness is steadily escalating, so to say that stress has become an epidemic is putting it mildly. Stress has been shown to be the catalyst for numerous diseases.[1] Before disease itself is manifested, our bodies will display warning signals in the form of health problems. However, most of us ignore these signs. Or worse, we treat them as though they are the whole problem, overlooking where they come from.

Sleep deprivation, fatigue, mental fog, irritability, weight gain, and sugar, starch, and caffeine cravings are not in themselves problems but rather symptoms of a problem. They are, however, the red flags that alert us that our overall stress level is beyond a healthy range.

To use a driving analogy, if the oil light goes on, you may be tempted to simply put a piece of tape over it. If you give in to your urge, you will no longer be visually alerted to the problem. But of course, the problem hasn't gone away. The problem will worsen until the car's engine seizes. A lit oil light is the mechanical equivalent of sugar cravings: the first sign that a problem is materializing. And while eating refined sugar will provide relief from the symptom, it will be short-lived. And, of course, the cause will remain unchecked.

REDUCING NUTRITIONAL STRESS

So, what does stress have to do with nutrition? That's exactly what I wondered when I began preparing (I hoped) to embark on a career as a full-time athlete. My aspiration was to race Ironman triathlons profession-ally. And I was willing to do whatever it took to make it happen. One thing it took, as you might expect, was a lot of training.

Over the years I had trained diligently and my fitness improved. But the rate at which I was improving was beginning to slow. Of course, as a person becomes proficient at something—anything—the rate of improve-ment will decline. However, it got to where my rate of improvement wasn't just declining but slowing to a halt. I had hit a plateau.

For the extraordinary amount of time and energy I was spending on training, the return I was now receiving in terms of enhanced fitness was modest at best. I had to try something different. I had to find a way to break through and advance to the next level.

But if not more training, what was it going to take? After a lot of research, I came across truly useful material. As I delved deeper into my investigation, I discovered what I needed to do to break out of my modest-at-best-gains rut. I had to increase my rate of recovery, the speed at which cellular repair took place.

This became the focal point of my research and evolved into what would consume my next several years. I had become convinced that improved cellular regeneration after exercise would be my express ticket to success, and I began searching for ways to accelerate it. This was the key: I knew that quicker recovery would allow me to schedule workouts closer together and therefore to cram more training into a shorter amount of time.

And while I understood that food provides us with the fuel to move around, what I was just beginning to realize was that it also supplies us with the building blocks we use to reconstruct our bodies during the regeneration process. Cellular tissue is constantly dying and regenerating, but for those who break down muscle tissue at an extraordinary rate—athletes, for example, by way of exercise—nutritional building blocks enable the body to grow

> The body needs premium building blocks to regenerate in a timely and thorough manner.

back stronger than it was pre-workout. Overcompensation by the body—its ability to grow stronger as a result of being broken down—is the training effect at work. But the body needs premium building blocks to regenerate in a timely and thorough manner. Fortunately for me, I had just realized— and still at a young age—that there was no such thing as overtraining, only under-recovering.

And, as I learned, nutrition—whether good or bad—plays a significant role in the regeneration process. Nutrition can speed it or slow it, depending on the quality of food. Adding to the physical strain of training, low-quality nutrition imposes stress of its own. And unlike the stress of training, from which the athlete receives a benefit (a greater level of fitness), the stress incurred from poor diet brings no gains. "Reducing stress by way of improved nutrition" was, in fact, the premise of my first book, *The Thrive Diet*.

When the body doesn't get the "biological building blocks"—the nutrients—it needs to keep pace with cellular regeneration, it experiences *nutritional stress*. And the body reacts to nutritional stress just as it does to mental or physical stress. The typical symptoms of stress begin to develop. It became apparent to me that taking in greater amounts of nutrients was a logical way to mitigate overall stress and therefore its symptoms.

My solution at first was simply to eat as much as I could. But, as I quickly learned the hard way, food is not necessarily synonymous with nutrition, at least not in the world we live in today.

What I had done was make sure I was fed, but unfortunately being fed is not the same as being nourished. There's a big difference. And my situation was in no way unique. In fact, I had just become average.

We North Americans are now an overfed yet undernourished society. Undernourishment is the new norm, no longer an exceptional state affecting only a small fraction of the population.

Here's how it works. Hunger is an essential, primal signal dating back to our earliest ancestors. Originating in our brain, the desire to eat is in fact a chemical signal triggered by the body's need for nutrients. To get fuel for our brain and muscles, as well as the building blocks for the ongoing repair of our cells, we need to eat. When we take in nutrient-rich food, such as fruit, our brain recognizes that we have responded to the chemical request it sent and turns off the signal, knowing that nourishment has been received. In the days when food was synonymous with nutrition, the more we ate, the better nourished we became. It really was that simple. That being the case, there was no desire to eat more than what we biologically required.

But times have changed. The highly refined foods that compose a large percentage of most North Americans' diet don't contain the nutritional components that would make them worthy of our consumption. These "foods" provide mass without sustenance. To make matters worse, they retain the calories. And calories without nutrients, also known as *low nutrient-dense food,* are the prime ingredients for a nutritional stress stew.

In *The Thrive Diet* I described my first exposure to significantly elevated stress, before I had discovered the major impact nutrition could have on regeneration. In one triathlon season in particular, the amount of training I was doing simply overwhelmed my system. It could not regenerate quickly enough to support the pace at which I was breaking it down. A high amount of physical stress, coupled with nutritional stress, elevated my level of cortisol (a stress hormone I'll tell you more about later), which then remained high for an extended period. After about four months, my stress problem had become chronic. I displayed all the telltale signs, but not understanding the relationship between stress and hormones, I ignored what I could and masked the rest: the general fatigue, difficulty sleeping, irritability, mental fog, and cravings for sugar and starchy food. While these symptoms of rampant stress were bothersome, they weren't nearly as debilitating for me as an athlete as those that were about to unfold. I actually began to gain weight. And it was *fat.* I was getting fatter, even though I was training full-time. The "experts" I consulted assured me the solution was simple: "If you're gaining fat, there can be only one reason—you're simply taking in more calories than what you're burning. Cut back on the amount you're eating and you'll lose weight."

This seemed odd to me. True, I was consuming a lot of food. But I had a hard time believing that it was in excess of what I was burning to fuel my significant volume of training. However, I was confused and out of ideas. So I tried the "experts'" suggestion. I cut back on the amount I ate.

What was to develop next I did not expect. Unbelievably, I began gaining weight even *more* quickly. Imagine: there I was, training 35 to 40 hours every single week, hardly eating, and *gaining* fat. How could that be?

It wasn't until about a year and a half later that the answer was revealed. After speaking with an endocrinologist—a scientist with a deep understanding of the intricate relationship between stress and hormones—it became clear to me that I had placed too much strain on my adrenal glands, burned them out, and, as a result, was "hormonally injured." Elevated cortisol affects all other hormones, and as a result, balance—homeostasis—in the endocrine system falters. Additionally—and most significantly—when cortisol is elevated, it becomes nearly impossible to tone muscle or lose fat. In extreme cases, the body gains weight and loses muscle. And that's exactly what was happening in my situation.

> Second only to overconsumption, the greatest reason for obesity in North America is that we are simply inundated with more stress than our adrenal glands can deal with in a sustainable, healthy manner.

As I learned the hard way, when I restricted the amount I ate, I gained fat rapidly because I wasn't getting adequate nutrition. My body desperately craved more nutrients, to help it cope with stress. Depriving my body of the nutrients it needed created greater nutritional stress and therefore more overall stress. Increased stress drove my cortisol levels higher, as my body sought to alleviate its stress through its own means. The higher my cortisol levels, the faster the symptoms of stress developed, such as fat gain.

My stress had come from the demands of training—physical stress. But excessive exercise is not the only cause of adrenal fatigue, and full-time athletes are not the only people at risk for elevated cortisol. The demands of modern life have the same cortisol-raising potential as physical stress. And as such, most North Americans have elevated cortisol levels—many with adrenal burnout—which prevent the body from toning muscle and burning fat effectively, despite regular exercise. Affecting more than 85 percent of North America's population, stress has reached never-before-seen levels. Second only to overconsumption (brought about by nutrient-absent food, as I explain on page 37), the greatest reason for obesity in North America is that we are simply inundated with more stress than our adrenal glands can deal with in a sustainable, healthy manner. But nutrient-dense whole foods can alleviate a considerable amount of that stress. And with lower stress come less severe and fewer symptoms.

The reality is that work, family, and the other stressors in our lives are sometimes not within our control. Fortunately, what we choose to eat is. Therefore, we can have a commanding influence on our overall stress levels. Once we have lowered our overall stress by eating well, we can more easily address some of the other daunting issues causing us traditional stress. But nutrition is a good place to start, and plant-based nutrient-dense whole foods are the base. The recipes in Chapter 6 encompass all the nutritional building blocks to ensure that nutritional stress is the least of your worries.

NUTRITION'S INFLUENCE ON SLEEP

We all know sleep is important and that without enough our mental and physical performance deteriorates. Brain function sharply declines, decision-making becomes labored, and reasoning escapes us. Our bodies don't repair the previous day's cellular wear and tear—the broken-down body tissue—in a timely fashion if we don't get enough rest, and stiffness and weakness are the result.

Delta-phase sleep is possible only when one's cortisol level is low.

As reported by several studies, people who get at least eight hours of sleep a night are not as susceptible to contracting disease, have a greater ability to focus, and are less prone to developing depression than those who sleep less. It's such findings that prompt "experts" to suggest that most of us "should sleep more."[2] I, however, suggest it's not *more* sleep that most of us need but *better-quality* sleep. To be specific, most of us need more deep delta-phase sleep. This is the phase in which growth hormone is released, naturally triggering cellular repair and regeneration.[3] To sleep in the delta phase is to sleep efficiently, and if we have plenty of delta-phase sleep, we need less total sleep. Unfortunately, the vast majority of North Americans aren't able to slip into the delta phase. Their stress level, and therefore their cortisol, is simply too high to let them. Physiologically, it's impossible to enter the delta phase while cortisol is elevated. Realizing the full restorative properties of sleep requires lowering our cortisol to a level few North Americans can reach without first significantly altering their diet.

Delta-phase sleep is possible only when one's cortisol level is low.

Clearly this relationship is a vicious circle. Stress prevents quality sleep so that we then need more sleep to compensate for its poor quality. Or fatigue takes hold, which we remedy with stimulants in the form of caffeine or sugar. Yes, more sleep will indeed help people feel better rested and acquire the benefits listed above. However, lower cortisol will enable more *efficient* sleep, and therefore not as much will be required to facilitate quick regeneration and the benefits that come with it.

Speaking biologically, improving sleep efficiency will reduce the quantity requirement. That's good. Those extra waking hours give us more time to spend as we please—provided, of course, we're alert and functioning at a high level, both mentally and physically. In order to be high functioning, we need to be well rested. So, although sleep is a central component of health and wellness, it is not the duration that is of utmost importance, it's the quality.

As you've undoubtedly noticed, the line between being awake and being asleep has become blurred for many people. Once the fleeting stimulation of coffee and sugar wears off, these people spend their days in a state of mental haze. Even after eight hours of sleep, they wake up feeling both physically and mentally tired. And it shows. Have you ever had a conversation with someone and asked yourself, "Is this person awake?" The answer is probably "not completely." Because when this person was sleeping, he or she probably wasn't completely asleep either, at least not in the deep, refreshing delta phase. And because that person couldn't sleep deeply, because of high levels of overall stress, he or she also couldn't be fully focused, alert, and present when awake. Of course, it's in our best interest to sharply define that line. When we sleep, we want to be completely and deeply asleep so that when we're awake, we can function optimally.

Because of the undeniable restorative value of high-quality sleep, there's been a long-running debate as to what's more important for overall health: high-quality nutrition or high-quality sleep.

On the one hand, nutrition provides building material to replace aging cells with new, vibrant ones. A nutrient-dense diet also reduces nutritional stress. On the other hand, high-quality delta-phase sleep is when that repair actually takes place. And high-quality deep sleep can occur only when cortisol levels are low. Since nutrient-dense food reduces stress, a healthy diet improves cortisol levels and thus the quality of sleep. Better rested people do not crave sugary and starchy foods, since they simply do not require their stimulating energy. And in turn, high-quality sleep makes it easier to maintain a healthy diet.

So I think it's safe to say that neither high-quality sleep nor nutrient-dense food is *more* important to health but that they are complementary; the benefit of each is enhanced by the other. All the same, a person needs to have his or her nutrition needs met through quality food, or the delta-phase sleep can't be had at all.

High-quality sleep is imperative for high-quality living. And to achieve high-quality sleep we require high-quality, nutrient-dense food.

Deep delta-phase sleep is necessary for efficient cellular restoration. Nutrient-dense food is necessary for delta sleep.

Those who base their diet on low nutrient-dense foods have to eat more to become equally nourished, which means the consumption of more calories without an increase in nutrition. This pattern leads to chronic hunger and, most likely, weight gain.

NUTRIENT DENSITY: WHAT IT IS AND WHY IT MATTERS

Known as macronutrients, protein, carbohydrate, and fat together compose—in varying ratios—100 percent of the calorie content in all foods (sugar, fiber, and starch are types of carbohydrate). This being the case, eating anything edible is a guarantee that you'll obtain macronutrients.

Altering the ratio of macronutrients is a strategy used by athletes in pursuit of specific goals. For example, increasing the fat and protein ratios (eating more fat and protein) during endurance training will help improve fat metabolism and prevent muscle breakdown. More protein and starchy carbohydrates, such as sweet potatoes, will help to volumize and build muscles in strength athletes. A diet higher in sugar (in the form of fruit) will provide easily digestible sustenance to athletes in need of quick energy immediately prior to short intense workouts. In *The Thrive Diet*, I explain these athletic fueling strategies in detail.

However, the term "nutrient density" does not refer to macronutrients but to micronutrients. Unlike macronutrients, micronutrients contain *no* calories. There is also no guarantee that food will contain them. In fact, as I explain in detail in Chapter 3, since the quality of the soil in which the food is grown determines its micronutrient content—based on the soil mineral content—a large percentage of our modern food supply lacks micronutrients because of over-farming. That's a problem.

The World Health Organization agrees, referring to micronutrients as the "'magic wands' that enable the body to produce enzymes, hormones and other substances essential for proper growth and development."[4]

All vitamins, minerals, trace minerals, phytochemicals, antioxidants, and carotenoids are classified as micronutrients. Other than protein, carbohydrate, and fat, every nutritional component is that of a micronutrient. A complete list of known micronutrients (it's theorized there are many yet to be discovered), along with the role each plays, can be found in "Guide to Nutrients" on page 293. As you can see, the list is extensive. And since micronutrients are the backbone of nutrition itself, their dietary presence will dramatically reduce nutritional stress and therefore the symptoms that accompany it.

> A simple ratio, a food's nutrient density can be calculated by dividing the combined and averaged micronutrient content in any given food by the number of calories it contains.

In fact, in 2005 the United States Department of Agriculture (USDA) published a report suggesting that we in North America consume too many calories and too few micronutrients, and that as a consequence our health is steadily declining.[5] Recognizing that consuming food no longer guarantees nutrient intake, in an uncharacteristically progressive move, the USDA came up with a simple yet effective way to help people view food in a different light. Nutrient content is what we ought to seek from food, not calories. In fact, the report urged, the more nutrients we can get from our food while taking in fewer calories in the process, the better. "Nutrient density" emerged as *the* food attribute to seek. A simple ratio, a food's nutrient density can be calculated by dividing the combined and averaged micronutrient content in any given food by the number of calories it contains. Since micronutrients are expressed in several different measurements (milligrams, micrograms, and international units), each was converted into a percentage of the USDA's recommended daily intake (RDI), to establish a benchmark and provide a starting point for the calculation. And while I don't necessarily agree with the percentages of RDI put forth by the USDA, their established values serve nicely as a means by which to institute a consistent benchmark.

In the years that followed, Dr. Joel Fuhrman, the author of an excellent book entitled *Eat to Live,* took the nutrient density ratio one step further and factored in antioxidants (which include phytonutrients, or phytochemicals, compounds found naturally in plants), for which there are currently no USDA recommended daily intakes. Yet their role in obtaining optimal health is well established as vital; as such, Dr. Fuhrman wisely includes them when calculating nutrient density values for various foods.[6]

Additionally, the USDA's evaluation system falls short in that phytonutrients have been shown to contain disease-fighting and anti-inflammation properties. Clearly, these nutritional compounds have value that should be factored into nutrient density scoring.

Interestingly, plants grown with herbicides and pesticides—rendering them non-organic—no longer need to develop compounds to defend themselves from weeds, insects, and other pests, because added chemicals now fill that role.[7] The disadvantage of this development—besides the synthetic chemical residue on our food—is that the self-protecting compounds plants would have naturally produced to make them undesirable to insects are in fact powerful phytonutrients.

Overall, because Dr. Fuhrman includes antioxidants in his nutrient-density calculation method, I believe it to be of greater value than the one originally put forth by the USDA. For this reason, when we assess nutrient density in Chapter 3, I have chosen to use his system.

However, by applying the Guiding Principles (beginning on page 21), we remove complicated calculations from the equation and simplify the pursuit of nutrient-dense foods. In addition, when food choices are plant-based whole foods, there is simply no need to count, or even observe, calories or grams of macronutrients. Basing the diet on plant-based whole foods will ensure everything falls into place. But before moving on to the Guiding Principles, let's look at a couple of areas where my whole-food Thrive diet, based on nutrient density, and a conventional diet, based on calorie counting, diverge significantly: fat and portion control.

High-Quality Raw Fat

Conventional diets say that fat is bad. And since fat contains more calories than carbohydrate or protein, it's true that even the healthiest fatty foods tend to be slightly less nutrient dense. However, the extra calories from high-quality fat can provide a benefit that offsets their slightly lower nutrient density.

Sources of fat that are unrefined, plant-based, and raw offer benefits in the form of omega-6 and omega-3 "essential" fatty acids (EFAs). These fatty acids are termed "essential" because the body needs them for peak health but can't manufacture them on its own—they must be obtained through food.

Those who take in adequate amounts of essential omega-3 and omega-6 fatty acids through their diet and who have them in correct balance have a lower risk of developing cardiovascular disease, diabetes, and arthritis.[8] The good news is that a diet comprised of plant-based whole foods delivers plenty of omega-6 and omega-3 as a matter of course. The ideal ratio of omega-6 to omega-3, namely, 2:1 to 4:1 (two to four times more omega-6 than omega-3) will automatically fall into place. Smoother, more supple skin and more efficient metabolism of fat are two immediate benefits that the correct balance of omega-3 and omega-6 EFAs in the diet bring.

Of course, an efficient fat metabolism directly translates into a leaner frame and—for the athlete—greater endurance during training sessions and races lasting two hours or more, Most raw nuts and seeds are an excellent source of high-quality fat. Although omega-3 is more difficult to come by than omega-6, excellent sources are flaxseeds, hemp seeds, chia, and, in particular, a seed called sacha inchi (what I classify as a "next-level food"). I'll be telling you more about these foods in Chapter 5, "Nutrient-Dense Whole Foods to Thrive," and of course the recipes in Chapter 6 offer several ways to make them part of your diet, too.

Nuts and seeds tend to score marginally lower on the nutrient density scale only because of their slightly greater fat content, not because of a lack of micronutrient levels. Interestingly, while hemp seeds and flaxseeds are full of micronutrients, their pressed oil is not. It's pure fat with very few micronutrients. Despite that, their oil is a worthy addition to your diet for its high levels of health-promoting omega-3 and omega-6 essential fatty acids. Coconut oil is another example that delivers more than you might expect. While it's not high in essential fatty acids, it is packed with medium chain triglycerides (MCT), a premium source of non-adrenal-stimulating fuel. I use coconut oil for its high-octane energy in several of my pre-workout, sport-specific recipes. (See Chapter 5 for more on coconut oil and MCTs.)

In short, cold-pressed, unrefined, plant-based seed and coconut oils are the exceptions to the nutrient-density rule.

As you'll notice, most of the salad dressing and sauce recipes in Chapter 6 are based on these oils.

Portion Control

Since I'm an advocate of nutrient-dense whole foods, I am not a supporter of portion control. The two simply can't coexist. If we eat nutrient-dense food, our chemical hunger signal will turn off naturally, as it did for our earliest ancestors. To forcefully deny ourselves continued eating when we desire to consume more is a reaction and a testament to our nutrient-lacking food. Forced

> When there's a lack of micronutrients in what we eat, our hunger signal remains active; overconsumption and weight gain are likely the result.

portion control is nothing more than a symptom of a systemic problem that has been breed by our unhealthy farming system and our misguided focus on food volume production as opposed to food nutrient density (which I explain in detail, starting on page 16). By means of pesticides and genetic modification, an artificially created abundance of food volume leads to one thing: a poverty of micronutrients. And, as we know, when there's a lack of micronutrients in what we eat, our hunger signal remains active; overconsumption and weight gain are likely the result.

I grew up in a wooded area in North Vancouver, and each spring I had a front-row seat to watch what happened when black bears stopped eating a natural, nutrient-rich diet and began literally eating garbage. Black bears wandering down from the forest in search of easily obtainable food would feed on garbage left on the curb for pickup. Food scraps were what they sought out in the suburban curbside disposal bins. High in calories, easy to obtain, and addictive, the processed human food would hook the bears. And in becoming hooked, the bears, just like humans, became lethargic and began packing on the pounds. By summer, they were undeniably tired and fat.

The only difference between these suburban garbage eaters and their counterparts in the forest was what they were eating. Instead of their natural diet—nutrient-rich roots, buds, berries, dandelions, and fruit—the suburban visitors were eating one that mirrored ours, high in calories, low in nutrients. And bears aren't the only wild animals to undergo this transformation when they begin eating our calorie-laden, nutrient-absent food. Raccoons, coyotes, and mountain lions—none are immune to the effects of low-quality food. And before long, as with humans on this diet, they become portly and lethargic.

Our preoccupation with producing a greater volume of food, rather than more nutritious food (I discuss the implications of this blinkered view of food production in detail in Chapter 2), is ultimately the culprit. And portion control can't fix the problem of undernourishment.

GUIDING PRINCIPLES

The Guiding Principles I set out below are five fundamental nutritional ideas that build on the premise of eating nutrient-dense, plant-based whole foods. These principles go hand in hand with my nutritional philosophy and are here to guide you when you're faced with making choices about what to eat. Of course, all the recipes in this book adhere to these principles.

My Five Thrive Guiding Principles

1. Eliminate biological debt: acquire energy through nourishment not stimulation.
2. Go for high-net-gain foods: make a small investment for a big return.
3. Aim for a high percentage of raw and low-temperature-cooked foods.
4. Choose alkaline-forming foods.
5. Avoid common allergens.

1. ELIMINATE BIOLOGICAL DEBT: ACQUIRE ENERGY THROUGH NOURISHMENT NOT STIMULATION

Biological debt is the term I use to describe the unfortunate, energy-depleted state that North Americans frequently find themselves in. Often brought about by eating refined sugar or drinking coffee to gain short-term energy, biological debt is the ensuing energy "crash."

There are two types of energy: one obtained from stimulation, the other from nourishment. The difference between the two is clear-cut. Stimulation is short-term energy and simply treats the symptom of fatigue. Being well nourished, in contrast, eliminates the need for stimulation, because a steady supply of energy is available to those whose nutritional needs have been met. In effect, sound nutrition is a preemptive strike against fatigue and the ensuing desire for stimulants. With nutrient-dense whole food as the foundation of your diet, there's no need to ever get into biological debt.

Generally speaking, the more a food is *fractionalized* (the term used to describe a once-whole food that has had nutritional components removed), the more stimulating its effect on the nervous system. And of course there's also caffeine to consider, North Americans' second-favorite drug (next to refined sugar). By way of stimulation, fractionalized foods and caffeinated beverages boost energy nearly immediately. But almost as quickly, within only a few hours, that energy will be gone. It is a short-term, unsustainable solution to the *symptom* of our energy debt. Adrenal gland stimulation *always* carries a cost. Obtaining energy by way of stimulation is like shopping with a credit card. You get something you desire now but that doesn't mean you won't have to pay eventually. The "bill" will come. And with that bill comes incurred biological interest: fatigue. Again.

We tend to rely upon additional stimulation to deal with this second wave of weariness, which in turn simply delays the moment when we pay off our tab. But the longer we put off payment, the greater the debt we accumulate. To continue our debt/credit analogy, to simply continue to summon energy by way of stimulation is like paying off one credit card with another. All the while, the interest is mounting.

Stimulation is a bad substitute for nourishment for another reason. It makes demands on the adrenal glands, prompting the production of the stress hormone cortisol. Elevated cortisol is linked to inflammation,[9] which is a concern for the athlete (and for anyone who appreciates fluid movement). Higher levels of cortisol also weaken cellular tissue, lower immune response, increase the risk of disease, cause body tissue degeneration, reduce sleep quality, and are a catalyst for the accumulation of body fat.[10] As if that weren't enough, chronic elevated levels of cortisol *reduce* the effectiveness of exercise, activity that normally helps to keep cortisol in check. Too-high cortisol levels can actually break down muscle tissue, as well as prevent the action of other hormones that build muscle. As a result, muscles not only become more difficult to tone but strength is likely to decline rather than increase.

Not surprisingly, if we keep on overstimulating our overstressed body, without addressing the real problem behind our fatigue, things only get worse. The severity of the symptoms of stress increase so that our health declines little by little. We put ourselves at greater risk for serious disease.

Often the first symptom of adrenal fatigue is increased appetite, followed by cravings, commonly for starchy, refined foods; difficulty sleeping; irritability; mental fog; lack of motivation; body fat gain; lean muscle loss; visible signs of premature aging; and sickness.[11] If this cycle of chronically elevated cortisol levels is allowed to continue, tissue degeneration, depression, chronic fatigue syndrome, and even diseases such as cancer can develop.

> Energy derived from good nutrition—cost-free energy—does not take a toll on the adrenal glands and so doesn't need to be "stoked" with stimulating substances.

In contrast, when we use nutrient-dense whole food as our source of energy, rather than fleeting pick-me-ups, our adrenals will not be stimulated, and, simultaneously, our sustainable energy level will rise because of the acquired nutrients. Energy derived from good nutrition—cost-free energy—does not take a toll on the adrenal glands and so doesn't need to be "stoked" with stimulating substances. In fact, one characteristic of wellness is a ready supply of *natural* energy that doesn't rely on adrenal stimulation. People who are truly well have boundless energy with no need for stimulants, such as caffeine or refined sugar.

A cornerstone of my dietary philosophy is to break dependency on adrenal stimulation. As you might expect, we accomplish this by way of nutrient-dense whole foods. And not just by supplementing our diet with them but by *basing* our diet on them. This diet, along with proper rest through efficient sleep (efficient because of our reduced stress, thanks to nutrient-dense food), will address the *cause* of the problem, not just the *symptoms* of nutritional shortfalls.

2. GO FOR HIGH-NET-GAIN FOODS: MAKE A SMALL INVESTMENT FOR A BIG RETURN

High-net-gain foods deliver us energy by way of conservation as opposed to consumption. Here's what I mean by that: the digestive and assimilation process is in fact an energy-intensive one. At the onset of eating, we begin spending digestive resources in an effort to convert energy stored within food—also known as calories—into usable sustenance to fulfill our biological requirements. And, as we know, whenever energy is transferred from one form to another, there's an inherent loss. However, the amount of energy lost in this process varies greatly and depends on the foods eaten.

Highly processed, refined, denatured "food" requires that significantly more digestive energy be spent to break it down in the process of transferring its caloric energy to us:

net energy gain = energy remaining
once digestive energy has been spent.

While it's true that a calorie is a measure of food energy, simply eating more calories will not necessarily ensure more energy for the consumer. If there were such a calorie guarantee, people who subsisted on fast food and other such calorie-laden fare would have abundant energy. And of course they don't. This is a testament to the inordinate amount of digestive energy required to convert such "food" into usable fuel.

(By the way, it's no coincidence that the cultures that have their largest, heaviest meals for lunch are the same ones who have afternoon siestas. Digestion is tiring.)

In contrast, natural, unrefined whole food digests with a considerably lower energy requirement. Therefore, we can gain more usable energy from simply eating foods that are in a more natural whole state, even if they have fewer calories.

When I grasped this concept, I began viewing food consumption as though it were an investment of sorts. My goal became to spend, or invest, as little digestive energy as possible to acquire the greatest amount of micronutrients and maximize the return on my investment.

For that reason, I refer to foods that require little digestive energy but yield a healthy dose of micronutrients as high-net-gain foods:

high net gain = little digestive energy spent,
substantial level of micronutrients gained.

With this principle in mind, I shifted my prime carbohydrate sources from processed and refined carbs, such as pasta and bread, to fruit. Fruit is packed with carbohydrate in the form of easily assimilated sugar, considerably easier to digest than refined grain flour. And fruit returns higher micronutrient levels than these processed, refined carb sources.

3. AIM FOR A HIGH PERCENTAGE OF RAW AND LOW-TEMPERATURE-COOKED FOODS

Foods that have not been heated above 118 degrees Fahrenheit are termed raw. There are several advantages to eating a large quantity of raw food in place of its cooked counterpart. Ease of digestion and assimilation, which directly translates into additional energy by means of an increase in net gain, is the most significant. Enzymes that contribute to overall health and aid digestion are not present in cooked food; heating above

118 degrees Fahrenheit destroys them. Before the body can turn cooked food into usable fuel, it must produce enzymes to aid in the digestion process. A healthy person can create these enzymes, but it costs energy, which creates a nominal amount of stress. As well, as we get older, our enzyme production naturally slows down; if we are not getting enough enzyme-rich foods in our regular diet, our enzyme-production system will have to work even harder. Overtaxing the system can weaken it further and lead to major digestive problems. Including enzyme-rich foods in our diet on a regular basis will help safeguard our bodies' ability to manufacture enzymes. Interestingly, a person who cannot produce digestive enzymes and does not obtain them through food can acquire the same diseases as someone suffering from malnutrition.[12]

> Food cooked at a high temperature can cause inflammation.

At only slightly above 118 degrees Fahrenheit, enzyme quality in food will sharply drop off. The next significant quality decline comes when food reaches a temperature of about 300 degrees Fahrenheit. This is the point at which essential fatty acids convert into trans fats. Additionally, food cooked at a high temperature can cause inflammation. When sugar is heated to a high temperature with fat, it can create end products known as AGEs (short for *advanced glycation end products*), which the body perceives as invaders. The immune cells try to break down these end products by secreting large amounts of inflammatory agents. If the cycle continues, it can result in problems commonly associated with old age: less elastic skin, arthritis, weakened memory, joint pain, and even heart disease.

Most of the base ingredients (the most common being corn, soy, or wheat flour) in refined foods on the typical grocery store shelf have been heated to well above 300 degrees. Because of this, the fat within those flours has become denatured and therefore is not only unusable by the body but perceived by the immune system as a threat, which weakens the body's reserves and causes a rise in cortisol.

As you will notice, many of the Thrive recipes in Chapter 6 are raw (look for the raw icon:). About 80 percent of my own diet is raw, and I find that percentage works well.

4. CHOOSE ALKALINE-FORMING FOODS

The measure of acidity or alkalinity is called pH, and maintaining a balanced pH within the body is an important part of achieving and sustaining peak health. If our pH drops, our body becomes too acidic, adversely affecting health at the cellular level. People with low pH are prone to many ailments and to fatigue.

The body can become more acidic through diet and, to a lesser extent, stress. Since our bodies are equipped with buffering capabilities, our blood pH will vary to only a small degree, regardless of poor diet and other types of stress. But the other systems recruited to facilitate this buffering use energy and can become strained. Over time, the result of this buffering is significant stress on the system, which causes immune function to falter, effectively opening the door to a host of diseases.

> Minerals are exceptionally alkaline-forming, so foods with a greater concentration of micronutrients—greater nutrient density—will inherently have a greater alkaline-forming effect.

Low body pH can lead to the development of kidney stones, loss of bone mass, and the reduction of growth hormone, which results in loss of lean muscle mass and increase in body fat production. And since a decline in growth hormone production directly results in lean muscle tissue loss and body fat gain, the overconsumption of acid-forming foods plays a significant role in North America's largest health crisis. However, food is not the only thing we put in our bodies that is acid-forming. Most prescription drugs, artificial sweeteners, and synthetic vitamin and mineral supplements are extremely acid-forming.

Low body pH is also responsible for an increase in the fabrication of cell-damaging free radicals and a loss in cellular energy production. Free radicals alter cell membranes and can adversely affect our DNA.

So what can we do to prevent all this? The answer is to consume more alkaline-forming foods and fewer acid-forming ones.

Minerals are exceptionally alkaline-forming, so foods with a greater concentration of micronutrients—greater nutrient density—will inherently have a greater alkaline-forming effect.

Another factor that significantly raises the pH of food and, in turn, the body, is chlorophyll content.

Responsible for giving plants their green pigment, chlorophyll is often referred to as the blood of plants. The botanical equivalent to hemoglobin in human blood, chlorophyll synthesizes energy. Chlorophyll converts the sun's energy that has been absorbed by the plant into carbohydrate. Known as photosynthesis, this process is responsible for life on Earth. Since animals and humans eat plants, we too get our energy from the sun, plants being the conduit. Chlorophyll is prized for its ability to cleanse our blood by helping remove toxins deposited from dietary and environmental sources. Chlorophyll is also linked to the body's production of red blood cells, making daily consumption of chlorophyll-rich foods important for ensuring the body's constant cell regeneration and for improving oxygen transport in the body and, therefore, energy levels. By optimizing the body's regeneration of blood cells, chlorophyll also contributes to peak athletic performance.

5. AVOID COMMON ALLERGENS

Causing, at the very least, symptoms such as mild nasal congestion, headache, and mental fog, sensitivities to certain foods are exceptionally common. Wheat, gluten (in wheat), corn, soy, and dairy are the most common of these allergens.

A sensitivity is an unpleasant reaction caused by eating food for which the body lacks the specific enzymes or chemicals to digest it properly. Unlike an allergic reaction, a sensitivity does not affect the immune system. Food allergies often become evident immediately upon consuming the food: there's no mistaking an allergic reaction, which comes on quickly and often violently, ranging all the way from abdominal cramps and vomiting or a tingling in the mouth, to life-threatening anaphylactic shock, with swelling of the tongue and throat and difficulty breathing. As serious as an allergic reaction may be, once the allergen is identified, the solution is straightforward: don't eat the food again! The symptoms of a specific food sensitivity, however, may not become evident for a few days or even a week after consumption, making its source difficult to trace. Food sensitivities, therefore, can be extremely difficult to immediately identify and eliminate, and in these cases, the strategy of eliminating common allergens from the diet is useful.

For several years, I had what I thought to be a bad case of hay fever. Each spring I would display the classic symptoms of pollen-caused allergies: dry eyes, sinus congestion, and mild flu-like symptoms. Thinking there was little I could do short of taking antihistamine drugs, which I wanted to avoid, I just put up with it. Since these symptoms flared up at the same time each year, it seemed obvious they must be related to environmental changes because of the onset of spring and the bursts of pollen in the air. Or so I thought.

When I began learning about sensitivities to food, I reanalyzed my diet. What I found intrigued me. Each year, as the winter turned to spring, I'd ramp up my cycling mileage in preparation for the coming triathlon season. And as the cycling ramped up, so did my consumption of the sports drink I sipped while logging the miles. My sports drink's base ingredient was maltrodextrin, which is an inexpensive form of carbohydrate, derived from corn. I got tested for food sensitivity, and sure enough, corn registered as one. Fresh, non-genetically modified corn on the cob didn't bother me, but corn in the highly processed state of maltrodextrin triggered an adverse reaction in my body.

From that point on, I began making my own sports drinks. You can find recipes for the original one I came up with, as well as some newer varieties, in the "Drinks" section in Chapter 6.

This discovery made me realize that many people have a food sensitivity—or several sensitivities—but they just don't know it. "Not feeling quite up to par" is their common description of their life in general. They rule out diet as the culprit since it's been a constant in their life—virtually unchanged—for years. Some people, as I did, blame environmental factors, such as dust or pollen. These annoying, low-grade symptoms, or sometimes a more general state of malaise, can go on or regularly recur for years; because the symptoms make just certain activities a bit more difficult without actually preventing them, the sufferer takes no action. But it is precisely the unchanging diet that is behind the symptoms.

If you suspect you may have a food sensitivity, try eliminating the common allergens—processed corn, wheat/gluten, dairy products, and soy—from your diet. Test by removing one food at a time for a period of 10 days so that you can isolate your reactions. If your symptoms subside when you are off the food, then you will know that it causes you problems and you'd be better off removing it from your diet. If you don't notice a change when you go off a food, then you can carry on eating it. It's that simple.

By the way, it's unlikely you'll find any processed, refined food at the supermarket that doesn't contain at least one of these common allergens. Corn and soy are particularly ubiquitous. They're cheap, shelf-stable, and take on other flavors well, so, in the eyes of the manufacturer, they're the perfect set of ingredients to increase the volume of processed foods, essentially being used as "filler," adding little to no nutritional value.

As described in this chapter, the implications of nutritional stress that results from a diet comprised of nutrient-absent foods can have a profound effect on us as individuals. In the next chapter, we'll look at the implications beyond the personal—to the effects on our environment of our food choices.

THRIVE AT A GLANCE

- A large percentage of overall stress is related to poor nutrition.
- If our diets do not supply adequate micronutrients, we will suffer nutritional stress and its debilitating symptoms.
- Nutrient density is the most comprehensive way to assess a food's value.
- High-quality deep sleep is dependent on high-quality nutrition.
- Energy gained by nourishment is sustainable; energy from stimulation is fleeting.
- By choosing easily digestible food, we reduce the amount of energy we need to expend on digestion.
- Less energy spent on digestion equates to more available energy.
- A diet high in alkaline-forming foods will reduce inflammation as well as the risk of disease.
- Foods that contain chlorophyll or are rich in minerals are particularly good alkaline-forming food sources.

2

EATING RESOURCES: The Environmental Toll of Food Production

Producing the vast amount of food required to support our population places a significant draw on our ecosystem, but it's a necessary exchange. Nonetheless, the divide between the resource requirements of plant-based food production and those of animal-based food production is impressive, greater than I ever could have imagined.

In the pages that follow, I explore the use of natural resources in relation to food production. The three natural resources required to produce food are arable land, water, and fuel (most of it fossil fuel). I also factor in the carbon dioxide, or CO_2, emissions that are released into the atmosphere through the burning of fossil fuel during the food production process. In addition, I consider other emission sources such as methane and nitrous oxide.

ARABLE LAND

Arable land refers to land that can be used for growing crops. It is undoubtedly one of our most precious natural resources and without it we couldn't produce even close to enough food to meet the demands of our rapidly growing population. While the land—the geographical space that croplands occupy—itself is vital, of equal importance is the quality of soil covering it.

Healthy topsoil is in fact a complex blend of elements that are vital to the growing process of nutritious food. Composed of a mixture of organic material, such as decaying plant matter, fungi, bacteria, and microorganisms, soil is much more than simply dirt. Just one acre can be home to 900 pounds of earthworms, 2400 pounds of fungi, 1500 pounds of bacteria, 133 pounds of protozoa, and 890 pounds of arthropods and algae.[1]

Topsoil also comprises a vast quantity of minerals that are the necessary catalysts for plants to produce vitamins, enzymes, antioxidants, a plethora of phyto-nutrients, and amino acids (protein), and that give them the ability to formulate hormones. Without minerals, none of these vital nutritional components can be constructed.

> Plants are the medium. They draw minerals from the soil, passing them on to us in a form we can assimilate.

And since we can't digest soil, plants come to our aid; they draw minerals into their stalks, leaves, and seeds, and then act as the delivery system through which the nutrition originating in the soil (the minerals) is passed on to us, through the food we eat.

Plants are the medium. They draw minerals from the soil, passing them on to us in a form we can assimilate.

Minerals are not only vital for good health but are in fact the base on which health is built. Low mineral intake is now accepted as a substantial increased risk factor for myriad conditions, including osteoporosis, type 2 diabetes, depression, and obesity. Mineral deficiencies have been shown to play a role in cardiovascular disease (CVD), which kills 910,000 people in the United States annually.[2] That equates to an American death every 35 seconds. And while nutrient-absent food isn't the only contributing factor, it's among the most significant. According to the World Health Organization, a major reason for the widespread incidence of CVD is that "people are consuming a more energy-dense, nutrient-poor diet and are less physically active."[3] So basically people are eating more calories than they need while getting too few micronutrients—the textbook definition of a low nutrient-dense diet. In fact, two-time Nobel prize–winning chemist Dr. Linus Pauling stated that "you can trace every sickness, every disease and every ailment to a mineral deficiency."[4] As mentioned earlier, nutritional stress is primarily a result of a lack of micronutrients, such as minerals and phytonutrients, in our food, and they cannot be present in the food if minerals are lacking in the soil. And as we know, stress is the root cause of most diseases. And the lack of micronutrients in food is the root cause of nutritional stress. So good health actually begins in the soil.

Today, the mineral content in our soil is drastically lower than it was even just a few decades ago, contributing to the strange paradox of a population overfed yet undernourished, an increasing and widespread health issue. According to the findings at the United Nations Conference on Environment and Development (UNCED), also known as the Earth Summit, held in 1992, North America "leads" the continents of the world in soil mineral depletion, having lost 85 percent of the mineral content in its soil over the past century (in comparison, South America's and Asia's minerals are 76 percent of what they were 100 years ago; Africa's, 74 percent; Europe, 72 percent; and Australia, 55 percent).[5] Sodium, calcium, potassium, magnesium, phosphorus, copper, zinc, and iron—all of them began their taper about 100 years ago, as population growth began to strain the land and its ability to produce an adequate supply of food.

For reasons such as population growth, the demand we place on our dwindling supply of arable land continues to increase. As the population increases, not only are there more mouths to feed, but housing more people also leads to suburban sprawl, which paves over arable land. So, as our demand for food continued to grow, the geographic space in which to grow it continued to shrink.

Conventional farming has reacted to this dilemma with a simple objective: to grow as much food as possible on as little land as possible. Mass and volume of production became paramount, but unfortunately the quality of the food grown was neglected in the process.

Since traditionally food has been understood to be synonymous with nutrition, the thinking traditionally has been that the more food we eat, the better nourished we'll be. But, not so, as I explained in Chapter 1; I learned this the hard way when I tried to better nourish myself in hopes of propelling my athletic performance. I ate more food to try to correct my lack of nourishment, which led to being overfed yet undernourished, a condition that unfortunately affects a significant percentage of the population. As I know now, I should have eaten not simply more food but more nutrient-dense food. The disconnect I and so many others have experienced is that though food once equalled nutrition, it's not necessarily so today. Not taking mineral content into account, conventional farming aims simply to produce more food on less land. Volume, mass, and calories are all that's considered when evaluating crop yield.

Because of heavy demands on the North American food system, genetic modification and widespread pesticide use have become the rule, not the exception. Despite the increasing awareness and popularity of organic food, its production worldwide pales in comparison to that of conventionally farmed and genetically modified organism (GMO) crops. According to Organic-World.net, only 0.8% percent of the world's total crops are grown without genetic modification and organically.[6] Clearly, a minuscule amount. Specifically looking to North America, Organic-World. net states that as of 2008 only 0.93 percent of Canadian crops were grown organically, only slightly better than the world average, while the United States' organic food crops registered at a dismal 0.6 percent. While genetic modification and chemical pest deterrents allow for a greater volume of food to be produced on less land, they may prove to cause serious health problems in years to come. And the decline of nutrient density is likely just one of the negative effects of producing our food this way. The possible complications of long-term GMO crops in our food supply are as yet unknown, and the effects on our health of chemical residue from pesticides are still to be determined.

Since there is a finite amount of each mineral in a given plot of land, the plants grown are limited by what is present in the soil. More plants doesn't mean more nutrition. The more food that's grown on a given plot of land, the less total amount of minerals each plant will contain, rendering each plant less nutrient dense. Clearly that's a problem.

Another factor is the type of crop. Each species of plant has a set limit as to how much of a given mineral it can draw into its cellulose. Here's how it works: when we eat food grown in over-farmed, mineral-depleted soil, to obtain the same amount of health-essential minerals, we need to consume more food. This of course equates to a greater intake of calories. In addition, the chemical hunger signal that I mentioned in Chapter 1 will remain active until it's satisfied that we've got the minerals—and the vitamins and the phytochemicals that the plant can produce only once it has drawn an adequate supply of the minerals from the soil—that we need to be healthy.

And if our body is not satisfied—and it won't be if the minerals aren't in the soil—it will urge us to eat more the only way it knows how: by making us hungry. So hunger doesn't necessarily mean we haven't eaten enough food. More often than not, today hunger means we haven't eaten enough *nutrient-rich* food.

Lack of minerals in the soil
↓
Micronutrient deficiency in our food
↓
Chronic hunger; tendency to overeat
↓
Weight gain and risk factor increased for
- Type 2 diabetes
- Hypertension
- Arthritis
- Osteoporosis
- Cardiovascular disease

Since the consumption of more calories with fewer minerals leads to a chronically active hunger signal, the tendency to overeat, followed closely by weight gain, will be the result. Being overweight will increase the risk factor for a wide range of diseases: type 2 diabetes, hypertension, arthritis, osteoporosis, and cardiovascular disease. So, simply producing a greater volume of food should not be our goal, but rather producing food that is of the highest nutrient density, which is governed, in large part, by soil quality.

Clearly the consumption of more calories and fewer nutrients equates to a diet of lower nutrient density, which can lead to health concerns and disease. And in fact, it's what the Standard American Diet is built on: lots of calories but very few micronutrients, such as minerals.

> Crop "yield" needs to be redefined and assessed as "nutrition contained within the food" as opposed to the total weight, volume, or caloric value of the food harvested.

Brian Halweil, a senior fellow at the Worldwatch Institute covering issues of food and agriculture, and the co-director of Nourishing the Planet, lays out the argument in a 2007 report titled *Still No Free Lunch: Nutrient Levels in U.S. Food Supply Eroded by Pursuit of High Yields.* Halweil describes the shortcomings of striving simply to obtain crop volume, mass, and calories from food. Instead, Halweil suggests, food producers ought to focus on the nutrient value of the food. Halweil eloquently concludes the report by saying:

> Yield increases per acre have come predominantly from two sources—growing more plants on a given acre, and harvesting more food or animal feed per plant in a given field. In some crops like corn, most of the yield increase has come from denser plantings, while in other crops, the dominant route to higher yields has been harvesting more food per plant, tree, or vine.
>
> But American agriculture's single-minded focus on increasing yields over the last half-century created a blind spot where incremental erosion in the nutritional quality of our food has occurred. This erosion, modest in some crops but significant in others for some nutrients, has gone largely unnoticed by scientists, farmers, government and consumers.[7]

Crop "yield" needs to be redefined and assessed as "nutrition contained within the food" as opposed to the total weight, volume, or caloric value of the food harvested.

Once I'd become familiar with the workings of this problem, the value of arable land was firmly impressed on me. I've never looked at rolling pastureland the same way since. What I learned next was hard to comprehend.

In 2008 I participated in the Students for Sustainability National Campus Tour, which was a joint initiative of the Canadian Federation of Students, the Sierra Youth Coalition, and the David Suzuki Foundation. We visited 22 campuses across Canada in 30 days, talking about the urgent need to create sustainable communities and campuses. While doing research for the speech I was preparing to deliver on the tour, I came across an extremely fascinating—and equally horrifying—piece of information. As unbelievable as it may seem, multiple reputable

Livestock production uses a *staggering* 70 percent of all arable land and 30 percent of the land surface of the planet.

sources, such as the United Nations and the Worldwatch Institute, corroborated that 70 percent of the food grown on—and drawing minerals from—our precious arable land was not in fact for us to eat but would serve instead as animal food, specifically, livestock feed.[8]

Animals being raised to be eaten—and animals being raised so that their by-products can be eaten, such as cows for milk and chickens for eggs—are collectively termed livestock. According to the United Nations, the raising of livestock is by far the single greatest anthropogenic use of land. Twenty-six percent of the ice-free surface of Earth is used as grazing land. Additionally, 33 percent of total arable land is used to grow food—primarily corn, soy, and wheat—to feed to animals. Factoring in grazing land and feedcrop requirements, livestock production uses a *staggering* 70 percent of all arable land and 30 percent of the land surface of the planet.[9]

While soil mineral depletion began approximately 100 years ago, the sharpest decline started about 30 years later, beginning in the 1940s. Not so coincidentally, this is when large-scale animal agriculture began commandeering the vast majority of our land.

Also, since 1940, cardiovascular disease has been on the rise.[10] And interestingly, its escalation has been in concert with widespread animal agriculture and the corresponding decline of mineral-rich soil. Coincidence?

Unfortunately, even those who don't eat meat are impacted by inefficient farming because of loss of nutrient-rich soil, since it affects the whole food supply.

Note that the lack of mineral content in animal feed is not a concern—in fact, it's desirable. As with humans, when animals eat nutrient-absent refined starch-based calories, they gain weight. The whole point of feeding animals this grain-based diet is to quickly increase their mass. And that it does. Of course, as I've mentioned, if there are no nutrients in the soil, then there are none in the animal, or in the person who eats it.

Raising animals for food is clearly an exceptional draw on arable land. But, to make matters worse, it's also an incredibly inefficient way of utilizing the dwindling supply that remains. For example, for every 16 pounds of plant matter (primarily composed of corn, wheat, and soy, most of which is genetically modified) fed to a cow, one pound of meat is returned.[11] Where do the other 15 pounds go? While a fraction is burned to fuel the cow's movement and some is lost into the atmosphere in the form of heat, the vast majority ends up as manure. So, we are literally turning one of our most valuable natural resources into manure.

Currently, the livestock population in the United States consumes more than seven times as much grain as the human population eats.[12]

According to the U.S. Department of Agriculture, the amount of grain fed to U.S. livestock would be able to feed about 800 million people (2.7 times the entire U.S. population) who followed a plant-based diet.[13]

A factory-farmed cow needs to consume 16 pounds of cattle feed
(wheat, corn, soy) to yield 1 pound of meat for human consumption.

And that estimate's based on people simply eating the crops grown for livestock. Now, if instead of growing GMO corn, soy, and wheat—destined to be fed to animals—we were instead to sow the fields with non-GMO, organic, nutrient-rich crops, such as hemp, flax, yellow peas, and kale, we would not only be able to feed more than double the U.S. population with ease but, most importantly, we'd be feeding them with nutrient-dense whole food. Far fewer people would be overfed yet undernourished, and disease would significantly decline.

Factory-farmed animals are fed wheat, corn, and soy, which are primarily carbohydrate based. Yet because of the sheer volume these animals are fed, they ingest six times as much protein as they yield in return.[14] And it takes 20 times as much land to grow beef for protein as it does to grow plants for protein.

What about grass-fed beef? If all who ate factory-farmed beef were to switch to grass-fed, there simply would not be enough meat to go around. Factory farming was created for a reason: to produce more food on less land. And it works. However, as already pointed out, it does so by yielding volume, mass, and calories rather than the true components of nutrition.

If beef eaters in the United Sates were to switch to exclusively grass-fed beef, one small steak about once every three weeks is all that would be available. There simply wouldn't be enough to meet demand. And if more grazing land were to be created, of course deforestation would be the result.

THRIVE AT A GLANCE

- Arable land with mineral-rich soil is vital to the production of micronutrient-rich food.

- Mineral-deficient food is a leading cause of disease in North America.

- More food doesn't necessarily mean more nutrition.

- As minerals decline, disease goes up.

- Calories without micronutrients cause overconsumption that leads to weight gain.

- Mineral content in our food has been declining since the 1940s, when industrialized farming began to take over the plains.

- Animal agriculture is an inefficient use of land and therefore depletes the soil of minerals.

- Seventy percent of food grown is destined to be fed to animals.

- If nutrient-dense food crops were planted in place of animal feed, a population over twice the size of North America's could be fed on the highest quality of food.

And as we know, it makes sense to "be the change we want to see in the world." So what if North Americans made the switch to eat grass-fed beef instead of factory-farmed meat? Would the result be a change we want to see? While one problem would be solved, another problem would arise—lack of grazing land. Clearly, swapping one problem for another will not give us a sustainable solution. We need to make the change—to be the change—that is a true, lasting solution.

FRESH WATER

Having grown up in North Vancouver and having rain fall from the sky for most of the year, learning to appreciate the value of water wasn't always easy. But in October 2008, while on the Students for Sustainability National Campus Tour, I met Maude Barlow. She is an author, an activist, and a world authority on water, and served as senior advisor to the United Nations. Understandably, I was honored to speak before her talk at Dalhousie University in Halifax, Nova Scotia.

After my speech, I sat back, watched, and listened. Barlow clearly articulated facts about water that I had never even considered. And as I began to see how the rest of the world lived—without water for proper sanitation or even adequate water to drink—I developed a newfound appreciation for it.

> Usable non-polluted fresh water is less than 1 percent of the Earth's total. And 70 percent of that 1 percent is primarily used for agriculture animal feed.

Referring to the much greater public concern about our looming energy crisis, Barlow stated that "no one ever died from a lack of energy."

That prompted me to delve deeper to find out more water-related facts. As nicely illustrated by the diagram on the next page, the United Nations research team concluded that fresh, non-polluted water should be considered one of our most valuable resources. While we are all acutely aware of the vital role water plays in keeping us alive, it wasn't until recently that the world's supply of drinkable water became a concern.

Using numbers put forth by the United Nations, 75 percent of the Earth is covered by water, yet 97.5 percent of that is salt water, which leaves just 2.5 percent as fresh water. However, 70 percent of that is frozen, leaving only 30 percent as ground water that we have access to. Unfortunately, a large amount of *that* water is polluted, so what we are left with as usable non-polluted fresh water is less than 1 percent of the Earth's total. And 70 percent of that 1 percent is primarily used for agriculture animal feed, and another 22 percent for industry. Therefore, what we have remaining is about 0.08 percent for domestic use.[15] Clearly, there is good reason to make a concerted conservation effort when using water around the house by using low-flow showerheads, not letting water run as we brush our teeth, and so on. But by far the most significant impact we can have on our conservation efforts is to reduce agriculture water usage. The 70 percent of non-polluted water that we use to irrigate fields is mostly going to animal feed crops. Not only are we using an inordinate amount of water in the production of animal feed, but upward of 85 percent of that feed is turned out as manure soon after consumption.[16] And that's the next concern. Farmed animals are not much more than a system for converting natural resources into manure, a by-product, which, to add insult to injury, is largely responsible for increasing the amount of water pollution so that our already dwindling supply of usable water becomes still scarcer.

And of course to transport the manure away, well, that takes energy too, from, of course, fossil fuel.

This depiction does not take into account the water damage that livestock contributes to. The livestock sector not only uses a vast amount

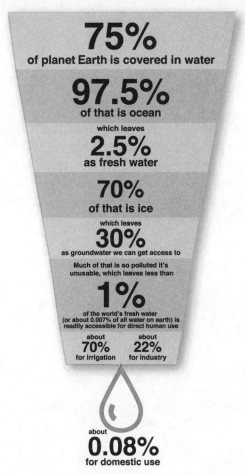

75%
of planet Earth is covered in water

97.5%
of that is ocean

which leaves
2.5%
as fresh water

70%
of that is ice

which leaves
30%
as groundwater we can get access to

Much of that is so polluted it's
unusable, which leaves less than

1%
of the world's fresh water
(or about 0.007% of all water on earth) is
readily accessible for direct human use

about
70%
for irrigation

about
22%
for industry

about
0.08%
for domestic use

Of the usable water we have, most is used to irrigate feed crops for
animals, leaving very little water for direct human consumption.
(Adapted from the original at treehugger.com. Used with permission.)

of water directly but, according to the United Nations, is responsible for significant water pollution. This is from the executive summary of a U.N. report, speaking of livestock production:

> It is probably the largest sectoral source of water pollution, contributing to eutrophication, "dead" zones in coastal areas, degradation of coral reefs, human health problems, emergence of antibiotic resistance and many others. The major sources of pollution are from animal wastes, antibiotics, and hormones, chemicals from tanneries, fertilizers and pesticides used for feedcrops, and sediments from eroded pastures.[17]

Water requirements to produce 1 pound of beef, not taking into consideration the water pollution caused by manure are as follows:

Factoring in irrigation for feed crops, it takes at least 2500 U.S. gallons of water to yield 1 pound of beef[18] (some sources estimate it can take as much as 12,000 gallons[19]). That's 59.5 standard bathtubs' worth of water to get a couple of sirloin steaks or a big helping of prime rib to your table.

In contrast, only about 60 gallons of water, enough to fill 1.36 bathtubs, are needed to produce a pound of sweet potatoes.[20]

About 100 gallons of water, the capacity of 2.3 standard-size bathtubs, are needed to grow one pound of hemp seed.[21]

In short, the amount of water it takes to produce a pound of beef is 25 times more than needed to grow a pound of hemp seed, and about 42 times more than needed to produce a pound of sweet potatoes.

FOSSIL FUEL

Oil, coal, and natural gas are collectively known as fossil fuel and abundantly overused in North America. In fact, 85 percent of energy produced in North America is derived from the burning of these carbon-rich deposits.[22] Essentially, fossil fuel is made up of prehistoric plant matter that, millions of years ago, extracted and "quarantined" carbon from the ancient atmosphere. As with plants today, they were made

> Plants are in fact solidified sunlight and, as such, the energy of the sun will be released back into the atmosphere when these ancient plants are burned or decay.

up of a combination of cellulose and sunlight. A plant is primarily the result of the process of photosynthesis, in which chlorophyll converts sunlight into carbohydrate for the plant to "eat." Cellulose is the plant matter created. Therefore, plants are in fact solidified sunlight and, as such, the energy of the sun will be released back into the atmosphere when these ancient plants are burned or decay. As Thom Hartmann eloquently puts it in his excellent book *The Last Hours of Ancient Sunlight,* fossil fuel is our savings account for the sun's energy that shone on Earth millions of years ago. *All* energy originates from the sun. Plants, of course, cannot exist without sunlight, and animals, of course, cannot exist without plants. We, as humans, are solar-powered too. Eating plants that, through photosynthesis, trap sunlight and use it to fabricate vitamins, grow cellulose, and draw minerals from the soil pass on to us the energy that originated from the sun.

Since our population began to grow considerably, and since our demand for energy paralleled this expansion, we got to a point where we could no longer rely simply on the current sunlight—solidified into trees—as a form of fuel (firewood) to keep pace with our escalating energy needs. So, we had to break open our savings account. Coal was the first of the fossil fuels to be burned since it was easily accessible and needed no refinement. Initially it was used in place of wood as a means to heat a room. The warmth and light from sunlight that had shone down on the Earth millions of years ago, combined with cellulose and carbon dioxide, was now being released back into the atmosphere. Enter the age of artificial climate change.

Soon after came the discovery of oil and the realization that it could serve as a dense source of energy, at that time primarily used for heating. The ability we accrued to refine and utilize it changed our path of evolution. It allowed us to grow our population more rapidly and has led to our ability to sustain its swift growth rate. Since the first chunk of coal was ignited, we have grown ever more dependent on our savings account, primarily consisting of two sources of oil.

The United States gets its oil from Canada and the Middle East. Neither source is without its share of challenges.

The Alberta oil sands, often referred to as the tar sands, supply a vast amount of the world's oil. Canada is now the number-one supplier of oil—ahead of Saudi Arabia—to the United States.[23]

The rich carbon deposits found in the oil sands are in the form of a thick, heavy type of oil called bitumen. Geologically, bitumen has not been "cooked" long enough to reach the light viscosity of coveted Middle Eastern crude.

The energy required to take bitumen from the earth and transform it into usable carbon products is extraordinary. In fact, some estimates suggest that for every unity of energy we obtain from the oil sands, we have used an equal amount of energy to extract and process it.[24] And since that energy comes from the burning of fossil fuel, our net energy gain would be zero. Clearly, this cancels out the oil sands as a solution to our energy shortfall. In addition, for every barrel of oil extracted and processed, three times that amount of water is needed, 90 percent of which is then deemed polluted.[25]

But the oil sands are great for the economy. Within our current structure, inefficiency is at the root of a robust economic system. The more people we can put to work doing things—and being paid well for it—the better. Alberta leads the way; it is the only Canadian province with a monetary surplus, so the argument is that that province must be doing something right. Or is it inefficiently?

The crude oil imported from the Middle East is much easier to extract from the earth and therefore takes much less energy and involves a less labor-intensive process. Plus, turning it into a usable product, such as gasoline, is considerably less involved, so that the process again requires less energy to be burned and fewer hands to make it happen.

There are clearly issues with both sources of oil. While less energy is required to extract and process crude oil form the Middle East, it must be transported to North America to be of use. As you can imagine, an immense amount of energy must be spent to load millions of gallons of oil on a freighter and have it shipped thousands of miles to North America. Also, it is not desirable that the United States be dependent on Middle Eastern nations for its energy, since its relationship with some of these nations is tumultuous. Should those nations decide to "turn off the tap," they would bring the United States, at its current rate of oil consumption, to its knees within a month.

Energy independence starts with reduced demand. And being more efficient is something we can all begin immediately. From there, new energy solutions can be developed domestically to sever our reliance on foreign oil.

And then there's the question of remaining supply. "How much more oil is left?" With deposits to our energy savings account being made only every few million years, how long can we continue to spend without earning before we're broke?

That's exactly the question being asked by many prominent scientists, many of whom subscribe to the "peak oil" theory. Defined as "the point at which maximum global oil production is reached, after which the rate of production goes into terminal decline," the "peak oil" hypothesis has become widely accepted.[26] Most who study the subject agree that the Earth will at some point simply run out of oil. But when exactly that time will come is hotly debated. Many scientists who study the subject believe we are on the cusp of peak oil now, or may have already experienced it, dating the start of the decline around 2009. But even the optimistic peak oil theorists warn that the peak is *likely* to happen within our generation, beginning in 2020. From there, as peak oil believers point out, a steep and terminal decline will abruptly follow.

A minority within the scientific community feels the decline will occur much later. If we were to run out of oil, they theorize, it would be centuries, or even millennia, from now. At which point, according to this minority, it is assumed that by then we'll have developed the technology to harness alternative forms of energy from the sun, wind, and the ever-changing sea tide.

But I think it's safe to say that no one really knows when that point will be. Or, as some see it, if it will come at all.

I should point out that there are in fact a handful of prominent scientists who dismiss the peak oil theory altogether. These people, subscribers to the abiotic hypothesis, as it is called, believe oil regenerates itself more quickly than what is classically believed, because, they theorize, it's not actually fossil fuel at all.[27] Therefore, it is not subject to the multi-million-year process of turning ancient decaying plants and dinosaurs into crude. The abiotic hypothesis posits that oil is a natural product produced in the mantel of the Earth. Therefore, as the theory goes, millions of years are not needed to create oil; the Earth manufactures and spews it out in a timely fashion.

> Since the combustion of oil creates toxic gas and releases pollution into the air that significantly increases the risk of disease, the less oil we need to burn, the better off we'll be.

So let's assume, against the vast scientific majority, that we will not reach peak oil in our generation—and maybe never. There's still an undeniable problem with our level of oil consumption: emissions. Best known as greenhouse gases, these emissions are being blamed for some pretty big problems. And some observers (the United Nations, for example) suggest those emissions are what cause climate change. (I expand on this topic in the "Air Quality" section on page 50.)

Regardless, I think we can all agree that, since the combustion of oil creates toxic gas and releases pollution into the air that significantly increases the risk of disease (including cardiovascular disease),[28] the less oil we need to burn, the better off we'll be.

As humans, we obtain our energy from food, which we produce by burning fossil fuel. So, essentially, we are trading the stored energy of fossil fuel (which originated from the sun) for caloric energy that's of biological use to us.

And each time we do this, as with any energy transfer, there is a loss. But when we pass that energy through an animal, the loss becomes significantly greater.

Growing food to feed to animals, which will in turn feed us, requires significantly more arable land than growing plants for our direct consumption. The reason so much food needs to be grown to feed these animals is that very little of the food energy passed on to the livestock is returned through their meat.

To give an idea of how energy is lost in the production of animals, here's a breakdown:

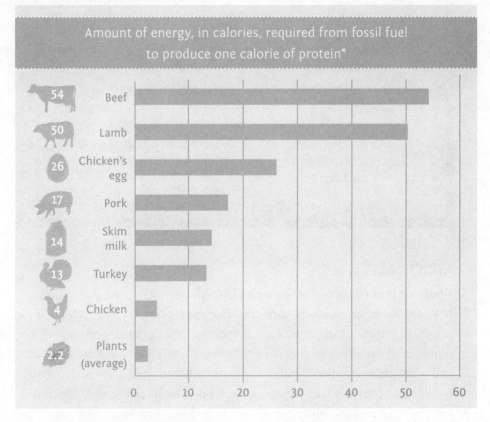

Amount of energy, in calories, required from fossil fuel to produce one calorie of protein*

Beef	54
Lamb	50
Chicken's egg	26
Pork	17
Skim milk	14
Turkey	13
Chicken	4
Plants (average)	2.2

On average, in the United States it takes 28 calories of fossil fuel energy to yield one calorie of animal protein for human consumption.[29]

In contrast, to obtain one calorie from plant-based protein, 2.2 calories of fossil fuel need to be burned.[30] This figure takes in account all fuel expenditures, including the energy it takes to produce fertilizer to grow the food and to run the farm machinery that harvests it.

Source of data: Cornell Science News, "U.S. Could Feed 800 Million People with Grain that Livestock Eat ..."[31]

*Of course, meat is inherently higher in protein than plants, so to make this comparison fairer, I've compared only the calories from protein. Contrasting the total number of calories produced from animal sources and those produced from plant sources would give an even more dramatically lopsided picture.

AIR QUALITY

When it's burned, fossil fuel provides us with energy by releasing stored heat and light from "ancient sunlight." However, along with the release of energy from prehistoric rays of sun comes the unwelcome by-product of sequestered carbon from prehistory, known as emissions. As dinosaurs and other ancient animals exhaled, plants collected their wayward carbon dioxide and over millions of years reduced it to fossilized sludge, which we know as coal and oil. When we burn coal and oil, the combustion releases this long-quarantined and ancient animal breath into our modern world. Believe it or not, each time you start your car, the emissions from the tailpipe are, in part, dinosaur breath being set free.

While plants may consume carbon dioxide, it's poisonous to us oxygen-breathing creatures. And because of this, reducing the amount of emissions set free into our atmosphere is in our best interest. Besides being deadly to inhale, carbon dioxide and other greenhouse gas emissions are being blamed for one of the most feared threats of modern times: artificial climate change. While most scientists agree that the Earth has experienced, and will continue to experience, a natural fluctuation in temperature change, a growing body of evidence suggests that emissions created by the combustion of fossil fuel—known as greenhouse gases—are playing a significant role in manipulating this cycle. Commonly referred to as global warming, the average temperature of the Earth is continuing to rise.

As the sun's warm rays shine on the Earth, some of their heat is absorbed, while some is reflected back into the atmosphere, escaping into space. However, greenhouse gases trap atmospheric heat from the sun's rays within the Earth's atmosphere, thereby causing temperature on Earth to rise. This is aptly known as the greenhouse effect. As the amount of greenhouse gases increases, so too does the average temperature of the Earth. Even a small temperature change can have a profound negative effect on ecosystems. And there have been significant changes already.

Former U.S. vice president Al Gore's 2006 Oscar-winning documentary, *An Inconvenient Truth,* shone a spotlight on the human hastening of this dire phenomenon. Citing evidence from the leading scientists on the subject, the film put forth the hypothesis that we are indeed to blame for the rapidly increasing average temperature rise of the Earth's atmosphere. Corroborated by a glut of scientific heavy-hitters, the film captured the world's attention and in doing so sounded the alarm bell. Its message was simple: we are to blame and we must change our ways before it's too late.

In one of the most compelling scenes of the film, computer-generated images showed areas, such as Florida, Shanghai, Calcutta, and Manhattan, slipping beneath the ocean surface as the polar ice caps melted because of an increase in global temperature, causing sea levels to rise. According to the film, if we continue on our current path, the polar ice caps will melt enough "in the near future" to cause a 20-foot rise in sea level, which would trigger the destruction of major coastal cities and result in "one hundred million environmental refugees."

> The volume of emissions created by operating cars, trucks, buses, airplanes, and ships, while considerable, pales in comparison to the volume of CO_2-equivalent greenhouse gases released through raising animals for food.

The film blamed our seemingly insatiable thirst for energy. Required to fuel our modern lifestyle and derived from burning fossil fuel, our energy needs and "wants" create emissions. A lot of emissions. And as a result an extraordinary amount of greenhouse gases are produced. *Everything* we do requires energy. And almost all of it is obtained by the burning of fossil fuel.

While the film certainly captured the world's attention, it drew criticism for offering few solutions to a problem it so passionately described. Adamantly making a case for human-hastened climate change and impending global disaster as a consequence, it left many feeling helpless. Other than suggesting, "Drive hybrid cars and replace incandescent light bulbs with compact fluorescents," the film left viewers wanting to know how they could mitigate the part they might be playing in the advancement of the crisis.

Coincidentally, also in 2006, a United Nations report was released stating that animal agriculture is one of the greatest contributing factors to anthropologic climate change because of its inordinate energy demands. The report went on to say that animal agriculture is a larger producer of greenhouse gas emissions, and therefore a larger contributor to artificial climate change, than all modes of transportation. *Combined.* The volume of emissions created by operating cars, trucks, buses, airplanes, and ships, while considerable, pales in comparison to the volume of CO_2-equivalent greenhouse gases released through raising animals for food. According to the U.N. report, *Livestock's Long Shadow,* a whopping 18 percent of these emissions result from raising livestock for food. In comparison, 13 percent of total greenhouse gas emissions (measured in CO_2 equivalent) can be attributed to the transportation sector.[32] As you can imagine, when the information was released that raising animals for food created 5 percent more greenhouse gas emissions than did all of transportation, people were in a state of disbelief.

Those best-intentioned carpooling hybrid drivers who stopped at the drive-thru to pick up their daily ham and egg sandwiches were caught off guard. Now there was confusion.

Here's the reason: greenhouse gas emissions comprise three types of gases: CO_2 (carbon dioxide), nitrous oxide, and methane. While CO_2 is the most common in terms of volume (it's what's expelled from a car's exhaust pipe), its global warming potential, or GWP, is 23 times lower than that of methane.[33] Emitted primarily from ruminants, most notably cows, methane has been fingered by the United Nations as a major contributor to anthropologic climate change. When cows and other ruminants eat grass, their digestive system breaks it down, mechanically, through chewing, and chemically, through fermentation. And a natural by-product of fermentation is gas, in this case, methane.

Yet, even worse, the vast majority of cows in North America are "produced" in factory farms where grass is not on the menu. Wheat, corn, and soy are popular feed choices since they "encourage" cows to reach their slaughter weight sooner. Cows have four stomachs, designed for the assimilation of grass, so when cows eat wheat, corn, and soy, digestive issues "erupt." And as a direct result, excessive gas is created and released into the atmosphere, contributing to the containment of atmospheric heat. And then there's the manure. Nitrous oxide has a global warming potential 296 times greater than that of CO_2, and 65 percent of the world's nitrous oxide emanates from livestock manure.[34]

So, clearly, the greatest single thing we can do as individuals to reduce the amount of greenhouse gases we contribute to producing is to eat foods that create fewer emissions in their production.

But what about those who dismiss the theory that we humans (and our farmed animals) have anything to do with climate change? Those who believe that the Earth is going through a natural cycle and we are just along for the ride and are in no way responsible for the average warming of the Earth? They are going against the overwhelming scientific majority and contradicting the findings by United Nations scientists, but to be fair, there is still a small scientific community that doesn't believe we are to blame, at all.

However, even if this small group of scientists turns out to be spot on, this is undeniable: when fossil fuel (and oil, if it turns out not to be fossil fuel after all) is converted into energy, it must be burned. And burning fossil fuel creates toxic gases, most commonly referred to as pollution.[35] And there's a direct correlation between air pollution levels and mortality rates.

Beginning in 1974, researchers at Harvard University set out to conduct a long-term study in hopes of revealing the correlation between air pollution and our number-one killer, cardiovascular disease. It had long been thought that air pollution could not be a good thing, but a long-term study on its health effects hadn't yet been conducted. And while we knew that the lack of minerals in our food is a major contributing risk factor for CVD, what about the microscopic particulates that we breathe in—air pollution? For this investigation, known as the Six Cities Study, the researchers collected data— over the course of 14 to 16 years—on 8000 people. The results showed what a rational person might suspect; that the greater amount of pollution in the air, the greater the risk of death from cardiovascular disease.[36] But the unmistakable correlation between the *amount* of air pollution and the cardiovascular disease mortality rates was striking.

The pollution in rural areas is generally created by food production.

The researchers concluded that there was a clear connection.

While it's true that most of the thick layer of pollution over the cities is attributable to automobiles, the pollution in rural areas is generally created by food production. Fewer people live in those areas, so perhaps, we might think, fewer people are directly affected by the pollution, but this is also the geographical space where our food is grown. As we know, plants quarantine carbon dioxide. When those plants are food crops, having them take in pollution becomes a problem. The pollution literally becomes part of our food. And then us.

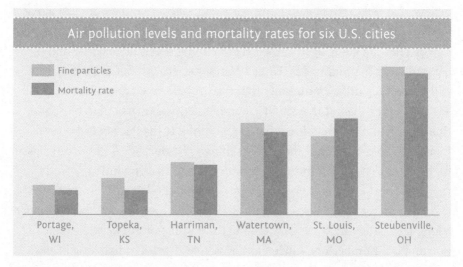

Air pollution levels and mortality rates for six U.S. cities

Fine particles
Mortality rate

Portage, WI · Topeka, KS · Harriman, TN · Watertown, MA · St. Louis, MO · Steubenville, OH

As you can see, the correlation between pollution and risk of death from CVD is striking.

Data adapted from Harvard Six Cities Study.

WHAT ABOUT EATING LOCAL TO REDUCE EMISSIONS?

Eating food that's been produced locally—commonly understood to be within a 100-mile radius—makes sense. Logically, the less distance food needs to travel to reach those who will consume it, the fewer emissions will be created.

In fact, this simple concept began gaining acceptance, and then even popularity, in 2005. Some early adopters were hailing it as our solution to climate change by reducing those dreaded "food miles" (the distance food must be transported from producer to consumer).

The thought of air-shipping food all over the world—bananas from Ecuador, pineapple from Hawaii, coconut from Thailand—seemed excessively decadent. Food miles were mounting, adding hundreds of thousands of pounds of CO_2 emissions to our atmosphere each year. But what was most disconcerting to those making an effort to eat local was our desire for out-of-season fruit—pears from Australia, peaches from South America, grapes from Chile—that, if we were patient, could be obtained locally in a few months.

Several books were published on the subject and, as the movement continued to progress, pop culture embraced it.

In fact, the word "locavore" was coined to describe proponents of the movement, and in 2007 the term not only made its way into the *New Oxford American Dictionary* but was named that dictionary's "word of the year."[37] The dictionary defined *locavore* simply as "a person who endeavors to eat only locally produced food." The locavores had arrived.

While I subscribe to the concept, grow a bunch of my own food, and shop at farmers' markets weekly, I couldn't help but wonder: of all the greenhouse gas emissions created by the production and delivery of food, how much of those were as a result of transportation (food miles) as opposed to production? Well, as I found out, I wasn't the only one with this question. Thankfully, some of those who shared my curiosity happened to be research scientists at Carnegie Mellon University in Pittsburgh. And with the means to get an answer, Christopher L. Weber and H. Scott Matthews conducted a study to determine the actual amount of CO_2-equivalent gases emitted by transportation of food, compared to the amount emitted by its production. The results were fascinating. Weber and Matthews published their findings in the prestigious journal *Environmental Science & Technology.*[38]

In the report the authors state, "Our analysis shows that despite all the attention given to food miles, the distance that food travels only accounts for around 11 percent of the average American household's food-related greenhouse gas emissions, while production contributes to 83 percent." A significant divide to say the least: 7.5 times more greenhouse gas emissions are created in the production of food than by its delivery. Specifically, nitrous oxide and methane, inadvertently produced by fertilizers (for animal feed crops), manure, and gas expelled during the animals' digestion account for a large portion of the CO_2-equivalent gases created during production.[39]

> 7.5 times more greenhouse gas emissions are created in the production of food than by its delivery.

The findings shed light on what many others and I had suspected: while eating local is sensible, environmentally speaking, *what* we eat is of greater importance. Significantly greater importance.

THE GREATEST EMISSION CREATOR

In July 2007, the compelling results of a study conducted by a team at the National Institute of Livestock and Grassland Science in Tsukuba, Japan, were published, and reported on in the U.K. press.[40] *New Scientist* magazine immediately picked up the story, and within hours the hard-to-believe numbers had gripped the scientific community's attention.[41] The study found that, when all factors were taken into consideration, the amount of greenhouse gases released into the atmosphere from the production of 2.2 pounds of beef was the equivalent of 80.08 pounds of carbon dioxide. To put that latter figure into perspective, it's the equivalent in CO_2 emissions of a midsize car that gets 26 miles to the gallon being driven 160 miles.

According to the U.S. Department of Agriculture, the average Canadian eats about 68.4 pounds of beef per year.[42] The CO_2 equivalent from its production would be equal to roadtripping in a midsize car from Vancouver to Toronto, and 90 percent of the way back: a total distance of 4976 miles.

The U.S. numbers are even higher, considering the average American eats 96.4 pounds of beef per year.[43] Producing that beef would create

1595 kilograms of CO_2, which equates to driving 7008 miles, or a road trip from New York City to Los Angeles *and* back, and then one-way from NYC to Miami. That's one person, for one year.

Not that this is at all realistic, but what *if* everyone in the United States stopped eating beef for one year? With a population of 307 million, the savings in CO_2 equivalent would be the same as what would be conserved by not driving that midsize car the distance of over 2 trillion miles (2,151,501,902,400, to be exact). Considerable, to say the least. And, in fact, since the average distance to the moon is 238,857 miles,[44] if everyone in the U.S. stopped eating beef for one year, that would be like not driving a distance equivalent to 9,007,489 trips to the moon. Think about it: the driving distance of over 9 *million* trips to the moon.

> If everyone in the U.S. stopped eating beef for one year, that would be like not driving a distance equivalent to 9,007,489 trips to the moon.

And if Canadians abstained from eating beef for one year? We'd save in emissions as much as if we'd chosen to not drive our cars a distance equal to 693,961 trips to the moon.

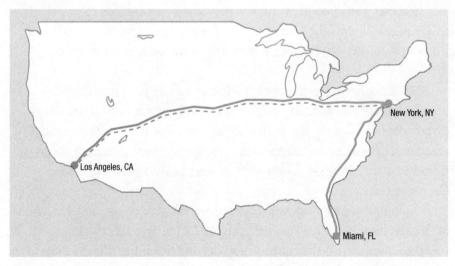

Driving equivalent in CO_2-equivalent savings if the average American did not eat factory-farmed beef for one year.

The same study also noted that the energy needed to produce the amount of fertilizer required to grow the feed to be consumed by that cattle is significant. For each kilogram of beef, 160 megajoules of energy is burned in fertilizer production. That's the same amount of energy that would be used to light a 100-watt bulb for 20 days.

So, to produce enough beef for the average Canadian to consume in one year would require the same amount of energy as keeping one 100-watt bulb lit for 622 days, or 14,928 hours.

By not eating beef for a year, an average American would conserve enough energy to light the bulb for 876 days.

Cows are not the only problem. According to a U.K. government agency called the Department for the Environment, Food and Rural Affairs (or DEFRA), while ruminants, such as cows and sheep, produce the most methane, other animals are not completely free from blame. The agency states that for every unit of meat from a pig that's produced, five times that amount of greenhouse gas is released into the atmosphere. For chickens, it's four times.[45] So while ruminants are the greatest offenders, energy ineffi-ciency, and therefore excessive emission production, is a factor in raising all livestock. While raising chickens produces less CO_2 emissions than raising any other farmed animals, the level of emissions is still double that of even the most inefficient plant crops.

For every one pound of chicken produced, four pounds of CO_2-equivalent emissions is released.[46] Therefore, the production of about 2.2 pounds of chicken releases as much CO_2-equivalent emissions as driving 14.6 miles.

Since the average amount of chicken eaten per person in Canada is 66.22 pounds[47] and for every pound of chicken produced four pounds of carbon dioxide is released into the atmosphere, that would add an annual amount of carbon dioxide to the atmosphere equivalent to the emissions from a midsize car being driven 439.4 miles (the distance from Prince George to Calgary).

And for Americans it's worse since they eat, on average, 102.3 pounds of chicken meat per year,[48] contributing 409.2 pounds of carbon dioxide into the atmosphere—the equivalent of driving 679 miles (the distance from Chicago to Washington, D.C.).

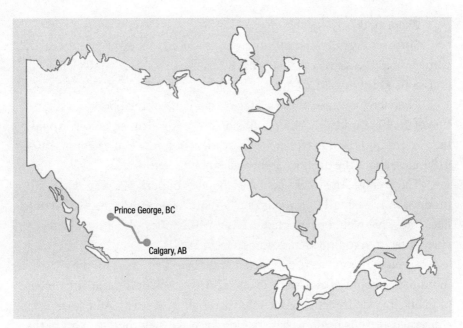

Driving equivalent in CO_2-equivalent savings if the average Canadian did not eat chicken for one year.

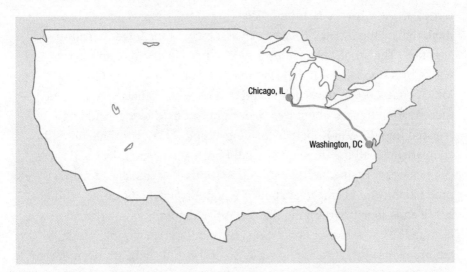

Driving equivalent in CO_2-equivalent savings if the average American did not eat chicken for one year.

As for pork?

Since for every 2.2 pounds of pork produced, about 11 pounds of carbon dioxide is released, that's the same amount of CO_2-equivalent emissions as driving 18.25 miles.

Therefore, because of the average Canadian's annual pork consumption of 50.49 pounds,[49] 114.5 pounds of CO_2-equivalent emissions would be released into the atmosphere. This works out to the equivalent of driving 418 miles, about the distance from Saskatoon to Lethbridge.

The average American eats 62.25 pounds of pork per year,[50] its production releasing 148 pounds of CO_2-equivalent emissions, which translates into what would be emitted driving 540.2 miles, about the distance from Pittsburgh to Providence, Rhode Island.

As more reports were conducted and as studies emerged, the inordinate amount of emissions created by livestock become an article of public knowledge (among the informed public, at least). And as such, concerned citizens begin wisely looking for ways they could be part of the solution. Those seeking to mitigate climate change and address other issues surrounding emission production began entertaining creative ways to do so.

In August 2007, *The Sunday Times* newspaper in the United Kingdom ran an article with the headline: "Walking to the shops damages planet more than going by car."[51] As we can assume was the intent, it captured attention. The article, penned by a staff writer, consisted of an interview with Chris Goodall, author of *How to Live a Low-Carbon Life*. Quoting Mr. Goodall's book, the article stated that based on official government fuel emission figures, about two pounds of carbon dioxide would be released into the atmosphere by driving a typical U.K. car three miles. In comparison, walking three miles would burn about 180 calories. If those calories were obtained by eating beef, the emissions associated with its production would come in at about 7.9 pounds, nearly four times as much as the emissions from driving the car. What about obtaining the calories from milk? It would take about 1¾ cups of milk to match the caloric requirements of the three-mile walk. So, if the milk came from a modern dairy farm, 2.6 pounds of carbon dioxide would be released, which is 25 percent more than would be released by the journey by car.

And as you may imagine, this article garnered attention, most of it from those who mistakenly assumed Chris was advocating we forgo

Driving equivalent in CO_2-equivalent savings if the average Canadian did not eat pork for one year.

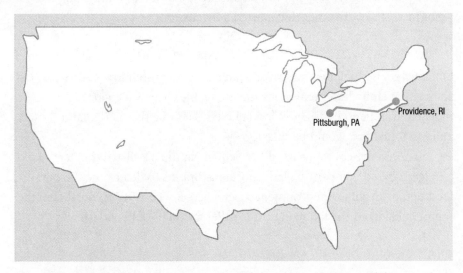

Driving equivalent in CO_2-equivalent savings if the average American did not eat pork for one year.

walking and increase our use of the automobile. Of course, this wasn't Goodall's point. By simply drawing our attention to the fact that all energy originates somewhere else, in my opinion, Goodall powerfully made his point: nothing is free. Even if we "generate" the energy ourselves, it had to come from somewhere, in his examples, from food. And, as we know, food production is a major energy draw. What I appreciate about this example is its articulation of energy origin, transfer, and use.

> A food source that emits fewer emissions during its life cycle is a good start for a more environmentally friendly food choice.

Of course, there are several factors not addressed in his example. How fit was the person? As we know, the fitter the person, the fewer calories he or she will burn, since a trait of greater fitness is improved efficiency. As efficiency improves, the fuel requirements (food in this case) to travel a given distance will decline. And by how much did this fictitious person reduce his or her risk of osteoporosis, type 2 diabetes, and cardiovascular disease by walking to the store? And what about the significant amount of energy used to manufacture the car? But that wasn't the point.

Certainly Chris Goodall did an excellent job in communicating his point in a "sticky" manner that started a discussion.

WHAT CONSTITUTES ENVIRONMENTALLY FRIENDLY FOOD CHOICES?

Obviously, simply not eating isn't a viable long-term solution. Clearly, a food source that emits fewer emissions during its life cycle is a good start for a more environmentally friendly food choice. And one that isn't a land, water, and fossil-fuel hog would be nice too.

In 2008 my colleagues and I at Sequel Naturals enlisted the help of a Calgary-based company called Conscious Brands to determine the best food options from an environmental perspective. Focusing on breakfast, the report considered the following three different types of breakfasts:

TRADITIONAL AMERICAN BREAKFAST

Consisting of two eggs, two slices of bacon, two links of sausage, one slice of toast, and 5.3 ounces of hash browns, the traditional breakfast scored 2.9 pounds of carbon dioxide.

LIGHT AMERICAN BREAKFAST

The light American breakfast comprised ½ cup of cereal, 1 cup of cow's milk, 1 cup of yogurt, and half a banana; it came in considerably lower, at just 12.3 ounces of carbon dioxide.

PLANT-BASED WHOLE FOOD SMOOTHIE BREAKFAST

The fully plant-based smoothie option, made up of a dry weight of 2.3 ounces of hemp protein, yellow pea protein, brown rice protein, flaxseed, maca, and chlorella came in at only 1.2 ounces of carbon dioxide. That's 10 times fewer emissions than that of the light American breakfast, and 38 times fewer than the traditional American breakfast. If we wanted to blend the smoothie with half a banana, it would come in at 2.1 ounces, still 22 times lower than the traditional American breakfast, and 5.8 times lower than the light American breakfast.[52]

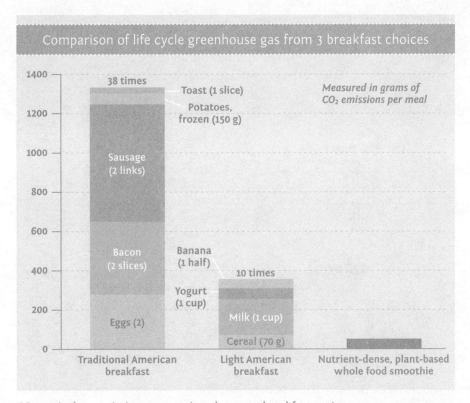

CO_2-equivalent emissions comparison between breakfast options.

So, with these numbers in hand, we can consider some options. For example, if one person were to switch his or her traditional American breakfast for a plant-based whole food option, the CO_2-equivalent savings would be equal to driving a midsize car from Vancouver, B.C., to Tijuana, Mexico: the whole length of the Western United States.[53]

Now, if everyone in the United States swapped his or her traditional breakfast for the plant-based option, the amount of emissions saved would be the equivalent to those created driving over 409 billion miles (409,853,744,250 miles, to be exact). That's equal in distance to over 1.7 million trips to the moon.

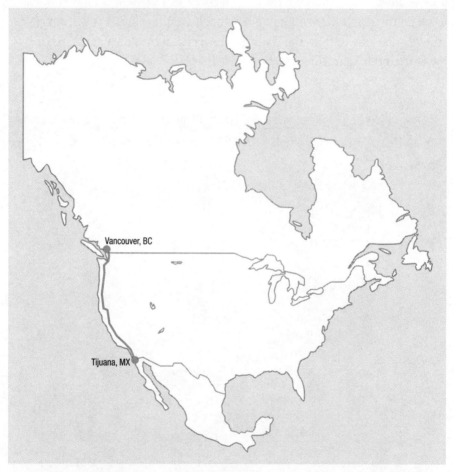

Driving equivalent of CO_2-equivalent savings if the average American swapped his or her traditional American breakfast option for a plant-based smoothie for one year.

In reality, not everyone in the United States eats a traditional American breakfast. Some eat lighter. Some eat heavier. But even if we use the light American breakfast and compare it to the greatly more nutritious plant-based option, the emission savings is impressive. In fact, it's with this information that we were able to fine tune my Vega Complete Whole Food Health Optimizer blender drink formula so that its production would require the least amount of each natural resource (more information on page 343).

This report was all very encouraging to me. It nicely illustrated that even though a substantial amount of natural resources are spent in obtaining our required nutrition, each of us has the ability to measurably reduce environmental strain by simply making informed food choices. *What we eat is paramount.*

But simply spending natural resources—and incurring the environment's cost associated with their use—is not the only consideration when selecting food. As I discussed in detail in Chapter 1, micronutrients are the most valuable component when assessing a food's nutritional worth.

And realizing that an inordinate amount of natural resources are required to produce food in this country, as well as appreciating the bearing on overall health that micronutrients have, I grew curious to see if there was a system that took both related issues into account when producing food, a system with a simple mandate: gain the highest levels of each micronutrient, expend the fewest resources to do so. I couldn't find one. In my search I did, however, come across carbon labeling—the displaying of a given food product's CO_2-equivalent emissions during its life cycle—and I did find an unrelated nutrient labeling system that indicated the nutrient density of select foods. But I found nothing that worked in concert to marry the two. This seemed to me to be a blind spot. A blind spot with ubiquitous, varied, and expensive consequences: excessive soil depletion, extreme fresh water consumption, fossil fuel gluttony, exorbitant greenhouse gases belching, to name just a few of the environmental issues. The

> If we knew that a food company was using less of each natural resource to produce more nutrient-dense food, we could, appreciating the importance of this, then choose to "vote" for that company by buying its products.

health issues associated with this blind spot were also rampant: constant hunger; overeating co-existing with malnourishment; general fatigue, difficulty sleeping, and dependence on stimulants, such as coffee and refined sugar; as well as significantly increased risk for many diseases, including cardiovascular disease. Could the solution be a simple ratio displayed on food labels, one part the amount of each natural resource expended, one part the micronutrients within the food that are gained in return? I believe that such a label would add another layer of transparency and completeness to the food system as a whole. It would clearly give us, as consumers, vital information upon which to base our buying decisions. If we knew that a food company was using less of each natural resource to produce more nutrient-dense food, we could, appreciating the importance of this, then choose to "vote" for that company by buying its products. And of course, in doing so, we would facilitate its growth, based on our values aligning with its values.

AN APPETITE FOR CHANGE:
Environmental and Health Solutions Through Food

3

As you've read in earlier chapters, when we factor in the natural resources required in their production, some foods are easier on the environment than others, and some stand out as exceptionally resource-intensive to produce.

As you'll read below, the Carbon Trust in the United Kingdom has introduced carbon labels to help British consumers understand the environmental implications of their food selections. The labels inform them as to how much carbon they will be responsible for "emitting" if they choose to buy certain foods.

That information's important, but it's only half of the equation.

Since micronutrients are the true measure of a food's value, in this chapter I look at ways to gain the most micronutrients while expending the fewest natural resources. I've put together a system that takes both environmental strain *and* micronutrient yield into consideration; I call it the "nutrient-to-resources ratio," and will explain it in detail later in the chapter. Foods with the highest nutrient-to-resources ratio are therefore the ones I suggest become the base of our diet, not only for peak health but also to put the least amount of strain on our natural resources.

I also compare the monetary cost of different foods. While it's undoubtedly less expensive to gain calories from highly processed food (and food-like substances) than from natural whole food, does the argument that "it's too expensive to eat healthy" hold up when we assess the cost of nutrition (micronutrients) as opposed to calories? How much highly refined food would we have to consume to match the nutrition we could get from a small amount of food in its natural state, and what would the cost difference be?

THE U.K.'S LOW CARBON DIET

In March 2007 the British government put forward a draft bill called the Climate Change Bill, which stated that the countries of the United Kingdom would be required to reduce their carbon emissions by 80 percent (based on 1990 standards) by the year 2050.[1]

A bold move and, as you might expect, a heavily criticized one by some industries. But despite the expected opposition, the bill was passed in November 2008, becoming law and fueling hopes that Britain would evolve into a low-carbon economy. The United Kingdom was now the only country in the world to have imposed such a law upon itself. But whether its ambitions will be fulfilled ultimately rests upon the actions of its people. Britons have responded to the challenge and, according to polls conducted, are actively seeking ways to reduce their carbon output. Through U.N. reports, most notably *Livestock's Long Shadow*, the British population has responded to the message that the most significant difference an individual can make is by way of informed food choices.

Industry's response to consumer concern was swift. Playing to consumers' interest (that could perhaps ultimately become a market demand), a nonprofit company called the Carbon Trust sprang up. Targeting companies wanting to estab-lish a "carbon footprint" for their products, the Carbon Trust offered a CO$_2$-assessment service and a label to indicate corporate environmental stewardship. Measuring the carbon dioxide emitted during

> With the aid of the labels, each person can now establish his or her individual "carbon budget," which presumably will make it easier for that individual to stay on a low-carbon diet.

the life cycle (farming, manufacturing, packaging, distribution, and disposal) of a given product, the Carbon Trust offered suggestions as to how carbon dioxide emissions could be reduced, and then would grant a carbon footprint label based on the current carbon dioxide production during the life cycle of the product. The company would make the label part of its packaging. Total "cradle to grave" emissions, measured in grams of CO_2e (carbon dioxide equivalent) emitted as a result of the product and packaging life cycle, were put in plain view.[2]

Further, to meet the Climate Change Bill's target of an 80 percent CO_2-emission reduction by 2050, each individual, says the Carbon Trust, must produce no more (on average) than 8.3 tons of CO_2e per year—which equates to 50 pounds per day. Of that 50 pounds, it's suggested that each person keep the emissions total from his or her food down to no more than 6.5 pounds of CO_2e per day. But wait—what? Clearly, these are abstract numbers, and without a context they have no value. However, thanks to the Carbon Trust, these otherwise unrelatable numbers are put into perspective and given relevance for the average person. With the aid of the labels, each person can now establish his or her individual "carbon budget," which presumably will make it easier for that individual to stay on a low-carbon diet. For example, Tesco, a large grocery store chain in the United Kingdom, enlisted the services of the Carbon Trust to help give its customers a sense of the CO_2-equivalent emissions they might release into the atmosphere should they use particular food products. Its house brand of orange juice, for example, displays a label stating that the production of a one-cup serving of the juice will be directly responsible for releasing 12.5 ounces of carbon dioxide emissions throughout its life cycle.[3] The idea is that with the labeling, those British citizens mindful of their carbon diet will be able to determine if they want to "spend" their limited carbon points to indulge in a glass of orange juice. (By the way, and interesting to note, since carbon labeling is not in itself law in the United Kingdom, only manufacturers of plant-based foods are taking part. That isn't surprising, since products containing meat and dairy would score horribly and presumably scare off those aspiring to stay carbon-lean. The carbon labeling would lift the veil of inefficiency and sales of those products with higher carbon footprints would, presumably, fall.)

I unequivocally commend the proactive stance the U.K. government has taken. However, the assessment services and carbon label of the Carbon Trust, though a step in the right direction, speaks to emissions with no reference to land, water, or fuel consumption. Its main shortcoming is that the emissions score is in no way tied to the nutritional component of the food on which it's affixed. What health-boosting nutrients do we get for our carbon expenditure? How do we decide if we want to "spend" our carbon on food when what we gain in return is not revealed?

To produce our food, we've clearly gone on a resource-spending binge. And, in the haze of our indulgence, what we seek in return has become unclear.

Since the inefficiency of translating natural resources into food has become strikingly apparent, the goal of conscientious food scientists has become to reap more calories from the land while expending fewer resources in doing so. And while I appreciate their effort to maximize return on natural resources' expenditure in food production, I don't feel the calorie is the best point of reference. A calorie is, indeed, a measure of food energy, so at first glance it may seem logical, when converting a precious unit of fossil fuel energy into food energy, to go for the highest possible yield of calories. But as I mentioned in Chapter 1, it's not a lack of calories that makes us sick. In fact, it's quite the opposite. Though the food-producing community may be conscientiously trying to reap more calories from the land for ostensibly environmental reasons, I believe that the health community has valuable insight to share regarding what, specifically, we ought to strive to pull—nutritionally speaking—from the land. And that is nutrient density. As I described in Chapter 1, nutrient density is expressed by a ratio of micronutrients delivered to calories consumed, and the higher the micronutrients and the lower the calories, the better. Nutrient density is a simple and effective way to rate true nutritional value. The goal, for the sake of North Americans' health, should be to draw more micronutrients from our land, not more calories. This rebalance would play a major role in preventing disease and obesity. The odd paradox of modern food production that enables us to be both overfed and undernourished would be immediately eradicated.

WHOLE FOODS MARKET: "HEALTH STARTS HERE" PROGRAM

In the spring of 2007, I had the opportunity to spend a long weekend at John Mackey's ranch. Just outside of Austin, Texas, about 40 like-minded people from the natural food industry converged on the Whole Foods Market CEO and visionary's 780-acre ranch. At this event, billed as sports weekend, it didn't take long to gather that Mackey wasn't a typical CEO. Sharing my view of food, John made it clear that he, too, strongly believed that we could lower our rate of disease and health problems in North America simply by making informed food choices.

And because of that simple preventative medicine philosophy, Whole Foods Market has taken the lead in North America by creating a program called "Health Starts Here."[4] As one of the program's proactive measures, nutrient density scores for almost all of its produce were displayed. Initiated by John and developed by Dr. Joel Fuhrman, the nutrient-density scores have dramatically increased sales in the highest-ranking foods. Health-conscious shoppers just needed to know which foods to buy more of, and Whole Foods Market was happy to tell them. Greater sales for Whole Foods Market and better health for the people who shop there: it was a win-win situation.

As you know, nutrient-density scores reflect micronutrient level in relation to calories, which is helpful when shopping for healthy choices. But those digits in no way tie in the environmental expenditure or reflect what resources were required to yield those micronutrients in the food.

> The goal is to get as high a level of health-boosting micronutrients from food, while expending the smallest amount of each natural resource to do so.

So while the British are drawing attention to and quantifying a food's carbon dioxide emissions, and while Whole Foods Market is calculating and displaying its nutrient density, neither system relates one value to the other.

Having had the opportunity to tour and work with both environmental and health experts, I feel as though the members of both groups have excellent perspectives—on their own set of values. Many in the nutrition industry are unfamiliar with the exceptional amount of resources it takes to create sustenance. And the environmental community is unacquainted with the value of nutrient density and therefore with what our resources ought to be providing in exchange for their consumption. And it's certainly not calories.

But in merging the two perspectives, we have the complete picture.

For this reason, I feel it's in our best collective interest to pursue a simple goal: get as high a level of health-boosting micronutrients from food, while expending the smallest amount of each natural resource to do so. That's it.

THE NUTRIENT-TO-RESOURCE RATIO: WHAT IT IS AND WHY IT MATTERS

In an effort to achieve this melding of perspectives, I've put together what I call the "nutrient-to-resource ratio." It takes into consideration both micronutrient gain and natural resource expenditure in the food production process.

When applied, the nutrient-to-resource ratio reveals how misguided our resource expenditure is in our failed attempt to nourish the country.

In Chapter 2, I showed the significant divide between the resources it takes to acquire food from animals and the amount it takes to acquire food from plants. That comparison was based on the conventional calorie-to-calorie or pound-to-pound measure, neither of which addresses the true nutritional makeup of the foods being compared. When we take micronutrients into account as well, that significant divide becomes almost hard to comprehend. Clearly, plant-based whole foods require—by far—the least amount of each resource to produce health-boosting micronutrients.

> I believe it should become standard practice to include a food's nutrient-to-resource ratio on all nutrition labels.

At this time, the information we'd need to calculate the ratio of nutrition from different foods to their draw on land, water, and fossil fuel during their production is not publicly available. I'm working to obtain these numbers so that a comprehensive evaluation system can be developed. Considering the health and environmental benefits of this information, I believe it should become standard practice to include a food's nutrient-to-resource ratio on all nutrition labels. The ratio would give consumers a broader and more comprehensive scope as to what products are, in fact, part of the solution.

However, emission (CO_2e) numbers are available for most foods. With that information, I calculated the nutrient-to-emission ratio for a number of foods, using CO_2e emissions as a measure of exchange for micronutrients.

The first step in calculating the ratio is to establish the amount of CO_2e produced to yield a single calorie of a given food. I then divided the nutrient density of the food by the emissions per calorie, and the result is the nutrient-to-emission ratio. You can find the formula for the calculation of emissions per calorie in "Calculating the Numbers" on pages 310–311, along with the sources of nutrition data I used to determine the nutrient density.

Here is a sample of nutrient-to-emission ratios I've calculated. Keep in mind that the higher the number, the greater the amount of micronutrients is delivered in relation to the amount of CO_2e released.

Plant-Based Food	Nutrient-to-Emissions Ratio
Almonds	1266
Lentils	238
Steamed vegetables (combination of carrots, broccoli, asparagus)	272

Animal	Nutrient-to-Emissions Ratio
Wild fresh local coho salmon	61
Baked chicken breast	11.6
Poached eggs	6.7
Farmed fresh local salmon	6.6
Domestic cheddar cheese	4
Beef tenderloin steak	1.01

As for the other resources, limited information is available with which to calculate the nutrient-to-resource ratios, so by necessity my assessment is limited too. However, I have worked out the ratios for a few foods. Full calculations for all the numbers in the next sections are given in "Calculating the Numbers" on page 310. There, I've presented the information in snapshot form to give you a quick sense of the draws on each resource that various plant and animal food sources demand. For each resource, I look at foods in pairs to contrast the draw each food makes on the particular resource. Comparing foods illustrates the substantial impact our food choices can have on the environment.

NUTRIENT-TO-ARABLE-LAND RATIO

Wheat, Corn, and Soybeans Versus Beef

As you will see from the calculations in "Calculating the Numbers" on page 310, significantly more arable land is needed to harvest an equivalent amount of micronutrients from beef as from wheat, corn, and soybeans:

These foods are common feed crops for animals. If we were to use these crops as food for the human population instead, we would gain 23.4 times the amount of micronutrients from the same amount of land.

Hemp Seed Versus Beef

However, if we were to plant hemp as food for humans in place of wheat, corn, and soybeans to feed beef cattle, we would gain 51.9 times the amount of micronutrients on an equal amount of land, compared to what we would obtain from beef.

To explain how I arrive at this figure, by weight, 5.33 times as much hemp seed can be produced as beef raised on the same amount of land (880 pounds of hemp seed per acre[5] compared with 165 pounds of beef[6]). And since beef has a nutrient density of 20 and hemp seed registers at 65 (3.25 times more), for every calorie you get from hemp seed, you'd have to eat 3.25 calories from beef to match the micronutrient level.

Since pound for pound, hemp seed contains about three times more calories than beef, to gain the equivalent in micronutrients from beef as from hemp seed would require 51.9 times more land.

To produce enough beef to match the per-calorie micronutrient level of hemp seed, 51.9 times more arable land would be needed.

Kale Versus Beef

Now, what if instead of growing wheat, corn, or soybeans we grew kale? I realize that comparing kale—among the most efficient and nutrient-dense crops—to beef—one of the worst in both categories—provides something of an extreme case, but the difference *is* impressive.

By weight, 232 times more kale than cattle can be produced on the same amount of land (38,400 pounds of kale per acre[7] compared with 165 pounds of beef). And since beef has a nutrient density of 20 and kale registers at 1000, which is 50 times greater, for every calorie you get from kale, you'd have to eat 50 from beef to match the micronutrient level.

Since beef has about four times the amount of calories per pound as kale, to gain the equivalent in micronutrients from beef as from kale would require 2900 times more arable land.

To gain an equal amount of micronutrients from beef as from kale, 2900 times more arable land would be needed,.

NUTRIENT-TO-WATER RATIO

Sweet Potatoes Versus Beef

As mentioned earlier, to produce a pound of beef, a minimum of 2500 gallons of water is required. In contrast, only 60 gallons of water is needed to produce a pound of sweet potatoes, a difference of a little over 41 times. Since beef has a calorie density of 2990 calories per kilogram and sweet potatoes register at 900 calories per kilogram—a difference of 3.3—we can determine that each calorie of beef requires just over 12 times more water to produce than a calorie from sweet potatoes. And since sweet potatoes have a nutrient density of 83, while beef registers at 20—a difference of 4.15—we can determine that a little over 52 times more water is required to gain an equal amount of micronutrients from beef as can be acquired by an equal number of calories from sweet potatoes.

It takes 52 times more water to produce an equal amount
of micronutrients from beef than from sweet potatoes.

NUTRIENT-TO-FOSSIL-FUEL COMPARISON

Plant Protein Versus Animal Protein

As noted in Chapter 2, the process of acquiring protein from animal sources rather than plant sources uses a great deal more fossil fuel. When we compare micronutrients, that divide becomes even greater. For example, lentils have a nutrient density of 100, whereas the average nutrient density of the animal products listed earlier in this chapter is 25.3. On average, you'd have to eat 3.95 calories from these animal products to obtain the same amount of micronutrients delivered in 1 calorie of lentils. And as mentioned earlier, on average, obtaining protein from animal products requires that

26.9 calories from fossil fuel energy be spent to obtain 1 calorie of protein. And since only 2.2 calories of energy needs to be spent to obtain 1 calorie of protein from lentils, choosing the plant protein equates to an energy savings of 12.22 times (26.9 ÷ 2.2 = 12.22). Multiplying the energy factor by the calorie factor (12.22 × 3.95) tells us that it takes 48.3 times more energy to obtain the same amount of micronutrients from the average animal product as from lentils.

Assuming, for comparison, that the protein amount is equal when factoring in micronutrient levels, 48.3 times more fuel would need to be burned to get an equal amount of micronutrients from animal products as from plants.

Even chicken—the most energy-efficient meat to produce, and a source that is relatively nutrient dense—fares poorly when compared with basic brown rice, which has an average nutrient density, since rice yields more micronutrients while requiring less fossil fuel to produce.

Since rice has a nutrient density of 41 and chicken has a nutrient density of 27, you'd have to eat 1.5 calories of chicken to obtain the same amount of micronutrients as found in 1 calorie of rice. Chicken requires 1.8 times more energy to produce than rice. If you multiply this energy factor (1.8) by the calorie factor (1.5), you'll see that 2.7 times more energy is needed to yield an equal amount of micronutrients.

Since the nutrient density of kale is 10 times that of lentils, 483 times as much fuel would need to be burned to obtain equivalent micronutrients from animal products as from kale.

Since the nutrient density of kale is 10 times that of lentils, 483 times more fuel would need to be burned to obtain equivalent micronutrients from animal products as from kale.

Cutting out the intermediary and going directly to the source uses significantly less fossil fuel and is therefore notably more efficient.

NUTRIENT-TO-EMISSION RATIO

In the following examples I use calories as the method of comparison, as opposed to weight or volume. Comparing the amount of emissions required to produce an equal number of calories from a variety of sources gives, I believe, the fairest comparisons. (However, when the ratio is calculated by using weight instead, the discrepancies are even greater. I've listed them below each example.) The complete calculations are in "Calculating the Numbers" on page 310.

Lentils Versus Chicken

Producing meat in the form of a chicken creates the fewest emissions as far as animal agriculture goes. Yet calorie for calorie, chicken emits 5.57 times more CO_2e in its production than lentils. And since lentils are 3.7 times more nutrient dense than chicken (lentils have a nutrient density of 100, and chicken has a nutrient density of 27), 20.6 times more greenhouse gases (5.57×3.7) will be released into the atmosphere to obtain the same amount of micronutrients from chicken as from lentils.

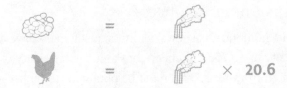

When comparing weight instead of calories, pound for pound, the divide would actually be slightly larger, registering at 28.8.

Steamed Vegetables Versus Baked Salmon

Steamed vegetables have a nutrient density average of 304, and salmon has a nutrient density of 39, which means that you'd need to produce 7.8 times more salmon to yield the same nutrient levels of steamed vegetables. Therefore, to obtain the equivalent amount of micronutrients from salmon as from steamed vegetables would mean 41.44 times more greenhouse gases being released into the atmosphere.

When comparing an equal weight of vegetables to salmon—pound for pound as opposed to calorie to calorie—an even greater divide is revealed. There would be 113.55 times more carbon dioxide released into the atmosphere to gain the equivalent micronutrient content from a pound of coho salmon as from a pound of vegetables.

Nuts Versus Domestic Cheese

Since raw nuts (almonds, cashews, walnuts, pistachios) have an average nutrient density of 36.75, and cheese has a nutrient density of 10, for every calorie obtained from nuts, 3.68 must be obtained from cheese to match the micronutrient content. There would be 2.49 grams of CO_2e released into the atmosphere in the production of 1 calorie from cheese, compared to only 0.03 grams to create a calorie from nuts.

304.73 times more CO_2e would be released into the atmosphere
to gain the equivalent micronutrient content from "American cheese"
as from raw nuts.

Since pound for pound nuts have more calories than cheese, cheese production would emit only 209 times as much carbon dioxide as nut production if the foods were compared on the basis of weight.

Kidney Beans Versus Beef Tenderloin

Just for fun, here's an extreme example: standard factory-farmed beef contrasted with beans.

Kidney beans emit 116 times (nutrient-to-emissions ratio of beef tenderloin: 19.72 ÷ 0.17 nutrient-to-emission ratio of kidney beans) less CO_2 than the production of beef tenderloin. Since kidney beans are 5 times more nutrient dense than beef, 580 times (116 × 5) more greenhouse gasses will be released into the atmosphere to gain an equal amount of micronutrients from beef tenderloin compared to kidney beans.

To obtain the same amount of micronutrients from beef as from beans, 580 times
more greenhouse gases would need to be released into the atmosphere.

When comparing equal weight of beans to steak rather than calories, the divide is even greater: 1560 times more carbon dioxide would be released into the atmosphere through the production of beef tenderloin than through the production of beans to produce an equal amount of micronutrients.

THE LOW COST OF HIGH NUTRITION

What about cost? We know that food requiring fewer resources to produce comes at a substantially lower environmental cost. And we know that those same nutrient-rich plant-based whole foods greatly reduce the risk of disease, which will likely translate into a cost savings at both an individual and a societal level. We also know that basing our diet on nutrient-dense whole foods has been shown to reduce sick days per year and enhance mental clarity; hence, a boost in productivity may result, which some may choose to put a dollar value on.

> It's certainly true that more calories and a greater mass of food (or food-like substances) can be had per dollar from highly processed refined sources, but does *nutrition* really cost more?

But what about the upfront monetary cost of healthier food—is it really more expensive? Yes. And no. It's certainly true that more calories and a greater mass of food (or food-like substances) can be had per dollar from highly processed refined sources, but does *nutrition* really cost more? I wondered, since micronutrients are the true measure of a food's worth, what the cheapest way to obtain them might be. Through conventional "standard" foods with lower price stickers? Or by way of plant-based whole foods priced a little higher? Bear in mind that the following examples are based on the retail (government-subsidized) prices for the meat option

The following sections describe what I found out.

Black Beans Versus Eggs

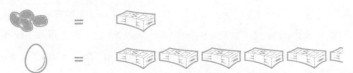

It costs 5.24 times more to gain nutrition from eggs as it does from black beans.

Eggs cost about $3.99 a dozen (about $2.99 per pound), and based on a caloric density of 1430 calories per kilogram, we can determine that each 100 calories costs 46 cents. Since eggs have a nutrient density of 27, their nutrient-to-cost ratio is 58.7 (27 ÷ $0.46). In contrast, black beans cost about $1.59 per pound, and based on a caloric density of 1320 calories per kilogram, each 100 calories from black beans costs 27 cents. Factoring in a nutrient density of 83, black beans have a nutrient-to-cost ratio of 307.41, 5.24 times greater than eggs. Therefore, you would have to eat six eggs, at a cost of $2, to match the micronutrient content you would acquire from 4 ounces of black beans, at an expenditure of 40 cents, a cost difference of over 5 times. ($1.59 [price per pound] ÷ 4 [price per 4 ounces] 0.40 × 5.24 = $2 worth of eggs) (40 cents × 5.24 = $2 [9.5 ounces of chicken])

Lentils Versus Chicken

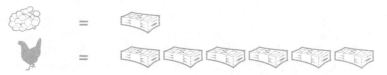

It costs 6 times as much to gain nutrition from chicken breast as it does from lentils.

Chicken costs about $4.99 per pound, so based on a caloric density of 1650 calories per kilogram, each 100 calories costs 67 cents. Since chicken has a nutrient density of 27, its nutrient-to-cost ratio is 40.30 (27 ÷ $0.67). In contrast, lentils cost about $1.99 per pound, and based on a caloric density of 1060 calories per kilogram, each 100 calories costs 41 cents. Factoring in a nutrient density of 100, lentils have a nutrient-to-cost ratio of 243.9, six times greater than that of chicken. Therefore, you would have to eat 9.5 ounces of chicken, at a cost of $3, to match the micronutrient content you would acquire from 4 ounces of lentils, at a cost of 50 cents, a cost difference of 6 times. ($1.79 [price per pound] ÷ 4 [price per 4 ounces] 0.45 × 41.67 = $18.75 worth of salmon) (50 cents × 6 = $3 [9.5 oz of chicken])

Chicken is 2.5 times more expensive by weight: 2.5 ÷ 1.55 = 1.6. Therefore, calorie for calorie, chicken is 1.6 times more expensive than lentils. Lentils are 3.7 times more nutrient dense than chicken: 3.7 × 1.62 = 6.

Therefore, you would have to spend 6 times more to gain equal nutrition from chicken as from lentils.

Flax Versus Coho Salmon

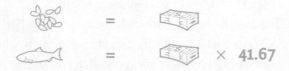

It costs 41.67 times more to gain nutrition from salmon as it does from flax.

Salmon costs about $15.99 per pound, and based on a caloric density of 1900 calories per kilogram, each 100 calories from salmon costs $1.85. Since salmon has a nutrient density of 39, its nutrient-to-cost ratio is 21.08 (39 ÷ $1.85). Flaxseed costs about $1.79 per pound, and based on a caloric density of 5340 calories per kilogram, each 100 calories costs $0.074. Factoring in a nutrient density of 65, flaxseed has a nutrient-to-cost ratio of 878.38, 41.67 times greater than that of salmon. Therefore, you would have to eat 1.17 pounds of salmon, at a cost of $18.78, to match the micronutrient content you would acquire from 4 ounces of flaxseed, at a cost of 45 cents, a cost difference of over 41 times.

WHY DOES A HAMBURGER COST MORE THAN AN APPLE?

Have you ever wondered how it is that meat and other highly processed food—all of which consumes an inordinate amount of land, water, and oil to produce—costs relatively little in dollars? How is it that a hamburger at a fast food restaurant can cost under $2 when a locally grown organic apple that uses a fraction of the energy to produce it costs the same or more? The short answer: government subsidies. In the United States, these happen through legislation known as the Farm Bill—the government's chief food policy tool. While it was introduced with the best of intentions, the Farm Bill has veered off course and no longer benefits the people as a whole. And certainly not the planet.

The original aims of the bill were achieving food security and helping struggling farmers make enough money to stay in business. However, the subsidies were misdirected. They began to flow mostly to the meat and dairy industry. Even with strong sales, the production cost of meat and dairy was simply too high (because of its substantial resource cost, in turn because of inefficiency) to enable farmers to make a profit. Had the government not stepped in and bailed out this exceptionally inefficient industry with monetary subsidies, the cost of meat and dairy would be prohibitive to almost everyone. Instead, subsidies turned these products into cheap commodities, causing consumption to skyrocket—and with it, pollution and disease. To rub salt in the wound, we're paying for it. Government money comes from the taxpayers. So, even those of us who don't support these inefficient industries by patronizing them prop them up by way of the current tax system. And now, to reopen the wound and dump more salt in, we also highly subsidize our reactive medical systems, which pours money into treating symptoms caused by poor-quality diets made possible, and encouraged, by subsidies. It truly is a vicious cycle.

> "Pay now for healthier food, or pay later to treat the symptoms of disease as a result of poor nutrition."

It's estimated that the health costs from poor diet are $250 billion per year in the United States alone, and the U.S. National Institutes of Health (NIH) has predicted that obesity will lower Americans' life expectancy by up to five years over the next few decades.

In 2006, I had an opportunity to speak to the U.S. Congress about exactly this issue: the Farm Bill. I've learned that when speaking to such governmental groups, it's important to focus on one thing: money. So I did. My suggestions were along the line of preventative medicine. I presented information suggesting that if subsidies were to be lifted, people simply would not be able to afford resource-gobbling food; they would have to eat simple plant-based foods. The most nutrient-dense food would then be the cheapest. Its lower cost would, of course, give people an economic incentive to eat better food. And, as a direct result, the disease risk factor of poor nutrition would be much reduced and the overworked health-care system would be much relieved of its strain. While nothing of significance developed as a result of my presentation, the nodding heads in the audience indicated that the simple logic of "Pay now for healthier food, or pay later to treat the symptoms of disease as a result of poor nutrition" seemed to strike a chord.

Considering the immense resource draw of meat production, it seems reasonable to say that if meat were sold at a fair market value, consumption of it would plunge. People simply wouldn't be able to afford it. Estimates on the fair market price for a typical fast-food burger, costed without subsidies, range from $35 all the way up to $200. The $35 calculation is based on the fossil fuel cost to produce it, while the $200 figure includes the land and water costs. In addition, the $200 burger price factors in the delivery of the food to the feedlots, and the cleanup of the water systems polluted by the cattle's manure. But, as it stands, with taxpayer-funded subsidies, this same hamburger at a fast-food restaurant will go for about $2. Sometimes as low as $1—if you select the "value menu" and agree to buy a carton of fries and a soft drink.

THE GREAT ENVIRONMENTAL DIVIDE: MEAL PLAN COMPARISONS

Following are three sample one-day meal plans to further illustrate the environmental impacts of our food choices. In each I've calculated and compared the amount of CO_2-equivalent emissions each meal is responsible for releasing during its production.

The first meal plan reflects the Standard American Diet, subsistence in the form of processed food-like substances, which relies heavily on red meat and dairy, with very little food in its natural state.

Second, I look at the "healthy" American diet. The go-to diet for many Americans once they've begun to experience health decline from years of eating the Standard American Diet, the "healthy" American diet is only marginally less processed and far from a long-term health solution. And, as I found, the amounts of CO_2-equivalent emissions created by its production are enough to be of major environmental concern.

The third meal plan is devised following the nutritional philosophy of this book: it uses plant-based whole foods with a high nutrient-to-resource ratio.

The numbers next to each meal are the grams of CO_2-equivalent emissions (here, CO_2e for short) released into the atmosphere in the production of the food composing each meal.

STANDARD AMERICAN DIET

Breakfast: Omelet with meat and cheese; cereal and milk; 12 oz coffee with cream and sugar. CO_2e: 3101 g

Lunch: Cheeseburger and fries; 12 oz soft drink. CO_2e: 3116 g

Dinner: Beef stir-fry; 12 oz beer; milk chocolate bar for dessert. CO_2e: 3387 g

TOTAL CO_2e: 9604 g

"HEALTHY" AMERICAN DIET

Breakfast: Buttermilk pancakes; scrambled eggs with cheese; latte. CO_2e: 2661 g

Lunch: Baked farmed salmon; 5 oz wine. CO_2e: 1396 g

Dinner: Chicken Caesar salad; baked potato with sour cream; 12 oz water. CO_2e: 1317 g

TOTAL CO_2e: 5374 g

WHOLE FOODS TO THRIVE SUGGESTION

Breakfast: Nutrient-dense plant-based smoothie (containing hemp protein, pea protein, rice protein, and flaxseeds) blended with banana and blueberries. CO_2e: 130 g

Snack: Seasonal fruit and nuts. CO_2e: 104 g

Lunch: Vegetable stir-fry and lentils. CO_2e: 361 g

Snack: Oven-roasted potatoes. CO_2e: 84 g

Dinner: Beans and rice with grilled vegetables and green salad. CO_2e: 308 g

Snack: Raw vegetables and hummus. CO_2e: 212 g

TOTAL CO_2e: 1199 g

So what do these numbers actually mean? That's what I wondered. To make them relatable, I've compared the CO_2-equivalent emissions released in food production with the CO_2 emissions created from driving. What I found was striking.

The chart on the next page shows the grams of CO_2-equivalent emissions that each diet would produce over various periods: a day, a week, a month, a year. I then compare these CO_2 levels with what's produced by different driving distances.

For a midsize car that averages between 26 and 28 miles per gallon, about half a pound (227 g) of carbon dioxide is released into the atmosphere to travel one mile.[8] Knowing that, we can directly compare driving with eating and its impact on the environment.

However, because the numbers for the consumption of arable land, water, and fossil fuel are not yet available for all food products, in the examples I use only CO_2-equivalent emissions to compare food's impact on the environment. Needless to say, the environmental toll is far greater when all factors are considered.

CO_2-equivalent emissions Diet by time period compared to driving distance		
Emissions from diet		**Emissions from driving**
STANDARD AMERICAN DIET		
Day: 9604 grams	=	42.4 miles
Week: 67,240 grams	=	296 miles
Month: 291,961 grams	=	1,289 miles
Year: 3,505,460 grams	=	15,476 miles
"HEALTHY" AMERICAN DIET		
Day: 5374 grams	=	23.7 miles
Week: 37,618 grams	=	166 miles
Month: 163,397 grams	=	720 miles
Year: 1,961,510 grams	=	8,650 miles
WHOLE FOODS TO THRIVE SUGGESTION		
Day: 1199 grams	=	5.28 miles
Week: 8393 grams	=	37 miles
Month: 36,469 grams	=	160.65 miles
Year: 437,635 grams	=	1928 miles

Sources of data: www.falconsolution.com/co2-emission/
227 g (1/2 lb) CO_2 to travel 1 mile in midsize car
Driving distances are for a midsize car that averages between 28 and 30 miles per gallon.

The contrasts among the diets are dramatic, to say the least.

But headlining just about every "what you can do to save the planet" list is the suggestion that we abstain from driving, or at least greatly reduce the amount we do. "Ride your bike, take transit, or—if you must—carpool." To many, driving has become a sign of environmental disregard. And while I agree that using alternative means of transportation and depending less on the automobile makes sense, are cars really deserving of so much attention?

Clearly, with automobile transportation we have a hands-on relationship. We regularly need to fill our vehicles with gasoline. This gives us an idea as to how much fuel we directly consume. Besides, filling up costs money, which for the average person reinforces the value of fossil fuel. That's good. Most of us appreciate the value of gas because of its high and ever-increasing cost and have a sense of the environmental damage its combustion is causing. We see the emissions spewing from tailpipes as we drive around. And sitting in rush-hour traffic, if we roll down the window, we experience directly how CO_2 emissions are adversely affecting air quality. Over cities such as Los Angeles, we can see a thick layer of smog engulfing the buildings as a direct result of tailpipe exhaust. And as such, we have a close, tangible, and therefore relatable relationship with our vehicle and its fossil fuel–burning, emission-creating ways.

> Making wise food choices can have a more significant impact on environmental preservation than eliminating driving altogether.

In contrast, we don't see the fuel needed in the production of food, and the emissions it releases are not apparent to the average person. But it's real nonetheless. And, as I mentioned earlier when I cited the 2006 U.N. report, the level of these emissions is considerably higher than that released by the automobile.

Toward the bottom of "planet-saving" suggestions, eating locally may sometimes be mentioned. And *maybe* switching to more plant-based options will round out the list. Maybe. But is the weighting of these suggestions reflective of the biggest environmental offenders?

We all have to eat, and making wise food choices can have a more significant impact on environmental preservation than eliminating driving altogether.

Difference in emissions created (equivalent to miles driven) from eating a Standard American Diet compared with Whole Foods to Thrive suggestions	
Day	37.12 miles
Week	259.84 miles
Month	1129 miles
Year	13,549 miles

In just one day, anyone who switched from the Standard American Diet to Whole Foods to Thrive would conserve the equivalent in emissions of the grams released in driving a little over 37 miles. Just one week of eating the Thrive way rather than the Standard diet would conserve as much CO_2e as is released driving 250 miles, the equivalent of traveling by car from Boston to New York City.

Spanning a year, the CO_2e savings would equal the grams emitted in driving from Los Angeles to New York City four-and-a-half times.

Since the average distance driven by each American is 12,500 miles per year,[9] switching from a Standard American Diet to a Whole Foods to Thrive way of eating would prevent more CO_2-equivalent emissions from entering the atmosphere than would abstaining from driving altogether.

Therefore, an average American can do more to mitigate climate change by changing what he or she eats than by completely cutting out driving.

And as for the "healthy" American diet?

Aware of the immense amount of methane and nitrous oxide that raising ruminants, such as cows and sheep, produces, some looking to reduce their carbon footprint are turning to meat from other animals. As you can see from the "healthy" American meal plan, it contains no meat from ruminants, yet its production still releases 338 percent more CO_2e than the suggested plant-based Whole Foods to Thrive meal plan. The environmental culprits in the "healthy" American meal plan? Fish, chicken, and dairy. While emitting less total CO_2e than the Standard American Diet, the "healthy" diet's switch to "white" meat is clearly not the best solution for environmental sustainability. And as mentioned in Chapter 1, such a diet's micronutrient level is considerably lower than we can obtain directly from plants.

4 EIGHT KEY COMPONENTS OF GOOD NUTRITION

While micronutrient density is the best gauge of a food's true nutritional worth, the eight components described in this chapter point to the specifics of what comprise healthy food. Six of the eight are directly related to nutrient density; the other two (whether a food contains essential fats or is a raw food) are not but do contribute to what constitutes a healthy choice.

As you can see, I've assigned an icon to represent each element of nutrition. In the next chapter, "Nutrient-Dense Whole Foods to Thrive," you can look for these icons as a quick key to identify the foods that are the top sources of each element.

ALKALINE-FORMING FOODS

The measure of acidity or alkalinity is called pH, and maintaining a balanced pH is an important part of reaching and sustaining peak health. The body can become more acidic through diet and, to a lesser degree, through stress. Since minerals are exceptionally alkaline-forming, the pH of any food is largely dependent on mineral content. Returning to the subject of soil quality, even greens—which are highly alkaline-forming due to their chlorophyll content—will not be as alkaline-forming if they are grown in mineral-depleted soil.

Alkaline-forming foods help to balance the body's pH. An acidic environment adversely affects health at the cellular level; people with low body pH are prone to fatigue and disease. And because acidity is a stressor, it raises cortisol levels, which results in impaired sleep quality.

To help your muscles recover and to lower your cortisol levels, consume highly alkalizing foods, such as those rich in chlorophyll, soon after exercise. Chlorophyll is the green pigment that gives leaves and green vegetables their color.

BEST SOURCES
- All green vegetables
- Seaweed
- Algae

BENEFITS
- Improves bone strength
- Reduces inflammation
- Improves muscle efficiency
- Reduces risk of disease

ANTIOXIDANTS

As mentioned in Chapter 2—in the first section, "Arable Land"—plants are only capable of producing antioxidants if they have drawn an adequate amount of minerals from the soil. For plants to develop their full antioxidant potential, they must be grown in mineral-rich soil.

> Antioxidant compounds found in fruits and vegetables cancel out the effects of the cell-damaging free radicals by slowing or preventing the oxidative process.

When our body's activity level rises, we use extra oxygen, which causes cellular oxidation. Oxidation can create free radicals, which reduce cell lifespan and cause premature cell degeneration. Damage done by free radicals has been linked to cancer and other serious diseases and to premature skin aging. Free radicals occur naturally in the body, with small amounts being produced daily, but stress can increase their presence. A reduction of stress through better nutrition combats the oxidative process and therefore free radical production. Antioxidants in foods also help to rid the body of free radicals by escorting them out of the body.

Because of the increased oxygen consumption associated with regular strenuous physical activity, it creates an abundance of free radicals. We therefore need to combat this negative side effect of exercise. Antioxidant compounds found in fruits and vegetables—vitamin C, vitamin E, selenium, and the carotenoids (compounds that give vegetables their orange color)—cancel out the effects of the cell-damaging free radicals by slowing or preventing the oxidative process. I noticed a clear improvement in how fast I recovered between workouts once I regularly began eating antioxidant-rich foods.

BEST SOURCES

- Organic berries
- Organic dark-colored fruit in general
- Organic colorful vegetables
- Green tea

BENEFITS

- Protects cellular health
- Speeds physical recovery
- Reduces risk of disease
- Improves skin's appearance and elasticity

CALCIUM

Since calcium is a mineral, the amount present in food has been declining over the years. Again, food quality is dependent on soil quality.

For most people, building, strengthening, and repairing bone is calcium's major role. Active people, however, have another important job for the mineral: muscle contraction and rhythmic heartbeat coordination. About 95 percent of the body's calcium is stored in the skeleton, but it's the remaining share that is the first to decline. Calcium in the bloodstream is lost in sweat and muscle contractions, so active people need more dietary calcium. Another micronutrient, vitamin D, maximizes calcium absorption. Vitamin D comes from the sun, so regular exposure to daylight will help your body absorb calcium and therefore help with bone maintenance.

Over the course of about the last 15 years, North Americans have been losing bone density and developing osteoporosis at a younger age than ever before in history. Initially, this loss was thought to be due to inadequate dietary calcium. Advertisements in magazines and on TV tried to convince people over the age of 40 to take calcium supplements. Unfortunately, the body doesn't properly absorb the inorganic forms of calcium found in supplements, so we'd need to consume a very large amount for supplementary calcium to have even a small impact on bone health. The net-gain principle suggests that the consumption of inorganic calcium is a poor use of energy. In fact, it's not uncommon for people who take calcium supplements to notice an energy dip within an hour or so after taking them.

Plants take inorganic calcium from the soil and convert it into an organic form of calcium that the human body can efficiently and completely make use of. Consuming an adequate supply of organic calcium from such sources as leafy green vegetables ensures that our bones stay strong and that muscle contractions remain smooth and efficient.

We must also make sure we don't remove the calcium that already exists in our bodies, so it's important to avoid acid-forming foods, which deplete our stores of calcium and so weaken the bones.

BEST USABLE PLANT SOURCES
- Dark leafy greens, such as spinach, kale, collard greens
- Unhulled sesame seeds

BENEFITS
- Improves muscle function and efficiency
- Increases bone strength
- Reduces the risk of osteoporosis

ELECTROLYTES
Electrolytes are electricity-conducting salts drawn from the soil. Calcium, chloride, magnesium, potassium, and sodium are the chief electrolyte minerals. Electrolytes in body fluid and blood regulate or affect the flow of nutrients into cells and of waste products out of cells, and are

essential for the regulation of muscle contractions, heartbeats, fluid levels, and general nerve function. When too few of these minerals are ingested, we may suffer muscle cramps and heart palpitations, light-headedness and trouble concentrating. In severe cases, lack of electrolytes leads to loss of equilibrium, confusion, and inability to reason.

You may have noticed salt-like crystals forming on your face when you perspire heavily. Those are electrolytes—what's left when the water component of sweat has evaporated—and they have to be replenished through food and drink. But not just any drink. When we consume too much fluid that does not contain electrolytes, it can flush out the remaining electrolytes from our body, referred to as water intoxication. While it isn't common among the general population, people who perform strenuous physical activity, especially in a warm environment, are susceptible.

Most commercial sports drinks contain unnecessary refined sugar and artificial flavor and color. Soon after my own experience with water intoxication, I developed my own formula for a natural, healthy, electrolyte-packed drink, my Lemon-Lime Sports Drink (the recipe is on page 130).

BEST SOURCES
- Coconut water
- Molasses and molasses sugar
- Seaweed (dulse and kelp in particular)

SECONDARY SOURCES
- Bananas
- Tomatoes
- Celery

BENEFITS
- Helps maintain hydration
- Improves the fluidity of muscle contractions
- Increases the heart's efficiency, lowers heart rate, improves endurance
- Boosts mental clarity

ESSENTIAL FATS

Essential fatty acids (EFAs) are an important dietary component of overall health. The word *essential* in the name means the body cannot produce these fatty acids—they must be ingested. There are two families of EFAs, omega-3 and omega-6.

EFAs support the function of the cardiovascular, immune, and nervous systems. Studies suggest that including omega-3, in particular, in the diet can benefit those who suffer from a wide range of ailments, including high blood pressure, high cholesterol, heart disease, diabetes, rheumatoid arthritis, osteoporosis, depression, bipolar disorder, schizophrenia, attention deficit disorder, skin disorders, inflammatory bowel disease, asthma, colon cancer, breast cancer, and prostate cancer.[1] These studies also suggest that an adequate supply of omega-3 may help reduce the risk of developing these ailments in the first place.

> EFAs play an integral role in the repair and regeneration of cells and therefore in keeping the body biologically young.

Responsible in part for the cells' ability to receive nutrition and eliminate waste, EFAs play an integral role in the repair and regeneration of cells and therefore in keeping the body biologically young. A balance of omega-3 and omega-6 EFAs will keep skin looking and feeling supple. EFAs also help fight infection and reduce inflammation. In addition, EFAs are linked to healthy and efficient brain development in children.

From an active person's perspective, when combined with proper endurance training, a diet with an adequate supply of EFAs can help improve endurance. Our bodies can store only a small amount of muscle carbohydrate. Once the body has burned all of its carbohydrate stores, it has to be refueled—as often as every 30 minutes during a long race or workout. However, once the body has adapted to a period of long, slow training (as I describe in my book *Thrive Fitness*), it becomes more efficient at burning body fat as fuel and thus is able to preserve its carbohydrate stores. This shift in metabolism is simply a trait of improved fitness, which therefore enables the body to burn less fuel to travel the same distance. This fuel shift means that refueling doesn't have to take place as often and endurance will be significantly improved. The fuel shift is facilitated by dietary EFAs, which need to be properly balanced between omega-6 and omega-3 to be effective. The ideal ratio is said to be 4:1. For every four parts omega-6 that's in your diet, you'll want to have one part omega-3. Fortunately, a plant-based whole food diet naturally provides that ratio.

BEST SOURCES
These all contain a balance of omega-3 and omega-6.
- Sacha inchi
- Chia
- Flaxseed
- Hemp

BENEFITS
- Improves endurance
- Increases the body's ability to burn body fat as fuel
- Improves the ability to stay well hydrated
- Improves joint function

IRON

Drawn into plant plasma from the soil, iron helps maintain blood cell health so that the heart can deliver oxygen-rich blood to the hardworking extremities—maximizing efficacy and therefore athletic performance. Iron also builds blood proteins essential for food digestion, metabolism, and circulation.

Iron is lost in sweat and is consumed during muscle contraction. The pounding impact of our feet on the ground during running can cause red blood cells to break down and thus lower their iron levels. People with low iron are at risk for anemia. Dietary iron helps counteract these problems.

About eight years ago, I went through a stage of reduced energy and poorer performance. I had a blood test to find out what was wrong. It showed that my iron level was low—not so low that I couldn't train at all, but certainly low enough to hinder my progress. I had borderline anemia. Because my active lifestyle consumed a lot of iron and because I did not eat animal products, which are higher in iron than plant-based foods, my doctor suggested I begin taking iron in tablet form. I knew a few people who had experienced stomach problems and even constipation when they began their iron supplementation program, so I wanted to see whether I could get all the iron I needed just from food. I found there are many good plant-based sources of iron. For me, a combination of about ¼ cup raw pumpkin seeds and a green salad daily did the trick. Within a few months my iron levels were back to optimal and have remained there ever since.

BEST PLANT-BASED SOURCES

- Pumpkin seeds
- Leafy greens (especially kale)
- Vega Complete Whole Food Health Optimizer (contains 100 percent of the recommended daily allowance)

BENEFITS

- Improves blood's oxygen-carrying ability
- Increases physical stamina
- Boosts energy

PHYTONUTRIENTS

Since phytonutrients are a specific form of antioxidants, they, along with other antioxidants, can only be produced if their host plant is grown in mineral-rich soil.

Phytonutrients are plant compounds that offer health benefits independent of their nutritional value. They are not essential for life, but they can help improve vitality and quality of life.

For example, a phytonutrient found in tomatoes improves blood vessel elasticity and thereby enhances blood flow through the heart. Tomatoes can thus lower the risk of developing cardiovascular disease and enhance athletic performance. The heavy processing of fruit and vegetables reduces the amount and effectiveness of phytonutrients, so these foods are best eaten raw. Every type of fruit and vegetable has at least a few phyto-nutrients, so simply eating many servings on a daily basis will boost health and performance.

- Colorful and green vegetables

BENEFITS

- Improves heart health
- Reduces the risk of cardiovascular disease
- Improves blood vessel elasticity, thereby improving circulation

RAW FOOD

As I noted earlier, eating a large percentage of raw food makes sense on several levels. High-temperature cooking and processing of food destroys enzymes and nutrients needed for efficient digestion. Before the body can make use of cooked food, it must produce enzymes to aid in the digestion process. That takes work, which of course is an energy draw and therefore creates a nominal amount of stress. In addition, food containing both sugar and fat cooked at a high temperature can provoke an immune response that causes inflammation. However, minerals are not damaged by heat, so cooking food will not lower mineral content. Vegetables that contain a high amount of starch, such as potatoes and sweet potatoes, are best eaten cooked.

BEST SOURCES

- Fruit
- Nuts
- Seeds
- Most vegetables

BENEFITS

- Improves digestibility of most (non-starchy) foods
- Maintains higher vitamin content in most foods
- Allows for higher net gain and therefore more energy

As you'll see, these beneficial traits are a large component of the foods that are the base ingredients for the recipes in Chapter 6.

5 | NUTRIENT-DENSE WHOLE FOODS TO THRIVE

The whole foods I'll tell you about in this chapter are among the prime ingredients in the recipes to follow in Chapter 6. And, as you might expect, they are among the most nutrient-dense foods there are while requiring less of each natural resource to produce than more traditional foods. As you can see, I've used the icons representing the eight key components of nutrition to indicate when a particularly high amount of a certain component is present in the food. These icons serve as a quick visual guide as to which foods are richest in these elements. These foods are essential components of daily meals in the Whole Foods to Thrive plan so I call them "Pantry Essentials"; you should make sure you have them on hand at all times.

WHOLE FOODS TO THRIVE PANTRY ESSENTIALS

Green Vegetables

Alkaline-Forming Antioxidants Calcium Electrolytes Essential Fats Iron Phytonutrients Raw

Because of their chlorophyll content, green vegetables are an excellent way to help alkalize the body, which, as I mentioned before, reduces inflammation and helps maintain bone health.

Chlorophyll also cleanses and oxygenates the blood, making it a true performance enhancer. More available oxygen in the blood translates into better endurance and an overall reduction in fatigue. In their raw state, chlorophyll-containing plants also possess an abundance of live enzymes that promote the quick rejuvenation of our cells. The consumption of green foods after exercise has been shown to help speed cellular regeneration. The consumption of chlorophyll-rich leafy green vegetables combined with moderate exercise is the best way to create a biologically younger body. Ounce for ounce, dark greens are also an excellent source of iron and calcium.

> The consumption of chlorophyll-rich leafy green vegetables combined with moderate exercise is the best way to create a biologically younger body.

You may not crave a plate full of fibrous, leafy green vegetables immediately after exercise, and they'd take up room in your stomach needed for other post-recovery nutrition. An easy way to ingest greens immediately after exercise is to mix a greens powder, such as chlorella or spirulina, into a fruit-based post-exercise recovery drink (the "Drinks" recipes start on page 128). Later in the day, a big salad is an ideal way to load up on more leafy greens.

All leafy greens are nutrient-dense and an excellent conduit for minerals in the soil; here are some of the more readily available ones:

- Beet greens
- Butter lettuce
- Collards
- Dandelion greens
- Dinosaur kale
- Mustard greens
- Red leaf lettuce
- Romaine lettuce
- Spinach
- Swiss chard

Fibrous Vegetables

Also mineral-rich, these vegetables are high in both soluble and insoluble fiber:

- Asparagus
- Beets
- Bok choy
- Broccoli
- Carrots
- Celery
- Cucumbers
- Daikon
- Green beans
- Green peas
- Green onions
- Sugar snap peas
- Watercress
- Zucchini

Starchy Vegetables

SQUASH

Antioxidants   Calcium   Iron   Phytonutrients

There are several types of squash, all with distinctly different flavors and textures; butternut, spaghetti, acorn, carnival, banana, zucchini, delicata, and kabocha are among the most popular and therefore the most common.

The most nutrient-dense form of starch-based food, squash is an excellent addition to the diet. Especially of value for those who aspire to pack on muscle, squash—combined with the correct workout—will contribute to the process of muscle building.

Many varieties of squash are grown in North America and can be found at most farmers' markets, especially in the autumn.

Other starchy vegetables include

- New potatoes
- Parsnips
- Pumpkin
- Sweet potatoes
- Yams

Sea Vegetables

Alkaline-Forming　Calcium　Electrolytes　Iron　Phytonutrients　Raw

Sea vegetables, often referred to as seaweed and less commonly as wild ocean plants, have been a staple of many coastal civilizations for thousands of years. Most notably, Asian cultures have long since embraced sea vegetables as an important part of their diet.

Sea vegetables are among the most nutritionally dense foods. Containing about 10 times the calcium of cow's milk and several times more iron than red meat, sea vegetables are easily digestible, chlorophyll rich, and alkaline forming. Packed with minerals, sea vegetables are the richest source of naturally occurring electrolytes known.

Dulse, nori, and kelp are the most popular sea vegetables in North America. Available in dried and ready-to-use form in most health food stores, their addition to many recipes is easy. Dulse provides the perfect mineral balance in a natural form and so is a superior source of the minerals and trace elements we need daily for optimal health.

Other, less common, sea vegetables are agar, arame, kombu, and wakame.

Pseudograins

Commonly referred to as grains but technically seeds, pseudograins are naturally gluten free and contain more protein (20 percent to 25 percent by volume) and are higher in micronutrient density than grains. You can use pseudograins in most recipes calling for rice.

BUCKWHEAT

Alkaline-Forming Phytonutrients Raw

Buckwheat is not actually wheat; it is a seed in the rhubarb family. Containing eight essential amino acids, including quite high amounts of the often-elusive tryptophan, buckwheat is a good source of protein. Tryptophan is a precursor to the neurotransmitter serotonin; having an adequate amount of tryptophan in the diet can be important to help enhance mood and mental clarity. Buckwheat is also high in vitamins E and B, calcium, and especially manganese.

Since buckwheat is gluten free, it is considerably more alkaline forming than gluten-containing grains. It is also a slow-release carbohydrate. Combined with a simple carbohydrate, buckwheat becomes one of the best endurance fuels available. Sprouted buckwheat digests and burns even more effectively because the sprouting process converts the complex carbohydrate into sugar, which the body can burn more efficiently than starch. But since the protein, fat, and fiber remain, this sugar will not cause an insulin spike and subsequent crash. Raw buckwheat can also be substituted for seeds in recipes to reduce fat content.

QUINOA

Alkaline-Forming Phytonutrients Raw

With a light, fluffy texture and mild earthy taste, quinoa balances the texture of other, heavier grains when combined with them.

Nutritionally similar to amaranth, quinoa consists of about 20 percent protein; it is high in lysine and is a good source of iron and potassium. High levels of B vitamins, in part responsible for the conversion of carbohydrate into energy, are also found in quinoa.

The preparation of quinoa is particularly important since it is naturally coated in a bitter resin called saponin. Thought to have evolved naturally to deter birds and insects from eating the seed, saponin must be removed by thorough rinsing to make quinoa palatable. Most of the saponin will have been removed before the quinoa is shipped to the store, but there will likely be a powdery residue.

Cook quinoa like rice, at a 1:2 quinoa-to-water ratio, for about 20 minutes. Quinoa can also be sprouted (Google "sprouting quinoa" for instructions).

WILD RICE

Alkaline-Forming Phytonutrients Raw

Wild rice is an aquatic grass seed rather than a true rice. High in B vitamins and the amino acid lysine, wild rice is much more nutritious than traditional grains. Native to the northern regions of the Canadian Prairie provinces, wild rice is seldom treated with pesticides since it thrives without them. (It is also grown as a domesticated crop in Minnesota and California.) Wild rice has a distinct, full-bodied flavor and a slightly chewy texture that complement many meals.

Cook like rice, at a 1:2 wild rice-to-water ratio, for about 30 minutes. It can also be sprouted (Google "sprouting wild rice" for instructions).

Seeds

FLAXSEED

Essential Fats Phytonutrients Raw

Grown mostly in the Canadian prairies, the seed of the blue-flowering flax plant is prized for its lignans and high omega-3 fatty acid content. The regular inclusion of lignans in the diet has been shown to reduce the risk of cancer. Flaxseed is also rich in fiber. However, it is its omega-3 essential fat content that makes flaxseed most valuable to athletes. As I noted earlier, aside from its ability to help reduce inflammation caused by movement, omega-3 plays an integral role in the metabolism of fat. A diet with a daily dose of 1 tablespoon of whole flaxseed will allow the body to more efficiently burn body fat as fuel. This is obviously a benefit to anyone wanting to shed body fat, but it is of major importance to athletes, who need to spare the energy stored in the muscles. As the body becomes proficient at burning fat as fuel (through training and proper diet), endurance significantly improves.

Whole flaxseed is high in potassium, an electrolyte responsible in part for smooth muscle contractions. Potassium is lost in sweat, so it must be replaced regularly to keep the body's levels adequately stocked. Potassium also helps to maintain fluid balance, assisting with the hydration process. Flaxseed is a whole food and a complete protein, and it retains its enzymes, allowing it to be absorbed and utilized by the body with ease, improving immune function.

HEMP SEEDS

Alkaline-Forming   Essential Fats   Phytonutrients   Raw

Hemp is available in three basic forms: seed, powder, and oil. Hemp seeds come straight from the plant and are rich in both omega-3 and omega-6 essential fatty acids. When pressed, the seed becomes hemp powder and oil. The powder, sometimes referred to as flour, is then milled finer to remove some of the starch. The result is hemp protein.

The protein present in hemp is complete, containing all 10 essential amino acids, which boost the immune system and hasten recovery. Hemp foods also have natural anti-inflammatory properties, key factors for speeding the repair of soft tissue damage caused by physical activity. Raw hemp products maintain their naturally high level of vitamins, minerals, high-quality balanced fats, antioxidants, fiber, and the very alkaline chlorophyll. Edestin, an amino acid present only in hemp, is considered an integral part of DNA. It makes hemp the closest plant source to our own human amino acid profile.

When it comes to protein, quality, not quantity, is paramount. I find hemp protein the easiest protein to digest. Since it is raw, its naturally occurring digestive enzymes remain intact. That and its relatively high pH allow it to be easily used by the body. As a result, the digestive strain placed on the body to absorb and utilize protein is reduced, making it a high-net-gain food. Top-quality, complete protein, such as hemp, is instrumental not only in muscle tissue regeneration but also in fat metabolism. Protein ingestion instigates the release of a hormone that enables the body to more easily utilize its fat reserves, thereby improving endurance and facilitating body fat loss.

PUMPKIN SEEDS

Alkaline-Forming   Essential Fats   Phytonutrients   Raw

Pumpkin seeds are rich in iron, a nutrient some people have trouble getting enough of, especially if they don't eat red meat. Anemia, a shortage of red blood cells in the body, is commonly caused by low dietary iron or by strenuous exercise. Iron is lost as a result of compression hemolysis (crushed blood cells due to intense muscle contractions). The more active the person, the more dietary iron she needs. Constant impact activity, such as running, reduces iron levels more dramatically than other types of exercise because of the more strenuous hemolysis. With each foot strike, a small amount of blood is released from the damaged capillaries. In time, this will lead to anemia if the runner doesn't pay close attention to her diet. Iron is also lost through sweat.

SESAME SEEDS

Alkaline-Forming   Calcium   Phytonutrients   Raw

Sesame seeds are an excellent, easily absorbable source of calcium. Calcium is in part responsible for muscle contractions—of particular concern to athletes, who will need to ensure that they maintain correct levels of calcium in the body. Calcium plays another important role in the formation and maintenance of bones and teeth. Athletes and people living in a warm climate will need extra amounts of dietary calcium since it is excreted in sweat.

I use a coffee grinder to grind sesame seeds into a flour, then store it in the refrigerator, for up to three months. I sprinkle the flour on salads, cereal, pasta, and soups. Some of the recipes in Chapter 6 call for sesame seed flour, to increase calcium content. When baking, it's possible to substitute sesame seed flour for up to one-quarter of the amount of regular, glutinous flour called for in the recipe. If the recipe calls for non-glutinous flour, the whole amount can be replaced with sesame seed flour. However, since sesame seed flour is slightly more bitter than most flours, you may want to experiment, gradually increasing the amount each time.

SUNFLOWER SEEDS

Phytonutrients   Raw

Made up of about 22 percent protein, sunflower seeds offer a good amount of dietary substance. Rich in trace minerals and several vitamins important for good health, sunflower seeds are a food worthy of regular consumption. Sunflower seeds are quite high in vitamin E and are antioxidant rich.

Legumes

Calcium   Iron   Phytonutrients   Raw

Legumes are plants that have pods containing small seeds. Lentils, peas, and beans are all in the legume family. Lentils and split peas are among the most commonly used legumes in this book's recipes for the simple reason that they don't need to be soaked before cooking.

Legumes in general have an excellent nutritional profile. High in protein, fiber, and many vitamins and minerals, a variety of legumes are part of my regular diet. Peas, and in particular yellow peas, have an exceptional amino acid profile. Also rich in B vitamins (in part responsible for converting food into energy) and potassium (an electrolyte needed for smooth muscle contractions), yellow peas are an excellent addition to an active person's diet. Because of peas' superior amino acid profile, manufacturers are now producing pea protein concentrates and isolates. This high-quality vegetarian protein is a good option for people with soy allergies.

Although some people avoid legumes because of their gas-producing reputation, legumes are no more a culprit than many other foods as long as they are prepared properly. After you have soaked beans and shelled peas in preparation for cooking, be sure to rinse them in fresh water. Rinse them again in fresh water after cooking. The water they soak and cook in will absorb some of the indigestible sugars that cause gas; rinsing it off will help improve their digestibility and minimize their gas production. Another way to improve the legumes' digestibility is to add seaweed to the pot when cooking them, to release the gas. A short strip of seaweed is enough for a medium-sized pot. As with all fiber-rich foods, legumes should be introduced slowly into the diet to allow time for the digestive system to adapt. Gradually increasing the amount of legumes you eat each day will ensure a smooth transition to a healthier diet.

Raw legumes are ideal for sprouting. Sprouting improves both legumes' nutritional value and their digestibility—enough so that they may be eaten raw. As well, sprouting allows the digestive enzymes to remain intact, eliminating gas production altogether.

These are the legumes I suggest for their nutritional value and taste:

- Adzuki beans
- Black beans
- Chickpeas
- Fava beans
- Kidney beans
- Lentils
- Navy beans
- Pinto beans
- Yellow and green split peas

Oils

Oils come in a wide assortment, each with a distinct taste and unique nutritional value. The key to keeping the flavors in your meals ever changing and your diet's nutrient value diverse is using various oils.

In the right amount, high-quality, cold-pressed, unrefined oils are among the healthiest of substances. My favorites are hemp, pumpkin, flaxseed, and, for cooking, coconut. Most oils contain the same nutrients as the plant seed they are from, just highly concentrated.

Not all oils are equal. Low-quality manufactured oil is one of the most damaging foods that can be consumed, eclipsing even refined carbohydrate. Many cheaper store-bought baked or fried products, such as muffins, chips, and cakes, contain trans fat, a near-poisonous substance unusable by the body. Trans fat, also known as trans-fatty acid, is added to many mass-produced commercial products to extend their shelf life, improve moisture content, and enhance flavor.

As for the oils used in the Whole Foods recipes, it's helpful to know which can be heated safely and which are best consumed raw. I never fry with hemp, flaxseed, or pumpkin seed oil because of their low burning point—the temperature point at which oil becomes molecularly damaged. Exceeding the burning point can convert healthy oils into trans-fatty acids. When baking with ingredients that contain fatty acids, such as flaxseed and other milled seeds, it is important that the temperature not exceed 350 degrees Fahrenheit. I rarely bake anything at temperatures above 300 degrees Fahrenheit, to ensure the fatty acids retain their nutritional value. For stir-frying, when the temperature is likely to exceed 350 degrees Fahrenheit, I use only coconut oil.

COCONUT OIL

Raw

Coconut oil is produced by pressing the meat of the coconut to remove the fiber. This is the only fat I use for frying. Sometimes called coconut butter since it's solid at a temperature below about 80 degrees Fahrenheit, coconut oil can be heated to a high temperature without converting to a trans fat. Surprisingly, coconut oil does not have a strong coconut taste, and it has almost no smell. When used in cooking, any remaining hint of the coconut taste leaves, making it a versatile oil.

Coconut oil is rich in medium-chain triglycerides, or MCTs. MCTs are unique in that they are a form of saturated fat yet have several health benefits. The body utilizes them differently from fat that does not contain MCTs. Their digestion is near-effortless and, unlike fat that does not contain MCTs (which gets stored in the cells), MCTs are utilized in the liver. Within moments of MCTs being consumed, they are converted by the liver to energy. It's for these properties that I include coconut oil in my energy gel recipes, beginning on page 275. They provide easily digestible energy, ideal during activity such as cycling or hiking.

EXTRA-VIRGIN OLIVE OIL

Raw

"Extra-virgin" means that the oil is from the first pressing of the olive. The subsequent pressing is referred to as virgin, the one following that produces regular olive oil. With a light taste and color, extra-virgin olive oil is a healthy addition to sauces, dips, and dressings. Although extra-virgin olive oil is a healthy oil, it delivers only minimal amounts of omega-3.

FLAXSEED OIL

Essential Fats   Raw

As you would expect, flaxseed oil is obtained by pressing flaxseed. Milder in taste than hemp and pumpkin seed oils, flaxseed oil contains the highest amount of omega-3 in comparison to omega-6, at a 5:1 ratio.

HEMP OIL

Essential Fats   Raw

Obtained by pressing hemp seed, hemp oil is one of the healthiest oils available. Dark green with a smooth creamy texture and mild, nutty flavor, hemp oil is an excellent base for salad dressings. Hemp oil is unique in that it has the ideal ratio of omega-6 and omega-3 fatty acids.

PUMPKIN SEED OIL

Essential Fats   Raw

Pumpkin seed oil is a deep green color with a hint of dark red. With a distinct, robust flavor, pumpkin seed oil is packed with essential fatty acids and has been linked to improved prostate health.

Nuts

ALMONDS

Antioxidants   Phytonutrients   Raw

The almond is one of the most popular nuts in North America. Almonds are resistant to mold without being roasted, making them a perfect nut to soak and eat raw. Particularly high in vitamin B2, fiber, and antioxidants, almonds have one of the highest nutrient levels of all nuts. That, combined with their high level of digestibility, especially when soaked, makes them a worthy addition to your diet. Although almonds don't need to be soaked, soaking makes them more nutritious—in this pre-sprouting state, their vitamin levels increase and the enzyme inhibitors are removed, making them even more efficiently digested.

MACADAMIA NUTS

Phytonutrients   Raw

Macadamia nuts contain omega-7 and omega-9 fatty acids. While these are nonessential fatty acids, meaning the body produces them, their inclusion in the diet has been linked to positive health benefits. Blending soaked macadamia nuts results in a creamy spread that makes for a healthy alternative to butter or margarine. Although soaked macadamia nuts are recommended for any of my recipes calling for macadamia nuts, they don't need to be soaked if you're short of time or unprepared.

WALNUTS

Electrolytes Phytonutrients   Raw

Walnuts are rich in B vitamins and possess a unique amino acid profile. Also rich in potassium and magnesium, walnuts can help maintain adequate electrolyte levels in the body, prolonging hydration. As with almonds and macadamia nuts, soaking improves their nutrition and digestibility. Walnuts complement many meals and snacks.

OTHER NUTS

The nuts listed below all offer high levels of nutrition in a compact form. These nuts can be substituted in recipes for the more common nuts, such as almonds and macadamia. Because of their diversity, incorporating them into your diet will ensure a greater variety of taste and nutrition. However, these nuts may not be readily available in grocery stores.

- Brazil nuts
- Cashews
- Filberts
- Hazelnuts
- Pecans
- Pine nuts
- Pistachios

Hazelnut trees grow wild in Europe and Asia. A staple in early humans' diet, hazelnuts have been eaten for thousands of years. Filberts are a variety of hazelnut that are cultivated and are often produced larger than wild hazelnuts to increase crop yield. Wild hazelnuts and filberts are nutritionally similar; both are excellent sources of the minerals manganese, selenium, and zinc.

Grains

As for true grains—as opposed to pseudograins—brown (or whole-grain) rice is my first choice. It's gluten free and offers considerably more micronutrients than other grains.

BROWN RICE

Phytonutrients

A staple of many countries, rice is one of the most consumed foods in the world by volume. Since brown rice has been unaltered over the years, the possibility of it causing an allergic reaction is low. Brown rice has a mild, nutty flavor.

The processing of brown rice is far less extensive than that of white rice, making it nutritionally superior to its white counterpart. Since only its outermost layer, the hull, is removed, brown rice retains its nutritional value. Brown rice is very high in manganese and contains large amounts of selenium and magnesium. It is a good source of B vitamins as well.

Purple sticky rice, or Thai black rice, is a nice alternative to standard brown rice. It can be substituted for brown rice at a 1:1 ratio.

Cook at rice-to-water ratio of 1:2. Put rice and water in a pot. Cover and bring to a boil. Once it's boiling, reduce heat to a simmer; simmer for 45 minutes. Remove from the stove and stir. Let cool.

Fruit

Pretty much all fruit is good. As mentioned in Chapter 1, fruit as a source of carbohydrate is considerably easier to digest than refined flour products and offers a significantly higher micronutrient density, hence rendering it a high-net-gain food. Therefore, making fruit the prime carbohydrate source instead of grain products will translate into greater usable energy.

Dates

Alkaline-Forming   Raw

Dates are nearly pure glucose, which, in its natural form, is a valuable type of sugar for people who are active. Glucose is rapidly converted to glycogen in the liver. Maintaining an adequate glycogen supply in both the muscles and the liver is imperative for sustained energy. For that reason, dates are best consumed shortly before, during, or immediately following exercise. Chlorophyll-rich foods also convert to glycogen, but not as quickly as glucose, therefore making the easily digestible, alkaline-forming date the ideal snack to fuel activity.

I use dates as the base ingredient for my whole food energy bar recipes starting on page 271.

Ginger

Phytonutrients   Raw

Fresh ginger is a worthy addition to any diet. Ginger can help the digestion process and ease an upset stomach. I use it in many recipes. Ginger has anti-inflammatory properties and so aids in the recovery of soft-tissue injuries and helps promote quicker healing of strains. I load up on ginger as my mileage increases to ensure inflammation is kept under control.

Green Tea

Alkaline-Forming   Antioxidants   Phytonutrients

While green, or incompletely fermented, tea leaves do contain a form of caffeine, it differs significantly from the form found in coffee beans. Theophylline causes a slow, steady release of energy over the course of several hours. Therefore, it does not cause caffeine jitters and places less stress on the adrenal glands. Green tea is also rich in chlorophyll and antioxidants.

However, since green tea is classified as a stimulant, it is something that I suggest drinking only before physical exercise. Green tea can help improve the level of intensity a person can reach during a workout or on race day. This leads to better, faster results. Theophylline has also been shown to help improve focus and concentration and to calm nerves. Before a big race, being able to relax and focus are valuable traits.

NEXT-LEVEL FOODS

These foods are sometimes harder to source; not all grocery stores will carry them, but most health food stores will. While these foods typically score among the top in terms of nutrient-to-resource ratio, they are not essential for healthy living. They will, however, provide a boost of nutrition in a concentrated form.

Kombucha

Alkaline-Forming   Antioxidants   Electrolytes   Phytonutrients   Raw

A popular health elixir in Asia, kombucha is a fermented tea, rich in organic acids, active enzymes, amino acids, and antioxidants. Its fermentation results in "good" bacteria content that helps improve the body's digestive strength, enabling it to metabolize and utilize nutrition more quickly. Additionally, kombucha acts as a natural muscle relaxant, helping muscles move with greater fluidity and ease, resulting in less energy expenditure and, ultimately, enhanced endurance. This is why I include a small amount of kombucha in my pre-workout drink. Also a liver detoxifier, kombucha helps speed cellular recovery. You'll find a basic kombucha recipe on page 133.

Sacha Inchi

Antioxidants Essential Fats Phytonutrients

With a green star-shaped fruit that yields a highly nutritious seed, the sacha inchi plant is native to the Peruvian Amazon. Also know as the "Incan peanut," the seed has several health properties, including easily digestible protein. High levels of the amino acid tryptophan are also present, making the seed a natural mood enhancer that lowers the risk of brain chemistry imbalances and depression. An exceptionally good source of omega-3 and omega-6 essential fatty acids, sacha inchi is among the most concentrated sources of omega-3 in the plant kingdom, coming in at 48 percent by volume. Also high in vitamin A and E, sacha inchi is rich in antioxidants.

Palm Nectar

Phytonutrients

Also known as coconut palm sugar, palm nectar is a low-glycemic form of natural sugar. Made from the sap of the date or coconut palm tree much as maple syrup is tapped from the maple tree, it is an ideal source of sustainable energy. It can be used as a low-glycemic sweetener, or as a premium, easily digestible fuel source that's ideal before a workout. Combining it with a high-glycemic sugar, such as that from dates or sprouted rice, I use palm nectar to balance the glycemic profile and slow the rate at which the sugar enters the blood stream.

Açaí Berries

Antioxidants   Electrolytes   Essential Fats   Phytonutrients   Raw

Açaí berries are the small, purple fruit of palm trees that grow in marshy areas in Central and South America, where it has been eaten by native people for centuries.

The berries are exceptionally rich in antioxidants and contain essential fatty acids and amino acids. Because they are also easy to digest, açaí are a high-net-gain food that can speed recovery after exercise.

In North America, açaí can be bought in most health food stores either frozen whole or freeze-dried in powdered form. The frozen berries are handy for making a smoothie. And you can mix the powder into recipes such as energy bars as an easy way to boost nutritional content.

Chlorella

Alkaline-Forming   Antioxidants   Essential Fats   Phytonutrients   Raw

Nutritionally speaking, chlorella—a fresh-water green algae—is a true superfood, comprising 67 percent protein; essential fatty acids; and a plethora of vitamins, minerals, and enzymes. Chlorella contains vitamin B12, which is difficult for vegetarians and vegans to find in forms other than laboratory-created tablets. Chlorella also possesses all 10 of the essential amino acids—the ones that must be obtained through diet for peak health. These amino acids, in conjunction with naturally occurring enzymes, are the most easily absorbed and utilized form of protein available. Many other complete proteins are much more energy-intensive to digest; by comparison, chlorella is a particularly high-net-gain food. Spirulina is also an excellent form of fresh water algae. While its protein content and B vitamin levels are lower than that of chlorella, it is still highly alkaline forming.

Coconut Water

Electrolytes   Raw

Coconut water is the nearly translucent fluid inside the coconut (not to be confused with coconut milk, which is a combination of coconut water blended with coconut meat). It has a light, sweet flavor. It is fat-free and contains high levels of simple carbohydrates, making it an ideal fluid to boost muscle glycogen without causing the stomach to become bogged down with digestive duties.

Packed with electrolytes, coconut water is the original sports drink. It has been used for decades to properly hydrate people who sweat profusely in tropical regions.

Maca

Phytonutrients

Maca, a root vegetable with medicinal qualities, is native to the high Andes of Bolivia and Peru. Known as an adaptogen, maca curtails the effects of stress by aiding the regeneration of the adrenal glands, making it an ideal food for the modern world. It helps lower cortisol levels, which will improve sleep quality. Of course, better-quality sleep directly translates into more waking energy. And maca increases energy by means of nourishment, not stimulation. I have found that I am better able to adapt to physical stress when I add maca to my diet.

Maca is a rich source of steroid-like compounds found in both plants and animals that promote quick regeneration of fatigued muscle tissue. During the off-season, I make a concerted effort to build strength and muscle mass in the gym. I've recently experienced exceptional strength gains by adding maca to my recovery drink. I can lift more weight than in previous years and I recover faster. It has enabled me to perform more high-quality workouts, thereby advancing my progress.

Published human clinical studies of maca used the gelatinized form of the vegetable. Gelatinization removes the hard-to-digest starchy component of the maca root. The result is an easily digestible, quickly assimilated, and more concentrated form of maca. Gelatinized maca has a pleasant, nutty taste and dissolves more easily than regular maca. When selecting maca, be sure to choose the gelatinized form for best results.

Chia

Antioxidants   Calcium   Electrolytes   Essential Fats   Iron   Phytonutrients   Raw

Chia seeds are small and round and look like white poppy seeds. Grown in the Amazon basin in Peru, chia laps up the nutrients in the rich, fertile soil and passes them on to the consumer. With a unique crunchy texture, chia is gaining in popularity in North America.

Particularly high in magnesium, potassium, calcium, and iron, chia can effectively replenish minerals used in muscle contractions and lost in sweat. Chia is truly one of the top foods for active people. And because it is high in both soluble and insoluble fiber, which helps to sustain energy and maintain fullness, chia is a true high-net-gain food. Packed with antioxidants and containing about 20 percent high-quality protein, chia is an ideal food to help speed recovery after exercise.

Aztec warriors were rumored to eat chia before going into battle to give them a nutritional boost and thereby improve their endurance. They were also said to have carried it with them when they ran long distances to be used as their body's primary fuel source. Since chia is nutritionally well rounded and complete, this may well have been the case. It is ideal to help maintain energy level during a workout, and the seeds are remarkably easy to digest.

Yerba Maté

Alkaline-Forming   Antioxidants   Electrolytes   Phytonutrients

Yerba maté is a species of holly native to subtropical South America. The leaves are rich in chlorophyll, antioxidants, and numerous trace minerals, and help aid digestion. However, since yerba maté does contain a form of caffeine, I suggest drinking it in a similar fashion to green tea, before exercise or when you really need extra short-term energy. Yerba maté is one of the healthiest forms of stimulation, yet any kind of stimulation will take its toll on the adrenal glands eventually.

However, after drinking yerba maté, it's important to make sure the adrenals are well nourished to help speed recovery. I include maca in my post-workout drink for this reason.

When sourcing yerba maté, it's of course best to avoid products from plantation-style farms that have cleared old-growth forest. Be sure to choose a brand that is wild harvest or has been grown with the jungle, not instead of the jungle. By making the harvesting of wild yerba maté economically viable for the producers, you will help prevent clearance of old-growth rain forest for the farming of animals. Before yerba maté rose to popularity outside of South America, it was common for large plots of land to be clear-cut for cattle-grazing land. While this is still a problem, in those areas with an abundance of yerba maté growing within the jungle, the yerba maté can be harvested without any alteration to the forest canopy; therefore, using the jungle to grow yerba maté enhances the "value" of the natural foliage so that in many cases there is more incentive to preserve it rather than to convert it to cattle pasture.

Stevia

Alkaline-Forming Phytonutrients

Stevia is a herb native to Paraguay. The intense sweetness of its leaf
is stevia's most celebrated feature. About 30 times as sweet as sugar,
dried stevia leaf contains no carbohydrates and so does not affect the body's
insulin levels. Stevia has been shown to help equalize blood sugar levels
raised by other sugars and starch consumed at the same time. Stevia, as you
might expect, is quickly gaining popularity as a natural sugar substitute
among those in pursuit of a leaner body. Improved digestion is another of
stevia's benefits. An excellent alternative to manufactured artificial sweeteners, stevia leaf is a whole food, just dried and ground into powder. When
choosing stevia, be sure you read the ingredients; it's commonly combined
with maltodextrin and other fillers. Selecting a stevia brand with stevia as
the sole ingredient is ideal. I add it to many of my foods. Its ability to help
regulate blood sugar levels is important for sustained energy. I even add
stevia to my sports drink to improve its effectiveness.

6

WHOLE FOODS RECIPES

Sunflower Seed Hemp Milk (and Chocolate Variation)

I usually make a week's supply of Sunflower Seed Hemp Milk at a time, which for me is about 8 cups. Hemp milk is a good substitute for cow's milk on cereal. For a tasty change, try using the chocolate version on your morning cereal. Hemp milk also adds a subtle creaminess and flavor to smoothies.

Time: 2 minutes • Makes about 3 cups

2 ½ cups water
½ cup hemp seeds
½ cup sunflower seeds

3 large pitted dates (fresh, or dried dates soaked overnight)
1 tbsp roasted carob powder or cocoa powder (for chocolate variation)

- In a blender, combine all ingredients and blend until smooth. To make into chocolate milk, you may add 1 tbsp of either roasted carob powder or cocoa powder before blending. Be aware that cocoa, unlike carob, does contain a bit of caffeine.
- Keep refrigerated, for up to 2 weeks.

Sacha Inchi Milk (and Chocolate Variation)

Soaking sacha inchi seeds doesn't enhance their nutritional value, but it does allow them to be blended into a smooth milk.

Time: 2 minutes • Makes about 3 cups

2 ½ cups water
1 ½ cups soaked sacha inchi seeds
3 large pitted dates (fresh, or dried dates soaked overnight)
2 tbsp roasted carob powder or cocoa powder (for chocolate variation)

- In a blender, combine all ingredients and blend until smooth. To make into chocolate milk, you may add 2 tbsp of either roasted carob powder or cocoa powder before blending. Be aware that cocoa, unlike carob, contains a bit of caffeine.

Ginger Pear Smoothie with Sunflower Seed Hemp Milk

The riper the pear, the sweeter the smoothie. If you'd like it even sweeter, add one or two fresh or soaked dried dates. Since ginger is a natural anti-inflammatory, this is an ideal choice for a post-workout snack.

Time: 2 minutes • Makes about 3 cups (2 servings)

1 banana
½ pear, cored
1 cup water
1 cup Sunflower Hemp Seed Milk (see p. 126)
1 tbsp ground flaxseed
1 tbsp hemp protein powder
1 tbsp peeled, grated ginger

• In a blender, combine all ingredients and blend until smooth.

....

Chocolate Almond Smoothie with Sacha Inchi Milk

Rich in protein and omega-3, this smoothie will keep you going for hours with sustainable, non-stimulating energy.

Time: 5 minutes • Makes about 3 ½ cups (2 large servings)

1 banana
2 fresh or presoaked dried dates
1 cup water
1 cup Sacha Inchi Milk (or chocolate variation) (see p. 126)
¼ cup almonds (or 2 tbsp raw almond butter)
1 tbsp ground flaxseed
1 tbsp hemp protein powder
1 tbsp roasted carob powder

• In a blender, combine all ingredients and blend until smooth.

Brendan's Original Lemon-Lime Sports Drink

Dates are high in glucose, which will enter the blood stream almost instantly. The sugar in the coconut water will enter the blood stream more slowly, spreading out the energy over a longer period.

Fresh dates are ideal. You can also soak dried dates for four hours beforehand to rehydrate.

Time: 5 minutes active • Makes 2 cups (1 large serving)

1 cup coconut water
1 cup water
2 dates (pitted fresh or presoaked dried)

Juice from ½ lemon
Juice from ¼ lime
Sea salt to taste

- In a blender, combine all ingredients and blend. If you prefer your drink smooth, strain out any pulp left from the lemon and lime.

....

The Brazier

THRIVE JUICE BAR, WATERLOO, ONTARIO

Jonnie Karan, co-founder of Thrive Juice Bar, named this one after me since he figured I'd like it. He was right. It's certainly one of my favorites.

Time: Under 5 minutes prep • Makes 2 cups

1 tsp raw cocoa nibs
1 tbsp organic cocoa powder
6–7 raw cashews
2 tsp blue agave nectar
¼ tsp vanilla extract
1 scoop or 1 ¾ ounces organic dairy-free
vanilla gelato

1 tbsp almond butter
⅓ ripe banana
⅓ cup rice milk
½ cup coconut water
2 or 3 ice cubes

- In a blender, combine all ingredients, including ice, and blend until smooth.
- It's best served immediately but can be kept up to 2 days in the fridge.

Thai Avocado Smoothie

THRIVE JUICE BAR, WATERLOO, ONTARIO

A unique-flavored and filling smoothie.

Time: Under 5 minutes • Makes 2 cups

½ ripe avocado
3 tsp dark raw agave
1 tsp vanilla extract
½ tsp chopped lemongrass

½ cup coconut water
⅓ cup rice milk
2 or 3 ice cubes

- In a blender, combine all ingredients, including ice, and blend until smooth.
- It's best served immediately but can be kept up to 2 days in the fridge.

Coconut Thai Lime Leaf Smoothie

THRIVE JUICE BAR, WATERLOO, ONTARIO

Lime leaf adds extra chlorophyll and a whole bunch of extra flavor in this exotic-tasting smoothie

Time: Under 5 minutes • Makes 2 cups

2 small scoops or ½ cup non-dairy
   coconut gelato
2 tsp blue agave
½ tsp chopped lemongrass
3 medium Thai lime leaves (available,
   usually frozen, in Asian markets)

Zest of half a lime
⅓ cup rice milk
½ cup coconut water
2 or 3 ice cubes

- In a blender, combine all ingredients, including ice, and blend until smooth.
- It's best served immediately but can be kept up to 2 days in the fridge.

Coconut, Mango, Yellow Curry Smoothie

THRIVE JUICE BAR, WATERLOO, ONTARIO

Straight-up delicious with anti-inflammation properties as a bonus from the curry.

Time: Under 5 minutes • Makes 2 cups

½ cup frozen mango
1 fresh orange, juiced
½ tsp lime, juiced
⅛ tsp yellow curry powder
2 tsp blue agave

¼ cup rice milk
2 ounces organic pure mango juice
⅓ cup coconut water
2 or 3 ice cubes

- In a blender, combine all ingredients, including ice, and blend until smooth.
- It's best served immediately but can be kept up to 2 days in the fridge.

La Belle Verte (The Beautiful Green One) Smoothie

CRUDESSENCE, MONTREAL, QUEBEC

Sweet, chlorophyll-rich goodness.

Time: Under 5 minutes • Makes 2 cups

¾ banana, frozen (or fresh banana
   + 2–3 ice cubes)
4 chunks pineapple
2 large leaves fresh kale
¼ cup parsley, chopped (firmly packed)

1 or 2 dates
2 tbsp shelled hemp seeds
1 pinch sea salt
1 ½ cups water

- Place all ingredients in a blender and add water to 16-ounce level. Blend until texture is like a smoothie, without lumps.
- It's best served immediately but can be kept up to 2 days in the fridge.

Kombucha Mojito

Delicious, nutritious, and refreshing. Can be served solo, or with a meal to aid in digestion.

Time: 5 minutes • Makes 2 servings

½ cup chopped mint
3 tbsp fresh squeezed lime juice
2 tbsp orange juice
3 tbsp palm sugar

½ tsp lime zest
Ice
2 cups kombucha
Mint leaves, for garnish

- Muddle the mint, lime juice, orange juice, palm sugar, and lime zest together to release the flavor of the mint leaves.
- Strain and pour into 2 glasses filled with ice. Top each glass with kombucha, stir, and top with a fresh mint leaf for garnish.

Green Mango Dessert Smoothie

Is it a smoothie or is it a dessert? The Vega Sport Performance Protein Powder with its 20 grams of protein, high-quality ingredients, and smooth texture pretty much rocks in this recipe, but another vanilla hemp or rice protein will work too. You can increase the protein powder to two servings and add a little extra water for a less dessert-like, more traditional smoothie.

Time: 5 minutes • Makes 2 servings

2 heaping cups frozen mango chunks
½ cup hemp milk
½ cup water
1 scoop vanilla-flavored Vega Sport
   Performance Protein Powder

Touch of white stevia, to taste
   (optional)
2 tbsp shredded coconut (optional)

- In a blender, combine mangoes, hemp milk, water, and protein powder and blend until completely smooth.
- If desired, boost sweetness with a touch of stevia, to taste, and blend again. Serve in a bowl and top with shredded coconut.

Pomegranate Smoothie

This is a simple, refreshing smoothie.

Time: 5 minutes • Makes about 3 ½ cups (2 large servings)

1 banana
1 date
2 cups cold water (or 1 ½ cups water plus 1 cup ice)
1 cup pomegranate seeds (the amount from 1 pomegranate)
1 tbsp ground flaxseed
1 tbsp hemp protein
1 tbsp hemp oil
½ tsp cayenne pepper

- In a blender, combine all ingredients, and blend until smooth.
- Can be kept up to 3 days in the fridge.

Tropical Pineapple Mango Smoothie

A good smoothie when you are on the go or feeling fatigued. A great tasting energy booster.

Time: 5 minutes • Makes about 3 ½ cups (2 large servings)

1 banana
2 fresh or soaked dried dates
2 cups cold water (or 1 ½ cups water plus 1 cup ice)
½ medium papaya
½ cup pineapple
1 tbsp ground flaxseed
1 tbsp hemp protein
tbsp coconut oil

- In a blender, combine all ingredients, and blend until smooth.
- Can be kept up to 3 days in the fridge.

Chocolate Goodness Smoothie

A full-flavor, nutrient-dense, chlorophyll-rich super smoothie that's filling enough to take the place of a meal.

Time: 5 minutes • Makes 1–2 servings

1 banana
1 cup frozen blueberries
1 cup unsweetened hemp milk
1 scoop Chocolate Vega Complete Whole Food Health Optimizer
1 tsp wheatgrass powder
2 tbsp raw cocoa powder
1 tsp mesquite powder
Stevia powder, to taste (substitute a touch of palm sugar
    or maple syrup if desired)
2–3 cups water
Handful of ice

• In a blender, combine all ingredients, including ice, and blend until smooth.

Toasted Chia Ginger Pear Cereal

With significant amounts of omega-3 and ginger, this cereal is considerably less acid-forming than standard ones. To make it even more nutritious, top with an energy bar cut into small pieces (recipes starting p. 266).

Time: 10 minutes prep; 1 hour to bake • Makes 4 cups (about 5 servings)

½ pear, diced
1 cup oats (or cooked or sprouted quinoa, to make cereal gluten free)
½ cup diced almonds
½ cup chia seeds
½ cup hemp protein
½ cup unhulled sesame seeds
½ cup sunflower seeds
¼ tsp ground stevia leaf
¼ tsp sea salt
¼ cup hemp oil
¼ cup molasses
2 tbsp apple juice
1 tbsp grated ginger root
Coconut oil

- Preheat oven to 250°F.
- In a large bowl, combine pear, oats, almonds, chia seeds, hemp protein, sesame seeds, sunflower seeds, stevia, and sea salt. In a small bowl, blend together hemp oil, molasses, apple juice, and ginger root. Add wet ingredients to dry ingredients, mixing well.
- Spread on a bake tray lightly oiled with coconut oil. Bake for 1 hour. Let cool, then break into pieces.
- Keeps refrigerated for up to 2 weeks.
- Eat with Sunflower Seed Hemp Milk (see p. 126) or Sacha Inchi Milk (see p. 126).

Sacha Inchi Baked Apple Cinnamon Cereal

With high-quality protein and omega-3 and omega-6 essential fatty acids, this cereal is filling. A great way to make it even more nutritious is to top it with an energy bar cut into small piece (recipes start on p. 266).

Time: 10 minutes prep; 1 hour to bake • Makes 4 cups (about 5 servings)

½ apple, diced
1 cup oats (or cooked or sprouted quinoa, to make cereal gluten free)
½ cup diced sacha inchi seeds
½ cup ground flaxseed
½ cup hemp protein
½ cup unhulled sesame seeds
½ cup sunflower seeds
1 ½ tsp cinnamon
¼ tsp nutmeg
¼ tsp ground stevia leaf
¼ tsp sea salt
¼ cup hemp oil
¼ cup molasses
2 tbsp apple juice
Coconut oil

- Preheat oven to 250°F.
- In a large bowl, combine apple, oats, sacha inchi, flaxseed, hemp protein, sesame seeds, sunflower seeds, cinnamon, nutmeg, stevia, and sea salt. In a small bowl, blend together hemp oil, molasses, and apple juice. Add wet ingredients to dry ingredients, mixing well.
- Spread on a baking tray lightly oiled with coconut oil. Bake for 1 hour.
- Let cool, then break into pieces.
- Keeps refrigerated for up to 2 weeks.
- Eat with Sunflower Seed Hemp Milk (see p. 126) or Sacha Inchi Milk (see p. 126).

Breakfast Blueberry Chia Pudding

Since chia rapidly absorbs fluid and takes on gelatinous properties when soaked, it makes an ideal nutrient-dense pudding base.

Time: 5 minutes active; 20 minutes total • Makes about 1 ½ cups (1 serving)

2 tbsp chia

¾ cup water

⅓ cup cashews

2–3 fresh pitted dates, or dried
  pitted dates soaked in water
  overnight to rehydrate

Pinch of cinnamon

Pinch of sea salt

Fresh or frozen blueberries

• Soak chia in water for 15 minutes. In a blender, combine with the rest of the ingredients, except the blueberries, and blend until smooth. Transfer to serving bowl and top with blueberries.

Chocolate Raspberry Chia Pudding

A delicious and filling breakfast.

Time: 5 minutes active; about 35 minutes total
Makes about 2 cups (1 large serving)

4 tbsp chia seeds

½ cup water

1 cup Sunflower Seed Hemp Milk (see p. 126)

1 tbsp cocoa

1 tbsp maple syrup

Raspberries, fresh or frozen

• Soak chia in water for 15 minutes. In small bowl, mix chia, water, and Sunflower Seed Hemp Milk together. Stir for 1–2 minutes or till consistent texture is reached. Add cocoa and maple syrup. Stir and then let sit for 10–15 minutes. As an option, it can be heated on the stovetop for about 2 minutes on low. Stir again, top with raspberries, and serve.

New Potato Pancakes

These whole food carbohydrate-rich pancakes will supply hours of sustainable energy, ideal a few hours before a long hike or bike ride. Use as many kinds of potatoes as possible, such as blue/purple potatoes, fingerlings, round white potatoes, and red potatoes.

Time: 15 minutes • Makes 8 medium-sized pancakes

1 pound mixed, unpeeled, new potatoes
3 tbsp ground flaxseed
2 tbsp brown rice flour
¼ tsp sea salt
1 carrot, shredded
About 1 tbsp coconut oil

- Using a hand grater or food processor, shred the potatoes (if using a food processor, pulse a couple of times using the S-blade after shredding, to make sure the potato shreds are not too long).
- In a small bowl, blend the flaxseed, flour, and salt.
- In a large bowl, toss the shredded potatoes and carrots together, then add in the dry mixture and combine. Use your hands to form 8 palm-size flat patties, about ½-inch thick. Set the patties aside on paper towels to absorb any excess moisture.
- In a large frying pan, heat about a tablespoon of coconut oil over medium high heat. When the oil is hot, add 4 patties. After a few minutes, when the patties are golden brown on the underside, flip them over and cook until the second side is crispy. When patties are cooked, transfer them back to paper towels to remove any excess oil until you are ready to serve them.
- Add a bit of new oil to the pan, and repeat with the remaining patties.

Buckwheat Banana Pancakes

Lightly flavored with cinnamon and nutmeg, these pancakes taste just like traditional pancakes.

Time: 15 minutes • Makes 2 large servings

1 cup buckwheat flour
¼ cup ground flaxseed
¼ cup hemp flour
2 tsp baking powder
1 tsp cinnamon
½ tsp nutmeg
1 banana
2 cups water
½ cup barley flakes (or buckwheat, sprouted or cooked)

- In a bowl, mix buckwheat flour, flaxseed, hemp flour, baking powder, cinnamon, and nutmeg.
- In a food processor, process the banana and water while slowing adding the dry ingredients until mixture is smooth. Stir in the barley flakes with a spoon or spatula.
- Lightly oil a pan with coconut oil and heat over medium heat. Pour in pancake batter to desired pancake size and cook for about 5 minutes or until bubbles begin to appear. Flip and allow to cook for another 5 minutes.

Blueberry Pancakes

Packed with taste and nutrition, these pancakes are a breakfast favorite.

Time: 15 minutes • Makes 2 large servings

2 fresh or soaked dried dates
1 cup blueberries
1 cup hemp milk
¾ cup water
½ cup buckwheat flour
½ cup sprouted or cooked quinoa
1 tsp baking powder
1 tsp baking soda
Sea salt to taste

- In a food processor, process all ingredients until smooth
- Lightly oil a pan with coconut oil and heat over medium heat. Pour in pancake batter to desired pancake size and cook for about 5 minutes or until bubbles begin to appear. Flip and allow to cook for another 5 minutes.

HEMP MILK

3 ½ cups water
1 cup hemp seeds
2 tbsp agave nectar

- In a blender, combine all ingredients. Keep refrigerated for up to 2 weeks. Makes about 4 cups.

Wild Rice Yam Pancakes

This is a heartier mixture than traditional pancakes, one that will give you a sense of fullness for several hours.

Time: 15 minutes • Makes 2 large servings

2 cups water
1 cup cooked or sprouted quinoa
1 cup mashed cooked yam
½ cup sprouted or cooked wild rice
¼ cup ground flaxseed
¼ cup ground sesame seeds
2 tsp baking powder
½ tsp black pepper

- In a food processor, process all ingredients until smooth.
- Lightly oil a pan with coconut oil and heat over medium heat. Pour in pancake batter to desired pancake size and cook for about 5 minutes or until bubbles begin to appear. Flip and allow to cook for another 5 minutes.

In general, since salads tend to be lower in carbohydrate, yet higher in minerals, they are usually best eaten later in the day, when the body doesn't require as much fuel (carbohydrate) but does need nutrition to rebuild and repair from the day's activities. For this reason I have one salad a day, always as dinner, or as part of dinner.

Thai Salad

MATTHEW KENNEY

A raw twist on a classic.

Time: 15 minutes   •   Makes 2 servings

2 handfuls mixed greens
½ cup finely diced pineapple
½ cup soaked, finely sliced sun-dried tomatoes
1 avocado, sliced
Sea salt
Freshly ground black pepper
½ red bell pepper, cut into long, thin strips
½ cup thinly sliced young coconut meat
½ cup chopped cashews
½ cup Creamy Thai Dressing (see p. 209)
Cilantro leaves, for garnish

- Place a handful of mixed greens in the center of each plate. Top with pineapple, sun-dried tomatoes, and avocado.
- Season with salt and pepper to taste. Top with red bell pepper, coconut, and cashews.
- Drizzle Creamy Thai Dressing generously over top just before serving. Garnish with cilantro leaves.

Beet Salad with Lemon Herb Cream Cheese

RAVENS' RESTAURANT, MENDOCINO, CALIFORNIA

Rich, earthy, full-flavored goodness. The chefs at Ravens' use a variety of beets from their garden, including Chioggia (also called candy-stripe) or yellow beets.

Time: 15 minutes active; 30 minutes to cook • Makes 4 servings

4 small garden beets
1 bunch of frisée or favorite greens (mesclun, arugula)

- Preheat oven to 450°F. Wash beets, then wrap in aluminum foil and roast in oven for 30 minutes. Remove from oven and allow to cool.
- When cool to the touch, slip off skins and slice.

DRESSING

3 tbsp Dijon mustard
¼ cup finely chopped shallots
½ cup white balsamic vinegar

2 cups olive oil
½ cup agave nectar
Salt to taste

- Mix all ingredients together.

LEMON HERB CREAM CHEESE

½ cup raw walnuts
Juice from ½ lemon (reserve second
    half, zest removed)

Lemon zest (add to taste)
1 clove garlic
Pinch of salt

- Place all ingredients in a food processor. Process until mixture resembles a coarse cream cheese. Add additional lemon juice if necessary and lemon zest for flavor.

To serve
- Arrange about one-quarter of the beets around the rim of a plate. Place one-quarter of the frisée in the center of the plate, then drizzle dressing over frisée and beets.
- Take two tablespoons: with one, scoop out some Lemon Herb Cream Cheese and with the other, form it into an ovoid shape and place on frisée.

Roasted Beet and Fennel Salad with Belgium Endive

MILLENNIUM RESTAURANT, SAN FRANCISCO, CALIFORNIA

A great winter salad. The rich dressing pairs well with the sweetness of the beets and bitterness of the endive.

Time: 15 minutes • Makes 2 servings

2 cups small diced beets
2 cups small diced fennel bulb
2–3 tsp olive oil
2 tsp balsamic vinegar
Juice of ½ Meyer lemon (optional)
Salt to taste
Black pepper to taste

- Blanch the beets and fennel for 3–4 minutes. Drain and toss with a small amount of olive oil and the balsamic vinegar.
- Roast on a non-stick baking mat or a parchment-lined sheet pan until al dente and glazed.
- Toss with Meyer lemon juice, if using. Adjust salt and pepper to taste.

To serve
6 Belgian endive spears per salad
Roasted beets and fennel mixture
Garlic-Green Peppercorn Dressing (see p. 206)
Grapefruit or mandarin orange segments, for garnish
Fresh tarragon and flat leaf parsley, for garnish

- For each salad, arrange endive around the perimeter of the plate. Fill each spear with the beet and fennel mixture.
- Drizzle the plate with 1–1 ½ ounces Garlic-Green Peppercorn Dressing.
- Garnish with grapefruit or mandarin orange segments, and a sprinkling of fresh tarragon and parsley leaves.

Roasted Vegetable Salad with Roasted Garlic Dressing

CANDLE 79, MANHATTAN, NEW YORK

Abigael Birrell, a Candle 79 chef, invented this luscious salad, full of roasted seasonal vegetables. She likes to serve it as a starter to a festive holiday dinner. It's also a good main course salad.

Time: 15 minutes active; 35–40 minutes to cook • Makes 4 servings

1 fennel bulb, trimmed and cut into bite-sized pieces
2 cups fingerling or new potatoes, cut into bite-sized pieces
1 cup baby turnips, peeled and cut into bite-sized pieces
2 medium-sized beets, peeled and cut into bite-sized pieces
2 medium apples, cored and sliced
1 tsp sea salt
Freshly ground black pepper
3 tbsp extra-virgin olive oil
Roasted Garlic Dressing (see p. 205)
2 bunches arugula, rinsed, trimmed, and stemmed
Toasted walnuts or pecans (optional)

- Preheat oven to 400°F.
- Toss the vegetables and apples with the salt, pepper, and olive oil in a large mixing bowl.
- Spread in a single layer on a baking sheet and bake until just tender, 35–40 minutes.
- Toss the vegetables with about 2 tbsp of Roasted Garlic Dressing to lightly coat the vegetables, and set aside.
- To serve the salad, arrange the arugula on 4 plates, then top with equal amounts of the warm vegetable mixture. Sprinkle with toasted walnuts or pecans, if desired, drizzle with a bit more dressing, and serve at once.

Shaved Zucchini and Sacha Inchi Salad

Refreshing yet filling.

Time: 10 minutes • Makes 4 servings

DRESSING

⅓ cup hemp oil (or Vega Antioxidant EFA oil)
2 tbsp fresh lemon juice
1 tsp coarse sea salt
½ tsp ground black pepper
¼ tsp dried crushed red pepper

- Mix oil, lemon juice, salt, black pepper, and crushed red pepper in a bowl. Set aside.

SALAD

2 pounds medium zucchini, trimmed
½ cup coarsely chopped fresh basil
¼ cup chopped sacha inchi
Salt and pepper to taste

- Using vegetable peeler, slice zucchini into ribbons, working from top to bottom of each zucchini. Put ribbons in large bowl.
- Add basil and chopped sacha inchi, then the dressing; toss to coat. Add salt and pepper, as much as desired.

Spicy Lentil Salad

Raw

BEETS CAFÉ, AUSTIN, TEXAS

Filling and protein-rich.

Time: 10 minutes • Makes 2 servings

2 tbsp minced onions
1 small clove of garlic, minced
¼ cup cilantro, finely chopped
2 medium tomatoes, finely chopped
½ tsp jalapeño, finely minced (or more for desired spice)
2 tsp lemon juice
1 cup sprouted lentils (½ cup before sprouting)
2 tsp apple cider vinegar
2 tsp olive oil
¼ tsp sea salt or to taste

- Place onions, garlic, cilantro, tomatoes, jalapeño, and lemon juice into a medium bowl and mix well.
- Add lentils, apple cider vinegar, and olive oil and mix. Add salt to taste.
- Store in the refrigerator.

Cumin-Style Cabbage Salad with Tart Green Apple

Raw

Unique and flavor-packed.

Time: 1 hour presoak; 15 minutes active • Makes 4–6 servings

½ medium green cabbage, sliced
   into thin strips
½ cup cashews, soaked in water
   for 1 hour
½ cup reserved cashew soak water
1 tbsp lime juice
½ tbsp balsamic vinegar

½ tsp ground cumin
1 clove garlic, peeled and
   mashed in a garlic press
1 tart green apple (such as
   Granny Smith variety),
   cut into matchsticks

- Place the cabbage in a large bowl and set aside.
- In a blender, combine the cashews, ½ cup soak water, lime juice, vinegar, cumin, and garlic and blend.
- Once the mixture is a smooth sauce, pour over the cabbage. Use clean hands to massage the sauce into the cabbage for a minute to help soften the cabbage slightly, into a slaw. Toss in the apple and serve.

....

Good Roots Salad with Coconut-Cumin Dressing

Surprisingly filling and energizing, this salad is packed with easily digestible carbohydrate. Be sure to peel all the vegetables before dicing and shredding.

Time: 10 minutes • Makes 4 servings

2 cups shredded beets (about
   2 medium)
2 cups shredded carrots (about
   4 medium)
4 cups diced jicama

2 tbsp pumpkin seeds
2 cups onion sprouts (about
   4 ounces in weight)
½ recipe Coconut-Cumin Dressing
   (see p. 209)

- Combine all ingredients and toss.

Raw

Dilled Spinach Salad with Avocado

Fresh tasting, filling, high in iron, and flavorful. If it's available, use fresh dill instead of dried. Add to taste.

Time: 10 minutes • Makes 2–4 servings

1 package (5 ounces) baby spinach leaves
1 beet, peeled and cut into large matchsticks
1 avocado, peeled and chopped
⅓ cup red onion, diced
2 tbsp hemp oil
1 tbsp balsamic vinegar
2 cloves garlic, pressed
1 tbsp dried dill
Salt and pepper to taste

- In a large bowl, combine the spinach, beet, avocado, and red onion.
- In a small bowl, whisk together the hemp oil, vinegar, pressed garlic, and dill. Add salt and pepper to taste.
- Just before serving, add the dressing to the salad and toss.

New Caesar Salad

With the distinctive flavor of a traditional Caesar salad, but with no cholesterol.

Time: 15 minutes • Makes 4 servings (1 cup of dressing)

DRESSING

¼ cup cashews
¼ cup water
¼ cup olive oil
2 ½ tbsp red wine vinegar
3 tbsp lemon juice
3 large cloves garlic
1 ½ tsp miso paste
3 tbsp wakame flakes
1 tsp Dijon mustard

• Blend all ingredients together.

SALAD

3–4 whole romaine lettuce hearts, torn into bite-sized pieces
Red Onion Flatbread (see p. 180) or flax crackers
½ ounce dulse strips, cut into small pieces
Freshly cracked black pepper, to taste

• Toss the lettuce with several spoonfuls of dressing in a large bowl (use as much dressing as desired). Place dressed greens on a plate along with a couple of pieces of Red Onion Flatbread or flax crackers and sprinkle with a few dulse strips. Generously adorn with some freshly cracked pepper.

Summertime Chef Salad

Raw

Fresh and light, this is a simple summertime staple.

Time: 10 minutes • Makes 4–6 servings

4–6 large handfuls mixed baby greens or chopped romaine lettuce

2 cups fresh white corn kernels (about 2 ears)

1 ½ cups grape tomatoes, halved

2 cups cucumber, thinly sliced

½ cup red onion, thinly sliced

4 cups jicama, cut into large matchsticks

1–2 avocados, cut into chunks

Dressing of choice

- For each serving, create a bed of greens, then place each of the other vegetables in a small mound around the plate for a decorative presentation. Drizzle with dressing of choice.

Summer Chopped Salad

TAL RONNEN

This is super-easy—a foolproof recipe—but you should make it right before you serve it. Chopped salads can get soggy if they sit around. Kids go crazy for this because of all the great flavors and textures.

Time: 20 minutes • Makes 4 servings

¼ pound green beans, cut into 1-inch pieces
5 radishes, finely diced
Dash of agave nectar
¼ English cucumber, finely diced
12 red and yellow cherry tomatoes, quartered
Kernels from 2 ears raw sweet corn
1 avocado, diced
1 cup baby arugula
1 shallot, minced
1 tsp minced fresh basil
1 tsp minced fresh oregano
Vinaigrette (see p. 208)
1 tsp freshly squeezed lemon juice

- Blanch the green beans in boiling water for 30 seconds, then chill in an ice bath. In the same boiling water, blanch the radishes for 20 seconds, then chill in an ice bath sweetened with a dash of agave nectar.
- Place all of the ingredients except for the Vinaigrette and lemon juice in a large bowl.
- Drizzle with the Vinaigrette and toss to coat. Sprinkle the lemon juice on top just before serving.

Watercress Salad with Roasted Beets

Unique and original flavor sets this nutrient-packed salad apart.

Time: 10 minutes; 1 hour to cool • Makes 4 starter salads

GINGERED BEETS

3 beets, washed and peeled

1 cup apple cider vinegar

2 tsp whole black peppercorns

2-inch piece of peeled ginger, cut into thin slices

¼ cup palm sugar

SALAD

2 bunches watercress, washed and de-stemmed

1–2 tbsp sesame seeds (black or white)

¼ cup Sweet Mustard Dressing (see p. 207)

Sesame seeds, for sprinkling

- Slice the beets into thin rounds and place in a large canning or heatproof jar.
- Pour the vinegar, peppercorns, ginger, and palm sugar into a small saucepan, and bring to a boil.
- Pour the hot mixture over the beets, seal jar, and place in the refrigerator. Beets may be enjoyed as soon as they have cooled (about 1 hour) or will keep in a closed, refrigerated container for up to 1 month.
- To prepare the salad, toss the watercress with the Sweet Mustard Dressing and divide onto serving plates.
- Place the beets on top of the greens, and sprinkle with sesame seeds.

Mexican Salad Bowl

Traditional Mexican salad.

Time: 10 minutes • Makes 4 servings

2 large handfuls baby salad greens
 or chopped romaine lettuce
1 cup cooked black beans
1 cup jicama, cubed
2 cups shredded carrots (about
 3 carrots)

1 cup sweet corn kernels
 (about one ear)
¼ cup chopped green onion

- In a large bowl, toss the vegetables with the dressing of your choice and serve.

Wilted Chard Salad with Lima Beans

Can be served as a meal.

Time: 10 minutes • Makes 2–4 servings

1 tbsp coconut oil
1 leek, white and light green parts,
 sliced thin
2 cloves garlic, minced
¼ tsp sea salt, or to taste
1 large bunch Swiss chard, stems
 removed, sliced into ½-inch strips

2 tsp balsamic vinegar
1 cup fresh or frozen lima beans,
 blanched for 2 minutes
¼ tsp red pepper flakes

- Heat the oil in a large skillet over medium heat until melted. Add the leek and the garlic and sauté for 2 minutes, stirring often, until the leeks have softened.
- Add the salt, Swiss chard, balsamic vinegar, lima beans, and red pepper flakes, and toss to combine.
- Cover the pan, reduce heat to medium-low, and cook for 4–5 minutes, until the chard has wilted. Remove from heat and serve immediately.

South of the Border Coleslaw

A fresh take on coleslaw.

Time: 15 minutes active; cashew soak time • Makes 4–6 servings

½ cup cashews, soaked in water for 1–2 hours
½ cup cashew soak water
2 tsp apple cider vinegar
2 large Medjool dates, pits removed
½ packed cup fresh cilantro, plus more for garnish
½ tsp sea salt
8 cups finely shredded green cabbage (shredded as thinly as possible)
1 cup shredded carrots

- When cashews are soaked and soft, place in a blender with water, vinegar, dates, cilantro, and salt. Blend until smooth and creamy (this may take a couple of minutes).
- In a bowl, mix together the cabbage and carrots. Pour the blended mixture on top and toss until well coated.
- Garnish with additional cilantro, if desired.

Dandelion Salad with Sun-Dried Tomatoes and Lentil Dressing

Flavorful sun-dried tomatoes combined with the bitterness of the dandelion greens give this salad special qualities. Lovers of bitter greens can substitute the mixed greens for additional dandelion.

Time: 5 minutes active; tomato presoak time • Makes 2–4 servings

2 heaping handfuls dandelion greens (about ½ bunch), trimmed and
   cut in half lengthwise
2 heaping handfuls baby mixed greens
⅔ cup sun-dried tomatoes, soaked 30 minutes in warm water until soft,
   sliced into ¼-inch strips
1 tbsp minced shallot
½ recipe Lentil Dressing (see p. 208)
⅓ cup cooked green lentils
Freshly cracked pepper

- In a large bowl, combine the greens, sun-dried tomatoes, and shallot. Pour the lentil dressing on top and toss thoroughly.
- To serve, top with ⅓ cup lentils and freshly cracked pepper.

Hemp Seed Kale Salad

Classic nutrient-rich, alkaline-forming salad at its best.

Time: 10 minutes • Makes 2–4 servings

1 large bunch of curly or latigo kale
3 green onions, minced (white parts only)
½ cup diced crimini mushrooms (optional)
1 tbsp red wine vinegar
1 tbsp miso paste
2 tbsp hemp oil
¼ tsp garlic powder
⅓ cup hemp seeds
1 red bell pepper, finely diced

- Wash and dry the kale thoroughly. Strip the stems away from the kale leaves and discard. Place the kale leaves in a large bowl, tearing apart any large pieces.
- Add the green onions, mushrooms, red wine vinegar, miso paste, hemp oil, and garlic powder.
- Use your hands to massage the ingredients into the kale for about 2 minutes, or until kale and mushrooms have softened slightly.
- Add the hemp seeds and bell pepper and toss thoroughly.

Quick Kale Avocado Salad

CHAD SARNO

This is a wonderful way to enjoy the mighty kale. Many are unfamiliar with raw kale, but working with this method of softening the kale with the other ingredients makes it not only much easier to digest but also incredibly delicious. Serve this recipe with your favorite cooked whole grain and a handful of raw or toasted seeds/nuts for a great protein-packed alkalizing meal.

Time: 10 minutes • Makes 2 servings

1 head kale, shredded (any variety is great)
1 large tomato, or red bell pepper, diced
1 ½ avocado, chopped
3 tbsp flaxseed oil
2 tbsp red onion, green onion, or leeks, finely diced
1 lemon, juiced
1 tsp sea salt
Diced fresh chilies or pinch of cayenne (optional)

- In large mixing bowl toss all ingredients together, squeezing as you mix to "wilt" the kale and cream the avocado. Serve immediately. This dish is also great if you want to use chard, collards, broccoli leaves, spinach, or any combination of these instead of kale.

Chinese Chopped Salad

A healthy makeover of a Chinese chicken salad, this fresh preparation offers substantial amounts of protein of the plant-based variety.

Time: 15 minutes • Makes 4 servings

DRESSING

1 ½ tbsp yacon syrup or agave nectar

1 ½ tbsp ume plum vinegar

1 ½ tbsp apple cider vinegar

1 tbsp + 1 tsp fresh ginger, grated

1 whole red jalapeño pepper, minced (with or without seeds)

¼ cup hemp oil or olive oil

2 tbsp fresh squeezed orange juice

¼ cup sesame seeds

- Combine all the dressing ingredients, except 2 tbsp of sesame seeds, in a blender, blending until as creamy as possible. Stir in the remaining sesame seeds by hand and refrigerate until ready to use.

SALAD

5 cups Chinese cabbage, shredded

5 cups romaine lettuce, shredded

¼ cup scallions, white and green parts

1 cup snow peas or sugar snap peas

2 cups mung bean sprouts

¾ cup chopped roasted sacha inchi seeds (or almonds)

½ cup dried goldenberries (optional)

- In a large bowl, chop and combine all salad ingredients, except for the sacha inchi seeds and goldenberries.
- Toss with dressing just before serving. Sprinkle with sacha inchi seeds and goldenberries.

Chopped Garden Salad

This is a really flexible recipe. Pick a dressing of your choice or just do simple oil and vinegar.

Time: 15 minutes • Makes 2 very large dinner salads or 4–6 side salads

1 large head romaine lettuce

2 carrots

1 large cucumber (if organic, do not peel)

2 large radishes

1 stalk celery

1 cup onion sprouts (or use another kind of sprout)

¼ cup sunflower seeds

2 tbsp hemp seeds (optional)

Salad dressing of choice

- Chop the lettuce, carrots, cucumber, radishes, and celery finely.
- Toss together in a large bowl with sprouts, sunflower seeds, and hemp seeds.
- Add dressing, if desired, just before serving.

Raw

Asian Vegetable Noodle Salad

A raw twist on an ancient classic.

Time: 10 minutes • Makes 2–4 servings

DRESSING

1 tbsp + 1 tsp hemp oil or olive oil

2 tsp sesame oil

2 tsp miso paste

2 ½ tbsp yacon syrup

2 tbsp balsamic vinegar

- Combine all the ingredients in a jar and mix well.

SALAD

6 large zucchinis, peeled with a vegetable peeler into long strips

2 carrots, peeled with a vegetable peeler into long strips

1 package of kelp noodles, drained (optional)

2 green onions

½ jalapeño, deseeded and minced

¼ cup chopped sacha inchi or chopped almonds

Cilantro leaves, for garnish (optional)

- Toss together the salad vegetables in a large bowl, mix in the dressing, and top with sacha inchi or almonds and cilantro.

Asian Carrot Avocado Salad

A quick and easy salad with an exquisite balance of Asian-influenced flavor. An all-time favorite recipe!

Time: 10 minutes • Makes 3–4 servings

DRESSING

2 tbsp flaxseed oil
3 tbsp lime juice
2 tsp ume plum vinegar
1 tbsp fresh ginger, grated
1 tbsp ground coriander
1 tbsp agave nectar

- In a food processor, blend oil, lime juice, miso, ginger, coriander, and agave nectar.

SALAD

4 cups grated carrots
½ cup cilantro, chopped
½ cup parsley, chopped
1 green onion, sliced (white and 1 inch of green parts)
½ cup raw sesame seeds, with 2 tbsp set aside
1 medium avocado, chopped

- Grate carrots and toss in a large bowl with cilantro, parsley, green onion, and sesame seeds (reserving 2 tbsp of seeds). Set aside.

To serve
- Pour dressing over salad, and toss well. Gently fold in avocado, and sprinkle reserved sesame seeds on top before serving.

Candied Grapefruit Salad

AMANDA COHEN

This is our way to make salad more fun. Start by preparing the grapefruit so the candy glaze can dry while you work on the rest of the recipe. You'll need a candy thermometer to test when the palm sugar's ready for dipping, and a sturdy piece of floral foam or Styrofoam heavy enough to hold the skewers upright while the grapefruit dries.

Time: 1 hour • Makes 4 servings

8 grapefruit segments with the peel removed, but as much of the pith (the white skin beneath the peel) left on as possible

8 eight-inch bamboo skewers
3 cups palm sugar
½ cup water

- Push the skewers through the bottom of the grapefruit segments until they're about halfway in.
- In a heavy stockpot, bring the sugar and water to 250°F. Insert the thermometer just to check temperature, then remove.
- Dip each skewered piece of grapefruit in the hot sugar and coat each thoroughly.
- Stick the bottom ends of the skewers into the foam and let them dry until hard.

DRESSING

¼ cup grapefruit juice
2 tbsp grapefruit zest
2 tbsp lemon juice
1 tbsp finely minced shallot

½ tsp Dijon mustard
¾ cup extra-virgin olive oil
2 tsp salt
¼ tsp pepper

- Blend everything but the oil in a blender, and then slowly stream in the oil. Add salt and pepper to taste.

SALAD

4 cups mixed greens
¼ ripe avocado, cubed
3 tbsp toasted sliced almonds

- Mix the greens with the salad dressing, and toss in the avocado and almonds. Divide among 4 plates. Put two grapefruit skewers on each plate.

Cream of Asparagus Soup

TAL RONNEN

Tal says, "When I lived in Virginia, asparagus was one of the only locally fresh vegetables you could find in spring. This recipe is versatile: If you can't find nice asparagus, use broccoli to make cream of broccoli instead. As in a lot of my recipes, cashew cream stands in for dairy here and makes for an equally rich, delicious dish."

Time: 1 hour, 15 minutes prep • Makes 6 servings

Sea salt

3 tbsp extra-virgin olive oil

1 large bunch asparagus, ends trimmed,
    cut into 2-inch pieces

2 stalks celery, chopped

1 large onion, chopped

2 quarts faux chicken or vegetable stock
    (try Better Than Bouillon brand)

1 bay leaf

1 cup Thick Cashew Cream
    (see p. 203) + 6 tsp, for garnish

Freshly ground black pepper

2 cups fresh baby spinach

Microgreens, for garnish

- Place a large stockpot over medium heat. Sprinkle the bottom with a pinch of salt and heat for 1 minute. Add the oil and heat for 30 seconds, being careful not to let it smoke. This will create a non-stick effect.
- Add the asparagus, celery, and onion and sauté for 6–10 minutes, until the celery is just soft. Add the stock and bay leaf, bring to a boil, then reduce the heat and simmer for 30 minutes. Add the Thick Cashew Cream and simmer for an additional 10 minutes. Remove and discard the bay leaf. Season to taste with salt and pepper.
- Working in batches, pour the soup into a blender, cover the lid with a towel (the hot liquid tends to erupt), and blend on high. Add the spinach to the last batch and continue blending until smooth. Pour the soup into a large bowl and stir to incorporate the spinach batch. Ladle into bowls. Garnish each bowl with microgreens and a teaspoon of Thick Cashew Cream.

Consommé: Tomato Water, Merlot-Pickled Onions, Avocado, and Mint

CHAD SARNO

Elaborate but ideal if you want to be fancy. If you plan to use one of the garnish options to dress up the meal, begin its preparation the night before or a day ahead so it's ready in time.

Time: 20 minutes active; 1 hour sit time for tomatoes • Makes 6 servings

12 vine tomatoes, chopped (mixed heirloom tomatoes preferred)
2 tsp sea salt
2 cloves garlic, sliced
1 tsp fresh chili, chopped
2 tbsp merlot vinegar (or aged sherry vinegar as substitution)
Mint leaves, torn, for garnish

- To release the natural water from the tomato, toss the chopped tomatoes with the sea salt, then massage the tomatoes to release water.
- Add sliced garlic, chopped chili, and merlot vinegar to the tomatoes. Allow to sit for ½–1 hour or so for flavors to marry in room temperature. During this time the tomato water will begin to drain off—help the process by regularly massaging the tomatoes.
- Saving the liquid for the broth, strain off tomato liquid with either a fine mesh strainer or a sprouting bag. (For a clear water, allow the mixture to hang in the sprouting bag so that gravity releases the clear tomato water ... for faster production, use the fine mesh strainer instead.) Store the tomato pulp for a future dish.
- Serve chilled and garnished with Wine Pickled Onions (recipe on the next page), torn mint leaves, and avocado balls. Alternatively, garnish with Cucumber-Cress Sorbet (recipe on the next page).

WINE PICKLED ONIONS

2 red onions, peeled, and sliced paper-thin on mandolin
½ cup red wine or merlot vinegar
3 tbsp agave syrup
Pinch of coarse sea salt
Pinch of cracked black pepper

- Toss all ingredients well, and gently massage. Allow to pickle for a few hours to overnight. Store in a jar and refrigerate; they will keep for up to 2 weeks.

CUCUMBER-CRESS SORBET (OPTIONAL SERVICE SUGGESTION)

½ cup cashews, soaked
1 avocado, removed from skin
½ cup watercress
½ cup cucumber, peeled
2 tbsp agave syrup
1 tbsp lemon juice
1 clove garlic
1 tsp sea salt
¼ cup water

- In high-speed blender, blend all ingredients until smooth. Either pour it into your choice of sorbet maker, following manufacturer's instructions, or line a square container with plastic wrap, then pour in blended mixture.
- Freeze overnight, pop the frozen block out the following day, slice it in strips, and put them through a single- or double-gear juicer, using the solid plate instead of the juicing screen, for a delicious sorbet!
- Serve a small quenelle with each bowl of consommé.

Black Bean Soup

CANDLE 79, MANHATTAN, NEW YORK

The day before you want to make this soup, soak your beans overnight in the refrigerator. It cuts down on the cooking time and makes them more digestible.

Time: 20 minutes prep; 45 minutes to cook • Makes 6 servings

2 tbsp extra-virgin olive oil
1 cup diced celery
1 cup diced yellow onion
¼ cup sliced leeks, trimmed and cleaned
1 cup diced zucchini
2 cloves garlic, minced
1 dried chipotle pepper
2 cups black beans, soaked overnight
1 bay leaf
1 tsp sea salt
12 cups filtered water
4 tbsp fresh cilantro, chopped
1 tsp fresh oregano, chopped
Chopped tomatoes, sliced avocado, tofu sour cream for garnish

- In a 4-quart pot, heat olive oil on medium heat. Add celery, onion, leeks, zucchini, garlic, and chipotle pepper. Sauté vegetables for 10–15 minutes or until they become translucent and very soft.
- Add black beans, bay leaf, salt, and filtered water. Cover pot, reduce heat to low, and allow to cook for 45 minutes or until the beans are cooked and all the vegetables have almost disappeared into the soup. The beans should be creamy in texture.
- Remove the bay leaf and divide the soup. Allow it to cool slightly and purée half in blender, beans too. Be very careful not to place the hot soup in the blender or else it will end up everywhere!
- Add puréed mixture to reserved soup. Stir in cilantro and oregano, and return to heat for about 5–10 minutes to reheat.
- Garnish with chopped tomatoes, sliced avocado, and tofu sour cream, or enjoy as is!

Tomato Soup

A plant-based twist on a classic creamy soup.

Time: 5 minutes (once Marinara/Pizza Sauce and UnMotza Macadamia Cheese are made) • Makes 1 serving

4 tbsp tomato sauce or Marinara/Pizza
  Sauce (see p. 200)
1–2 tbsp UnMotza Macadamia Cheese
  (see p. 200)
4 tbsp chopped tomato
2 tbsp chopped cucumber, red bell
  pepper, zucchini
1 tbsp red onion
2 tbsp sunflower sprouts

½ clove of garlic
1 tsp each sea salt and dried oregano
1 tbsp extra-virgin olive oil
1 tsp apple cider vinegar
Pinch of black pepper
1 cup of warm or hot water
Optional: 1 tsp nutritional flakes,
  chopped cilantro, or parsley,
  or all three

- Combine all ingredients in a bowl and eat as is, or combine in a blender and blend until smooth.

Curry Soup

Rich and filling.

Time: 10 minutes (once UnMotza Macadamia Cheese is made) • Makes 1 serving

4 tbsp UnMotza Macadamia Cheese
  (see p. 200)
2 tbsp chopped red bell pepper
4 tbsp chopped zucchini
1 tbsp red onion
3 tbsp chopped celery
½ clove of garlic
2 tbsp sunflower sprouts

1 tsp each sun dried sea salt and
  curry powder
¼ tsp turmeric powder
1 tbsp first cold pressed coconut oil
1 tsp apple cider vinegar
Pinch of black pepper
1 cup warm or hot water
Optional: 1 tsp nutritional flakes

- In a blender, combine all ingredients, except UnMotza Macadamia Cheese, and blend until smooth.
- Pour soup into a bowl and pour UnMotza Macadamia Cheese on top.

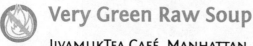

Very Green Raw Soup

JivamukTea Café, Manhattan, New York

Raw

This soup is like a green juice, with all the whole food fiber benefits.

Time: 5 minutes prep • Makes 1 serving

Bowl of mixed lettuce greens
4 cherry tomatoes
½ apple
1 tsp spirulina powder (optional)

1 tbsp olive oil
1 tsp lemon or lime juice
Salt and pepper to taste
Approx 1 cup water

- Place all ingredients into a blender or food processor. Add just enough water to blend ingredients into a creamy consistency and serve cold.

Spinach Cream Soup

Crudessence, Montreal, Quebec

Raw

Creamy and filling, this soup is chlorophyll-rich.

Time: 5 minutes • Makes 4 servings

¼ cup shelled pistachios
⅛ cup pine nuts
1¼ cups spinach
⅔ cup avocado
¼ cup lemon juice
3 cups water

⅓ cup red onions
1 tsp sea salt
½ tsp ground black pepper
1 small clove garlic
1 tbsp dried rosemary

- Place all ingredients into a blender and blend thoroughly until the texture is smooth and creamy. Serve cold.

Raw

Chilled Summer Greens and Avocado Soup
with Truffled Cashew Sour Cream

MILLENNIUM RESTAURANT, SAN FRANCISCO, CALIFORNIA

Sophisticated and nutritious.

Time: 10 minutes • Makes 4 servings

3–4 medium avocados (ripe)
2 scallions
1 cup diced seeded cucumber
2 cups watercress or arugula

Juice of 2 lemons
1 ½ cups water
Salt and pepper to taste

- In a blender, combine all ingredients and purée. Adjust salt and pepper to taste.
- Serve or chill covered with wrap on top of the soup to keep it from oxidizing. Do not store more than 2 hours.
- Serve each portion with fresh herbs of choice, a drizzle of a peppery-grassy olive oil (Posolivo, Scabicia Tuscan or Sevillano, or Arbequenia varietal oil), and 2 tsp of Cashew–Truffle Oil Cream (see p. 202).

Raw

Cool Cantaloupe Soup

Simple summer refreshment. Almost an applesauce-like texture.

Time: 10 minutes • Makes 2–4 servings

4 cups cantaloupe flesh (about 1 medium)
2 tbsp fresh lemon juice
2 tbsp fresh lime juice

- In a blender, combine all ingredients and blend until smooth. Chill before serving.

Chilled Cream of Beet Soup

Sweet beets combine with creamy avocados for this alluring, satisfying, and not to mention stunningly colored soup. This soup has great healthy benefits too. It is vegan, gluten free, and cholesterol free.

Time: 10 minutes prep; 1 hour to roast   •   Makes 4 servings

4 medium beets
1 avocado, chopped
1 lime, juiced
2 cups water
3 tbsp hemp seeds
1 tbsp ground coriander
¼ tsp sea salt
Fresh cilantro leaves and black pepper, for garnish (optional)

- Heat the oven to 350°F. Trim the beets and remove stems and end. Individually wrap each beet in tin foil. Roast for 1 hour and allow to cool completely.
- Using a paper towel, rub off the beet skins. Chop coarsely.
- Place the cooled beets, avocado, lime juice, water, hemp seeds, coriander, and sea salt in a blender. Blend until completely smooth.
- Place soup into refrigerator, and allow to chill for a minimum of 30 minutes.
- Pour into serving bowls and sprinkle with cilantro and black pepper, if desired.

Chilled Cucumber Avocado Soup

HORIZONS, PHILADELPHIA, PENNSYLVANIA

Time: 15 minutes prep; 1 hour to chill • Makes 6 servings

8 cucumbers, peeled and seeded
¼ cup chopped onion
1 bunch fresh cilantro, leaves only
½ cup fresh mint leaves (packed)
2 limes, juiced
1 ripe avocado
1 clove garlic
2 tbsp olive oil
2 tsp salt
2 tsp pepper
1 tsp organic palm sugar
¼ cup vegan mayo (preferably veganaise)
½ tsp Dijon mustard
Toasted pumpkin seeds, for garnish
Cumin oil (olive oil with ground cumin, lightly heated then cooled)

- Purée all ingredients in a food processor, except pumpkin seeds and cumin oil, adding in enough water to reach the consistency you like. It should be fairly thick, but adjust to your liking. Let chill for at least 1 hour.
- Serve chilled and garnish with toasted pumpkin seeds and cumin oil.

Lentil Soup with Wilted Spinach

A filling protein- and chlorophyll-rich soup.

Time: 40 minutes • Makes 4 servings

1 tbsp coconut oil

2 large cloves garlic, minced

2 cups vegetable broth

1 cup dry lentils, rinsed thoroughly

1 bay leaf

2 tbsp fresh lemon juice

2 cups packed baby spinach leaves

Salt and pepper to taste

- In a large soup pot, heat the coconut oil over medium heat. Add the garlic, and cook for 1 minute until garlic is fragrant and begins to turn golden.
- Pour in the vegetable broth, and add the lentils and bay leaf. Raise the heat to bring to a boil, then reduce heat to maintain a simmer.
- Cook, uncovered, for about 30 minutes, until lentils are just tender, adding more water if necessary to maintain a broth around the legumes.
- Stir in the lemon juice and the spinach and cook for 2 minutes longer, or until spinach has turned bright green and has wilted. Season with salt and pepper to taste and serve warm.

Chilled Carrot Ginger Soup

The ginger in this recipe is very light. Ginger lovers can easily up the quantity.

Time: 5 minutes • Makes 2–4 servings (4 cups)

3 cups carrot juice

1 avocado (peeled and pitted)

1 tbsp minced fresh ginger

2 tsp miso paste

- Combine all ingredients in a blender and blend until smooth. Serve chilled or slightly warmed.

Young Coconut Soup

Delicately seasoned. Add more jalapeño or ginger for a spicier taste.

Time: 10 minutes • Makes 2–4 servings (4 cups)

2 cups young coconut meat (about
   2–4 young coconuts)
2 cups young coconut water
4 dice-sized cubes of peeled fresh ginger

2 tsp fresh lime juice
1 tsp yacon syrup
1 jalapeño pepper, de-seeded
¼ tsp sea salt

- Combine all ingredients in a blender and blend until smooth and creamy. Serve chilled or at room temperature.

Fresh and Creamy Tomato Basil Soup

This soup is best when made during the peak tomato season—summer—for maximum flavor. If tomatoes are not in their prime or simply not sweet enough, add a touch more yacon syrup to balance the acidity.

Time: 10 minutes active; 1–2 hours presoak; 30 minutes to chill • Makes 2 ½ cups

3 medium tomatoes
½ cup raw cashews, presoaked in water
   (1–2 hours)
1 ½ tbsp yacon syrup
2 tbsp olive oil

1 ½ tbsp grated white onion
¼ cup shredded fresh basil + extra
   for garnish
¼ + ⅛ tsp sea salt
Fresh black pepper to taste

- Cut the tomatoes in half. Remove and discard all the seeds, then coarsely chop the remaining tomato and place in blender.
- Drain the cashews, reserving ½ cup water, and place in a blender.
- Combine all remaining ingredients in a blender, including the reserved soak water, and blend until smooth and creamy. If necessary, add more water to achieve desired texture.
- Allow soup to chill in refrigerator for 30 minutes or longer before serving.

Purée of Turnip Soup

A comforting holistic soup made from the delicious yet often forgotten turnip.

Time: 40 minutes • Makes 4 servings

2 tbsp coconut oil
1 medium yellow onion, chopped
2 large cloves garlic, minced
3–4 turnips (about 1 pound), chopped into 1-inch pieces
1 medium sweet potato (about ½ pound), peeled and chopped into 1-inch pieces
2 ½ cups vegetable broth
1 cup unsweetened almond milk
¼ tsp salt, or to taste
½ tsp black pepper

- Heat a large saucepan over medium heat. Melt the coconut oil, and add the onions and garlic. Sauté for 3–5 minutes or until onions begin to turn translucent.
- Add the turnips and the sweet potato and sauté for 2 minutes longer. Pour in the vegetable broth and almond milk. Bring to a boil, then reduce heat and simmer for 20–25 minutes, or until turnips and potatoes are tender.
- Transfer soup contents to a blender, and purée until completely smooth (blend in a couple of small batches, if necessary).
- Pour soup back into pot and reduce at a low simmer for 5 minutes longer. Add salt to taste and black pepper.

Puréed White Bean Soup

Simple and warming.

Time: 25 minutes • Makes 4–6 servings

1 tbsp coconut oil
2 leeks, white and light green parts sliced thin
1 tsp fresh thyme
3 tbsp lemon juice
30 ounces cooked white beans (fresh or canned)
2 ½ cups water
¼ cup cashews
1 tsp palm sugar
Salt and freshly ground black pepper, to taste
Fresh thyme leaves and freshly cracked pepper, for garnish

- In a large saucepan, heat the coconut oil, leeks, and thyme over medium-high heat for 2 minutes, or until the leeks are just softened.
- Transfer mixture to a blender and add lemon juice, beans, water, cashews, sugar, a pinch of salt, and a generous amount of black pepper. Blend on high for several minutes until completely smooth.
- Transfer back to the saucepan and simmer at medium-low heat for about 15 minutes, stirring occasionally. Do not overcook or soup will become too thick—add additional water if needed.
- Season with additional salt if desired, and garnish each bowl lightly with fresh thyme leaves and freshly cracked pepper before serving.

Stuffed Mushrooms

Raw

Karyn's Fresh Corner Café, Chicago, Illinois

Time: 15 minutes active; 6 hours to soak; 1–2 hours to sit • Makes 6 servings

2 cups almonds (soaked)
20–30 button mushrooms
Bit of olive oil and salt (to massage on mushrooms)
½ yellow or green pepper
½ red pepper

2 stalks celery
3 cloves garlic
¼ cup olive oil
1 tsp salt

- Soak the almonds for 6 hours in purified water, drain, and rinse. Set aside.
- De-stem the mushrooms. Massage the caps very gently with a bit of oil and salt and let sit for 1–2 hours.
- Purée the almonds, peppers, celery, garlic, olive oil, and salt in food processor. Stuff pâté mixture into the mushroom caps.

Arame Sesame Brown Rice

Ravens' Restaurant, Mendocino, California

A Japanese macrobiotic-inspired classic.

Time: 5 minutes (once rice is cooked) • Makes 4 servings

2 tbsp sesame oil
½ cup bell peppers, diced
3 tbsp toasted arame

4 cups cooked rice
2 scallions, chopped, for garnish
2 tbsp sesame seeds, for garnish

- In saucepan, heat the sesame oil, then add the peppers and arame and sauté for 1 minute on medium-high heat.
- Add rice and stir well until heated thoroughly.
- Garnish with scallions and sesame seeds.

Mashed Kabocha Squash with Toasted Coconut

This delicious recipe makes a great side dish.

Time: 10 minutes • Makes 4 servings

¼ cup shredded coconut
1 pound cooked kabocha squash
3 tbsp light coconut milk
Salt and pepper, to taste

- In a pan over medium-low heat, toast the coconut for 1–2 minutes, stirring constantly, until golden brown. Coconut burns easily, so remove from heat immediately after cooking.
- Mash the squash and coconut milk together with a fork or handheld blender. Season with salt and pepper if desired, and top with toasted coconut.

....

Garlic Thyme Sweet Potato Oven Fries

Delicious and very filling.

Time: 5 minutes active; 35 minutes to cook • Makes 2 servings

2 medium sweet potatoes
2 cloves garlic
2 tbsp coarsely chopped pumpkin seeds
1 tbsp thyme

1 ½ tbsp coconut oil
½ tbsp basil
Sea salt to taste

- Preheat oven to 300°F.
- Cut the sweet potatoes into wedges or chunks. In bowl, combine the garlic, pumpkin seeds, thyme, coconut oil, basil, and sea salt.
- Add the sweet potatoes, stirring with your hands to make sure all the pieces are covered with the mixture.
- Spread the sweet potatoes on a baking tray lightly oiled with coconut oil; bake for about 35 minutes.
- If you prefer the potatoes crispier, leave in the oven for an extra 5–10 minutes.

Kitchari

JIVAMUKTEA CAFÉ, MANHATTAN, NEW YORK

This is based on a traditional Indian porridge-like dish. It is very grounding and stabilizing without being too filling. It is often used in Ayurveda as a detox fast.

Time: 2 minutes active; 1 hour to cook • Makes 4 servings

2 cups red lentils

1 cup short-grain brown rice or white basmati rice

8 cups water

1 tbsp salt

- Place the lentils and rice in a large soup pot; wash and rinse with cold water 3 times.
- Add the water and bring to boil.
- Boil for 5 minutes, then turn heat to medium-low and cook for 1 hour.
- Add salt, stir well, and serve.

Spirulina Green Millet

JIVAMUKTEA CAFÉ, MANHATTAN, NEW YORK

Offers carbohydrate and chlorophyll-rich simplicity.

Time: 15 minutes active; 20 minutes to cook • Makes 1 serving

1 cup dry millet

2 cups water

4 tbsp flaxseed oil

4 tbsp powdered spirulina

1 tbsp Braggs Liquid Aminos (or soy sauce)

- Boil the millet in the water, reduce heat, and cook until all water is absorbed (about 20 minutes).
- Wait for the millet to "dry out" (about 10 minutes).
- Transfer the millet to large bowl. Add the oil and, using a large fork, mix well.
- Sprinkle spirulina over the millet and continue to mix well.
- Finally, add Braggs, mixing well until the millet is bright green.

Red Lentil Dal

THE GREEN DOOR RESTAURANT, OTTAWA, ONTARIO

The Indian classic.

Time: 10 minutes, 30–45 minutes to cook • Makes 8–10 servings

2 cups red lentils
5 cups water
2 tbsp olive oil
1 tsp whole cumin seeds
¼ tsp asafetida (hing)
½ tsp garam masala
½ tsp ground cumin
¼ tsp ground cardamom
¼ tsp cayenne or 1 hot pepper, diced
2 cloves garlic, crushed
1 tsp salt, or to taste
¼ cup chopped fresh cilantro, for garnish
1 small tomato, diced, for garnish

- Wash the lentils well by rinsing and pouring off 3 changes of water. Cook the lentils in the water until well done (30–45 minutes). Do not drain. Set aside.
- Put a medium-sized pot on low heat. Add the oil and whole cumin seeds. Fry until they smell toasted.
- Add the rest of the spices and cayenne or hot pepper, and toast lightly. Add the garlic and sauté until light brown.
- Add the red lentils and salt. Cook on low heat for 10 minutes. Garnish with chopped cilantro and tomato.
- This will make a fairly thick dal. Add water for a thinner consistency.

Grilled Spinach with Salsa Rustica
(Italian Tomato Relish)

HORIZONS, PHILADELPHIA, PENNSYLVANIA

Bursting with flavor.

Time: 15 minutes • Makes 4 servings

SPINACH

2 tbsp olive oil
1 tsp minced garlic
1 pound washed baby spinach
Salt and pepper

- In a large skillet, heat the olive oil and garlic and toss the spinach until just past half-cooked.
- Season with salt and pepper.

SALSA

1 cup diced plum tomatoes
1 tsp capers
1 tsp chopped black olives
1 tsp chopped fresh oregano or thyme or both
2 tsp minced red onion
1 tsp olive oil
½ tsp sherry vinegar
Pinch of salt
¼ tsp black pepper

- Toss all of the salsa ingredients together and serve over the spinach.

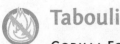

Tabouli

GORILLA FOOD, VANCOUVER, BRITISH COLUMBIA

A raw twist on a classic.

Time: 10 minutes • Makes 2 servings

2–3 medium tomatoes, diced
⅓ medium cucumber, seeded and diced
10 sprigs parsley
1 tbsp lemon juice
¼ tsp salt
1 tbsp olive oil

- Mix the diced tomatoes and cucumbers. With a knife, hand-mince parsley and stir into the cucumber and tomato mix.
- Add the lemon juice, salt, and olive oil and mix well.

Couscous Flower

CRUDESSENCE, MONTREAL, QUEBEC

Raw, fiber-rich, and delicious.

Time: 5 minutes • Makes 8 servings

8 cups cauliflower, shredded in food processor
3 ½ ounces zucchini, diced
1 ounce onions, finely sliced
3 ½ ounces parsley, chopped
5 ounces tomatoes, diced
½ cup currants

3 tbsp lemon juice
3 tbsp olive oil
1 tsp sea salt
¼ tsp garlic purée
3 tbsp cumin
¼ tsp ground black pepper

- Mix all ingredients together by hand in a large salad bowl.

Almond Cakes with Mango Tartar Sauce

LIVE ORGANIC FOOD BAR, TORONTO, ONTARIO

Raw

Flavorful and filling, these cakes may be served as a side or a complete meal.

Time: 40 minutes; 12–18 hours to soak and dehydrate
Makes 6 servings (about 24 cakes and 2½ cups of sauce)

ALMOND CAKES

¼ bunch of celery

2 cups + 1 tbsp of soaked almonds

1 tbsp olive oil

Juice of 3 ½ limes

Water to blend

¼ cup of dulse flakes

2–3 green onions, thinly sliced with
   some green + some for garnish

1 ¼ inch of ginger, grated

⅛ cup of Nama Shoyu soy sauce

Celtic sea salt if needed

Fresh dill, for garnish

- In a food processor, start by pulsing the celery into pulp. Place in a medium-size mixing bowl. Set aside.
- In the food processor, add the almonds, olive oil, lime juice, and as much water as needed to blend into a fine, smooth mixture. Add this blended mixture to the bowl with the celery. Mix in remaining ingredients by hand. Season with Celtic sea salt if needed.
- Take 2 tbsp of dough and form into croquette shapes (mini patties) and dehydrate for at least 12–18 hours at 115°F until crisp on outside but moist in the middle.

MANGO TARTAR SAUCE

½ mango, peeled, pitted, and diced

¼ piece of red onion, diced

¼ English cucumbers, diced

½ daikon radish, peeled and diced

1 tbsp capers

2 tbsp chopped dill

½ tbsp rice wine vinegar

1 tbsp agave nectar

- In a medium mixing bowl, mix all ingredients. Chill for at least 30 minutes in a refrigerator before serving. This mix will last for at least 3–4 days in a refrigerator.

To serve
- Top each Almond Cake with 1 tsp Mango Tartar Sauce. Garnish with additional green onions and fresh dill.

Bok Choy Couscous with Sacha Inchi

Toasting the millet helps give it better texture and flavor.

Time: 10 minutes prep; 30 minutes to cook • Makes 2 servings

¾ cups dry millet

2 cups water

1 tbsp miso paste

1 tbsp sesame oil

2 tsp brown rice vinegar

½ tsp ume plum vinegar

1 tbsp minced shallot

2 cloves garlic, minced

2 tsp peeled and grated ginger

1 ½ pounds baby bok choy, leaves separated from the stems

2 tbsp yacon syrup

⅓ cup coarsely chopped sacha inchi

- Toast the millet in a frying pan over medium high heat for 3–4 minutes, or until the millet is fragrant and begins to sound as if it is popping.
- In a medium saucepan, combine the water, toasted millet, and miso paste. Bring to a boil, then reduce heat to a simmer.
- Cook uncovered until the water is absorbed (about 30 minutes), let stand for 5 minutes, then fluff with a fork.
- Once the millet has finished cooking, prepare the vegetables. Heat the sesame oil over medium-high heat. Add the shallot and garlic, and cook for 30 seconds until slightly carmelized. Add the baby bok choy and the prepared ginger–yacon mixture. Toss to combine. Cover and cook for 3–5 minutes, or until the bok choy is bright green and tender crisp (add a little water to the pan to prevent burning, if needed).
- Remove from heat. Place the millet in a large serving bowl and top with the contents of the cooked vegetable pan. Mix well. Sprinkle the sacha inchi seeds on top and serve warm.

Rustic Sweet Onion Flatbread

Satisfying like bread but light and with intense flavor. If you like you can use a dehydrator for this recipe instead of the oven.

Time: 10 minutes active; cooking time varies • Makes 8 servings

¾ cup tomato, chopped

⅓ cup raisins

1 tbsp miso paste

1 red onion, chopped (about 2 cups)

1 large garlic clove, chopped

¾ cup flaxseed powder

½ cup hemp seeds

2 tbsp chia seeds

- Preheat the oven to 250°F and line a baking sheet with parchment paper.
- In a food processor, blend the tomato, raisins, and miso paste together until smooth.
- Add the onion, garlic, and flaxseed powder and blend again as smooth as possible. Mix in the hemp seeds and chia seeds by hand.

Dehydrator method (preferred)

- Spread the mixture onto several teflex sheets, and dehydrate at desired temperature until dried out into a flatbread. For best results, flip over after 8–10 hours to allow even dehydration.

Oven method

- Pour the mixture onto the lined baking sheet. Use a spatula to spread into a thin layer, forming about a 10 × 14-inch rectangle (spread as evenly as possible to ensure even cooking).
- Bake for 1 hour, then remove from oven and place another sheet of parchment paper on top of the spread. Holding the two layers together, flip the whole spread over so that the new parchment is on the bottom. Carefully peel away the top layer of parchment from the spread and discard.
- Return to the oven for 15–20 minutes, or until the edges begin to brown slightly. Turn off the oven, and leave the tray inside until the oven has cooled, about 30 minutes. Break into larger pieces for flatbread, into smaller pieces for crackers, or crumble into croutons.

Roasted Garlic Quinoa

One of the healthiest and tastiest ways to infuse quinoa with flavor.

Time: 5 minutes; 30 minutes to cook • Makes 3–4 servings

1 whole head garlic, roasted
2 tbsp melted coconut oil + some for drizzling
1 tsp lemon juice
4 cups cooked quinoa
½ tsp sea salt, or to taste
½ tsp fresh thyme leaves, for garnish

- To roast the garlic, heat the oven to 400°F. Cut off the top of the head of garlic, about ¼–½ inch down, and place it inside a small piece of aluminum foil.
- Drizzle with coconut oil, close the foil loosely, and bake for 30 minutes. Open foil after removing from the oven.
- Once the garlic has cooled, squeeze the roasted garlic cloves out of the peel and into a small bowl.
- Add the coconut oil and lemon juice and mix well. Pour the resulting sauce into a medium saucepan and heat on medium-low heat.
- Add the quinoa and sea salt and cook, stirring, for 2–3 minutes until quinoa is heated through and the flavors have incorporated. Garnish with fresh thyme before serving.

Superfood Gomashio

Sprinkle on top of salads, rice, and quinoa for a flavorful salty condiment. You can also skip the toasting step and use raw.

Time: 5 minutes • Makes ⅔ cup

4 sheets nori
¼ cup hemp seeds
¼ cup sesame seeds

1 tsp sea salt
1 tsp sesame oil

- Rip or crumble the nori into small flakes over a bowl. Slowly spoon into a running food processor.
- Add the hemp seeds, a little at a time, stopping the machine if needed to scrape down the sides and re-incorporate ingredients. Add the sesame seeds, sea salt, and sesame oil.
- Heat a small frying pan over medium heat. Add the mixture, and toast for about 1 minute, stirring constantly, just until the nori is crisp and fragrant and begins to shrink in size.
- Remove from heat quickly to prevent burning. This will keep for several weeks in an airtight container.

Baby Zucchini and Avocado Tartar

MATTHEW KENNEY

This recipe calls for ring molds.

Time: 10 minutes • Serves 4

2 firm avocados, finely diced
4–5 baby zucchini, finely diced
2 tbsp lemon juice
1 tbsp olive oil + drizzle, for garnish
2 tbsp micro basil (or finely minced basil)

1 tbsp chives, minced + some
   for garnish
1 tsp agave
2–3 tsp coarse salt
Freshly ground pepper, for garnish

- Toss all ingredients until well combined.
- To serve, divide into four servings and press into ring molds. Garnish with chive, freshly ground pepper, and a drizzle of olive oil.

Simple Italian Zucchini Noodles

A classic example of the Italian mastery of letting beautiful ingredients shine.

Time: 15 minutes active prep • Makes 4 servings

4 medium yellow zucchini
1 tsp sea salt
1 tbsp coconut oil
5 large cloves garlic, minced
2 tbsp hemp oil
¼ tsp red pepper flakes

- Trim the ends of the zucchini. Using a hand-held vegetable peeler, carefully strip the zucchini, layer by layer, into noodle-like pieces and gather into a colander (for best results, discard the watery center section that holds the seeds).
- Toss the zucchini strips with the sea salt, then place the colander over a large bowl to catch excess moisture. Leave the zucchini to rest for 30 minutes.
- After 30 minutes, wash the zucchini thoroughly with warm water to remove any excess salt and let drain for 5 more minutes.
- Heat the coconut oil over medium heat, and add the garlic. Cook for 1–2 minutes until the garlic begins to turn golden.
- Add the zucchini and toss well, cooking for 1 minute longer or until heated through.
- Remove from heat and toss with hemp oil and red pepper flakes, adding more sea salt to taste. Serve warm.

Almond Noodles with Carrots and Wakame

This recipe takes noodles with peanut sauce to the next level.

Time: 20 minutes • Makes 4 servings

NOODLES

2 packages kelp noodles (24 ounces)
3 tbsp almond butter
1½ tbsp yacon syrup
1 tbsp brown rice miso
1 tsp wasabi powder
¼ tsp garlic powder
1 tsp ginger powder
¼ cup water, or more as needed
1 tsp sesame oil
2 medium carrots, grated

GARNISH

1 scallion, sliced thin (white and green parts)
2 tbsp almonds, chopped

- Rinse the kelp noodles thoroughly and place in a pot of warm water. Let soak for 10 minutes to soften.
- While the noodles are soaking, whisk the almond butter, yacon syrup, miso, wasabi, garlic, ginger, and water in a bowl or blender until smooth. Set aside.
- Gently heat the sesame oil over medium heat in a large frying pan or wok. Add the kelp noodles and carrots and stir fry for 2–3 minutes, tossing to ensure even cooking. Add the almond sauce and cook for 2 more minutes, or until the sauce has thickened.
- Transfer to a serving bowl and top with scallions and chopped almonds.

Variation: This may also be served as a raw dish. Simply reduce the amount of water in the sauce to 1–2 tbsp, and omit the sesame oil entirely. Instead of cooking, just toss the ingredients together and serve at room temperature.

Roasted Broccoli with Avocado Pesto

Fiber and flavor rich—delicious.

Time: 20 minutes • Makes 4 servings

BROCCOLI

3 pounds broccoli, cut into florets (including some stem)
2 tbsp melted coconut oil
Salt and freshly ground pepper

SAUCE

1 medium avocado
4 cloves garlic
3 packed cups fresh parsley leaves, + 2 tbsp for garnish
¼ cup lemon juice
2 heaping tbsp packed dulse
½ tsp sea salt

- Preheat the oven to 475°F. Line a baking sheet with aluminum foil for easy clean-up.
- Toss the broccoli with the coconut oil and sprinkle with salt and pepper. Spread out on the baking sheet and roast for 15 minutes or until tender-crisp, tossing the broccoli halfway through.
- While the broccoli is cooking, make the sauce. Mash the avocado in a small bowl and place in a food processor. With the machine running, add the garlic, parsley, lemon juice, dulse, and sea salt and process until a smooth paste has formed.
- Toss the avocado mixture with the hot broccoli in a large bowl, then transfer to a serving plate and sprinkle with reserved 2 tbsp parsley.

Fresh Pasta Puttanesca

A vegetable-in-place-of-pasta twist on a classic.

Time: 15 minutes prep; 1 hour to soak and sit   •   Makes 4 servings

PASTA

4 large zucchini
1 ½ tbsp Vega Antioxidant EFA Oil Blend (or use hemp oil or virgin olive oil)
Pinch of sea salt

SAUCE

4 sun-dried tomatoes
1 ½ cups cherry tomatoes, quartered
1 tsp pressed garlic
4 tsp dulse flakes
3 tbsp Vega Antioxidant & EFA Oil Blend (or use hemp oil or flaxseed oil)
½ cup kalamata olives, pitted
¼ cup fresh parsley, chopped
½ tsp red pepper flakes
¼ tsp sea salt

- To begin the sauce, place the sun-dried tomatoes in enough water to cover them. Set aside and let soak for 45 minutes or until soft.
- To make the "pasta," use a vegetable peeler to dispose of green skin on the zucchini, then peel long fettuccini-like strips with the rest. Toss strips in the oil and sea salt until well coated, and set aside for 15 minutes to soften.
- To make the sauce, put sun-dried tomatoes (once soaking is complete) along with the remaining sauce ingredients into a food processor. Pulse until just finely chopped, about 5 or 6 times.
- Toss sauce with pasta and serve, sprinkling with extra parsley if desired.
- The sauce also works as a cooked recipe. Simmer on low heat for 3–4 minutes until just heated through, and then serve with the pasta.

Optional: If you have a dehydrator and a little patience, spread sauce on a teflex sheet before mixing with pasta, and dehydrate for an hour—deliciousness will result. Stir before combining with pasta, and serve warm.

Cheesy Broccoli Bowl

I like to serve this recipe as is (uncooked), enjoying the maximum nutritional potential of these great superfoods. If you're craving a warm dish though, no worries—simply steam the florets lightly for a few minutes, then combine with the sauce. Either way, the delicious cheesy flavor and addictive broccoli crunch will have you coming back for a healthy second round every time.

Time: 10 minutes • Makes 4 servings

2 tbsp nutritional yeast

½ tsp paprika

¼ tsp garlic powder

¼ cup raw tahini

1 tsp brown rice or chickpea miso paste

3 tsp lemon juice

2 tbsp coconut oil, melted

3 tbsp water, if needed

1 ½ tsp ume plum vinegar

8 cups finely chopped fresh broccoli florets

2–3 tbsp hemp seeds

- In a small bowl, mix the nutritional yeast, paprika, and garlic powder together. Stir in the tahini and brown rice or chickpea miso paste.
- Pour in the lemon juice, coconut oil, water, and ume plum vinegar, and whisk thoroughly. If a thinner sauce is desired, add extra water, a tablespoon at a time.
- Put the broccoli into a larger bowl and pour sauce on top. Toss until evenly coated (for best results, use your hands!).
- Sprinkle with hemp seeds and serve.
- Keeps refrigerated for several days.

Traditionally, spreads are spread on bread. But in sticking with my dietary philosophy, I use them mostly on raw vegetables.

Zucchini Hummus

GORILLA FOOD, VANCOUVER, BRITISH COLUMBIA

A lighter version than the traditional chickpea hummus, with zucchini as its base.

Time: 5 minutes • Makes 1 ½ cups

1 cup sesame seeds
2–3 medium cloves garlic
10 sprigs parsley
5 tbsp lemon juice
⅓ tsp salt
3 medium zucchini, cut into 1-inch pieces

- Grind the sesame seeds with the S-blade in a dry food processor until it is a creamy tahini. Set this paste aside in a bowl.
- Finely mince the garlic with the S-blade in the food processor.
- Add parsley, lemon juice, salt, and zucchini and blend until nearly puréed.
- Add the tahini back in and blend all ingredients together until smooth and creamy.

Zucky Hummus

CRUDESSENCE, MONTREAL, QUEBEC

With sunflower seeds and zucchini as its base, this hummus's flavor is further enhanced by the ginger.

Time: 10 minutes active; 8 hours presoak • Makes 4 ½ cups

1 ½ cups of sunflower seeds (soaked for 8 hours)
2 ¼ cups coarsely chopped zucchini
¼ cup lemon juice
¼ cup water
¼ cup olive oil
1 tsp ginger juice
1 tsp sea salt
2 tsp cumin
3 cloves garlic
Pinch cayenne powder

- Soak the sunflower seeds for 8 hours.
- Rinse the soaked seeds well before use. Mix all ingredients together in a blender until creamy and smooth.

Pecan and Dill Pâté

CRUDESSENCE, MONTREAL, QUEBEC

You can use this pâté to replace the rice in sushi rolls. Just make them as usual, with a lot of finely chopped veggies and avocado to replace the fish ... simply amazing!

Time: 10 minutes active; 8 hours presoak • Makes 6 cups

3 cups sunflower seeds (soaked for 8 hours)
2 cups pecans
¼ cup apple cider vinegar
¼ cup lemon juice
½ cup sunflower oil
¼ cup water
1 cup fresh parsley, chopped coarsely (packed firmly)
½ cup fresh dill chopped coarsely (packed firmly)
3 cloves garlic
2 tsp sea salt

- Soak the sunflower seeds for 8 hours.
- Rinse the soaked seeds well before use. Using a food processor, grind them into a powder.
- Add the rest of the ingredients and mix until a creamy texture without lumps is obtained.

Walnut Chili Pâté

Gorilla Food, Vancouver, British Columbia

Filling, yet easy to digest and packed with vegetables, this pâté has an intense flavor.

Time: 10 minutes active, 6–8 hours presoak • Makes 1 cup

½ cup sunflower seeds (soaked overnight)
⅓ cup walnuts (soaked overnight)
2 stalks celery
¼ medium zucchini
¼ bunch cilantro
5 sprigs parsley
½ tbsp lemon juice
¼ tsp salt
¹⁄₁₆ tsp cayenne chili powder
¼ tsp chili blend powder
¹⁄₁₆ tsp cumin
¹⁄₁₆ tsp coriander

- Soak the seeds and nuts for 6–8 hours or overnight and rinse well before using.
- In a food processor with the S-blade, coarse-grind the sunflower seeds and walnuts. Put in a medium-size bowl.
- With the food processor and the S-blade, purée celery, zucchini, cilantro, parsley, lemon juice, salt, cayenne chili powder, chili blend powder, cumin, and coriander.
- Add this purée into the seed and nut mix and mix everything well by hand.

Sacha Inchi Sunflower Seed Pâté

This mild pâté is a great accompaniment to flavored crackers.

Time: 10 minutes • Makes about 2 cups

2 cloves garlic
1 cup sacha inchi seeds
1 cup sunflower seeds
½ cup walnuts

⅓ cup hemp oil
¼ cup orange juice
1 tsp sea salt

- In a food processor, process all ingredients together until smooth. Keep refrigerated for up to 2 weeks.

Creamy Cilantro Tahini Pâté

BLOSSOMING LOTUS, PORTLAND, OREGON

Smooth and flavorful, this pâté is ideal for vegetable dipping or to add to a salad for a nutritional boost.

Time: 15 minutes active; 2 hours presoak • Makes about 4 servings

2 cups sunflower seeds (soaked 2 hours)
2 cups rough chopped fresh cilantro
1 ½ peeled garlic cloves
1 ½ tbsp chopped jalapeño
3 tbsp small diced yellow onion
6 tbsp fresh lemon juice

2 tsp agave
6 tbsp raw tahini
3 tbsp olive oil
1 ½ tbsp ground cumin
1 tsp salt

- Soak the sunflower seeds for 2 hours, rinse, and drain.
- Clean the cilantro and chop thoroughly.
- Blend everything except sunflower seeds in a blender, put into large container, add sunflower seeds, and stir. Blend in batches in food processor. Adjust salt to taste.

Sacha Inchi Butter

A delicious omega-3-rich, nut-free alternative to peanut butter.

Time: 10 minutes • Makes ¾ cup

1 cup sacha inchi seeds
2 tsp hemp oil
Dash of sea salt

- Using a food processor quipped with the S-blade, process all ingredients for 5–10 minutes, or until desired creaminess is reached. Remove with rubber spatula.

Sacha Inchi and Sunflower Seed Chocolate Spread

A delicious, healthy, omega-3-rich, nut-free alternative to Nutella.

Time: 10 minutes active; 8 hours presoak

½ cup sacha inchi seeds
1 cup sunflower seeds
2 tbsp roasted carob powder
1 large pitted date (fresh or soaked for 8 hours)

- Using a food processor quipped with the S-blade, process all ingredients for 5–10 minutes, or until desired creaminess is reached. Remove with rubber spatula.

Seasonal Fig Jam

Great on toasted sprouted bread, or atop a warm morning porridge. The riper/ softer the figs, the sweeter the jam. Make when figs are at their peak ripeness.

Time: 5 minutes • Makes about 1 cup

1 cup chopped fresh, very ripe figs (such as black mission)
3–4 large Medjool dates, pits removed (amount depends on
    how naturally sweet the figs are)
1 tbsp lemon juice
⅓ cup apple juice
2 tbsp chia seeds

- Place all ingredients in a blender and process until smooth.
- Cover and refrigerate mixture for 30 minutes before serving to allow the chia seeds to thicken the mixture. It will keep for 3–4 days, refrigerated.

Fresh Fruit Strawberry Jam

Excellent on crispy crackers or as a dessert spread/sauce.

Time: 10 minutes active; 30 minutes to chill • Makes about 1 cup

1 ½ cup diced strawberries
3 large Medjool dates, pits removed
1 tbsp lemon juice
1 tbsp chia seeds

- Use a blender to blend *half* of the strawberries with the dates, lemon juice, and chia seeds until smooth.
- Stop machine and add the remaining strawberries. Pulse blender several times until berries are incorporated yet still chunky.
- Cover and refrigerate mixture for 30 minutes before serving to allow the chia seeds to thicken the jam and the flavors to develop. It will keep for 3–4 days, refrigerated.

Guacamole

GORILLA FOOD, VANCOUVER, BRITISH COLUMBIA

Classic, simple guacamole.

Time: 10 minutes • Makes 1 cup

2 medium avocados
1 ½ tbsp lemon juice
½ tsp salt
⅛ small onion

• Mash the avocado, lemon juice, and salt together with a fork. With a knife, finely mince the onion and mix into the avocado mash-up.

....

Ranch Dipping Sauce

Serve as a dip for homemade root fries or carrots, or even as a salad dressing.

Time: 10 minutes • Makes about 1 cup

1 cup young coconut meat (about 2 coconuts)
⅓ cup young coconut water (add additional if needed for blending)
2 tbsp apple cider vinegar
1 ¼ tsp onion powder
½ tsp sea salt
¼ tsp dried dill
1 medium clove garlic

• Use a blender to blend all ingredients together until creamy and smooth. Add additional coconut water if needed to blend properly.

Lemon Tahini Dipping Sauce

Ideal with carrots, celery, and cucumber.

Time: 5 minutes • Makes about ½ cup

3 tbsp tahini
3 tbsp lemon juice
3 tbsp water
¼ tsp garlic powder
1 minced scallion (white and light green parts only)
1 tbsp minced parsley

- Mix all the ingredients together in a jar. Refrigerate when not in use—it will keep for several weeks.

Salsa

GORILLA FOOD, VANCOUVER, BRITISH COLUMBIA

Flavorful fresh raw salsa can be added to a salad or can be eaten simply with an avocado.

Time: 10 minutes • Makes about 1 cup

1 fresh date
⅛ tsp sea salt
¹⁄₁₆ tsp cayenne chilies
1 tbsp lemon juice

⅛ medium yellow onion
2 medium tomatoes
¼ bunch cilantro

- In a food processor with the S-blade, blend the date, salt, and cayenne chilies. Slowly add the lemon juice to blend the dates until fully clear.
- Add the onions and tomatoes to the food processor and pulse-chop everything into a saucy but still chunky consistency. Pour into a bowl.
- With a knife, hand-mince the cilantro and stir well into the mix.

Sacha Inchi Black Bean Lime Salsa

Protein, micronutrient, and EFA rich, this salsa is a nutritional powerhouse.

Time: 10 minutes active; 2–3 hours to soak • Makes about 2 cups

Juice of ½ lime
2 cloves garlic, finely chopped
1 tomato, diced
½ onion, diced
1 cup black beans
1 cup coarsely chopped or torn cilantro

1 tbsp balsamic vinegar
1 tbsp hemp oil
½ cup chopped sacha inchi seeds
½ tsp cayenne pepper
¼ tsp sea salt

- In a bowl, combine all ingredients. Allow the salsa to sit for a few hours at room temperature so that the flavors infuse.
- Keep refrigerated for up to 1 week.

Hemp Black-Eyed Pea Cayenne Salsa

This is an ideal salsa to add protein-packed spicy flavor to your vegetables.

Time: 10 minutes active; 2–3 hours to sit • Makes about 2 cups

Juice of 1 lemon
1 tomato, diced
½ onion, diced
1 cup black-eyed peas
1 cup coarsely chopped or torn cilantro
1 tbsp balsamic vinegar

1 tbsp hemp oil
2 tbsp hemp seeds
½ tsp cayenne pepper
½ tsp chili flakes
¼ tsp sea salt

- In a bowl, combine all ingredients. Allow the mixture to sit for a few hours at room temperature so that the flavors infuse.
- Keep refrigerated, for up to 1 week.

Yellow Bell Pepper Pine Cheese

LIFE FOOD GOURMET, MIAMI, FLORIDA

Flavor-packed and full of nutrition, thanks to the nutritional yeast.

Time: 5 minutes • Makes 1 cup

1 yellow bell pepper
½ cup pine nuts
3 ½ tbsp nutritional yeast
1 tsp sea salt
1 tsp tarragon

1 tbsp olive oil
1 tsp apple cider vinegar
1 tsp agave nectar
1 small clove garlic

- Place all ingredients in a Vita-Mix or regular blender and blend thoroughly.

Live Cashew Cheese Base

BLOSSOMING LOTUS, PORTLAND, OREGON

Ideal as a base for other recipes or any time you need a cheese-like substitute. The rejuvelac (a liquid made from fermented grains) is rich in healthy bacteria that aid in digestion.

Time: 5 minutes active; 8 hours to sit • Makes about 2 ½ cups

2 cups soaked cashews
¾ cup rejuvelac

- Soak cashews overnight. Then rinse well, drain, and blend cashews in food processor. Slowly add the rejuvelac until the mixture is thick and creamy.
- Let the mixture sit at room temperature overnight to ferment. Don't fully cover, so that gases can escape.
- Refrigerate when fermenting time is over.

Live Seasoned Cashew Cheese

BLOSSOMING LOTUS, PORTLAND, OREGON

Ideal as a topping for steamed broccoli, cauliflower, and carrots.

Time: 5 minutes (once Live Cashew Cheese Base is made)
Makes about 5 servings

2 ½ cups Live Cashew Cheese Base (see p. 198)
Pinch sea salt
Pinch black pepper
1 tsp extra-virgin olive oil (can substitute hemp oil)

- Put all ingredients into a blender. Blend until smooth.

Live Cashew Sour Cream

BLOSSOMING LOTUS, PORTLAND, OREGON

"Live," in that it's made with soaked (germinated) nuts, this recipe is creamy yet easy to digest.

Time: 5 minutes (once Live Seasoned Cashew Cheese is made)
Makes about 10 servings

1 ¼ cups Live Seasoned Cashew Cheese (see above)
4 tsp fresh lemon juice
½ tsp sea salt
Pinch black pepper
4 tsp extra-virgin olive oil (can be substituted with hemp oil)
2 ½ tbsp filtered water

- Blend all ingredients in food processor.

Marinara/Pizza Sauce

Light and flavorful, this is a versatile sauce.

Time: 5 minutes active; 15 minutes to soak • Makes 4 cups

1 ½ sun-dried tomatoes

2 Roma tomatoes

¼ cup olive oil

1 ½ tbsp Italian seasoning

1 tbsp oregano

1 handful fresh basil

2 tbsp palm nectar granuals

1 tsp apple cider vinegar

1 ½ tsp sun-dried sea salt

1 tbsp anise seeds (optional)

- Soak sun-dried tomatoes for about 15 minutes. Blend all ingredients except sun-dried tomatoes. Add sun-dried tomatoes and blend again.

UnMotza Macadamia Cheese Sauce

For a lighter recipe, substitute the macadamia nuts with pumpkin seeds. Always use a rubber spatula to scrape the sides of the processor and get an even mixture.

Time: 5 minutes • Makes about 3 cups

1 ½ cups macadamia nuts

1 ½ cups Irish moss gel

½ cups nutritional yeast

2 tsp sun-dried sea salt

¼ cup nut milk

- Place nuts in a food processor with an S-blade and grind them into a thick paste. Add the remaining ingredients and pulse a few more times.

Sun-dried Tomato Sauce

CRUDESSENCE, MONTREAL, QUEBEC

This flavor-packed versatile sauce can be used on pizzas or as a tasty accompaniment for roasted vegetables.

Time: 10 minutes • Makes about 5 cups

2 ½ cups sun-dried tomatoes
5 medium fresh tomatoes, quartered
1 carrot, sliced coarsely
1 tbsp garlic paste
1 tbsp olive oil
½ cup Mediterranean spices
½ tsp cayenne powder
½ cup onion, sliced coarsely
½ cup onion, finely sliced
½ cup parsley, finely chopped

- Rehydrate the sun-dried tomatoes in a bowl of water while you gather the other ingredients.
- In the blender, process the rehydrated tomatoes (without the soaking water) with the rest of the ingredients except the onions and parsley.
- Do not fill the blender more than ⅔ full at a time. You may have to repeat the process a few times, placing the fresh tomatoes in the bottom of the blender and adding the other ingredients on top.
- Pour the tomato mixture into a bowl, and fold in the coarsely sliced and finely chopped onions and the chopped parsley.

Note: Mediterranean spices can be bought as a mix. The spices include annatto, cumin, oregano, sweet and hot paprikas, rosemary, and saffron.

Live Avocado Goddess Sauce

BLOSSOMING LOTUS, PORTLAND, OREGON

Can be used as a chip or vegetable dip, or as an addition to a salad.

Time: 15 minutes • Makes about 3 cups

2 ripe avocados
2 ½ tbsp apple cider vinegar
2 tsp minced garlic
1 tbsp red onion
½ tbsp dried parsley
½ tsp salt

2 tsp ground cumin
½ tsp pepper
1 ½ tbsp nutritional yeast
2 ½ tbsp fresh lemon juice
1 cup filtered water

- Put all ingredients in a large container and blend with an immersion wand until smooth with a medium thickness. Add more salt, apple cider vinegar, and lemon juice to taste.
- Store in the refrigerator for up to 10 days.

Cashew–Truffle Oil Cream

MILLENNIUM RESTAURANT, SAN FRANCISCO, CALIFORNIA

Tastes as fancy as it sounds.

Time: 10 minutes active; 1 hour presoak

4 tbsp raw cashews, presoaked in warm
  water 1 hour
2 tsp nutritional yeast
Water as needed

1 tsp balsamic vinegar
Black truffle oil to taste
Salt to taste

- Rinse and drain the cashews. In a blender, blend the cashews with the yeast and enough water to just cover the nuts.
- Slowly add more water while blending until the consistency is that of thickened cream.
- Add the balsamic vinegar, truffle oil, and salt to taste.

Regular, Thick, and Whipped Cashew Cream

TAL RONNEN

To make the regular cashew cream into thick cream, reduce the amount of water in the blender so that the water just covers the cashews.

Time: 10 minutes active; overnight presoak
Makes about 3 ½ cups regular cream or 2 ¼ cups thick cream

2 cups whole raw cashews (not pieces, which are often dry), rinsed very well under cold water

- Put the cashews in a bowl and add cold water to cover them. Cover the bowl and refrigerate overnight.
- Drain the cashews and rinse under cold water. Place in a blender with enough fresh cold water to cover them by 1 inch.
- Blend on high for several minutes until very smooth. (If you're not using a professional high-speed blender, such as a Vita-Mix, which creates an ultra-smooth cream, strain the cashew cream through a fine-mesh sieve.)

WHIPPED VARIATION

Lighter and possibly more delicious than cashew cream.

Time: 5 minutes active (once Thick Cashew Cream is made); 2 hours to chill
Makes about 2 cups

1 cup Thick Cashew Cream (see above)   ¼ cup water
¼ cup light agave nectar   ⅔ cup refined coconut oil, warmed
½ tsp vanilla extract   until liquid

- Place the Thick Cashew Cream in a blender and add the agave nectar, vanilla, and ¼ cup water. Blend until thoroughly combined.
- With the blender running, slowly drizzle the liquid coconut oil in through the hole in the blender lid. Blend until emulsified.
- Pour into a bowl and chill in the refrigerator, covered, for 2 hours. Stir before serving.

Spiced Sesame Sauce

GORILLA FOOD, VANCOUVER, BRITISH COLUMBIA

Ideal on raw, steamed, or stir-fried vegetables.

Time: 10 minutes • Makes 1 cup

⅓ cup sesame seeds
1 ½ tbsp lemon juice
⅔ cup water
⅛ tsp salt
½ tsp cumin

¼ tsp paprika
½ date
1/16 tsp cayenne
5 sprigs parsley

- In a blender, combine all ingredients except the parsley and blend until smooth. With a knife, mince the parsley and place in a bowl.
- Pour the sauce in with the parsley and stir together.

Miso Mushroom Gravy

LIVE ORGANIC FOOD BAR, TORONTO, ONTARIO

Great on your favorite nut loaf.

Time: 10 minutes • Makes about 4 cups

½ cup of olive oil
Juice of 1 lemon
3 stalks of celery, chopped
⅛ cup Nama Shoyu soy sauce

3 cloves garlic
1 ¾ cup miso (brown rice if possible)
¾ cup pitted dates
1 ¾ cup sliced button mushrooms

- In a blender, mix all ingredients except for the sliced mushrooms and blend until smooth.
- In a medium bowl, add the mushrooms to the gravy. This keeps for a few days in a cold refrigerator.

Roasted Garlic Dressing

CANDLE 79, MANHATTAN, NEW YORK

The chefs at Candle 79 say they always keep a good amount of Roasted Garlic Dressing on hand. Not only is it excellent on salads, but they also serve it with roasted or steamed vegetables, rice, and grains.

Time: 15 minutes active; 35–40 minutes to cook • Makes 2 cups

1 cup garlic cloves, peeled
1 cup extra-virgin olive oil
½ cup filtered water
¼ cup balsamic vinegar
¼ cup sherry wine vinegar or red wine vinegar
1 tbsp white miso
2 tbsp minced fresh thyme or 1 tsp dried oregano
Pinch of ground nutmeg
1 tsp sea salt
2 tsp freshly ground black pepper
¼ cup dried currants or cranberries (optional)

- Preheat oven to 350°F.
- Put the peeled garlic cloves in a baking dish and cover with extra-virgin olive oil. Cover the dish with foil and roast approximately 25 minutes or until golden brown. When cool enough to handle, remove the garlic with a slotted spoon and transfer to a blender.
- Reserve the roasted garlic oil for another use.
- Add the water, vinegars, miso, thyme or oregano, nutmeg, salt, and pepper to the garlic, and blend until smooth. Add a bit more water if necessary. Add currants or cranberries, if desired. The dressing will keep in the refrigerator, covered, for 1 week.

Note: The reserved roasted garlic oil will keep in a covered container for 1 week. Use it to drizzle over pasta, vegetables, and salad greens. It's also excellent when used in sautés and stir-fries.

Maple Cayenne Tahini Dressing

This is a full-flavored dressing with a bit of bite. The tahini offers a good amount of calcium, and the cayenne pepper helps get the blood flowing.

Time: 10 minutes • Makes about 1 ½ cups

½ clove garlic
½ cup balsamic vinegar
½ cup hemp oil
¼ cup water
2 tbsp tahini

½ tbsp dill
½ tsp cayenne pepper
¼ tsp maple syrup
Sea salt to taste

- Put all ingredients into a blender. Blend until smooth.

Garlic–Green Peppercorn Dressing

MILLENNIUM RESTAURANT, SAN FRANCISCO, CALIFORNIA

Smooth, with a bite of pepper, this dressing adds full-bodied flavor to any salad.

Time: 15 minutes • Makes 2 ½ cups

1 ½ cups Regular Cashew Cream
   (see p. 203)
1 tsp fine-ground green peppercorn
¼ tsp coarsely ground green peppercorn
¼ cup extra-virgin olive oil
Salt as needed

1 clove minced garlic
¼ cup cider vinegar
2 tsp nutritional yeast
1 tbsp minced fresh dill
1 tbsp minced fresh tarragon

- Blend the Regular Cashew Cream through the green peppercorn. Slowly emulsify in the oil.
- Adjust salt. Whisk in remaining ingredients.

Chili-Lime Dressing

BLOSSOMING LOTUS, PORTLAND, OREGON

A full-flavored, healthy version of the Mexican-inspired classic.

Time: 15 minutes • Makes about 10 servings

2 ½ peeled garlic cloves

1 tsp deseeded New Mexico pepper

Pinch cayenne pepper

2 tsp fresh orange juice

5 tbsp fresh lime juice

1 tbsp raw agave syrup

1 tsp cumin powder

2 tsp medium diced red onion

½ cup loosely packed fresh cilantro

½ avocado

½ cup filtered water

1 ½ tsp sea salt

12 tbsp olive oil or hemp oil

- In a blender, process the garlic, peppers, orange juice, and lime juice together until smooth.
- Add in the agave syrup, cumin, onion, and cilantro.
- Add the avocado, water, and salt. Blend and drizzle in oil while blender is on low.

....

Sweet Mustard Dressing

A twist on a traditional honey mustard dressing, this is ideal as a vegetable dip or salad dressing.

Time: 5 minutes • Makes about ⅔ cup

⅓ cup yacon syrup

¼ cup Dijon mustard

2 tbsp apple cider vinegar

4 tsp almond butter

½ tsp sea salt, or to taste

- Mix all ingredients together in a jar and keep refrigerated till ready to use.

Vinaigrette

Tal Ronnen

A simple classic.

Time: 5 minutes • Makes 2 servings

1 tbsp white wine vinegar
½ tsp agave nectar
3 tbsp extra-virgin olive oil
Sea salt and freshly ground black pepper

- Place the vinegar and agave nectar in a small bowl, then, whisking constantly, slowly pour in the oil in a thin stream. Season with salt and pepper to taste.

Lentil Dressing

Flavorful, protein-rich salad dressing.

Time: 5 minutes • Makes about ¾ cup

⅔ cup cooked green lentils, unsalted
6 tbsp olive oil
2 tsp miso paste
2 tsp Dijon mustard
4 tsp balsamic vinegar
Freshly cracked pepper

- Use a blender or food processor to blend the lentils, oil, miso, mustard, and vinegar into a semi-smooth mixture. Add pepper to taste.

Coconut-Cumin Dressing

Delicious creamy texture and a West Indian–inspired flavor. This dressing uses young Thai coconuts, which are white on the outside, unlike the more commonly found brown "hairy" ones, otherwise known as "old coconuts." Young Thai coconuts can be found in Whole Foods Markets and most health-food stores, as well as in Asian markets.

Time: 10 minutes • Makes about 1 cup

1 cup young coconut meat (about 2 coconuts)
2 tbsp young coconut water
1 ½ tbsp balsamic vinegar
2 tbsp lime juice
1 ¼ tsp cumin powder
1 medium clove garlic
2 tsp palm sugar
¼ tsp sea salt

• Blend all ingredients in a blender until smooth.

Creamy Thai Dressing

Matthew Kenney

Smooth, filling, and delicious, this is an ideal dressing for any salad.

Time: 10 minutes • Makes 2 ½ cups

¾ cup sesame oil
½ cup Nama Shoyu soy sauce
¼ cup olive oil
¼ cup lime juice
1 tbsp maple syrup
3 tsp red chili flakes
1 tsp sea salt
¼ cup chopped cashews

• Blend all ingredients in a blender until smooth.

Raw

Pasta with Marinara Sauce

KARYN'S FRESH CORNER CAFÉ, CHICAGO, ILLINOIS

Raw and light zucchini "pasta" with a fully-flavored sauce.

Time: 10 minutes prep; 30 minutes to marinate • Makes 4 servings

2 large zucchini
½ tbsp sea salt
¼ cup extra-virgin olive oil
1 ½ cups sun-dried tomatoes, soaked in water
4 cups fresh sweet tomatoes
2 cloves garlic
1 ½ tbsp agave nectar
¼ cup green olives
¼ cup black olives
1 cup basil
⅛ cup tamari

- Peel the zucchini (optional) and shred into long spaghetti strands with spiralizer or peeler.
- Marinate zucchini with the salt and 2 tbsp olive oil for 30 minutes.
- Gently squeeze out excess water.
- In a blender, process the sun-dried tomatoes with the soaking water, 2 cups of fresh tomatoes, garlic, and agave nectar. Pour in a large bowl.
- Slice or chop the olives, dice the remaining fresh tomatoes, chiffonade the basil and add them to bowl.
- Add the remaining olive oil and the tamari. Stir gently.
- Add drained zucchini to sauce and toss to coat.

Asian Noodle Bowl with Kelp Noodles and Almond Chili Sauce

Raw

Cru, Los Angeles, California

A unique and filling sauce on light sea-vegetable noodles

Time: 10 minutes prep • Makes 2 servings

ALMOND CHILI SAUCE

½ cup raw almond butter
¼ cup tamari
¼ cup olive oil
⅛ cup lemon juice
⅛ cup water
¼ cup agave nectar
1 tsp salt
1 pinch chipotle powder
1 ½ tbsp chili flakes

• Mix everything together in a bowl until well combined.

NOODLES

½ cup shredded cabbage
1 cup kelp noodles
½ cup cucumbers, sliced

GARNISH

Cilantro, finely chopped
Radish sprouts
Cashews, chopped
Chilies, chopped

• Toss noodle ingredients with 1–2 ounces of the Almond Chili Sauce. Garnish with cilantro, radish sprouts, cashews, and chilies.

Pumpkin Gnocchi with Jerusalem Artichoke Purée

MILLENNIUM RESTAURANT, SAN FRANCISCO, CALIFORNIA

This is a wonderful dish. It is a bit more complex to make than some of the other recipes, but it's well worth the effort. You might want to first try this recipe on a weekend!

Prep time: 40 minutes • Makes 4 servings

GNOCCHI

2 large yellow Finn potatoes, baked
2 cups cooked pumpkin, skin removed
⅔ cup of all-purpose flour, plus more as needed
   (can use quinoa or buckwheat flour to make gluten free)
Salt to taste

- Remove the skin from potatoes. Either pass the potatoes and pumpkin through a potato ricer or mash the potatoes and pumpkin until smooth.
- Slowly mix in the flour until you form a soft dough.
- Cut off ¼ of the dough and roll into a 1-inch-thick rope. Use the remaining dough to make 3 more ropes.
- Cut each rope into ½-inch segments, and pinch the sides of each piece of dough so it looks like a bow tie. Place the finished gnocchi on a floured pan.
- Freeze the gnocchi for 1 hour.
- Near serving time, bring at least 1 gallon of salted water to a boil. Add half of the gnocchi.
- Cook for 5–6 minutes until the gnocchi float to and remain on the surface for 2 minutes.
- With a slotted spoon, remove gnocchi to a plate, and coat with a little extra-virgin olive oil. Repeat with the remaining gnocchi.

JERUSALEM ARTICHOKE PURÉE

1 leek, cleaned medium dice
4 tbsp olive oil
1 cup peeled Jerusalem artichokes, medium dice
1 cup peeled diced parsnip
Vegetable stock to cover
2 tsp nutritional yeast
Salt and pepper to taste
Fruity extra-virgin olive oil, truffle oil, or nut oil to taste
Fresh grated nutmeg to taste
Fried sage leaves
Chopped parsley

- In a saucepan, sweat the leek in the oil over medium-low until soft.
- Add the artichoke and parsnip. Cover with stock by 1 inch over the vegetables.
- Add the yeast. Simmer, covered, until the vegetables are soft.
- Purée the sauce until smooth in a blender. Adjust salt and pepper.
- Toss the gnocchi in the sauce. Place on a serving plate and drizzle with olive, truffle, or walnut oil.
- Grate nutmeg over the top, crumble fried sage leaves over that, and sprinkle with parsley.

Raw Zucchini and Carrot Lasagna with Almond "Ricotta"

RAVENS' RESTAURANT, MENDOCINO, CALIFORNIA

A true gourmet recipe, ideal for entertaining and showcasing perfectly balanced flavors.

Time: 25 minutes prep; 8 hours to marinate • Makes 6 servings

Note: Start this recipe the day before serving or early in the morning: the vegetables need to marinate at least 8 hours. Nutritional yeast, often used in vegan cooking, has a nutty, cheesy flavor. It's available at health food stores.

6 zucchini (about 2 ¾ pounds), trimmed and thinly shaved lengthwise into strips

2 large carrots, trimmed and thinly shaved lengthwise into strips

5 tbsp lemon juice (from about 2 lemons)

5 tbsp olive oil, divided

1 ¾ tsp kosher salt, divided

¾ tsp freshly ground black pepper, divided, + more to taste

1 ½ cups raw slivered almonds

3 cloves garlic

2 bunches kale (about 2 pounds), ribs removed, leaves roughly chopped

1 cup lightly packed basil leaves (about 1 bunch)

2 tbsp nutritional yeast (optional)

3 cups baby spinach, cut into thin strips

3 medium tomatoes, cored and roughly chopped

3 medium beets, preferably Chioggia (also called candy-striped) or yellow, peeled and very thinly sliced

Fennel bulb, halved lengthwise, cored, and very thinly sliced

Chopped chervil, or another favorite herb, for garnish

- In a large bowl, toss the zucchini and carrots with the lemon juice, 1 tbsp oil, ¾ tsp salt, and ¼ tsp pepper. Cover and chill. Marinate, gently tossing 2 or 3 times, 8 hours or overnight; drain well, reserving marinade. Meanwhile, put almonds in a large bowl, cover with 2 inches water, and set aside to soak 8 hours or overnight; drain well.
- In a food processor, purée the almonds, garlic, and 2 tbsp oil until smooth; transfer to a large bowl and set aside. In the same processor (no need to clean it), working in batches as needed, pulse the kale and basil, scraping down the sides often, until very finely chopped; transfer to the bowl with the almond mixture. Add the yeast, if using, ½ tsp salt, and ¼ tsp pepper, stirring to combine; set aside.
- Toss the spinach with the reserved zucchini marinade, squeezing with your hands to wilt it; wring dry, and discard marinade. Purée the tomatoes in a blender until smooth; season with ½ tsp salt and ¼ tsp pepper.
- To assemble lasagna, alternate layers of zucchini and carrots with kale-almond mixture and spinach, making 4 layers of each, in a 9" × 13" dish. Arrange beets on top, then scatter fennel over beets.
- Serve immediately, or chill for a few hours first. It can be eaten cold or at room temperature. To serve, pour some of the tomato purée onto each plate, then top with a piece of lasagna. Garnish with chervil and pepper to taste, and drizzle with remaining 2 tbsp oil.

Summertime Succotash with Creamy Rosemary-Garlic Sauce

One of the most delicious ways to get full, this recipe is ideal for after a long bike ride, when hunger is at its peak.

Time: 40 minutes prep • Makes 4–6 appetizer servings

4 cups zucchini, diced (about 2)

½ tsp sea salt

1 cup shelled fresh or frozen lima beans, blanched for 2 minutes

1 tbsp olive oil or hemp oil

2 tsp lemon juice

1 tsp miso paste

1 garlic clove, mashed

6 sprigs roasemary, for garnish

Black pepper, for garnish

- Using sharp knife, slice the summer squash lengthwise as thinly as possible to form long, flat noodle-like strips. Discard the core area with the seeds as it contains too much moisture. Cut each prepared squash strip into two pieces so that when folded in half, the squash resembles a small ravioli in size.
- Toss the strips with the sea salt. Place in a colander over the sink or a large bowl. Let sit for 30 minutes, rinse, then pat dry.
- To make the filling, place the blanched lima beans, oil, lemon juice, miso paste, and garlic in a food processor and purée until smooth. Set aside until needed.

CREAMY ROSEMARY-GARLIC SAUCE

¾ cup cashews

½ cup water

1 tbsp lemon juice

1 tsp red wine vinegar

¾ tsp minced fresh rosemary

1 large garlic clove

¼ tsp sea salt

- Place the cashews and water in a blender, and process until a thick and smooth cream has formed. Add the remaining ingredients and blend until incorporated.
- To assemble the ravioli, place a small spoonful of the filling mixture on one side a squash strip, then fold over the other side. Pat lightly to seal. Repeat to make the remaining ravioli.
- On each serving plate, pour a small bed of Creamy Rosemary-Garlic Sauce, then place a few ravioli on top. Garnish with a rosemary sprig and black pepper and serve.

Beet Ravioli with Basil Macadamia Ricotta

A fancier dish that is best served as an appetizer. You can also use striped heirloom beets for a stunning presentation, but regular red beets will work as well.

Time: 20 minutes prep • Makes 4–6 appetizer servings

3–4 medium yellow beets, peeled and trimmed
1 cup macadamia nuts
½ tsp sea salt
2 tbsp tahini paste
2 tsp lemon juice
2 tbsp chopped fresh basil
Hemp oil, for garnish
Basil, shredded, for garnish
Freshly cracked black pepper, for garnish

- Shave the beets very thinly into circular rounds by using a mandolin or a sharp knife. Set aside.
- Place the macadamia nuts and sea salt in a food processor, and process until finely ground. Transfer to a small bowl and mix in the tahini, lemon juice, and basil to form a "ricotta."
- For each ravioli, spoon a small portion of ricotta into the center of a beet round, and top with a second beet round, like a sandwich. Once the ravioli have been plated, drizzle with a little hemp oil, basil, and pepper.

Zucchini Pasta with Chunky Tomato Sauce

This zucchini has a classic-style sauce, but the addition of Brazil nuts or walnuts gives it a non-traditional crunch.

Time: 15 minutes active prep • Makes 4 servings

4 medium zucchini
1 tsp + ½ tsp sea salt
20 sundried tomatoes, soaked in warm water until soft (about 30 minutes)
½ cup tomato soak water
2 Roma tomatoes, chopped
1 medium clove garlic
2 heaping tbsp chopped fresh basil
1 heaping tbsp chopped fresh oregano
1 tbsp raisins
1 tbsp hemp oil
½ cup chopped walnuts or chopped Brazil nuts
Pinch red pepper flakes

- Trim the ends of the squash. Using a hand-held vegetable peeler, carefully strip the squash, layer by layer, into noodle-like pieces and gather into a colander (for best results, discard the watery center section that holds the seeds).
- Toss squash strips with 1 tsp of sea salt and place the colander over a large bowl to catch excess moisture. Let rest for 30 minutes.
- After 30 minutes, wash the squash thoroughly with warm water to remove any excess salt, and let drain for 5 more minutes.
- In a food processor, blend together the sundried tomatoes, ½ cup soak water, fresh tomatoes, garlic, basil, oregano, raisins, hemp oil, and ½ tsp of sea salt into a chunky paste.
- Add the nuts and pulse a few times to chop the nuts finely (but do not blend).
- Toss the sauce with the zucchini strips and sauté over medium-low heat for 1–2 minutes to warm through.

Variation: Skip the sauté step and serve at room temperature as a delicious raw dish.

Spaghetti Squash with Three-Step Marinara

Trade in that nutrient-void pasta slopped with a jar of some name-brand red stuff for this fresh and flavorful natural meal. Mesquite and the sweet herb stevia lend a mild sweetness to balance out the acid undertones from the tomatoes, and plentiful fresh herbs bring this dish to life. Served with spaghetti squash, this is a feel-good delicious dinner through and through.

Time: 15 minutes prep; 25 minutes to cook • Makes 2–3 servings as a main dish

2 pounds spaghetti squash (about 1 medium)
1 cups chopped onion
3 medium cloves garlic, minced
2 tbsp coconut oil
3 cups deseeded and chopped tomatoes
　　(about 5 medium)
⅛ tsp green stevia or ½ tsp agave nectar

½ tsp mesquite powder (optional)
3 tbsp chopped fresh oregano
2 tbsp parsley, + 2 tbsp chopped,
　　for garnish
¼ tsp sea salt
½ tsp black pepper

- Cut the squash in half, lengthwise. Place cut side down in a baking pan, and add ½ inch of water to the pan. Bake at 350°F for 40 minutes.
- While squash is cooking, make the sauce. In a large saucepan, sauté the onions and the garlic for about 5 minutes, or until the onions begin to become translucent.
- Stir in all the other ingredients and bring to a boil. Reduce heat to low, and simmer for 20 minutes.
- Place the sauce in a blender and purée until fairly smooth. Return the sauce to the pan, and keep at a low simmer until ready to be served.
- When the squash has finished baking, remove from the oven and allow to cool for a couple minutes. Flip the squash over, and scrape the flesh with a fork to extract the spagetti-like strands from the shell.

To serve
- Place a mound of spaghetti squash on a plate, top with sauce and garnish with parsley. Or combine the squash with the sauce directly in the pan and sauté for 2–3 minutes to reheat and remove any excess moisture.

Note: Do *not* substitute white stevia for the green in this recipe 1:1. White stevia is much sweeter than green. If you choose to use white stevia, add only a tiny pinch to avoid over-sweetening.

Summer Squash Fettuccini with Lemon Pepper Cream

If you make one recipe, make this one: amazingly gourmet tasting, so simple to make. Zucchini can be substituted for the yellow squash. Don't worry about the high quantity of salt—most of it will be washed away.

Time: 15 minutes prep and blend; 1–2 hours to soak cashews; 30 minutes to soak squash; 3 minutes to cook • Makes 6 servings

4–5 medium long yellow summer squash
1 tsp + ¼ tsp sea salt, divided
⅔ cup cashews, soaked in water for 1–2 hours
¾ cup water
1 tsp lemon zest
1 tbsp apple cider vinegar
½ tsp freshly ground black pepper, or more to taste
1 tbsp coconut oil
1 leek, thinly sliced

- Trim the ends of the squash. Using a hand-held vegetable peeler, carefully strip the squash, layer by layer, into noodle-like pieces and gather into a colander (for best results, discard the watery center section that holds the seeds).
- Toss the squash strips with 1 tsp of sea salt and place the colander over a large bowl to catch excess moisture. Let rest for 30 minutes. After 30 minutes, wash the squash thoroughly with warm water to remove any excess salt and let drain.
- Place the soaked, soft cashews in a blender with the water, lemon zest, vinegar, freshly ground pepper, and remaining ¼ tsp sea salt. Blend for several minutes on high until a very smooth cream has formed.
- Heat the coconut oil in a large saucepan over medium heat, then add the leeks. Cook for 2–3 minutes until the leeks have softened. Reduce heat to low and add the blended cream to the pan.
- Stirring constantly, gently warm the mixture for about 1 minute. Add the prepared squash noodles to the pan and toss with the sauce for a just moment until to heat the squash—about 30 seconds to 1 minute.
- Remove from heat and serve immediately, garnishing with additional freshly cracked pepper, if desired.

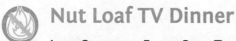

Nut Loaf TV Dinner

Raw

LIVE ORGANIC FOOD BAR, TORONTO, ONTARIO

A filling and flavorful loaf, but you'll need a dehydrator for this recipe.

Time: 15 minutes prep; 12–18 hours to dehydrate • Makes 4 servings

3–4 cloves garlic
1 red onion
3 ½ cups walnuts, soaked
2 cups pumpkin seeds, soaked
2 cups and 1 tbsp sunflower seeds, soaked
Water to blend
2–3 stalks celery
1 inch ginger, finely chopped
⅛ bunch fresh parsley, chopped
⅛ bunch fresh rosemary, de-stemmed and chopped
¼ tsp cumin seeds
3 tbsp olive oil
¼ tsp Celtic sea salt

- In a food processor, blend the garlic until pasty. Blend the onion until finely chopped.
- Blend the walnuts, pumpkin, and sunflower seeds with a touch of water to blend until smooth, but not too watery; it should be thick. Transfer into a medium-size bowl.
- Finely chop the celery. Add to mixture with remaining ingredients. Adjust salt to taste.
- Form into loaf shape about 4 inches wide and 8 inches long. Cut into ¼-inch slices.
- Lay out onto dehydrator (it is not necessary to use teflex sheets). Dehydrate for 12–18 hours at 115°F until outside is crispy and inside is slightly moist.
- Serve with cauliflower mash potatoes and top with Miso Mushroom Gravy (see p. 204).

Thai Vibe

Live Organic Food Bar, Toronto, Ontario

Raw

This Thai-inspired noodle salad contains kelp noodles, which are available at local health stores and come ready to eat. Kelp is seaweed high in minerals and vitamins and balances the body's pH levels, among other benefits. Maca will give you fuel for the day. This sauce can be used on pasta or rice as well. Enjoy!

Time: 10 minutes prep • Makes 2–4 servings

MIXED VEGETABLES

½ packages of enoki mushrooms, picked apart
1 bunch of dandelion greens, chiffonade
1 red pepper, julienne
1 red onion, julienne
1 bunch of cilantro, lightly chopped
1 bag kelp noodles
Sunflower sprouts or almonds, for garnish

SAUCE

1 cup of almond butter
2 tbsp chopped ginger
5 cloves of garlic
½ red onions
1 ½ red peppers
2 small carrots

1 and ½ Thai chilies or (jalapeños)
¼ cup of rice vinegar
½ cup of lime juice
2 tbsp of agave nectar
1 tbsp of Maca
Celtic sea salt to taste

- Blend all sauce ingredients together in a blender until smooth.
- In a large bowl, combine all vegetables and the kelp noodles. Add the sauce and toss and serve.
- Garnish with sunflower sprouts or almonds.

Roasted Maple Glazed Brussels Sprouts
with Chestnuts

Delicious and surprisingly filling, this is an ideal side dish to compliment a holiday meal.

Time: 10 minutes prep • Makes 6 servings

¼ cup shallots, minced

2 tbsp coconut oil

4 cups Brussels sprouts, halved lengthwise

1 cup chestnuts, roughly chopped, roasted, and peeled
  (fresh, jarred, or vacuum-packed)

1 tsp chopped fresh thyme

¼ tsp each sea salt and black pepper, to taste

1 ½ tbsp maple syrup (grade B)

1 tsp Dijon mustard

1 tbsp apple cider vinegar

- Preheat oven to 425°F.
- In a large roasting pan or ovenproof sauté pan over medium heat, cook the shallots in the oil until soft, about 2 minutes.
- Add the Brussels sprouts, chestnuts, thyme, salt, and pepper and stir well.
- Transfer the pan to the oven and roast for 15 minutes. Remove from the oven and mix in the maple syrup, mustard, and vinegar with the vegetables. Toss well. Return the pan to the oven and continue roasting until the Brussels sprouts are tender, about 10 minutes more.
- Transfer to a serving dish and serve immediately.

Shanghai Rice Bowl

Fresh, Toronto, Ontario

The Shanghai is one of the daily special items on rotation at Fresh. This is a tasty rice bowl that always hits the spot and is easy to prepare.

Time: 10 minutes; 20 minutes for the rice • Makes 2 servings

⅓ cup water
4 baby bok choy, cut in half lengthwise
6 tbsp olive oil
6 tbsp tamari
3 cups shiitake mushrooms, stems removed and halved if large
4 cups cooked brown basmati rice
½ cup Tahini Sauce (see the next page)
2 tsp Mixed Herbs (see p. 227)
2 cups sunflower sprouts
2 tbsp hulled hemp seeds
1 cup cooked or canned chickpeas
2 lemon wedges, for garnish

- Put the water in a wok or skillet over high heat. Add the bok choy halves and cover. Steam 5 minutes until bok choy is almost tender. When water evaporates, add 2 tbsp olive oil, 2 tbsp tamari, and the shiitake mushrooms. Sauté 5 minutes until bok choy and mushrooms are tender. Set aside.
- Divide the cooked rice between 2 large rice bowls, and drizzle both with Tahini Sauce, 4 tbsp olive oil, and 4 tbsp tamari. Sprinkle with Mixed Herbs.
- Place the sautéed bok choy and shiitake mushrooms on the rice, and top with sunflower sprouts, hemp seeds, and chickpeas.
- Garnish with lemon wedges and serve.

TAHINI SAUCE

This is Fresh's most versatile sauce; it tastes great on everything. Tahini, or sesame butter, is made from ground sesame seeds and is high in protein and a good source of essential fatty acids. Tahini is the traditional accompaniment for falafel but can also be used in any kind of stuffed sandwich, as a salad dressing, or as a sauce with rice bowls or noodles.

Time: 5 minutes prep • Makes 4–6 servings

2 cloves garlic, minced
½ cup chopped parsley
½ tsp sea salt
2 tbsp lemon juice
⅔ cup filtered water
½ cup tahini

- In a blender, process the garlic, parsley, salt, and lemon juice until smooth.
- Add the water and tahini, and process until smooth. You may need to add a bit more water if your raw tahini is especially thick. Add water a tablespoon at a time until you get a pourable consistency.
- Store in a sealed container in the fridge for up to 4 days.

Temple Rice Bowl

FRESH, TORONTO, ONTARIO

If we acknowledged and treated our bodies with the same care and consideration as we do a place of worship, we would all be a lot healthier. This is a modest rice bowl that will make you feel good. The hummus, you will discover, goes wonderfully on brown rice.

Time: 10 minutes prep; 20 minutes for the rice • Makes 2 servings

⅓ cup water

4 baby bok choy, cut in half lengthwise

7 tbsp olive oil

4 tbsp tamari

8 wedges tomato

4 cups cooked brown basmati rice

2 tsp Mixed Herbs (see the next page)

1 cup Hummus (see the next page)

2 cups sunflower sprouts

2 tbsp hulled hemp seeds

4 slices red onion, chopped

2 lemon wedges, for garnish

- Put the water in a wok or skillet over high heat. Add the bok choy halves and cover. Steam until the bok choy is almost tender. When water evaporates, add 1 tbsp olive oil, 1 tbsp tamari, and the tomato wedges. Sauté 1 minute until the bok choy is tender, then set it aside.
- Divide the cooked rice between 2 large rice bowls. Mix the remaining 6 tbsp olive oil and 3 tbsp tamari together, then drizzle it over the rice. Sprinkle with Mixed Herbs.
- Place the Hummus in the middle of each rice bowl, and arrange the bok choy around the edge. Top with sunflower sprouts, hemp seeds, and red onion.
- Garnish with lemon wedges and serve.

HUMMUS

This is Fresh's basic hummus recipe. It can be adjusted in countless ways, depending on your preference. You could add parsley or cilantro for an herbal twist, some toasted cumin and cayenne for a bit of spice, or some roasted red peppers, black olives, or sun-dried tomatoes for a Mediterranean taste. If you prefer a milder garlic flavor, roast the garlic before using it.

Time: 5 minutes prep • Makes 8 servings

2 cups canned or cooked chickpeas
3 cloves garlic
2 tbsp tahini
4 tbsp lemon juice
½ tsp sea salt
1 tbsp filtered water

- In a food processor, purée all ingredients. Add more water if necessary to get the consistency you like.

MIXED HERBS

This combination of dried herbs adds flavor to many of the recipes at Fresh. The mixture will last forever; just give it a little rub between your fingers before using to release the flavors and aromas. Mixed Herbs are great sprinkled on salads, pastas, noodles, or rice bowls, and keep indefinitely in a sealed jar.

Time: 5 minutes prep • Makes 6 tablespoons

1 tbsp dried oregano
1 tbsp dried basil
1 tbsp dried marjoram
1 tbsp dried dill
1 tbsp dried thyme
1 ½ tsp dried rosemary
1 ½ tsp dried sage

- Combine all ingredients in a bowl. Mix well.

Millet Bowl with Greens and Toasted Sunflower Seeds

A macrobiotic-inspired dish with a perfect balance of flavors.

Time: 45 minutes prep • Makes 4–6 servings

1 cup millet, uncooked
3 cups water
1 tbsp miso paste
¼ cup hulled sunflower seeds
1 tbsp coconut oil
1 leek, white and light green parts only, thinly sliced
1 clove garlic, minced
1 tbsp fresh lemon juice
2 cups packed baby spinach
2 cups packed Swiss chard, stems removed and cut into ½ inch strips
Pinch red pepper flakes
Sea salt and black pepper, to taste

- In a dry sauté pan over medium heat, toast the millet for 2–3 minutes, until fragrant. Bring the water to a boil, stir in the miso paste. Reduce heat to a simmer, then add the millet. Cook, uncovered, until water has absorbed, about 30 minutes. Fluff with a fork and keep warm.
- Toast the sunflower seeds over medium heat in a dry skillet for 2–3 minutes, until golden brown and fragrant. Stir constantly to ensure even cooking and to prevent burning. Remove from pan and set aside.
- Heat the coconut oil in a large skillet and add the leek. Cook for 2 minutes to soften, then add the garlic and cook for 30 seconds longer. Add the lemon juice, spinach, Swiss chard, red pepper flakes, and salt and pepper to taste. Reduce heat to medium low, cover the pan, and cook for 5 minutes or until vegetables have wilted and are bright green.
- To serve, place the vegetables on top of the millet and sprinkle with toasted seeds.

Italian Garden Stir Fry

Herb-rich flavors give this stir fry a burst of freshness.

Time: 15 minutes prep • Makes 4–6 servings

1 tbsp coconut oil
½ onion, chopped
2 cups chopped zucchini
1 cup chopped mushrooms
2 cloves garlic, minced
15 ounces cooked garbanzo beans
1 large tomato, puréed (about ½ cup liquid)
1 tsp palm sugar
1 lemon, cut into wedges
2 tbsp chopped fresh mint
1 tbsp chopped fresh oregano
2 tbsp chopped fresh parsley, plus more for garnish
Salt and pepper, to taste
Cooked brown rice or quinoa for serving

- Heat the coconut oil over medium heat in a large cooking pan. Add the onion, zucchini, and mushrooms, and cook for 3–4 minutes until softened.
- Add the garlic and cook for 1 minute longer, then add the beans, tomato purée, and sugar, and squeeze in the juice from half of the lemon wedges.
- Bring to a simmer, then reduce heat to medium low and mix in the fresh herbs. Cook until the liquid has reduced, about 5–10 minutes.
- Add the remaining juice from remaining lemon wedges, remove from heat, and season with salt and pepper, to taste. Serve over a warm bed of rice or quinoa.

Eggplant Rollatini

To save time, make the filling and sauce while the eggplant is cooking (sauce is optional). Other lentils may be used in the place of red lentils—the red lentils are simply more elegant looking. Very filling.

Time: 1 hour prep • Makes 4 servings (1 cup of sauce)

2 eggplants
½ tsp sea salt

FILLING

1 tbsp coconut oil
2 cups cooked red lentils (unsalted; see p. 175)
1 tbsp miso paste
3 tbsp tahini
2 tbsp fresh minced oregano
2 tbsp fresh minced basil
2 tbsp fresh squeezed lemon juice
2 cloves garlic, minced

SAUCE

1 cup sundried tomatoes, soaked in warm water for 20 minutes
1 cup chopped tomatoes
¼ tsp salt
2 garlic clove
2 tbsp chopped fresh basil
1 date
2 tbsp shallot, minced

- Cut off the ends of the eggplant. With a flat end down on the cutting board, slice into ¼-inch strips. Place in a bowl and toss with ½ tsp sea salt to help extract some of the bitterness. Let rest for 10 minutes, then wash thoroughly and pat dry.
- Heat oven to 350°F. Lay the eggplant strips flat on a baking sheet and brush with coconut oil. Alternatively, a coconut oil cooking spray may also be used. Bake for 15 minutes, remove from oven, and let rest until cool enough to handle.
- Meanwhile in a medium bowl, mix together all the filling ingredients. Set aside.
- Blend all sauce ingredients together in a blender until fully combined.

To serve

- Spread 1 cup of the sauce on a baking dish. On a separate work surface lay each eggplant strip flat and spread a heaping tablespoon of filling across each. Roll up into a cylinder and place atop of the bed of sauce.
- Pour the remaining sauce on top and cover with aluminum foil. Bake for 30 minutes, until heated through and the sauce is bubbly.

Quinoa Pilaf with Swiss Chard and Lemon

Macrobiotic-inspired with the additional flavor of garlic.

Time: 5 minutes prep; 20 minutes to cook and sit • Makes 4–6 servings

4 cups water

2 cups quinoa

1 large bunch Swiss chard (about 1 pound)

2 tbsp coconut oil

2 large cloves garlic, minced

2 tbsp minced shallot

1 ½ tbsp lemon zest (grated lemon peel)

½ tsp sea salt, to taste

- Bring the water to a boil and add the quinoa. Cover and cook for about 15 minutes until the water has evaporated. Uncover, fluff with a fork, and let stand for 5 minutes.
- Meanwhile, trim the stems off the chard. Roll leaves together into a large cigar shape and cut into ½-inch slices.
- Heat the oil over medium heat. Add the garlic and shallot, and cook for 1 minute, stirring constantly. Turn the heat to low, add the chard and toss well, then cover and cook for 2 minutes.
- Remove from heat and add the mixture to the quinoa. Add the salt and lemon zest and stir to combine. Serve warm.

Wild Rice with Kabocha Squash and Sage Butter

This is one of my top five recipes, period. It's the perfect fall meal.

To save time, make the rice and butter while the squash is cooking. Yams may also be used in place of the kabocha.

Time: 1 hour prep; 30–45 minutes for the rice • Makes 4 servings

1 pound kabocha squash (about ½ medium squash)
3 tbsp melted coconut oil + 1 tbsp, divided
½ cup wild rice
½ cup brown rice
2 cups water
½ tbsp chopped fresh sage, packed
1 tbsp minced shallots
½ tsp salt

- Preheat the oven to 400°F. Cut the squash in half, then scoop out and discard the seeds.
- Use 1 tbsp coconut oil to lightly brush the cut areas of the squash, and place cut side down on a baking sheet. Bake for 40–45 minutes or until soft when pierced with a fork.
- When cool enough to handle, cut into 1-inch chunks (skin may be left on for extra flavor and nutrition or disposed of). Keep warm.
- To make the rice, combine the rices and water in a saucepan. Bring to a boil, reduce heat to low, and let simmer, covered, until done.
- Meanwhile, in a food processor, blend 3 tbsp coconut oil, sage, shallots, and salt until smooth.

To serve
- In a large pan, heat the sage butter mixture over medium-low heat for 1 minute. Add the rice and toss to combine, and cook for 1 minute longer while stirring constantly. Remove from heat and carefully fold in the squash.

Spicy Black Bean Chili

BLOSSOMING LOTUS, PORTLAND, OREGON

The ideal winter meal.

Time: 15 minutes prep • Makes about 6 2-cup servings

3 tbsp sunflower oil
1 cup diced red onion
¼ green bell pepper
½ cup small diced delicata squash
5 tbsp minced garlic
½ cup whole coriander seeds, toasted
1 tomato, crushed
1 tomato, diced
1 quart water
1 ½ tbsp unrefined cane sugar (optional or can use agave nectar)
2 tsp sea salt
2 ½ tsp apple cider vinegar
½ cup cooked sweet corn kernels
2 ½ tbsp ground cumin powder
1 tsp chipotle powder
3 cups cooked black beans
3 tbsp nutritional yeast

- In very large soup pot, heat the sunflower oil on high; add the onion, bell peppers, squash, and minced garlic. Continue to cook 3–5 more minutes, stirring as needed.
- Toast the coriander and grind in blender. Add both cans of tomatoes to soup pot, along with the water, sugar, salt, vinegar, corn, cumin, chipotle, and coriander. Reduce heat to medium.
- Add cooked beans and nutritional yeast. Turn off heat and mix well. Blend 1.5 quarts of chili in blender and mix back in rest of chili.

Superfood Chia Chili

LIFE FOOD GOURMET, MIAMI, FLORIDA

Raw

A nutrient-packed twist on a traditional dish.

Time: 10 minutes; 6 hours to soak the quinoa • Makes 4 servings

¼ cup chia seeds

1 cup water

4 tbsp tomato sauce (optional but highly recommended)

1 handful fresh cilantro, finely chopped

1 handful fresh parsley, finely chopped

¼–½ cup sprouted red quinoa or sprouted buckwheat

2–4 tbsp each celery, tomato, bell pepper, red onion, chopped

2–4 tbsp cold pressed olive oil

1–3 tbsp tomato powder

1–4 tsp chili powder

1–2 tsp cumin

½–1 tsp turmeric

1–3 tsp apple cider vinegar

½ tsp jalapeño flakes (optional)

½–1 tsp of sun dried sea salt

- Combine the chia seeds and ½ cup of warm water in a bowl and mix thoroughly. Add the remaining ingredients, including the remaining ½ cup of water.

West African Yam and Bean Patties

Just as filling as most burgers, and better tasting too.

Time: 30 minutes prep • Makes 8 patties

1 tbsp coconut oil + extra for frying

1 onion, diced

1 pound yams, diced

1 carrot, grated

4–5 cloves garlic, minced

½ tsp ginger powder

2 tsp paprika

¼ tsp cayenne pepper

1 can pinto beans, drained and rinsed

1 cup cooked brown rice

¼ cup quinoa flakes

¼ cup finely chopped almonds

Sea salt and pepper, to taste

Lime wedges, for garnish

- Heat 1 tbsp of coconut oil in a skillet over medium-high heat. Add the diced onion, reduce the heat to medium, and cook until the onions are soft and translucent, about 3–5 minutes. Stir in the yams and add a pinch of sea salt. Cover and cook until the yams are completely tender, about 5 minutes, stirring occasionally. Add the garlic, ginger powder, paprika, and cayenne and cook until fragrant, about 30 seconds. Remove from heat.
- Empty the pinto beans into a large bowl, and use a fork to mash them. Add the cooked vegetable mixture along with the cooked rice, quinoa flakes, and almonds. Stir to combine and then add salt and pepper, if desired. Hand-shape the mixture into 8 patties.
- Heat a small spoonful of coconut oil in a pan over medium-low, then set a couple of patties in the hot pan. Cook the patties for about 6 minutes, then flip them over—you should see a nice crust on the cooked side. If they break apart during the flipping, just reshape them with the spatula—they'll hold together once the second side is cooked. Cook the second side for another 6 minutes. Repeat until all patties are cooked.
- Serve the patties on your bun of choice or atop a bed of greens à la "protein style," with a lime wedge on the side for garnish.

Tip: Chop the yams coarsely, then use a food processor to pulse them into a fine dice.

Fajita Patties

With a bit of a Mexican bite, these "burgers" are a departure from the usual.

Time: 35 minutes active prep • Makes 4–6 servings

1 medium red bell pepper, shredded
1 medium green bell pepper, shredded
2–3 tbsp coconut oil, for frying
1 medium onion, diced (about 1 cup)
2 cloves garlic, minced
2 tsp cumin powder
½ tsp chili powder
¼ tsp cayenne pepper
½ tsp sea salt
1 cup cooked quinoa
¼ cup finely chopped walnuts
¼ cup sprouted flax powder (or finely ground flaxseeds)
2 tsp palm sugar
Lime wedges

- Place the red and green bell peppers in a colander, and squeeze firmly to remove excess liquid (try to get the peppers as dry as possible for best results). Set aside.
- In a medium frying pan, heat 1 tbsp coconut oil over medium-high heat. Add the onion, reduce heat to medium, and sauté for 3–5 minutes or until onions begin to turn translucent. Add the peppers, garlic, cumin, chili powder, cayenne, and sea salt, then cook for 2 minutes longer, stirring frequently. Remove from heat and transfer to a large bowl.
- To the cooked vegetable mixture add the quinoa, walnuts, flax powder, and palm sugar, and mix well. Hand-shape into 4–6 patties.
- Heat a small amount of coconut oil in a large frying pan over low heat. The lower the heat the better, as it will allow the patties to cook longer and get a better crust. Place the patties on the pan carefully, and cook until browned underneath, about 10 minutes. Flip the patties over and repeat on the other side.
- Just before serving, drizzle with fresh lime juice from the lime wedges.

Homestyle Lentil Patties

The classic lentil burger.

Time: 30 minutes active; 50 minutes total • Makes 4–6 servings

2–3 tbsp coconut oil, for frying
1 cup chopped onions (about 1 medium)
1 carrot, shredded
1 celery stalk, minced
1 ½ tbsp balsamic vinegar
1 cup cooked red lentils (see p. 175)
2 tbsp sunflower seeds
¼ cup sprouted flax powder (or finely ground flaxseeds)
Sea salt and black pepper, to taste

- In a medium frying pan, heat 1 tbsp coconut oil over medium-high heat. Add the onions, carrot, and celery, and sauté until the onions are translucent and the vegetables are soft. Mix in the vinegar, remove from heat, and transfer to a large bowl.
- To the vegetable mixture, add the remaining ingredients and mix well, mashing the lentils slightly with a fork. Adjust salt as needed. Place in the refrigerator for 20 minutes to allow the flaxseed to help firm the mixture. Hand-shape into 4–6 patties.
- Heat a small amount of coconut oil in a large frying pan over low heat. The lower the heat the better, as it will allow the patties to cook longer and get a better crust. Place the patties on the pan carefully, and cook until brown underneath, about 10 minutes. Flip the patties over and repeat on the other side. Serve warm.

Indian-Spiced Lentil Hemp Patties

Nutrition-packed and with a unique flavor, these burgers will satisfy a large appetite.

Time: 30 minutes prep • Makes 4–6 patties

2–3 tbsp coconut oil, for frying
1 cup diced yellow onion (about 1 medium)
1 stalk celery, minced
1 cup diced red pepper
6 medium cloves garlic, minced
¼ tsp ginger powder
1 tsp coriander powder
1 tsp cumin powder
½ tsp turmeric
½ cup hemp seeds
3 tbsp sprouted flax powder (or finely ground flaxseeds)
1 ¼ cups cooked red lentils (see p. 175)
½ tsp sea salt, or to taste

- In a medium frying pan, heat 1 tbsp coconut oil over medium-high heat. Add the onion, celery, and red pepper, and sauté until the onion is translucent and the vegetables are soft. Add the garlic and sauté for 1 minute longer. Remove from heat and transfer to a large bowl.
- To the vegetable mixture, add the remaining ingredients and mix well, mashing the lentils with a fork. Adjust salt as needed. Hand-shape into 4–6 patties.
- Heat a small amount of coconut oil in a large frying pan over low heat. The lower the heat the better, as it will allow the patties to cook longer and get a better crust. Place the patties on the pan carefully, and cook until brown underneath, about 10 minutes. Flip the patties over and repeat on the other side. Serve warm.

Chia Bean Patties

Loaded with omega-3-rich chia and mineral-rich wakame sea vegetables, these burgers are a nutritional powerhouse. They can also be cooked, then gently warmed as needed.

Time: 10 minutes prep (once beans are cooked, or use canned); 20 minutes to soak
Makes 8 burgers

2 cups cooked black beans (unsalted)
1 ½ cups cooked brown rice
⅓ cup chia seeds
¼ cup nutritional yeast
1 ½ tsp fresh oregano, minced (or ½ tsp dried)
1 tsp palm sugar
1 stalk celery, minced
1 carrot, minced
½ cup minced yellow onion
3 tbsp brown rice miso
¼ cup quinoa flakes

- With a potato masher or the back of a fork, mash the black beans in a large bowl into a chunky purée. Mix in the remaining ingredients, one at a time. (Alternatively, pulse ingredients together in a food processor, allowing some chunks to remain.) Form into 8–10 patties and refrigerate for 1 hour.
- Heat a spoonful of coconut oil in a large non-stick frying pan over medium-low heat. Add a few of the patties and cook for several minutes on each side until both sides are brown (the finished patties will be crispy on the outside and soft on the inside).

To serve
- Place burgers on sprouted-grain buns with your favorite toppings or "protein-style" in a collard-green leaf.

Variation: Soak 2 tbsp dried wakame flakes for 20 minutes. Drain and mash into the mixture before cooking. Wakame adds an extra-strong punch of trace minerals and a light savory flavor.

Black Bean Sliders

Try serving on a crispy cracker or onion bread with a wedge of avocado.

Time: 10 minutes active; 30 minutes to sit
Makes 2 servings (8–10 small patties)

1 tbsp olive oil
3 large cloves garlic, minced
1 cup diced yellow onion (about ½ medium onion)
1 cup diced green bell pepper (about ½ pepper)
15 ounces cooked black beans (made fresh or canned)
3 tbsp chia seeds
2 tbsp chopped walnuts (optional)
1 cup cooked brown rice
¼ cup quinoa flakes
2 tbsp pumpkin seeds
2 tsp chili powder
1 tsp cumin powder
¼ tsp cayenne powder
2 tbsp nutritional yeast
1 ½ tbsp yacon syrup
Salt, to taste

- Heat the oil over medium heat. Add the garlic, onion, and bell pepper.
- Cook until the vegetables have softened, about 5 minutes. Remove from heat.
- Mix in remaining ingredients. With a fork or simply by hand, mash down the beans to form a chunky mixture.
- Place in the refrigerator, covered, for 30 minutes to allow chia seeds to swell and absorb the excess moisture.
- Form the mixture into 8–10 small patties (sliders). Coat a frying pan with a small amount of oil (olive oil or coconut). Over medium heat, cook the patties for about 5 minutes on each side or until browned.
- Alternatively, place sliders in a pan and cook in the oven with the setting on broil for 5–8 minutes or until just beginning to brown.

Portobello Patties

Full of flavor, these plant-based burgers are extremely versatile.

Time for cooked method: 10 minutes prep; 2–3 minutes to cook   •   Prep time for raw method: 10 minutes; 10–12 hours to dehydrate   •   Makes 4 burgers

½ cup ground flax seeds

¼ cup hemp seeds

2 tbsp fresh parsley, chopped

½ tsp fresh thyme, chopped

1 tbsp EFA oil

2 tbsp brown rice miso paste

1 tsp pressed garlic

½ tsp black pepper

2 tbsp nutritional yeast

2 tsp Dijon mustard

¼ cup chopped walnut pieces

2 medium Portobello mushrooms,
   chopped into large cubes

⅓ cup quinoa flakes (for cooked
   method only; omit for raw method)

- In a food processor, combine all ingredients except walnuts and mushrooms, and process until well combined. Turn off the processor and add in the walnuts and mushrooms.
- Pulse just 3 or 4 times until the mushrooms are diced, but do not blend. Remove mixture from the processor and transfer to a medium bowl.

Raw method (best)

- Shape the mixture into 4 balls and flatten into patties on a teflex or nonstick sheet. Dehydrate for 10–12 hours at 115°F, then flip and transfer to a mesh sheet and "cook" for 2 more hours.
- These keep for several days wrapped and refrigerated, and can be reheated as needed.

Cooked method

- Mix the quinoa flakes into the mushroom mixture and let stand for 20 minutes.
- Sauté the chopped mushrooms in a dry pan for 2 or 3 minutes to cook out some of the moisture. Follow the rest of the recipe to form burgers.
- To cook, shape the mixture into 4 balls (press firmly to compact) and form into patties. Over very low heat, oil the pan with 1 or 2 tbsp coconut oil.
- Add the patties, and cook approximately 10 minutes on each side. The patties will remain soft on the inside but utterly enjoyable.

To serve

- Try "protein style" on a salad with a mustard vinaigrette, in a collard green with avocado and vegetables, or in a wrap or bun of choice with all the trimmings.

Pizza

The following pizzas are unique in that their crusts are made from legumes and vegetables as opposed to flour; therefore, their crusts are protein and nutrient packed.

Time: 15 minutes prep; about 45 minutes to cook
Makes 1 large pizza (4 servings)

Follow this procedure for all the pizza recipes

- Preheat oven to 300°F. In a food processor, process all crust ingredients until mixture starts to ball up. Lightly oil the baking tray with coconut oil. Spread mixture on tray to about ¼-inch thick (it can be thicker or thinner if you prefer).
- Choose a pizza sauce from the "Spreads, Dips, Sauces, and Dressings" section.
- Bake for 45 minutes. (This will vary slightly depending on the moisture content of the vegetables and the desired crispness of the pizza.)

Buckwheat Sunflower Seed Carrot Pizza

The crust of this pizza is lighter tasting, with a distinct carrot flavor. Choose a pizza sauce from the "Spreads, Dips, Sauces, and Dressings" section.

Time: 15 minutes prep; about 45 minutes to cook
Makes 1 large pizza (4 servings)

CRUST

1 cup ground sunflower seeds
1 cup raw buckwheat
1 cup grated carrot
¼ cup coconut oil
½ tsp parsley
Sea salt, to taste

TOPPING

1 tomato, sliced
½ Spanish onion, diced
1 cup chopped celery
½ cup chopped fresh basil
½ cup grated carrot
½ cup chopped green onions

- Preheat oven to 300°F.
- In a food processor, process all crust ingredients until mixture starts to ball up.
- Lightly oil the baking tray with coconut oil. Spread mixture on tray to about ¼-inch thick (it can be thicker or thinner if you prefer).
- Top with your choice of sauce.
- Spread toppings evenly on the crust.
- Bake for 45 minutes.

Black Bean Chili Pizza

Fiber rich and nutrient dense, this one will end hunger. Choose a pizza sauce from the "Spreads, Dips, Sauces, and Dressings" section.

Time: 15 minutes prep; about 45 minutes to cook
Makes 1 large pizza (4 servings)

CRUST

1 ½ cups cooked brown rice
1 ½ cups black beans
¼ cup coconut oil
1 tbsp chili powder
1 tsp chili flakes
Sea salt to taste

TOPPING

1 tomato, sliced
½ onion, diced
1 cup chopped bell peppers (any color)
½ cup grated beet
½ cup chopped green onions
1 tsp oregano (or 1 tbsp fresh)
1 tsp thyme (or 1 tbsp fresh)

- Preheat oven to 300°F.
- In a food processor, process all crust ingredients until mixture starts to ball up.
- Lightly oil the baking tray with coconut oil. Spread mixture on tray to about ¼-inch thick (it can be thicker or thinner if you prefer).
- Top with your choice of sauce.
- Spread toppings evenly on the crust.
- Bake for 45 minutes.

Quinoa Falafels

Nutrient and flavour packed, these falafels are a divergence from the traditional. You can make the lemon tahini sauce while the falafels are cooking.

Time: 30 minutes prep • Makes about 1 dozen

3 tbsp flax powder
⅓ cup water
2 cups cooked quinoa
3 tbsp nutritional yeast
2 tsp miso paste
1 tsp tahini
1 tsp onion powder
½ tsp garlic powder
¼ cup minced parsley
1 tsp minced fresh oregano
2 tbsp melted coconut oil
Lemon Tahini Dipping Sauce (see p. 196)

- Mix together the flax powder and water in a small bowl and let rest for 10 minutes to form a gel.
- Meanwhile, heat the oven to broil. In a medium bowl, combine quinoa, nutritional yeast, miso paste, tahini, onion powder, and garlic powder, and mix well. Add the flax gel, along with the parsley and oregano, and combine thoroughly.
- To make the falafels, roll a heaping tablespoon of the mix into a ball and flatten into a 2" circle.
- Line a baking tray with aluminum foil, and brush with 1 tbsp coconut oil where the falafels will sit. Place the falafels on the tray, and brush the tops with the remaining tablespoon of melted coconut oil. Put the tray in the oven, and broil for 8 minutes.
- Remove from the oven, flip the patties with a spatula, and return to the oven for another 5–7 minutes, or until patties have formed a dark brown crust.
- Let cool for 1 minute before serving. Enjoy with Lemon Tahini Dipping Sauce on the side or drizzled on top.

Live Falafels

BLOSSOMING LOTUS, PORTLAND, OREGON

Raw

Enzyme packed, filling, and delicious, but you will need a dehydrator for
this recipe.

Time: 15 minutes prep; overnight to soak nuts and dehydrate
Makes about 4 servings

6 tbsp Brazil nuts (soaked overnight)

1 ½ tbsp raw tahini

½ cup dry walnuts

1 scallion, chopped

1 tbsp fresh lemon juice

½ of 1 medium-size jalapeño pepper, seeded and rough chopped

½ cup cilantro

½ cup parsley

1 tbsp dried oregano

1 tsp ground cumin

Salt and pepper, to taste

3 tbsp water, as needed

- Soak the Brazil nuts overnight.
- In blender, purée all ingredients except nuts, and transfer to large bowl.
- In food processor, chop the walnuts into fine mixture and add to the bowl with
  the puréed ingredients. Drain the Brazil nuts and chop in the food processor.
  Add to the bowl.
- Thoroughly mix all ingredients. Use a small scoop to make round balls and
  place on dehydrator sheets.
- Dehydrate for 1.5 hours at 115°F, then turn down to 90°F and dehydrate
  overnight.

Gorilla Food Green Tacos

Gorilla Food, Vancouver, British Columbia

Raw

A filling and satisfying raw taco.

Time: 5 minutes prep (once Walnut Chili Pâté, Salsa, and Guacamole are made)
Makes 4 tacos

4 Romaine lettuce leaves
1 ¼ cup Walnut Chili Pâté (see p. 191)
1 ¼ cup Guacamole (see p. 195)
16 tbsp Salsa (see p. 196)

- Cut just the very bottom off the romaine lettuce bunch so that the leaves fall apart.
- Wash the leaves and pat dry to be used as your taco shells.
- Spread a wide base of Walnut Chili Pâté down the center of each leaf.
- Top the pâté evenly with Guacamole, leaving the pâté visible on the sides.
- Top the Guacamole with Salsa, leaving both the guacamole and pâté visible on the sides.
- Plate and enjoy eating these with your hands.

Collard Green Buckwheat Wrap

This nutrient-dense wrap is surprisingly filling. For the dressing, choose any in the "Spreads, Dips, Sauces, and Dressings" section.

Time: 10 minutes prep • Makes 1 large serving

1 avocado
2 Roma tomatoes
1 cucumber
1 large carrot
2 strips dulse (about ¼ cup, tightly packed)
1 cup sprouted or cooked buckwheat
1 leaf collard green
3 tbsp salad dressing

- Peel and cube the avocado, slice the tomatoes and cucumber, and grate the carrot. Along with the dulse and buckweat, place on a collard green.
- Drizzle salad dressing over top. Roll up, tucking the ends in so the wrap is secure. Cut into pieces if desired.
- A collard green leaf also serves as a good wrap for guacamole combined with quinoa, buckwheat, or brown rice.

Variation: To serve as a complete meal, add ½ cup black-eyed peas and ½ tsp cayenne pepper to the mixture to spice it up.

Spicy DLT Sandwich

DLT = dulse, lettuce, tomato

Strips of dulse sea vegetables makes this a flavor- and nutrient-packed sandwich. You can use rice bread for a gluten-free option.

Time: 10 minutes prep • Makes 1 serving

2 slices sprouted bread (such as Ezekiel Sesame), lightly toasted if desired
½ avocado, mashed
1–2 tbsp fresh salsa (amount depends on spice level)
¼ cup dulse strips (or 1 ½ tbsp dulse flakes)
Lettuce leaves
Tomato slices

- Toast the bread if desired. Spread the avocado on one side and the salsa on the other.
- Distribute the dulse evenly atop the avocado, stack with lettuce and tomato, and put the sides together.

Butternut Squash Soup

Karyn's on Green, Chicago, Illinois

Can be served as a creamy and delicious meal, or in a smaller portion as a complement to a salad. Ideal in the fall.

Time: 5 minutes prep; about 90 minutes to cook • Makes 4 servings

SOUP

2 whole butternut squash (about 5 pounds)
2 cans (about 28 ounces) of coconut milk
Salt, to taste

- Cut tops off the squash and discard. Cut squash in half, lengthwise, and de-seed, saving excess squash for stock. Place halved squash, cut side down, on sheet trays lined with parchment paper. Roast in 350°F oven for about 1 hour or until largest squash is soft throughout.
- Cool to room temperature. Remove the skin from the roasted squash and put the squash in a large sauce pot. Combine coconut milk with squash.
- Simmer squash and coconut milk, stirring occasionally until mixture is heated throughout.
- Remove from heat and use a blender to process mixture on high until smooth. Season with salt to taste, then strain through fine-mesh sieve.
- Pour about 1 cup of soup into each of 4 bowls. Top with some chickpeas and garnish with paprika.

CHICKPEAS

¼ cup canned chickpeas, strained   ½ tsp paprika + some for garnish
Oil   Salt to taste

- In fryer at 325°F (or in sauce pot with 325°F oil), fry chickpeas until they begin to brown. Strain excess oil off chickpeas, then toss with paprika and salt in mixing bowl.

Caramelized Brussels Sprouts with Mustard Vinaigrette

KARYN'S ON GREEN, CHICAGO, ILLINOIS

An ideal complement for soup and a salad.

Time: 5 minutes prep; 20 minutes to cook • Makes 4 servings

MUSTARD VINAIGRETTE

1 shallot, minced
2 sprigs of tarragon, chopped
¼ cup champagne vinegar
1cup whole-grain mustard
Lemon juice and salt to taste

- Whisk all vinaigrette ingredients together. Thin with water if desired.

CARMELIZED BRUSSELS SPROUTS

1 pound blanched Brussels sprouts (halved or quartered, depending on size)
Olive oil for sautéing
Salt to taste

- Blanch Brussels sprouts in salted boiling water until al dente. Shock in a bowl of ice water. Drain and pat dry with paper towels. Sear Brussels sprouts in a very hot sauté pan with olive oil until well caramelized. Fold in mustard vinaigrette and season with salt, to taste.

Summer Squash Roast

HORIZONS, PHILADELPHIA, PENNSYLVANIA

Ideal served with a salad, in the summer.

Time: 5 minutes prep; 20 minutes to cook   •   Makes 6 servings

2 pounds assorted summer squash, seed centers removed
¼ cup of olive oil
1 tsp minced fresh garlic
Salt and pepper
1 cup kalamata olives, pitted and quartered
1 tsp olive oil
1 packed cup fresh basil leaves
1 tbsp vegan sour cream
Pinch salt and pepper

- Preheat oven to 450°F.
- Cut the squash into 1- to 2-inch chunks and toss with ¼ cup of the olive oil, the garlic, salt, and pepper. Roast in the oven for 8 minutes or until just tender.
- Toss the olives in 1 tsp of olive oil and roast in the oven for 12 minutes.
- Purée the basil and sour cream in the food processor. Toss the roasted squash in the basil mixture and garnish with the roasted olives. Add salt and pepper to taste.

Heavenly Baked Delicata Squash

Simple, flavorful squash.

Time: 10 minutes prep; 45 minutes to cook • Makes 4 servings

4 Delicata squash (preferably organic so that you
   can eat the skin and the flesh)
2 tbsp + 1 tbsp virgin coconut oil, divided
1 tbsp dried oregano
½ tsp sea salt or to taste
Freshly cracked pepper, to taste
½ cup water

- Preheat the oven to 400°F.
- Cut off the ends of the squash and slice in half lengthwise. Remove all seeds with a spoon and discard. Rub 2 tbsp of coconut oil equally on the insides of the squash, and sprinkle evenly with oregano, sea salt, and pepper.
- Place squash cut side down on a large baking tray, and pour ½ cup water in the bottom of tray. Loosely cover the tray with aluminum foil or another tray, and bake for 45 minutes or until squash is very tender when pierced with a fork.
- To serve, place squash face-up on serving plate, and drizzle with remaining tablespoon of coconut oil.

Note: Coconut oil is solid at room temperature but melts quickly. It does not matter whether the oil is solid or liquid in this recipe, though it may be easier to distribute when softened or melted.

Turnip Masala

The Green Door, Ottawa, Ontario

A new twist on an Indian classic.

Time: 15 minutes prep; 35 minutes to cook • Makes 6–8 servings

8 cups rutabaga, cut into 1-inch cubes
Olive oil and salt for roasting the rutabaga
4 tbsp olive oil
¼ tsp asafetida
½ tsp ground cardamom
½ tsp ground cumin
½ tsp ground turmeric
½ tsp curry powder
1 tsp garam masala
1 tsp ground coriander
3 cloves garlic
1 fresh hot chili pepper (optional)
1 medium onion, diced
2 cups water
1 tsp salt
1 cup diced tomato
½ cup chopped fresh cilantro, for garnish

- Preheat oven to 350°F.
- Toss the cubed rutabaga in a little olive oil and salt. Spread on a baking sheet and bake for 20–30 minutes or until lightly browned and very close to being fully cooked.
- Put a medium-size pot on low heat. Add the oil and all the spices. When they are lightly bubbling, add the garlic and fresh hot pepper. Sauté for 1 minute, then add the onion.
- When the onion is nicely sautéed, add the water. Turn the heat up to high, add the salt and tomato, and bring to a boil.
- Add the roasted rutabaga. Cook for a few minutes, then remove from heat.
- Garnish with fresh cilantro.

Roasted Parsnip with Coconut Fennel Sauce

THE GREEN DOOR, OTTAWA, ONTARIO

An ideal complement to soup.

Time: 10 minutes prep; 35 minutes to cook • Makes 6–8 servings

4 cups parsnip (about 3 regular-size parsnips)
Olive oil and salt for roasting the parsnips
1 tbsp olive oil
2 tbsp freshly ground fennel seed
1 can coconut milk (14 ounces)
Juice of 3 limes
Zest of 1 lime
1 tsp salt, or to taste
Chopped fresh parsley, for garnish

- Preheat oven to 350°F.
- Wash the parsnips. Cut them into chunks averaging 1-inch long, and of roughly equal size so that they roast evenly.
- Drizzle the parsnips with olive oil and lightly sprinkle with salt. Bake for 20 to 30 minutes, or until done.
- Place oil in a large saucepan, add the fennel seed, and roast gently at low heat for 2–3 minutes. Add the coconut milk, lime juice and zest, and salt. Heat slowly, stirring with a wooden spoon.
- When hot, add the roasted parsnip. Adjust salt to taste. Garnish with parsley.

Roasted Cauliflower

Simple, nicely flavored cauliflower.

Time: 5 minutes prep; 45–50 minutes to cook • Makes 2 servings

1 head cauliflower
2–3 tbsp coconut oil, melted
Sea salt
½–1 tsp paprika

- Preheat oven to 375°F.
- Line a baking sheet with parchment paper for easy clean-up.
- Cut the cauliflower into golf ball-size chunks. Toss with the coconut oil, along with the salt and paprika, to taste.
- Spread out evenly on top of prepared sheet and place in the oven for about 40–50 minutes, tossing once halfway through cooking time.

Curried Cauliflower "Rice"

Shredding the cauliflower makes it cook faster and results in a bowl of fluffy vegetable "rice." Simply grate the head of cauliflower with a standard grater using a medium setting.

Time: 10 minutes • Makes 2–4 servings

4–5 cups shredded cauliflower
2 tsp curry powder blend (use your favorite brand or mix)
½ tsp sea salt
1 tbsp coconut oil
1 tbsp apple cider vinegar

- In a bowl, mix together the shredded cauliflower, curry blend, and sea salt.
- In a large skillet, heat the coconut oil over medium heat until melted. Reduce heat to medium and add the cauliflower. Cook, stirring constantly, for about 5 minutes, or until cauliflower is cooked through. Stir in the apple cider vinegar. Remove from heat, and serve.

Mashed Potatoes and Turnips

Like mashed potatoes but more nutritious.

Time: 30 minutes   •   Makes 6–8 servings

2 quarts water
1 tsp sea salt + additional to taste
1 ½ tbsp palm sugar
3 cups turnips, cut into 1-inch cubes (about 2 medium)
5 cups red potatoes, cut into 1-inch cubes (about 5–6 medium)
1 tbsp coconut oil
2 tbsp nutritional yeast

- Pour the water, sea salt, and palm sugar in a large pot, and bring to a rolling boil.
- Add the turnips and potatoes and cook for 20 minutes, or until tender when pierced with a fork. Drain thoroughly and transfer to a large bowl.
- Add coconut oil and nutritional yeast, and mash into a chunky purée. Serve warm.

Parsnip Oven Fries

Ideal as an autumn meal addition or as a snack, parsnip's "earthy" flavor is brought out by the higher heat of roasting.

Time: 5 minutes prep; 30 minutes to cook   •   Makes 4 servings

2 pounds of parsnips (about 2–3 large roots)
2 tbsp coconut oil, melted
Salt and pepper to taste

- Preheat oven to 450°F.
- Peel the parsnips and trim the ends. Cut in half lengthwise, then slice into ½-inch sticks. Spread out onto a large baking pan and toss with oil, salt, and pepper.
- Bake for 25–30 minutes or until golden brown, tossing parsnips once or twice during baking to ensure even cooking.

Spicy Cocoa-Hazelnut Zouzous

CRUDESSENCE, MONTREAL, QUEBEC

Zouzous are the Crudessence energy balls.

Time: 10 minutes • Makes about 20 zouzous

3 ½ ounces Brazil nuts

3 ½ ounces hazelnuts

7 ounces (slightly less than ½ pound) sunflower seeds, ground
to a fine powder with food processor or coffee grinder

¾ cup cocoa powder

⅛ cup agave nectar

2 tbsp maca powder

2 tsp mesquite

1 ½ tbsp carob powder

¼ cup cayenne powder

1 ⅓ cups date paste (make by blending about 2 cups of lightly
packed pitted dates in a food processor)

- Using the food processor, grind the Brazil nuts and hazelnuts just till they
  are reduced to a powder, not a butter. Transfer into a large bowl and add the
  other ingredients, mixing by hand.
- With a small ice cream scoop, form the mixture into balls of approximately
  ⅛ cup or 2 tbsp per zouzou.

Salt and Vinegar Kale Chips

Crispy, nutrient dense, and flavor packed. These are among the most nutritious and delicious snacks you can make.

Time: 15 minutes prep; cook time varies • Makes 2 medium-size servings or 1 large

½ cup raw sunflower seeds
1 tbsp apple cider vinegar
½ tbsp balsamic vinegar
Water for blending
½ tsp sea salt, to taste
1 bunch curly kale

- Combine the sunflower seeds, both vinegars, and sea salt in a blender. Blend for several minutes until a chunky paste has formed, adding a tablespoon of water into the blender as needed to assist with blending. (The more water that is added, the longer the chips will take in the oven.)
- Strip off the kale leaves into a bowl and discard the stems. Tear up any large pieces roughly, and pour the creamed mixture on top of the kale. Using clean hands, massage the mixture into the kale for one minute to evenly coat the leaves.

Oven method
- Heat the oven to 200°F. Place a piece of parchment paper on top of a baking sheet, then spread out the kale chips evenly over the surface to ensure even cooking time.
- Bake for about 2 hours (time varies according to relative humidity), or until kale has dried out and is crispy. Keep a close eye on the kale at the end of its cooking process to make sure it does not burn.
- Enjoy immediately or keep in an airtight container for up to 2 weeks.

Dehydrator method
- Warm the dehydrator to 115°F. Spread out the kale onto 4 mesh dehydrator sheets, and dehydrate for 10–12 hours, or until crispy (time may vary depending on relative humidity).

Cool Coconut Orange Squares

Easy to make and liked by pretty much everyone.

Time: 45 minutes • Makes about 1 ½ dozen small squares

½ cup cashews
½ cup macadamia nuts
2 large Medjool dates, pits removed
1 tsp orange zest
¾ tsp vanilla extract
½ cup + 2 tbsp shredded dried coconut
¼ palm sugar
Pinch sea salt

- In a food processor, grind the cashews into a coarse flour. Add the macadamia nuts and process until finely chopped. Add in the dates, orange zest, and vanilla extract and mix until dates are combined. Add the remaining ingredients and process until a sticky dough has formed.
- Place a piece of parchment paper or wax paper on a flat surface. Spoon out the dough onto the paper in a mound, and place another piece of paper on top. Use a rolling pin to flatten the dough into a ½-inch-thick layer. Freeze for 30 minutes or longer, till mixture is firm, before cutting into small squares. Best served frozen.

Nori Crisps

Crunchy and fast to make. Since nori burns easily, try baking a small practice batch in the oven to get your timing down.

Time: 10 minutes • Makes about 50 crisps (2–4 servings)

½ tbsp sesame oil
½ tbsp melted coconut oil
½ tsp rice vinegar
1 tbsp palm sugar, + extra for sprinkling
¼ tsp sea salt, + extra for sprinkling
5 nori sheets

- Preheat oven to 350°F.
- In a small glass or cup, mix together both oils, vinegar, 1 tbsp palm sugar, and ¼ tsp sea salt.
- With a sharp knife or scissors, cut each sheet of nori in half, then cut each half into 5 strips. Lay the strips on a non-stick baking sheet, then lightly brush the surface of the nori with the oil mixture.
- Place in the oven for 3 minutes or until the nori has turned a dark green. Nori burns very easily, so keep a close eye on the time while baking.
- Remove the baking sheet from the oven and lightly sprinkle the strips with additional palm sugar or a bit of sea salt, if desired.

Sour Cream and Onion Kale Chips

Flavor-bursting, nutrient-dense treat.

Time: 35 minutes prep; cook time varies
Makes 2 medium-size servings or 1 large

½ cup cashews, presoaked 35 minutes
⅓ cup water
1 tbsp onion powder
¼ tsp garlic powder
1 ½ tbsp apple cider vinegar
¼ tsp sea salt
2 tbsp minced fresh parsley
1 bunch kale

- Combine the soaked cashews, water, onion powder, garlic powder, vinegar, salt, and parsley in a blender and process until smooth, stopping the machine and scraping down the sides if needed (this may take several minutes). Set aside.
- Wash the kale to remove any grit, then carefully dry the leaves. Strip the leaves into a bowl, roughly tearing any large pieces. Discard the stems. Pour the creamed mixture on top of the kale, using a small spatula or spoon to remove the mixture from the blender. Using clean hands, massage the mixture into the kale for 1 minute to evenly coat the leaves.

Oven method

- Heat the oven to 250°F. Line two baking sheets with parchment paper, then spread the kale chips over the sheets, as evenly and flatly as possible, to ensure even cooking time. Bake between 1 ½ and 2 hours (time varies according to the dryness of the kale), tossing halfway through the baking, until the kale has dried out and is crispy but not burnt. Keep a close eye on the kale at the end of its cooking process, and remove any premature crispy chips from the batch if needed. Enjoy immediately or keep in an airtight container for up to 2 weeks.

Dehydrator method

- Warm the dehydrator to 115°F. Spread the kale onto 4 mesh dehydrator sheets, and dehydrate for 10–12 hours, or until crispy (time will vary depending on relative humidity).

BBQ Red Bell Pepper Kale Chips

Adjust the red pepper flakes to add more or less heat.

Time: 25 minutes prep; cook time varies
Makes 2 medium-size servings or 1 large

1 bunch kale

1 cup red bell pepper, chopped

½ cup water

1 tbsp olive oil

½ cup sunflower seeds

¼ tsp salt

½ teaspoon chipotle powder

1½ tsp onion powder

¼ tsp garlic powder

2 tbsp raisins

- Strip the kale leaves into a bowl, discarding the stems. Tear up any large pieces and set aside.
- Combine all ingredients except for the kale in a food processor or single-serving blender and process until smooth, stopping the machine and scraping down the sides if needed (this may take several minutes).
- Pour the blended mixture on top of the kale, using a small spatula or spoon to remove it from the food processor. Using clean hands, massage the mixture into the kale for 1 minute to evenly coat the leaves.

Dehydrator method (preferred)
- Warm the dehydrator to 115°F. Spread out the kale onto 4 mesh dehydrator sheets, and dehydrate for 10–12 hours, or until crispy (time may vary depending on relative humidity).

Oven method
- Heat the oven to 200°F. Place a piece of parchment paper on top of a baking sheet, and spread the kale chips over the surface, as evenly as possible, to ensure even cooking time.
- Bake for about 75–100 minutes (time varies according to relative humidity) or until kale has dried out and is crispy. For best results, toss the kale several times. Keep a close eye on the kale at the end of its cooking process to make sure it does not burn.
- Enjoy immediately or keep in an airtight container for up to 2 weeks.

Red Raspberry Frozen Fruit Pops

Using stevia is a terrific way to cut down on using sugars. Its sweetness varies from brand to brand, so you'll have to experiment and use it to taste. In this recipe, stevia extends the sweetness of the fruit sugars from date syrup and raspberries without adding a single extra calorie. If you're not a fan of stevia, use another natural sweetener, such as palm sugar or a little extra date syrup.

Time: About 10 minutes; 3 hours to freeze
Makes about 10–12 oz of mixture (servings depend on size of molds)

½ cup unsweetened almond milk
½ tsp vanilla extract
1 cup frozen raspberries
White stevia powder (or liquid), to taste
2 tbsp date syrup

- Get vessels ready for frozen pops—use either a plastic mold or kit, or small cups with sticks.
- In a medium bowl, stir together the almond milk and vanilla. Add the frozen raspberries, and use a fork to "mash" them into the liquid—the milk will begin to freeze into a slush around the raspberries, which is the objective.
- Mix until chunky, but not blended. Add a tiny, tiny dash of stevia and mix well. Taste the mixture and add more stevia if needed, mixing after each addition. The overall taste should be quite sweet, as freezing will bring the sweetness down a notch.
- Drizzle the date syrup into the mixture. Stir once or twice *only*—just enough to incorporate the syrup into the raspberry mix but allowing large date syrup swirls to remain.
- Carefully spoon the mixture into the molds. Freeze for 3 hours or until mixture is completely frozen through. Thaw for a minute or two just before serving.

energy bars and gels

In sharp contrast to conventional energy bars, these bars are true high-net-gain food.

They provide fast and sustained energy, and are easy and quite quick to prepare—no cooking involved. I like to make a big batch once a month, wrapping the bars individually in plastic wrap and freezing them, so that I always have a variety on hand. Because these bars maintain a supple and chewy consistency even when frozen (unlike commercial bars, which freeze rock-solid), you don't need to wait for them to thaw to enjoy them.

Follow this procedure for all the energy bar recipes, unless otherwise stated.

- In a food processor, process all ingredients until desired texture is reached. If you prefer a uniformly smooth bar, process longer. If you would rather a bar with more crunch and texture, blend for less time.
- Generally, if I'm making the bars specifically to be eaten during physical activity, such as long training rides, I'll blend the mixture until it is smooth, as this will reduce the amount of chewing required. However, for variety, I'll also be sure to make a few batches at the same time that are crunchier, to eat as a regular snack.
- Remove mixture from processor and put on a clean surface. There are two ways to shape the bars: you can roll the mixture into balls or shape it into bars.
- To shape into balls, use a tablespoon or your hands to scoop the mixture (however much you like to make one ball), then roll it between the palms of your hands.
- To shape as bars, flatten the mixture on the clean surface with your hands. Place plastic wrap over top, then, with a rolling pin, roll mixture to desired bar thickness. Cut mixture into bars.
- Alternatively, form mixture into a brick, then cut as though slicing bread. As the bars dry, they become easier to handle.

Spiced Açaí Energy Bars

Time: 45 minutes total; 15 minutes active • Makes 8–10 bars

¾ cup raw almonds

¾ cup pitted Medjool dates
(about 6 large)

3 tbsp açaí powder

2 tbsp raisins

¼ cup dried apricots

1 tbsp chia seeds

½ tsp ginger powder

½ tsp cinnamon powder

¼ tsp vanilla extract

Tiny pinch of salt (optional)

- Mix all the ingredients together in a food processor just until a dough has formed (allowing some almonds to remain coarsely chopped).
- Place a sheet of plastic wrap on a cutting board and spill the dough out on top. Use your hands to press and form it into a 1-inch-thick rectangle.
- Cover the rectangle of dough with plastic wrap and place it in the freezer for 30 minutes, then cut into 8–10 bars. These are best eaten cold, since they're stickier when they're warm.

Green Energy Bars

Chlorophyll-packed, these green bars taste much better than they look.

Time: 15 minutes • Makes 8 bars

1 cup raw cashews

1 cup pitted Medjool dates (about 8 large)

2 tsp wheatgrass powder

¼ cup hemp seeds

- Mix the cashews, dates, and wheatgrass powder together in a food processor just until a coarse dough has formed (allowing some cashews to remain coarsely chopped). Add the hemp seeds and pulse several times until combined.
- Place a sheet of plastic wrap on a cutting board and spill the dough out on top. Use your hands to shape the dough into rectangle about 1-inch thick, then cut into 8 pieces. Wrap and keep in the freezer for long-term storage.

Chocolate Sacha Inchi Blueberry Energy Bars

High in antioxidants and flavonoids, these bars help reduce free radical damage in the body and improve cellular recovery after workout.

Time: 10 minutes • Makes about a dozen 1 ¾ oz bars

1 cup fresh or soaked dried dates
¼ cup sacha inchi seeds
¼ cup blueberries
¼ cup roasted carob powder
¼ cup ground flaxseed
¼ cup hemp protein
¼ cup unhulled sesame seeds
1 tsp fresh lemon juice
½ tsp lemon zest
Sea salt, to taste
½ cup sprouted or cooked buckwheat (optional)
½ cup frozen blueberries
¼ cup chopped sacha inchi seed

- In a food processor, process all ingredients except the buckwheat, blueberries, and chopped sacha inchi seeds. Knead the buckwheat, blueberries, and sacha inchi seed into the mixture by hand.

Walnut Cranberry Energy Bars

High in antioxidants and flavonoids, these bars help reduce free radical damage in the body and improve cellular regeneration.

Time: 10 minutes • Makes about a dozen 1 ¾ oz bars

1 cup fresh or soaked dried dates
¼ cup walnuts
¼ cup fresh or frozen cranberries
¼ cup ground flaxseed
¼ cup hemp protein
1 tbsp coconut palm sugar
¼ cup unhulled hemp seeds
Sea salt to taste
½ cup sprouted or cooked buckwheat (optional)
¼ cup copped walnuts
½ cup frozen cranberries

- In a food processor, process all ingredients except the buckwheat, walnuts, and cranberries. Once the ingredients are well blended, knead in the buckwheat, walnuts, and cranberries into the mixture by hand.

These gels contain coconut oil, which provides direct energy to the liver and dramatically improves endurance when combined with a carbohydrate source. And they contain chia. Chia provides sustained nutrients in an easily digestible whole food form. When the energy from the glucose contained in the dates begins to wear off, the slower-release energy from the maple syrup is activated, followed by the ultra-slow sustained energy release of chia.

The gels take only minutes to make, and can be carried in a standard two-ounce gel flask, available in most running stores.

Carob Energy Gel

2 large Medjool dates
1 tbsp agave nectar
1 tbsp ground chia
1 tbsp coconut oil
1 tsp lemon zest

1 tbsp fresh lemon juice
1 tsp cocoa nibs (or substitute carob powder)
Sea salt, to taste

- Blend all the ingredients together into a gel-like consistency. For an extra kick, add 1 tsp of ground yerba maté.

Lemon Lime Energy Gel

2 large Medjool dates
1 tbsp agave nectar
1 tbsp ground chia
1 tbsp coconut oil

½ tbsp lemon zest
½ tbsp lime zest
½ tsp dulse
Sea salt to taste

- Blend all the ingredients together into a gel-like consistency. For an extra kick, add ½ tsp of ground yerba maté and ½ tsp of green tea.

Raw

Raw Berry Parfait

KARYN'S FRESH CORNER, CHICAGO, ILLINOIS

Simple, raw, and delicious.

Time: 5 minutes active; 6 hours presoak • Makes 6–8 servings

2 cups cashews
6 cups purified water
¼ cup agave nectar or sweetener of choice
1 cup ground flaxseed or granola
1 ½ cups thinly sliced strawberries

- Soak cashews for 6 hours in 4 cups purified water. Drain and rinse.
- Blend cashews, 2 cups water, and agave nectar to a creamy consistency.
- In a large bowl or among individual parfait glasses, spoon in a dollop of the cashew cream, cover with a thin layer of granola or flax, then arrange a thin layer of strawberry slices.
- Repeat with one or two more rounds of layers. Best served chilled.

Raw

Banana Nut Bread

BEETS LIVING FOOD CAFÉ, AUSTIN, TEXAS

Making this "bread" requires some planning and a dehydrator, but it's worth the extra effort.

Time: 10 minutes active; 12–16 hours to dehydrate • Makes 12 servings

6 fresh bananas (approximately 20 ounces by weight)
2 cups cored and chopped apple
1 ½–2 cups zucchini, peeled and chopped
¼ cup dried raisins or cranberries
1 cup frozen blueberries or cranberries
2 tsp lemon juice
½ tsp vanilla
1 cup ground chia seeds
½ cup soaked, dehydrated walnuts, chopped

- Place the bananas, apples, zucchini, raisins, blueberries, lemon juice, and vanilla in a food processor outfitted with the S-blade. Process until smooth and pour into a large bowl.
- Add the ground chia seeds. Combine thoroughly with a whisk, making sure to break up any clumps.
- Use a scoop or a ¼-cup measure to "plop" mixture onto a paraflex sheet on dehydrator tray lined with the mesh sheet. Spread lightly into desired shape and sprinkle with chopped walnuts.
- Dehydrate for 6–8 hours.
- Invert onto a different tray. Peel back the paraflex and continue to dry for an additional 6–8 hours.

Key Lime Pie

CRUDESSENCE, MONTREAL, QUEBEC

Raw

A delicious, creamy, raw version of the classic Key Lime Pie.

Time: 20 minutes • Makes 1 pie (serves about 8)

CRUST

1 cup macadamia nuts
1 ½ cups coconut, shredded
⅛ cup date paste

⅛ tbsp sea salt
¼ tbsp vanilla extract

- Using a food processor, process macadamia nuts to a butter. Add the other ingredients and process until the mixture is uniform. Spread over the bottom of a pie pan.

FILLING

6 cups avocado flesh
1 ½ cup lemon juice
1 cup coconut butter

1 cup agave nectar
½ tsp vanilla extract

- Place all ingredients in a blender and blend until well combined. Empty the mixture on top of the crust and spread evenly.

ICING

¼ cup lemon juice
2 cups macadamia nuts
3 tbsp agave nectar
1 tsp vanilla extract

Pinch of sea salt
¾ cup water
3 tbsp coconut oil, melted

- Place all ingredients—*except* coconut oil—into the blender and mix until a smooth uniform liquid without lumps is obtained. Add the melted coconut oil while blender is on; blend till oil is mixed throughout. Carefully spread the icing on top of the filling. Beginning at the center of the pie, create two spiral lines of icing. With a fork, draw the shape of a spider's web. Refrigerate before serving.

Banana Crème Pie
MATTHEW KENNEY

Raw

Delicious raw and liked by all, this pie can be safely served to traditional dessert eaters.

Time: 30 minutes active; 30 minutes to sit • Makes 1 pie

CRUST

1 ½ cups macadamia nuts

½ cup coconut flakes

½ tsp salt

3 tbsp agave syrup

1 tbsp coconut oil

1 tsp vanilla extract (or 1 bean)

- In a food processor, process the macadamia nuts, coconut flakes, and salt until they are crumbly flour. Add the agave, oil, and vanilla, and lightly pulse until all ingredients are well combined but mixture only sticks together when pressed between your fingers.

BANANA CRÈME FILLING

3 cups soaked cashews

2 cups mashed banana

1 cup agave syrup (or ½ cup, with
   ½ cup honey)

2 tsp vanilla extract

1 tbsp lemon juice

¼ tsp salt

½ cup coconut oil

- Blend the first six ingredients until smooth. Add the coconut oil and blend until combined.

COCONUT CRÈME

1 ½ cups soaked cashews

1 ½ cups coconut milk

½ cup agave

1 tbsp vanilla

1 tsp lemon juice

1 cup coconut oil, liquefied

Pinch of salt

1 sliced banana for layering

- Blend the first five ingredients until smooth. Then add coconut oil and salt and continue to blend until completely combined. Chill in the refrigerator for a few minutes before using in order to let it set.

To serve
- Press crust into a 9-inch tart pan with a removable bottom. Pour in banana crème filling. Top with one thinly sliced banana. Top with coconut crème. Let set in the refrigerator for at least 30 minutes before serving.

Cherry Tartlets with Vanilla Crème

Chad Sarno

Serve this dessert when you want to be fancy and impress.

Time: 15 minutes active; 1 hour to sit • Makes 12–15 tartlets

CRUST

1 cup raw macadamia nuts

1 cup unsweetened, dried, fine shredded coconut

¾ cup agave nectar

3 tbsp lemon zest

⅛ tsp sea salt

- In a food processor, process all ingredients until finely ground. The mixture should form into a ball when pressed. Using small 2- or 3-inch tartlet shells, line each with plastic wrap, press 2 tbsp crust mixture into each tartlet shell, then remove the tart from the shell and chill.

VANILLA CRÈME

1 cup cashews (presoaked in water 4 or more hours to soften)

¼ cup agave nectar

3 tbsp coconut butter

½ cup orange juice

1 whole vanilla bean seeds scraped

1 tbsp lemon zest

1 tbsp lime zest

- In a high-speed blender, blend cashews, agave, coconut butter, juice, vanilla bean, and lemon and lime zest. Chill blended mixture until ready to serve, at least 1 hour.

GARNISH

1 cup pitted whole cherries

Mint sprigs or edible flowers (optional)

LAVENDER SYRUP (OPTIONAL)

¼ cup dried lavender flowers

1 cup agave nectar

1 cup hot water

- Steep the lavender flowers in hot water for 20 minutes, then strain the water and whisk in the agave.

To serve

- Set one tartlet shell in the center of each plate and top with 2 tbsp vanilla crème. Add a little lavender syrup, if you want to be really fancy. Garnish with cherries and mint or edible flowers.

Chocolate-Covered Sacha Inchi Bananas

A nutrient-rich twist on a simple, delicious classic.

Time: 10 minutes active; 5–6 hours to freeze • Makes 8 servings

¼ cup sacha inchi, chopped fine
4 bananas, peeled and halved, frozen for a minimum of 1 hour
1 batch Simple Chocolate Sauce (see below)

- Place the sacha inchi on a plate. Dip the cold banana halves into the sauce, covering as much surface area as possible, then roll in sacha inchi pieces.
- Place the banana piece on a plate and repeat with remaining bananas. Cover the coated bananas loosely and place the plate in the freezer until bananas are frozen through, about 5 hours.

Simple Chocolate Sauce

Use this sauce as a perfect ice cream topping, a dip for cold fruits, or drizzled on top of other desserts.

Time: 5 minutes • Makes ⅔ cup

⅓ cup coconut oil
½ cup cocoa powder
¼ cup agave nectar or maple syrup

- In a small saucepan, melt the coconut oil over low heat until liquid. Add the cocoa powder and agave nectar and whisk to form a smooth sauce.

No-Bake Double Chocolate Chip Maca Cookies

Soft and sweet. The secret is in the toasting of the coconut, which gives these the taste of "baked" cookies. For the chocolate, use the darkest chocolate you can find.

Time: 15 minutes • Makes about 1 ½ dozen

¼ cup shredded coconut

⅓ cup coconut flour

2 tbsp cocoa powder

2 tsp maca powder (gelatinized)

2 tbsp palm sugar

Pinch sea salt

2 tbsp almond butter

½ tsp vanilla extract

¾ cup Medjool dates, pits removed (about 6 large)

1 ¾ ounces dark chocolate, chopped into small chunks

- Heat a small frying pan over low heat and add the coconut. Toast the coconut until it has turned an amber color (about 1 minute), stirring constantly as coconut will burn easily. Remove from heat.
- Place the toasted coconut, coconut flour, cocoa powder, maca, palm sugar, and sea salt into a food processor and start the machine. Add the almond butter and vanilla extract, then, one at a time, add the pitted dates and process until a crumbly dough has formed.
- Stop the machine and check the consistency: the dough should stick together easily between two fingers when pinched. If too wet, add additional coconut flour. If too dry, add water about ¼ tsp at a time until correct consistency is reached.
- Transfer to a bowl and mix in the dark chocolate chunks. Using clean hands, grab a small amount—about a tablespoon—and squeeze and roll to form a tight ball. Flatten the ball using your palm or the back of a glass to form a small cookie.

Mint Chip Ice Cream

With a fresh coconut and strong mint flavor, this ice cream is a favorite.

Time: 10 minutes active; about 6 hours to freeze • Makes about 1 pint (serves 4)

½ cup young coconut meat (about 1 coconut)
1 ¼ cup raw cashews
1 cup water
¾ cup vanilla unsweetened almond milk
⅓ cup fresh mint leaves
⅓ cup + 1 tbsp agave nectar
12–15 drops mint extract
⅓ cup chopped dark chocolate

- Blend the coconut meat, raw cashews, water, almond milk, fresh mint, and agave together into a smooth cream.
- Carefully add the mint extract—a little goes a long way, so take care not to spill. Blend again, taste, and add more extract if desired.
- Transfer to a bowl or Tupperware container and freeze, covered, for 1 hour until chilled. Add in the chopped dark chocolate and mix thoroughly. Return to the freezer and enjoy when frozen through—about 4-6 hours. Let defrost for a couple of minutes to soften slightly before serving.

Vanilla Protein Ice Cream with Caramel Swirl

Time: 10 minutes prep; about 6 hours to freeze • Makes 4–6 servings

ICE CREAM

1 cup cashews

1 cup water

1 banana

1 scoop Vanilla Vega Sport Protein Powder

2 tbsp agave nectar

1 tsp vanilla extract

- Blend the cashews, water, banana, protein powder, agave nectar, and vanilla in a blender until smooth. Transfer to a bowl or Tupperware container, cover, and place in the freezer.

CARAMEL SWIRL

2 tbsp yacon syrup

2 tsp almond butter

Pinch of sea salt

- In a small glass, mix the yacon, almond butter, and a pinch of sea salt together to form a sauce. After the ice cream has frozen for about 1 hour and begun to thicken, fold the sauce into the ice cream—mixing just enough to form caramel swirls (do not fully incorporate). Place the ice cream back in the freezer until frozen—about 5 hours.

Chocolate Chip-Maple Maca Ice Cream

If you have an ice cream maker, feel free to put it to use for a fluffier texture.

Time: 10 minutes active; 6–8 hours to freeze • Makes about 1 pint

1 cup cashews
1 cup water
½ banana
¼ cup maple syrup
2 tbsp gelatinized maca powder
Pinch sea salt
1 tsp vanilla extract
⅓ cup chopped dark chocolate

- Blend all ingredients—except the chopped dark chocolate—in a blender until completely smooth. Transfer to a bowl or Tupperware container and freeze for 30 minutes.
- Mix in dark chocolate chunks into the cold ice cream, then continue freezing, covered, until frozen through—about 6–8 hours.
- Let defrost for 5 minutes to soften before serving.

Cantaloupe Ginger Ice

The ginger flavor is very delicate. Only use fresh. Ginger lovers can add more.

Time: 15 minutes active; about 4 hours to freeze • Makes 4 servings

4 cups cantaloupe flesh (about one medium)
1 ¼ tsp fresh grated ginger
2 tbsp agave nectar (or use white stevia, to taste)

- Combine all ingredients in a food processor, and purée until mostly smooth. Transfer mixture to a shallow bowl or pan, cover, and place in the freezer. About once an hour, use a fork to break up any frozen chunks into a fluffy snow. Repeat until entirely frozen, about 4 hours.

....

Watermelon Lemon Granita

Granita is an Italian ice. Very refreshing.

Time: 15 minutes active; about 4 hours to freeze • Makes 4 servings

4 cups watermelon flesh, seeds removed
½ tsp lemon zest
2 tbsp lemon juice
1 tbsp agave nectar (or use white stevia, to taste)

- Combine all ingredients in a food processor, and purée until mostly smooth.
- Transfer mixture to a shallow bowl or pan, cover, and place in the freezer. About once an hour, use a fork to break up any frozen chunks into a fluffy snow.
- Repeat until entirely frozen, about 4 hours.

Indigo Granita

The antioxidant powerhouses of blueberries and açaí combine forces to produce a colorful, healthy, and refreshing frozen berry dessert. Serve in a wine glass with a sprig of mint for the full esthetic effect.

Time: 15 minutes active; about 4 hours to freeze • Makes 2–4 servings

1 ½ cups frozen blueberries, partially thawed
¼ cup lemon juice
1 tbsp maple syrup
1 tbsp agave nectar
2 tbsp açaí powder
Touch of stevia, if desired
2 tbsp water + more if needed to blend

- Combine all ingredients in a food processor, and purée until mostly smooth.
- Transfer mixture to a shallow bowl or pan, cover, and place in the freezer. About once an hour, use a fork to break up frozen chunks and fluff into a snow. Repeat until entirely frozen, about 4 hours.

....

Bodacious Berry Cookie Crumble

Great crumbled on top of ice creams or use as a flavorful pie crust.

Time: 5 minutes • Makes 8- or 9-inch pie crust, depending on thickness

1 cup cashews
8 large Medjool dates, pitted
2 tbsp Vega Shake & Go Bodacious Berry Instant Smoothie Mix
1 tbsp coconut oil

- Grind the cashews into a flour in a food processor. With the machine still running, add the pitted dates, one at a time, until fully incorporated. Add the Smoothie Mix and the oil and process until a dough has formed.

Summery Simple Peach Tart

Best made in the summer with in-season peaches.

Time: 5-10 minutes • Makes 6 tarts

1 cup almonds
¼ cup ground flaxseeds
4 dates
2 tbsp palm sugar
2 tsp vanilla extract
Pinch sea salt
2–3 peaches, chopped
Cinnamon, for garnish
Vanilla or maca ice cream, for serving

- To make the crust, grind the almonds into a flour in a food processor. With the machine still running, add the flax seeds, dates, sugar, vanilla, and sea salt and process until fully incorporated. Press the dough into a tart pan to form a crust and top with peaches and cinnamon.

Variation: Place peaches in individual serving bowls and simply crumble the almond mixture on top. Serve with any ice "cream" recipe from this book.

Yacon Dessert Ravioli

The mild taste and satisfying crunch of reconstituted yacon is complemented by a delectable sweet citrus-spice filling. Serve with a good non-dairy vanilla ice cream, or make the raw version below and keep this recipe 100 percent raw. An impressive gourmet dessert with gorgeous flavor notes that are not to be forgotten!

Time: 10 minutes active; about 6 hours to freeze • Makes 12 ravioli (serves 2–4)

2 cups orange juice, or enough to immerse yacon slices
24 large yacon slices (try to pick larger, rounder slices)
½ cup walnuts
¼ cup raisins
1 ½ tbsp yacon powder
¼ tsp nutmeg
¼ tsp cinnamon
½ tsp orange zest (grated orange rind) + some for garnish
One recipe Vanilla Protein Ice Cream with Caramel Swirl (see p. 279)
Yacon syrup, for garnish

- In a large bowl, pour orange juice over yacon slices until submerged. Allow to reconstitute for 1 hour.
- In a food processor, combine walnuts, raisins, yacon powder, nutmeg, cinnamon, and orange zest. Process until a sticky filling has formed. Remove the mixture from the processor and knead to combine ingredients completely, if necessary.
- Remove yacon slices from orange juice and blot on a paper towel to remove any extra juice. To make the ravioli, place a small teaspoon of the filling in the center of a yacon slice. Top with another yacon slice, and pat the edges slightly to form a ravioli. Repeat until all yacon slices are used.

To serve
- Place 3 or 4 ravioli on a dessert plate, and top with a scoop of Vanilla Protein Ice Cream. Drizzle with yacon syrup and dust with orange zest, if desired.

White Chocolate Raspberry Cheesecake
with Chocolate-Almond Crust

Cacao butter is the secret ingredient to making this phenomenally smooth and sophisticated non-dairy cheesecake. Not only does the cocoa butter perfume the mixture with white chocolate luxury, but it also helps solidify the cheesecake. The cocoa paste has the same effect in the dark chocolate almond crust, which retains a satisfying chocolate crunch because of the hint of cocoa nibs. Mesquite powder enhances the chocolate flavor even further, serving as a natural flavor match for cocoa.

Time: 20 minutes active; 30 minutes to freeze • Makes 1 9-inch cheesecake or 6 5-inch mini cheesecakes

CRUST

1 cup raw almonds

1 cup Medjool dates, pits removed (about 8)

2 tsp mesquite powder

3 tbsp melted cocoa paste (approximately ¼ cup unmelted shavings; melt using a double boiler method)

½ tsp almond extract

Pinch sea salt

1 tbsp cocoa nibs

- Use a food processor to blend the almonds, dates, mesquite, cocoa paste, almond extract, and sea salt together into a crumbly dough. Add the cocoa nibs and pulse several times until just combined but not blended through, to retain a bit of crunchy texture.

CHEESECAKE

2 pints fresh raspberries

2 cups raw cashews

3 ½ ounces melted cocoa butter (about 1 cup dry shavings;
   melt using a double boiler method)

½ cup agave nectar or maple syrup

2 tbsp lucuma powder

4 tbsp vanilla extract

2 tbsp fresh lemon juice

Simple Chocolate Sauce (see p. 276; optional)

Fresh raspberries, for garnish

- Use a blender to purée the raspberries. (If you wish, you may strain the berries to remove all the seeds but retain the liquid, then place the juice back in the blender. If you don't mind a few seeds in the cheesecake, you can skip this step.) To the raspberries, add the remaining ingredients and blend until very smooth (this may take a few minutes).

To serve

- Use a 9-inch springform pan and press the crust evenly across the bottom to form a dense, flat layer. Pour the cheesecake mixture on top.
- Place in the freezer for at least 30 minutes to set; freeze until ready to serve. To serve, drizzle with Simple Chocolate Sauce if you wish, then decorate with fresh raspberries.

Variation: Make mini cheesecakes instead. When the cake components are ready, line six 5-inch ramekins tightly with plastic wrap, allowing extra wrap to flow over on all sides of the ramekin to ease later removal (alternatively, simply use mini-springform molds). Make the mini cakes just as you would the larger one. When ready to serve, simply lift the cheesecakes out of the ramekins by pulling up on the plastic wrap.

Apple-Pear Energy Tartlets

Of course, simply doubling the recipe and making one large tart works too.

Time: 10 minutes • Makes 6 3-inch tartlets

1 medium sweet crunchy apple (Fuji or Gala work well)
1 medium soft pear (Bartlett or Bosc)
½ tsp pumpkin pie spice, or cinnamon
1 tbsp yacon syrup (optional)
3 Vega Whole Food Vibrancy Bars, Original Flavor

- Chop the apple and the pear into a fine mince. Toss with pumpkin pie spice and yacon syrup until well combined. Set aside.
- Using a standard-size muffin tin or tartlet molds, tuck a layer of plastic wrap into one of the cups to use as a mold.
- Break the energy bars into halves. Take one of the bar pieces and press it firmly into the lined cup, coaxing the bar "dough" evenly along the bottom and up the sides about ½ inch with your fingers to form the tartlet base.
- Use the plastic wrap to remove the packed base from the muffin tin, then peel away the plastic. Use the plastic again to form the remainder of the tartlets.
- Liberally spoon in the apple-pear filling evenly into each of the tartlet crusts and serve.

GUIDE TO NUTRIENTS

MICRONUTRIENTS

Phytonutrients

Also referred to as a phytochemical, a phytonutrient is a plant compound that, by boosting the immune system, offers health benefits independent of its nutritional value. As you can see, for "Best sources" I specify "organic" foods. That's because phytonutrients are often the plant's natural pest deterrent, whereas if chemical pesticides are used, the plant won't produce phytonutrients. It's suspected that a vast quantity of phytonutrients—with benefits we aren't even aware of—have yet to be discovered. But it's believed that they will most probably be present in the same foods that contain known phytonutrients.

Best sources: organic vegetables, organic seeds, organic fruit, organic nuts, green tea, yerba maté.

Antioxidants

Antioxidants are naturally occurring compounds, including vitamins C and E and the mineral selenium. Antioxidants are most prized for their ability to protect cells. Helping rid the body of free radicals, antioxidants are credited with helping to maintain cellular health and to promote cellular regeneration. If not for antioxidants, cellular damage caused by stress would advance quickly and likely lead to disease.

Best sources: organic colorful fruits and vegetables, organic berries, organic sweet peppers.

VITAMINS

Vitamin A

Vitamin A helps the body resist infection, which it is more prone to after physical exertion, and allows the body to use its reserves for repairing and regenerating muscle tissue (instead of fighting infection)—leading to quicker recovery. Vitamin A helps support growth and repair of muscle and maintains red and white blood cells—crucial for performance.

Best sources: orange and dark green vegetables, including carrots, pumpkin, sweet potatoes, winter squash, broccoli, kale, parsley, and spinach; apricots, mango, papaya, cantaloupe.

Vitamin B1

Vitamin B1 helps the body convert carbohydrate into energy. Maintaining high energy levels depends in part on maintaining adequate vitamin B1 in the diet. People who eat healthy food rarely have a problem getting enough vitamin B1; it's plentiful in many foods. Also, because active people expend more energy than the average person, they need more vitamin B1. Again, this is usually not a problem, since with increased activity comes increased appetite.

Best sources: legumes, pseudograins, nuts, brown rice, nutritional yeast, and blackstrap molasses.

Vitamin B2

Vitamin B2 helps break down amino acids (protein) for the body to use. Utilization of amino acids is a key factor in quick muscle recovery and regeneration after exertion. Like vitamin B1, B2 helps the body convert carbohydrate into energy.

Vitamin B2 aids in the formulation of growth hormones, a primary factor in muscle health and development. It also contributes to healthy red blood cell production. Red blood cells are the carriers of oxygen to working muscles, making them an integral part of performance

Best sources: legumes, pseudograins, nuts, brown rice, nutritional yeast, and blackstrap molasses.

Vitamin B3

Vitamin B3 is essential for the body's breakdown and utilization of carbohydrate and protein. As with other B vitamins, vitamin B3 plays an integral part in the conversion of food into energy. Vitamin B3 has an important role in keeping the digestive system healthy as well. A healthy digestive system will allow the body to get more out of its food, reducing hunger and the amount of food needed. Also, a healthy digestive system will extract trace minerals from food, essential for performance.

Best sources: beets, sunflower seeds, nutritional yeast.

Vitamin B5

As with other B vitamins, vitamin B5 helps the body convert food into energy. As well, vitamin B5 facilitates the production of steroids—an integral part of the regeneration process after exertion. This vitamin is found in a wide variety of healthy foods, and deficiency is uncommon.
Best sources: seeds, pseudograins, avocados.

Vitamin B6

As a B vitamin, B6 too participates in the release of energy from food and in the formation of red blood cells. Vitamin B6 aids in the production of antibodies—essential for warding off infection and maintaining the ability to recover from exertion quickly. Vitamin B6 contributes to cardiovascular health, helping the heart efficiently circulate blood in a greater volume as demanded by the active person.
Best sources: pseudograins, greens, bananas, brown rice, walnuts, avocados, oats.

Folate (Folic Acid)

Folate is a B vitamin that is found naturally in foods; when in supplement form, it is called folic acid. Folate works in tandem with vitamin B12 to help produce oxygen-carrying red blood cells. Folate plays an integral role in helping the body make use of dietary protein, facilitating muscle repair. The heart relies on folate, in part, to help it maintain a smooth, rhythmic, efficient beat—and gives it a higher tolerance for physical activity.
Best sources: leafy green vegetables, legumes, pseudograins, orange juice, nutritional yeast.

Vitamin B12

Vitamin B12 is essential for a healthy nervous system, aiding in coordination and smooth muscle movement. As with other B vitamins, B12 plays a role in the production of red blood cells and conversion of food to usable energy. Unlike other B vitamins, B12 is not plentiful in foods. Special attention must be paid to ensure dietary B12 needs are met, particularly if the diet doesn't contain animal products and exercise level is moderate to high.
Best sources: chlorella, miso, nutritional yeast, kombucha.

Vitamin C

Vitamin C is a powerful antioxidant, meaning it plays an integral role in reducing damage to body tissue and muscle done by physical activity; it is therefore essential for active people. Cellular damage that occurs as a result of environmental factors, such as pollution, will be minimized by daily ingestion of vitamin C. The ability to minimize environmental stress will greatly improve the body's ability to ward off infection and allow it to recover from physical activity considerably quicker. Iron absorption is improved when iron is ingested at the same time as vitamin C–rich foods.
Best sources: most vegetables and fruits (especially citrus fruits).

Vitamin D

Vitamin D allows the body to absorb calcium more efficiently—a key factor for proper bone formation (and healing) and smooth muscle contractions.
Best sources: nutritional yeast, exposure to sunlight.

Vitamin E

Vitamin E, like vitamin C, is a powerful antioxidant. Active people need higher levels of vitamin E than sedentary people, as vitamin E, in concert with other vitamins, reduces the constant stress exercise places on the body.

Promoting cardiovascular health by maintaining an optimal ratio of "good" to "bad" cholesterol is another role of vitamin E. The ability to maintain the ideal ratio is a key factor for proper growth hormone production—the cornerstone of muscle rejuvenation post-exertion. Vitamin E also combats the effects of harmful free radicals produced by physical activity.
Best sources: flaxseed oil, hemp oil, pumpkin seed oils, and especially raspberry seed, cranberry seed, and pomegranate seed oils; nuts, avocados.

Vitamin K

Vitamin K plays a significant role in blood clotting. It also provides the heart with nutrients it needs for optimal function.
Best sources: leafy green vegetables, pine nuts.

Biotin

Biotin works in concert with the B vitamins as a converter of food into usable energy.

It is also necessary for cell growth and helps the production of fatty acids and the metabolism of fats and amino acids.

Best sources: nuts, nutritional yeast.

Carotenoids

Carotenoids are fat-soluble phytochemicals, meaning that they are stored in fat cells and not excreted in urine. Therefore, they remain in the body longer. A class that includes over 600 naturally occurring pigments synthesized by plants, algae, and photosynthetic bacteria, carotenoids are easily spotted since they are yellow, orange, deep green, and red. Many of them also have antioxidant properties and include

- Alpha-carotene
- Beta-carotene
- Cryptoxanthin
- Lutein
- Lycopene
- Zeaxanthin

Best sources: carrots, sweet potatoes, squash, spinach, kale, collard greens, tomatoes, red and yellow sweet peppers. Most colorful fruits and vegetables.

MACROMINERALS

Calcium

For most people, bone strength and repair is calcium's major role. Active people, however, have another important job for the mineral: muscle contraction and ensuring a rhythmic heartbeat. Upward of 95 percent of the body's calcium is stored in the skeleton, and a decline in calcium levels may take years to manifest as osteoporosis. But a decline will be noticeable as an irregular heartbeat and muscle cramps—the responsibilities of that remaining few percent. Since calcium in the bloodstream is lost in sweat and muscle contractions, a higher dietary level for active people is recommended.

The body orchestrates the effective combination of calcium and vitamin D to maximize calcium absorption.

Best sources: leafy green vegetables, unhulled sesame seeds, tahini.

Magnesium

Critical for muscle function, magnesium helps the heart beat rhythmically by allowing it to relax between beats, which allows all other muscles to relax. Magnesium also assists in calcium's bone production.

Best sources: leafy green vegetables, string beans, legumes, pseudograins, bananas, nuts, avocados.

Phosphorus

Critical in the maintenance of the body's metabolic system, phosphorus allows the body to use food as fuel. Phosphorus works with calcium in the production, repair, and maintenance of bones.

Best sources: pseudograins, most tropical fruit.

Potassium

Potassium, an electrolyte, helps the body maintain fluid balance and therefore hydration. Being properly hydrated is essential for efficient movement. Proper hydration will maintain the blood's light viscous flow, increasing the amount the heart can pump and improving performance. Smooth, concise muscle contractions are one of potassium's responsibilities. Nerve impulse transmission and cell integrity also rely, to a degree, on potassium. As a result, smooth motor function, heartbeat efficiency, and the ability to strongly contract a muscle are dependent on adequate potassium intake. As with other electrolytes, potassium is lost in sweat, so active people need more.

Best sources: leafy green vegetables, most fruits (especially bananas and kiwis).

Sodium

Sodium is a vital component of nerves as it stimulates muscle contraction and helps to keep calcium and other minerals soluble in the blood. With too little, muscle stiffness and cramping can be the result.

TRACE MINERALS

Boron

Important for bone health and strength, boron has also been linked to a reduction in free radical production during the conversion of food into usable energy. Deficiency may result in reduced motor skills and difficult learning.
Best sources: fruits, vegetables, nuts and legumes.

Chloride

Vital in the production of hydrochloric acid, chloride is secreted from the parietal cells of the stomach in preparation for digestion
Best sources: sea vegetables, sea salt.

Chromium

Chromium works with other vitamins and minerals to turn carbohydrate into usable energy.
Best sources: pseudograins, nuts, nutritional yeast, black pepper, thyme.

Cobalt

A component of vitamin B12, cobalt is a nutritional factor necessary for the formation of red blood cells.
Best sources: chlorella, miso, nutritional yeast, kombucha.

Copper

Like vitamin C, copper assists iron absorption in the body. With iron, copper plays a role in the transport of oxygen throughout the body—imperative for optimal performance. As a member of the body's defense network, copper works in concert with antioxidants to reduce the effects of environmental and physical damage, providing the body with a strong platform to regenerate and build strength.

Best sources: legumes, seeds, pseudograins, raisins, nuts.

Iodine

Iodine is integral to thyroid hormone production. Thyroid hormone assists the cells in the fabrication of protein and the metabolism of fats—essential for energy maintenance. High levels of iodine are lost in sweat, making active people's requirements higher than those of less active people.

Best sources: sea vegetables (especially dulse).

Iron

The main role of iron is to fabricate hemoglobin to facilitate red blood cell health. An adequate iron level is of paramount importance for the active person. A well-maintained iron level ensures the body is able to deliver oxygen-rich blood to the hard-working extremities, maximizing efficacy. Also used to build blood proteins needed for food metabolism, digestion, and circulation, dietary iron is essential for proper functionality.

Best sources: spinach, legumes (especially split peas), pumpkin seeds.

Manganese

As an activator of antioxidant enzymes, manganese contributes to an expedited process of recovery, essential to all those who are physically active. Manganese is a cofactor in energy production, metabolizing protein and fats.

Best sources: leafy green vegetables, legumes, pseudograins, nuts, rice.

Molybdenum

A trace mineral, molybdenum's chief role is as a mobilizer, moving stored iron from the liver into the bloodstream—of particular significance to active people. An aid in the detoxification processes, molybdenum helps the body rid itself of potentially toxic material, minimizing stress.

Best sources: legumes, pseudograins, nuts.

Selenium

In concert with vitamin E, selenium preserves muscle tissue elasticity, allowing fluent, supple movement. A trace mineral, selenium combines with other antioxidants to shield red blood cells from damage done by physical exertion. It also improves immune function. As with other antioxidants, selenium offers protection from environmental stress encountered by most people on a regular basis.

Best sources: Brazil nuts, walnuts, whole-grain rice, nutritional yeast.

Zinc

Zinc's major role is to allow the body to use dietary protein as building blocks, for the regeneration of muscles. As well, zinc plays an integral role in the preservation of proper immune function.

Best sources: pseudograins, pumpkin seeds, nutritional yeast.

MY FAVORITE RESTAURANTS AND CAFÉS

BEETS LIVING FOODS CAFÉ (Austin)

5th Street Commons
1611 W 5th Street, Suite 165
Austin, TX 78703
www.beetscafe.com

Beets Living Foods Café features an upscale raw-food dining experience.
Owner/Operator Sylvia Heisey is passionate about healthful living and
sharing it with the world; the chefs of Beets Café create uncommonly good
food that is alive with flavor and nutrition and prepared with love. The
menu utilizes seasonal vegetables, fruits, nuts and seeds, and the chefs' first
choice is always local and organic, supporting farmers with sustainable
practices that preserve the Earth. Meal selections, packaged in eco-friendly
containers, are also available for take-out.

BLOSSOMING LOTUS (Portland)

1713 Northeast 15th Avenue
Portland, OR 97212
(503) 228-0048
www.blpdx.com

Located in the historical Irvington district on Portland's East Side,
Blossoming Lotus is open seven days a week to provide customers with a
wide array of organic, vegan, sustainable meals.

CANDLE 79 (New York City)

154 East 79th Street at Lexington Avenue
New York, NY 10021
(212) 537-7179
www.candle79.com

Candle 79 is the upscale, elegant sister restaurant of the famous Candle
Café, whose famed vegan, organic cuisine is presented in a different setting:

a two-floor fine dining oasis with an organic wine and sake bar for the conscientious yet sophisticated eater. Candle 79 has twice been rated as Zagat's #1 vegetarian restaurant, and its reputation is known internationally. It is the restaurant choice of neighborhood regulars, celebrities, politicians, and corporate tycoons.

Combining creative, beautifully presented healthy, and delicious cuisine with gracious and knowledgeable service in a gorgeous duplex space, Candle 79 is at the forefront of a movement to bring elegance to vegetarian cuisine, and to bring the concepts of "local," "seasonal," "organic," and "vegan" into the culinary mainstream.

CRU (Los Angeles)
1521 Griffith Park Blvd
Los Angeles, CA 90026
(323) 667-1551
www.crusilverlake.com

In the Silverlake area of Los Angeles, Cru serves up delicious, seasonal, organic, raw food that is nutrient-rich and perfectly flavor-balanced.

CRUDESSENCE (Montreal)
105 Rachel Street West
Montreal, QC H2W 1G4
(514) 510-9299
www.crudessence.com

A healthy oasis in the heart of the city, Crudessence serves vegan, organic, and local living food in an inspired atmosphere. The Crudessence team emphasizes respect for all living things: a healthy body in a clean environment and a strong, sustainable ecosystem. A full range of catering services and an accompanying culinary academy are integral parts of Crudessence's mandate to instill in others the happiness of conscious eating and support for local agriculture.

FRESH AT HOME (Toronto)

www.freshrestaurants.ca

Fresh on Bloor

326 Bloor Street West
Toronto, ON M5S 1W5
(416) 531-2635

Fresh on Crawford

894 Queen Street West
Toronto, ON M6J 1G3
(416) 913-2720

Fresh on Spadina

147 Spadina Avenue
Toronto, ON M5V 2L7
(416) 599-4442

With a comprehensive menu, and three locations, Fresh offers a wide variety of nutrient-dense healthy fast food for eat in or take away. Also, their menu of fresh juices is impressive.

GORILLA FOOD (Vancouver)

101-436 Richards Street
Vancouver, BC V6B 2Z3
(604) 684-3663
www.gorillafood.com

Started by an organic, vegan, raw food enthusiast, Gorilla Food is passionate about creatively and consciously providing raw, organic, vegan foods to people who care for themselves and the world. Gorilla's kitchen and eat-in space is at 436 Richards Street in the bustling core of downtown Vancouver. It also provides a line of vegan raw food ingredients, available at select retail environments. Finally, Gorilla's convenient delivery service brings the best locally sourced raw foods directly to the customer, connecting the urban consumer with organic farmers who are growing natural foods in the most sustainable ways.

THE GREEN DOOR (Ottawa)

198 Main Street
Ottawa, ON K1S 1C6
(613) 234-9597
www.thegreendoor.ca

Since 1988, the daily practice in the kitchen and bakery of the Green Door Restaurant reflects a desire to create delicious and wholesome food that is as close to its natural state as possible. Recipes are kept simple; there are no fillers, processed ingredients, or preservatives in any of the dishes. There is no set menu; every morning, the team assesses the day's supply of fresh vegetables, and creates the menu based on what ingredients are in season or in plentiful supply—although some popular dishes are produced daily. The Green Door specializes in high-quality foods made from ingredients that are locally produced.

HORIZONS (Philadelphia)

611 South 7th Street
Philadelphia, PA 19147-2103
(215) 923-6117
www.horizonsphiladelphia.com

Over the years, Horizons has become a major contributor to the dining scene in Philadelphia. In 2006, Horizons was awarded "3 Bells" by *Philadelphia Inquirer* restaurant critic Craig LaBan. That same year, it was named "Restaurant of the Year" by *VegNews* magazine. Since then, Horizons continues to be celebrated on a local and national level as one of the pioneers of vegan cooking. Horizons has been recognized by *Philadelphia Magazine* as one of the Top 50 Restaurants in Philadelphia and by the *New York Times* as one of Philadelphia's Best New Restaurants. Horizons made culinary history on November 4, 2009, by cooking the first vegan dinner at the prestigious James Beard House in New York City.

JIVAMUKTEA CAFÉ (New York City)

2nd Floor, 841 Broadway
New York, NY 10003
(800) 295-6814
www.jivamuktiyoga.com/cafe/index.html

Located within Jivamukti Yoga School, JivamukTea Café's mission is to
facilitate the elevation of consciousness and support the practice of ahimsa
(non-harming) through delicious, nourishing, vegan meals that use organic
foods free from genetic modification and by employing sustainable practices
that support the continued viability of all precious life forms on our shared
planet. It provides a peaceful and educational environment for all to enjoy
food, drink, art, music, and community events that support and expand
upon this vision.

KARYN'S ON GREEN (Chicago)

130 South Green Street
Chicago, IL 60607-2625
(312) 226-6155
www.karynsongreen.com

With delicious food and a vibrant bar scene, Karyn's on Green is making
vegan sexy in the midst of Chicago's traditional Greektown neighborhood.
It serves lunch, dinner, and drinks seven days a week in a chic atmosphere,
providing an earth-friendly vegan approach to contemporary dining. The
diverse menu features classic American dishes reinterpreted with innovative
presentations but without the meat, fish, chicken, or dairy—and highlights
a host of satisfyingly bold flavors and imaginative combinations.

KARYN'S FRESH CORNER CAFÉ (Chicago)

1901 North Halsted Street
Chicago, IL 60614
(312) 255-1590
www.karynraw.com/Raw-Cafe

Karyn's Fresh Corner Café is a one-of-a-kind holistic wellness center.
Founder Karyn Calabrese is committed to using only whole fresh foods

(organic) whenever possible. Fresh Corner creates fresh, delicious, high-quality meals that are high in fiber and low in fat. Consistency is the key to making healthy lifestyle choices, and the Café's Living Food Menu helps consumers maintain that consistency. The Café also features classes and workshops throughout the year on the importance of food choices and how to prepare tasty, nutritious meals at home.

LIFE FOOD GOURMET (Miami)

1248 SW 22nd Street
Miami, FL 33145-2936
(305) 856-6767
www.lifefoodgourmet.com

Life Food Gourmet offers fresh and creative raw meals and snacks, from simple to extravagant.

LIVE ORGANIC FOOD BAR (Toronto)

264 Dupont Street
Toronto, ON M5R 1V7
(416) 515-2002
www.livefoodbar.com

Live is proud to offer an array of gourmet, gluten- and sugar-free vegan dishes to invigorate and cleanse the body, mind, and soul. Many different cultural foods and ingredients are combined in handmade organic dishes. Each dish is prepared and served by experienced staff and guided by Live's founders, internationally known Raw Chef and restaurateur Jennifer Italiano and her financial partner and brother, restaurateur Chris Italiano. With their deep love of food and service, they ensure a life-changing culinary experience.

MILLENNIUM (San Francisco)

580 Geary Street
San Francisco, CA 94102-1650
(415) 345-3900
www.millenniumrestaurant.com

Millennium creates a gourmet dining experience out of vegetarian, healthy, and environmentally friendly foods, striving to make vegetarian dining fun and exciting. Nestled in the heart of a food lovers' city, Millennium is committed to keeping that tradition alive. The cuisine is influenced by the flavors and styles of many cultures and all dishes are completely animal-free. Millennium is dedicated to supporting the essential earthly concepts of organic food production: small farms, sustainable agriculture, recycling, and composting. The kitchen cooks with fresh produce delivered every day, and chooses organic whenever possible; the restaurant is completely free of genetically modified foods.

PURE FOOD & WINE (New York City)

54 Irving Place
New York, NY 10003
(212) 477-1010
www.oneluckyduck.com/purefoodandwine

Even though I haven't used any recipes from Pure Food & Wine in this book, I want to mention and highly recommend this restaurant. It's gourmet raw food at its finest. My friend Sarma (who has also written a couple of excellent books, *Raw Food/Real World* and *Living Raw Food*) is the founder and chef.

RAVENS' RESTAURANT (Mendocino)

44850 Comptche Ukiah Road
Mendocino, CA 95460-9007
(707) 937-5615
www.ravensrestaurant.com

Located at Stanford Inn Eco-Resort on the Mendocino coast, Ravens'
Restaurant is haute contemporary vegetarian and vegan cuisine blended
with ecologically responsible paradigms. Dishes are based on locally
harvested products from seaweed to morels; 99 percent of all ingredients
are organic, and most produce is from regional organic growers or the
Inn's California-Certified Organic Farm. The wine list comprises primarily
wines produced from certified organic vineyards, biodynamic vineyards, or
those using sustainable, traditional farming practices. All food wastes are
composted and the compost dug back into garden beds; all glass, papers,
and cardboard are sent to recycling. Ravens' Restaurant is independently
owned and lovingly cared for by Jeff and Joan Stanford.

THRIVE JUICE BAR (Waterloo)

105-191 King Street South
Waterloo, ON N2J 1R1
(519) 208-8808
www.thrivejuicebar.com

Named after my first book *Thrive*, Thrive Juice Bar was started by a big
supporter of the plant-based whole food philosophy that I put forth in my
book. Now a friend, founder Jonnie serves up some of the best-tasting, most
nutrient-dense smoothies you'll ever drink.

CALCULATING THE NUMBERS

2   EATING RESOURCES

The Greatest Emission Creator

Since a midsize car emits ½ pound of CO_2 to travel 1 mile (www.falconsolution.com/co2-emission/), we can calculate the following:

CANADIAN BEEF EMISSIONS

256 km (distance needed to drive a mid-size car to produce the equivalent in CO_2e as the production of 1 kg of beef) × 31.1 kg (average annual beef consumption of each Canadian) = 7961 km. converted to miles: 4976

4976 × 33,311,400 (Canadian population) = 166,757,526,400 miles ÷ 238,857 miles (average distance to the moon) = 693,961 trips to the moon

Driving distance from Vancouver to Toronto: 2719 miles (4375 km)

U.S. BEEF EMISSIONS

256 km (distance needed to drive a midsize car to produce the equivalent in CO_2e as the production of 1 kg of beef) × 43.8 kg (average annual beef consumption of each American) = 11,278 km. converted to miles: 7008

7008 × 307,006,550 (U.S. population) = 2,151,501,902,400 miles ÷ 238,857 miles (average distance to the moon) = 9,007,489 trips to the moon

CANADIAN CHICKEN EMISSIONS

23.5 km (distance needed to drive a midsize car to produce the equivalent in CO_2e as the production of 1 kg of chicken) × 30 kg (average annual chicken consumption of each Canadian) = 703 km (439.4 miles)

U.S. CHICKEN EMISSIONS

23.5 km (distance needed to drive a mid-size car to produce the equivalent in CO_2e as the production of 1 kg of chicken) × 46.5 kg (average annual chicken consumption of each American) = 1086 km (679 miles)

CANADIAN PORK EMISSIONS

Since for every 1 kg of pork produced, 5 kg of CO_2 is released, that's the same amount of CO_2e as driving 18.25 miles.

29.3 km (distance needed to drive a midsize car to produce the equivalent in CO_2e as the production of 1 kg of pork) × 22.9 kg (average annual pork consumption of each Canadian) = 672 km (418 miles)

U.S. PORK EMISSIONS

29.3 km (distance needed to drive a midsize car to produce the equivalent in CO_2e as the production of 1 kg of pork) × 29.6 kg (average annual pork consumption of each American) = 867 km (538 miles)

What Constitutes Environmentally Friendly Food Choices?

1335 miles conserved × U.S. population = 409,853,744,250 miles ÷ 238,857 miles average distance to the moon = 1,715,895.98 trips to the moon

For the light breakfast: emission savings would be 26% (light is 10 ÷ 38 of traditional). Therefore, savings from eating the plant-based option over the "light" option would conserve as much CO_2 as would be emitted by driving the equivalent distance as making 446,133 trips to the moon.

3   AN APPETITE FOR CHANGE

Nutrient-to-Emission Ratio

Formula: CO_2e, in grams, emitted in the production of 1 kg of given food divided by the number of calories per 1 kg of given food. The nutrient density of the given food divided by the sum of the above calculation yields the nutrient-to-emission ratio.

$$\frac{CO_2e, \text{ in grams, emitted in the production of 1 kg of food}}{\text{number of calories per 1kg of food}} = \begin{array}{l} CO2e, \text{ in grams,} \\ \text{emitted to produce} \\ 1 \text{ calorie of food} \end{array}$$

$$\frac{\text{nutrient density of the food}}{CO_2e, \text{ in grams, emitted to produce 1 calorie of food}} = \text{nutrient-to-emission ratio}$$

Since the eatlowcarbon.org suggested serving size is 4 oz (¼ pound), you can find the amount of CO_2e that is emitted in the production of 1 kg of each food by multiplying the amount of CO_2e emitted by 4 oz by 4, which will give you the amount of CO_2e emitted by one pound of the food. To convert that to 1 kg, multiply by 2.2.

You can use Nutri-Facts.com to find out how many calories are in each 100 g of that food and then multiply that number by 10 to give you the amount in 1 kg.

Here's an example using beef:

7641 g of CO_2e per 4oz.* 7641×4 (to yield amount of CO_2e for 1 lb) = $30,564 \times 2.2$ (to convert to 1 kg) = $67,240 \div 3410$ (calories per 1 kg[†]) = 19.7 grams of CO_2e needed to produce 1 calorie from beef. 20 (nutrient density of beef[‡]) ÷ 19.7 = 1.02.

Beef tenderloin steak has a nutrient-to-emissions ratio of 1.02.

* eatlowcarbon.org
[†] Nutri-Facts.com
[‡] eatrightamerica.com/nutritarian-lifestyle/Measuring-the-Nutrient-Density-of-your-Food

ALMONDS
$20 \times 4 = 80 \times 2.2 = 176 \div 5810 = 0.03$. $38 \div 0.03 = 1266$
Almonds have a nutrient-to-emissions ratio of 1266.

LENTILS
$56 \times 4 = 224 \times 2.2 = 492.8 \div 1160 = 0.42$. 100 (nutrient density) ÷ 0.42 = 238.
Lentils have a nutrient-to-emissions ratio of 238.

STEAMED VEGETABLES
$83 \times 4 = 332 \times 2.2 = 730.4 \div 650 = 1.1$. $300 \div 1.1 = 272$
Steamed vegetables (combo of carrots, broccoli, asparagus) have a nutrient-to-emissions ratio of 272.

FRESH WILD LOCAL COHO SALMON

$101 \times 4 = 404 \times 2.2 = 888.8 \div 1390 = 0.64.\ 39 \div 0.64 = 61$

Fresh wild local coho salmon has a nutrient-to-emissions ratio of 61.

BAKED CHICKEN BREAST

$436 \times 4 = 1744 \times 2.2 = 3836 \div 1650 = 2.32.\ 27 \div 2.32 = 11.6$

Baked chicken breast has a nutrient-to-emissions ratio of 11.6.

POACHED EGGS

$661 \times 4 = 2644 \times 2.2 = 5816 \div 1420 = 4.01.\ 27 \div 4.01 = 6.7$

Poached eggs have a nutrient-to-emissions ratio of 6.7.

FARMED FRESH SALMON

$1203 \times 4 = 4812 \times 2.2 = 10{,}585 \div 1780 = 5.9.\ 39 \div 5.9 = 6.6$

Farmed fresh salmon has a nutrient-to-emissions ratio of 6.6.

DOMESTIC CHEDDAR CHEESE

$1140 \times 4 = 4560 \times 2.2 = 10{,}032 \div 4030 = 2.49.\ 10 \div 2.49 = 4$

Domestic cheddar cheese has a nutrient-to-emissions ratio of 4.

BEEF TENDERLOIN STEAK

3410 calories per 1000 g.

7641 in CO_2e per 4 oz = 7641×4 (lbs) $\times 2.2$ (converting to kg) = $67{,}240 \div 3410$ (calories per 1 kg) = 19.7 grams of CO_2e needed to produce 1 calorie from beef. 20 (nutrient density) $\div 19.7 = 1.02$.

Nutrient density: 20

Beef tenderloin steak has a nutrient-to-emissions ratio of 1.02.

Nutrient-to-Arable-Land Ratio

Please note that in "Calculating the Numbers," I've rounded to two decimal places. Inconsequential discrepancies in calculations will occur if you round to more or fewer decimal places in your calculations.

CATTLE FEED VERSUS BEEF

Since it takes 16 pounds of cattle feed to yield 1 pound of beef,* we can determine how much land would be needed to yield the equivalent micronutrient level from cattle feed itself, compared to beef.

Looking at a traditional cattle feed makeup (half corn and the other half composed of half wheat and half soybeans), we get a nutrient density of 41.25.

Since beef has a nutrient density of 20, I divided 41.25 by 20 to get the difference in nutrient density.

41.25 ÷ 20 * = 2.06 (difference in nutrient density)

Multiplying the 16 (multiple of how much more cattle feed is needed than beef to produce the same weight of food) × 2.06 (difference in micronutrient level between cattle feed and beef) = 33. That means that 33 times more land would need to be used to gain the equivalent micronutrient levels from beef as from wheat/corn/soybeans.

To compare calories to calories (as opposed to weight to weight), we can find out the amount of calories in 1 kg of each of the foods, then find the difference. I use Nutri-Facts.com to do this.

Since beef has 1.41 times more calories than an equal weight of wheat, corn, and soy[†] (in the combination mentioned: ½ corn, ¼ soybeans, and ¼ wheat), we can divide 33 (times more land needed to gain micronutrients from beef as opposed to an equivalent weight of cattle feed) by the amount that beef is more calorie dense than cattle feed, which is 1.41. That equals 23.4. Therefore, calorie for calorie, we'd need 23.4 times more land to gain the same amount of nutrition from beef as compared to the nutrition we'd gain from the food we fed to the cow.

*using eatrightamerica.com
[†] using nutritional data from Nutri-Facts.com

16 (multiple of cattle feed, by weight, that is fed to a cow as opposed as to what is returned in food) × 2.06 (nutrient density of cattle feed, 41.25, divided by nutrient density of beef, 20) = 2.06 × = 33 (32.96 rounded).

33 ÷ 1.41 (times more calorie dense beef is than cattle feed is) = 23.4

HEMP VERSUS BEEF

880 (pounds of hemp grown per acre) divided by 165 (pounds of beef produced on an acre) = 5.33 × 3.25 (how many times more nutrient dense hemp is than beef) = 17.33 × 3 (times by which hemp is more calorie dense than beef) = 51.9

KALE VERSUS BEEF

38,400 (pounds of kale grown per acre) divided by 165 (pounds of beef produced on an acre) = 232 × 50 (how many times by which kale is more nutrient dense than beef) = 11,600 divided by 4 (times by which beef is more calorie dense than kale) = 2900

Nutrient-to-Water Ratio

BEEF VERSUS SWEET POTATOES

Formula: Difference in nutrient density (in multiples) × difference in amount of water (in multiples) required to produce each crop on the same amount of land. Divide that number by the number of multiples the animal food is more calorie dense than the plant food is (if not more dense, then multiply those two numbers).
Nutrient density of beef: 20
Nutrient density of sweet potatoes: 83
Amount of water required to produce 1 pound of beef: 2500 gallons (minimum)
Amount of water required to produce 1 pound of sweet potatoes: 60 gallons
Beef has 3.3 more calories than sweet potatoes for an equal weight (900 calories per 1 kg of sweet potato compared with 2990 calories per 1 kg of beef)
83 ÷ 20 = 4.15 × 41.66 (2500 ÷ 60, difference in amount of water required, in multiples) = 172.88 ÷ 3.3 = 52.3

Nutrient to Fossil Fuel Comparison

LENTILS VERSUS ANIMAL PROTEIN

25.29 average nutrient density of animal products

26.9 average number of calories of energy required to produce 1 calorie of protein from animals

2.2 number of calories of energy required to produce 1 calorie of protein from lentils

100: nutrient density of lentils

Nutrient-to-Emission Ratio

Formula: By weight, amount of plant food divided by animal food that can be grown on an equal amount of land. Multiply that number by how many times greater in nutrient density the plant food is over the animal food. Take that number and divide it by how many times greater in calorie density the animal food is over the plant food. (Occasionally, the plant food will be more calorie dense, in which case you'll multiply those two numbers.) This will tell you how much more land (in multiples) will be needed to produce, calorie for calorie, the equivalent in micronutrient content for the two foods that you compared.

Difference in CO_2e (in multiples) that it takes to produce an equal weight of the two foods, divided by the multiple by which animal food is more calorie dense than plant food. (Occasionally, the plant food will be more calorie dense, in which case you'll multiply the two numbers.) Take that number and multiply it by the difference (in multiples) that the plant food is more nutrient dense.

Here's an example:

LENTILS VERSUS CHICKEN

CO_2e

Lentils: 1 lb = 224 g CO_2 emission equivalent. 224 ÷ 2.2 (to kg) = 101.82

Baked chicken: 1 lb = 1744 g CO_2 emission equivalent. 1744 ÷ 2.2 = 792.73

792.73 ÷ 101.82 = 7.79

7.79 times more CO_2e is released into the atmosphere to produce the same amount (in weight) of chicken as lentils.

To calories:

1160 cal per 1 kg lentils

1650 cal per 1 kg chicken breast

$1650 \div 1160 = 1.4$ more calories for equal weight of chicken

$7.8 \div 1.4 = 5.57$ times CO_2 equivalent calorie to calorie

Nutrient density of lentils = 100

Chicken = 27

3.7 times more nutrient dense $\times$ 5.57

3.7×5.57 times CO_2 equivalent calorie to calorie = 20.6

STEAMED VEGETABLES VERSUS BAKED SALMON

Calorie to calorie:

Steamed vegetables (cauliflower, carrots, broccoli): 1 lb = 333 g CO_2e

Baked farmed coho salmon: 1 lb = 4812 g CO_2e

$4812 \div 333 = 14.45$ times more CO_2 will be released into the atmosphere to produce an equal weight of coho salmon compared to steamed vegetables

1780 calories per 1 kg of coho salmon

650 calories per 1 kg of cauliflower, carrots, broccoli

$1780 \div 650 = 2.74$

Average nutrient density of cauliflower, carrots, broccoli = 304

Nutrient density of salmon = 39

$304 \div 39 = 7.8$ times more nutrient dense

$14.45 \div 2.74 = 5.27 \times 7.8 = 41.44$

NUTS VERSUS DOMESTIC CHEESE

Calorie to calorie:

Raw nuts: 1 lb = 80 g CO_2e

American cheese: 1 lb = 4560 g CO_2e

$4560 \div 80 = 57$ times CO_2e

5810 cal per 1 kg (equal combination of raw almonds, cashews, walnuts, and pistachios)

4030 cal per 1 kg American cheese

$5810 \div 4030 = 1.44$

Average nutrient density of nuts (almonds, cashews, walnuts, pistachios) = 36.75

American cheese = 10

36.75 ÷ 10 = 3.7 times more nutrient dense nuts are than American cheese

57 × 1.44 = 82.02 × 3.7 = 304

KIDNEY BEANS VERSUS BEEF TENDERLOIN

Calorie to calorie:

Beef tenderloin: 1 lb = 30,564 g CO_2e

Beans: 1 lb = 96 g CO_2e

30,564 ÷ 96 = 318.37 times more CO_2e is released into the atmosphere to produce the equivalent in weight of beef as of kidney beans

Nutrient density:

Nutrient density of kidney beans = 100

Nutrient density of beef = 20

100 ÷ 20 = 5

Conversion to calories from weight:

Since there are 2.69 times more calories in 1 kg of beef than there are in kidney beans, to compare calories to calories, we divide by 2.69

1270 calories per 1 kg

3410 calories per 1 kg

3410 ÷ 1270 = 2.69

318.38 × 5 = 1591.85 ÷ 2.69 = 591.76

Pound for pound:

Beef tenderloin: 1 lb = 30,564 g CO_2e

Beans: 1 lb = 96 g CO_2e

30,564 ÷ 96 = 318.37 times more CO_2e is released into the atmosphere to produce the equivalent in weight of kidney beans and beef

Nutrient density of kidney beans = 100

Nutrient density of beef = 20

100 ÷ 20 = 5

318.37 × 5 = 1591

The Low Cost of High Nutrition

Lentils to chicken
Lentils: 1060 cal per 1 kg
Chicken: 1650 cal per 1 kg
1650 ÷ 1060 = 1.55
Lentils cost per lb: $1.99
Chicken cost per pound: $4.99
$4.99 ÷ $1.99 = 2.5
Lentils nutrient density: 100
Chicken nutrient density: 27
100 ÷ 27 = 3.7
65 ÷ 39 = 1.66
2.5 ÷ 1.55 = 1.6 × 3.7 = 5.96

Flaxseed to salmon:
Flaxseed: 5340 cal per 1 kg
Salmon: 1900 cal per 1 kg
5340 ÷ 1900 = 2.8
Flaxseed cost per pound: $1.79
Salmon cost per pound: $15.99
15.99 ÷ 1.79 = 8.9
Flaxseed nutrient density: 65
Salmon nutrient density: 39
65 ÷ 39 = 1.66
2.8 × 8.9 × 1.66 = 41.37

Black beans to eggs:
Black beans: 1320 cal per 1 kg
Eggs: 1430 cal per 1 kg
1430 ÷ 1320 = 1.08
Black beans cost per pound: $1.59
Eggs cost per dozen: $3.99
Eggs cost per pound (9 eggs): $2.99
12 ÷ 9 = 1.33
3.99 (cost of 12 eggs) ÷ 1.33 = 3.98 (close to 3.99, cost of 12 eggs)
2.99 ÷ 1.59 = 1.88

To gain calories from black beans would cost 1.88 times less than to get them from eggs

Black beans nutrient density: 83

Eggs nutrient density: 27

$83 \div 27 = 3.07$

$1.88 \div 1.08 = 1.74 \times 3.07 = 5.37$

For more details on the calculations in this section, you may visit www.brendanbrazier.com.

.

NOTES

1 Health's Dependence on Nutrition

1. Centers for Disease Control and Prevention, *Stress ... at Work, DHHS (NIOSH) Publication Number 99–101* (National Institute of Occupational Safety and Health [NIOSH], 1999), NIOSH Publications and Products webpage, www.cdc.gov/niosh/docs/99-101/.

2. WGBH Educational Foundation and the Harvard Medical School Division of Sleep Medicine, "Why Sleep Matters: Benefits of Sleep," Healthy Sleep Web Site, http://healthysleep.med.harvard.edu/healthy/matters/benefits-of-sleep.

3. C. Gronfier, R. Luthringer, M. Follenius, N. Schaltenbrand, J.P. Macher, A. Muzet, G. Brandenberger, "A Quantitative Evaluation of the Relationships Between Growth Hormone Secretion and Delta Wave Electroencephalographic Activity During Normal Sleep and After Enrichment in Delta Waves," *Sleep* 19, no. 10 (1996): 817–24, www.journalsleep.org/Articles/191010.pdf.

4. World Health Organization (WHO), "Nutrition Topics: Micronutrients," WHO website, www.who.int/nutrition/topics/micronutrients/en.

5. U.S. Department of Health and Human Services (HHS) and U.S. Department of Agriculture (USDA), "Chapter 2: Adequate Nutrients Within Calories Needs," in *Dietary Guidelines for Americans 2005,* updated July 9, 2008, USDA webpage, www.health.gov/dietaryguidelines/dga2005/document/html/chapter2.htm.

6. Joel Fuhrman. Website, eatrightamerica.com/nutritarian-lifestyle/Measuring-the-Nutrient-Density-of-your-Food. See also Fuhrman, *Eat to Live: The Amazing Nutrient-Rich Program for Fast and Sustained Weight Loss,* Rev. edn. (Little, Brown, 2011).

7. Sarah Burns, "Nutritional Value of Fruits, Veggies Is Dwindling; Chemicals That Speed Growth May Impair Ability to Absorb Soil's Nutrients," *Prevention*, updated July 9, 2010, www.msnbc.msn.com/id/37396355.

8. University of Maryland Medical Center, "Omega-6 Fatty Acids" (2006), www.umm.edu/altmed/articles/omega-6-000317.htm.

9. Arthur C. Guyton, M.D., and John E. Hall, M.D., *Textbook of Medical Physiology,* Ninth edn. (W.B. Saunders, 1996), 963.

10. Ibid.

11. Diana Schwarzbein, M.D., *The Schwarzbein Principle II, The Transition* (Health Communications, Inc., 2002), 114.

12. J.C. Waterlow, "Enzyme Changes in Malnutrition," *Journal of Clinical Pathology* 4 (1970): 75–79, www.ncbi.nlm.nih.gov/pmc/articles/ PMC1176288.

2   Eating Resources: The Environmental Toll of Food Production

1. Preston Sullivan, *Sustainable Soil Management: Soil System Guide*, publication (Fayetteville, AR: National Sustainable Agricultural Information Service, National Center for Appropriate Technology, May 2004), http://attra.ncat.org/attra-pub/PDF/soilmgmt.pdf.

2. Mary Carter, "Heart Disease Still the Most Likely Reason You'll Die," CNN, November 1, 2006, www.cnn.com/2006/HEALTH/10/30/heart.overview/ index.html; "About the American Heart Association ...," Fiscal Year 2005–06, South Carolina, Fact Sheet, American Heart Association, www. americanheart.org/downloadable/heart/11453667380862005-2006%20 SC%20Fact%20Sheet.doc.

3. World Health Organization "Noncommunicable Disease Prevention and Health Promotion, Global Strategy, Facts Related to Chronic Disease, Fact Sheet," 2003, www.who.int/hpr/gs.fs.chronic.disease.shtml#:%20 WHY%20IS%20THIS%20HAPPENING?

4. "Linus Pauling—Biography," Nobelprize.org, January 26, 2011, http:// nobelprize.org/nobel_prizes/chemistry/laureates/1954/pauling-bio.html.

5. PhysicalNutrient.net, "Soil Mineral Depletion: Can a Healthy Diet Be Sufficient in Today's World?" www.physicalnutrition.net/soil-mineral-depletion.htm.

6. Organic-world.net, "Table: World: Organic Agriculture by Country: Organic Agricultural Land, Share of Total Agricultural Land, Producers 2008," www.organic-world.net/statistics-world-area-producers.html.

7. Brian Halweil, *Critical Issue Report: Still No Free Lunch: Nutrient Levels in U.S. Food Supply Eroded by Pursuit of High Yields* (The Organic Center, September 2007), www.organic-center.org/reportfiles/Yield_Nutrient_ Density_Final.pdf.

8. *Livestock's Long Shadow: Environmental Issues and Options* (Food and Agriculture Organization of the United Nations, Rome, 2006), www.fao. org/docrep/010/a0701e/a0701e00.htm.

9.   Ibid.

10. R.I. Levy and J. Moskowitz, "Cardiovascular Research: Decades of Progress, A Decade of Promise," *Science* 217 (1982): 121–29.

11. Mark Gold, *The Global Benefits of Eating Less Meat: A Report by Compassion in World Farming Trust*, Foreword by Jonathon Porritt (Hampshire, UK: CWF Trust, 2004), 22, http://awellfedworld.org/PDF/CIWF%20Eat%20 Less%20Meat.pdf.

12. D. Pimentel, "Livestock Production and Energy Use," in R. Matsumura, ed., *Encyclopedia of Energy* (San Diego, CA: Elsevier, 2004): 671–76.

13. U.S. Department of Agriculture, *Agricultural Statistics* (Washington, DC: U.S. Department of Agriculture, 2001).

14. David Pimentel and Marcia Pimentel, "Sustainability of Meat-based and Plant-based Diets and the Environment," *American Journal of Clinical Nutrition* 78 (September 2003): 660S–663S, www.ajcn.org/cgi/content/ full/78/3/660S#top.

15. Hillary Mayell, "UN Highlights World Water Crisis," National Geographic News, June 5, 2003, http://news.nationalgeographic.com/ news/2003/06/0605_030605_watercrisis.html.

16. *Livestock's Long Shadow*.

17. Ibid.

18. John Robbins, "2,500 Gallons All Wet?" EarthSave Foundation website, www.earthsave.org/environment/water.htm.

19. Cornell Science News, "End Irrigation Subsidies and Reward Conservation, Cornell Water-Resources Study Advises ...," Cornell University, Ithaca, NY, January 20, 1997, www.news.cornell.edu/releases/ Jan97/water.hrs.html.

20. Cornell Science News, "U.S. Could Feed 800 Million People with Grain that Livestock Eat, Cornell Ecologist Advises Animal Scientists; Future Water and Energy Shortages Predicted to Change Face of American Agriculture," Cornell University, Ithaca, NY, August 7, 1997, www.news. cornell.edu/releases/aug97/livestock.hrs.html.

21. Anndrea Hermann, M.Sc., P.Ag- V.P. Canadian Hemp Trade Alliance 2010, email interview by author, July 28, 2010.

22. U.S. Department of Energy, "Fossil Fuels," www.energy.gov/energysources/ fossilfuels.htm.

23. GlobalPost.com, "Top 7 Suppliers of Oil to the U.S.," July 28, 2010, www.globalpost.com/dispatch/100726/top-7-us-oil-importers.

24. "The Oil Drum: Net Energy. Discussion About Oil and Our Future," www.theoildrum.com/node/3839.

25. *Petropolis: Aerial Perspectives on the Alberta Tar Sands: A Film by Peter Mettler*, dir. Peter Mettler (Greenpeace Canada), www.petropolis-film.com/#/tarsands.

26. "Peak Oil," Wikipedia.org, http://en.wikipedia.org/wiki/Peak_oil.

27. Jerome R. Corsi, "Discovery Backs Theory Not 'Fossil Fuel,'" WorldNetDaily, posted February 1, 2008, www.wnd.com/?pageId=45838.

28. United States Environmental Protection Agency, *Latest Findings on National Air Quality, 2001 Status and Trends,* EPA Publication No. EPA 454/K-02-001 (Research Triangle Park, North Carolina: U.S. Environmental Protection Agency Office of Air Quality Planning and Standards Emissions, Monitoring, and Analysis Division, 2002), www.epa.gov/air/airtrends/aqtrnd01/summary.pdf.

29. Cornell Science News, "U.S. Could Feed 800 Million People ..."

30. Ibid.

31. Ibid.

32. Food and Agriculture Organization of the United Nations, Agriculture and Consumer Protection Department, "Spotlight: Livestock Impacts on the Environment," November 2006, www.fao.org/ag/magazine/0612sp1.htm. Full report: *Livestock's Long Shadow* (see n. 820).

33. *Livestock's Long Shadow* (see n. 820), xxi.

34. Ibid.

35. *Latest Findings on National Air Quality, 2001 Status and Trends.*

36. National Institute of Environmental Health Sciences, National Institutes of Health, "Air Pollution & Cardiovascular Disease," www.niehs.nih.gov/health/impacts/cardiovascular.cfm. Full report (executive summaries and commentary): *Reanalysis of the Harvard Six Cities Study and the American Cancer Society Study of Particulate Air Pollution and Mortality: A Special Report of the Institute's Particle Epidemiology Reanalysis Project* (Health Effects Institute, 2000), http://pubs.healtheffects.org/getfile.php?u=273.

37. "Oxford Word of the Year: Locavore," OUP blog, November 12, 2007, http://blog.oup.com/2007/11/locavore.

38. Carnegie Mellon University, "Carnegie Mellon Researchers Report Dietary Choice Has Greater Impact on Climate Change Than Food Miles," News webpage, April 17, 2008, http://www.cmu.edu/news/archive/2008/April/april17_foodmiles.shtml. Full report: Christopher L. Weber and H. Scott

Matthews, "Food Miles and the Relative Climate Impacts of Food Choices in the United States," *Environmental Science & Technology* 42 (2008): 3508–13, http://pubs.acs.org/doi/pdf/10.1021/es702969f.

39. Ibid.

40. Ian Sample, "Meat Production 'Beefs Up Emissions,'" July 19, 2007, www.guardian.co.uk/environment/2007/jul/19/climatechange. climatechange. Full article: Michele Fanelli, "Meat Is Murder on the Environment," *New Scientist,* July 18, 2007, www.newscientist.com/article/mg19526134.500.

41. Ian Sample, "Meat Production 'Beefs Up Emissions,'" July 19, 2007, www.guardian.co.uk/environment/2007/jul/19/climatechange.climatechange. Full article: Michele Fanelli, "Meat Is Murder on the Environment," *New Scientist,* July 18, 2007, www.newscientist.com/article/mg19526134.500.

42. United States Department of Agriculture, Foreign Agricultural Service, "Beef: Per Capita Consumption Summary Selected Countries, Kilograms Per Person [table]," in *Livestock and Poultry: World Markets and Trade, 2006,* www.fas.usda.gov/dlp/circular/2006/06-03LP/bpppcc.pdf.

43. Ibid.

44. "Lunar Distance: Astronomy," Wikipedia entry, http://en.wikipedia.org/wiki/Lunar_distance_(astronomy).

45. Department for Environment Food and Rural Affairs (DEFRA), Science and Research Projects webpage, "Impacts of Food Production and Consumption—EV02007," http://randd.defra.gov.uk/Default.aspx?Menu=Menu&Module=More&Location=None&Completed=0&ProjectID=14071. Full report: C. Foster, K. Green, M. Bleda, P. Dewick, B. Evans, A. Flynn, J. Mylan, *Environmental Impacts of Food Production and Consumption: A Report to the Department for Environment, Food and Rural Affairs,* Manchester Business School (London: DEFRA, 2006), http://randd.defra.gov.uk/Document.aspx?Document=EV02007_4601_FRP.pdf.

46. Ibid.

47. "Beef: Per Capita Consumption Summary Selected Countries ..."

48. Ibid.

49. Ibid.

50. Ibid.

51. Dominic Kennedy, "Walking to the Shops 'Damages Planet More Than Going by Car,'" *The Times*, August 4, 2007, www.timesonline.co.uk/tol/news/science/article2195538.ece.

52. "Vega for Sustainability and Reducing Your Carbon Foot Print," http://myvega.com/sustainability. Values calculated using Falcon Solution's CO_2 Emissions Calculator (www.falconsolution.com/co2-emission).

53. Ibid.

3 An Appetite for Change: Environmental and Health Solutions Through Food

1. Department of Energy and Climate Change [U.K.], "Legislation: Climate Change Act 2008," www.decc.gov.uk/en/content/cms/legislation/cc_act_08/cc_act_08.aspx.

2. Carbon Trust, "The Carbon Reduction Label," www.carbon-label.com.

3. *Case Study CTS055: Working with Tesco: Product Carbon Footprinting in Practice* (London, UK: Carbon Trust, 2008), www.carbontrust.co.uk/publications/pages/publicationdetail.aspx?id=CTS055. www.carbon-label.com/casestudies/Tesco.pdf.

4. Whole Foods Market, "Health Starts Here" webpage, www.wholefoodsmarket.com/nutrition.

5. Anndrea Hermann, email interview (see chap. 2, n. 33).

6. Cornell Science News, "U.S. Could Feed 800 Million People ..." (see chap. 2, n. 20).

7. North Williamette Research and Extensive Center, Oregon State University, "Commercial Vegetable Production Guides: Collards and Kale," last revised April 23, 2002, http://nwrec.hort.oregonstate.edu/collards.html; Pimentel and Pimentel, "Sustainability of Meat-based and Plant-based Diets and the Environment" (see chap. 2, n. 26).

8. Values calculated using Falcon Solution's CO_2 Emissions Calculator (www.www.falconsolution.com/co2-emission). 2.837 tonnes of carbon dioxide.

9. U.S. Department of Energy, *Transportation Energy Data Book,* Edition 29 (June 30, 2010), http://cta.ornl.gov/data/index.shtml.

4 Eight Key Components of Good Nutrition

1. University of Maryland Medical Center, "Omega-3 Fatty Acids" (2006), www.umm.edu/altmed/articles/omega-3-000316.htm.

GLOSSARY

Abiogenic theory
Proponents theorize that oil is not fossil fuel but rather originates from deep carbon deposits created during the formation of the earth. Abiogenic theorists point to the discovery of methane on Saturn's moon Titan as evidence supporting the hypothesis that hydrocarbons can form without biology, and they don't believe we are in danger of running out of oil.

Aerobic exercise
Aerobic means with oxygen. Any exercise that requires the constant breathing of oxygen to maintain pace is considered aerobic. Running at a moderate pace is an aerobic form of exercise, while sprinting full-out is not. Sprinting is classified as anaerobic, that is, without oxygen.

Anthropogenic
Of, relating to, or resulting from human influence on the natural world. For example, *anthropogenic climate change* refers to changes in the climate attributed to human activity.

Antioxidants
Antioxidants are the name given to several naturally occurring compounds—vitamin C, vitamin E, selenium, and carotenoids—prized for their cell-protection and cell-regeneration attributes. They help remove body-aging and cancer-causing free radicals from the body.

Arable land
Land fit for the growing of food. It must be in a climate conducive for agriculture with soil with sufficient nutrients to sustain crops.

Biological age
Biological age refers to the time that has passed since the body's most recent round of cellular regeneration. Biological age can be reduced by speeding the regeneration process of the body. Complementary stress, such as exercise and high-quality food, reduces biological age, while uncomplementary stress and refined foods increase it.

Biological debt

Biological debt refers to the state of fatigue the body goes into after energy from stimulation has dissipated. It is often brought about by eating refined sugar or drinking coffee to gain short-term energy.

Carbon dioxide (CO_2)

The gas emitted from the combustion of fossil fuel, carbon dioxide is known as a greenhouse gas.

Catabolic

A metabolic state in which a "breaking down" rather than a "building up" occurs in body tissues is referred to as catabolic. This state is most commonly precipitated by stress and therefore by the release of cortisol.

Celiac disease

Celiac disease is the intolerance of gluten-containing foods, such as wheat. A celiac who consumes gluten risks damaging the small intestine.

Climate change

Generally known as an altering of weather patterns over the course of several years, the term *climate change* is most commonly used to refer to an anthropological shift in climate (one that his caused by humans). However, climate change can be used to refer to both natural and anthropological shifts in climate. Global warming is the best known form of climate change.

CO_2e

CO_2e is an internationally accepted abbreviation for carbon dioxide equivalent. It's a measure that expresses the amount of all greenhouse gases released into the atmosphere, not just carbon dioxide. Other greenhouse gases, such as methane and nitrous oxide, have a global warming potential (GWP) considerably greater than that of CO_2 (carbon dioxide), therefore CO_2e takes their potency into account to allow for comparisons to be drawn between multiple sources that emit different forms of gas.

Electrolytes

Electrolytes are electricity-conducting salts. Throughout our body tissue, fluid, and blood, electrolytes conduct charges that are essential for muscle contractions, heart beats, fluid regulation, and general nerve function. Chloride, calcium, magnesium, sodium, and potassium are the chief minerals in electrolytes. A diet too low in these minerals can cause muscle cramps and heart palpitations. When we drink too much liquid that does not contain electrolytes, it can flush out the body's remaining electrolytes, causing muscle cramping and heart palpitations.

Emissions

Emissions generally refer to gases (such as carbon dioxide) created by the burning of fossil fuel. Methane and nitrous oxide are also examples of emissions.

Empty food

This term is usually assigned to foods that are heavily processed or refined. With little if any nutritional value, such foods still have plenty of "empty" calories, and usually starch and sugar, all of which can lead to quick weight gain and a feeling that hunger is never being satisfied.

Energy independent

Achieving energy independence—the desirable state of not requiring energy (oil, for the most part) to be imported from other countries—has been a long-standing goal of the United States, for both economic and national security reasons. It's widely accepted that energy independence will not be possible until the United States dramatically reduces its demand and develops alternative forms of energy—for example, solar and wind power, and biofuels from sustainable sources such as algae—on a large scale.

Essential fatty acids (EFAs)

EFAs are fats that cannot be produced by the body but must be obtained through food in order to achieve peak health. Omega-3 and omega-6 are the two essential fatty acids.

Fractionalized

Food that is no longer whole is termed fractionalized. White flour, for example, is fractionalized, because the germ and the bran components of the wheat have been removed, leaving only the flour.

Free radicals

Free radicals are damaging compounds that alter cell membranes and can adversely affect our DNA. Occurring naturally in the body, free radicals are produced on a daily basis in small amounts. However, as stress increases, so too does the production of free radicals. If stress is allowed to persist in the body for an extended period, the damage done by free radicals can be significant; they have been linked to cancer and other serious diseases. Research has also shown that free radicals cause premature signs of aging when they remain in the system. Reducing stress through better nutrition is one way to combat free radical production. Specifically, antioxidants help rid the body of free radicals by helping it excrete them in urine and sweat.

Fructose

Also known as fruit sugar, fructose is naturally occurring in most fruits. Since it is very sweet, it is often extracted from fruits to sweeten other foods.

Genetically modified organism (GMO)

A GMO is an organism whose genetic makeup has been altered using genetic engineering techniques—usually in an effort to enable more food to be grown on less land. Genetic modification is most commonly performed on crops such as corn and soy to help them survive being sprayed with pesticides. The practice is a concern within the scientific community as well as among the general public because no long-term studies have been done to determine the safety of GMOs.

Global warming

The increase in Earth's average atmospheric and water temperature since the mid-twentieth century is known as global warming. Greenhouse gases are thought to be the prime cause.

Global warming potential (GWP)
GWP is a measure of the amount of greenhouse gas released into the atmosphere. Carbon dioxide is assigned a 1, meaning it's the benchmark against which other gases are measured. Methane, for example, is 23 times as potent as carbon dioxide in terms of its damaging environmental attributes; therefore, its GWP is 23.

Glucose
Glucose is a form of simple carbohydrate and the primary sugar found in the blood. The best Thrive source for glucose is dates.

Greenhouse effect
The trapping of the sun's warmth within the earth's atmosphere by greenhouse gases is known as the greenhouse effect. The greenhouse effect is considered to be the largest contributing factor to hastening global warming.

Greenhouse gases
Gases emitted into the atmosphere that prevent a large portion of the sun's heat from leaving the Earth's atmosphere, leading to global warming. Greenhouse gases include carbon dioxide, methane, and nitrous oxide, which are emitted from a variety of sources, including the combustion of fossil fuels and the raising of livestock for food.

Growth hormone
Growth hormone, often simply referred to as GH, is what stimulates muscular growth and cell reproduction. It is released during intense exercise and sleep.

High-net-gain nutrition
Net gain is the term I use to refer to the usable nutrition the body is left with once food is digested and assimilated. Food that is nutrient dense and requires little energy to digest and assimilate can be referred to as high-net-gain food. The higher the net gain of food, the more energy that can be garnered from it.

Lignans

Lignans are plant-derived compounds that combine with others to fabricate the cell wall of the plant. Lignans are regarded as one of the best compounds to help protect against cancer and reduce cholesterol levels. When we consume lignan-rich foods, friendly bacteria convert the plant lignans to mammalian lignans, thereby allowing the release of their therapeutic attributes within the body.

Livestock

Livestock are domesticated animals raised in an agricultural setting for the production of food. They include animals that will directly become food, as well as animals that produce food, such as milk and eggs.

Methane

A greenhouse gas, methane is released from livestock, primarily ruminants, such as cattle and sheep.

Muscular functionality

Muscles that are fit and move with ease while putting minimal strain on the cardiovascular system because of their superior efficiency can be said to have become functional. Strength equals efficiency, which culminates in muscular functionality.

Nitrous oxide

A greenhouse gas with a GWP considerably higher than methane and carbon dioxide, nitrous oxide is released from livestock waste.

Nutrient density

A ratio by which micronutrient content is averaged and divided by calories. Nutrient density is an attribute to seek when striving healthier food options.

Organic

Foods grown without the use of synthetic herbicides or pesticides are said to be organic. All food was organic until the mid-twentieth century, when new synthetic chemicals were introduced to food crops in an effort to reduce damage done by weeds, insects, and rodents. The hope at that time was that this novel way of farming would allow the production of more food on less land. But since then people have become concerned about the effects on their health of ingesting chemical residues on food. Non-organic food is often referred to as "conventional."

Peak oil theory

This theory defines "peak oil" as the point at which global oil extraction has reached its maximum rate, after which the rate of production enters "terminal decline."

Phytonutrient

Also referred to as a phytochemical, a phytonutrient is a plant compound that by boosting the immune system offers health benefits beyond its nutritional value. Classified as micronutrients, phytonutrients are not essential for life but can help improve vitality and, in turn, the quality of life. See "Guide to Nutrients" on page 293 for more information.

Ruminant

A mammal that digests plant-based food by moistening it within its first stomach, then regurgitating the semi-digested mass, "cud," and re-chewing it. This process, "ruminating," breaks down plant matter and stimulates digestion. Cows, goats, and sheep are ruminants that are commonly farmed for their meat and by-products.

Simple carbohydrate

Sometimes referred to as simple sugar, simple carbohydrate is prevalent in most fruits. The body's most readily usable form of fuel, and therefore its first choice for fuel, simple carbohydrate is necessary for both mental and physical activity. If the body is not fed foods that contain simple carbohydrate, it will have to convert complex carbohydrates, but that conversion takes extra work and therefore is not a good use of energy. Glucose and fructose, the primary simple carbohydrates, are ideal fuels in that they are already in a form that the body can utilize. As well, digestive enzymes can break these simple carbs down more efficiently than they can their complex carbohydrate counterparts.

Sterols

Sterols are steroid-like compounds found in both plants and animals. Plant sterols have the ability to lower cholesterol and have been recognized as beneficial to heart health and in the fight against cardiovascular disease.

Strength-to-weight ratio

The amount of weight able to be lifted in comparison to body weight is known as the strength-to-weight ratio. It is particularly advantageous for endurance athletes to increase this ratio, because becoming stronger will garner no performance improvement if body weight also rises. This is because the extra weight of the muscle will offset the strength gains. Strength-to-weight ratio can be increased by building muscular strength (and therefore efficiency and practical function) while not increasing the size or weight of the muscle.

Trace minerals

Also known as microminerals, trace minerals have several important functions in the body that add up to optimal health. As the name suggests, these minerals are needed only in trace amounts, and a diet rich in a variety of foods will ensure their inclusion.

Trans fats

Also known as trans-fatty acids, these are a form of fat produced by heating oils to high temperature, thus altering their chemical compound and making these fats difficult for the body to process. They also inhibit the body's ability to efficiently burn healthy fats as fuel.

Whole food

Food that has had no nutrient—macro or micro—removed during processing.

RESOURCES

BOOKS

Environment

Barlow, Maude. *Blue Covenant: The Global Water Crisis and the Coming Battle for the Right to Water*

Hartmann, Thom. *Last Hours of Ancient Sunlight: The Fate of The World and What We Can Do About It Before It's Too Late*

Heinberg, Richard. *The Party's Over: Oil, War and the Fate of Industrial Societies*

Goodall, Chris. *How to Live a Low-Carbon Life: The Individuals Guide to Tackling Climate Change*

Suzuki, David, and David Taylor. *The Big Picture: Reflections on Science, Humanity, and a Quickly Changing Planet*

Health

Barnard, Neal D. *Dr. Neal Barnard's Program for Reversing Diabetes: The Scientifically Proven System for Reversing Diabetes without Drugs*

Campbell, T. Colin, and Thomas M. Campbell II. *The China Study: The Most Comprehensive Study of Nutrition Ever Conducted And the Startling Implications for Diet, Weight Loss, And Long-term Health*

Freedman, Rory, and Kim Barnouin. *Skinny Bitch*

Fuhrman, Joel. *Eat to Live: The Revolutionary Formula for Fast and Sustained Weight Loss*

Food and Recipes

Esselstyn, Rip. *The Engine 2 Diet: The Texas Firefighter's 28-Day Save-Your-Life Plan that Lowers Cholesterol and Burns Away the Pounds*

Melngailis, Sarma. *Living Raw Food: Get the Glow with More Recipes from Pure Food and Wine*

Morris, Julie. *Superfood Cuisine*

Phyo, Ani. *Ani's Raw Food Essentials: Recipes & Techniques for Mastering the Art of Live Food*

Ronnen, Tal. *The Conscious Cook: Delicious Meatless Recipes That Will Change the Way You Eat*

Silverstone, Alicia. *The Kind Diet: A Simple Guide to Feeling Great, Losing Weight, and Saving the Planet*

Steele, Jae. *Ripe from Around Here: A Vegan Guide to Local and Sustainable Eating (No Matter Where You Live)*

Wignall, Judita. *Going Raw: Everything You Need to Start Your Own Raw Food Diet and Lifestyle Revolution at Home*

Food Issues

Heintzan, Andrew, and Evan Solomon (eds). *Feeding the future: From Fat to Famine, How to Solve the World's Food Crises*

Joseph, John. *Meat Is for Pussies*

Moby and Miyun Park (eds). *Gristle: From Factory Farms to Food Safety (Thinking Twice About the Meat We Eat)*

Pollan, Michael. *In Defense of Food: An Eater's Manifesto*

Schlosser, Eric. *Fast Food Nation: The Dark Side of the All-American Meal*

Weber, Karl (ed). *Food Inc.: A Participant Guide: How Industrial Food Is Making Us Sicker, Fatter, and Poorer-And What You Can Do About It*

MOVIES

Blue Gold: bluegold-worldwaterwars.com
Flow: flowthefilm.com
Forks over Knives: forksoverknives.com
Fuel: thefuelfilm.com
Future of Food: thefutureoffood.com
King Corn: kingcorn.net
Supersize Me: super-size-me.morganspurlock.com

WEBSITES

Nutrition / Environment / Recipes / Blogs / Health

The Kind Life: Alicia Silverstone's blog includes writings on health, nutrition, style, and the environment. Also includes a forum and serves as an excellent plant-based recipe archive.
thekindlife.com

Crazy Sexy Life: Includes a daily blog and article posting by a large number of leaders in the know about plant-based health, nutrition, and style. Started by Kris Carr, from the film *Crazy, Sexy, Cancer.*
crazysexylife.com

Heather Mills: Information about her ongoing projects, including charity work and her plant-based V-Bites Café and food products. Includes regularly updated health, nutrition, and fitness advice.
heathermills.org

Georges Laraque: Former NHL player and (as of September 2010) deputy leader of the Green Party of Canada. Includes video interviews and a regularly updated blog about environmental issues and training as an elite athlete while eating a plant-based diet.
georgeslaraque.com

VegNews *magazine online:* The home of the popular magazine includes the latest information on plant-based nutrition, restaurants, and all things related to leading a plant-based life. Includes a blog and video content.
vegnews.com

John Mackey: The blog of the founder and CEO of Whole Foods Market always contains an intelligent, proactive, and well thought out applicable message, often including articles about Conscious Capitalism, a movement created by Mackey.
www2.wholefoodsmarket.com/blogs/jmackey

Vega Community: An active forum on all things related to nutrition, health, sports, training, and the environment. Visitors may create a profile and post questions as well as gain access to the latest information on topics of interest to them.
vegacommunity.com

Elizabeth Kucinich: Sound food, environmental, and political suggestions.
elizabeth.kucinich.us

Ecocentric: The official blog for sustainable food, water and energy programs of the GRACE Foundation.
ecocentricblog.org

Sustainable Table: Advocates for and celebrates local sustainable food, educates consumers on food-related issues, and works to build community through food.
sustainabletable.org

Meatless Monday: A non-profit initiative of The Monday Campaigns, in association with the Johns Hopkins Bloomberg School of Public Health. Its goal is to help you reduce your meat consumption by 15 percent in order to improve your personal health and the health of the planet.
meatlessmonday.com

The Eat Well Guide: Offers free resources, maps, and directories for finding fresh, locally grown, and sustainably produced food in the United States and Canada.
eatwellguide.org

The Meatrix films: Award-winning humorous satires (translated into more than 30 languages) that educate about the dark origins of industrial food production. themeatrix.com

H2O Conserve: An online source for tools and information—including the popular Water Footprint Calculator—which enable individuals to make water conservation part of their everyday lives. h2oconserve.org

Mind Body Green: Provides a wealth of information about nutrition, health, and the environment. Updated daily. mindbodygreen.com

*En*theos:* Philosophical information delivered in approachable, easy to grasp language. entheos.me

H Life Media: H Life is an online holistic-health lifestyle publication to redefine the concept of health, to inform us about what it truly means to lead a healthy lifestyle, and to empower us to take responsibility for our wellness and our lives. hlifemedia.com

G Living: The go-to site for the modern green movement. G Living showcases all that is fresh and green in the world of food, fashion, transportation, architecture, and more. gliving.com

Choosing Raw: A wealth of plant-based-nutrition information with near-daily updates, entertaining and informative blog posts, and recipes. choosingraw.com

No Meat Athlete: A comprehensive blog focusing on plant-based nutrition for athletes. nomeatathlete.com

Julie Morris: A regularly updated blog on food, health, and green living. It also includes excellent videos of plant-based recipe preparation. juliemorris.net

Olsen Haus: A source for high-fashion, environmentally friendly shoes. olsenhaus.com

Physicians Committee for Responsible Medicine: Doctors and laypersons working together for compassionate and effective medical practice, research, and health promotion. PCRM is a nonprofit organization. pcrm.org

T. Colin Campbell Foundation: A non-profit organization, the TCCF offers scientific and health information available to the public, without influence from industry or commercial interests.
tcolincampbell.org

Dr. Joel Fuhrman: A wealth of information on disease prevention through nutrient-dense whole foods.
drfuhrman.com

Centers for Disease Control and Prevention: Collaborating to create the expertise, information, and tools that people and communities need to protect their health through health promotion, prevention of disease, injury and disability, and preparedness for new health threats.
cdc.gov

World Health Organization: WHO is the directing and coordinating authority for health within the United Nations system. It is responsible for providing leadership on global health matters, shaping the health research agenda, setting norms and standards, articulating evidence-based policy options, providing technical support to countries, and monitoring and assessing health trends.
who.int

Organic Athlete: A non-profit organization whose mission is "to promote health and ecological stewardship among athletes of all ages and abilities by sharing information, building community, and inspiring through athletic example."
organicathlete.org

Environment / Policies

Carbon Trust: The Carbon Trust is a not-for-profit company with the mission to accelerate the move to a low carbon economy. It provides specialist support to help business and the public sector cut carbon emissions, save energy, and commercialize low carbon technologies.
carbontrust.co.uk

Carbon Reduction Label: The Carbon Reduction Label indicates the amount of CO_2 used to produce a product. The label also indicates that the company is actively working to reduce its carbon footprint.
carbon-label.com

David Suzuki Foundation: It works with government, business, and individuals to conserve our environment by providing science-based education, advocacy, and policy work, as well as acting as a catalyst for the social change that is demanded.
davidsuzuki.org

Worldwatch Institute: The Worldwatch Institute is an independent research organization recognized by opinion leaders around the world for its accessible, fact-based analysis of critical global issues. Its three main program areas include climate and energy, food and agriculture, and the green economy.
worldwatch.org

Post Carbon Institute: The Post Carbon Institute is leading the transition to a more resilient, equitable, and sustainable world through a shift away from energy derived from fossil fuel.
postcarbon.org

Department of Energy and Climate Change: The U.K.'s Department of Energy and Climate Change is responsible for all aspects of U.K. energy policy and for tackling global climate change on behalf of the nation.
decc.gov.uk

Institute for the Analysis of Global Security: IAGS is a non-profit organization that directs attention to the strong link between energy and security, and provides a stage for public debate on the various avenues to strengthen the world's energy security.
iags.org

Council on Foreign Relations: CFR is an independent, nonpartisan membership organization, think tank, and publisher.
cfr.org/about

Department for Environment Food and Rural Affairs: DEFRA is the department responsible in the U.K. government for policy and regulations on the environment, food, and rural affairs.
ww2.defra.gov.uk

New Energy Choices: Aims to promote policies that ensure safe, clean, and environmentally responsible energy options.
newenergychoices.org

United Nations: The United Nations is an international organization founded in 1945 by 51 countries and is committed to maintain international peace and security, develop friendly relations among nations, and to promote social progress, better living standards, and human rights.
un.org

Conscious Brands: A strategic planning organization that works with leaders to further expand and implement sustainability into the core of their businesses.
www.consciousbrands.com

The Ad Hatter: A search-engine optimization specialist who only takes on socially and environmentally responsible companies, specializing in nutrition sites.
theadhatter.com

Hemp History Week: A non-profit grassroots campaign that supports the legal growing of hemp for food and building materials in the United States. Advocating hemp's cultivation by presenting its health, economic, and social benefits to Congress, the backers of Hemp History Week aim to reinstate the growing of this versatile crop, thereby benefiting the American people. Hemp History Week falls in May of each year, but the group's work continues year-round.

hemphistoryweek.com

Canadian Hemp Trade Alliance: A wealth of information on everything hemp as it relates to the Canadian industry.

hemptrade.ca

Terracycle: Terracycle takes all types of garbage and turns it into consumer products with little or no waste. Upcycling, as Terracycle calls it, saves thousands of tons of garbage from going into landfills.

terracycle.net

Free Range Studios: Free Range Studios works to sell revolutionary ideas and products that build a more just and sustainable world. It is driven by a belief that the right stories told in revolutionary ways can transform society. Creators of *The Story of Stuff* and the *Meatrix*.

freerange.com

The Story of Stuff: A simple and effective 20-minute video focused on the lifecycle of consumer goods, from cradle to grave.

storyofstuff.com

Calculators

Carbon emissions from driving, air travel, and home:
terrapass.com/carbon-footprint-calculator

Water usage spanning household needs to industrial food production consumption:
www.waterfootprint.org

Automobile CO_2 based make and model:
falconsolution.com/co2-emission (used to calculate emissions in "The Great Environmental Divide: Meal Plan Comparisons" section on page 87)

Food miles:
fallsbrookcentre.ca/cgi-bin/calculate.pl

Nutritional information: Comprehensive information database that contains nutritional facts on over 6000 foods.
nutri-facts.com

Companies

Sequel Naturals: In 2003 Charles Chang (owner of his newly formed company, Sequel Naturals) and I partnered to produce a replica of the blender drink formula I had been making for myself since the early 1990s. I first began drinking it to help reduce recovery time between workouts, and it allowed me to train more and improve at a faster rate. I believe this was the defining factor that allowed me to have a career as a professional Ironman for seven years.

We launched Vega Complete Whole Food Health Optimizer in Canadian health food stores in September 2004, and the following year we began exporting it to the U.S. It's now available in most health food stores and GNCs in Canada and the U.S., and it's begun appearing in traditional supermarkets and pharmacies.

Of course, when formulating Vega products, I adhere to the nutritional and environmental ideas I write about in this book. They are nutrient-dense, high-net-gain, alkaline-forming, and plant-based, made by expending as few natural resources as possible to obtain their health boosting qualities. In fact, formulating Vega products is when I began to realize how tightly personal and environmental health was intertwined. The idea of striving to gain as many micronutrients while expending as few natural resources as possible—leading to the creation of the nutrient to resource ratio—came from my drive to develop a truly nutrient-dense, convenient, commercially available food product while limiting its strain on the environment.

The website includes recipes, informative videos, my recently published articles on health, nutrition, and environmental issues, and all things Vega.

myvega.com.

ZoN Fitness: In spring 2010, Greg Holmes of ZoN Fitness, a Chicago-based manufacturer and distributor of sporting goods, contacted me. He had read *Thrive Fitness* and felt that my approach to exercise—a tool to enhance overall quality of life—was in perfect alignment with his.

We became partners and in doing so, I created a comprehensive written and video downloadable fitness and nutrition program based on my books *The Thrive Diet* and *Thrive Fitness*. This program is free (visit zonfitness.com to access).

I was delighted to align myself with ZoN Fitness, having longed for a broad mainstream platform to disseminate what I consider to be vital information, information that I believe should be taught in school—the practical value of greater fitness and proper nutrition for long-term health. As the largest distributor of sporting goods in the world, ZoN Fitness serves as an outstanding conduit for information to an immense and diverse audience. In Canada, ZoN Fitness is available at all Canadian Tire stores, and in the United States at most sporting-goods and department stores.

Visit the website for complete information about my ZoN Fitness program, ZoN fitness equipment, and to get the latest information on health and fitness.

www.zonfitness.com.

INDEX

BESTSELLING BOOKS TO THRIVE

It's not a program. It's a lifestyle.

Begin the journey towards holistic health. Join Brendan Brazier's FREE online 30-day 'Thrive in 30' program.

Sign up today at: Thrivein30.com

Join Brendan Brazier as your personal guide as he walks you through the steps needed to achieve optimal performance by means of plant-based whole foods.

Based on Brendan's books (*The Thrive Diet* and *Thrive Fitness*), Thrive in 30 consists of three emails a week for four weeks, with each containing a short video and concise text component. The program is geared to help busy people seamlessly apply optimal nutritional habits and incorporate a high-return exercise program, ultimately improving their overall quality of life.

When you join Thrive in 30 you'll learn:

- How to combat stress by consuming plant-based whole foods and applying the principles of high net-gain nutrition
- Why whole food plant-based sources of protein and fat are optimal and how they can synergistically be combined with exercise to help you build a biologically younger body
- How you can alkalize your body's pH to combat disease, improve sleep quality, and shed fat
- What superfoods to eat that will balance hormones, detoxify your body, and lower cholesterol
- How to identify common foods that cause unexplained, mystery illnesses
- How to use nutrition to strategically fuel your body for greater endurance, maximize the return from workouts, and recover faster
- Why nutrition, exercise, and sleep are the secret to empowered mental health
- Strategies that will boost your health and make your gains permanent

 vegn **SIGN UP FOR FREE AT:** thrivein30.com